Progress in Soil Science

Series Editors

Alfred E. Hartemink, Soil Science, University of Wisconsin, Madison, WI, USA

Alex B. McBratney, Sydney Institute of Agriculture School of Life and Environmental Sciences, The University of Sydney, Sydney, NSW, Australia

Progress in Soil Science series publishes books that contain novel approaches in soil science in its broadest sense – books in the series should focus on true progress in a particular area of the soil science discipline. The scope of the series is to publish books that enhance the understanding of the functioning and diversity of soils in all parts of the globe. The series includes multidisciplinary approaches to soil studies and welcomes contributions of all soil science subdisciplines. Key themes: soil science - soil genesis, geography and classification - soil chemistry, soil physics, soil biology, soil mineralogy - soil fertility and plant nutrition - soil and water conservation - pedometrics - digital soil mapping - proximal soil sensing - soils and land use change - global soil change - natural resources and the environment. *Submit a proposal*: Proposals for the series will be considered by the Series Editors. An initial author/editor questionnaire and instructions for authors can be obtained from the Publisher, Dr. Robert K. Doe (robert.doe@springer.com).

U. C. Sharma · M. Datta · Vikas Sharma

Soil Acidity

Management Options for Higher Crop Productivity

Springer

U. C. Sharma
Jammu, Jammu and Kashmir, India

M. Datta
Kolkata, West Bengal, India

Vikas Sharma
Soil Science and Agricultural Chemistry
SK University of Agricultural Sciences
and Technology
Jammu, Jammu and Kashmir, India

ISSN 2352-4774 ISSN 2352-4782 (electronic)
Progress in Soil Science
ISBN 978-3-031-76356-4 ISBN 978-3-031-76357-1 (eBook)
https://doi.org/10.1007/978-3-031-76357-1

This Springer imprint is published by the registered company Springer Nature Switzerland AG
The registered company address is: Gewerbestrasse 11, 6330 Cham, Switzerland

Foreword

It is important to know just what is meant by acid soils having a global extent of 30% or nearly 3950 Mha of the land area developed under Udic or Ustic moisture regime both in cold and temperate climate, dominated by Spodosols, Alfisols, Inceptisols and Histosols and the southern tropical belt consisting largely of Ultisols, Alfisols and Oxisols. It gives me a sense of contentment to write the foreword of the Book on *Soil Acidity: Management Options for Higher Crop Productivity* authored by Drs U. C. Sharma, M. Datta and Vikas Sharma to move forward for addressing the issues of crop production scenario in the vast area under acid soils globally. The writers are all accomplished professionals with extensive backgrounds in working with acid soils.

It may be pertinent to mention researchers apprehension owing to presence of monomeric aluminium in soil solution in soils below pH 5.5 in acid soils of India as well as in other parts of the globe, causing aluminium toxicity and sometimes toxicity of iron and manganese to plants growing in acid soils and also resulting in poor crop growth. In acid soils, most microbial processes including the mineralisation of organic matter followed by nutrient cycling are tangibly reduced because of diminished growth and reproduction of soil microbes. Organic acids, having a large spatial heterogeneity in acid soils, are known to perform many functions in soils including root nutrient acquisition, mineral weathering, microbial chemotaxis and metal detoxification.

The plant behaviour under soil acidity stress is also dealt by the authors in this publication. In addition to naturally available acid soils in our globe, some agricultural practices, industrial processes as well as mining are also known to promote soil acidity. Proton (H^+) rhizotoxicity, in addition to dissolution to harmful elements, Al, Fe and Mn in soil solution, can also seize root growth and reduce nutrient availability. Under high activity of hydrogen ions, the ability of plant cell to maintain cytoplasmic pH remains disrupted obstructing the normal growth of plants and also water uptake in plants. Considering the low productivity of some crop plants, evolved to survive in low soil pH condition, there is a large variability in crop selection to have adequate growth and development in acid soils without any ameliorative measures.

As discussed by the authors, soil acidity along with low nutrient availability is the major constraint to crop production in acid soils. The detrimental effects of soil acidity are P sorption caused by clay mineralogy, oxides and hydrous oxides of Fe and Al. On the other hand, soil acidity converts available soil nutrients into unavailable forms. There is a shortage of available Ca, K, Mg, P and Mo on one hand along excess/toxicity of metallic ions. On highly weathered tropical soils, soil acidity followed by low CEC are the major constraints for crop production. Crops differ in their susceptibility to soil acidity. In acid soils, there may be loss in crop diversity, low crop yield, decline in fertiliser response and sometimes total crop failure is also noted. Production of some crops, barley, wheat and beans is extremely low even after application of optimum fertiliser doses on acid soils. Poor plant growth, loss of crop growth vigour, poor nodulation, stunted root growth, high disease incidence, luxuriant weed growth and distorted leaf colours are major symptoms with the increase in soil acidity.

In order to have proper soil management in acid soils, acid tolerant crops may be grown according to soil pH selectivity. Crops can be divided into the following groups, highly acid tolerant crops, moderately acid tolerant crops and slightly acid tolerant crops. Other ways to consider as an effective management are application of liming materials, application of organic amendments/compost/biochar, application of tree leaf, use of basic fertiliser/rock phosphate, soil/water management and increase in the efficiency of N and potash fertilisers.

So, the present publication on Acid soils in the Global perspectives with plethora on research information can be used effectively by Researchers, Policy Planners, Farmers and Students to unravel the mechanisms of soil acidity development, impact on crop growth and modules of acid soil management to triumph over the productivity constraints in acid soils. I am aware that the authors' new endeavour will significantly contribute to the solution of the issues associated with crop growth on acidic soils because they have extensive expertise working at the Indian Council of Agricultural Research and Agricultural University.

Dr. Himanshu Pathak
Secretary (Dare), Ministry of Farmers
Wefare and Agriculture
Director General, Indian Council
of Agricultural Research
New Delhi, India

Preface

It is a matter of delight leading to the process of erudition for assemblage of our efforts to write a book on soil acidity which affects nutrient availability, microbial activity, elemental toxicity, physical properties and finally crop/plant growth and development in a substantial manner. Over the years, the documentation in the domain of soil acidity being a world wide phenomenon has been attempted to a large extent but suffers constraints considering an extensive acceptability for the science and technology to be accentuated to overcome the impact of acidity problems.

Soil pH is important to life as the enzymes that catalyse the chemical reactions of life require a specific pH to function. On the other hand, it can be said that if pH is increased or decreased significantly, biochemicals like proteins can be denatured and become non-functional, causing a cell death. So, life on earth depends on appropriate pH levels in and around living organisms and human life requires a maintenance of proper pH level in the serum of about 7.4, a slightly alkaline range to survive. Any increase or decrease in the blood pH can lead to several disorders.

As an infant depends on mother for his or her food, so is the case of human on soils. The only difference is that as the infant grows, he or she gets rid of dependency for physical need on his or her mother but we even after million years of evolution depend on soils from birth to death. A man will die but the carbon will return back to the soil for the plant/crop to assimilate and with time, again and again on a cycle of plant and animal life. In addition to carbon, soils supply several major requirements such as an anchoring medium as well as water/nutrient supply and storage that are necessary to propagate crops being consumed by us.

Owing to our working periods in agriculture mostly on acid soils, we have undertaken the task of writing a book on *Soil Acidity: Management Options for Higher Crop Productivity* for covering the various topics under acidity arena. While introducing the topics, one should recognise soil degradation as a key factor in acid soils leading to poor agricultural productivity and affecting the livelihoods of farmers and their environments. The sustainable use of acid soils requires nutrient inputs, such as fertilisers, inorganic and organic amendments such as lime, biochar, compost, manure as well as soil health or fertility is a manageable soil property and its management is of utmost importance for optimising crop production to achieve a sustainable target.

Acid soils occupy approximately 30% or 3950 mha of the world's ice free land area and occur mainly in two global belts in northern area having cold and temperate climate and southern tropical a/subtropical regions. Sixty-seven per cent of acid soils support forests and woodlands and approximately 18% are covered by savanna, prairie and steppe vegetation. The remaining acid soil area is used to grow arable crops. In cold and temperate climate as well as in high altitude regions, the presence of decaying organic matter produces hydrogen ions responsible for soil acidity. In warm and humid environments, soil pH decreases over time through acidification due to leaching from high amounts of rainfall. Due to differences in chemical composition of parent materials, soils will become acidic after variable lenghths of time. Soils developed from granite materials are likely to be more acidic than soils developed from calcareous shale or limestone. Harvesting of crops has also its effects on soil acidity development due to large removal of lime like elements for crop nutrition. It is noted that high yielding forages can cause soil acidity to develop faster than with other crops. The use of fertilisers, especially those supplying nitrogen, has often been blamed as cause of soil acidity. High acidic condition is noticed in peat soils, being formed by the natural accumulation of partially decayed biomass and these soils are the largest reserves of terrestrial organic carbon in the ecosystem. So, a broad information on the variability in acid soils is reported in the book contained herein.

Chemistry and microbiology of acid soils are also interesting. There are different pools of soil acidity prevailing in acid soils present across the globe. Regarding pools of soil acidity, active acidity is the quantity of hydrogen ions present in soil water solution. The second pool, exchangeable acidity, refers to the amount of acid cations, aluminium and hydrogen occupied on cation exchange sites. On the other hand, residual acidity comprises of all bound aluminium and hydrogen in soil minerals. The solubility of aluminium increases as the soil pH decreases. Al^{3+} toxicity is a major constraint to crop production in acid soils where soil pH is lower than 5.5. During Al-hydrolysis, there is a release of H^+ and for every one mole of Al^{3+}, 3 moles of H^+ are released into solution. Soluble polymeric hydroxyl aluminium ions present in acid soils are also discussed exhaustively. In addition to elemental toxicity in acid soils, there is a deficiency of soil nutrients in acid soils present in different places. Most microbial processes, including the breakdown of organic matter and cycling of nutrients are reduced in acidic soil because growth and reproduction of soil microbes, primarily bacteria and fungi, are reduced.

Acid soils significantly limit crop production worldwide. The mechanisms and genes conferring tolerance to acid soil stress has been a focus of intense research interest over the past decade, thus having a huge a variability of tolerance/susceptibility of crops towards soil acidity. Various amendments are also used in reclamation or alleviating the impact of soil acidity on crops/plants and availability of amendments may play a vital role in soil acidity management. Managing acid soils is the most important for the long-term sustainability of farming systems and soil resources. There are three management approaches, such as minimise acidification, use of soil amendments and acid tolerant crops preferably suitable genotypes for a proper a crop production modules to be adopted.

We hope that the present book will enrich the knowledge and concerned information regarding acid soils, their characteristics and sustainable management for crop production to the readers including researchers and students and the technology will ultimately percolate to the planners worldwide for implementation by the farming community. Springer Publishers are profusely thanked for accepting our manuscripts for publishing in the form of a book.

Jammu, India — Dr. U. C. Sharma
Kolkata, India — Dr. M. Datta
Jammu, India — Dr. Vikas Sharma

Acknowledgements Due to their hazardous levels of metals like Al, Fe, and Mn content, as well as, deficiencies of other elements, particularly P, acid soils restrict crop productivity. Numerous scientists have produced excellent work on a variety of topics related to managing acid soils for increased agricultural yields. All of those esteemed writers, whose writings are referenced throughout this book, have our sincere gratitude and gratefulness.

Contents

Abbreviations

AD	Anno Domini
AER	Anion Exchange Resin
Al	Aluminium
ALMT	Aluminium-Activated Malate Transporter
ALMT1	Aluminium-Activated Malate Transporter 1
ALMT2	Aluminium-Activated Malate Transporter 2
ALS	Acetolactate Synthase
AMO	Ammonia Monooxygenase
ANC	Acid Neutralising Capacity
AOB	Ammonia Oxidising Bacteria
ASS	Acid Sulphate Soils
ASSOD	Soil Degradation in South and Southeast Asia
ATA	Agricultural Transformation Agency of Ethiopia
ATPase	Adenosine Triphosphatase
ATP-SR-Y	Acid Tolerant Population Streak Resistant Yellow
AVS	Acid Volatile Sulphide
BNF	Biological Nitrogen Fixation
CAAA	Clean Air Act Amendments
CCA	Canonical Correlations Analysis
CEC	Commission of the European Communities
CIAT	International Centre for Tropical Agriculture
CIESIN	Center for International Earth Science Information Network
CIMMYT	International Maize and Wheat Improvement Centre
CRISPR/Cas9	Clustered Regularly Interspaced Short Paliesdromic repeats—Associated Protein 9
CSIRO	Commonwealth Scientific and Industrial Research Organization
D'ed	Developed Countries
D'ng	Developing Countries
DNA	Deoxyribonucleic Acid
DS	Direct Seeded
DTZ	Digital Transition Zone in Roots

ECCE	Effective Calcium Carbonate Equivalent
ECEC	Effective Cation Exchange Capacity
EFB	Empty Fruit Bunches
ELD	Economics of Land Degradation
ENM	Engineered Nanomaterials
F1	First Generation Progeny
F2	Second Generation Progeny
FA	Fulvic Acid
FAO	Food and Agriculture Organization
FAO, UN	Food and Agriculture Organization of the United Nations
FYM	Farmyard Manure
GAEZ	Global Agro-ecological Zones
GAEZ	Global Agro-Ecological Zones GEF Global Environment Facility
GCA	General Combining Ability
GHG	Green House Gases
GIMMS	Global Inventory Modeling and Mapping Studies
GIS	Geological Information Systems
GL	Green Liquor
GLADA	Global Assessment of Land Degradation and Improvement
GLADIS	Global Land Degradation Information System
GLASOD	Global Assessment of Human-Induced Soil Degradation
GLII	Global Land Indicator Initiative
GLOBIO	Global Biodiversity Model
GPFLR	Global Partnership on Forest Landscape Restoration
GSP	Global Soil Partnership
Gt	Gigatonnes
GWAS	Genome-Wide Associated Studies
HA	Humic Acid/Hydroxy Apatite
HIV-Hydroxy	Interlayering in Vermiculite
HLPE	High Level Panel of experts
HPLC	High Performance Liquid Chromatography
HT	Humid Tropical Climate
HvAACTI	Aluminium Tolerance Gene (Barley) Through Citrate Exudation of MATE family
IAA	Indol Acetic Acid
IAEA	International Atomic Energy Agency
IASS	Institute for Advanced Sustainability Studies
ICAR	Indian Council of Agricultural Research
ICRISAT	International Crops Research Institute for Semi-Arid Tropics
ICSU	International Council for Science
IFA	International Fertilizer Association
IFPRI	International Food Policy Research Institute
INPW	Inland Pacific Northwest
INRM	Integrated Natural Resource Management

IPBES	Intergovernmental Platform on Biodiversity and Ecosystem Services
IPCC	Intergovernmental Panel on Climate Change
IRRI	International Rice Research Institute
ISRIC	International Soil Reference and Information Centre
ITPS	Information Technology Professional Solutions
IUSS	International Union of Soil Sciences
JRC	Joint Research Centre of the European Commission
LADA	Land Degradation Assessment in Drylands
LDN	Land Degradation Neutrality
LDSF	Land Degradation Surveillance Framework
MAS	Master of Applied Sciences
MATE	Multidrug and Toxic Compound Extrusion Transporters
MATE	Multidrug and Toxic Compounds Extrusion
MBC	Microbial Biomass Carbon
MDG	Millennium Development Goal
MDH	Malate Dehydrogenase
ME	Malic Enzyme
MEA	Millennium Ecosystem Assessment
MEA	Millennium Ecological Assessment
Mha	Million Hectares
MIDH	Mutant Isocitrate Dehydrogenase
mRNA	Messenger Ribonucleic Acid
Mt	Million Tonnes
NAAS	National Academy of Agricultural Sciences
NARS	National Agricultural Research System
NDVI	Normalised Difference Vegetation Index
NE	North East
NISF	National Institute for Soils and Fertilizers
NPP	Net Primary Production
NRCS	Natural Resources Conservation Service
OA	Organic Anion
OECD	Organisation for Economic Co-operation and Development
OHPP	Old Himalayan Picdmont Plains
OM	Organic Matter
OPVs	Open Pollinated Cultivars
PAGE	Pilot Assessment of Global Ecosystems
PASS	Prospective Acid Sulphate Soils
PBL	Netherlands Environmental Assessment Agency
PCA	Principal Component Analysis
PEP	Phosphoenolpyruvate Phosphatase
PEPCase	Phosphoenolpyruvate Carboxylase
PhoA	Alkaline Phosphatase Structural Gene
PLFA	Phospholipid Fatty Acids
PSM	Phosphate Solubilizing Microorganism

qRt-PCR	Quantitative Real Time Polymerase Chain reaction
QTL	Quantitative Trait Locus
RHB	Rice Husk Biochar
RNA	Ribonucleic Acid
ROS	Reactive Oxygen Species
RP	Rock Phosphate
rRNA	Ribosomal Ribonucleic Acid
RUE	Rain Use Efficiency
SAHH	S-Adenosyl-L-Homosysteinase
SbMATE	Sorghum Aluminium Tolerance Gene MATE Family
SCA	Specific Combining Ability
SDG	Sustainable Development Goal
SIS	Soil Information System
SLM	Sustainable Land Management
SOC	Soil Organic Carbon
SOLAW	State of the World's Land and Water Resources
SOTS	EU Soil Thematic Strategy
SSA	Sub-Saharan Africa
STAR	Steroidogenic Actute Regulatory Protein
TEA	Total Exchangeable Acidity
UN	United Nations
UNCBD	United Nations Convention on Biological Diversity
UNCCD	United Nations Convention to Combat Desertification
UNCOD	United Nations Conference on Desertification
UNEP	United Nations Environment Programme
UNESCO	United Nations Educational Scientific and Cultural Organization
USDA	United States Department of Agriculture
WAD	World Atlas of Desertification
WFS	World Food Security
WOCAT	World Overview of Conservation Approaches and Technologies
WRB	World Reference Base for Soil Resources
WWF	World Wide Fund for Nature
ZmMATE1	Maize MATE Genes
ZPC	Zero Point of Charge

Chapter 1
Introduction

Abstract Soils supply a wide range of ecological services required to suit human requirements. Provisioning services include things like food, wood, fibre, raw materials, and infrastructure support; regulating services include things like flood mitigation, nutrient and contaminant filtering, carbon storage and greenhouse gas regulation, detoxification and waste recycling, and pest and disease population control; and cultural services include things like recreation, aesthetics, heritage values, and cultural identity. Because of the acidity of the soil, many crops face growth and yield constraints worldwide. The basic problems with the chemical properties of highly acidic soils include not just acidity per se, but also the presence of toxic materials and components such soluble forms of Al, Fe, and Mn, nitrites, and various hazardous organic acids. The toxicity of aluminium (Al) is one of the primary barriers to agricultural productivity on acid soils, which are present on up to 40% of arable land globally. When Al is dissolved in soil solutions with an acidic pH, plants are at risk from this element, which is found in the third-highest concentration in the Earth's crust. Because leaching of these elements from the soil lowers pH and reduces the soil's capacity to act as a buffer, mineral acid soils are created by parent materials that are naturally acidic and lacking in basic cations (Ca, Mg, K, and Na). Because it dissolves in soil, aluminium (Al) ions—which are phytotoxic—become more prevalent in soil solutions. Organic acid soils generally have no Al buffering and can have pH values much below four. They are characterised by high quantities of humic acids and partially degraded plant debris.

Keywords Acid soils · Aluminium · Iron · Manganese · Crops · Phytotoxic · pH

1.1 General Description

Numerous crops around the world experience growth and yield limitations due to soil acidity. In addition to acidity itself, the presence of poisonous substances and components such soluble forms of Al, Fe, and Mn, nitrites, and numerous harmful organic acids are the fundamental issues with the chemical characteristics of more

U. C. Sharma et al., *Soil Acidity*, Progress in Soil Science,
https://doi.org/10.1007/978-3-031-76357-1_1

acidic soils. One of the main obstacles to agricultural yield on acid soils, which can be found on up to 40% of arable lands worldwide, is aluminium (Al) toxicity. Al, which makes up the third-highest concentration in the crust of the Earth, is hazardous to plants when dissolved in soil solutions with acidic pH levels (Kochian 1995).

According to von Uexkull and Mutert (1995), 3950 million hectares of global land have acidic soil, of which 15% are planted with annual and perennial crops. They further elaborated that these acidic soils make up around thirty percent of the total land surface on the planet. The northern belt of acid soils, which is comprised mostly of organic acid soils and may be found in the moist northern temperate zone, is responsible for the maintenance of coniferous forests. A band of mineral-rich and acidic soils can be found to the south of the tropical rainforests. These soils are most commonly found in the moist tropics, subtropics, and temperate forests. But many of the soils in other places are likewise acidic. Approximately two billion hectares, or 14%, of the world's total ice-free area, are made up solely of tropical acid soils. The percentage of acidic soils in the world's arable soils—those that can support agricultural growth—is about 40%. According to Uexkull and Mutert (1995), just 12% of cropland is acidic at the moment, but this percentage is rising globally. Figure 1.1 shows distribution of acid soils in the World (USDA/NRCS).

Mineral acid soils are produced by parent materials that are acidic and naturally deficient in the basic cations (Ca, Mg, K, and Na), or by the leaching of these elements from the soil, which lowers the pH and the soil's ability to act as a buffer. Aluminium (Al) is soluble in soil and becomes more abundant in the soil solution as phytotoxic aluminium ions. Al is present in most mineral soils in sufficient amounts to buffer the soil to a pH of about 4. Large concentrations of humic acids and partially decomposed plant matter make up organic acid soils, which often have no Al buffering and a pH that can drop well below 4 (Kidd and Proctor 2001). Depending on the soil type, a variety of factors influence acid soil toxicity. Al toxicity is the main factor limiting plant growth in acidic soils with a high mineral concentration. Under acidic conditions, Al is liberated from soil minerals such as $Al(OH)_2^+$, Al(OH), and

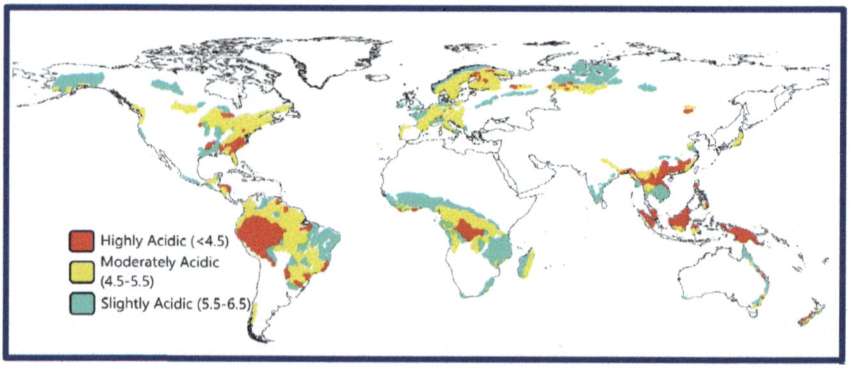

Fig. 1.1 World distribution of acid soils (USDA, NRCS, World soil resources, Washington, D.C.) (Soil Survey Staff, available at http://websoilsurvey.nrcs.usda.gov/)

$Al(H_2O)_3^+$, the latter of which is sometimes referred to as Al (Kinraide 1991). Al ions quickly impede root growth in the majority of agriculturally significant plants at micromolar doses.

Smith (1872) is credited with making the initial discovery of acidic deposition in England during the nineteenth century. After it was reported that acidic precipitation and surface water acidification had occurred in Sweden and the adjoining parts of Scandinavia in the late 1960s and early 1970s (Oden 1968), acidic deposition became a worry for the environment. This was due to the fact that acidic deposition is harmful to the ecosystem. Acid precipitation, in both its dry and its wet forms, makes its way to natural places, which may be terrestrial or aquatic in nature. The level of acid deposition can be determined using a scale known as "pH." The pH of a substance can be used to determine how acidic it is. A pH of 7.0 is found in pure water. After World War II, countries in Europe and eastern North America began significantly increasing their consumption of fossil fuels, which led to the beginning of the modern age of anthropogenic acid deposition in those regions. The United Nations Conference on the Human Environment, which took place in Stockholm, Sweden, in 1972, was the impetus for the beginning of international collaboration to combat air pollution and acid deposition. Both natural sources, such as volcanoes and decaying vegetation, and man-made sources, primarily emissions of sulphur dioxide (SO_2) and nitrogen oxides (NO_x) arising from the combustion of fossil fuels, are responsible for the production of acid deposition's predecessors, also known as the chemical forerunners of acid deposition. Since it was first established, this first agreement has had several protocols added to it, which increase its scope. The Clean Air Act of 1970 and its revisions in 1990 are largely responsible for lowering the amount of acid rain that falls on the United States. In the 1970s, work began on drafting a Memorandum of Intent between the United States of America and Canada to reduce air pollution and acid deposition. This was done in an effort to improve environmental conditions.

It is now feasible to examine how three decades of clean air legislation have affected the reduction of emissions, levels of air pollution, trends, and the chemical implications of acid deposition, as well as the recovery of ecosystems. It is also possible to think ahead to the proposed reauthorisation of the Clean Air Act and the subsequent national assessment of acidic deposition, which is scheduled to take place in the middle of 2001. It is essential to perform periodic reviews of the knowledge collected from long-term monitoring, process-level research, and modelling in order to facilitate the evaluation of regulatory programmes and the settlement of complex environmental challenges. The study that follows places a significant emphasis on the lessons learned since the passage of the Clean Air Act Amendments in 1990 and goes into considerable detail regarding their discussion. This act's primary concern was the recovery of ecosystems in the Northeastern region of the United States. It also addressed the effects of acid deposition on the forest vegetation, soils, and surface waterways of that region. If these inputs are converted by soil microbes to nitrate (NO_3^-), the atmospheric deposition of ammonium (NH_4^+), which is formed from NH_3 emissions, can help to contribute to the acidity of soil and water. Agricultural activities, however, have been recognised as the main cause of NH_3 emissions in the

US by local and regional studies (Jordan and Weller 1996). The biggest component is typically thought to be animal and poultry manure.

Understanding that a soil with a pH of 5 is 10 times less acidic than a soil with a pH of 4 is crucial because pH is scaled on a logarithmic basis. Acidic soils are those that have a pH of less than 7 (i.e. a neutral pH). Thus, soils with a pH of less than 5.5 may be classified as acid (i.e. those in which growth of many plant species is negatively affected); nevertheless, this has little practical significance because most plant species will only display reduced growth at significantly lower pH.

1.1.1 Soil's Function

Numerous ecological services that are necessary to meet human needs are provided by soils. Food, wood, fibre, raw materials, and infrastructure support are examples of provisioning services; flood mitigation, nutrient and contaminant filtering, carbon storage and greenhouse gas regulation, detoxification and waste recycling, and pest and disease population control are examples of regulating services; and recreation, aesthetics, heritage values, and cultural identity are examples of cultural services (Dominati et al. 2014). Recognising the wide range of variation in these ecosystem services' attributes is also essential. One example of how regulation services often give humans longer-term, less immediately tangible benefits than provisioning services do is the regulation of carbon dioxide as a greenhouse gas through the stabilisation of carbon in soils as organic matter (Kopittkea et al. 2019). However, providing services usually assist people in a very concrete and very instantaneous way. There is an unprecedented demand for soils and the services they provide. The demand for soil to generate food, fibre, energy, and physical support for infrastructure is the greatest of these pressures, and it is rising quickly due to the rapid rise of the human population and our changing way of life. In fact, according to estimates, the number of people on the planet did not rise above 6 million until the advent of agriculture (Livi Bacci 2017), and it has since risen to around eight billion. By the year 1000, there would be 250 million people, 6.1 billion by the year 2000, and 9.8 billion by the year 2050 (UN 2017). Accordingly, between 2005 and 2050, food production must grow by 70% in order to attain global food security (ELD 2015). We must analyse how the growing demand for the provision of services being placed on soils is already leading to their unsustainable degradation, as this is of utmost relevance to our topic.

1.1.2 Surface Acidity

Soil acidification occurs when the pH of the soil's aqueous phase (its solution) falls into the acidic range, which is defined as less than 7.0. While any pH value less than 7.0 on the pH scale is considered acidic, values less than 5.5 indicate soil acidification, which may impair soil functions. Such effects may be significant for both soil horizons at the surface and those below. According to estimates, acid soils make up as much as 30% of the world's ice-free land area, or about 3950 Mha (von Uexkull and Mutert 1995).

The amount of precipitation and vegetation support the natural process of soil acidity. Acidity is generally produced through the breakdown of plant wastes and precipitation. As a result, over a long period of time, the process of soil weathering, which gradually eliminates basic cations (Ca, Mg, K, and Na) from the soil and raises the relative proportion of acidic cations (Al and Fe), results in soil acidification. Acidity is also increased by burning forests or removing basic cations from crops. Naturally acidic soils are more likely to exist in areas with old soils (those not affected by recent glacial events) and a humid climate. A significant fraction of the world's soils are acidified, especially in the wet tropics and northern boreal forests.

Significant acidity can be produced by the oxidation of sulphides in the soil. In places that were originally submerged by sea water, drained, and subsequently exposed to atmospheric oxygen, acid sulphate soils develop. The most typical situation would be coastal regions close to salt marshes, floodplains, and mangrove forests. When sulphide-bearing mine spoils from coal or metal mining operations are exposed to the atmosphere, acid mine soils develop. Such regions can experience severe off-site acidification with detrimental ecological effects due to acid mine drainage.

Other anthropogenic activities, such management techniques used in crop and soil production or the burning of fossil fuels and the ensuing acid deposition, can further hasten soil acidification. Due to the nitrification process, which turns ammonium into nitrate, the usage of fertilisers containing ammonium significantly contributes to acidity. This problem is exacerbated by the use of many of the most widely used inorganic fertilisers in the world, including urea, diammonium phosphate, monoammonium phosphate, and anhydrous ammonia. Acidity in the soil has three main detrimental effects on plants. Generally speaking, when the soil pH is between 5.5 and 7.5, plant nutrients in the soil are most readily available for plant absorption.

1. Acidification of the soil can reduce plant nutrient availability and potentially cause shortages. The main nutrients to be concerned about are calcium, magnesium, potassium, phosphorus, and molybdenum.
2. The presence of high concentrations of soluble Al and Mn in extremely acidic soils (pH 5.0), which can be hazardous to growing plants, is a second factor. Al toxicity is, in reality, a significant global constraint on crop productivity. At a pH of about 5.0, soluble Al levels often start to rise dramatically (this will vary slightly by soil); the free Al ions then go through a series of hydrolysis reactions that not only increase acidity further but also significantly raise the quantity of lime needed to neutralise it. In order to reduce the effects of soil acidity, regulating soil pH to stay above this crucial range is important. According to Sumner and Noble's estimation in 2003, Al phytotoxicity was a possibility in 67% of the world's acidic soils.

3. The third main adverse consequence is related to the microbes that fix biological N. These become blocked in extremely acidic environments, which restricts the amount of N that legumes can fix.

The ability of a soil to tolerate variations in soil pH, whether there be increases or decreases, is known as soil buffering capacity. It is a crucial factor that affects a soil's propensity to become acidic as well as how easily acidity may be later rectified with lime. Buffering capacity is positively connected with organic matter content, clay content, having basic minerals as the soil's parent material, and calcium carbonate content. This implies that in soils containing these components, it would require more acidity to lower the pH. In other words, compared to a soil with a finer texture and a larger amount of organic matter, a sandy soil with low levels of organic matter would have a low buffering capacity. When it comes to soils that are susceptible to acid-producing processes, anthropogenic influences on soil acidity are most noticeable. This is the case, for example, in regions like eastern Canada and eastern Europe that get acid deposition from the United States and western Europe, respectively. Acidic soils support native flora that is adapted to the environment. Therefore, the amount of acid soils that need to be managed will be reduced by efforts to limit deforestation in the tropics and at northern latitudes, as well as through efforts to lessen the conversion of savannah and grasslands into agricultural production. When the natural flora is disturbed, difficulties with acid soil typically get worse because erosion and soil C oxidation both rise. In affluent nations, a decrease in N and S emissions to the atmosphere has also been effective in lowering acid precipitation.

The best way to correct soil acidity for soils previously used for agriculture is to use lime that contains calcium carbonate ($CaCO_3$). The soil pH must be raised to a specific level, often at least above the pH at which Al toxicity is no longer an issue, by adding enough lime to overcome the soil's buffering capability. The expense and the availability of lime are the main drawbacks of this method. Additionally, small amounts of pelletised lime may be directly mixed in with the seeds or transplants. Liming might not be practical if lime is not readily available nearby. Although each would need to be assessed separately, a range of different materials can have some liming benefits. Gypsum ($CaSO_4{\cdot}2H_2O$) applications, which are particularly effective for addressing subsurface acidity because the gypsum is fairly soluble and can drain beyond the surface soil horizon, can help reduce Al toxicity without altering the soil pH. The P combines with soluble Al and helps reduce Al toxicity, providing a narrow zone of soil with better growing conditions for roots. Subsurface band applications of P fertilisers can also be beneficial.

Utilising crops and/or cultivars within a specific crop that are acid-tolerant is another tactic. Using N fertilisers without ammonium, such as calcium nitrate, as a prophylactic measure may be preferable than using ammonium-containing N fertilisers, which may eventually require lime to balance the acidity they generate. When it comes to minimising the overall extent of acid soils that have an adverse effect on ecosystem functioning at a specific place, efforts to minimise land use conversion are likely to have the most impact. Additionally, developing nations must be included in the effort to minimise N and S emissions to the atmosphere. Acidic agricultural soils will require ongoing efforts to maintain or enhance pH values in order to stop additional production reductions.

1.1.3 Soil as a Source of Energy

Not only are there more demands on soils now and in future to ensure global food security, but there are also more demands on soils from other uses. The demand for biofuels has increased significantly since the start of the twenty-first century, and global ethanol production has increased from about 20 GL from 2000 to about 120 GL as of 2018 (OECD/FAO). Currently, the majority of these biofuels are made from cereals in temperate regions and sugarcane in tropical ones. As a result, 120 Mt more of cereal production—primarily maize—has been needed over the past ten years for biofuel generation. According to recent estimations, 32 million ha of agricultural land are needed for this biofuel production (Langeveld et al. 2013). In comparison, the total cropping area worldwide is 1600 million ha (FAO 2018), or around 0.24% of the world's ice-free soil and 2% of all cropland. Future increases in biofuel production are anticipated to be less significant than those in food production, with global ethanol output rising to approximately around 120 GL now produced will increase to 130 GL by 2027 (OECD/FAO 2018). There is little evidence that the production of biofuels is significantly displacing food production relative to the total area of crops, and present and predicted volumes only require a tiny amount of soil (Langeveld et al. 2013). However, considering the negative effects of agricultural output on soil quality and soil degradation, particularly the concomitant emission of greenhouse gases from soils as a result of agricultural production, the use of soils for biofuels needs to be carefully considered.

1.1.4 Soil Degradation Due to Unsustainable Agriculture Intensification

The capacity of soils to offer their various ecosystem services is altered by anthropogenic changes in how they are used, notably for the intensification of agricultural output to achieve food security. Therefore, decision-making procedures must clearly take into account how the advantages of boosting agricultural production intensity compare to the value loss in other services. According to estimates by the FAO and ITPS (2015), 33% of soils are currently moderately to highly deteriorated as a result of erosive processes, salinisation, acidification, pollution, or compaction, and 52% of agricultural land has already experienced mild to severe soil degradation (ELD 2015). According to predictions, land degradation over the next 25 years may result in a 12% decline in global agricultural productivity and a 30% increase in food prices (ELD 2015).

1.1.5 Loss of Soil OM and Corresponding GHG Emissions

It is essential to ensure that there is an adequate quantity of organic matter present in the soil in order to preserve its fertility. It is necessary for the soil to have organic matter in order to fulfil all of the fundamental requirements of the crop plant in order to maximise yield potential. These requirements include enhancing the cation exchange capacity of the soil, raising the water-holding capacity of the soil, and improving the soil's infiltration rate, in addition to acting as a slow-release storage for macro- and micronutrients. On the other hand, agricultural production is to blame for the mineralisation of the soil, which led to the loss of organic matter in the soil. According to the findings of Kopittke et al. (2017), agricultural land for an extended length of time resulted in a decline of soil organic carbon (OC) stores of between 30 and 60%. This is significant given that OC accounts for approximately 80% of soil organic carbon and 58% of organic matter. This loss of OM is the result of the removal of organic materials (harvestable products) from the soil, in addition to the increased microbial activity brought on by higher soil temperatures and soil profile oxygenation. Both of these factors contribute to increase soil temperatures. As one might expect, cultivating land results in a loss of organic matter (OM) in the soil, which in turn reduces the fertility of the land in its natural state. Another role of organic matter (OM) in soil fertility is the preservation of water in a form that may be used by plants. The water that is stored in soils, commonly referred to as "green water," is responsible for 65% of all freshwater that exists on Earth (Sposito 2013). This "green water" supplies 90% of the water that is used for agricultural production around the world. However, it is well known that cultivating land and the resultant loss of organic matter in the soil both lead to decreased agricultural yields. This is because the amount of water that is available to plants decreases as a result of cultivating land. Long-term cropping promotes the mineralisation of organic matter (OM), which in turn releases carbon dioxide, a gas that contributes to global warming. In addition, OM is necessary for the maintenance of the natural fertility of the soil. Because they can store carbon, soils are an important component in the C cycle that occurs on a global scale. According to Jobbágy and Jackson (2001) and Scharlemann et al. (2014), the top three metres of the ocean contain a total of 2344 Gt of organic carbon. This amount is greater than the total amount of organic carbon found in the atmosphere (870 Gt C in 2018) and in vegetation (600 Gt C). Because long-term cropping causes 30–60% of the soil organic matter (OC) to mineralise, a considerable amount of carbon dioxide is emitted as a result of agricultural production. This results in a significant amount of greenhouse gas emissions. According to Smith et al. (2015), the amount of carbon dioxide that was released from soils as a direct result of changes in land use is estimated to have been 50 Gt of C (186 Gt of CO_2) between the years 1860 and 2010. These releases were caused by the soils themselves. Because of this, the use of soils for agricultural production is a direct contributor to climate change because of the release of carbon dioxide that occurs as a result of the degradation of organic matter (OM) in the soil. The loss of old OC from soils, notably from permafrost soils and sediments that store significant OC stocks but are starting to thaw, is of particular concern as a result of temperature rises linked with climate change (Schuur et al. 2015). So how to boost agricultural output globally while diminishing soil fertility and releasing greenhouse gases as a result of the loss of OM is a significant concern.

1.1.6 Soils' Function in Agriculture's Sustainable Intensification

Although soil is a limited resource, the demand for the services it provides (such as food, fibre, energy, and physical support) is rising quickly in tandem with the world's population. But as agricultural practices are currently being intensified, soils are already degrading in an unsustainable way, making them less useful to humans in the long run, harming the environment, and making it harder for them to adapt to other environmental stresses.

1.1.7 Enhancing Agricultural Practices and Processes

A number of techniques have been highlighted for enhancing agricultural systems and agronomic techniques. The following are a few of the most crucial actions:

1. Apply fertiliser repeatedly during the growth season or use slow-release fertilisers to more carefully match the N supply to the plant's N demand (Ju et al. 2009). This lessens the degree of soil acidification, lessens the already significant nitrous oxide emissions from soils, and lessens other negative off-site environmental effects.
2. Marginal soils should not be used for productive agriculture; these soils frequently suffer from the worst deterioration and offer the lowest productivity.
3. Practise conservation agriculture (such as reduced tillage) to promote water infiltration and storage, decrease erosion, and decrease the rate at which soil organic matter (OM) is lost from soils. Due to improved soil structure and fertility, this helps to sequester carbon in soils and reduces erosion.
4. To lessen the negative effects brought on by the loss of soil OM, organic compounds should be returned to the soil and superfluous plant products should be removed (Somasundaram et al. 2017).
5. Avoid soil-contaminating practices and clean up already-contaminated areas, according to Wang et al. (2019).
6. Enhancing irrigation water quality and timing to lessen the effects of salt-affected soils (Letey et al. 2011).

1.1.8 Definitions of Soil in Different Systems of Soil Classification

With the exception of living things, regions having perpetual ice that are not covered by anything else, and areas of water deeper than 2 m, the World Reference Base for Soil Resources (FAO 2014) defines soil as any substance that comes into contact with the atmosphere and is less than 2 m of the Earth's surface. This definition does not include water bodies deeper than 2 m. The term "soil" refers to a natural body that may be found on the surface of the Earth, has the capacity to fill space, and can be recognised by either one or both of the following qualities. (i) Solids (minerals and organic stuff), liquids, and gases are the three components that make up soil. (ii) Horizons or layers that can be differentiated from the core material and are in a position to support plants with rooted systems (also called strata). According to the Russian Classification System (Shishov et al. 2004), soil is a natural-historical entity that is solid-phase and has a network of connected horizons which collectively constitute a genetic profile. Soil is the result of the topmost layer of the lithosphere being altered due to the presence of soil-forming processes. Soil is a solid-phase entity with a network of connected horizons that collectively constitute a genetic profile. Some methods of soil categorisation take into account urban soils, such as those that are "sealed" by concrete or asphalt, strata of composts or other fertile materials that are utilised in the production of lawns and gardens, surface layers, mine debris, or garbage heaps (Rossiter 2007). In addition to fine Earth patches within hard rock desquamation fractures, such as those seen in Antarctic endolithic soils (Goryachkin et al. 2012) and underwater soils (Demas 1993), the idea of soils as natural entities also includes very thin films in caves. This is because caves sometimes have very little humidity, which makes it difficult for fine Earth patches to form. As a result, the definition of soil is greatly expanded. Even extrapolating it to other planets has been suggested by soil scientists (Targulian et al. 2010).

1.1.9 Soils, Land Use, and Management Interactions

Many soils have been disturbed by humans either directly or indirectly. However, it is not always easy to distinguish between natural influence and direct and indirect human influence (Smith 2005). Some human endeavours do, however, clearly have a direct effect. These include mining, land management, soil sealing, land degradation, and land use change. The level of land usage has a significant effect on soils as well. Indirect effects of human activity on soils include acid deposition (for instance, sulphate and nitrogen) and heavy metal contamination.

1.1.10 Soil Degradation and Shifts in Land Use

Biomass is removed and soils are disturbed as a land's cover changes, such as from forest or natural grassland to grazing or crops. This in turn causes changes in soil characteristics and biodiversity as well as losses of soil carbon and other nutrients. Increases in soil carbon and nutrients can come about as a result of some land cover conversions, such as afforestation following the abandonment of farmland. Even when a change in cover does not occur due to land use, such as increased grazing intensity or the harvesting and regeneration of forests, the characteristics of the soil might still deteriorate. Approximately 24% (35 million km^2) of the world's land area is degraded. According to Bai et al. (2008), rangeland makes up 20–25% of degraded land, 19% of it is needle-leaved forest, and 23% is broadleaved forest. The type of land cover change, temperature, and mode of vegetation removal (such as land clearing fires, mechanised harvest) all have a significant impact on the size and form of the changes.

1.1.11 Policies for Future Soil Development

Governments around the world are obliged to implement a variety of progressive policies and programmes. Such regulations should expressly acknowledge the significance of regulatory services and cultural services given by soil, such as the cycling of greenhouse gases, in addition to the short-term provisioning services of soil. Such regulations would comprise the following:

1. Promoting soil-friendly agronomic techniques is important. Even while many of these methods are rather well recognised and frequently simple to put into practice, doing so is typically not financially feasible. The degradation of soils is a result of this failure to provide soils an economic value (ELD, Economic of Land Degradation 2015).
2. Limit or avoid turning native ecosystems' soils into agricultural land (ELD 2015).
3. Recognising the importance of the myriad services the soils give will help you control competing uses of soils, such as the urbanisation of prime agricultural soils.
4. Create financial incentives to support strategies that acknowledge the importance of soil ecosystem services (ELD 2015).
5. Encourage more interaction between educators, landowners, and fundamental and applied scientists, especially in the era of "post-truth" (Bouma 2018).
6. Increase public awareness of the importance of soils (McBratney et al. 2014). Society will not treat soils, the resource that supports our very existence, with the respect it merits if it does not recognise the significance of soils.

1.2 Al Toxicities in Acid Soils

The root apex is the main site of Al poisoning. The complexity of how aluminium impacts cellular processes makes it difficult to pinpoint the main effect(s) of Al poisoning. The primary root becomes stunted and the growth of lateral roots is prevented when exposed to Al. Due to the inhibition of cell elongation and cell proliferation, affected root tips are short and stubby. Due to the limited root system that results, the plant is more vulnerable to drought stress because its ability to absorb nutrients and water is hindered. Plants vulnerable to Al toxicity have significantly decreased crop quality and yield (Jovanovic et al. 2006, 2007; Samac and Tesfaye 2003). The grain production of maize in tropical populations has increased with a shift towards cultivars that were more resistant to the acidic conditions of the soil. According to Garvin and Carver (2003), the majority of breeding efforts that were undertaken to increase production in acidic soil focussed on building tolerance to Al toxicity.

Aluminium resistance mechanisms fall into one of two groups (Taylor 1991; Heim et al. 1999; Kochian et al. 2005; Zhou et al. 2007). Both the root exclusion of aluminium and the plant's detoxification of aluminium ions fall under these categories. The creation of a pH barrier in the rhizosphere by the plant itself is one of the exclusion mechanisms that are seen in plants. Another exclusion mechanism is the release of al-chelating chemicals from the roots of the plant. Citric acid and malic acid, as stated by Delhaize et al. (1993), have the capability to chelate aluminium and decrease or avoid the harmful effects of aluminium at the cellular level. The enzyme activity within the plant is shielded from the potentially detrimental effects of aluminium. There is evidence to suggest that organic acids have a role in the process of aluminium exclusion. This role may be played either through the release of aluminium from the root or by the detoxification of aluminium in the cytoplasm.

Increasing crop productivity in the tropics with minimal environmental impact and cost may be possible through genetic adaptation of plants to Al toxicity. This strategy is particularly intriguing for maize because there is germplasm that is tolerant to Al that may be used for selection and genetic research. There are conflicting findings from the numerous investigations that have been conducted to clarify the genetic control of Al tolerance in maize. But as agreed upon by the authors, the trait is quantitatively inherited and is governed by a small number of genes (Lima et al. 1995). The majority of genetic investigations on maize aluminium tolerance have used seminal root development in nutrient solution as a screening tool. The ability to distinguish between tolerant and susceptible maize genotypes using nutrient solutions with high aluminium concentrations has been demonstrated (Martins et al. 1999; Cançado et al. 1999). Even though much research have been done, it is still unclear what genetic factors and molecular processes underlie the genetic heterogeneity in maize Al tolerance.

1.2.1 Root Growth and Aluminium Toxicity

High Al concentrations make it more challenging to evaluate physiological reactions. A significant amount of Al in the nutritional growth medium may become inert through complexation, polymerisation, or precipitation (e.g., with phosphate). As a result, the concentration of free Al that promotes toxicity in plant metabolism may be significantly lower than the level found in the growth media (Mengel and Kirkby 1987). In tolerant genotypes of Zea mays L., low levels of Al can also stimulate root development. According to Tabuchi and Matsumoto (2001), the reduction in root growth is the most severe effect of Al exposure on plant growth in non-accumulators plant species when soil pH is low (Marschner 1995). The monomeric form of aluminium (Al^{3+}) can easily polymerise into the considerably more phytotoxic polymeric form, which is found in maize.

1.3 The Development of Acid Soils

Two of the most common ways that acidic soils can be created are through the use of fertilisers and the removal of crop waste. Because our soils are notoriously lacking in a variety of nutrients, including nitrates and phosphates, amongst others, it is common practice to supplement the soil with fertilisers that contain these elements. In the process known as nitrification, microorganisms in the soil convert ammonium NH_4^+ from fertilisers such ammonium nitrate (NH_4NO_3) or soil organic matter into nitrate and hydrogen ions. Nitrate and hydrogen ions are both necessary for plant growth. The amount of hydrogen ions that are introduced into the soil is variable and is determined by the type of fertiliser used. If nitrate is not absorbed by plants, it has the potential to leach out of the root zone. This is done to ensure that the electrical balance is maintained. Sodium, potassium, or calcium are the cations that frequently drain out of the soil. This is due to the fact that hydrogen ions are held by the soil more weakly. Because of this, there is an increase in hydrogen ions everywhere throughout the system, which results in acidity. If a negatively charged nitrate ion is taken up by the roots of the plant, then the plant must expel an equally negatively charged hydroxide ion in order to maintain the correct level of electrical equilibrium. When the hydrogen ion unites with the hydroxide ion in the soil to generate water. Gazey and Davies (2009) stated that the hydrogen ion will no longer contribute to the acidity of the soil because it will no longer be contributing to the formation of water. The usage of these fertilisers, particularly in excessive amounts, contributes to acidification.

The pH of most plant matter is somewhat on the alkaline side, and plant roots have the ability to take in both positively and negatively charged anions (such as nitrate, phosphate, or sulphate) as well as cations (such as ammonium, potassium, calcium, or magnesium). Plants take in a greater number of cations than anions, and in order to keep their electrical systems in proper working order, they must release a positively charged hydrogen ion into the surrounding soil each time they take in a cation that has a positive charge. There is a net export of alkalinity and residual hydrogen ions are left in the soil when plant material is removed from the soil by grazing, harvesting, or the concentration of dung into stock camps rather from being returned to the soil. This can happen when the plant material is not returned to the soil. This results in a negative alkalinity balance. As the farm's crops are taken year after year, the soil ages and turns acidic (Gazey and Davies 2009). The sheer act of harvesting cereal grains and cutting hay for livestock feed can increase the amount of hydrogen ions in the soil, which over time will cause the soil to become more acidic.

1.3.1 Decay of Organic Matter

Although it takes a long time for decaying leaves, needles, bark, and grass clippings to contribute acid to soil, they are able to do so over the course of several years. As organic matter decomposes, carbon dioxide is released, and when this gas combines with soil moisture, a small quantity of carbonic and other weak acids is produced.

1.3.2 Parental Content

The parent material is what makes up the actual soil. Typically, it's a form of stone, similar to how sand is a crushed rock. Shale and limestone are more alkaline than acidic, so the soil that develops from these rocks tends to have a high pH. Granite soil, on the other hand, will produce more acidic soil because of its acidic chemical composition. This is among the factors that make regions close to rocky shorelines unsuitable for cultivating plants or commodities.

1.3.3 Leaching and Precipitation

Organic material is removed from the soil as rainwater permeates it. A place is on the verge of becoming acidic if it receives more than 30 inches of precipitation annually. This is a very slow process that can take more than a century to complete, although rain-deposited chemical and particulate pollution might hasten it. Although acid rain occurs naturally, most of it is caused by anthropogenic pollution from cars and industry.

1.3.4 High-Yielding Plants

The lime or alkaline elements in neutral to alkaline soil are depleted by high intensity agriculture. To feed themselves, plant roots extract this substance from the soil. If fields aren't allowed to lie fallow and recover, or if they aren't altered with lime and fertilisers, high-yield crops grow more quickly and take even more nutrients from the soil, leaving it very acidic. This is the quickest way to produce acidic soil, yet it's also the simplest to fix with sensible gardening techniques.

1.3.5 Woodland Soils

Soils in forests are often quite acidic, rich in organic matter, and have a low potential for nutrient and chemical accumulation. When the soil quality is poor, the role of ecosystem management will be even more crucial for maintaining their long-term sustainability because organic matter in the topsoil is always more susceptible to change than mineral phases. Directly, in the event of a significant shift in the amount and distribution of rainfall, or even in temperature; however, more subtly and indirectly, through the intricate web of interactions and feedbacks. Climate change can add considerable limits to the bioavailability of nutrients. In this chapter, the unique characteristics of forest soils are analysed and contrasted with those of agricultural soils. This article highlights the dangers that face forest soil.

The application of fertilisers and the elimination of byproducts of crop production are two of the most prevalent ways that acidic soils can be generated. Microorganisms in the soil are responsible for a process that is known as nitrification. Because there are more negatively charged surface sites on soil constituents than positively charged surface sites, nitrate ions, which have a charge of NO_3^-, can be easily removed from the vast majority of agricultural soils. In contrast to positively charged surface sites, negatively charged surface sites are more prevalent. A cation with a positive charge is lost during the leaching process of nitrate. This is done to make sure that the correct electrical balance is kept at all times. Cations such as sodium, potassium, and calcium are the most common elements that are lost when soil is drained. This is because the soil has a less strong ability to hold onto hydrogen ions than it does other ions. As a consequence of this, there is an increase in the number of hydrogen ions all throughout the system, which causes the system to become more acidic. If the roots of the plant are able to take in an ion with a negative charge that is nitrate, then the plant will need to release an ion that also has a negative charge and is hydroxide in order to maintain the correct amount of electrical balance. Gazey and Davies (2009) claim that once the hydrogen ion in the soil combines with the hydroxide ion in the soil to make water, the hydrogen ion will no longer contribute to the acidity of the soil since it will no longer be contributing to the formation of water. The utilisation of these fertilisers, particularly in quantities that are considered excessive, leads to the acidity of the environment.

1.3.6 Soil Deterioration

The process of soil acidification causes soil pH to drop over time, and the issue frequently goes undetected. Acidification can happen naturally over thousands of years, particularly affecting locations with high rainfall. However, with intense agricultural practices, fast acidification can happen over a few years. Either the subsurface alone or the surface soil can be affected by acidification. Surface acidity can be effectively treated and has a significant positive impact on plant development and output. Correcting subsurface acidity is challenging and expensive. Farmers in high-risk locations should find the issue as soon as they can. Soil acidification has increased and developed into a major environmental problem due to ongoing nitrogen (N) enrichment and sulphur (S) deposition. Due to the influencing elements, a complete knowledge of the overall pattern of ecosystem belowground processes in response to soil acidity is still difficult. Meng et al. (2019) used 304 observations from 49 separate studies for a meta-analysis of the effects of soil acidification on belowground processes, focusing primarily on soil cations, soil nutrients, respiration, root, and microbial biomass. The findings demonstrated that adding acid considerably decreased soil pH by 0.24 on average, with forest ecosystems experiencing a smaller pH decrease than non-forest ecosystems. These suggested that soil acidification affects soil carbon processes. According to a meta-analysis, soil acidification has a significant negative effect on below-ground processes and has the potential to reduce soil carbon emission (Meng et al. 2019).

1.3.7 Threatened Areas

The coastal regions with higher rainfall that are used for intensive agriculture are the most lighter-textured sands and loams with low quantities of organic matter. Sands and loams with higher quantities of organic matter are also at risk. The most resilient clay soils are those that range from neutral to alkaline, such as the brigalow soils and the black clay soils that may be found in the Darling Downs and Central Queensland.

The following are some factors that hasten acidification:

- Using ammonium-based nitrogen fertilisers at rates higher than what plants need on naturally acidic soils.
- Leaching of nitrate nitrogen from the root zone, which was initially applied as fertilisers based on ammonium.
- Routine removal from the paddock of waste materials and products produced by plants and animals. Alkalinity is removed from the soil by growing plants. The soil turns more acidic as harvested goods are removed. Where a significant amount of material is removed, as in the production of silage, hay, and sugarcane, the impact is highest.

1.3.8 Impacts

Other than decreases in crop and pasture output, which may be dramatic in acute situations, there are no outward signs of soil acidity. Some nutrients may become less accessible as soils become more acidic, while other components may accumulate to dangerous levels.

The following issues may arise in acidic soils, some or all of them:

- A decrease in the amount of nutrients recycled by soil microorganisms (e.g., a possible decrease in the supply of nitrogen).
- The amount of phosphorus in the soil that plants can access may decrease.
- Substantial deficiency of calcium, magnesium, and molybdenum,
- Plants' capacity to utilise subsurface moisture can be constrained.
- The soil may emit aluminium, which is poisonous to plants and microbes.
- Manganese levels could become poisonous.
- Cadmium (a heavy metal pollutant) uptake by pastures and crops may rise.

According to Kochian et al. (2004), soil acidity is one of the primary causes of land degradation and affects around fifty percent of the potentially arable soils across the globe. Soils frequently become acidic as a result of a natural process known as carbonic acid leaching, which occurs when CO_2 is dissolved in rainwater. The acidification process will continue until there is a balance between the amount of compounds being removed and those being added back in. Basic cations in the environment, such as calcium (Ca) and magnesium (Mg), are taken from the environment as a result of crop harvesting and leaching. On the other hand, these bases are simultaneously supplied as a result of the decomposition of organic waste and the weathering of minerals (Abebe 2007; Sanchez 1977). According to the principles of geology, the acidity of the soil increases as the amount of precipitation does. According to studies conducted in 2015 by Behera and Shukla and in 2014 by Fageria and Nascente, agricultural management techniques, the removal of organic matter, the continuous use of acid-producing fertilisers, contact exchange between exchangeable hydrogen on root surfaces and the bases present in exchangeable form on soils, and microbial production of nitric and sulfuric acids can all be factors that contribute to soil acidity. As a result of soil acidification, increasing N: P and N: K ratios appear to have a negative impact on the abundance of species that are threatened with extinction, as stated by Roem and Berendse (2000).

Soils in tropical regions with high humidity tend to become acidic as a result of the natural leaching of basic cations when there is an abundance of precipitation. According to Agegnehu et al. (2019), when the pH of the soil solution drops below 5, aluminium ion becomes the predominate ion. According to Sivaguru and Horst (1998), excessive aluminium causes harm to the root apex and hinders the expansion of the root system in acidic soils. Crops produced on acid soils are limited by the lack of nutrients and water availability due to the poor root growth, which also results in lower water and nutrient uptake. Reduced crop growth and yield is the overall result (Marschner 2011; Wang et al. 2006). Ethiopia's growing scope and level of soil acidity severely restrict crop production. For instance, farmers have switched to growing oats instead of wheat and barley in several barley, wheat, and faba bean-growing regions of the central and southern Ethiopian highlands (Haile and Boke 2009). Oats are more resistant to soil acidity than wheat and barley.

Climate-related elements like rainfall, temperature, topographic features, and morphological characteristics are the key soil-forming factors that contribute to an increase in soil acidity (Abebe 2007; Brady and Weil 2016). The main soil types that are dominated by soil acidity are nitisol/oxisol regions. In general, non-calcareous parent materials result in naturally acidic soils.

In the humid tropics, managing acidic soils is the main area of concern. However, identifying and describing a problem area does not warrant a significant study effort. It is necessary to conduct a thorough assessment of our current understanding of soil processes related to soil acidity and the management of acid soils (Fageria and Baligar 2008). Even though there has been a lot of research on managing acid soils in South America, Africa (Eswaran et al. 1997b; Tully et al. 2015), Asia, and Australia (Bai et al. 2008; Eswaran et al. 1997c), there is still a lack of more in-depth knowledge and understanding of the issue surrounding managing acid soils and available management options. The development of appropriate solutions for the sustainable management of soil and water resources in acid soil agroecosystems should be the main objective of management of acid soils (MAS). The resource base is being weakened by indiscriminate clearing, improper land use, and poor management of soil and water resources. Lower productivity, deteriorating soil structure, and diminishing soil fertility are all consequences of losing topsoil (Bronick and Lal 2005; Lal 2015). In such situations, intensive agriculture provides a practical and desirable alternative to farming the marginal hillsides. These methods must concentrate on acid soil reclamation, highland erosion management, and organic matter depletion in acid soil. To evaluate the possible effects of new technology on productivity and the environment, it is necessary to analyse the linkages between policies and land use strategies (Lal 2015; Thomas 1995).

Increasing agricultural production to fulfil the demands of a growing population while preserving or enhancing the natural resource base ought to be MAS's ultimate goal. Although reducing biophysical restrictions on soil and water resources is the main goal, applying new technology calls for the creation of suitable socioeconomic and policy frameworks as well as the active involvement of farmers. Exploiting degraded areas that are incapable of supporting crop production due to soil acidity is one of the top priorities in the effort to boost agricultural growth. However, the research and development methods employed up to this point paid little attention to this important issue and were unable to create an all-encompassing solution to slow its progression. Since the problem's scope is now known, it is important to focus on reducing its negative effects and fostering its contribution to food production and natural resource management (Agegnehu et al. 2019).

1.3.9 Acidity Level of the Soil

The capacity of the soil to carry out its tasks in a sustainable way is referred to as soil quality (Lal 2015). The biophysical and chemical characteristics of soil evolve throughout time, and management techniques hasten these changes. One of the main causes of land deterioration in the world is acidity of the soil. According to Sumner and Noble (2003), the world's ice-free soils are acidic to a degree of about 30%. Soil acidity primarily affects locations with warm climates, including subtropical and tropical areas. According to Eswaran et al. (1997a), 32% of all arable land worldwide is acidic. Ultisols, Entisols, and Oxisols make up about two-thirds of all acidic soils on Earth (Rengel 2011). About 3.75 million km^2 or 14.3% of Africa's total land area is covered with Oxisols, also known as ferralsols. From little to high fixation, 10.76 million km^2 or 35% of Africa's total land area is characterised by P fixation, and of this, 8.23 million km^2 is typical of high P fixation (Eswaran et al. 1997c). Similar to tropical American soils, which are mostly acidic and have few plant-available nutrient reserves. The delicate soils and natural resources are under a lot of stress because of the urgent need to expand agricultural production. These soils are prone to compaction, have hazardous levels of Al and Mn, and are quickly eroded (Thomas 1995). Land degradation has a significant negative influence on biodiversity in addition to reducing the productive potential of agricultural land, rangelands, and forest resources (Akhtar et al. 2011). It has a negative impact on the ecological productivity and integrity of vast tracts of land or landscapes that are used by humans.

1.3.10 Types of Acidity in Soil

There are primarily two types of acidity in acidic soils:

Active acidity: This results from the soil solution's high concentration of H^+ ions, which can be attributed to hydrolytic acid salts, water-soluble organic acids, and carbonic acid (H_2CO_3). The pH level of a water suspension or soil extract can be used to calculate it. It has a direct impact on how plants and soil microbes flourish.

Exchange acidity: This refers to the amount of H and Al ions that are adsorbed on soil colloids. Ions in adsorption and soil solution are in balance (active and exchange acidity), allowing for a smooth transition from one form to another (Fig. 1.2). Because adsorbed H and Al ions seep into the soil solution, soil acidity is also known as adsorbed, potential, or reserve acidity. Such an equilibrium situation is quite useful in practice since it serves as the foundation for the soil's buffering capacity or resistance to pH shift. The active acidity in the soil solution is where bases (such as lime) initially react. The acid is then progressively released into the active form by the reserve acidity pool (Somani 1996).

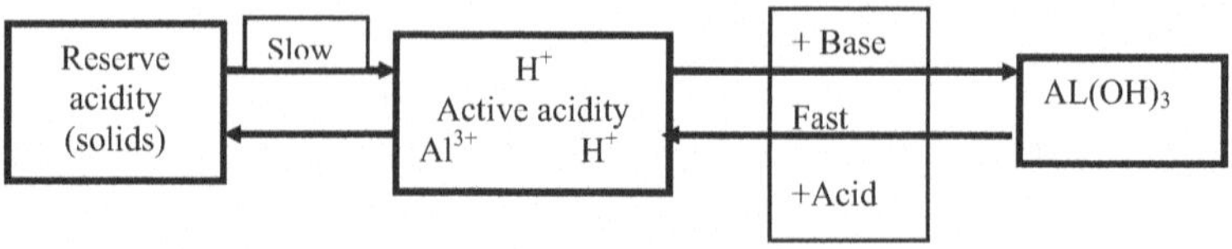

Fig. 1.2 Relationship of equilibrium between the acidity of the exchange medium (reserve) and the acidity of the solution (active). *Source* Somani (1996)

1.4 Deposition of Acid

According to research that has been carried out over the course of the last many years (Driscoll et al. 2001), acidic deposition has an effect on forested landscapes as well as aquatic ecosystems in North America, Europe, and Asia. This difficult issue is an indication of a new type of environmental concerns that have a scale that spans multiple regions and defies regulation in an easy-to-understand manner. Acidic deposition is a potential consequence of air pollution on a global scale and has the potential to have a large effect on a regional scale. It differs greatly from one place to another.

1.4.1 Aquatic-Terrestrial Linkages

There exists a clear correlation between the pace of ecosystem development of acid neutralising capacity (ANC) and the rate at which acidifying compounds are deposited from the atmosphere. Therefore, a lot of the impacts of acid deposition depend on how quickly compounds that cause acidification accumulate in the environment. Mineral weathering, cation exchange, and the immobilisation of SO_4^{2-} and N are some of the key terrestrial processes that produce ANC, a measurement of the ability of water or soil to neutralise inputs of strong acid (Charles 1991). Mineral weathering is just one example of a terrestrial activity. The neutralisation of acid occurs during the solution phase of the process, and the rate at which this happens has a direct correlation with the amount of water that moves through terrestrial and aquatic ecosystems. The effects of acidic deposition on surface waters might vary depending on the season and the stream velocity. Surface waters typically reach their most corrosive state in the springtime, following the occurrence of precipitation and snowmelt events. As a consequence of precipitation, which also results in the thawing of snow, the water table may rise from the subsoil into the higher soil layers, moving upwards. When compared to the subsurface, the success rate of acid-neutralising processes like mineral weathering and cation exchange is typically lower at these horizons. Both chronic and episodic acidification can be caused by strong inorganic acids, which can come from air deposition, naturally occurring processes, or a combination of the two.

1.4.2 Soil Sulphur Buildup

According to watershed input–output budgets that were built in the 1980s for northeastern forest ecosystems (Rochelle and Church 1987), the quantity of sulphur that was exported by surface waters (mainly as SO_4^{2-}) was virtually identical to the amount of sulphur that was deposited from the atmosphere. According to the findings of those investigations, putting restrictions on emissions ought to make atmospheric S deposition lessen while concurrently cutting down the quantity of S that gets into surface waters. The Clean Air Act Amendments of 1970 led to a reduction in the amount of sulphate that was deposited into the atmosphere, which in turn led to a decrease in the concentration of SO_4^{2-} in surface waters throughout the Northeast (Likens et al. 1990; Stoddard et al. 1999). These results were published in the journal Environmental Science and Technology. Recent research on the mass balance of watersheds in the Northeast (Driscoll et al. 1998) found that the loss of SO_4^{2-} from watersheds is greater than the deposition of S in the atmosphere.

1.4.3 Ecosystem Recovery

The acidic deposition causes a change in the chemical makeup of both the forest and the aquatic ecosystems, which is detrimental to both. The amount of sulphur that has been deposited from the atmosphere in the northeastern United States has increased by more than a factor of five in the past 150 years (Charles 1991). As a result, the majority of acid-sensitive ecosystems have been subjected to significant quantities of strong acids for an extended period of time. Some aquatic ecosystems in the Northeast are showing evidence of chemical recovery as a result of the considerable drop in atmospheric S deposition that has occurred since the 1970 Clean Air Act Amendments (CAAA) (Stoddard et al. 1999). This decline in S deposition has been attributed to the CAAA. It would appear that the passage of important chemical thresholds is scheduled to coincide with the beginning of any negative consequences that may have an impact on biotic resources. For instance, according to Cronan and Schofield (1990) and Cronan and Grigal (1995), if the molar Ca:Al ratio in the soil water is less than one and the soil base saturation percentage is less than twenty percent, this implies that acidic deposition poses a hazard to the forest plants. In addition, acidic deposits in surface water pose a threat to aquatic biota when the pH is lower than 6.0, the ANC is lower than 50 eq L^{-1}, or the concentration of inorganic monomeric Al is higher than 2 mol L^{-1} (MacAvoy and Bulger 1995). The technique will be broken up into two stages, notwithstanding the scarcity of information and expertise in acidification recovery, particularly at the ecosystem level. This is because the approach will be separated into two parts. First, a phase of chemical recovery in forest and aquatic ecosystems will receive assistance from reduced acidic deposition as a result of emission restrictions since these limits will help reduce acidic deposition.

The length of time it will take for ecosystems to recover after the initial phase will vary greatly from one another and be determined by the following factors:

- How much atmospheric deposition has decreased.
- The regional decrease of base cation exchangeable soil pools.
- The regional rate of mineral weathering and base cation inputs from the atmosphere.
- The quantity and rate of S and N released from soil pools as $SO_4{}^{2-}$ or $NO_3{}^{-}$ to drainage waters (Galloway et al. 1983).

1.4.4 Acid Deposition and Acid Rain

Acid deposition from the atmosphere can occur in two ways: either as a dry coating on dust, smoke, or other aerosols (small, microscopic airborne particles) or as a wet deposition in precipitation (rain, snow, or fog). Both in Europe and the USA, acid rain was a big problem in the last few decades of the twentieth century. The issue in these places has significantly decreased because of strict emission control rules. However, acid rain is still a significant issue in several emerging nations, particularly China.

1.4.5 What Causes Acid Rain?

The origins of the acidic materials are sulphur dioxide (SO_2), ammonia (NH_3), nitrogen oxides (NO_x), and acidic particles that are discharged into the atmosphere when fossil fuels are burned in power plants and cars. The combustion of fossil fuels, especially coal, in electric power plants is responsible for around two-thirds of the SO_2 and one-fourth of the NO_x emissions in the US. Acid rain is the consequence of these gases reacting with oxygen, water, and other elements in the atmosphere to produce various acidic chemicals. The outcome is a moderate mixture of nitric acid and sulfuric acid. Prevalent winds carry sulphur dioxide and nitrogen oxides across state and international borders, often across hundreds of kilometres, when they are emitted from power plants and other sources. The majority of acid rain is deposited away from power plants. Many of these are in the midwest of the USA, where acid rain is widespread both there and along the entire east coast. The amount of acid rain has decreased, as regulations on power plant emissions have increased, from 28.8 million metric tonnes in 1978 to 17.8 million metric tonnes in 1998.

1.4.6 *Rain's Acidity*

Due to fast industrial development, acid rain has since the middle of the twentieth century grown to be a significant global environmental issue (Blank 1985; Duan et al. 2016). According to research done by Zhao et al. (2009) and Yang et al. (2012), the primary sources of acidic and acidifying pollutants are the combustion of fossil fuels, which releases sulphur dioxide (SO_2) and nitrogen oxides (NO_x), and agricultural operations, which produce ammonia (NH_3). According to research by Gao et al. (2018), even while SO_2 and NO_x emissions have fallen in Europe and North America, a rise in those emissions is being caused by the burning of coal in many developing countries. The natural process of soil acidification has been accelerated (Grieve 2001; Kunhikrishnan et al. 2016) as a result of increases in nitrogen and sulphur deposition that were caused by human activity. Both Stevens et al. (2010) and Liang et al. (2018) state that the acceleration of soil acidification by humans has a negative impact on biogeochemical cycles and the operation of ecosystems. For this reason, it will be essential to have an understanding of the broad patterns in ecosystem processes associated with acid deposition across a variety of habitats in order to accurately predict the behaviour of ecosystems in response to future global change. However, there hasn't yet been a thorough worldwide analysis of how acid deposition affects ecosystem processes. Ecosystems are impacted by acid deposition in a variety of ways, particularly for underground processes. First of all, increased soil H^+ intake as well as $SO_4{}^{2-}$ and $NO_3{}^-$ induced by acid deposition may have a direct impact on microbial activity (Kuperman and Edwards 1997). Second, base cations (such as K^+, Mg^{2+}, and Ca^{2+}) will compete with H^+ from acid deposition for replacement, increasing base cation leaching out of the soil. This further decreases the capacity of the soil to buffer acid (Driscoll et al. 2003) and the availability of nutrients (Likens et al. 1996). Third, increased soil acidification caused by ongoing acid deposition may cause free Al^{3+} and Fe^{3+} to be mobilised and released into the soil solution. This toxic element buildup in the topsoil may eventually hinder root development and microbial activity, which would decrease soil respiration (Godbold et al. 1988; Kochian 1995; Poschenrieder et al. 2008).

1.4.7 *Precipitation*

Precipitation acidity is quantified in pH units, where

$$pH = \log(H^+)$$

where H^+ is the number of hydrogen ions that are dissolved in a watery weak solution. The pH value determines how acidic or basic precipitation is, with a lower pH being more so. Pure water has a pH of 7.0, but pure rain has a pH of 5.6 due to carbon dioxide dissolving in water to produce the weak acid carbonic acid, H_2CO_3.

$$CO_2 + H_2O = H_2CO_3$$

In extreme circumstances, the pH of precipitation from highly contaminated air can be less than 2. Precipitation typically has a pH between 4.4 and 5.8. The capacity of soils to neutralise acid has a significant impact on the way acid rain affects the environment. The most efficient materials are limestone and other calcium carbonate-containing rocks and soils. Calcium carbonate reacts with acids to form neutral chemicals and carbon dioxide. For instance, calcium carbonate reacts with sulfuric acid and hydrochloric acid:

$$H_2CO_3 = CO_2 \text{ gas} + H_2O$$
$$CO_2 + CaCl_2 + H_2O = 2HCl + CaCO_3$$

If the soil is primarily composed of quartz, like sand, it will resist weathering and lack the bases necessary to neutralise acid. If the soil contains very little base, such as limestone, the acid is only mildly neutralised or, over time, not at all. Acids in rain can also be neutralised by mature forest soils. Such soils contain a lot of humic acids and are acidic. According to Krug and Frink (1983), mature soils in forests in New England or Norway have a pH of 3.8 and may hold as much acid as would fall in 1000 years of rain at 1 m/year with a pH of 4.3. These soils have a high level of acid rain resistance. Thin alpine soils devoid of carbonates and humic acid, as well as the granite rocks that they are above, cannot neutralise acid rain.

In many parts of North America and northern Europe, soil and water have become more acidic, and this is generally attributed to acid rain. However, the same variables that have long been recognised to significantly acidify soils through the natural processes of soil formation are also frequently thought to make landscapes sensitive to acidification by acid rain. Regions that are experiencing acidification have also recovered from extensive and reckless land usage. There is proof that long-term acidity brought on by changes in land use and the ensuing vegetation succession is superimposed by acid rain. In order to evaluate the benefits anticipated from suggested reductions in emissions of oxides of sulphur and nitrogen, it is necessary to thoroughly analyse the interactions between acid rain, acid soil, and plants on a watershed basis.

1.4.8 Nutrient Leaching

Acids remove soil's nutrients. They deprive plants of vital nutrients including calcium, potassium, magnesium, and trace metals by being taken deeper into the Earth or into streams. Metals are typically affixed to clay particles created by the weathering of rocks. Ion exchange is the process through which the acid's H^+ anion swaps out the metal ions in the clay. The metals show up as sulphates and nitrates in groundwater. One Ca^{2+}, Mg^{2+}, or two K^+ anions are replaced by two H^+ anions.

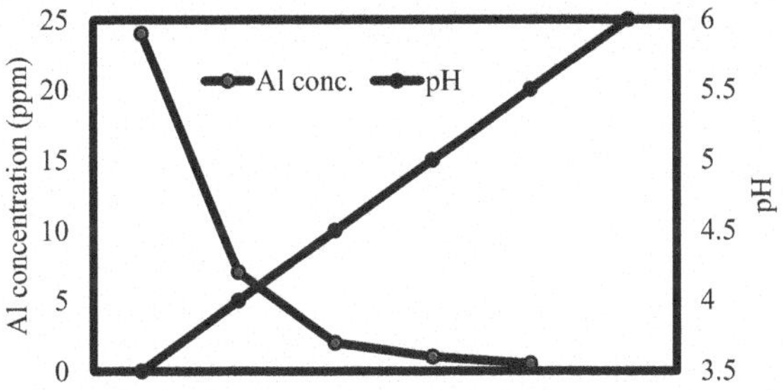

Fig. 1.3 Relationship between pH and Al concentration in soil

Table 1.1 Soil pH (in water) as affected by cropping practices

Depth	Soil pH of natural vegetation area	Soil pH of cropped area
Up to10 cm	5.6	4.5
10–20 cm	5.1	4.3
20–30 cm	5.0	4.2
30–50 cm	4.8	4.6
50–70 cm	4.5	4.5

1.4.9 Al and pH Relationships

Soil pH drops to a point where aluminium dissolves. Soluble aluminium can increase significantly in response to even a smallpH drop (Fig. 1.3). This kind of aluminium restricts access to nutrients and water and hinders the growth of roots. Grain size decreases, production declines, and poor crop and pasture development are all results of inadequate water and nutrients. The effects of aluminium poisoning on crops are usually most noticeable during seasons with a dry finish because plants have limited access to water stored in the subsoil for grain filling. The primary effects of agricultural practices on soil pH are indicated in Table 1.1.

1.4.10 Acidification of the Soil's Main Effects

A slightly acidic soil pH is ideal for plant growth because it promotes the activity of soil microbes and makes plant nutrients more accessible. In farming systems, acid conditions will lead to a decrease in plant yield. Due to this, any sort of productive agriculture will be less profitable and more dependent on fertilisers. The availability of vital nutrients will increase if soil pH is adjusted to a more optimum pH range. When the pHCa is 4.8 or lower, nutrient poisoning can happen in acidic soils (Slattery et al. 1999). Aluminium (Al) and manganese (Mn) toxicity are the two main toxicities in acid soils (Slattery et al.1999). Aluminium and manganese are toxic to plant roots when they are more readily available in the soil solution in strongly acidic soils (pHW

4.3). Root stunting is the most typical plant sign of aluminium toxicity on acidic soils (Slattery et al. 2001). Reduced root growth obstructs the uptake of nutrients and water, which lowers yield. Some plants can tolerate higher quantities of Al in the soil solution than others, but as the pH decreases, so do the farming possibilities. If this occurs, soil degradation may progress to an irreversible state. These methods will result in a worse soil resource, one that cannot be restored to its original condition and is expected to maintain present crop productivity. It is obvious that there is a need to maintain this precious resource as much as possible, and one tactic is to avoid having very acidic soils. Acidic soils make it difficult for productive plants like lucerne, phalaris, canola, and barley to establish and develop. Low pH and toxic aluminium both have a permanent negative impact on lucerne establishment (Yokota and Ojima 1995). As a way to halt the acidification process, deep-rooted perennial pastures are being grown (Ridley et al. 1990). If certain plant species cannot grow because the soil pH is too low, nitrate leaching will persist, speeding up acidification and increasing aquifer recharge, both of which would worsen the salinity issues on dry land. Recent research by Slattery et al. (1998) has demonstrated that permitting soils to stay at a pHCa of 4.0 or lower for a lengthy period of time (10 years) has serious consequences for soil clay loss from primary clay minerals. By allowing weathered soils to become acidic over the course of one agricultural generation, there appears to be little damage done to the mineral structure of the soil. It is evident that if acidification is allowed to continue unchecked for prolonged periods of time, the soil will continue to lose clay and increase its silica content. To stop this irreversible soil damage, liming materials must be used to keep the soil from becoming excessively acidic. In all of Australia's major grain-producing regions, the extent of acidity is developing in the topsoil layers of cropping soils (Coventry 1991; Slattery and Burt 1997). Although the subsurface soil layers in many of these areas are frequently alkaline, they can also be acidic and hostile because they can be sodic, saline, or both. The rich, nutritious soil environment for plant roots is provided by the surface soil layers, and they must be safeguarded from irreparable clay loss owing to soil acidity. The possibility of developing acidity in previously neutral subsoil is another issue if surface soil acidity persists without abatement. Due to the slow flow of neutralising minerals like lime and base-rich organic matter down the soil profile, subsurface acidity is difficult to remediate (Ridley and Windsor 1992; Myers and De Pauw 1995). Because they are typically found in recharge zones, some granitic soils already have severe issues with subsoil acidity, which prevents plant growth and exacerbates dry land salinity. Despite the fact that agriculture has made these granite subsoils' pH worse, recharge control on these soils will need special consideration.

1.5 Production of Food

There has long been discussion and concern about the alleged limits to feeding a growing world population. Tertullian, a respected church leader, brought up the subject as early as the third century AD. Following Malthus, the discussion gained steam in the late eighteenth century, and more recently, with the publication of Paul Ehrlich's Population Bomb. However, global food production increased more quickly than both per capita consumption and population growth. From 2.5 billion in 1950 and 3.7 billion in 1970, the population climbed to about 8.0 billion in 2022. According to the UN population forecasts, which were taken from the medium version of the 2008 release used here, there may be 9.15 billion people on the planet in 2050 (Alexandratos and Bruinsma 2012). Thus, over the next 28 years, we anticipate a growth of 1.15 billion. Because of this slowdown, global agriculture will increase at a slower rate than in the past. After factoring for waste, animal feed, and non-food uses, the average amount of food available per person around the world in 2005–2007 increased to 2770 kcal/person/day. Therefore, in theory, there is enough global aggregate food consumption to feed almost everyone. Despite this, 2.3 billion people live in nations with a per capita energy intake of less than 2500 kcal, 0.5 billion in countries with a per capita energy intake of less than 2000 kcal, and 1.9 billion people live in countries with an energy intake of more than 3000 kcal. The causes are generally acknowledged: Agricultural poverty, which has various characteristics but is associated with inadequate access to food produced in other nations and inability to grow agriculture in many low-income countries (Alexandratos and Bruinsma 2012).

1.5.1 Soil as a Source of Nutrients and Fibre

The biggest strain on soils is caused by the fast rising need for food and fibre. This is attributable to an expanding human population as well as an increase in per capita calorie consumption brought on by rising prosperity and dietary changes. For instance, the daily calorie intake per person in China went from 1400 kcal in 1963 to 3100 kcal in 2013 (FAO 2018). Soils are inextricably tied to food production. According to estimates, humans get 98.8% of their daily calories from soil (2849 kcal per person) and only 1.2% from aquatic sources (35 kcal per person, FAO 2018). Soil pressure increases since there is a greater need for food and fibre. The rate of land use change has substantially accelerated in the previous 300 years, and in the beginning, increases in food production were mostly achieved by expanding the area under cultivation. Currently, permanent meadow and pasture make up around 3300 million ha (or 25% of ice-free soil), while croplands make up about 1600 million ha (or about 12% of the total ice-free soil) (FAO 2018). However, since the middle of the twentieth century, it has been much easier to feed the world's rapidly expanding population by intensifying production (here, "agricultural intensification" is defined as an increase in crop yield production per unit of area), as opposed to by continuing to expand the land area. In fact, despite the fact that the world's population expanded by over three times between 1950 and 2015 (from 2.5 to 7.3 billion), the percentage of ice-free soil used for crops only slightly, from 9.2 to 12.2% (Fig. 1.4). The "Green Revolution," or intensification of agriculture, has involved the use of better genotypes, fertilisers, pesticides, water, and other agronomic practices.

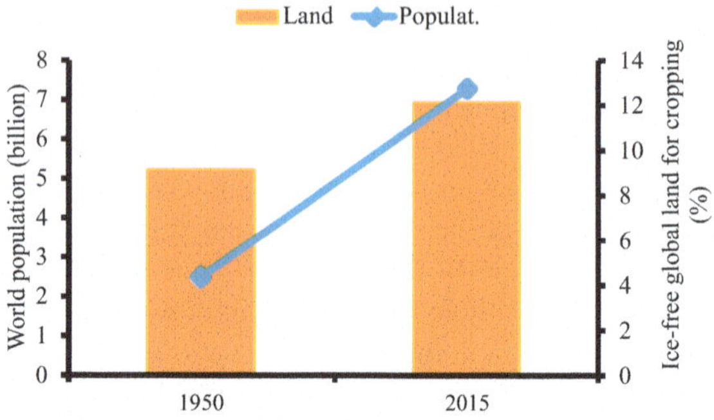

Fig. 1.4 Increase in World population and land utilised for cropping between 1950 and 2015. *Source* OurWorldInData.org/land-use

The majority of our advancements in cereal production have been made possible by yield improvements. Since 1961, the typical cereal yield has increased by 175%. In comparison to 1961, the world can now produce roughly three times as much cereal from a given amount of land. In certain areas, this growth has been considerably more pronounced (Ritchie 2017). However, it should be remembered that during the Green Revolution, environmental effects (such as excessive fertilisation and pesticide leaching) were mainly disregarded. In future, it is anticipated that yield increases, rather than area expansion, will continue to account for the majority of the increase in food and fibre production needed to attain global food security.

Since the increase in cereal output has not been very significant, area expansion is likely to remain crucial. Intensification puts more strain on the soil and frequently results in degradation. This affects the long-term provision of food and fibre necessary for achieving global food security as well as the soils' capacity to offer their numerous ecosystem functions. It is also crucial to take into account land use changes for food production in terms of ecological repercussions and planetary boundaries (Steffen et al. 2015).

Figure 1.5 shows the trends in increase percent of cereal production, crop yield, population, and land used for cereals over base year 1961.

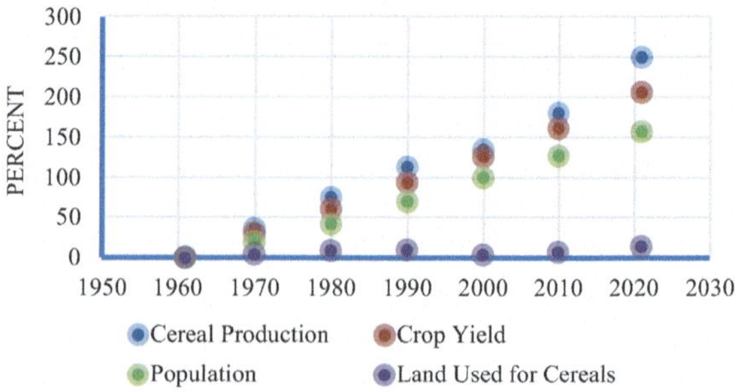

Fig. 1.5 Trends in increase percent of cereal production, crop yield, population, and land used for cereals over base year 1961. *Source* OurWorldInData.org/crop-yields

1.5.2 Population and Income Are the Primary Determinants

A slower rate of growth in agricultural consumption is anticipated as a result of the anticipated decline in world demographic growth during the ensuing 30 years (down from 1.6% between 1963 and 2007 to 0.75% per year between 2005 and 2050). It is crucial to remember that the slowdown in global population growth is a combination of certain countries' continued rapid development and others' slowing down or declining. The majority of nations with fast-growing populations in future are those that already have high rates of undernourishment and insufficient food intake. For example, according to the FAO (2006a, b) revision of the UN Population Prospects from 2002, Sub-Saharan Africa will have 1557 million people, or 17% of the world's population, in 2050. In 2050, it is anticipated to reach 1753 million, or 19% of the global population. The region's anticipated population in 2050 has been increased further in the 2010 revision (UN 2011), reaching 1960 million or 21%. The expected improvements in global food security can be considerably changed by such dramatic changes in several food-insecure nations.

1.5.3 Changing Dietary Patterns: Towards Repletion and Over-Nutrition

From 2005–2007–2050, the total demand for agricultural products is projected to increase at 1.1% year, down from 2.2% annually in previous decades. The key forces behind such anticipated changes are population expansion, increases in per capita consumption, and dietary changes that result in a greater consumption of cattle products. There won't be much room for additional rises in per capita consumption after significant portions of the world's population have reached those levels. While most emerging nations, including India, have not yet fully shifted to levels of meat consumption characteristic of western diets, most industrialised nations have largely finished the transition to livestock-based diets. Thus, even after taking into account increases in per capita consumption and dietary changes, the rise of global food production required to meet the growth of demand will be less than in the past. Worse, considerable portions of the population are affected by the obesity epidemic while undernourishment is still extensively widespread in a number of emerging countries with low national averages. These nations are dealing with a double burden of malnutrition, which presents new difficulties and strains for their healthcare systems.

While it is anticipated that developing nations would consume more livestock products, there may still be significant gaps between them and wealthy nations in terms of their consumption of meat and milk. That is, the adoption of western-style livestock-based diets in many emerging nations will be gradual. Some significant nations have advanced quickly in that direction, including China and Brazil. However, as they increase their consumption levels, they will inevitably slow down.

This tendency will be strengthened for aggregate demand by the likelihood that both countries will begin to experience demographic decline starting in 2020.

It won't be any simpler to achieve these output increases than in the past; in fact, the opposite is frequently true for a number of reasons. Crop yield growth has slowed down significantly, and concerns have been raised that the trend may not change. The question is not whether yields would increase at the previous high rates, as they most likely would not, independent of the various nations and crops. Instead, the question is whether the reduced growth potential and slight increases in arable land are adequate to meet the higher needs. Furthermore, there is a significant possibility that climate change will have a detrimental impact on the ability of agricultural resources to provide food in many parts of the world. The sustainability of the food production system is under scrutiny generally. The likelihood of continuing to use high levels of external inputs in production, increase the proportion of animals in total output, increase cultivated land and irrigation, and transport goods over vast distances is questioned. The necessity for "sustainable intensification" of output is frequently argued for (Royal Society 2009; Nature 2010; Godfray et al. 2010). Will the anticipated production volumes be viable to reach? We'll outline certain combinations of projected yield growth, water use, and land use that could support the production estimates. Figure 1.6a, b shows the per capita consumption and demand of cereals in the developing, developed countries and the world, between 1970 and 2020 and expected from 2020 to 2050.

1.5.4 Yields Are Decreasing

While yield growth is slowing, resource restrictions on agricultural output are more severe than in the past (Fig. 1.7). This is the main cause of people's expressed concerns that the world's food output may not be sufficient to feed a growing population and maintain food security for everyone. In this regard, it is important to keep in mind that the ability of the globe to produce food is only distantly related to food security, if at all. This is true for at least two reasons:

(a) There are enough extra resources for food production in some regions of the world that are just waiting to be put to use if only institutional and economic frameworks would allow it.
(b) Production constraints exist and frequently result in Malthusian situations at the local level because they frequently have a negative impact on not only the possibility of increasing food supplies but also serve as the main drivers of overall development and the emergence of poverty traps. Production limitations continue to play a significant role in determining food security.

It is not a good metric for assessing future prospects to argue that providing food security for a growing population will become more challenging due to the fact that there is currently less untapped land and water and less room for yield development

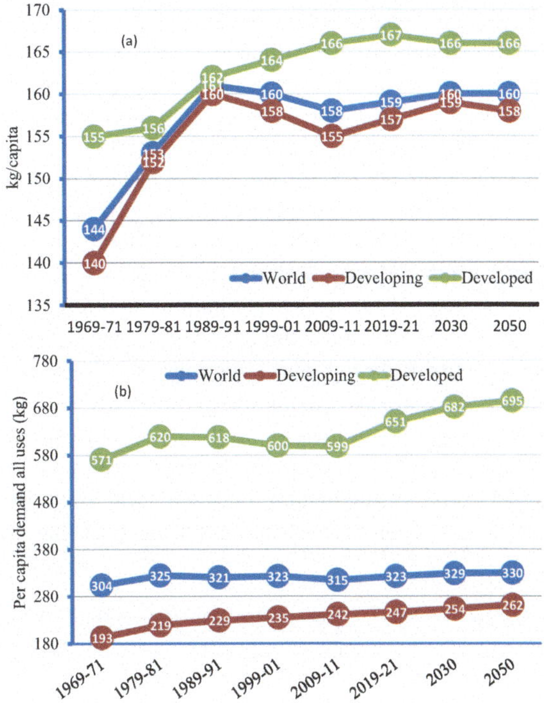

Fig. 1.6 **a** Per capita consumption of cereals in the World, developing (D'ing) and developed (D'ed) countries, **b** per capita demand of cereals for all uses. *Source* Modified over, Alexandratos and Bruinsma (2012)

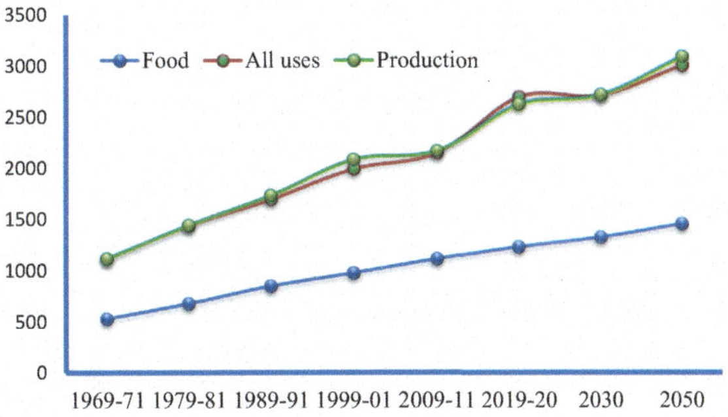

Fig. 1.7 Demand for food (human consumption, in blue), all uses (red) and cereal production (green) in the World. *Source* Alexandratos and Bruinsma (2012)

than in the past. Instead, the question is whether there are enough resources to meet future demands, which will increase at a considerably slower rate than in the past. Studies on the suitability of land for crop production conducted by the International Institute for Applied Systems Analysis and FAO for the Global Agroecological Zones study (GAEZ) and updated version (Fischer et al. 2002, 2011) indicate that there are 7.2 billion hectares of land globally that have the potential to produce rainfed crops. Of this land, 1.6 billion ha are currently used for crop production, including irrigated crops. According to the findings of the GAEZ review, approximately 75 million acres of land that is now being utilised have been deemed unsuitable (Table 1.2). A fraction of this type of inappropriate terrain that is now in use is comprised of irrigated desert. The remaining area is estimated to be 2.3 billion ha. However, not all of it should be seen as having the potential to be useful for the production of crops for two reasons. There are approximately 220 million hectares of such land in use today, of which 47 million hectares are irrigated. However, out of the remaining 2.9 billion hectares, there are 1.5 billion hectares that are of poor quality for rainfed agricultural production and are classified as marginally suitable or very marginally suitable. To begin, there are 2.8 billion hectares of land that are covered by forests, protected areas, or other non-agricultural uses that have the potential to grow in future. These uses include infrastructure, habitations, and other similar purposes.

The globe had a plentiful supply of grains up until around 2006, but yield growth was steadily declining. This is demonstrated by the real price of grains having been on the fall since at least the middle of the 1980s and then remaining relatively constant until the significant increase in the price index in the years 2007–2008. Worldwide averages represent a compilation of numerous predictions for the various nations and cereal crops, distinguishing between yield gains from rainfed and irrigated areas. Information about important variables is provided in Table 1.3.

Table 1.2 Global potential area for rainfed crops (million ha)

Land class	Rainfed	Irrigated	Spare usable	Forest/protected/built-up	Total
Best land	381	61	349	524	1315
Good land	682	135	1064	1299	3180
Marginal land	177	47	1522	992	2738
Not suitable	43	32	–	–	75

Source Chapter 4, (from the GAEZ). Global Agroecological Zones

Table 1.3 Important variables during different periods

	2005/2007	2050	2080	2100
Population, UN 2010 revision	6574	9300	9960	10,125
kcal/person/day	2770	3072	3202	3230
Cereals (kg/capita)	157	159	160	161
Cereals, different uses (kg/capita)	312	329	337	337
Meat (kg/capita)	38.67	49.3	55.1	55.0
Meat Production (Mt)	258	455	524	564
Cereal production (Mt)	2067	3008	3183	3210
Cereal yield (kg/ha, rice paddy)	3320	4300	4830	5250
Arable land area (Mha)	1590	1660	1632	1615

Source Alexandratos and Bruinsma (2012), Sharma (2023)

Table 1.4 Undernourishment in developing nations (%)

Region	1990–1992	2005–2007	2015	2030	2050
Developing countries	19.7	15.9	11.7	7.9	4.1
Excluding SE Asia	19.1	13.5	9.8	6.9	4.1
Africa (Sub-Saharan)	33.6	27.6	21.4	14.5	7.1
N. Africa/Near East	6.0	7.4	6.0	4.7	3.4
S. America and Caribbean	12.2	8.5	6.3	4.1	2.5
South Asia	21.5	21.8	16.1	10.5	4.2
East Asia	19.2	11.0	6.8	4.2	2.8

Source Alexandratos and Bruinsma (2012)

1.5.5 The Frequency of Malnutrition

According to the FAO assessment (FAO 2010), 827 million people in developing nations (16% of their population) were estimated to be undernourished in 2005–2007, when daily food consumption was 2620 kcal on average (Table 1.4). The three-year average of 1990–1992 was 810 million; this estimate is not materially different from that figure, which was higher at 20% of their overall population at the time. The 1996 World Food Security (WFS) set the goal of reducing the number of undernourished people in developing countries by 2015 at the latest using the 3-year average of 1990/1992 as a base.

1.5.6 Demographics

According to the low, medium, and high predictions, the world's population will increase from the 6.6 billion in 1950 to 8.0, 9.15, and 10.5 billion in 2050, respectively (Fig. 1.8). A moderate prediction suggests that a very abrupt decrease in global demographic growth is likely. By the final decade of our forecasts, 2040–2050, further slowing will reduce it to 0.4% each year. The world population is predicted to reach a peak of 9.4 billion people in the year 2075 and then start steadily declining to 9.2 billion people by the year 2100, according to the Medium Variant estimate.

Even if the growth rate has drastically decreased, the absolute annual increments are still substantial. In the decade that ended in 2010, 79 million people were added to the world's population annually. This number is expected to stay over 50 million each year through the middle of the 2030s. By 2050 (Fig. 1.9), a quicker drop should reduce the annual increase to 27 million. Almost all of these increases will occur in emerging nations. In the late 2040s, the developed world's population will begin to shrink. There will be more diversity among the developing nations. In the early 2040s, East Asia will have switched to negative demographic growth. Past 2050, South Asia and Latin America will both have negative growth, with the Near East and North

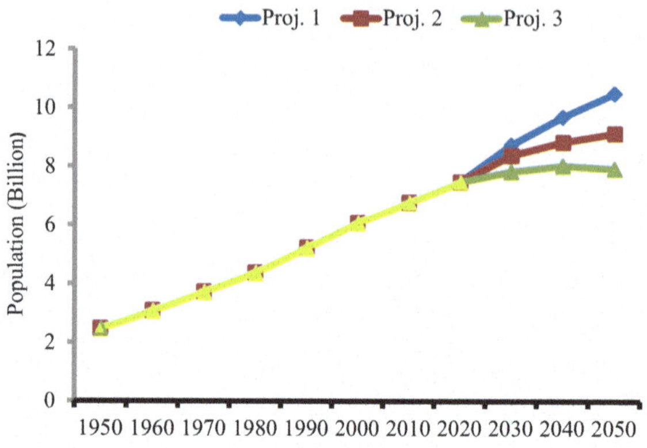

Fig. 1.8 World population projection between 1950 and 2050. *Source* UN (2009)

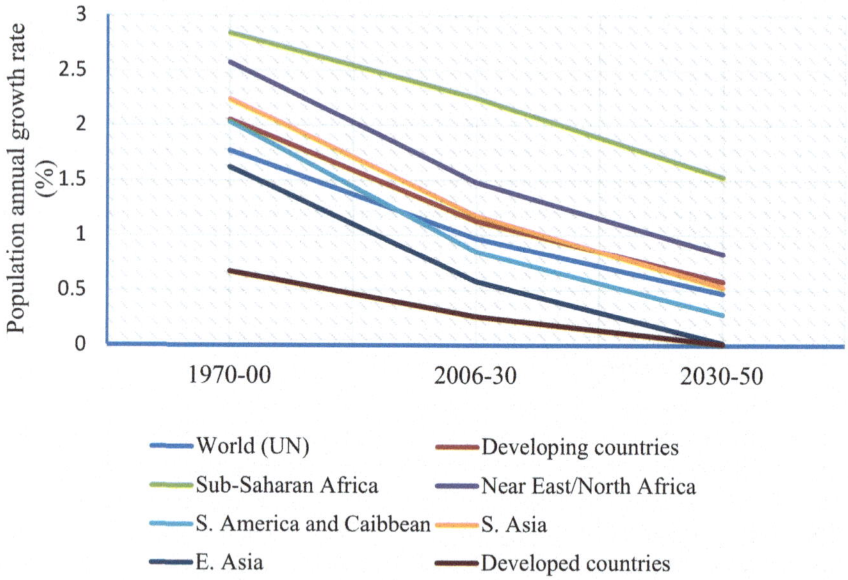

Fig. 1.9 Population growth rate in different regions of the world. *Source* Alexandratos and Bruinsma (2012)

Africa region moving in the middle of the 2080s. Sub-Saharan Africa, which will have 2.2 billion people and is still adding about 11 million people annually, would essentially be the only region that is still experiencing population growth by the year 2080.

Demographic predictions for some nations, mostly in Africa, indicate that their populations in 2050 would be very sizable multiples of their present populations (Fig. 1.10). This possibility raises the important question of whether significant advancements in food consumption per person and nutrition could be made in the

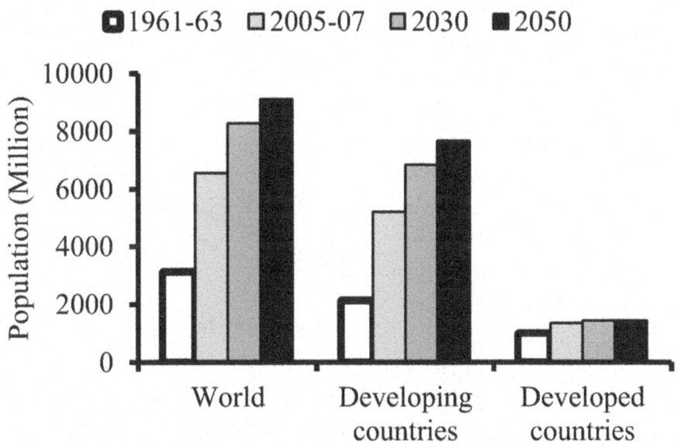

Fig. 1.10 Population increase in the world over time. *Source* Alexandratos and Bruinsma (2012)

near future, especially in nations that heavily rely on agriculture for employment, income, and food supplies as well as those with limited agricultural resources.

1.5.7 Global Food Demand and Population Growth

Figure 1.11 illustrates the annual per-capita consumption of cereal in China, India, and the developing world. Inter-country variations in population growth rates and levels of per capita food consumption have an impact on the consequences of global population expansion on the demand for food. According to current forecasts, many nations with high food per capita have slow (and eventually negative) population growth rates, which limits the potential for additional rises. Extremely large variations in population growth rates and per capita consumption could have the effect of slowing down global demand growth and decreasing global consumption. These traits will slow the increase in world food demand. Figure 1.12 shows a comparison of cereal production between developing and industrialised nations.

1.5.8 Changes in Dietary Habits

Major changes in eating habits are taking place all throughout the world, even in the intake of everyday staples in favour of more varied diets. There have been significant negative health effects associated with these changes in food intake on a global and regional scale (Kearney 2010). Nutritional shift affects populations in those nations undergoing fast transition. This transition's diversity may be the result of variations in sociodemographic parameters and other consumer traits. The policies of trade liberalisation over the past more than three decades have implications for health because they have facilitated the "nutrition transition," which is linked to rising rates of obesity and chronic diseases like cardiovascular disease and cancer.

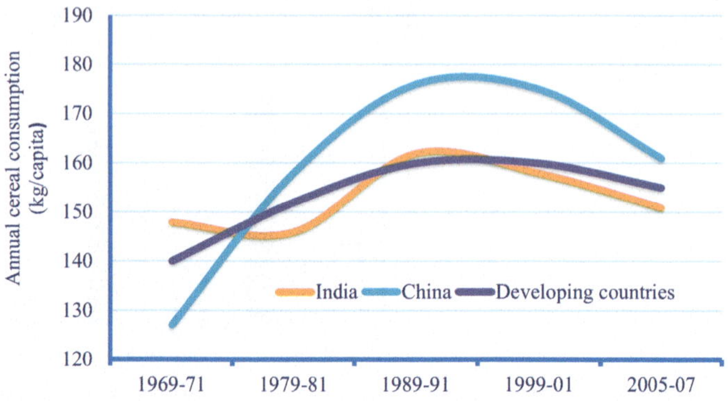

Fig. 1.11 Annual cereal per capita consumption in India, China, and the world. *Source* Alexandratos and Bruinsma (2012)

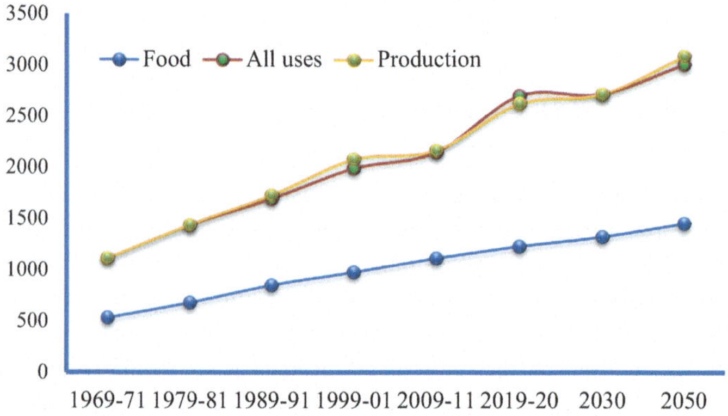

Fig. 1.12 Trends in cereal production in the world (million metric tonnes). *Source* Alexandratos and Bruinsma (2012)

The urbanisation of the world and the marketing of the food business are additional variables that affect health. Due to improvements in agricultural practices during the past 50 years, including better productivity, a wider variety of food, and a reduced reliance on seasonality, the world's capacity to feed its population has increased. Food availability has expanded along with rising earnings. This has resulted in major changes in food consumption during the last 60 years.

1.5.9 Trends of World Food Consumption

According to Alexandratos (2006), over the past four decades, there has been a considerable improvement in rising food intake per person, with an increase of about

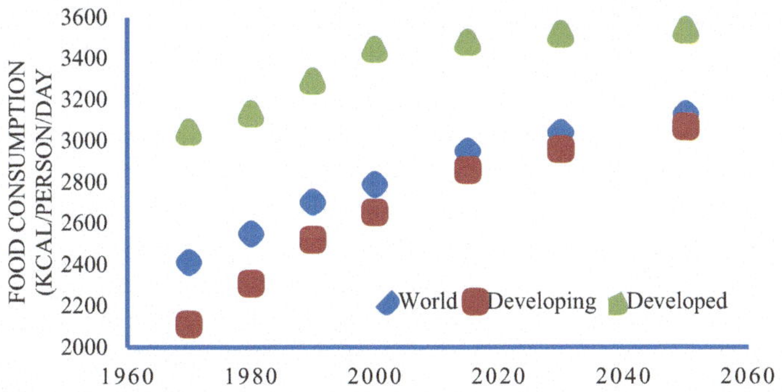

Fig. 1.13 Food consumption (kcal per person per day) (based on Alexandratos 2006) (D'ing C, Developing countries; D'ed C, Developed countries)

400 kcal per person per day, from 2411 to 2789 kcal (Fig. 1.13). This represents an increase from 2411 to 2789 kcal in the total amount of daily calories consumed by an individual. As a direct consequence of this, the number of people who are undernourished has significantly dropped. However, there are still a number of developing countries (primarily in Sub-Saharan Africa, such as Somalia, Burundi, Rwanda, and Kenya) in which the average amount of food consumed by a person has actually declined even further from the already extremely low level that it was at.

Changes in the diet's composition have been shown to go hand in hand with the sharp increase in food energy that has been observed internationally. Such dietary adjustment appears to go through a pattern that involves two key stages.

The recent pattern of emerging nation economies growing much more fast than developed country economies is likely to continue in future, analysts say. The gap in relative income would be much smaller by 2050. However, given the huge gaps in absolute per capita income that exist today, the absolute differences would still be noticeable and might even get wider. Furthermore, inequities within countries and between areas in the wealthy world of today would probably become more pronounced. Deprivation and undernourishment would still be widespread in less restrictive circumstances, despite the fact that it would be much less frequent than it is now. Food demand is anticipated to increase as a result of these modifications. Production would have to nearly triple in the developing countries in order to achieve this. This suggests that the production of several key commodities will dramatically increase. In order to meet demand by 2050, agricultural output would need to more than double in South Asia and Sub-Saharan Africa.

1.5.10 Future Trends and Constraints

The rate of output growth, on the other hand, may be more difficult to keep up with than it was in the past due to a number of factors. These factors include climate change, the demand that humans place on natural resources, a lack of investment in agriculture, and technological deficiencies. There is reason to believe that rapid technological improvement and innovation will enable us to sustainably satisfy the needs for food in the foreseeable future. However, this can only be achieved through the implementation of astute and smart public policies, increased funding, and public–private partnerships that make the most of the opportunities to maintain the current levels of production, grow yields in a way that is environmentally friendly, and reduce poverty and food insecurity. Through deforestation, greenhouse gas emissions, and nitrate pollution of aquatic bodies, agriculture has also harmed the larger environment (FAO 2011). The average annual increase in cereal yield has decreased since the 1990s (Fig. 1.14). Cereal yield growth rates below 1% year would be a concerning sign because the significant additional amounts of food required in the coming decades will be mostly produced through yield gains rather than massive expansion of the farmed land. Crop yields across rich and poor nations also differ significantly. Wheat and rice yields in poor countries are very low (about half) compared to highly rich countries. Annual growth rates of wheat, rice, and maize over time have been shown in Fig. 1.15.

The world's population is ageing.

Since virtually the beginning of the twenty-first century, annual rates of population increase have been on a downward trend everywhere across the world. In the late 1960s, total fertility rates (TFR) reached their highest levels in history, peaking at 4.5. The total fertility rate dropped to 2.5 in 2015, which resulted in yearly population growth rates of 1.2% across the globe. Until recently, absolute yearly increments continued to rise despite the fact that worldwide population growth rates were slowing down; but, since then, they have started to significantly fall. The absolute annual population gains are hovering well about 80 million people. By 2050, the medium version anticipates a little fall in absolute numbers to a little more than 55 million people, and a further decline to 15 million each year by the end of the century (Fig. 1.16). This prediction is based on the assumption that fertility rates will remain relatively stable.

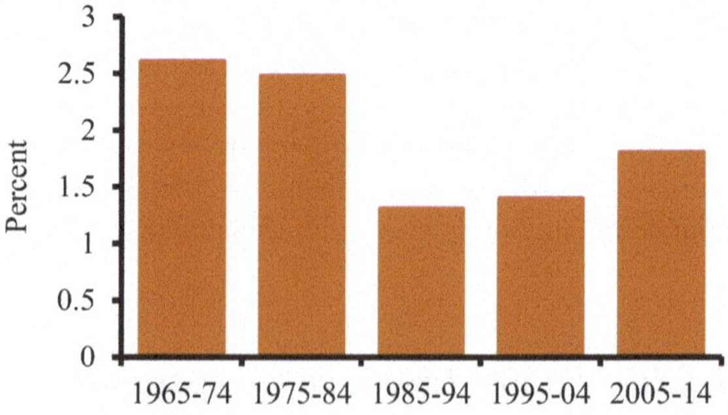

Fig. 1.14 Mean annual yield growth rates of cereals. *Source* FAO (2016)

Fig. 1.15 Annual growth rate (%) of wheat, paddy, and maize in the world. *Source* FAO (2016)

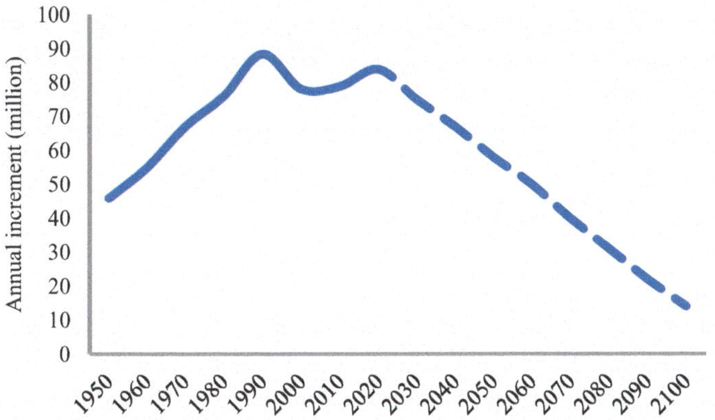

Fig. 1.16 World annual population increments, observed (linear line) and expected (dashed line) between 1950 and 2100.

1.5.11 The Nutritional Change is Accelerating Due to Rapid Urbanisation

The world's population was largely rural for many years. More than 60% of the population resided in rural areas 35 years ago. Since then, there has been a significant shift in the ratio of urban to rural areas, with metropolitan areas now making up about 54% of the world's population. In 2050, 35 years from now, more than two-thirds of the world's population may reside in urban regions (UN 2015). Urbanisation is being supported by changes in agriculture, particularly in terms of technical development and the introduction of labour-saving technologies. The effects of urbanisation have also and will probably continue to have an impact on, agriculture, food, and nutrition. Figure 1.17 depicts the ratio of rural to urban population throughout time.

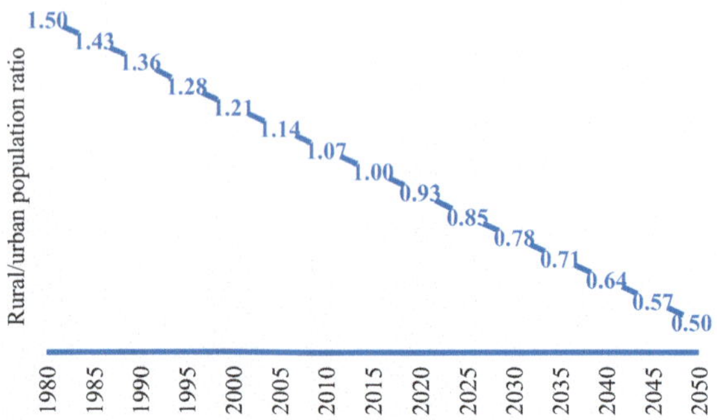

Fig. 1.17 Transition in rural/urban (*R*/*U*) world population ratio over a period of 70 years. *Source* Based on UN (2015)

Global urbanisation might result in a net 2.4 billion people being added to towns and cities by 2050, which is more than the 2.2 billion increase in the world's population. This implies that there could be a net loss of roughly 200 million people in rural areas. The net decline in rural populations is due to a number of variables, most notably greater mortality rates and shorter life expectancies in rural areas, rather than just an emigration from rural to urban areas. The lower urban fertility rates are more than countered by these elements.

1.6 Losses and Wastes of Food

From production to consumption, almost 30% of all food produced on Earth is wasted (HLPE 2014). The losses and waste in the food system over time are not accurately estimated. Evidence to date, however, points to the loss or waste of 1.3 billion tonnes annually, or one-third of the edible portion of food that was initially meant for human use, with 670 million tonnes in high-income countries and 630 million tonnes in low- and middle-income nations. Food losses and waste frequently lead to decreased profitability for farmers and other members of the food business as well as higher prices for consumers, both of which have an effect on food insecurity by restricting access to food for underprivileged individuals. If food losses and waste were decreased, food availability would increase and global food security would be increased.

Major food losses occur upstream, during harvest, and during post-harvest handling in low-income countries as a result of poor infrastructure, old equipment, a lack of knowledge, and production investment. Additionally, food losses are frequently caused by management and technical issues with harvesting, storing, transporting, processing, packaging, and marketing. The industries that produce and process fisheries and agriculture on a small- and medium-scale incur the greatest losses. Uncertainty around the weather, the status of the market, and subpar institutional frameworks all have an impact on losses. Each year, post-harvest procedures lose almost 13 million tonnes of grains, or more than 15% of the continent's entire yield. Every region, with the exception of South and Southeast Asia, loses or wastes more than 30% of the food that was meant for human consumption. Approximately 5.9% of food losses and waste in Sub-Saharan Africa—where they are particularly high at 36%—occur at the retail and consuming stages, whereas more than 30% do so during the harvest, post-harvest, and processing phases.

1.6.1 Using Acid Soils to Fill the Food Scarcity Gap

The health of the world's soils is critically important to ensuring that people will always have access to food. On the other hand, only 10–12% of the natural soils are capable of supporting agricultural activities. As a consequence of this, food insecurity is a problem that affects people all around the world. More than seventy-five percent of the world's population is deficient in one or more of the necessary mineral components, which results in the intrinsic malnutrition of micronutrients, often known as the hidden hunger. Therefore, in order to keep food production at its optimal level, appropriate land use regulations must to be formulated taking into consideration the salinity event that is gradually eroding the ecosystem of the soil (Gebrehiwot 2022).

Toxic consequences, nutritional imbalance, and decreased microbial activity are the main limiting factors linked to acidic soils. Reduced availability of nutrients such as calcium, magnesium, phosphorus, nitrogen, boron, and molybdenum as well as the poisoning of metals like manganese and aluminium are caused by acidity in the soil. According to Thakuria et al. (2016), it also slows down biological processes in the rhizosphere, especially those that are symbiotic or mutualistic between plants and beneficial fauna and flora. Both directly and indirectly, these elements influence plant growth. Severe water stress brought on by constrained root growth in the subsurface horizon is another significant limitation related to acidic soils (Adams 1984). These factors collectively severely restrict the potential for raising crop productivity. On 67% of the nation's acidic soils, aluminium toxicity and the resulting phosphorus deficit pose significant crop production challenges (Eswaran et al. 1997a).

In locations where soil acidity is a problem, crop productivity is low and fast declining. With a focus on annual and perennial crops, rainfed farming is primarily used in humid and subhumid regions where acid soils are prevalent (Dixon et al. 2019). The majority of the forests and woodlands that once covered the acid soil areas have been fast vanishing, and they have been replaced by a variety of small-holder farmers using techniques ranging from shifting cultivation to more intense fallow systems. Permanent cultivation methods have developed in areas of central and east Africa where population pressure has been the greatest. Therefore, in order to decrease the expansion of agricultural land on underperforming lands, it is important to close production disparities through long-term accretion, cropping efficiency, and preventing food waste (Foley et al. 2011). By lowering the yield and quality of cash crops including coffee, tea, pineapple, oil palm, rubber, and sisal, which are significant sources of foreign income for various African nations, soil acidity has also been having an impact on large-scale farms (Vlek et al. 2010). One of the key tactics for achieving food and nutrition security is acid soil management. On the other hand, managing acid soils continues to be a significant concern that necessitates funding. To create, adopt, and scale up an appropriate acid soil management technique, in-depth research and knowledge are necessary.

Liming is a widespread activity that is used to enhance pH levels, reduce surface soil acidity, and mitigate the toxic effects of aluminium. Liming can be accomplished with any type of liming chemical that can reduce the acidity of the soil. These materials are often compounds of calcium or magnesium, such as oxides, hydroxides, and carbonates of calcium and magnesium. These liming chemicals, when applied to acidic soils, substitute hydrogen and aluminium ions in the colloidal phase of the soil solution, which results in a decrease in the acidity of the soil. It is recommended that the pH levels of the surface soil should be 5.5, and the pH levels of the subsoil should be 4.8. In order to reduce the acidity of the soil, it is recommended that liming materials be applied on a consistent basis so that the pH can be raised to a safe level (pH > 5.5). When the pH is more than 4.8, the harmful effects of aluminium in the subsoil are mitigated. The amount of lime that is applied will vary depending on factors such as the pH profile of the soil, the kind of soil, the agricultural system, the amount of rainfall, and the quality of the lime. Other methods include: Utilising Al-tolerant cultivars is one method that has the potential to be an effective remedy for the problems that are created by the toxicity of aluminium in acidic soils. In order to accomplish this objective, genotypes must be developed that have an elevated capacity for withstanding acidic soils and high concentrations of Al. Together with the procedure of liming, this has the potential to boost agricultural productivity in soils that are acidic. It is possible that the problems caused by the acidity of the subsoil can be solved by applying lime and gypsum combined.

References

Abebe M (2007) Nature and management of acid soils in Ethiopia. Haramaya University, Haramaya, Ethiopia

Adams F (1984) Crop response to lime in the southern United States. In: Adams F (ed) Soil acidity and liming. American Society of Agronomy Inc., Madison, WI, pp 211–265

Agegnehu G, Yirga C, Erkossa T (2019) Soil acidity management. Ethiopian Institute of Agricultural Research (EIAR). Addis Ababa, Ethiopia

Akhtar Schuster M, Thomas RJ, Stringer LC, Chasek P, Seely M (2011) Improving the enabling environment to combat land degradation: institutional, financial, legal and science-policy challenges and solutions. Land Degrad Dev 22:299–312

Alexandratos N, Bruinsma J (2012) World agriculture towards 2030/2050: the 2012 revision. ESA Working Paper No. 12–03. FAO, Rome

Alexandratos N (ed) (2006) World agriculture: towards 2030/50, interim report. An FAO perspective. Earthscan, London, UK; FAO, Rome, Italy

Bai ZG, Dent DL, Olsson L, Schaepman ME (2008) Proxy global assessment of land degradation. Soil Use Manage 24:223–234

Behera SK, Shukla AK (2015) Spatial distribution of surface soil acidity, electrical conductivity, soil organic carbon content and exchangeable potassium, calcium and magnesium in some cropped acid soils of India. Land Degrad Dev 26:71–79
Blank LW (1985) A new type of forest decline in Germany. Nature 314:311–314
Bouma J (2018) The challenge of soil science meeting society's demands in a "posttruth", "fact free" world. Geoderma 310:22–28
Brady N, Weil R (2016) The nature and properties of soils. Pearson Education, Columbus, EUA
Bronick CJ, Lal R (2005) Soil structure and management: a review. Geoderma 124:3–22
Cançado GMA, Loguercio LL, Martins PR, Parentoni SN, Borém A, Paiva E, Lopes MA (1999) Hematoxylin staining as a phenotypic index for aluminum tolerance selection in tropical maize (Zea mays L.). Theor Appl Genet 99:747–754
Charles DF (ed) (1991) Acidic deposition and aquatic ecosystems. Regional case studies. Springer-Verlag, New York
Cronan CS, Grigal DF (1995) Use of calcium/aluminum ratios as indicators of stress in forest ecosystems. J Environ Qual 24:209–226
Cronan CS, Schofield CL (1990) Relationships between aqueous aluminum and acidic deposition in forested watersheds of North America and Northern Europe. Environ Sci Technol 24:1100–1105
Coventry DR (1991) The injection of slurries of lime, associated with deep tillage, to increase wheat production on soils with subsoil acidity. In: Wright RJ (ed) Plant–soil interactions at low pH. Kluwer Academic Publications, Dordrecht, the Netherlands, pp 437–445
Delhaize E, Craig S, Beaton CD, Bennet RJ et al (1993) Aluminum tolerance in wheat (Triticum aestivum L.) (I. Uptake and Distribution of Aluminum in Root Apices). Plant Physiol 103:685–693
Demas GP (1993) Submerged soils: a new frontier in soil survey. Soil Survey Horizons 34:44–46
Dixon J, Garrity DP, Boffa J-M, Williams TO, Amede T, Auricht C, Lott R, Mburathi G (2019) Farming systems and food security in Africa: priorities for science and policy under global change. Routledge, London
Dominati E, Mackay A, Green S, Patterson M (2014) A soil change-based methodology for the quantification and valuation of ecosystem services from agro-ecosystems: a case study of pastoral agriculture in New Zealand. Ecol Econ 100:119–129
Driscoll CT, Likens GE,Church MR (1998) Recovery of surface waters in the northeastern U.S. from decreases in atmospheric deposition of sulfur. Water Air Soil Pollut 105:319–329
Driscoll CT, Lawrence GB, Bulger AJ, Butler TJ, Cronan CS, Eagar C, Lambert KF, Likens GE, Stoddard JL, Weathers KC (2001) Acidic deposition in the Northeastern United States: sources and inputs, ecosystem effects, and management strategies. BioScience 51:180–198
Driscoll CT, Driscoll KM, Mitchell MJ, Raynal DJ (2003) Effects of acidic deposition on forest and aquatic ecosystems in New York State. Environ Pollut 123:327–336
Duan L et al (2016) Acid deposition in Asia: emissions, deposition, and ecosystem effects. Atmos Environ 146:55–69
ELD (2015) Report for policy and decision makers: reaping economic and environmental benefits from sustainable land management. Economics of Land Degradation (ELD) Initiative, Bonn, Germany
Eswaran H, Reich P, Beinroth F (1997a) Global distribution of soils with acidity. In: AC Moniz, (ed.), Plant Soil Interactions at Low pH, Brazilian Soil Science Society, pp 159–164
Eswaran H, Almaraz R, van den Berg E, Reich P (1997b) An assessment of the soil resources of Africa in relation to productivity. Geoderma 77:1–18
Eswaran H, Reich P, Beinroth F (1997c) Global distribution of soils with acidity. In: Moniz AC (ed) Plant-soil interactions at low pH. Brazilian Soil Science Society, Sao Paulo, pp 159–164
Fageria N, Baligar V (2008) Ameliorating soil acidity of tropical Oxisols by liming for sustainable crop production. Adv Agron 99:345–399
Fageria NK, Nascente AS (2014) Management of soil acidity of South American soils for sustainable crop production. Adv Agron 128:221–275
FAO (2006) Soil biodiversity portal. http://www.fao.org/ag/agl/agll/soilbiod/index_en.stm

FAO (2006) World reference base for soil resources 2006. A framework for international classification, correlation and communication. World Soil Resources Reports, 103. Rome, Italy, 128 p

FAO (2018) FAO statistical databases. http://apps.fao.org/. Food and Agriculture Organization of the United Nations (FAO)

FAO (2010) The state of food insecurity in the world 2010. http://www.fao.org/publications/sofi-2010/en/

FAO (2011) Land tenure, climate change mitigation and agriculture. Mitigation of climate

FAO (2014) World reference base for soil resources 2014. World Soil Resources Report 106. Rome, 120 pp

FAO (2016) FAO food change in agriculture (MICCA) programme, June 2011. Rome. Price index [Website] (available at www.fao.org/worldfoodsituation/foodpricesindex). Accessed 15 Nov 2016

FAO and ITPS (2015) Status of the world's soil resources (SWSR)—main report. Food and Agriculture Organization of the United Nations and Intergovernmental Technical Panel on Soils, Rome, Italy

Fischer G, van Velthuizen H, Shah M, Nachtergaele F (2002) Global agro-ecological assessment for agriculture in the 21st century: methodology and results. RR-02-002, IIASA, Laxenburg and FAO, Rome. http://www.iiasa.ac.at/Research/LUC/SAEZ/pdf/gaez2002.pdf

Fischer G, van Velthuizen H, Nachtergaele F (2011) GAEZ v3.0—global agroecological zones model documentation, (mimeo). IIASA, Luxemburg. http://www.iiasa.ac.at/Research/LUC/GAEZv3.0/gaez2010-Flyer_1final1.pdf

Foley JA, Ramankutty N, Brauman KA, Cassidy ES, Gerber JS, Johnston M, Mueller ND, Connell CO' RD, Paul CWP, Balzer C et al (2011) Solutions for a cultivated planet. Nature 478:337–342

Galloway JN, Norton SN, Church MR (1983) Freshwater acidification from atmospheric deposition of sulfuric acid: a conceptual model. Environ Sci Technol 17:541A-545A

Gao Y, Ma M, Yang T, Chen W, Yang T (2018) Global atmospheric sulfur deposition and associated impaction on nitrogen cycling in ecosystems. J Clean Prod 195:1–9

Garvin DF, Carver BF (2003) The role of the genotype in tolerance to acidity and aluminium toxicity. In: Rengel Z (ed) Handbook of soil acidity. Marcel Dekker, New York, pp 387–406

Gazey C, Davies S (2009) Soil acidity: a guide for WA farmers and consultants Bulletin 4784 (ISSN 1833-7246). The Department of Agriculture and Food, Western Australia, 52 p

Gebrehiwot K (2022) Soil management for food security. In: Natural resources conservation and advances for sustainability, pp 61–71

Godbold DL, Fritz E, Huttermann A (1988) Aluminum toxicity and forest decline. Proc Natl Acad Sci USA 85:3888–3892

Godfray HCJ, Beddington JR, Crute IR, Haddad L, Lawrence D, Muir JF, Pretty J, Robinson S, Thomas SM, Toulmin C (2010) Food security: the challenge of feeding 9 billion people. Science 327:812

Goryachkin S, Cherkinsky A, Mergelov N, Shorkunov I (2012) Endolithic (bio)weathering and rock varnish in East Antarctica as early Earth analogs of 'protosoils.' Mineral Mag 76(6):1777

Grieve IC (2001) Human impacts on soil properties and their implications for the sensitivity of soil systems in Scotland. CATENA 42:361–374

Haile W, Boke S (2009) Mitigation of soil acidity and fertility decline: challenges for sustainable livelihood improvement: evidence from southern region of Ethiopia. In: Sustainable land management and poverty alleviation. Ethiopian Development Research Institute, Addis Ababa, pp 131–143

Heim A, Luster J, Brunner I, Frey B, Frossard E (1999) Effects of aluminium treatment on Norway spruce roots: aluminium bindings forms, element distribution, and release of organic substances. Plant Soil 216:103–116

Jobbágy EG, Jackson RB (2001) The distribution of soil nutrients with depth: global patterns and the imprint of plants. Biogeochemistry 53:51–77

Jordan TE, Weller DE (1996) Human contributions to the terrestrial nitrogen flux. Bioscience 46:655–664

Jovanovic Z, Djalovic I, Komljenovic I, Kovacevic V, Cvijovic M (2006) Influences of liming on vertisol properties and yields of the field crops. Cereal Res Commun 34:517–520

Jovanovic Z, Djalovic I, Tolimir M, Cvijovic M (2007) Influence of growing sistem and NPK fertilization on maize yield on pseudogley of Central Serbia. Cereal Res Commun 35:1325–1329

Ju XT, Xing GX, Chen XP, Zhang SL, Zhang LJ, Liu XJ, Cui ZL, Yin B, Christie P, Zhu ZL, Zhang FS (2009) Reducing environmental risk by improving N management in intensive Chinese agricultural systems. Proc Natl Acad Sci 106:3041–3046

Kearney J (2010) Food consumption trends and drivers. Phil Trans R Soc B 365:2793–2807. https://doi.org/10.1098/rstb.2010.0149

Kidd PS, Proctor J (2001) Why do plants grow poorly on acid soils: are ecologist missing the obvious. J Exp Bot 52:791–799

Kinraide TB (1991) Identity of the rhizotoxic aluminium species. Plant Soil 134:167–178

Kochian LV (1995) Cellular mechanisms of aluminium toxicity and resistance in plants. Ann Rev Plant Physiol Plant Mole Biol 46:237–260

Kochian LV, Hoekenga OA, Pineros MA (2004) How do crop plants tolerate acid soils? Mechanisms of aluminum tolerance and phosphorous efficiency. Annu Rev Plant Biol 55:459–493

Kochian LV, Pineros MA, Hoekenga OA (2005) The physiology, genetics and molecular biology of plant aluminum resistance and toxicity. Plant Soil 274:175–195. https://doi.org/10.1007/s11104-004-1158-7

Kopittke PM, Dalal RC, Finn D, Menzies NW (2017) Global changes in soil stocks of carbon, nitrogen, phosphorus, and sulphur as influenced by long-term agricultural production. Glob Chang Biol 23:2509–2519

Kopittkea PM, Neal Menziesa NW, Wangb P, McKennaa BA, Lombic E (2019) Soil and the intensification of agriculture for global food security. Environ Int 132:105078

Krug EC, Frink CR (1983) Acid rain on acid soil: a new perspective. Science 221(4610):520–525

Kunhikrishnan A et al (2016) Functional relationships of soil acidification, liming and greenhouse gas flux. In: Sparks DL (ed) Advances in agronomy, vol 139. Elsevier, New York, pp 1–71

Kuperman RG, Edwards CA (1997) Effects of acidic deposition on soil invertebrates and microorganisms. In: Reviews of environmental contamination and toxicology. Springer, New York, pp 35–138

Lal R (2015) Restoring soil quality to mitigate soil degradation. Sustainability 7:5875–5895

Langeveld JWA, Dixon J, van Keulen H, Quist-Wessel PMF (2013) Analyzing the effect of biofuel expansion on land use in major producing countries: evidence of increased multiple cropping. Biofuels Bioprod Biorefin 8:49–58

Letey J, Hoffman GJ, Hopmans JW, Grattan SR, Suarez D, Corwin DL, Oster JD, Wu L, Amrhein C (2011) Evaluation of soil salinity leaching requirement guidelines. Agric Water Manag 98:502–506

Liang G, Zhang D, Lu Y, Feng X, Luo G, Zhao Z (2018) Response of soil C:N: P stoichiometry to long-term simulated acid rain in a monsoon evergreen broad-leaved forest at Dinghushan nature reserve. Ecol Environ Sci 27:844–851 (In Chinese with English abstract)

Likens GE, Bormann FH, Hedin LO, Eaton JS, Driscoll CT (1990) Dry deposition of sulfur: a 23-year record for the Hubbard Brook Forest ecosystem. Tellus 42B:319–329

Likens GE, Driscoll CT, Buso DC (1996) Long-term effects of acid rain: response and recovery of a forest ecosystem. Science 272:244–246

Lima M, Miranda, Filho JB, Furlani PR (1995) Diallel cross among inbred lines of maize differing in aluminum tolerance. Braz J Genet 4:579–584

Livi Bacci M (2017) A concise history of world population, 6th ed. Wiley/Blackwell, Hoboken, NJ

MacAvoy SE, Bulger AJ (1995) Survival of brook trout (Salvelinus fontinalis) embryos and fry in streams of different acid sensitivity in Shenandoah National Park, USA. Water Air Soil Poll 85:439–444

Marschner H (1995) Mineral nutrition of higher plants, 2nd edn. Academic Press, New York, p 889

Marschner H (2011) Marschner's mineral nutrition of higher plants. Academic Press, London
Martins PR, Parentoni SN, Lopes MA, Paiva E (1999) Eficiĕncia de indices fenotĭpicos de comprimento de raiz seminal na avaliaĉão de plantas individuais de milho quanto ă tolerăncia ao aluminio. Pesquisa Agropecuăria Brasileira 34:1897–1904
McBratney A, Field DJ, Koch A (2014) The dimensions of soil security. Geoderma 213:203–213
Meng C, Dashuan Tian D, Zeng H, Li Z, Yi C, Niu S (2019) Global soil acidification impacts on belowground processes. Environ Res Lett 14:074003
Mengel K, Kirkby EA (1987) Principles of plant nutrition. International Potash Institute, Bern, Switzerland
Myers RJK, De Pauw E (1995) Strategies for the management of soil acidity. In: Date RA, Grundon NJ, Rayment GE, Probert ME (eds) Plant-soil interactions at low pH: principles and management. Kluwer Academic Publishers, Dordrecht, The Netherlands, pp 729–741
Nature (2010) How to feed a hungry world. Nature 466(7306), 29 July 2010
Oden S (1968) The acidification of air precipitation and its consequences in the natural environment. Bulletin of Ecological Research Communications NFR. Translation Consultants Ltd., Arlington (VA)
OECD-FAO (2018) Agricultural outlook 2017–2026. OECD Publishing, Paris, France. OECD/FAO, 2018
Poschenrieder C, Gunse B, Corrales I, Barcelo J (2008) A glance into aluminum toxicity and resistance in plants Sci. Total Environ 400:356–368
Rengel Z (2011) Soil pH, soil health and climate change. In: Soil health and climate change, pp 69–85
Ridley AM, Windsor SM (1992) Persistence and tolerance to soil acidity of phalaris and cocksfoot in north-eastern Victoria. Aust J Exp Agric 32:1069–1075
Ridley AM, Helyar KR, Slattery WJ (1990) Soil acidification under subterranean clover (Trifolium subterraneaum L.) pastures in North-Eastern Victoria. Aust J Exper Agric 30:195–201
Rochelle BP,Church MR (1987) Regional patterns of sulfur retention in watersheds of the eastern US. Water Air Soil Poll 36:61–73
Roem WJ, Berendse F (2000) Soil acidity and nutrient supply ratio as possible factors determining changes in plant species diversity in grassland and heathland communities. Biol Conserv 92:151–161
Rossiter DG (2007) Classification of urban and industrial soils in the world reference base for soil resources. J Soils Sediments 7(2):96–100
Royal Society (2009) Reaping the benefits: science and the sustainable intensification of global agriculture. RS Policy document 11/09, Royal Society, London
Samac DA, Tesfaye M (2003) Plant improvement for tolerance to aluminium in acid soils. Plant Cell, Tissue Organ Cult 75:189–207
Sanchez PA (1977) Properties and management of soils in the tropics. Soil Sci 124:1–187
Scharlemann JPW, Tanner EVJ, Hiederer R, Kapos V (2014) Global soil carbon: understanding and managing the largest terrestrial carbon pool. Carbon Manage 5:81–91
Schuur EAG, McGuire AD, Schädel C, Grosse G, Harden JW, Hayes DJ, Hugelius G, Koven CD, Kuhry P, Lawrence DM, Natali SM, Olefeldt D, Romanovsky VE, Schaefer K, Turetsky MR, Treat CC, Vonk JE (2015) Climate change and the permafrost carbon feedback. Nature 520:171
Shishov L, Tonkonogov V, Lebedeva I, Gerasimova M (2004) Classification and diagnostics of soils of Russia. Smolensk, Oekumena, 341 pp. [in Russian]
Sivaguru M, Horst WJ (1998) The distal part of the transition zone is the most aluminum-sensitive apical root zone of maize. Plant Physiol 116:155–163
Slattery MC, Burt TP (1997) Particle size characteristics of suspended sediment in hillslope runoff and stream flow. Earth Surf Proc Land 22:705–719
Slattery WJ, Edwards DG, Bell LC, Coventry DR, Helyar KR (1998) Soil acidification and the carbon cycle in a cropping soil of north-eastern Victoria. Soil Res 36(2):273–290

Slattery WJ, Conyers MK, Aitken RL (1999) Soil pH, aluminium, manganese and lime requirement. In: Peverill KI, Sparrow LA, Reuter DJ (eds) Soil analysis: an interpretation manual. CSIRO Publishing, Collingwood, pp 103–128

Slattery JF, Coventry DR, Slattery WJ (2001) Rhizobial ecology as affected by the soil environment. Austr J Exp Agric 41:289–298

Smith RA (1872) Air and rain. Longmans, Green, London

Smith P (2005) An overview of the permanence of soil organic carbon stocks: influence of direct humaninduced, indirect and natural effects. Eur J Soil Sci 56:673–768

Smith P, House JI, Bustamante M, Sobocká J, Harper R, Pan G, West PC, Clark JM, Adhya T, Rumpel C, Paustian K, Kuikman P, Cotrufo MF, Elliott JA, McDowell R, Griffiths RI, Asakawa S, Bondeau A, Jain AK, Meersmans J, Pugh TAM (2015) Global change pressures on soils from land use and management. Glob Chang Biol 22:1008–1028

Somani L (1996) Crop production in acid soils, 1st ed. Agrotech Publishing Academy, New Delhi

Somasundaram J, Reeves S, Wang W, Heenan M, Dalal R (2017) Impact of 47 years of no tillage and stubble retention on soil aggregation and carbon distribution in a vertisol. Land Degrad Dev 28:1589–1602

Sposito, G., 2013. Green water and global food security. Vadose Zone J. 12.

Steffen W, Richardson K, Rockström J, Cornell SE, Fetzer I, Bennett EM, Biggs R, Carpenter SR, de Vries W, de Wit CA, Folke C, Gerten D, Heinke J, Mace GM, Persson LM, Ramanathan V, Reyers B, Sörlin S (2015) Planetary boundaries: guiding human development on a changing planet. Science 347

Stevens CJ, Thompson K, Grime JP, Long C, Gowing DJG (2010) Contribution of acidification and eutrophication to declines in species richness of calcifuge grasslands along a gradient of atmospheric nitrogen deposition. Funct Ecol 24:478–484

Stoddard JL, Jeffries DS, Lükewille A, Clair TA, Dillon PJ, Driscoll CT, Forsius M, Johannessen M, Kahl JS, Kellogg JH (1999) Regional trends in aquatic recovery from acidification in North America and Europe. Nature 401:575–578

Sumner ME, Noble AD (2003) Soil acidification: the world story. In: Rengel Z (ed) Handbook of soil acidity. Marcel Dekker, New York, pp 1–28

Tabuchi H, Matsumoto H (2001) Changes in cell wall properties on wheat (Triticum aestivum) roots during aluminium-induced growth inhibition. Physiol Plant 112:353–358

Targulian V, Mergelov N, Gilichinsky D, Sedov S, Demidov N, Goryachkin S, Ivanov A (2010) Dokuchaev's soil paradigm and extraterrestrial 'soils'. In: Gilkes RJ, Prakongkep N (eds) Proceedings of the 19th world congress of soil science; soil solutions for a changing world. Division symposium. Astropedology, IUSS, Australia, Brisbane, pp 1–4

Taylor GJ (1991) Current views of the aluminum stress response: the physiological basis of tolerance. Curr Top Plant Biochem Physiol 10:57–93

Thakuria D, Hazarika S, Krishnappa R (2016) Soil acidity and management options. Indian J Fertilisers 12:40–56

Thomas RJ (1995) Management of Acid soils (MAS): an interdisciplinary and multi-institutional collaborative undertaking. In: "Proceedings of a DSE/IBSRAM International workshop on soil, water and nutrient management research". IBSRAM, Bangkok, DSE:Zschortau, pp 71–86

Tully K, Sullivan C, Weil R, Sanchez P (2015) The state of soil degradation in Sub-Saharan Africa: baselines, trajectories, and solutions. Sustainability 7:6523–6552

Von Uexkull HR, Mutert E (1995) Global extent, development and economic impact of acid soils. Plant Soil 171:1–15

United Nations (2009) World population prospects: the 2008 revision

United Nations (2011) World population prospects: the 2010 revision

UN (United Nations) (2015) World population prospects: the 2015 revision [Website] (available at https://esa.un.org/unpd/wpp). Accessed Nov 2016

UN (2017) World Population Prospects: the 2017 revision, key findings and advance tables. Working Paper No. ESA/P/WP/248. United Nations, New York, NY

Vlek PL, Le QB, Tamene L (2010) Assessment of land degradation, its possible causes and threat to food security in Sub-Saharan Africa. In: Food security and soil quality. CRC Press, Boca Raton, p 57–86

Wang J, Raman H, Zhang G, Mendham N, Zhou M (2006) Aluminium tolerance in barley (Hordeum vulgare L.): physiological mechanisms, genetics and screening methods. J Zhejiang University Sci 7:769–787

Wang P, Chen H, Kopittke PM, Zhao F-J (2019) Cadmium contamination in agricultural soils of China and the impact on food safety. Environ Pollut 249:1038–1048

Yang Y et al (2012) Significant soil acidification across northern China's grasslands during 1980s–2000s. Glob Change Biol 18:2292–2300

Yokota S, Ojima K (1995) Physiological response of root tip of alfalfa to low pH and aluminium stress in water culture. In: Plant-soil interactions at low pH: principles and management. Springer, Dordrecht, pp 329–331

Zhao Y, Duan L, Xing J, Larssen T, Nielsen CP, Hao J (2009) Soil acidification in China: is controlling SO_2 emissions enough? Environ Sci Technol 43:8021–8026

Zhou LL, Bai GH, Carver B, Zhang DD (2007) Identification of new sources of aluminum resistance in wheat. Plant Soil 297:105–118

Chapter 2
Global Status and Extent of Acid Soils

Abstract Acid soils cover 3950 m ha, or one-third of the world's ice-free territory. These soils are the product of either ustic or udic moisture regimes and are mostly found in two global regions. The northern belt (cold and moderate climate) is dominated by spodosols, alfisols, Oxisols, and histosols, whereas the southern tropical belt is dominated by Ultisols and Oxisols. Around 18% of acid soils are made up of savanna, grassland, and steppe flora in addition to 67% of acid soils supporting forests and woods. Just 4.5% of the total land (179 m ha) is utilised for crop cultivation due to corrosive soil. Tropical perennial crops are grown on an additional 33 million hectares. Whatever efforts are made to develop acid soils for agriculture and agroforestry in tropical climates should concentrate on the deforested and abandoned areas with damaged acid soils. This will be difficult without a large initial expenditure and the right technologies. A three-pronged development approach is proposed to help prevent or lessen the yearly devastation of over 5 million hectares of tropical forests caused by "untraditional shifting cultivators." It would help protect the fragile natural ecosystems on acidic tropical soils, which are currently believed to be necessary for the survival of life as we know it. Acid sulphate soils are considered the "nastiest soils on earth" due to their high acidity, increased mobility of potentially dangerous components, and limited nutrient absorption. Although they account for a very small percentage of all issue soils worldwide, they often have detrimental consequences on local aquaculture, agriculture, and the environment. They are often seen in heavily populated locations along the beaches of several developing countries. Acrisols and ferralsols, two types of tropical and subtropical soils, are found in older, densely wooded environments. They are usually yellowish to reddish in colour, heavily worn, and severely leached. The hydrolysis of silicate minerals and the quick removal of weathering products lead to low pH and low amounts of weathered product in the soil solution. More ferralsols and acrisols are found on older land surfaces in humid tropical settings. On gently sloping Tertiary and Pleistocene erosion surfaces that have experienced several cycles of weathering, erosion, and deposition, they can be discovered. The principal habitat of ferralsols is the wet tropical regions, covering an estimated 745 million hectares worldwide. Mostly, they are found on the continental shields of Central Africa and South America. Only areas

U. C. Sharma et al., *Soil Acidity*, Progress in Soil Science,
https://doi.org/10.1007/978-3-031-76357-1_2

with easily weathered basic rock and hot, humid climates outside of continental shields, such as Southeast Asia and a few Pacific Islands, are home to ferralsols.

Keywords Acrisols · Ferralsols · Extent of acid soils · Global distribution of acid soils · Acid sulphate soils

2.1 General Description

According to von Uexkül and Mutert (1995), acidic soils are most frequently found in wet tropical and temperate locations, which are both typically rich in water and heat that are necessary for plant growth. This suggests that acid soils offer enormous potential for productivity. However, aluminium (Al) toxicity and nutrient deficiency together with acid soil acidity are the main factors that limit plant productivity (Zhao et al. 2014). The combination improvement of plant Al tolerance and nutrient-use efficiency is essential for increasing agricultural output in acidic soils. The most prevalent mineral ingredient needed by plants is nitrogen (N). Globally, the availability of soil N has a significant impact on the growth and development of crops (Gutiérrez 2012). Agriculture uses a lot of N fertilisers to grow the crops that supply food for a growing portion of the world's population. According to Erisman et al. (2008), 4 billion individuals have been born since 1908, or about 27% of the world's population over the past century. Ineffective nitrogen utilisation leads to excessive nitrogen fertilisation, which in turn contributes to a number of environmental problems, including water eutrophication, greenhouse gas emissions, loss of nitrate (NO_3^-), acid rain, and soil acidification (Ju et al. 2009). It is essential for today's agriculture to have both large yields and efficient use of nutrients. As a result, it is essential to have an understanding of how plants respond to nitrogen in order to improve the efficiency with which plants utilise nitrogen (Kant et al. 2011).

A third of the world's ice-free land, or 3950 m ha, is made up of acid soils. These soils are mostly found in two worldwide belts and are the result of udic or ustic moisture regimes. Spodosols, Alfisols, Inceptisols, and Histosols dominate the northern belt (cold and moderate climate), whereas Ultisols and Oxisols predominate in the southern tropical belt. In addition to supporting forests and woods in 67% of the acid soils, savanna, grassland, and steppe vegetation make up around 18% of the acid soils. The area with acid soil that is used for growing crops is only 4.5% (179 m ha). 33 million more hectares are used for tropical perennial crops.

Tropic and wetland forests play an essential part in maintaining the harmony of local, regional, and global ecosystems as well as in protecting water supplies, flora, and fauna. Large tracts of anthropogenic savannas on severely eroded and degraded acid soils have replaced the more than 250 million ha of tropical forest lost to destructive logging and abusive modern shifting farming in the second part of this century. While the northern belt's acid soils are getting more protection and reforestation.

According to von Uexkül and Mutert (1995), the deforested and abandoned areas with damaged acid soils should be the focus of any efforts to develop acid soils for agriculture and agroforestry in tropical regions. Without a substantial initial investment and proper technology, this will be challenging. It is suggested that a three-step development strategy could assist stop or avoid the annual destruction of more than 5 Mha of tropical forests by "untraditional shifting cultivators." It would aid in preserving the delicate natural ecosystems on tropical acid soils, which are currently thought to be essential for the continuation of life on Earth.

Particularly when the soil pH drops below 5, aluminium (Al) in the soils is soluble into ionic forms. These ionic forms of Al have been found to be extremely harmful to plants, initially inhibiting root extension by damaging the root apex's cell structure and negatively affecting the roots' ability to absorb water and nutrients. As a result, plant growth and development are severely hampered. On the other hand, phosphorus (P) is readily fixed by clay minerals found in acidic soils, such as kaolinite and other iron oxides, making it inaccessible to roots. As a result, P insufficiency and Al toxicity are seen as the two main obstacles to crop production in acidic soils.

Without a doubt, soil is essential to the world since it provides nearly all of the food and fibre needed to support the human population as well as essential ecosystem functions. Although 1.35 billion hectares of arable land are available worldwide, only 0.20 ha are available per person, which is not allocated equally (Anderson et al. 2010). For instance, 71% of the world's population, 46% of the world's arable land, and the majority of the world's low-quality, unproductive land are all found in Asia and Africa. In addition to having more acreage per person, the industrialised nations of North America and Europe also have more resources for soil protection. The preservation of all lands is crucial. The problem of land degradation is considerable, especially in locations with fragile, low-quality lands, despite significant advancements made with contemporary practices such as conservation tillage (Anderson et al. 2010). Virtually all life on Earth depends on soil, according to Professor Dennis Greenland, a British soil scientist who has worked in both rich and developing nations. He argues that it is crucial to preserve our soils and boost their production. Failure to regulate soil degradation caused the demise of numerous ancient civilizations.

The "nastiest soils on earth" have been dubbed as acid sulphate soils because of their extreme acidity, enhanced mobility of potentially hazardous components, and low bioavailability of nutrients. They only make up a small portion of all problem soils in the globe, yet they frequently have serious negative effects on local agriculture, aquaculture, and the environment. Along the coasts of many emerging nations, they are frequently situated in densely populated areas. Because of this, individuals who are least prepared to deal with the low agricultural yields and decreased water quality that can come from acid sulphate soil disturbance may have major consequences (Ljung et al. 2009). The value of ecosystems for human health and well-being was stressed in the Millennium Ecosystem Assessment, which was commissioned by the United Nations in 2000. These include the services they offer as food and water supplies, for the prevention and treatment of disease, and for the cultural services that ecosystems give. Many efforts to control acid sulphate soil have been focused on issues relating to agriculture, aquaculture, and the environment, but little attention has been paid to the connection to human health (Ljung et al. 2009). Kofi Annan, the Secretary-General of the United Nations, called for the Millennium Ecological Assessment (MEA) in 2000 to investigate the effects that changes in the environment have on people's quality of life and to provide a foundation of information to support improved choices for sustainable development (Millennium Ecosystem Assessment 2003; Corvalan et al. 2005). The MEA, which comprised approximately 1400 specialists from around the world between 2001 and 2005, produced a cutting-edge analysis of the state of ecosystems today, their patterns, and the services they offer. Provisioning, regulating, and cultural ecosystem services were recognised as the three main types of ecosystem services necessary for human health and well-being. A variety of additional ecosystem activities were also identified as being required to maintain their production. It was acknowledged that compared to any previous time in human history, the last 50 years have seen humans impact ecosystems at a faster rate. Although the use of irrigation systems and changes to land use have greatly improved human health and produced positive net economic gains, more than 60% of the services they provide are still being used in an unsupportable manner (Corvalan et al. 2005). It should be highlighted that the environmental effects of ASS in general have been sufficiently documented (Dent 1986; Dent and Pons 1995; Wilson et al. 1999); the focus is on the human health repercussions of ecosystem change. Sulphuric acid is produced when sulphide minerals in soil oxidise, creating ASS. This process causes the soil pH to decrease, and once the soil buffering capacity is reached, ASS with pH values frequently below 4 is formed (Indraratna et al. 1995). It is well known that ASS is created by the oxidation of the iron sulphide mineral pyrite (FeS_2), which is the end product of processes involving the reduction of massive amounts of sulphate in saltwater (Bronswijk et al. 1995; Simpson and Pedini 1985). In particular, ASS is known to be produced by the oxidation of the iron sulphide mineral pyrite (FeS_2). As a consequence of this, potential and actual ASS are commonly detected in close proximity to both existing and formerly existing beaches, in addition to further inland (McBride 1994; Sammut 2004). The difference between actual and potential acid soil stress lies in whether or not the soil has gotten acidified as a result of oxidation (referred to as actual ASS), or whether or not the soil material is still unoxidised but has the potential to become acidified if it comes into contact with air (referred to as potential ASS).

2.2 Soil Charter of the World

The foundation of life on earth is soil.

Though there may be fewer people than ever before who would comprehend the reality of this statement, we currently know more about soil than ever before. Since less human effort has been used to cultivate the soil over the previous century, less direct touch with the soil has been experienced in the majority of places. However, the entire human society is now more dependent than ever on both the concrete products of the soil and the more intangible services it offers for the upkeep of the biosphere (FAO 2015). There is need to elucidate crucial relationships between human welfare and soil health and to establish a standard by which to gauge our success as a society in protecting this priceless resource.

Although soils are essential to life on Earth, human pressure on soil resources is beginning to reach critical levels. One key component of sustainable agriculture is careful soil management, which also offers a useful lever for climate regulation and a route for preserving ecosystem services and biodiversity. The World Soil Charter offers a list of nine guiding principles that condense our current knowledge of soil, the various functions it plays, and the dangers it faces from these roles. Thus, the nine principles serve as a brief yet thorough introduction to this paper.

2.2.1 Preamble

Preamble

1. Although soils are essential to life on Earth, human pressure on soil resources is beginning to reach critical levels. One key component of sustainable agriculture is careful soil management, which also offers a useful lever for climate regulation and a route for preserving ecosystem services and biodiversity.
2. In the final report of the United Nations Conference on Sustainable Development, titled "The Future We Want," which was published in June 2012 in Rio de Janeiro (Brazil), the economic and social significance of good land management, including soil, is acknowledged, in particular its contribution to economic growth, biodiversity, sustainable agriculture and food security, the eradication of poverty, the empowerment of women, addressing climate change, and improving water accessibility.

2.2.2 Principles

Principle 1: Despite the fact that soils are necessary for life to exist on Earth, human strain on soil resources is starting to become severe. Careful soil management is essential to sustainable agriculture because it helps regulate the temperature, preserves ecosystem services, and promotes biodiversity. The World Soil Charter provides a summary of our present understanding of soil, the different roles it plays, and the threats these roles provide in the form of a list of nine guiding principles.

Principle 2: Soils vary in shape, characteristics, and the degree of ecosystem services they offer because they are the product of complex actions and interactions of processes over time and space. In order to eradicate poverty and achieve food security, good soil governance necessitates understanding these soil capacities and encouraging land use that respects the variety of capabilities.

Principle 3: Sustainable soil management preserves or enhances the cultural, regulating, and provisioning services that soil offers without seriously affecting biodiversity or the soil functions that underlie those functions. Maintaining equilibrium between the soil's regulating functions for water quality and availability, the composition of the atmosphere's greenhouse gases, and the supporting and provisioning functions for plant growth is especially crucial.

Principle 4: Decisions about soil management are often implemented locally and in a variety of socioeconomic settings. It frequently takes multi-level, interdisciplinary initiatives by numerous stakeholders to produce specific metrics suitable for acceptance by local decision-makers. It's crucial to have a strong commitment to using indigenous and local knowledge.

Principle 5: The range of chemical, biological, and physical qualities found in a soil determine its ability to perform various tasks in major part. To attain sustainability, it is imperative to understand the current status of those properties, their function in soil functions, and the impact of change—both natural and human-induced—on them.

Principle 6: Global biodiversity, which includes everything from microorganisms to plants and fauna, is primarily stored in soils. This biodiversity is essential to maintaining the functions of the soil and, by extension, the ecosystem goods and services provided by the soils. Soil biodiversity must be preserved in order to protect these functions.

Principle 7: Whether actively managed or not, all soils offer ecosystem services important to the control of water at several scales and the regulation of the global climate. These benefits of soils as a global commongood can be diminished by changing how land is used. Only in the context of global evaluations of the contribution of soils to important ecosystem services can the impact of local or regional land use transitions be properly assessed.

Principle 8: Inherently, soil functions and their capacity to supply ecosystem services necessary for human well-being are reduced or eliminated by soil deterioration. To retain the services offered by all soils, major soil deterioration must be minimised or eliminated. This is also far more cost-effective than rehabilitation of degraded soils.

Principle 9: By using the right rehabilitation procedures, it is sometimes possible to restore the fundamental functions and contributions of degraded soils to ecosystem services. This expands the area that can be used for service delivery without requiring a change in land use.

2.2.3 Functions of Soil and Ecosystem Services

It becomes necessary to explicitly identify the tasks that the soil performs as a result of assessments of risks to those functions. Seven soil functions are acknowledged by the proposed Soil Framework Directive (CEC 2006) of the European Union as being at risk from disturbances to the soil:

1. Biomass production, which includes forestry and agriculture.
2. Preserving, filtering, and converting nutrients, chemicals, and water.
3. The reservoir of biodiversity, including ecosystems, species, and genes.
4. The natural and cultural surroundings for people and human activity.
5. Raw material source.
6. Serving as a pool for carbon.
7. Geological and archaeological heritage archive.

Simultaneously launched in 2000 were the European Union's Soil Thematic Strategy (SOTS) and the United Nations' Millennium Ecosystem Assessment (MEA 2005). The European Union launched the SOTS, while the United Nations created MEA. The MEA aimed to evaluate how ecosystem change affects human well-being and develop the scientific basis for policies that promote ecosystem preservation and sustainable use. Costanza et al. (1997) and Daily et al. (1997) established the ecosystem services framework upon which the MEA was built. The MEA was built on top of this structure.

2.2.4 Global Perspective of Arable Soils

According to the World Fact Book 2009, there were 6.790 billion people living in the world in July 2009. Asia accounts for 32% of the world's arable land, followed by North America (17%), and Africa (14%). The lowest amount of arable land per person is found in Africa and Asia, which together account for 46% of all arable land and 71% of the world's population. According to estimates, soil erosion causes 0.3–0.8% of the world's arable land to become unusable for agricultural production each year. Wind and water erosion are to blame for 84% of this deterioration. Around the world, 45% of arable soils are currently impacted in some way (Lal 2007).

2.3 Status and Extent of Acid Soils

2.3.1 Acrisols and Ferralsols

Tropical and subtropical soils called Acrisols and Ferralsols are found in older, heavily forested settings. They are typically very leached, highly worn, and yellowish to reddish in hue. A significant portion of low-activity clay minerals, like kaolinite and sesquioxides (goethite, gibbsite, and haematite), are present in the majority of the soils. Low pH and low quantities of weathered product in the soil solution result from hydrolysis of the silicate minerals and fast elimination of weathering products. In older land surfaces in humid tropical climates, acrisols and ferralsols are more prevalent. They can be found on gently undulating Tertiary and Pleistocene erosion surfaces that have undergone numerous cycles of weathering, erosion, and deposition.

2.3.2 Extent and Location in the World

Acrisols, which are mostly found in the equatorial tropics on ancient, badly worn land surfaces in Southeast Asia, West Africa, and the central region of South America, occupy around 1 billion hectares (Table 2.1; Fig. 2.1). Ferralsols are distributed over around 745 million hectares worldwide, with the wet tropical regions being their primary habitat (Fig. 2.2). They are mostly found on the continental shields of South America and Central Africa. Ferralsols can only be found in regions like Southeast Asia and a few Pacific Islands that have easily weathered basic rock with a hot, humid climate outside of continental shields.

Land use

Tropical rainforests naturally cover acrisols and ferralsols. Acrisols and ferralsols are suitable for use in agriculture as pasture or arable land, although their sustained usage depends much on good land management. Subsistence farming is practised across sizable tracts of the Acrisol and Ferralsol ecosystems, in part through shifting cultivation.

Production constraints

Extremely acidic and nutrient-deficient, Acrisols and Ferralsols frequently contain hazardous quantities of exchangeable aluminium. When there are dry spells during the rainy season, the high levels of exchangeable aluminium may operate as a strong barrier to deep rooting of the plants, causing physiological dryness of crops. Due to their high potential for phosphorus fixation, acrisols, and ferralsols also exhibit low phosphorus availability. For crop productivity, lime, macronutrients, and frequently micronutrients (such as molybdenum), must be provided. The poor cation exchange

Table 2.1 Extent (million ha) of soil groups by continent

Soil group	Africa	Australasia	Europe	North America	North and Central Asia	South and Central America	South and SE Asia	Total
Acrisols	92.7	32.5	4.2	114.8	148.2	341.2	263.0	996.6
Arenosols	462.4	193.2	3.8	25.5	3.5	119.0	94.5	901.9
Calcisols	171.2	113.9	56.7	114.7	95.2	24.3	220.1	796.1
Ferralsols	319.3	0	0	0	0	423.3	0	742.6
Histosols	12.3	1.2	32.8	93.4	99.5	9.2	24.8	273.2
Leptosols	381.5	48.8	64.8	83.3	710.9	246.6	48.8	1584.7
Podzols	11.3	8.5	213.6	220.8	21.8	5.5	6.0	487.5
Solonchaks	48.6	16.6	2.3	0.1	46.9	24.3	48.5	187.3
Solonetz	13.8	38.2	7.9	10.7	30.0	34.6	0	135.2
Vertisols	106.2	90.0	5.9	9.1	11.8	38.0	76.3	337.3

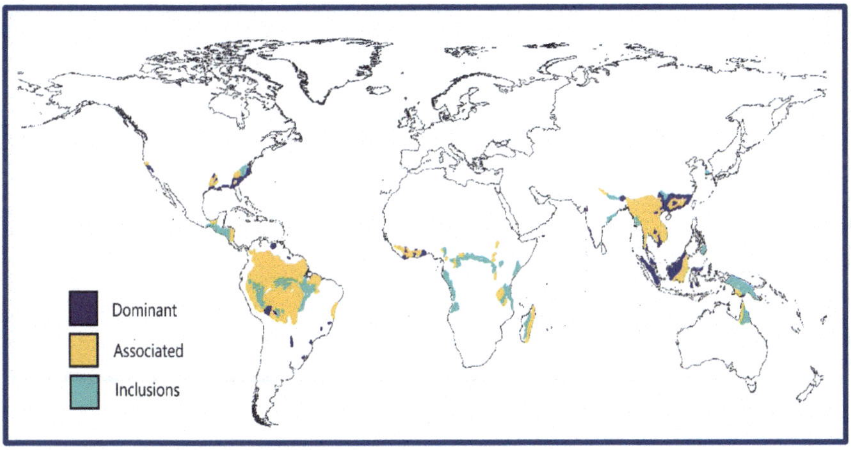

Fig. 2.1 Acrisols in the world. Based on WRB and FAO/UNESCO soil map of the world

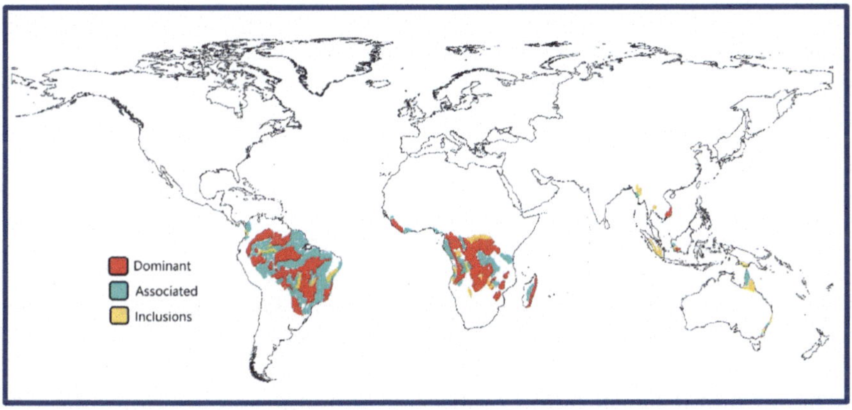

Fig. 2.2 Ferrasols in the world. *Source* FAO-GIS (1998)

capacity of the clay minerals in these soils makes the exchange capacity of the organic matter extremely significant; nevertheless, the organic matter is quickly mineralized in the hot tropical climate.

Nitisols

The Food and Agriculture Organisation classifies soil into 30 different soil types, including nitisol. Nitisols, which make approximately 1.6% of the planet's land area, are primarily found in eastern Africa at higher altitudes, along the coast of India, in Central America, and on tropical islands like Cuba, Java, and the Philippines. They are possibly the most naturally fertile of the tropical soils due to their high nutrient concentration and deep, porous structure. For plantation farming, they are extensively

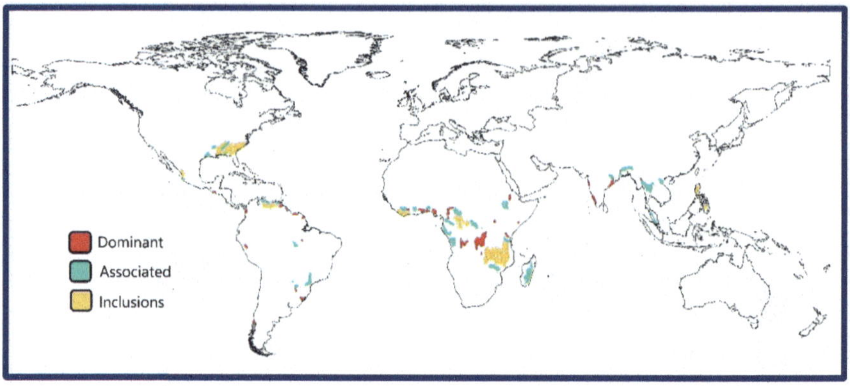

Fig. 2.3 Location of Nitisols in the world

used. By a considerable buildup of clay, nitisols are technically classified. The soil structure is thought to be largely influenced by iron oxides and high water content. The biological activity of the soil has a significant impact on nitisols as well, which causes the upper part of the soil profile to homogenise (Fig. 2.3).

2.3.3 Podzols

A spodic horizon with a buildup of organic matter, iron, and aluminium within 200 cm of the soil surface distinguishes podzols from other soil types. The presence of an albic horizon beneath a spodic horizon frequently denotes a translocation and accumulation process. Underneath the original plant is a cover layer of acid organic litter, followed by a distinct bleached (albic) horizon caused by the leaching out of humus, iron, and aluminium. More humus can form and clay minerals can dissolve in the presence of strong leaching and severely acidic conditions. The results of biological decomposition are also transmitted from the eluvial layers to the spodic horizon together with iron and aluminium. The soil types known as podzols are found in northern, cool-temperate areas with coniferous forests. Podzols can also be found all over the tropics, where they flourish on sand layers of nearly pure quartz.

Worldwide extent and location

An estimated 487 million hectares of podzols are found globally (Table), primarily in the northern hemisphere's temperate and boreal zones. A total of 32 million hectares or less are covered by tropical podzols; the Malesian region (Kalimantan, Sumatra, Irian), northern Australia, and the Rio Negro and Guyanas of South America are notable locations. There, they are primarily found in old, alluvial quartz sands and residual sandstone weathering in humid areas. Podzols soils are infertile because of their sandy texture, low pH level, and low nutritional status. As a result, Podzols' usage in agriculture is restricted. Normally, they are set aside for coniferous forests, light grazing, or being left fallow. However, if irrigation, liming, and fertilisers are used properly, reasonable yields of arable crops could be obtained.

Principal production limitations

Podzols are extremely sandy, acidic, and nutrient-poor. Crops grown on Podzols commonly experience phosphate insufficiency, aluminium toxicity, and nitrification restriction. A thin iron pan that is impermeable may also result from the leaching out of iron and subsequent buildup in lower layers. Then, they might exhibit hydromorphic characteristics. For arable crops, podzols are intrinsically sterile. These soils must first be limed, fertilised, and cultivated in order to grow crops. If created, the iron pan must be broken in order to enhance the drainage of the soil. If fertilisers and lime are consistently applied, podzols can then be fruitful.

2.3.4 Sandy Soils

Sandier soils are those where issues with agricultural land use are most prevalent and a coarse texture predominates in the context of agricultural problem soils. The following sections emphasise the use of arenosols as the principal Reference Group for real "sandy" soils. Since sandy soils are typically dry and hot, their delayed chemical weathering results in only weakly formed soils with limited profile horizon formation. In reaction to extreme temperature changes, physical weathering predominates. Most erosion is caused by wind. In a desert sandy soil, the A-horizon is only marginally developed and difficult to perceive, but it gets more pronounced in semi-arid sandy soils. The structure of soil is exceedingly flimsy and fragile. Sandalloid soils have relatively low organic matter concentrations because of the absence of vegetation cover, which has a very minimal impact on organic material formation. However, the rate and duration of chemical reactions, the creation of organic matter, and biological activity are all governed by the actual soil moisture content in sandy soils. Sand-filled areas can be broken down into two main groups: residual sands, which are produced by the weathering of older, typically quartz-rich soil or rock (typically in tropical climates), and shifting or recently deposited sands, such as those found in deserts and beach areas where sand is accumulated by the selective action of wind and water. Arenosols can be found in arid to (per)humid climates, as well as in severely cold and hot temperatures. Related land formations range from recently formed dunes, beach ridges, and sandy plains to very old plateaux covered in light forests.

Global extent and location

Arenosols are found across approximately 900 million hectares (Table), the majority of which are located in the dry zone. This zone encompasses the southern Sahara, Southwest Africa, and western Australia. Arenosols that have been highly leached can be found across several million hectares in the perhumid tropics, mainly in South

America and certain locations of Southeast Asia. It's possible that there are a few teeny-tiny enclaves of arenosols all over the surface of the globe.

Land use

There are several unused wastelands with sandy soils. Large-scale grazing is possible on sandy soils in the dry zone, although irrigation also makes it viable to grow crops for use in agriculture. Temperate regions employ mixed grazing and arable farming, but during dry spells, supplemental (sprinkler) irrigation is frequently required. Sandy soils are chemically depleted and extremely vulnerable to erosion in the perhumid tropics, thus they must be carefully managed if they are to be used for agriculture.

Main production constraints

The low water-holding capacity and rapid infiltration rate caused by sandy soils' coarse texture serve as the primary production restrictions. Normal low nutrient levels and nutrient retention lead to low intrinsic fertility for agricultural output. The soil readily loses nutrients to leaching. The soils are extremely vulnerable to wind erosion because of their low soil structure. But when proper management is used and water is readily available for crop cultivation through irrigation or due to a shallow water table, sandy soils can be quite productive.

According to von Uexkül and Mutert (1995) and Dalovic et al. (2012), the two global bands with the highest prevalence of acidic soils are the northern belt, which features a climate that is cold, damp, and temperate, and the southern tropical belt, which features a climate that is warm, humid, and tropical. The acidic soils of the world can be broken down geographically as follows: the Americas make up 40.9% of the total, followed by Asia (26.4%), Africa (16.7%), Europe (9.9%), Australia and New Zealand (6.1%). According to von Uexkül and Mutert (1995), woods account for around 67% of the area covered by acidic soil. This is followed by 18% of savanna and prairie vegetation, 4.5% arable land, and less than 1% of perennial tropical lands. According to Nwago et al. (2004), acid soils cover 75% of arable land in Cameroon but just 13% of all land in Kenya (Kisinyo et al. 2014). According to Venter et al. (2001), approximately 11 million hectares of South Africa's soils are moderately acidic, compared to 5 million ha that are highly acidified. 85% of the soils in KwaZulu-Natal have a pH under 5, and around 50% of them have an acid saturation level of greater than 10% (Roberts and Smeda 2001).

In the humid tropics, the vast majority of the soils are acidic and infertile. The two-thirds of the humid tropical land mass are made up of Oxisols and Ultisols, and the remaining 7% is made up of psamments and spodosols (Table 2.2). On a continental scale, tropical South America has the highest concentrations of Oxisols and Ultisols, with Africa and Asia having lower concentrations (Szott et al. 1991).

The main factors limiting plant growth on these soils are aluminium toxicity and insufficient nutrient availability. According to Sanchez et al. (1989), the soils of the humid tropics have 980 (66%), 850 (57%), 565 (38%), 270 (18%), 255 (17%), and 155 (11%) million ha that are, respectively, low in nutrient reserve, toxic to aluminium, high in P fixation, acidic but not toxic to aluminium, highly erodibility with poor drainage, and have low CEC. It has been proposed that agroforestry systems

Table 2.2 Distribution of acid soils in the humid tropics

Order	Humid America	Tropical	Humid Africa	Tropical	Humid America	Tropical	Total tropics	Humid
	Mha	Percent	Mha	Percent	Mha	Percent	Mha	Percent
Oxisols	332	50	179	40	14	4	525	35
Ultisols	213	32	69	16	131	35	413	28
Psamments	6	1	67	15	17	4	90	6
Spodosols	10	2	3	1	6	2	19	1

Source Sanchez et al. (1989)

are the most appropriate ways to manage soils in the humid tropics in a sustainable, productive way (Nair 1984). But there aren't enough facts to back up this claim (Sanchez 1979, 1987). These systems include Latin American coffee and chocolate production systems (Russo and Budowski 1986), home gardens in Asia and Africa (Michon et al. 1986; Fernandes et al. 1984), and alley cropping (Kang et al. 1990).

Agroforestry is seen to be particularly appropriate to marginal soils in the tropics, despite the relevance of such preferred systems on fertile soils. The ability to achieve the three primary goals of agroforestry—nutrient recycling, preservation of soil organic matter, and defence against erosion and runoff—in acidic soils, however, remains to be seen. If this is the case, agroforestry could be a significant alternative to slash-and-burn farming (Szott et al. 1991).

The predominant farming method in the area is shifting cultivation, which is distinguished by the heavy use of labour and extensive utilisation of land. Most farm holdings are fewer than 20 ha, of which 1 to 2 ha are always in production and the remaining portion is fallow secondary forest. Because of the region's present high rates of migration as well as its limited availability and access to land, production is additionally hampered by declining soil fertility brought on by shorter fallow periods and low organic matter.

2.3.5 Distribution of Acidic Soils

Ferralsols and Acrisols, as well as to a lesser extent Plinthisols, Alisols, and Nitisols, are considered acidic soils (IUSS Working Group WRB 2015). Acrisols occupy 87.8 million ha, or 2.9%, of Africa's total land area, whereas ferralsols occupy 312.4 million ha, or 10.3% (Tully et al. 2015). In Africa, phosphorus (P) fixation ranges from slight to high over an area of over 10.8 million km^2, or 35% of the continent's total land area (Eswaran et al. 1997). Of this, 8.2 million km^2 have high soil acidity-induced P fixation. The scope and severity of the acidification of the soil in SSA are both on the rise. According to the Agriculture Transformation Agency (ATA) (2014), soil acidity affects 43% of Ethiopia's arable land, of which 28% is extremely acidic (pH 4.1–5.5) (Abebe 2007). In Ethiopia, soils formed on non-calcareous parent materials like

Nitisols and Acrisols are naturally acidic. In contrast to many African nations where Acrisols and Ferralsols predominate, the primary acidic soil associations in Ethiopia are Dystric Nitisols and Orthic Acrisols with Inclusions of Leptosols (Abebe 2007). Nitisols have excellent agricultural potential because of their solid structure and large water storage capacity. Even immediately following precipitation or during the dry season, these soils are workable (IUSS Working Group WRB 2015). They typically have very low available P and low CEC (Regassa and Agegnehu 2011; Abebe 2007). Because of Ca, Mg, P, and Mo deficiency as well as potential Al and Mn toxicity, strongly acidic soils are infertile (Fageria and Nascente 2014). Iron (Fe), manganese (Mn), and aluminium (Al) are highly concentrated in acidic soils. Al becomes the dominating ion in the soil solution when the pH falls below 5.0 and is soluble in water. Excess Al damages the root apex and prevents root growth in acidic soils (Sivaguru and Horst 1998). Crops cultivated on acid soils are confined with poor nutrients and water availability, which leads to reduced growth and yield of crops (Wang et al. 2006; Marschner 2011). This is because of the poor root growth that reduces water and nutrient intake. Acidic soils like acrisols and ferralsols, which have good physical qualities but low nutrient availability, carry dangerous levels of Al and Mn (IUSS Working Group WRB 2015).

Vertisols are a particular class of fine-textured, poorly drained clay soils that are typically discovered on sedimentary plains, both on level ground and in depressions. On hillslopes and the piedmont plains, vertisols are found in smaller areas. They live in environments with distinct dry and wet seasons, from semi-arid and subhumid tropical to subhumid temperate and mediterranean. The seasonal rainfall ranges from 150 to 2000 mm; however, it often ranges from 500 mm to a thousand mm year. Due to the predominance of swelling clay minerals, particularly montmorillonite, vertisols have unique characteristics. They exhibit a high degree of homogeneity in their physical features, which are mostly determined by the high percentage of clay (40–80%) and the unique mineralogy of the clay. Wide and deep fissures appear in the earth during the dry season. When the clays swell following the first rains, these fractures seal. Churning or pedoturbation, or internal mass motions in the soil, are brought on by stresses created by the swelling. Due to this, a distinctive structure forms with big, planar soil blocks lower in the solum and wedge-shaped structural aggregates in the surface soil. Leaching of soluble weathering products is severely constrained in vertisol locations due to poor drainage, pH is above 7, and there is a lot of readily available calcium and magnesium. These circumstances favour the production of clay minerals of the smectite group, particularly montmorillonite. Free carbonates and gypsum accumulations are frequent in semi-arid regions. Under irrigation, saline and sodic vertisols may form, but this is uncommon in the wild. The reproductive status of most vertisols is moderately high but uneven. Without fertiliser or manure, vertisols may sustainably, albeit modestly, produce crops year after year. This is as a result of the constant surface elevation of subsoil caused by pedoturbation. Any nutritional shortage issues are easily resolved because nitrogen and frequently phosphate levels are also too low. Tricalcium phosphate fixation may take place in small amounts, but it is far less common than in the acidic tropical soils (Oxisols, Ultisols) in wetter areas. The response to potassium varies. Micronutrients and secondary elements are frequently lacking. It has been demonstrated that meticulous fertilisation can double or treble the output of crops like sugar cane and cotton. The surface features of vertisols vary, and these greatly affect how they respond to soil tillage operations. There are two major categories: the vertisols that self-mulch and crusty vertisols. During the dry season, these feature a fine (granular or crumb) surface soil that is 2–30 cm deep. Dessication and soil shrinkage result in the formation of this fine tilth. When such soils are ploughed, the clods break apart after being repeatedly wet and dried. In the dry season, seedbed preparation is hardly ever necessary when this mulch is well-developed. Crusty podsols are big, hard clods that form when they are ploughed and may last for two to three years before they collapse enough to allow the creation of a healthy seedbed (Dudal 1965). If such soils are to be farmed, mechanical tillage is necessary. They have a thin, hard coating. The soil's tensile tension has an impact on whether the characteristic of self-mulching or crusting occurs. Soil texture is one of the elements that affect this stress. While crusty vertisols in the same region were more sandy and less clayey (e.g. 50% clay; 35% sand), it was discovered that vertisols in the Sudan were self-mulching when they had clay levels of 60–80%. When calcareous concretions of a size comparable to sand are present in significant quantities, which appear to disrupt the consistency of the clayey soil material, soils are also significantly self-mulching. According to Jewitt et al. (1979), where rainfall exceeds 500 mm, the surface mulch of vertisols in the Sudan is not well-developed. Other observations have demonstrated that vertisols typically lack calcareous concretions when there is greater precipitation. Extremely high salt concentrations encourage the development of a hard surface crust. Regional distribution of Vertisols is shown in Table 2.3.

Table 2.3 Regional distribution of vertisols (based on data from FAO/UNESCO soil map of the world, length of growing period according to FAO Agroecological zones project, Rome)

Region	Area (T* ha)	Area (T ha) per length of growing period**			
		< 90 days	90–180 days	180–300 days	> 300 days
Africa	107,000	38,700	42,300	22,800	3200
Near and Middle East	5700	1600	3100	1000	–
Asia and Far East	57,800	14,600	38,700	3900	600
Latin America	26,900	2700	6700	10,100	7400
Australia	48,000	–	–	–	–
North America	10,000	–	–	–	–
Europe	5400	–	–	–	–
World total	258,000				

*T (000' ha)

**The growth season is defined as the number of days (in a year) when evapotranspiration is greater than 50% of total precipitation, plus the time needed to evapotranspire an estimated 100 mm of water from excess precipitation (or less if not accessible) stored in the soil profile (FAO 1978). Data on the growth cycle as reported by the FAO Agroecological Zones Project, Rome

2.3.6 Constraints

Vertisols are very plastic-like and sticky when wet, yet they have a very rigid consistency when dried. Vertisols rarely stay in the ideal soil moisture range for tillage (wet soil with a friable consistency), which is a restricted range. However, tillage activities can also be carried out during the dry season with the aid of large gear. In the wet season, mechanical tillage severely compacts the soil. Really soggy terrain is impassable. With a surface mulch or fine tilth, dry vertisols have a high infiltration rate. However, when the soil is wet, the clays swell and close the surface macropores and fissures. Thus, the soil becomes completely impervious. Since the soil has reached its field capacity, there is essentially no water flow in vertisols due to their extremely poor hydraulic conductivity. The amount of soil in which water is held is little because of the restricted ability of soil to absorb moisture. In regions with heavy rainfall, flooding can be a significant issue since standing water can make tillage operations difficult. Open drains can be utilised to drain surface water, or crops can be cultivated on ridges with the excess water being directed to a main collector drain via the intervening furrows. Drainage from moles is essentially impossible. Vertisols have little structural stability, making them extremely susceptible to water erosion. Therefore, slopes more than 5% should not be used for arable cropping. Instead, contour cultivation with a groundcover crop is advised on gentler slopes, or the land can be used for pasturage. To prevent slumping, adequate surface drainage must be given during terracing.

2.3.7 Problem Soils

Table 2.4 and Fig. 2.4 indicate the distribution of problematic soils. Around 151 million hectares of land are covered by planosols and other acidic, seasonally wet soils, mostly in subhumid and wetter climates with marked wet and dry seasons. The largest regions of planosols (767 million hectares) are found in Latin America. They are most frequent on almost level areas of the landscape above ordinary river levels, though they can be found on river plains, such as those of the Brahmaputra (Bangladesh) or the Luena (Zambia). These soils are also found in less strongly seasonal climates, where low hydraulic conductivity of the subsoil or substratum combined with a nearly flat land surface results in periodic waterlogging. The surface horizons of planosols have a lower clay concentration than the deeper, increasingly permeable layers inside them. In general, albeit less obviously, other acidic, seasonally wet soils exhibit the similar pattern. In general, topsoil has lower clay activity (cation exchange capability and moisture retention) than deeper strata. In such severe circumstances, the upper soil horizons have very little structural stability. The sandy topsoil may be extremely compact when dry, but it is not cemented.

Table 2.4 Problem soils in the world (million hectares)

Region	Vertisols	Peat soils	Acid sulphate soils	Planosols	Saline and sodic soils
Africa	105.0	12.2	3.7	15.2	69.5
Near and Middle East	5.7	–	–	–	53.1
Asia and Far East	57.8	23.5	6.7	2.7	19.5
Latin America	26.9	7.4	2.1	67.2	59.4
Australia	48.0	4.1	–	49.3	84.7
North America	10.0	117.8	0.1	12.3	16.0
Europe	5.4	75.0	–	4.0	20.7
World	258.8	240.0	12.6	121.4	322.9

FAO/UNESCO Soil Map of the World (1971–1979)

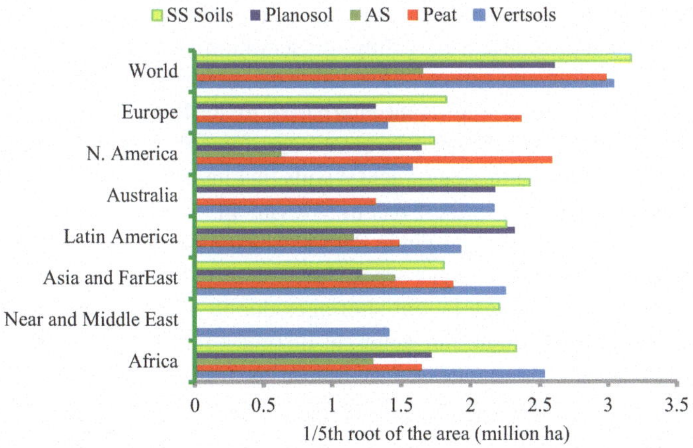

Fig. 2.4 Area (1/5th root) under problematic soils in the World (million ha). SS, Saline and sodic' AS, Acid sulphate (based on data from FAO/UNESCO Soil Map of the World 1971–1979)

2.4 Agriculture on Acid Soils

Acidic soils have in the past defied long-term agriculture. Man has a tendency to settle on high-base quality soils throughout civilization (Sanchez 1976). Al saturation can be the cause of poor crop development in acidic soils (Coleman and Thomas 1967). Generally speaking, pH has a direct impact on plant growth, excluding severe circumstances. However, the hydrogen ion concentration at a relatively low pH of 4.2 may prevent or even reverse cation uptake by plant roots (Jackson 1967). Acid soils' poor fertility and productivity are typically caused by one or a combination of:

- Al, Fe and/or Mn toxicity

- P, K, Ca/or Mg deficiency
- Physical considerations, such as; low water-holding capacity (Oxisols, Spodosols, Psamments), prone to crusting and compaction (Oxisols, Ultisols).

The lack of available nutrients and the toxic effects of aluminium are the primary factors that inhibit plant growth in these soils. Agroforestry systems have been suggested as the most suitable ways to manage soils in humid tropics in a productive and sustainable manner (Nair 1984). This is because agroforestry systems are the most suitable ways to manage soils in humid tropics. But there aren't enough facts to back up this claim (Sanchez 1979, 1987). Furthermore, the agroforestry systems that are most frequently used as examples are situated in regions with a high concentration of naturally fertile base-rich soils like Alfisols and Andosols. These systems include Latin American coffee and chocolate production systems (Russo and Budowski 1986), home gardens in Asia and Africa (Michon et al. 1986; Fernandes et al. 1984), and alley cropping (Kang et al. 1990). Agroforestry is seen to be particularly appropriate to marginal soils in the tropics, despite the relevance of such preferred systems on fertile soils. The ability to achieve the three primary goals of agroforestry—nutrient recycling, preservation of soil organic matter, and defence against erosion and runoff—in acidic soils, however, remains to be seen. If so, agroforestry could be a significant alternative to slash-and-burn farming. Kang et al. (1990) noted the urgent need to assess agroforestry systems in humid tropical settings with a predominance of acidic soils that have high levels of extractable aluminium and low fertility in their recent assessment on alley cropping. A lack of workforce is currently limiting output in Peru's Amazon basin by prohibiting farmers from clearing and caring for larger areas. Additionally, farmers are unable to hire more workers to address these labour shortages due to a lack of savings, credit, or options for off-farm jobs that can generate income (Torres et al. 1983). Because of the region's present high rates of migration as well as its limited availability and access to land, production is additionally hampered by declining soil fertility brought on by shorter fallow periods. The study team identified three agroforestry systems as potential solutions to get around these production barriers: (1) alley cropping systems; (2) modified leguminous fallows to speed up the restoration of soil fertility and minimise the duration of time spent in fallow; and (3) sequential cropping systems that include fruit trees and annual crops. According to Raintree (1987), the applicability of these technologies, which collectively constitute a number of different agroforestry options, will be contingent on the relative availability of land, labour, and capital.

Even while it is possible to find woody species that are able to flourish in acidic soils and can be coppiced, the annual crop advantages that these plants provide are significantly reduced as a result of competition from hedges. The majority of farmers who practise shifting cultivation eventually give up their farms because of the increasing challenges of weed control and/or the deteriorating fertility of the soil. Farmers would have a better chance of increasing their crop:fallow ratio as well as their production per unit of time if they used managed fallows rather than wild secondary vegetation since managed fallows are more effective at weed control and restoring soil fertility. These concepts were investigated by Szott (1987), Szott et al. (1987), and Palm and Szott (1989) during the period of 4.5 years of managed leguminous fallow growth in an abandoned shifting cultivation area.

In many areas of tropical and subtropical Africa, acid soils are common. Traditional slash-and-burn methods and the usage of crop varieties that have been suited to the local environment have given subsistence farmers access to a low but sustainable level of production. However, the shortened fallow periods brought on by population pressure are insufficient to re-establish soil fertility. Technologies created to address these issues should combine crop cultivars resistant to acidity and drought with lime and fertiliser inputs, as well as composting or other conventional methods (Spaargaren 1989). Additionally, technologies should be adapted to the socioeconomic context and expectations of farmers.

According to Coleman and Thomas (1967), aluminium saturation may be to blame for the poor crop development that occurs in acidic soils. At a pH that is quite low, 4.2, there is a possibility that the concentration of hydrogen ions will impede or even reverse the cation uptake by plant roots (Jackson 1967). In addition to nutritional factors, the productivity of many acid soils is affected by physical characteristics such as a low capacity for storing water (Spodosols, Oxisols, and Psamments) and a propensity for crusting, erosion, and compaction (Oxisols, Ultisols). The temperate forest zones of the northern hemisphere contain the second-highest concentration of potentially arable acid soils, behind the humid tropics, which have the highest concentration overall. Large swaths of traditionally corrosive soil have been significantly improved in the majority of highly industrialised nations, especially in western Europe, North America, and the Far East, as a result of the massive use of lime and (phosphate) fertiliser. This is especially true in the Far East, North America, and the United States. On the other hand, dangerously high levels of acidity have been found in the soils of various European countries that are still covered by forest cover (Tamm and Andersson 1989). This was caused by excessive depositions of nitrogen and sulphur from the atmosphere. Acidification of intensively farmed soils is a significant issue in many regions of the world, including the United States of America (Jackson and Reisenauer 1984) and the territory that was formerly known as the Soviet Union (Breburda 1990). The acidity of soils caused by pasture management practices that emphasise the use of legumes is an issue in parts of Australia that are intensively managed (Williams 1980, 1990). Some management techniques that affect soil cation levels include the effects of acid-forming (N) fertilisers on soil acidification, the removal of basic cations from harvested crops, and increased leaching and runoff losses as a result of extended periods when the cropped soil lacks the proper vegetation cover. Other practices include harvesting crops, which remove basic cations, and the removal of basic cations from harvested crops. Most of the wood used in the world comes from forests that thrive in acidic soils. The majority of crops planted in acidic soils include palm oil, tea, coffee, pineapple, rambutan, mangosteen, and jackfruit, as well as several tuber crops like cassava, yams, and sweet potatoes, rubber, chincona, and some other medicinal plants. A little under 1% of the approximately 11 million ha of wooded land that is cleared each year is turned into highly profitable agricultural, while the majority of the remaining area is turned into unproductive grassland. The health of the entire human population ultimately depends on the stability of the global climate and the biodiversity that results from the natural flora that grows on acidic soils. Many unproductive acid soils can now be converted into very productive land thanks to modern science and technology (where climate and topography permit). However, it would not be a good idea to overly employ our current expertise and knowledge to cultivate forested acid soils, as we might need them in the long run more for ecological stability and the preservation of biodiversity than for a quick increase in food production.

Only soils with a pH_2O at their surface zones below 5.5 are regarded to be acidic soils in this context. A total of 3950 million ha (Tables 2.5 and 2.6), or about 30% of the world's ice-free territory, are covered by acid soils. Table 2.7 displays the total area of acid soils by taxonomic orders.

Table 2.5 Global extent of acid soils by soil group in the World

Soil group	Area (million ha)	Soil group	Area (million ha)
Fluvisols	49.8	Podzols	415.2
Gleysols	401.8	Planosols	15.3
Regosols	293.2	Acrisols	731.0
Arenosols	280.3	Nitosols	117.9
Rankers	60.9	Ferralsols	726.6
Andosols	33.9	Histosols	270.2
Cambisols	299.5	Total	3950.5
Podzoluvisols	254.9	% of area	30

Source von Uexkül and Mutert (1995)

Table 2.6 Global extent of acid soils by region in the world

Region	Area (million ha)	% of total area	Region	Area (million ha)	% of total area
Africa	659.4	22	SE Asia and Pacific	314.6	63
Australia and New Zealand	239.3	30	North America	662.4	30
Near East	4.8	1	Central America	37.0	35
Europe*	391.3	37	South America	916.9	14
Far East	212.4	12	North and Central Asia**	512.4	57

Source von Uexkül and Mutert (1995)
*Including former Soviet Union, **Asian parts of former Soviet Union

Table 2.7 World acid soils area as per soil Taxonomy Orders (von Uexkül and Mutert 1995)

Order	Area	
	Million ha	%
Ultisols	864	21.8
Entisols	824	20.9
Oxisols	727	18.4
Inceptisols	561	14.2
Spodosols	415	10.5
Histosols	270	6.8
Alfisols	255	6.5
Andisols	34	0.9
Total	3950	100.0

Table 2.8 Total land area and arable land (million ha) in the world (US Report 1967)

Region	Total land	Crop land	Arale land
Australia/New Zealand	820	32	153
Europe	480	154	174
North America	2110	239	465
South America	1750	77	681
Africa	3010	158	737
Former USSR	2240	227	353
Asia	2740	519	637
Total	13,150	1406	3190

The resources of possibly arable land were evaluated by the US Presidents' Science Advisory Committee (US Report 1967) (Table 2.8). Potentially arable land was defined as an area having soils that were cultivable and reasonably productive for food crops that were suitable for the environment (Abrol 1982). This assessment was backed by FAO (1969) and Buringhetal. It is believed that tropical acidic soils have the greatest potential for agricultural growth (Dudal 1988).

2.5 Acid Soils' Regional Distribution

There are two major geographic areas in the world where udic and ustic moisture regimes led to the development of acid soils. According to the FAO World Soil Resources Map (FAO 1991b), the majority of acid ferralsols, acrisols, and nitosols (Oxisols, Ultisols) were created in the south under tropical or subtropical conditions. In contrast, the majority of acid gleysols, histosols, and fluvisols are found in the north belt. About 40.9%, or 1617 million hectares, of the world's acid soils are found in the American continent (Table 2.9). With 1044 million ha (26.4%) of the world's

Table 2.9 World acid soils—present use (million ha) by main regions (FAO 1991a, b; Sanchez 1987; von Uexkül and Mutert 1995)

Region	Arable and permanent cropped	Forest and woodland	Permanent pastures	Other land	Total area under acid soils
America	63	1080	279	195	1617
Europe	40	311	28	12	391
Africa	11	450	195	3	659
Asia	90	725	117	112	1044
Oceania	8	55	80	96	239
World	212	2621	699	418	3950
% of total	15	65	21	10	30

acid soils, Asia has the second-highest percentage. The continent of Australia and New Zealand has the least amount of acidic soils in the world, covering 6.1% of its total surface, or 239 million ha.

2.5.1 Acidic Soils in Africa

In large portions of western, central, eastern, and southern Africa, acid soils are present. They can be found in practically any climate and frequently bear little resemblance to the underlying rocks. These soils are typically maintained using the "slash-and-bum" method, which involves clearing the vegetation and burning it, then cultivating the soil that has been enriched by the ashes. Local crop types that can withstand the harsh soil conditions have been evolved in numerous places. They often don't produce much, and external factors like a lack of rain, pests, and illnesses can greatly diminish their output. In many nations, population pressure has caused the traditional management system's required fallow period to renew soil fertility to be shortened or even abandoned. As a result, there has been soil erosion, yield decline, vegetation change, and even desertification. For African small farmers to successfully cultivate their lands, there is an increasing requirement for novel management tools. The majority of the research undertaken has focused on a specific aspect of the production system, such as suitable cultivars, reducing soil acidity, or fertiliser responsiveness. Due to the intricacy of the issue, integrated study hasn't often been done on all of the elements and how they interact. However, a diverse research effort will be required if new technologies are to be implemented and used successfully. At the very least, soil science, climatology, agronomy, agroeconomy, and socioeconomics should be included in this multidisciplinary approach.

The upland soils in Africa with acidic soils have a pH (H_2O) of less than 5.5 (pH (KCl) is often less than 4.8) and are generally more or less well drained. In the Oxisol, Ultisol, and Inceptisol orders of the Soil Taxonomy, they cover a wide variety of soils

(Soil Survey Staff 1975). Although they are often very deep, somewhat deep soils that frequently have a stone line predominate in western Africa. Negative chemical and physical characteristics of these soils include a low pH and a high aluminium saturation (often more than 60%), a low amount of available macro- and micronutrients, an imbalance between the different nutrients, a low nutrient holding capacity, phosphorus fixation, frequently low organic matter content, a low water-holding capacity due to coarse-textured surface layers and/or microaggregation, and negative surface characteristics like vulnerability to erosion. These soils also have a low amount of available macro- and micronutrients. The high aluminium saturation and low pH of these soils are well-known characteristics. Under natural circumstances, topsoil pH is typically the highest and aluminium saturation is typically the lowest, not exceeding 10–20%. This radically alters in the subsoil, where an aluminium saturation is sometimes measured to be greater than 60% at a depth of 30 to 50 cm. This sharp growth results in shallow roots, which makes plants extremely sensitive to even brief dry episodes. Exchangeable aluminium is not high in an absolute sense, because of the typically modest cation exchange capabilities. But as research has demonstrated, it appears to be quite challenging to transfer the lime downward to address subsurface acidity. Liming studies conducted in northern Zambia found that 8–10 t ha^{-1} of lime were required to noticeably affect the aluminium saturation at a depth of 60 cm in soils that had an exchange capacity of slightly more than 10 cmol(p+)/kg clay. The tests were carried out in soils that included a high concentration of clay.

The heavily worn African soils virtually never have enough micronutrients. While studies almost always demonstrate a response to sulphur, the majority of crop stands exhibit shortages in zinc, iron, or boron in one way or another. When fertiliser is added, the imbalance between the nutrients is caused by the absence of nearly all critical nutrients. If the type of lime used to reduce soil acidity is not carefully chosen, an imbalance between calcium and magnesium may quickly manifest, harming the crop. In many African soils, particularly the heavily worn soils of the Central African Plateau, phosphorus is also in poor supply. These regions often have thin topsoils with available P concentrations of 5–10 ppm (Bray technique) and subsoils with available P concentrations of less than 5 ppm. These numbers tend to be greater in the shallower soils of western Africa, in soils with a noticeably high organic matter content, and in soils that are frequently fertilised with organic manure. Acidic soils that support forest savanna-style plants often have thin surface layers that are no thicker than 10 cm and have an organic matter (OM) concentration that is no more than 2–3%. After cultivation, the OM concentration often decreases by a third, leaving just 2% to 1%, and commonly even less, in the surface layer. Many African soils exhibit microaggregation, which has a significant impact on both the soil's ability to retain water and how water moves through it. On the Central African Plateau, it is common to come across soils that, when farmed, are known locally as developing a "hardpan." In actuality, these soils lack a pan and are more accurately described as "hard-setting" (Mullins et al. 1987).

Ethiopia, one of eastern Africa's developing nations, is seriously challenged by issues with soil degradation. One of the major issues limiting and preventing lucrative and long-term agricultural productivity in many regions of the world is soil acidity.

According to Negene (2019), soil acidity affects about 40.9% of Ethiopia's total arable land, of which 27.7% has moderately to weak acids (pH 5.8–6.7) and 13.2% has strongly to moderately acidic soils (pH less than 5.5). About 28.1% of Ethiopia's soils are strong acid soils (pH 4.1–5.5), which make for about 43% of the country's arable land, according to Ethio SIS (2014). The absence of phytotoxic chemicals, such as soluble aluminium (Al) and manganese (Mn), as well as nutritional deficits, such as those in phosphorus (P), calcium (Ca), and magnesium (Mg), are the two main reasons that limit the fertility of acidic soils. Numerous acid soil management strategies have been established across the nation to address these issues. According to Edwards and Bell (1989), calcium and molybdenum deficiencies are more frequent than aluminium or manganese toxicity when it comes to limiting plant growth in acid soils of tropical Australia. However, adding soluble fertilisers has the potential to make many acidic soils hazardous for aluminium. The use of lime to reduce acid soil's infertile restrictions on the growth of agriculturally significant plant species is the cornerstone of managing soil acidity in tropical Australia.

2.5.2 Acidity of the Soil's Effects on Crop Production

Acidity causes intricate interactions between elements that limit plant growth and involve the physical, chemical, and biological characteristics of the soil (Robarge 2008). Major physical obstacles to growing crops on tropical soils include soil erosion and poor water retention. On acidic soils, aluminium toxicity and calcium, magnesium, and phosphorus deficiency or scarcity are thought to be the main chemical restrictions on plant growth (Menzies 2003). When it comes to biological characteristics, soil acidity has a negative impact on the activities of helpful microorganisms. This has a significant impact on the decomposition of organic matter, the mineralization of nutrients, and the immobilisation, uptake, and utilisation of nutrients by plants. As a result, crop yields are negatively impacted (Fageria 2009).

In Australia, places with significant rainfall (> 1500 mm yr^{-1}) and a brief dry season are home to acid tropical soils. These regions are located in northeast Queensland's wet tropics. The area consists mostly of slender coastal plains that are bordered by mountains that reach heights of up to 1600 m inland. This area experiences less rainfall overall, evaporation that is only approximately 60% of northern areas, less noticeable seasonality in the rainfall, and significantly lower temperatures than northeast Queensland. The principal crop grown in the Australian wet tropics is sugarcane, which is primarily grown in coastal lowlands and takes up roughly 0.126 million ha. On the elevation of Atherton Tableland, other crops grown include potatoes, peanuts, and maize. In northeast New South Wales and southeast Queensland, the pattern of land use on the acidic soils is also varied. According to Hetherington et al. (1988), sugarcane has a long history of receiving large fertiliser inputs and is a crop that is very tolerant of acidic soil. In comparison with other countries' land use patterns, the wet tropics of north Queensland are almost unique in that they lack any farming practices even remotely resembling shifting cultivation, rainfed paddy rice production, and practically any plantation crops like cacao, rubber, oil palm, or coconuts. The requirement for food crop production is rising as a result of population pressure in Southeast Asian nations; this need is being met by increasing the use of land for multiple cropping and by intercropping food crops with young plantation crops like rubber and oil palm after replanting. In contrast to the situation in Southeast Asia, land use concerns in north Queensland are compelling the creation of more open areas for recreational activities associated with the region's rapidly expanding tourism industry.

2.5.3 *Acidic Soils of Thailand*

Acid soils cover at least 5.5 million hectares of Thailand. This region is used extensively for crop production, although with mixed results. A greater economic benefit from using this property more effectively would be felt throughout the nation. The studies have been conducted to highlight acid soil features in Thailand and research initiatives to enhance the use of these soils across the nation (Parkpian et al. 1991).

Thailand has a vast variety of soil types due to its unique physiography and climate. In Thailand, there are a lot of locations with acidic soil. These soils are used for many of the major cropping systems and are thought to make up a total of 40% of Thailand's agricultural land (19.7 million hectares). These soil types are used to grow rice on about 3 million hectares, though frequently with little success. Improvements in the use of these soils would therefore have significant economic implications. In Thailand, the prevalence of acid soils is more likely to be growing than declining. Farmers are striving to cultivate regions that can no longer support agricultural production without additional adaptations as a result of growing populations and increasing pressure on the available land resources. Large regions that were once in forests are turning into wastelands that cannot support agriculture and where forests cannot regrow because measures are rarely undertaken.

2.5.4 *Acid Soils in India*

Approximately 48 to 49 Mha of India's 142 Mha of arable land are covered by acid soils, of which 25 Mha have a pH below 5.5 and 23 Mha have a pH between 5.6 and 6.5 (Mandal 1997). According to Maji et al. (2012), 6.24 (1.9%) and 24.41 (7.4%) Mha, respectively, of the country's total geographic area, were covered by strongly acidic and moderately acidic soils. Nearly 65% of the soils in India's northeastern region have a pH of less than 5.5, which is considered to be severe acidity (Sharma and Singh 2002). Aluminium (Al) and manganese (Mn) toxicity, low pH in general, and shortages or low availability of some important elements, such as calcium (Ca), magnesium (Mg), phosphorus (P), boron (B), and molybdenum (Mo), are the main growth-limiting factors linked to acid soil infertility. These elements may directly inhibit the growth and development of both below-ground and above-ground plant parts, or they may indirectly do so by impeding the formation and operation of symbiotic or mutualistic relationships with advantageous soil biota communities, which would otherwise slow down biological activity in the rhizosphere. Water stress that occurs from the limiting of root expansion into acid subsurface layers may also be a barrier to plant growth in extremely acidic soils (Adams 1984). The fundamental issue limiting plant growth on acidic soils is well acknowledged to be Al toxicity (Kochian 1995). In order to increase crop productivity, it is necessary to address the low soil pH levels. State-wise distribution of acid soils in India is shown in Table 2.10.

2.5.5 *Acid Soil Infertility Factors*

Toxicities of aluminium and manganese, low pH in general, and shortages or low availability of several critical minerals, such as calcium, magnesium, phosphorus, and molybdenum, have all been linked to acid soil infertility as the main growth-limiting factors (Foy 1984). These elements may inhibit the growth of plants directly or indirectly by interfering with the formation and operation of symbiotic relationships with rhizobia, mycorrhizas, and actinomycetes. Water stress, which is caused by the restriction of root expansion into acid subsurface layers, may also be a barrier to plant growth in extremely acidic soils (Adams 1984). We'll now look at how calcium shortage, molybdenum deficiency, aluminium toxicity, and manganese toxicity affect plant growth in acidic soils of the Australian tropics and subtropics. Despite indications that soils used to cultivate sugarcane have become more acidic over the past 10 to 15 years (Chapman et al. 1981; Wood 1985), no field data have been collected to support the hypothesis that aluminium toxicity limits sugarcane development (Haysom et al. 1986). When grown in rotation with sugarcane or as a replacement crop in areas where sugarcane was once grown, other more susceptible crops, such as navybean, soybean, or maize, are more likely to experience issues Al toxicity (Pearson 1975). Teitzel and Bruce (1973a, b) were the first to hypothesise that pasture species' reactions to lime on soils from Queensland's wet tropics required the reduction of aluminium toxicity. Due to an overestimation of monomeric aluminium in the former, perhaps brought on by the inclusion of some soluble organic matter-aluminium complexes, soil solutions from surface soils had higher monomeric aluminium species activity than those from subsoils (Bruce et al. 1989). The following equation can be used to show how the activity of $A1^{3+}$ in subsurface soil solution depended on ionic strength and aluminium saturation:

$$\left(R^2 = 0.947\right)\mathrm{In}\left(A1^{3+}\right) = 6.97 + 1.96\ \mathrm{In\ Iss} + 0.077\ \mathrm{AI\ sat}\%.$$

In only one out of 18 subsoils, Isbell and Edwards (1988) expected that the toxicity of aluminium would prevent soybean roots from elongating. As well as perhaps in seven of the 18 surface soils from northeast Queensland's wet tropics for which soil

Table 2.10 State-wise distribution of acid soil in India ('000 ha)

State	Strongly acidic	Moderately acidic	Slightly acidic	Total area	Total geographical area	% of TGA
	pH < 4.0	pH 4.5–5.5	pH 5.5–6.5			
Andhra Pradesh	0.0	0.0	2827.5	2827.5	27,504.5	10.3
Arunachal Pradesh	4775.9	1742.7	268.8	6787.4	8374.3	81.1
Assam	23.5	2331.2	2332.7	4687.5	7843.8	59.8
Bihar	0.0	36.7	2324.9	2361.6	9416.3	25.1
Chhattisgarh	156.4	5930.1	4386.6	10,473.1	13,480.5	77.7
Goa	3.6	113.7	191.1	308.4	370.2	83.3
Himachal Pradesh	0.0	157.0	1620.6	1777.6	5567.3	31.9
Jammu & Kashmir	0.0	93.3	1480.1	1573.4	22,223.6	7.1*
Jharkhand	0.0	999.6	5772.1	6771.7	7971.4	84.9
Karnataka	0.0	61.4	3254.7	3316.1	19,179.1	17.3
Kerala	138.0	2789.6	753.2	3680.8	3886.3	94.7
Madhya Pradesh	0.0	1124.7	10,601.8	11,726.5	30,864.1	38.0
Maharashtra	0.0	240.0	4332.6	4572.6	30,771.3	14.9
Manipur	426.9	1437.2	325.1	2189.2	2232.7	98.1
Meghalaya	0.0	1186.3	1054.4	2240.7	2242.9	99.9
Mizoram	0.0	1267.6	777.3	2044.9	2108.1	97.0
Nagaland	118.9	1483.3	55.7	1657.9	1657.9	100.0
Odisha	0.0	261.6	8409.7	8671.3	15,570.7	55.7
Sikkim	278.9	323.4	2.8	605.1	709.6	85.3
Tamil Nadu	264.0	347.3	4294.5	4905.8	13,005.8	37.7
Tripura	56.6	749.0	237.2	1042.8	1048.6	99.5
Uttar Pradesh	0.0	0.0	337.5	337.5	24,104.6	1.4
Uttarakhand	0.0	1183.6	2300.6	3484.2	5336.5	65.3
West Bengal	0.0	555.6	4199.7	4755.3	8875.2	53.6
Other states	0.0	0.0	0.0	0.0	64,381.0	0.0
Total	8242.6	24,414.6	62,141.2	92,798.4	328,726.3	28.2
Area (%)	1.9	7.3	18.9	28.2	100.0	

Source Maji et al. (2012)

*Based on total area of Jammu & Kashmir, including Pakistan occupied area of the state

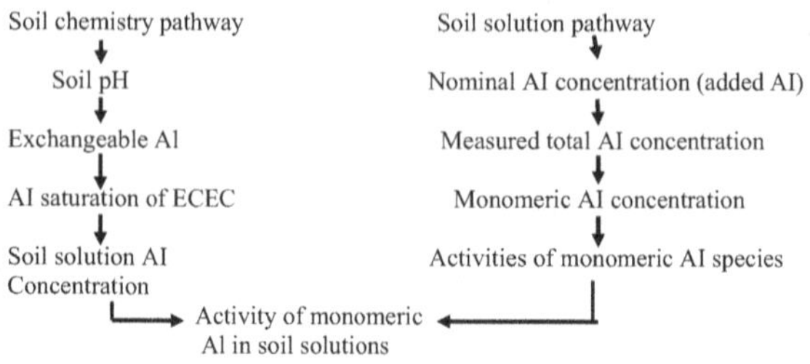

Fig. 2.5 Understanding aluminium toxicity. *Source* Edwards and Bell (1989)

solution studies were available. Figure 2.5 shows the aluminium toxicity limitations to plant growth (Edwards and Bell 1989).

2.5.6 Toxic/Nutritional Deficiencies

Lack of calcium

In acid soils in southeast Queensland (Lee and Wilson 1972; Jones 1982); pasture legumes have been found to be deficient in calcium. Additionally, the sugarcane cultivar Q90 produced in Queensland's wet tropics was where the first signs of calcium and magnesium shortage were discovered in the late 1970s. These discoveries were made a few years after the sugar industry switched to high analysis fertilisers, eliminating the modest calcium inputs that were previously present in the yearly applications of superphosphate.

Toxicity of manganese

There aren't many reports of manganese toxicity inhibiting plant growth in acidic soils in northern Australia, and those that do exist aren't very convincing. Fergus (1954) found that manganese toxicity was present in French beans that were cultivated in a krasnozem with a pH of 4.4 in the Mary River valley, which is located close to Gympie in southeast Queensland. The pH of the surface horizon of the virgin soil was initially 6.0, and regular applications of ammonium sulphate on pineapples added to the problem. However, Siman et al. (1971) claimed that manganese toxicity is a common restriction to growth of French beans on krasnozems in northeast New South Wales. Lovett and Johnson (1969) observed manganese toxicity in tobacco and beans cultivated on acid soils of the Mareeba-Dimbulah area in north Queensland. Although there have not been many instances of manganese poisoning in the field, it is certain that soil acidification will increase the frequency of manganese toxicity in the future.

Deficiency of molybdenum

In Australia's tropical and subtropical acid soils, molybdenum deficiency seems to be relatively common (Bruce and Crack 1978; Rayment 1978). Tropical pasture legumes growing on acid Queensland soils have been reported to respond to molybdenum in numerous instances. Generally, the responses to molybdenum have been found to be higher on acid soils that are of volcanic origin (krasnozems), as opposed to soils that were built on granite. The Orders Entisols, Oxisols, Ultisols, Alfisols, and Inceptisols of the United States are included in the list of soils above. In soils with high levels of accessible nitrogen and by applying lime, which enhances molybdenum availability, molybdenum insufficiency can be concealed (Kerridge 1978). Kerridge et al. (1972) have shown that the molybdenum shortage on some krasnozems (Oxisols) was so severe that reactions to the element happened even when additional lime was present. *Soil deterioration*

Throughout Asia's tropical and subtropical regions, particularly in Southeast Asia, some parts of East Asia, and some portions of South Asia, there are significant areas of acidic soils. In the coastal plains of Southeast Asia and southern China, acid sulphate soils are very common. There are approximately 112 thousand hectares of acid sulphate soils in China and 7.5 million ha altogether in Southeast Asia (Shamshuddin et al. 2014). According to Hicks et al. (2008), the soils in these areas are also vulnerable to exogenous acid, such as acid deposition. Due to their nature, the ferralitic, basaltic, and grey degraded soils that make up nearly one-third of Vietnam are very susceptible to acidification and degradation (NISF 2012). The improper and imbalanced application of chemical fertilisers is a major factor in the acidity of the soil in Vietnam. According to information from the International Fertilisers Association (IFA), the total amount of NPK chemical fertiliser used in Vietnam grew 31 times between 1961 and 2012 (IFA 2012). Increased and imbalanced fertiliser use results in imbalance of essential nutrients in the soil, detrimental to plant growth. The soil may become more acidic as a result of several fertiliser types, particularly organic fertilisers (Nguyen 2014). The existence of sulphate soil in Vietnam is another factor contributing to the process of acidity. The Red River delta's sulphate soil area has grown by 7000 ha, and the soil's S concentration has sharply increased. The delta of the Mekong River has lost 261,000 ha of land, and the amount of total exchangeable cations and total dissolved salts has also decreased. Climate change's detrimental effects on this process in Vietnam may also include temperature rise, precipitation changes, and sea level rise. Over half (55.8 million ha) of the 102.8 million hectares of acid highland soils in Indonesia (pH 5 and 50% base saturation) are suited for various forms of agriculture and plantation (Puslitbangtanak 2000). Technologies with relatively cheap costs are widely accessible to address the issue of soil acidity. These include the use of organic matter, balanced fertilisation, and the choice of crops that can withstand acidity. However, farming practices, such as the use of fertilisers that cause acidification, like ammonium sulphate, make the issue worse (Kamprath 1984).

Acid soils are primarily found south of the Yangtze River in tropical and subtropical areas of China. About 204 million acres, or 22.7% of the nation's total land area, are covered by acidic soils. Southern China's soil acidity has escalated during the past three decades as a result of severe acid deposition and the extensive use of fertilisers containing ammonium. Between the 1980s and 2000s, soil pH in subtropical areas of the nation declined on average by 0.23 units for cereal cropland and 0.30 units for cash agriculture (Guo et al. 2010). Ammonium-based fertilisers in China were mostly responsible for accelerating soil acidification on agriculture, whereas acid deposition was primarily to blame for acidification on forest soils. Although SO_2 emissions in China have decreased by 20% over the previous 10 years, acid deposition is still a significant issue, and its effects on soil acidification are not likely to be reversed very soon (Zhao et al. 2009). In subtropical areas of China, Japan, Southeast Asia, and parts of South Asia, tea is widely grown. In tea gardens, the soils have become significantly more acidic (Wang et al. 2010). The soils in vegetable greenhouses were acidified and salinized as a result of excessive ammonium-based fertiliser application and a lack of leaching (Guo et al. 2010).

2.5.7 Causes and Methods for Acidity

Acidification of the soil is mostly caused by:

(1) persistent rainfall that causes base cations to dissolve locally;
(2) sinking soils that may contain acidic sulphate;
(3) acid deposition, which results from the release of acidic materials into the air or water due to dredging, mining, urbanisation, or industrialization;
(4) excessive use of ammonium-based fertilisers (such as ammonium sulphate) in agricultural practices associated with intensive agriculture; and
(5) the removal of all harvested materials through deforestation and other land use practices, which frequently causes the pH of the topsoil to decrease.

Only the first of these five factors may be attributed to natural causes; the other four are entirely the consequence of the actions of humans. The degree to which the soil is acidic tends to rise with time in natural ecosystems. Older soils have consequently been subjected to more weathering and are more acidic than younger soils, soils from dry climates, or soils derived from more basic rocks, particularly those in humid climates or those formed from acidic rocks (Helyar and Porter 1989; von Uexkül and Mutert 1995). Younger soils have also been subjected to less weathering and are less acidic than older soils. The biggest worry for soil acidification is in soils with a low capacity to buffer pH changes and in soils with low pH levels, such as acid soils in tropical regions with extensive weathering (Harter 2007; Johnson et al. 1982). In assessing a soil's ability to act as a buffer and, thus, its susceptibility to acidification, soil texture, and organic matter concentration are key factors to consider (Helyar and Porter 1989; Steiner et al. 2007). The least protected against acidification are light sandy soils that are deficient in organic matter. Metal sulphides are found in acid sulphate soils, and when they are exposed to oxidation, they form sulfuric acid. Acid sulphate soils naturally form on land in aquatic ecosystems and as a result of modifications to the hydrology and use of the land brought about by humans. In addition to these indirect consequences, acid deposition harms plants, fish, and the chemistry of soil and water resources. Acid deposition is currently a major worry in countries that are fast growing, like China (Chen 2007). Land use and soil management are significant factors in defining the chemical characteristics of the soil. Barak et al. (1997) claim that intensive agricultural practices that involve significant nitrogen fertiliser inputs and significant product removal raise soil acidity. This is a problem in soils that are unable to absorb the rise in H^+, such as those with low levels of negatively charged organic matter, clay, and lime. In fact, the process that turns ammonium into nitrate releases hydrogen ions (H^+) into the soil solution, which may cause the pH of the soil to decrease. This is a concern in both agricultural and woodland ecosystems (Cavelier et al. 1999; von Uexkül and Mutert 1995). Crop harvesting and deforestation remove significant amounts of biomass. The removal of base cations from the soil during harvesting creates the potential for a rise in the acidity of the soil.

2.6 Tropics and Acid Soils

Light texture (sandy) soils are more common on the Northeast plateau and to a lesser extent on the Southeast coast and Peninsula. The most typical crops grown in these sandy soils include peanuts, cassava, sugar cane, black gram, and okra. Wetland rice is also widely farmed in the Northeast plateau. However, lateritic soils, one of the most important upland soils for field crops, make approximately 1.4 million hectares and are primarily found in the Central highlands. This kind of soil is used widely for the growth of maize and soybeans.

2.6.1 Classes of Land Quality and World Agriculture

The ability of the land to carry out a certain function in relation to use is the basis for land quality classes. The production of food, mostly grains, and the ability to react to management techniques that support sustainable land management are the two functions that are most crucial for human well-being. To assess current and future global food production, land quality classes based on grain productivity may be used. According to Blum and Eswaran (2004), just 2.4% of the world's land is classified as Class I, or prime land with few management-related restrictions and optimal soil temperature and moisture conditions for annual crops. 40% of the world's food and feed production comes from these lands, which are primarily found in tropical areas and are heavily utilised for grain production. About 9.5% of the world's land resources fall into land quality classes II and III, which have some usage restrictions. These lands are typically found in temperate areas with mild climates, and they react well to excellent management. As long as conservation tillage is used and fertiliser use is well controlled, there is little to no risk of soil degradation. Better fertiliser management must boost productivity just as much as increasing fertiliser applications. These fields may become much more sustainably farmed if no-till farming techniques are used.

2.6.2 Agriculture Development on Acidic Soils

Acid soils were mostly unexplored until the turn of the century, and only a small number of individuals who practised shifting agriculture found food there. Without major external inputs, traditional management systems could not support ongoing arable crops. Beginning in the late 1900s, the development of acid soils for agricultural use moved slowly until World War II. Only a limited part of deforested land can be maintained under productive and sustainable land use regimes, despite the fact that the majority of deforested land is converted to agricultural use. A number of reasons in the post-war era have led to an extensive invasion of acid soils. Stangel and Uexküll (1990) list the main contributing elements.

- The tropics' rapid population growth.
- The creation and application of more effective machinery for cutting down trees and building roads.
- Progress in breeding and selecting food crops that can grow on acidic soils.
- Rapidly expanding need for tropical timber.
- Growing demand for tropical plantation products like rubber, palm oil, coffee, and tea (Table 2.11).

When developing acid soils for agriculture and agroforestry, von Uexkül and Mutert (1995) suggested three steps, emphasising the need to focus on deforested and abandoned areas of deteriorated acid soil in the tropics. This would require enough funding and the right technological inputs. Following is the three-step development process:

Table 2.11 World arable crops on acid soils: area, yield, production (FAO 1991a; von Uexkül and Mutert 1995)

Crop	% on acid soils	Area (million ha)	Yield t ha^{-1}	Production (million tonnes)
Rice	13	18.60	1.10	20.46
Maize	20	25.82	1.80	46.47
Wheat	5	6.97	2.58	18.05
Barley	20	14.56	2.49	36.33
Oats	70	14.53	2.00	29.05
Rye	85	14.00	2.23	31.22
Soybean	35	20.27	0.55	11.12
Potatoes	60	10.61	15.10	160.21

- A preliminary stage in which leguminous cover crops are established, assisted by liming and P applications, to improve the biological and nutrient cycle of the topsoil and guard against compaction, erosion, and crusting. This phase is vast and "unproductive" in that little money is made from the materials that are gathered.
- A second phase aimed at enhancing and deepening the rooting zone for commercially viable crops to increase farm profitability. Investment is still necessary.
- A third phase of healthy eating and effective management to create a system that is both ecologically and economically successful.

2.7 Acid Sulphate Soils (ASS)

The term "acid sulphate soil" refers to sediments and soils that contain iron sulphides, the most prevalent of which being pyrite. It is possible to successfully handle acid sulphate soils. For the treatment of acid sulphate soils, several management strategies are available. The physical and chemical properties of the acid sulphate soil, the hydrological situation, and the site's environmental sensitivity will all play a role in choosing the best management strategy. Before any disturbance, it is crucial to do an acid sulphate soil assessment at each site to aid choose the best course of action (Das and Das 2015). Acid sulphate soil requires varied management strategies to increase agricultural yield due to the existence of diverse soil groups (Das and Das 2015).

Acid sulphate soils are those that have enough sulphides in them to become strongly acidic (pH 3) when drained and aerated enough for cultivation. It is also referred to as cat clay at times. Included are all soils where sulphuric acid may be generated and have an extended impact on key soil properties. Not all acid sulphate soils are a concern. But when the water table is lowered and the soils are drained or exposed to the air, the sulphides combine with oxygen to generate sulphuric acid. Iron, aluminium, and other heavy metals, especially arsenic, can also be liberated from the soil when this sulphuric acid is produced. After mobilisation, metals and acid can have harmful effects on soil and plants, including the death of vegetation, the extinction of fish and other aquatic life, the acidification of groundwater, the failure of steel and concrete structures, and the killing of fish and other aquatic animals.

Due to their ability to create sulphuric acid when disturbed or exposed to air, these soils are known as prospective acid sulphate soils (PASS). Acid sulphate soils are typically found at coastlines where saltwater inundates the ground. About 2 million acres of land in Thailand and India are covered by acid sulphate soil. There are about 390,000 ha of acid sulphate soils in India. These can be found in Tamil Nadu, West Bengal, Andhra Pradesh, Odisha, and Kerala. About 110 thousand ha of Kerala's land is made up primarily of highly organic sulphates; 26 thousand ha of them are salinity-affected. In West Bengal, the Sundarbans region has the largest concentration of the big group sulfaquents' 280 thousand ha acid sulphate soil. According to estimates, there are around 0.26 million hectares of acid sulphate soils in coastal locations,

Table 2.12 Global distribution of acid sulphate soils

Country	Area (M ha)
Australia	3.000
Vietnam	2.140
Venezuela	2.000
Brazil	1.111
Guyanas	1.246
Indonesia	0.409
Thailand	1.500
Malaysia	0.657
Central America	0.650
Madagascar	0.528
India	0.390

Source Attanandana et al. (1986), Angeloni et al. (2004)

including Kerala and the Andaman and Nicobar group of islands. Table 2.12 shows the area under acid sulphate soils (ASS) in various countries of the world.

2.7.1 What Are ASS Soils

All soils impacted by iron sulphide minerals are referred to as acid sulphate soils (ASS). These soils could either already contain sulphuric acid or have the capacity to do so when the sulphide minerals are exposed to oxygen, leading to deoxygenation or the release of pollutants. This broad definition incorporates:

(i) The wide genetic soil kinds of potential and actual (or active) ASS, first identified by Pons in 1973; Fanning (2002) recognised post-active ASS.
(ii) Acid sulphate soils with material that contains monosulphidic black ooze (MBO), an organically rich substance; material that is subaqueous or waterlogged and includes high concentrations of monosulphides that deoxygenate the air (Sullivan et al. 2002).
(iii) Acid sulphate soils under subaqueous, waterlogged, and drained wetland soil conditions.

Acid sulphate soils form in coastal, estuary, mangrove swamp, coastal back swamp/marsh, and inland salty wetlands (Fitzpatrick et al. 1996). These environments are favourable for the development of sulphide minerals, particularly iron pyrite (FeS_2), which is why acid sulphate soils develop there. One of the byproducts of the sulphate reduction process, which refers to the usage of SO_4^{2-} rather than O_2 during microbial respiration, is iron sulphide minerals. In almost all subaqueous soils in seas, estuaries, lakes, rivers, and wetlands, sulphate reduction takes place naturally. However, a number of variables affect how much sulfidic material will build up in a specific environment. High rates of sulphate reduction and sulphide buildup are mostly dependent on:

(i) elevated sulphate levels in surface or groundwater,
(ii) prolonged soil and sediment saturation that favours anaerobic conditions,
(iii) the presence of labile carbon to support microbial activity, and
(iv) the presence of iron minerals.

Boman et al. (2008) explored the sulphur dynamics of boreal brackish-water sediments rich in metastable iron sulphide, which, when oxidised, has a major impact on the chemistry of stream water. A location close to sea level and a location around 40 m above the latter in a zone of present isostatic land uplift (Finland) were used to collect two cores that extended from the upper oxidised layer (acid sulphate soil) into the underlying iron sulphide-bearing sediment (potential acid sulphate soil). Both of these locations were used to collect the cores. Instead of being generated by a natural elevation, as is commonly believed to be the case, the formation of these famed acidic soils was clearly demonstrated by the findings to have been caused by widespread ditching of farmlands. S and Ni are lost in large quantities above the depth of artificial drainage; however, below this depth, pyrite is widespread, metastable iron sulphide grows with depth, and the concentration of Ni remains rather stable. At one of the places, the elemental S has been preserved as a result of the dynamic processes that take place in the tight region that exists between the reduced and oxidised layers. It was discovered that the sulphur isotopic composition of acid sulphate soils for metastable iron sulphide and pyrite was clearly bimodal, roughly matching to a distribution in stream-water sulphate that had been previously reported from the same region. According to Boman et al. (2008), this indicates that pyrite, and not metastable iron sulphide, is the predominant source of sulphate and acidity in some of the damaged drains.

The Vietnam Mekong River Delta, a sizable low-lying river plain just above sea level that spans 1.6 million ha, contains one of the highest concentrations of acid sulphate soils in the entire world. Acid sulphate soils have large levels of iron (Fe), sulphates (SO_4^{-2}), and aluminium (Al), and when they are drained, they form sulfuric acid (H_2SO_4), which lowers the pH of the soil below 4 (Huu et al. 2022). These metals build up in the topsoil throughout the tropical dry season, which makes them some of the hardest soils to grow crops in since they are poisonous to plant root growth and development and reduce yields. However, the Vietnam Mekong Delta produces 50% of the nation's rice crop, 95% of the country's rice exports, 65% of its aquaculture production, 60% of its exported fish, and 70% of its fruit (Loc et al. 2021).

Andriesse and van Mensvoort (2006) estimated that there are approximately 17 million ha of acid sulphate (AS) soils around the world. These soils pose a risk to the environment due to the high level of acidity they contain as well as the large concentration of potentially transportable metals that they contain (van Breemen 1993; Astrom and Bjorklund 1997). The biggest occurrences of these soils in Europe may be found in Finland (Andriesse and van Mensvoort 2006). They are located on the western coastal plains of that nation and encompass an area that is between 43,000 and 130,000 ha (Yli-Halla et al. 1999). In addition to the coastal lowlands of Southeast

Asia, West Africa, eastern Australia, and Latin America, these soils can also be found in these regions. According to Roots and Åström (2006), the AS soils in Finland are notorious for spilling acidity and harmful metals into sensitive watercourses, which commonly results in the death of fish (Hudd 2000). The parent materials of the acid sulphate soils in Finland are the iron sulphide-containing sediments that were created during the Holocene in the brackish Litorina Sea, which is now known as the Baltic Sea. These sediments are also referred to as prospective acid sulphate soil or PAS soil. PAS and AS soils are typically found between 0 and 45 m above the current sea level (Palko 1994). This is because of the post-glacial isostatic land uplift that took place in the region and is still taking place in the central and northern parts of Scandinavia at a pace of 4–9 mm per year. Pyrite (FeS_2) typically accounts for 1–4% of the dry weight of PAS soils, while elemental sulphur, organic sulphur, and sulphate typically occur in only trace amounts (van Breemen 1973), and metastable iron sulphide (such as mackinawite and greigite) typically accounts for less than 0.01% of the dry weight of PAS soils. On the other hand, metastable iron sulphide is found in the PAS soils (brackish-water sediments) on both the Swedish (Georgala 1980; Sohlenius and Öborn 2004) and Finnish (Backlund et al. 2005) sides of the Gulf of Bothnia. This was discovered by Georgala in 1980 and Sohlenius and Öborn in 2004. This can be deduced from the dark coloration as well as the presence of a significant amount of acid volatile sulphide (AVS). The majority of prior oxidation-mechanism investigations have concentrated on the mineral pyrite since it is typically the primary iron sulphide species in these environments (Arkesteyn 1980; Ward et al. 2004).

Large portions of the coastal floodplains in east Australia are covered in acid sulphate soils (ASS). These hydromorphic pyritic sands release a lot of sulphuric acids when they oxidise. These soils may also come into touch with estuarine tidal waters that have a high ionic strength due to their geographic location. As a result, the acidity (pH) and cation concentrations in soil porewaters and nearby aquatic systems (such as rivers, estuaries, and agricultural field drains) are frequently highly variable. Montmorillonite, which is well known for its shrink-swell characteristics, is present in relatively high concentrations in acid sulphate soils, particularly in the unoxidised gelatinous deeper layers. Changes in the concentration of cations, such as H_3O^+, can have an impact on the interactions between montmorillonite platelets and may also have a major impact on the hydraulic conductivity of materials containing this clay (Le et al. 2008). In order to simplify the natural system, an investigation of the unique effects that various cations (H^+, Ca^{2+}, Fe^{2+}, and Na^+) had on a soil–water suspension while it was contained within a filtration cell unit was carried out. This allowed for the simplification of the natural system. The material coordinates theory was utilised in the process of carrying out the computation in order to ascertain the consolidation coefficient, hydraulic conductivity, and moisture ratio of the filter cakes that had been deposited.

Because of the rapid rate of development in Australia's coastal regions, there is a growing demand for a deeper understanding of the factors that influence the creation of acid and high dissolved concentrations of elements like iron and aluminium from so-called acid sulphate soils (ASS), as well as the movement of these elements. These elements include iron and aluminium. Sulfidic minerals were present in these ancient hydromorphic soils, which are often referred to as Holocene coastal sediments or marine muds. Pyrite (FeS_2) was the sulfidic mineral that was found most frequently. Sulphate, either unoxidised or with the potential to become acidic. It can be difficult to dewater potential acid sulphate soils (PASS) because these soils usually include a high percentage of swelling clays. According to van Oploo (2000) and Smith et al. (2003), extreme weather conditions or disturbances caused by humans, such as an extended drought, drainage, dredging, or excavation, make it possible for oxygen to oxidise sulphides in PASS, which results in the production of true ASS. In the end, the oxidation process creates acidic porewaters and, frequently, large concentrations of dissolved cations, particularly of Fe(II) and Al(III) (Glamore 2003). These acidic porewaters and high concentrations of dissolved cations have the potential to be exported into surrounding streams and waterbodies, where they can kill aquatic life and erode civil infrastructure. According to Tulau (1999), potential acid sulphate soils are characterised by the presence of sulfidic soil components that include pyrite but have not been subjected to an amount of oxidation that would bring the pH level down below 3.5. According to this explanation, the "partial" pyrite oxidation that occurs during PASS can result in significant amounts of ferrous iron and sulphuric acid in the soil porewater:

$$Fe^{2+} + H_2 + 2SO_4^{-2} \rightarrow FeS_2 + 7/2O_2 + H_2O$$

Despite the fact that the soil pH was as low as 2.81, the plant's defence system against the cellular damage induced by reactive oxygen species continued to function correctly. This was the case despite the fact that the pH of the soil had dropped as low as 2.81. According to Liua et al. (2009), this may be the reason why this particular plant species is able to withstand extremely acidic environments so well. ASS are widely dispersed in mining sites as well as coastal floodplains (Lin et al. 2004; Maddocks et al. 2004). Metals with a high potential for toxicity are released when sulphuric acid is produced as a result of the oxidation of sulphide minerals. Toxic chemicals generated from ASS may harm ecosystems in two different ways:

(1) soil deterioration that prevents or even eliminates plant growth removing all vegetation from the land surface (Cherry et al. 2001);
(2) acid runoff that taints aquatic ecosystems downstream (Lin et al. 2007).

Multiple hazardous chemicals found in acid sulphate soils have the potential to interact and have a negative impact on the surroundings they are exposed to. The combined toxicity of several toxicants on a particular organism may be less than, equal to, or greater than that anticipated from the total of the toxicity of each chemical alone, i.e. "antagonistic," "additive," or "synergistic," respectively. As a result, it is insufficient to analyse or forecast the effects of an ASS on ecosystems by simply analysing its chemical properties. Because it enables a clear connection to be made between ASS-borne pollutants and biological response, an eco-toxicological approach is preferred.

2.7.2 Formation of Acid Sulphate in Soils

Sulphides in soils oxidise, resulting in the formation of acid sulphate soils. Sulphide is oxidised to sulphate by a biological reaction when soil is drained and subsequently aerated, creating sulphuric acid. The severity of this is based on the amount of sulphides in the soil. The effects of acid sulphate soil leachate might be short-lived (peaking with the first rains following dry periods) or very long-lasting. Acid from acid sulphate soils that drained 100 years ago is still being produced in some areas of Australia. *Thiobacillus ferrooxidans* bacteria, which produce acid sulphate soil, are the most active oxidizers below pH 4.0. Seawater, groundwater, and surface water sulphate (dissolved salt) are converted into sulphide (another type of chemical containing sulphur) by the *Thiobacillus ferrooxidans* bacteria. The major constituents of acid sulphate soils, metal sulphides, are created when this sulphide combines with metals, particularly iron, in the soil sediments or water column.

2.7.3 Specifications of Acid Sulphate

A sulphuric horizon with a pH of less than 3.5 and sulphide content (yellow colour) is present in acid sulphate soil. Sulphuric horizon is formed of organic or mineral soil material that is at least 15 cm thick and has a pH of less than 3.5 due to sulphuric acid (1:1 soil:water by weight or in a minimum of water to permit measurement). This material provides proof that sulphuric acid is the cause of the low pH value. Additionally, it contains sulphidic substances with oxidisable sulphur compounds. They are either mineral or organic soil components with a pH higher than 3.5. The organic matter content of acid sulphate soils is often significantly higher than that of normal soil, and among acid sulphate soils, the organic matter content tends to be higher, the lower the pH of the surface soil. Sulphaquepts, Sulphaquents, Sulphihemists, and Sulphohemists are a great group concerned in acid sulphate soil. In lowland rice, hydrogen sulphide frequently forms and produces akiochi sickness, which hinders the roots' ability to absorb nutrients. This is due to the fact that, although having a high organic matter concentration, low pH inhibits the decomposition of organic matter and slows ammonification.

2.7.4 Infertility Issues

Acid sulphate soils are typically unproductive and lower agricultural output. Their low productivity may be brought on by the acidity, salinity, toxicity of aluminium, toxicity of iron, nutrient deficiency, low base status, and toxicity of hydrogen sulphide in the soil. The sulphuric acid lowers pH, which reduces the availability of several soil nutrients to plants. As a result of the acid dissolving iron and aluminium from the soil, plants may now access them in dangerous sulphuric by hydrogen ions directly, especially at pH values below 3.5 to 4. However, in this pH range, aluminium toxicity is probably more significant. Acid sulphate soils can have an impact on landscaping and plant development when used as waste. Acid sulphate soils affect animal productivity. Good pasture is discouraged by the acid. Grazing animals may consume excessive amounts of iron and aluminium from their diet of acid-tolerant plant species and acidic water. It generates so much sulphuric acid that no vegetation will grow, leaving the soil's surface looking naked and scorched.

According to earlier research that was based on scant mineralogical evidence, Fe^{2+}, which was present in the higher Loxton Sands, was oxidised and hydrolysed to produce Al (Harrington 2004):

$$Fe(OH)_3 + H^+ \rightarrow Fe^{2+} + 2H_2O + 0.5O_2 + e^-$$

In the low pH Lakes of western Australia, the hydrolysis of Fe2 + has also been proposed as the primary mechanism for acid production (McArthur et al. 1991). One of the most acid-producing processes in nature is the oxidation of pyrite. Since it was not found in the core material, it was not regarded as a potential source of acidity. The reaction is complicated and involves many phases, but it ultimately results in:

$$2FeOOH + 4SO_4 + 8H^+ \rightarrow 2FeS_2 + 15/2O_2 + 5H_2O$$

In Thailand, there are 896,000 ha of acid sulphate soils, which make up around 1.5 million hectares of total land area. The remaining areas (604,000 ha) are spread across the Peninsula, the Eastern Coast, and other locations close to the Thai coast (Rimwanich and Suebsiri 1983). Acid sulphate soils in Thailand can be split into two groups based on the origin of the materials (marine or brackish water): (a) prospective acid sulphate soils (Sulfaquents) and (b) acid sulphate soils (Sulfaquepts and Sulfic Tropaquepts). As previously indicated, potential acid sulphate soils are widespread and dispersed over coastal locations. These soils cannot be ploughed after rain since they are in mangrove vegetation and have a sticky nature. True acid sulphate soils, on the other hand, are those that have grown over deposits of brackish water, include jarosite mottles, and have a sulfuric horizon present at depth that indicates the acidity level. These acid sulphate soils also include a lot of organic content, are relatively easy to drain, and may be cultivated (Moorman 1963).

In the Lower Lakes of South Australia, which include Lake Albert, Lake Alexandrina, and the tributaries of Finniss River and Currency Creek, numerous types of acid sulphate soils are described. Due to the intense drought conditions in the river watershed, several locations in the lower reaches of the River Murray have recently seen declining and record low water levels. The exposure and drying of hypersulphidic materials has the potential to have negative environmental effects such as metal mobilisation, deoxygenation of water, development of malodors (H_2S, organo-S compounds), and soil and water acidification. These soils can have hazardous effects on aquatic biota and human health if they are not addressed (Fitzpatrick et al. 2010). They can also cause water bodies to become more acidic. At numerous areas in the Lower Lakes, CSIRO has carried out acid sulphate soil characterisation study (Fitzpatrick et al. 2008b, e). This research was done to find where acid sulphate soils are found and to understand how they are distributed. The specific type and extent of these soils could not be determined with certainty in the Fitzpatrick et al. (2008a, b, e) baseline research due to the reconnaissance scale character of those studies and the low sample density.

Acid sulphate soils are organic substrates (peat), sediments, or naturally existing soils that arise when land is flooded. These soils include iron sulphide minerals or their oxidation products, most commonly in the form of the mineral pyrite. The discharge of sulphuric acid from the soil may also cause the release of iron, aluminium, and other heavy metals, including arsenic. Once mobilised in this way, the acid and metals can have a number of negative effects, including the death of plants, the acidification of groundwater and other bodies of water, the extinction of fish and other aquatic life, and the eventual collapse of concrete and steel structures. The most vulnerable soils and sediments were created within the previous 10,000 years, following the last significant sea level rise. Sulphate in the ocean mingled with sediments from the land that included iron oxides and organic materials when the sea level rose and submerged the land. By reducing sulphate ions in seawater or groundwater to hydrogen sulphide, lithotrophic bacteria like *Desulfovibrio desulfuricans* are able to breathe in these anaerobic conditions. As a result, very fine-grained, highly reactive framboid crystals of iron sulphides, such as (pyrite), are created, which react with the dissolved ferrous iron. Warmer temperatures provide these bacteria with more ideal conditions up to a point, increasing the likelihood that iron sulphides will occur. As opposed to pyrite formed in more temperate climes, tropical wet environments, such as mangrove swamps or estuaries, may have larger concentrations of the mineral. The pyrite is stable up until it comes into contact with air, at which point it quickly oxidises and turns into sulphuric acid. Acid is still being released from some acid sulphate soils in Australia that drained 100 years ago (Sammut and Lines-Kelley 2000).

Millions of hectares worldwide are covered with fine-grained sulphide-bearing sediments, principally iron sulphides, according to Andriesse and van Mensvoort (2006). Many times, the exploitation of land on these soils has resulted in significant environmental problems. The sediments were formed during the Holocene in the old Baltic water under anoxic conditions, and they eventually emerged from the water as a result of isostatic land uplift (up to 9 mm per year). These sediments were later found in Finnish coastal locations. Currently, these soils can be found up to 100 m above sea level (Erviö 1975). Today's river and stream estuaries are still producing sulphide-containing sediments. The metal sulphides oxidise during the drainage of these sediments, releasing a sizable amount of sulphuric acid. As a result, the pH rapidly drops, and acid sulphate soils are created. The soil is typically defined by an abundance of secondary iron mineralization, such as schwertmannite, according to Österholm and Åström (2002), and the oxidation depth frequently lowers to 2 m. Large amounts of metals are released into the porewater under these extremely acidic conditions as a result of the breakdown of metals from oxidised sulphides and weathering silicate minerals. At high water flow circumstances, acidic porewater that is rich in metals, like copper, is more common. Significant ecological damage has been attributed to the discharge of metals like Al, Cd, Co, La, Mn, Ni, and Zn into recipient streams (Macdonald et al. 2007; Palko et al. 1985; Weppling 1993). Most of the metals emitted precipitate in river estuaries (Nordmyr et al. 2008). These fine-grained soils, which are mostly utilised in agriculture, are easy to cultivate if drained and limed due to their well-developed structure and high nutrient content. The majority of this soil oxidation and subsequent acidity of water courses has occurred over the past 50 years, as efficient subsurface pipe drainage has become more widely used (Saarinen et al. 2010).

2.7.5 Chemical Process

Soils containing pyrite (FeS_2), commonly called cat clays, can become extremely acidic (pH 4) due to the oxidation of pyrite into sulphuric acid (H_2SO_4) during drainage. In its most basic form, this chemical reaction is as follows (Acid Sulphate Soil Technical Manual 2003; Dent 1986).

$$2FeS_2 + 9O_2 + 4H_2O \rightarrow 8H^+ + 4SO_4^2 + 2Fe(OH)_3(\text{solid})$$

The result, $Fe(OH)_3$, iron(III) hydroxide (orange), precipitates as a solid, soluble mineral, immobilising the alkalinity component while the acidity in the sulphuric acid remains active. High levels of aluminium (Al^{3+}, released from clay minerals under the effect of the acidity) are formed as a result of acidification, which is bad for vegetation. Additional results of the chemical reaction include:

1. A pungent gas called hydrogen sulphide (H_2S).
2. Sulphur, a solid yellow substance.
3. Iron(II) sulphide (FeS), a solid that is dark grey or blue.
4. Red solid hematite (Fe_2O_3).
5. Brown mineral called goethite (FeO.OH).
6. Brown mineral known as Schwertmannite.
7. Iron sulphate substances, such as jarosite.
8. H-Clay (hydrogen clay, a stable mineral with a high proportion of adsorbed H^+ ions, but low in nutrients).

Fe^{2+}, the ferrous ion, and Fe^{3+}, the ferric ion, respectively, are the bivalent and trivalent forms of iron that can exist in the body. While the ferric form is insoluble, the ferrous form is soluble in water at any pH. The ferric forms predominate more when the soil becomes more oxidised. Colours ranging from black, brown, blue-grey, red, orange, and yellow can be seen in acid sulphate soils. By adding seawater, the hydrogen clay can be improved because the magnesium (Mg) and sodium (Na) in the seawater replenish the hydrogen that has been adsorbed.

2.7.6 Acid Sulphate Soil Materials

1. The Australian Soil Classification now refers to soil components as sulphuric materials (Isbell 1996). In essence, these are soil minerals with a pHw 4 due to sulphuric acid oxidation.
2. Sulfidic materials are components of the soil that are characterised by the presence of detectable sulphide minerals (this property is defined as having a sulfidic S concentration that is more than or equal to 0.01%). This phrase ought to be utilised in a descriptive manner (e.g. sulfidic soil material or sulfidic sediment), and it ought to adhere to the standard definitions utilised by other scientific fields, such as geology and ecology (e.g. sulfidic sediment). At the moment, the level of Cr-reducible sulphide that must be below in order for the procedure to be effective is 0.01%. Other techniques for identifying sulfidic materials include X-ray diffraction, visible Raman spectroscopy, infrared spectroscopy, and regional variability of subaqueous and terrestrial acid sulphate soils and their characteristics for the Lower Lakes region of South Australia.
3. When a layer between 2 and 10 mm thick is incubated aerobically at field capacity, hypersulphidic material is identified by experiencing a significant (overall decrease of at least 0.5 pH unit) drop in pH to 4 or lower (1:1 by weight in water, or in a minimum of water to permit measurement) (Sullivan et al. 2010). Hypersulphidic material is defined as sulfidic material with a field pH of 4 or higher.
4. Monosulphidic materials are described as organic-rich, subaqueous, or wet solids with measurable amounts of monosulphides by Sullivan and colleagues (2010). According to Sullivan et al. (2009), monosulphidic black oozes are unique substances that are distinguished by their consistency of gel.

2.7.7 Acid Sulphate-Containing Soils, Existence, and Magnitude

The bulk of tropical low-lying coastal regions with acid sulphate soils were originally home to mangrove swamps. Potential acid sulphate soils have a pH that is almost neutral, but after drainage and oxidation, the pH considerably rises. The tropics are home to about 10 million hectares (ha) of known and potential acid sulphate soils, with the global total most likely not exceeding 14 million ha.

The majority of dryland crops cannot thrive on acid sulphate soils due to the dangerous levels of aluminium and the scarcity of phosphorus. The two primary detrimental elements for wetland rice are low phosphorus and dangerously high amounts of dissolved iron. The main issues in acid sulphate soils with near-neutral potential (Sulfaquents, Sulfic Fluvaquents) include high salinity, poor bearing capacity, uneven land surface, and the possibility for significant acidification during droughts. In comparison with soils (Sulfic Tropaquepts, Sulfic Haplaquets) that have the pyritic substratum at larger depths, young acid sulphate soils (Sulfaquepts), which feature this horizon near to the surface, are frequently more acidic. Huge swamps with sulfaquepts that are highly acidic, like those in Vietnam, can produce acidic floodwater that could destroy crops grown on surrounding better land.

Even though they affect all of the earth's crust and only make up a small portion of the overall problem soils that the world's population must deal with, acid sulphate conditions are a short-term problem that plagues the people. However, a number of the other types of problem soils are typically found in remote locations far from dense populations of people and in areas with other serious issues. The formation of the acid sulphate soils would therefore be of urgent importance since they typically occur in good conditions for food production, frequently close to densely populated coastal areas or river plains. Area under different soil groups is shown through Fig. 2.6.

Acid sulphate soils can be found all over the world, in environments ranging from the cold and temperate to the hot and humid tropics. On the other hand, it would appear that tropical deltas are the places where they are found in the greatest numbers (van der Kevie and Yenmanas 1972). Every continent possesses acid sulphate soils, with the exception of Australia. The most extensive habitats can be found in the northeastern region of South America, West Africa, and South and Southeast Asia. It has been stated that large areas of Finland, Sweden, Holland, and the United States all have soils that are high in acid sulphate content. Both Kawalec (1972) and van Breemen and Pons (1978) documented the spread of acid sulphate soils across the entire planet. Kawalec studied the distribution of acid sulphate soils in South and Southeast Asia. In terms of their location, the most majority of acid sulphate soils are formed in coastal regions from sediments that are either very recent or only slightly recent. Most of the time, the areas where they have been able to generate marine and estuarine deposits are restricted to those that are situated in close proximity to an ocean. In general, you can find them in the following environments: (a) Coastal areas influenced by marine or brackish water; (b) Freshwater backswamps that were formerly brackish; (c) Lake bottoms that have been drained and polderised, like in the Netherlands; and (d) Inland continental environments with innate pyritic parent rock or locations where lignite and coal have been mined for a long time. Table 2.13 displays the distribution of acid sulphate soils that are present and potentially present throughout East and Southeast Asia. The fact that Indonesia, Thailand, and Vietnam

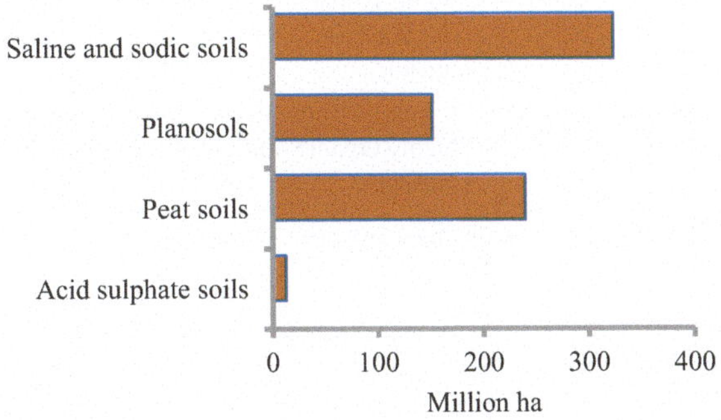

Fig. 2.6 Area under different soil groups. Adapted from Beek et al. (1980), based on data from FAO/UNESCO

account for two-thirds of the total five million hectares should be highlighted. More than one million hectares were reported by Moormann (1961) in South Vietnam, almost 200,000 by Watts (1969) in West Malaysia, and over 1.5 million by Pons and Kevie (1969) in Thailand. In the tidal marshes of Bangladesh, Myanmar, and India there is a possibility of finding certain acid sulphate soils, which are also referred to as "Sulfaquents." In Thailand, there are about 800 thousand hectares of acid sulphate soils, most of which are found in the lower Bangkok Plain's delta flat. The rest is dispersed along the peninsula's and the nation's east coast's coastal regions and estuaries. The majority of the delta flat, which makes up 35.6% of the total land area of the Bangkok Plain, is covered in the most significant acid sulphate soils in the nation. These soils are acidic and sulphate-rich. Both the northern and southern acid sulphate soils of the delta flat are between 5000 and 6000 years old, whereas the soil in the middle of the delta flat is between 2500 and 3000 years old.

Table 2.13 Distribution of acid sulphate soils in Southeast and East Asia (based on van Breemen and Pons 1978)

Country	Area (000' ha)	Soil classification
Bangladesh	400[a]	Sulfaquents, Sulfaquepts
China	67	Sulfaquepts, Sulfic Haplaquepts
India	(110 + 280[a])	Sulfaquepts that are highly organic, Sulfaquents that are partly (26,000 ha) influenced by salinity
Indonesia	2200	The majority of highly organic Sulfaquents, Sulquepts, and Sulfihemists
Japan	21	Sulfaquepts, Sulfic Haplaquepts, and a potentially acidic shallow sea bottom
Malaysia	160	Very organic sulfaquepts and sulfaque, acidified mangrove swamps caused by lobster mounds
Myanmar	180[a]	Sulfaquents
Philippines	7	Sulfic Tropaquepts, Sulfaquepts, highly organic Sulfaquepts
South Korea	3	Sulfic Haplaquepts, Sulfaquents
Thailand	670	Sulfic Tropaquepts (550,000 ha), Sulfaquents (10,000 ha), Sulfaquepts (50,000 ha), Sulfaquepts, Sulfaquents, Sulfaquepts, partially highly organic
Vietnam	1000	Primarily Sulfaquepts, with minor portions of Sulfic Tropaquepts and highly organic Sulfaquents. Partially highly organic Sulfaquents are also present

[a]These figures are probably gross overestimates

2.7.8 Chemistry of Aluminium in Acid Sulphate Soils

In the crust of the planet, aluminium is the metallic element that is most prevalent. It is therefore not unexpected that a lot of the features of acid soils are governed by the chemistry of aluminium as hydrolysis of the aluminium ion creates a moderately strong acidic environment. In fact, Jackson (1963) characterised aluminium bonding as the guiding principle of soil science. The oxidation of sulphides to sulfuric acid, which may also react with a number of aluminium compounds in soil, results in the production of further acidity in acid sulphate soils. Coleman and Thomas (1967) just provided an outstanding review on the age-old subject of soil acidity.

Hydrolysis

Chemically, it has been assumed that the first-stage hydrolysis reaction is how aluminium is hydrolysed in aqueous solutions:

$$\mathrm{AlOH^{2+} + H = Al^{3+} + H_2O}$$

Where because of convenience, the water of hydration is withheld. The thermodynamic equilibrium constant for this reaction, expressed as ion activity as negative logarithms, is given by:

$$\mathrm{AlOH + pH - pAl = pK}$$

Numerous measurements of this hydrolysis constant have been made (Sillen and Martell 1964, 1971), and the majority of researchers agree that pK is roughly 5.0.

We should first establish the problem or problems that need to be solved by this research before discussing the various types of research that ought to be conducted on acid sulphate soils. There are two key components to the acid sulphate issue:

(i) the development of hazardous circumstances for agricultural plants or fish as a result of the quick creation of acid, once or repeatedly, in excess of the soil's rapid buffering capability;
(ii) the inefficient or expensive removal process, or the transient or long-term neutralisation of the acid.

The potential acidity that may still be present in the subsurface and the substantial phosphate fixation that can be found in the surface and subsoil of many reclaimed acid sulphate soils are two examples of less serious but longer-lasting difficulties that may require an adjustment to the management practices that are used. These problems may require an adjustment to the management practices used. While research increases the likelihood of successfully modifying the vegetation and drainage status of big acid sulphate sites, it is important to understand and take into account the effects these changes will have not only in the immediate area but also in other parts of the world. The land itself, the land and water downstream (including coastal streams and the sea), the plant and animal life locally and downstream, as well as the people living in or around the region to be developed, could be impacted by these changes. For the purpose of making effective use of acid sulphate soils, it is required to conduct both fundamental and supportive forms of research.

2.7.9 How Acid Sulphate Soils Were Created

Pyrite production and pyrite oxidation are the two key processes involved in the creation of acid sulphate soils.

Pyrite formation

The formation of pyrite is the result of a number of peculiar circumstances that exist in tropical coastal locations. The sulphate in seawater, which is biologically converted to sulphide in the anaerobic mud, is the source of the sulphur in pyrite. The bacterial sulphate reduction process requires an energy source, and organic matter is typically easily accessible due to the extensive plant growth in coastal locations. Additionally, ferrous iron (Fe^{2+}) must be readily available; it is often produced by reducing insoluble ferric compounds that are a byproduct of clay's weathering. Pyrite is formed and accumulates in tropical coastal wetlands as a result of the interaction of sulphate from seawater, organic matter from plant development, anaerobic conditions brought on by the exclusion of air oxygen by the excess water, and the presence of Fe^{2+}.

$$FeS_2(s) + 8HCOa - (aq) + 4H_2O(J) = Fe_2Oa(s)$$
$$+ 4SO_4^{2-}(aq) + 8CH_2O_2^{+} + 1/2O_2(g)$$

This chain of events comprises reducing all sulphate to sulphide, which is then oxidised to disulphide ($S_2{}^{2-}$) using Fe (III) and oxygen as oxidants.

According to Pons et al. (1982), the direct precipitation of Fe^{2+} and $S_2{}^{2-}$ to form FeS_2 produces pyrite in a matter of days under favourable conditions (Goldhaber and Kaplan 1974). On the other hand, the solid–solid reaction of FeS and S to form FeS_2 is a time-consuming process that can take months or even years to produce measurable quantities of pyrite. When exposed to air, the fine-grained pyrite found in tidal sediments easily oxidises, releasing Fe (II) sulphate and sulphuric acid:

$$Fe^{2+} + 2SO_4^{2-} + 2H^+ \rightarrow FeS_2 + 7/2O_2 + H_2O$$

Depending on what happens to the iron, acid will be produced along with pyrite oxidation. Every "mole" of pyrite yields one equivalent of H^+ if all iron is oxidised and stays in solution as Fe^{3+}. When all iron is oxidised and hydrolysed into solid ferric oxide, acid production is at its peak (van Breemen 1982):

$$FeS_2 + 15/4O_2 + 7/2H_2O \rightarrow Fe(OH)_3 + 2SO_4^{2-} + 4H^+.$$

Pyrite is oxidised by dissolved Fe (III) more quickly than by oxygen.

$$15Fe^{2+} + 16H^+ + 2SO_4^{2-} \rightarrow FeS_2 + 14Fe^{3+} + 8H_2O$$

2.7.10 Ozone-Related Products

In the soil, the majority of the iron (II), hydrogen, and sulphate ions generated during pyrite oxidation go through further processes. Some of the significant oxidation products include the following.

Jarosite

Only acidic (pH 2 to 4) oxidised (Eh > 400 mv) conditions can produce jarosite. The majority of acid sulphate soils prominently display the pale yellow (2.5–5 Y 8/3–8/6) jarosite ($KFe_3(SO_4)_2(OH)_6$) mineral. It frequently takes the form of earthy void fillers or mottles in the soil matrix, and it always produces a clear X-ray diffraction pattern. The metastable jarosite in acid sulphate soils will eventually hydrolyse to goethite. The pale yellow mottling is so distinctive that it serves as a diagnostic criterion for identifying acid sulphate soils along with pH (USDA 1975). However, some acid sulphate soils, particularly those with significant levels of organic matter, lack jarosite (Kosaka 1971).

Metal Oxides

Iron from oxidised pyrite primarily manifests as Fe (III) oxides. The oxidation of dissolved Fe (II) sulphate generated during the pyrite oxidation can produce fine-grained goethite directly and quickly, or more slowly, via hydrolysing jarosite. Hematite may make up a portion of the Fe (III) oxides in the B horizon in the better drained, more fully formed acid sulphate soils, producing noticeable red mottles.

Gypsum

If the soils have a significant amount of a neutralising substance like $CaCO_3$. Acid sulphate soils' ability to produce gypsum is a sign that they are generally good for farming.

2.7.11 Limitations of Soils with Acid Sulphate

Acid sulphate soils typically yield low. One or more of the following unfavourable conditions, such as soil acidity, salinity, aluminium toxicity, iron toxicity, a lack of important nutrients, a low base status, and hydrogen sulphide toxicity, may be to blame for their low production.

pH of the soil

The documented negative effects of acid sulphate soils may be caused by hydrogen ions directly, especially at pH values below 3.5 to 4. However, in this pH range, aluminium toxicity is probably more significant.

Salinity

Salinity frequently has an impact on the acid sulphate soils found in tidal zones. According to Pasricha and Ponnamperuma (1976), salinity worsens other toxicities by weakening plants and possibly increasing the concentration of iron and aluminium in the sick solution. Additionally, in many young acid sulphate soils, the tidal electrolyte content rises to dangerous levels upon soil reduction (Ponnamperuma 1972).

Aluminium toxicity

Aluminium toxicity is one factor that can hinder some plant species' ability to flourish. High levels of Al impair phosphorus, calcium, and potassium uptake as well as cell division and the functioning of several enzyme systems (Rorison 1972). Al toxicity affects the majority of plants growing on acidic sulphate soils with pH values below 4. Al toxicity (68 ppm in soil solution) was found to be the cause of the death of rice plants raised on soil from Vietnam.

Iron toxicity

Rice becomes hazardous when dissolved iron concentrations exceed 300–400 ppm (Ponnamperuma et al. 1955). After six weeks of submersion, Fe II content in a Vietnamese soil reached 800 ppm, according to Nhung and Ponnamperuma (1966). Two weeks after being immersed in an acid sulphate soil, 5000 ppm of water-soluble Fe was discovered in a subsequent investigation (Ponnamperuma 1972). After 30 days of incubation, Tanaka and Navasero (1966) discovered 500 to 700 ppm of Fe in solutions taken from two Malaysian and Vietnamese acid sulphate soils.

Low nutrition

Content the most significant issue with acid sulphate soils, particularly the Sulfic Tropaquepts of the Bangkok Plain, is phosphorus shortage. This is because iron and aluminium toxicity and damaging salinity are not present. The phosphate reaction is boosted by nitrogen supply (Attanandana et al. 1981).

Reduced base status

Acid sulphate soils are created when bases are eliminated as sulphate and aluminium occupy the majority of the exchange complex. As a result, Ca and K deficiencies in acid sulphate soils are likely to occur. In the top 35 cm of a Thai acid sulphate soil, exchangeable Ca content of 3.5–5.0 me/100 gm and exchangeable Mg of 3.0–3.2 me/100 gm have both been reported (Sombatpanit 1970). The soils from lobster mounds had the highest exchangeable Ca value, with a value of 1.6 me/100 gm, according to Andriesse et al. (1972). According to Pham-Huu-Anh et al. (1961), the 0–35 cm, the cat clay horizon, and the reduced horizon each contained 3.0, 2.9, and 3.6 me/100 gm of exchangeable calcium; the equivalent values for magnesium were 4.1, 5.7, and 9.0 me/100 gm, indicating the soil's brackish (or marine) origin.

The dangers of hydrogen sulphide

It has been established that hydrogen sulphide is hazardous to rice plants due to its restriction of the roots' ability to oxidise. According to research by Moraghan and Patrick (1974), sulphate reduction started very slowly at pH 5 and accelerated as the pH rose to 7. Van Breemen (1975) demonstrated that sulphate reduction occurs when the pH rises above 4.5 to 5, resulting in sulphide, which then precipitates as FeS. H_2S and toxicity may be a possibility if the soil has little Fe^{2+}.

2.7.12 Kinds of Soils

The geographical distribution of possible acidic sulphate soil elements is explained in relation to climate zones and the physiography of coastal plains. This provides an explanation for the geographic distribution of these soil elements. An excess of pyrite relative to acid-neutralising components is referred to as potential acidity for this reason. The fundamental components and environmental factors that must exist in order for pyrite and acid-neutralising components to form, accumulate, or sediment are enumerated and explained in terms of current and historical physiographic settings, as shown by examples from the world around us. These conditions are favoured by low rates of sedimentation or land subsidence in relation to the sea level as well as a tropical humid climate. A relative rise in land elevation or an increase in sedimentation rate reduces the impact of tidal flushing and mangrove vegetation. This causes tidal creeks to rapidly silt up and closed shorelines to quickly accrete at levels much above mean sea level. The potential acidity of tidal deposits is thereby diminished.

Due to altered hydrology in the Murray–Darling Basin (MDB) system, subaqueous and marginal soils have been able to significantly accumulate sulfidic material. Sulfidic materials pose minimal to no risk of acidification if left undisturbed and submerged in water (Fitzpatrick et al. 2009). However, when exposed to air, the existing sulphides combine with oxygen to make sulphuric acid and frequently produce sulphuric materials (soil with a pH below 4 as a result of sulphide oxidation). Significant amounts of acidity may then be released into the water when these sulphuric compounds are covered with it. Since 2006, record low inflows and river levels have caused many wetlands, rivers, and lakes in the MDB to dry out, exposing sulfidic material and causing soil acidification (Fitzpatrick et al. 2009).

2.7.13 Acid Sulphate Soil's Characteristics

The primary soil properties of these soils may be affected by the presence of sulphuric acid in these soils or by the potential for sulphuric acid formation in these soils (Pons 1973; Dent 1986; Dent and Pons 1995). Additional potential concerns include the deoxygenation of soil or surface waters as well as the release of contaminants when the sulphide minerals are exposed to oxygen. Fanning (2002) asserts that ASS recognises the following three main genetic soil materials:

1. Sulphuric material with a pH of less than 4 that contains sulphuric acid (and maybe iron sulphide minerals) is found at shallow depths. Previously, these substances were referred to as genuine, active, or raw ASS substances.
2. Previously known as prospective or unripe ASS materials, sulphidic, hyper-sulphidic, or hyposulphidic materials contain mostly iron sulphide minerals (FeS_2).
3. Materials that are predominantly composed of monosulphide minerals (FeS) that are still wet. Monosulphide black ooze (MBO) was the former word used. One of the byproducts of the sulphate reduction process is the formation of iron sulphide minerals.

High rates of sulphate reduction and sulphide buildup are necessary for:

1. Large amounts of dissolved sulphate.
2. Soils and sediments that have been sufficiently saturated for an extended period of time to encourage reducing conditions (such as subaqueous and soggy soils; see glossary).
3. Labile carbon is available to support microbial activity.
4. Availability of minerals containing iron or iron in solution.

Hypersulphidic soil materials are sulphide-containing and capable of acidifying to sulfuric materials, which can be environmentally harmful if exposed to air through disturbance. Pyrite is oxidised by exposure, producing 4 mol of acidity (i.e. 2 mol of sulphuric acid) for every mole of pyrite. When oxidation results in a pH of 4 or less, this process converts hypersulphidic material to sulfuric material (Isbell 1996); take note that a sulfuric horizon has a pH of 3.5 or less according to the US definition in Soil Survey Staff (1999, 2003). Pyrite first begins to weather when it is exposed to oxygen (in air). If submerged in water under decreasing conditions, pyrite is often stable. Sulfuric acid is typically produced by bacteria that are present near the soil surface and the repeated wetting/drying cycles that fine-grained pyritic framboids are subjected to. Several authors have provided detailed descriptions of the intricate steps involved in the oxidation process and the creation of sulfuric compounds (Fanning and Fanning 1989; Nordstrom and Alpers 1999; Bigham et al. 2002). When highly acidic sulfuric materials with sulphides and low concentrations of acid-neutralising minerals (like calcite) are exposed to air by drainage or excavation, a major acid-drainage issue may arise. If the amount of acidity created surpasses the soil's ability to act as a buffer, acidification results.

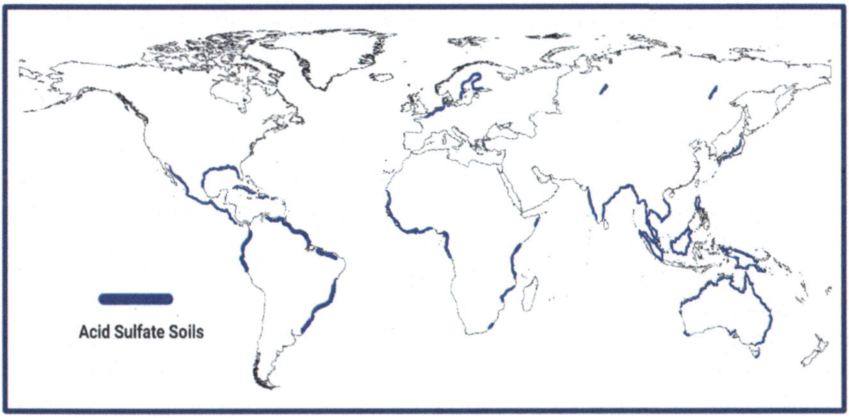

Fig. 2.7 Acid sulphate soils are distributed generally throughout the world (based on Simpson and Pedini 1985 and Fitzpatrick 2005)

Around coastal locations, acid sulphate soils are common (Fig. 2.7). They are also locally linked to freshwater wetlands and saline groundwater in some agricultural areas. According to Fitzpatrick et al. (2002), coastal acid sulphate soils in Australia cover an estimated 80,000 km^2 and are found beneath floodplains and estuaries close to where most Australians reside. Acid sulphate soil disturbance is frequently linked to dredging, excavation, dewatering, and marina construction activities. Potential acid sulphate soils (PASS) are acid sulphate soils that have not been disturbed. Actual acid sulphate soils (AASS) are acid sulphate soils that have undergone disturbance (Acid sulphate soils Technical Manual 2003).

2.8 Options for Acid Sulphate Soil Management

The main management option would be to alter the situation (to slow or stop the rate and extent of pyrite oxidation), even though the main cause of the formation of sulfuric materials is increased disturbance of hypersulphidic material (promoting sulphide oxidation). This can be accomplished by either rapidly drying sulfidic material to stop the biological processes that cause acid production or by maintaining sulfidic material in anaerobic conditions under saturated circumstances. Once oxidation has taken place, reversing the process by rewetting is challenging because this is the moment when acid dangers and metal mobilisation risks are at their peak. Because of this, it's crucial to characterise ASS landscapes and create reliable ASS risk or hazard maps at the proper sizes. To choose the best options for drainage and the most suitable management of the soils when they are drained, it is essential to understand the properties and processes of the soil as well as the hydrology of the water table. The management of ASS during development can enhance the quality of discharged water while safeguarding the environment and infrastructure. Applying low-cost land management techniques typically results in such gains (Dear et al. 2002). The following general concepts are followed by ASS management, listed in order of priority (Fitzpatrick et al. 2008b).

Since the 1960s, numerous watercourses in midwestern Finland have routinely experienced brief but intense spells of poor water quality during high water flow due to acidity and metals from acid sulphate soils. Due to this, there have been sporadic large-scale fish extinctions and sharp declines in fish populations. However, a low pH is a common feature of many watercourses in northern Europe that is brought on by naturally occurring organic acids, non-carbonate bedrock, and historically significant amounts of acid rain (Toivonen and Toivonen 2011). As a result, the significance of acid sulphate soils has not always been well discussed. However, in contrast to other acid sulphate soils tested in Finland, some of them even appear to be leaching significant amounts of Fe. These soils include Al, Cd, Co, Ni, and Zn to recipient streams. In rivers, streams, and lakes, patterns of acidity, sulphate, sulphur isotopes, organic carbon, and metals were discovered (Toivonen and Toivonen 2011).

(1) The greatest cause of fish mortality in the watershed is unquestionably acid sulphate soils, which are also the most significant source of metals in the lake.
(2) The load from dredging operations and the release of organic (humic) acids is negligible for the lake as a whole.
(3) Only 7% of the entire catchment's young near-field sediments provided for around 20–30% of the total acidic burden.
(4) The southernmost part of the acid/metal load entered the lake system.
(5) It appears that this load can be kept from spreading into the northern half of the lake system by controlling exits to the sea.

2.8.1 The ASS Millennium Ecosystem Assessment

According to the Millennium Ecosystem Assessment, human health and well-being are significantly dependent on ecosystem services. These services include those that provide sources of food, water, fuel, and livelihood; regulate the spread of contaminants and diseases; provide a variety of intangible benefits through cultural services; and provide support services that ensure the continued production of the aforementioned ecosystem services. There is little room for debate that significant disturbance of ecosystem processes has a negative impact on human well-being, given that human well-being is dependent on the resources in question. The degree of the adverse health consequences will vary depending on how dependent the population that is exposed is on the resource and ecosystem that is being disrupted, such as whether or not it is their only source of food or water. Since subsistence populations do not have the choice of purchasing food goods from other locations, their health is frequently very directly tied to the productivity of the ecosystems in their area. According to Corvalan et al. (2005), the connection between the productivity of ecosystems and human health in urban regions is not as clear, but its significance is not diminished. People who face negative effects on ecosystem services, such as diminished aquacultural and agricultural outputs, are also the people who have the least capacity to deal with them. This is due to the fact that ASS are usually located in high population density places inside developing nations. Even though ASS only make up a small portion of the world's problem soils, the difficulties connected with their cultivation and the environmental ramifications can thus have considerable effects on health, particularly on a local and individual basis. This is especially true in situations where there is a high concentration of ASS. In addition to the provision of food and a social safety net for rural communities, which contributes to an improvement in health (Corvalan et al. 2005), the MEA recognises the value of agriculture in terms of its ability to generate employment possibilities and economic benefits.

2.8.2 Availability of Clean Water Supply

Increased soil acidity causes a number of potentially hazardous metals to become mobilised, which can then be transmitted to neighbouring streams and contaminate drinking water sources, rendering local water supplies unfit for irrigation or household use (Minh et al. 1997). This can lead local water supplies to become unfit for use in home or agricultural settings. According to Howa and Cuong (2006), places that are classified as ASS have greater water concentrations of a range of metals than areas that are not classified as ASS. In a study that was carried out in the Mekong Delta (Buschmann et al. 2008), it was discovered that the arsenic and manganese concentrations in 71% of the investigated wells that were utilised for drinking water were found to be at levels that were higher than the standards set out by the WHO. As a consequence of the draining of arsenic-rich estuarine peat soil for the purpose of urban expansion in western Australia, it was discovered that the pH of the groundwater there can reach as low as 2.4, and arsenic concentrations can reach as high as 7300 g/L. Because it is essential for the production of food, the care of stock animals, drinking, personal hygiene, cleaning, and cooking, insufficient supply of freshwater can have significant detrimental consequences on the wellness of humans. Vorosmarty et al. (2005) found a significant correlation between financial investments in freshwater services and improvements in human health, which is evidence of the close relationship that exists between the availability of freshwater and one's state of well-being.

2.8.3 Assessment and Management

The environment's hydrological circumstances and the soil's physical characteristics play a major role in determining the best practices for managing water in acid sulphate soils. Application of water management strategies for reclamation on a wide scale may be severely constrained by labour, the volume of water required, and potential environmental effects (Tuong 1993).

Due to the high expenses necessary and the severe acidity, acid sulphate soil reclamation by chemicals (such as liming) is typically not an option. The only realistic strategy to manage these soils, according to various writers (Dent 1986; Sen 1988), is through appropriate water management. Because most acid sulphate soils experience adverse hydrological circumstances, such as waterlogging during the rainy season and a lack of water during the dry season, water management is crucial. As one of the initial steps in the reclamation process, it is not unexpected that several water management strategies have been tried out by farmers, governmental, and private organisations. Both success and failure have resulted from their efforts, and mistakes have frequently been made again. This demonstrates the uneven distribution of information on the reclamation of acid sulphate soils that is currently available.

2.8.4 Surface Water Management

It is well known that the rate of decrease in surface soils containing acid sulphate is sluggish (Ponnamperuma et al. 1973; Van Breemen 1976). Recent investigations have provided more proof that this is especially true for mature acid sulphate soils. Ritsema et al. (1991) found that extended submersion had no impact on the pH of genuine acid sulphate soils. In two out of three ripe acid sulphate soils, Satawathananont et al. (1991) found no discernible decrease in soluble aluminium after 6 months of submersion. Keeping these ripe soils submerged is obviously not a necessary precaution from the perspective of soil amelioration, albeit the presence of surface water will aid in reducing the capillary ascent of toxins to the root zone. Surface layers of unprocessed acid sulphate soils are frequently abundant in organic materials and iron. Even while the pH may have raised to higher than 4.4, continuous submersion will not always result in a reduction of water-soluble AI concentration to a non-toxic level (Satawathananont et al. 1991). Rice plants typically perform better in higher spots of wet fields than those in continually submerged locations in raw acid sulphate soils. The increase in dissolved iron I1 and/or H_2S and the acidity of the surface water are two possible "side effects" of submersion that could account for this (Van Breemen 1976). Hanhart and Ni (1991) suggested intermittent drainage as a middle ground. They discovered that intermittently drained rice developed stronger root systems, fewer empty grains, and greater weight per panicle. Continuous submersion of rice produced more tillers but also produced significant bronzing symptoms.

2.8.5 Leaching

Leaching of acid sulphate soils has been shown to have advantageous effects, according to Ponnamperuma et al. (1973), Kivinen (1950), and van Breemen (1976).

The majority of early research, however, used small, homogenised samples or small pots, where the water flow pattern and the physical characteristics of the soil were significantly altered from the actual field circumstances. This is problematic because such features determine the rate of leaching and, consequently, its usability in the field. Unaltered soil core samples and field tests show that acid sulphate soil leaching is a sluggish, water-intensive process. Aluminium can become toxic at concentrations as low as 2 ppm in solutions, making it the most likely threat to agricultural crops (Dent 1986). Aluminium leaching from undisturbed soils is extremely challenging. The aluminium concentration in the root zone could not be reduced to a level that was safe for plant growth using even one metre of water. According to field research, the leaching of undisturbed soils is a progressive process that is only allowed to occur at drainage canals' edges, where there is insufficient water exchange between the canal and the soil body, at a distance of about 10 m. Similar progressive leaching was observed naturally at Pulau Petak, Kalimantan, where there is an average annual rainfall of about 2300 mm, according to Konsten et al. (1990). They calculated that 1.3% of the total acidity measured in the upper 65 cm was leached annually. At this rate, it will take at least 80 years to get rid of all acidity. In areas with poor drainage, it can take longer.

According to Dent and van Mensvoort's prediction from 1993, a significant portion of the ASS would turn into wasteland because of growing production costs and declining water supplies. It was acknowledged that in order to prevent further issues, management must take place on a landscape size and concentrate on creating sustainable systems. Because the ASS processes are intricate, any evaluation or modelling of the consequences must take into account not just the soil environment but also the receiving bodies, the climate, and the chemical interactions between these and geographical factors (Kijne 2006; Wilson et al. 1999). Aspects of social, legal, policy, and economic management will also need to be taken into account (Powell and Martens 2005). In a simulation model examining the best water management options for ASS in an Indonesian rice-growing region, Bronswijk et al. (1995) discovered that abandoning the area, flooding it, and finding the farmers another suitable location was the best course of action.

The findings demonstrated that acid sulphate soils might be enhanced to increase soil productivity for rice farming. For these areas, liming, green manuring, and liquid organic fertilisers are suitable techniques. Sulphate soils with a slight acidity are simple to handle. With green manure and liquid organic fertilisers, a small amount of marl (499 kg/rai; a rai is comparable to 1600 m^2 or 0.16 ha) is sufficient. For soils with medium fertility and moderately acidic sulphate levels, a suitable rate of green manure and liquid organic fertilisers (1000 kg/rai) is sufficient. According to Maneewan and Sanguansubpayakorn (2007), in severely acidic sulphate soils with high fertility, an adequate rate of marl (1123 kg/rai) combined with green manure is sufficient. It is clear that fertilisation and liming are effective techniques for paddy crops. In irrigation areas, farmers are encouraged to use organic fertilisers like green manure, farm manure, compost, and liquid organic fertiliser that can produce high rice yields and high farmer incomes in order to reduce acidity, prevent acidification, and reclamation.

Potential acid sulphate soils have a high pyrite content and are poorly drained soils. In the field, the pH of the soil will be neutral or slightly acidic. According to Attanandana and Vacharotayan (1986), drainage causes the soil to become significantly more acidic, which adversely impacts plant growth by making aluminium and iron toxic while also decreasing the availability of phosphorus and other minerals. The term "para" or "pseudo" acid sulphate soils refers to soils where the acid has been neutralised or leached out to the point where root development and microbiological activation are no longer hindered. These soils still exhibit jarosite mottles, high soluble sulphate concentrations, and high percentages of Al saturation in the clay complex, but do not contain toxic levels of free or soluble AI.

2.8.6 Amelioration

Strong acidity is a major factor in why millions of hectares of acid sulphate soil in South and Southeast Asia are left fallow or are unsuccessfully farmed. The productivity might be enhanced by cultivating rice on these lands. Future food deficits in South and Southeast Asia could be lessened. Acid sulphate soils are among the most problematic soils and need to be improved using the proper techniques. Since the soils are lowland and suited for rice cultivation, Attanandana and Vacharotayan (1986) list the following as amelioration strategies for wetland rice:

1. Drainage and leaching.
2. Submergence.
3. Liming.
4. Addition of manganese dioxide.
5. Application of nitrogen, phosphorus, and potassium, as well as using rock phosphate as a phosphate source.
6. Application of resistant varieties.

2.8.7 Acid Sulphate Soil's Effects

Potentially acidic sulphate soil disturbance can be harmful to aquatic environments, fish, and plant life. Several effects may result from the discharge of acidic leachate into groundwater and surface waterways, according to Sammut and Lines-Kelley (2000).

- Ecological harm to riparian and aquatic environments, including iron precipitation, an increase in fish disease outbreaks, and fish mortality.
- Impacts on aquaculture and fisheries projects in estuaries (increased disease, loss of spawning grounds, etc.).
- The contamination of groundwater with heavy metals such as arsenic, aluminium, and others.

- A decline in agricultural productivity due to soil pollution with metals, primarily aluminium.
- Corrosion of steel and concrete pipes, bridges, and other subsurface assets causes damage to infrastructure.

2.8.8 Agriculture's Effects

Potentially acidic sulphate soils, sometimes known as cat clays, are frequently left uncultivated or, when they are, they are planted beneath rice to keep the soil moist and prevent oxidation. Normally, subsurface drainage of these soils is not advised. Because of climatic dry spells and a lack of irrigation water, acid sulphate soils cannot be kept constantly moist while grown; nevertheless, surface drainage may aid to eliminate the toxic and acidic compounds (produced in the dry spells) during rainy seasons. Acid sulphate soils may eventually be recovered with the aid of surface drainage. Thus, Guinea Bissau's native people have succeeded in improving the soils, although it has taken them many years of meticulous management and labour.

According to a study conducted in South Kalimantan, Indonesia, which has a perhumid environment, upland rice, peanuts, and soybeans can be successfully grown on acid sulphate soils with widely separated subsurface drainage systems. Older residents of the area had already established themselves here and were able to grow a range of products (including tree fruits) by using hand-dug drains that extended from the river through the land and into the back swamps. Although the crop yields were small, they were sufficient to support a respectable standard of living. Acid sulphate soils that have been reclaimed have a well-established soil structure and are very permeable, yet they are unproductive because of the leaching that has taken place.

2.8.9 Sediment Nature and Qualities

The size of the acid sulphate problem still has to be determined through some fundamental research. It is still necessary to test the overall geogenetic hypothesis of pyrite production in sediments. There are currently no widely applicable practical field methods for characterising sediments that include pyrite. It may not be as crucial as previously believed that the total levels of potential acid exceed the soil's capacity to neutralise them. But because they may coexist in close proximity, soils with and without such an excess of potential acid need to be thoroughly identified and documented. This may be accomplished most successfully once the parameters of sedimentation, vegetation, and pyrite accumulation have been determined, and the characteristics of the sediments and the processes of sedimentation have been thoroughly comprehended and accurately recorded for every estuary region. It is still necessary to identify and map using the existing or improved criteria and procedures, even though this basic information may not be sufficient in the future.

Acid sulphate soils can be used for a wide range of purposes, including:

(i) Protection against flooding, a wildlife reserve, or low-level agricultural water storage for all sorts of acid sulphate soils, including some nearby peat sites.
(ii) Oil palm, wetland rice, rubber for inland, freshwater environments.
(iii) Generation of biomass energy or the production of wood and bark for places closer to the sea, which are primarily covered with mangrove vegetation and some with palm vegetation.
(iv) Fishponds with brackish water in regions with a consistent dry season for the salty areas along the shore.
(v) Fish, shellfish, and other marine life nurseries for the saltwater edges. This brings us full round to the initial application that was mentioned: preserving the ecology.

Thus, the freshwater and saltwater zones can be distinguished as the two primary types of acid sulphate soil environments, each with distinct development options and extremely diverse socioeconomic ramifications.

2.8.10 Management of Acid Sulphate Soils

Numerous management strategies are employed to lessen the likelihood of acid sulphate soils and their impacts. The following general methods for reclamation are recommended:

1. Avoidance is the most effective management technique for managing acid sulphate soil and is to be taken into account at all locations. When kept under wet, undisturbed circumstances, acid sulphate soils are inert. Avoidance is frequently the least expensive and most environmentally friendly choice.
2. Disturbance Reduction: If acid sulphate soils must be used for crop cultivation, there should be as little disturbance as possible. To effectively minimise disturbance, a thorough acid sulphate soil research must be undertaken. The location can be properly defined before investigating measures to reduce disturbance.
3. Liming is the primary and most crucial method of neutralising acid sulphate soil. Alkaline and neutralising minerals must be physically incorporated into the soil. Due to its alkaline pH, lime buffers any acids created and brings the pH of the soil to a healthy level. Acidity created in disturbed acid sulphate soils can be neutralised by properly combining the right kind and quantity of lime. Lime requirements are reduced if acid sulphate soils are leached in the early stages of acidification.
4. Re-flooding: The goal of re-flooding is to slow the pace of pyrite oxidation and neutralise real acidity. Re-flooding depends on creating an environment where the reduction of Fe, Mn, S, and N may occur. The decrease of these substances is what causes acid soils to often see an increase in pH following waterlogging. Re-flooding is another method for controlling water table height that can be utilised to stop PASS or ASS from further oxidising.

5. Re-flooding with seawater has the potential to hold the current acid leachate and push it deeper into the soil profile, limiting the export of oxidation products and slowing diffusion into the tidally exchanged seawater. Even if further modification is later required, this strategy has the advantages of being straightforward, affordable, and improving the problem.
6. Sandalloid material that contains iron sulphides is appropriate for hydraulic separation. To hydraulically separate the sulphides from the sandy components, sluicing or hydrocycloning is used. This method works incredibly well in regions where the sediments are deficient in organic matter and include between 10 and 20% clay and silt. The method's isolated sulfidic material needs specific handling, which could involve strategic reburial or neutralisation.
7. Sulphate-reducing bacteria may be able to stop the oxidation of pyrite by reinstating reducing conditions inside the bunded area. As a result, the sulphide formation processes that take place in the mangrove soils outside the bund wall would be reinstated. Chemical changes are brought about by bioremediation in the soil, water, and any accumulating sediment. If there is sufficient in situ microbial creation of acid-neutralising capacity, bioremediation would be a natural process and cost-effective.
8. In situ soils can be covered with clean fill if earthworks do not alter groundwater levels. Undisturbed in situ potential acid sulphate soil can be covered with a sizable amount of clean fill. It is not possible to specify a minimum fill depth for developments that are either residential or commercial/industrial. Instead, the inquiry should be used to identify the severity and extent of acid sulphate soil at each location in order to estimate the appropriate depth of fill. Any accompanying infrastructure can be installed inside clean fill once a site has been completely covered, preventing excavation or dewatering from upsetting any in situ acid sulphate soil.
9. Flooding and intermittent drainage: In order to reduce the amount of acid sulphate soil, soils may be flooded (anaerobic) or submerged in water. With this technique, rice farming is essentially the only use for the region. But regrettably, drought does occasionally happen and quickly promotes soil acidity. When potential acid sulphate soils are flooded, the water frequently turns acidic, harming crops. Intermittent drainage led to better root systems, fewer empty grains, and greater weight per panicle in rice. Continuous submersion of rice produced more tillers but also produced significant bronzing symptoms.
10. Water table management: In acid sulphate soil, the sulphuric horizon may occasionally be covered by a non-acidifying layer. It is then feasible to empty the area such that just the sulfuric layer remains submerged (anaerobic). After damage from excessively intensive draining has been done, the soils can be repaired by raising the water table.
11. Deep soil mixing is done using a hollow-flight auger with a large diameter (one to three metres). It also uses special "paddles" to mix the dirt. The soil slurry is mixed with lime or cement and various binding agents as holes are drilled into the soft substrate, creating firm, supportive columns in the soil following cementation.

12. Crops that can be grown successfully include rice, which is the most ideal because of its high acid tolerance. The use of rice as a crop decreases the toxicity of iron and aluminium by raising the pH of the soil in acid sulphate soils. Upland rice, peanut, and soybean farming have shown promising results on acid sulphate soils with a widely dispersed subsurface drainage system.

2.8.11 Conventional Management Practices

Slash-and-burn shifting agriculture is a traditional management method that is still used extensively. There are two versions. To release nutrients and improve the pH of the soil, one form involves clearing an entire field and burning logs and vegetation; the other involves leaving trees intact and burning only branches. The termite mounds, which have a higher fertility than the surroundings, are occasionally effectively utilised by farmers. This is particularly true if the soils they operate on are very deficient. Systems based on burning grass have evolved in savanna settings where trees are uncommon. For instance, the "mala" method in the Congo entails mowing the grass and putting it in bunds. The grass inside these bunds is then burned before the topsoil is applied, and as soon as it begins to rain, corn or other crops are planted on the bunds. The multistorey, multiculture system is another type of management structure seen in Africa. In this strategy, the farmer keeps the huge, practical trees because they provide shade, protect the soil from erosion, or because they provide fruit. Plantains typically make up the second floor, or bananas, while annual crops, frequently a blend of maize, beans, groundnuts, and vegetables, are grown on the third tier at ground level. There is always some food and/or money available to the farmer. This method is frequently found in quite densely populated locations, leading one to wonder if it did not naturally evolve as a change from moving farming to a more stationary style of agriculture.

2.8.12 Newly Developed Management Methods

The best method of clearing land has historically been by hand. Mechanised clearance was first used when European settlers in Africa began to establish huge farms. When state farms, parastatals (organisations in Africa controlled by Government), or private businesses began operating in several African nations following independence, land development ministries frequently adopted this practice. This mechanical cleaning has frequently wreaked havoc on these delicate soils. Organic matter and nutrients are frequently concentrated in thin surface layers. This layer is likely to be destroyed via mechanised land clearance techniques that attempt to completely remove all vegetation, including the roots, destroying any remaining soil fertility in the process. The subsoil is additionally compacted as a result of it. For these reasons, developing acid soils shouldn't involve employing mechanised land clearing with bulldozers or mechanised land clearing with bulldozers and other heavy machinery.

A successful soil management approach requires more than just fertilisation and picking the suitable crop variety. Many farmers, particularly those who operate on a small scale, lack the resources or access to acquire lime and fertilisers. They require methods that are simple for them to learn and don't demand a lot of effort. This has been realised, and research with varying degrees of success has been focused on this strategy. Leguminous crops are used in rotation more and more, and crop leftovers are returned to the soil to repair or at the very least maintain appropriate amounts of organic matter. Additionally, innovative methods like agroforestry are becoming popular.

References

Abebe M (2007) Nature and management of acid soils in Ethiopia. Haramaya University, Haramaya, p 99

Abrol IP (1982) Reclamation of waste lands and world food prospects. In: Wither soil research. International Congress Soil Science, New Delhi, India, pp 317–337

Acid Sulfate Soil Technical Manual 1.2 (2003) CSIRO land & water, Australia

Adams F (1984) Crop response to lime in the southern United States. In: Adams F (ed) Soil acidity and liming, 2nd edn. American Society of Agronomy, Madison, Wisconsin, pp 211–265

ATA (2014) Soil fertility mapping and fertilizer blending. Agricultural Transformation Agency (ATA), Addis Ababa

Anderson D, Lafond G, Head I (2010) Global perspective of arable soils and major soil associations. Prairie Soils Crops J 3:1–8

Andriesse JP, van Breemen N, Blokhius WA (1972) The influence of mud lobster (Thallasina anomale) on the development of acid sulfate soils in Mangrove. swamps in Sarawak (East Malaysia). In: Proceedings of the international symposium on acid sulfate soils, 13–20 August 1972. Wageningen, Netherlands

Andriesse W, van Mensvoort MEF (2006) Acid sulfate soils: distribution and extent. In: Lal R (ed) Encyclopedia of soil science, 2nd ed. Taylor and Francis Group, New York, USA, pp 14–19

Angeloni J, Peek A, Appleyard S, Wong S, Watkins R (2004) Acid sulfate soils: distribution, impacts and regulation (A western perspective). Corros Prevent 106:1–13

Arkesteyn GJMW (1980) Pyrite oxidation in acid sulphate soils: the role of microorganisms. Plant Soil 54:119–134

Attanandana T, Vacharotayan S (1986) Acid sulfate soils: their characteristics, genesis, amelioration and utilization. Southeast Asian Studies 24:154–180

Attanandana T, Vacharotayan S, Kyuma K (1981) Chemical characteristics, and fertility status of acid sulfate soils of Thailand. In: Proceedings international symposium on add sulfate soils, January 18–30, 1981. Bangkok, Thailand

Backlund K, Boman A, Fröjdö S, Åström M (2005) An analytical procedure for determination of sulphur species and isotopes in boreal acid sulphate soils and sediments. Agric Food Sci 14:70–82

Barak P, Jobe BO, Krueger AR, Peterson LA, Laird DA (1997) Effect of long-term soil acidification due to nitrogen fertilizer inputs in Wisconsin. Plant Soil 197:61–69. https://doi.org/10.1023/A:1004297607070

Beek KJ, Blokhuis WA, Driessen P, van Breemen N, Brinkman R, Pons LJ (1980) Problem soils: their reclamation and management. ILRI Publ. 27:43–72

Bigham JM, Fitzpatrick RW, Schulze DG (2002) Iron oxides. In: Dixon JB, Schulze DG (eds) Soil mineralogy with environmental applications. Soil Science Society America Book Series No 7. SSSA, Madison, Wisconsin, pp 323–366

Blum WEH, Eswaran H (2004) Soils for sustaining global food production. J Food Sci 69:38–42. Published on Web 2/20/2004)

Boman A, Åström M, Fröjdö S (2008) Sulfur dynamics in boreal acid sulfate soils rich in metastable iron sulfide—the role of artificial drainage. Chem Geol 255:68–77

Breburda J (1990) Development of agricultural yield levels and soil K-status in Eastern and Western Europe. In: Development of K fertiliser recommendations. International Potash Institute, Switzerland, pp 17–35

Bronswijk J, Groenenberg J, Ritesma C, van Wijk A, Nughoro K (1995) Evaluation of water management strategies for acid sulfate soils using a simulation model: a case study in Indonesia

Bruce RC, Crack BJ (1978) Chemical attributes of some Australian acid tropical and subtropical soils. In: Andrew CS, Kamprath EI (ed) Mineral nutrition of legumes in tropical and subtropical soils. CSIRO, Melbourne, Australia, pp 59–73

Bruce RC, Warrell L, Bell L, Edwards D (1989) Chemical attributes of some Queensland acid aoils. I. Solid and solution phase compositions. Soil Res 27:333–351

Buschmann J, Berg M, Stengel C, Winkel L, Sampson ML, Trang PT, Viet PH (2008) Contamination of drinking water resources in the Mekong delta floodplains: arsenic and other trace metals pose serious health risks to population. Environ Int 34:756–764

Cavelier J, Aide TM, Dupuy JM, Eusse AM, Santos C (1999) Long-term effects of deforestation on soil properties and vegetation in a tropical lowland forest in Colombia. Ecotropicos 12:57–68

Commission of the European Communities (CEC) (2006) Communication from the Commission to the Council, the European Parliament, the European Economic and Social Committee and the Committee of the Regions. Thematic Strategy for Soil Protection. COM 231 Final, Brussels

Chapman LS, Haysom MBC, Chardon CW (1981) Checking the fertility of Queensland's sugar land. In: Proceedings of Australian society of sugarcane technologists, 1981 conference, pp 325–332

Chen J (2007) Rapid urbanization in China: a real challenge to soil protection and food security. CATENA 69:1–15

Cherry DS, Currie RJ, Soucek DJ, Latimer HA, Tren GC (2001) An integrative assessment of watershed impacted by abandonedmined land discharges. Environ Pollut 111:377–388

Coleman NT, Thomas WG (1967) The basic chemistry of soil acidity. Agron Monogr 12:1–41

Corvalan C, Hales S, McMichael A (2005) Ecosystems and human well-being: health synthesis, a report for the Millennium Ecosystem Assessment. World Health Organisation, Geneva

Costanza R, d'Arge R, de Groot R, Farber S, Grasso M, Hannon B, Limburg K, Naeem S, O'Neill RV, Paruelo J, Raskin RG, Sutton P, van den Belt M (1997) The value of the world's ecosystem services and natural capital. Nature 387:253–260

Daily GC, Matson PA, Vitousek PM (1997) Ecosystem services supplied by soil. In Daily G (ed) Nature's services: societal dependence on natural ecosystems. Island Press, Washington, DC, 412 pp, pp 113–132

Dalovic IG, Jockovic ĐS, Dugali´c GJ, Bekavac GF, Purar B, Šeremeši´c SI, Jockovi´c MĐ (2012) Soil acidity and mobile aluminum status in pseudogley soils in the ˇcaˇcak–kraljevo basin. J Serb Chem Soc, 77:833– 843

Das SK, Das SK (2015) Acid sulphate soil: management strategy for soil health and productivity. Popular Kheti 3:2–7. Available online at www.popularkheti.info

Dear SE, Moore NG, Dobos SK, Watling KM, Ahern CR (2002) Soil management guidelines. In: Queensland acid sulfate soil technical manual (version 3.8). Department of Natural Resources and Mines, Indooroopilly, Queensland

Dent DL (1986) Acid sulphate soils: a baseline for research and development. International Institute for Land Reclamation and Improvement Publication No. 39, Wageningen

Dent D, Pons L (1995) A world perspective on acid sulfate soils. Geoderma 67:263–276

Dudal R (ed) (1965) Dark clay soils of tropical and subtropical regions. FAO Agric Dev Pap No 83, Rome

Dudal R (1988) Changing patterns in the utilization of upland soils in the tropics and subtropics. In Management and Fertilization of Upland Crops in the Tropics and Subtropics. Nanjing Institute of Soil Science, Nanjing, PRC, pp 1–5.

Edwards DG, Bell LC (1989). Acid soil infertility in Australian tropical soils. In: Craswell ET, Pushparajah E (eds) Management of acid soils in the humid tropics of Asia. ACIAR Monograph No. 13 (IBSRAM Monograph No. 1), x + 118 p

Erisman JW, Sutton MA, Galloway J, Klimont Z, Winiwarter W (2008) How a century of ammonia synthesis changed the world. Nat Geosci 1:636–639

Erviö R (1975) Kyrönjoen vesistöalueen rikkipitoiset viljelysmaat. Cultivated sulphate soils in the drainage basin of river Kyrönjoki. J Scient Agric Soc Finland 47:550–561

Eswaran H, Reich P, Beinroth F (1997) Global distribution of soils with acidity. In: Moniz AC (ed) Plant-soil interactions at low pH. Brazilian Soil Science Society, Sao Paulo, pp 159–164

Ethio SIS (2014) Soil fertility mapping and fertilizer blending. Agricultural Transformation Agency (ATA) Report, Ethiopia soil information system (Ethio SIS). Ministry of Agriculture, Addis Ababa

Fageria NK (2009) The use of nutrients in crop plants. CRC Press, Boca Raton, Fl.

Fanning DS (2002) Acid sulfate soils. In: Lal R (ed) Encyclopedia of soil science. Marcel Dekker, New York, pp 11–13

Fanning DS, Fanning MCB (1989) Soil: morphology, genesis and classification. John Wiley & Sons, New York

FAO (1969) Provisional world plan for agricultural development. Italy, Rome

FAO (1978) Reports on the agro-ecological zones project. Methodology and results for Africa

FAO (1991a) Production yearbook, 1990, vol 44. Italy, Rome

FAO (1991b) World soil resources report 66. Italy, Rome

FAO (2015) Revised World Soil Charter (2015). Food and Agriculture Organization of the United Nations Viale delle Terme di Caracalla 00153 Rome, Italy

FAO/UNESCO (1971–1979) Soil map of the world. Volumes I-X. UNESCO, Paris

FAO-GIS (1998) World reference base for soil resources. World Soil Resources Report, #84. FAO, Rome, 88pp

Fageria NK, Nascente AS (2014) Management of soil acidity of South American soils for sustainable crop production. Adv Agron 128:221–275

Fergus IF (1954) Manganese toxicity in an acid soil. Queensland J Agric Sci 11:15–27

Fernandes ECM, O'Ktingati A, Maghembe J (1984) The Chagga homegardens: a multistoried agroforestry cropping system on Mt. Kilimanjaro (Northern Tanzania). Agro for Syst 2:73–86

Fitzpatrick RW, Davies PG, Thomas BP, Merry RH, Fotheringham DG, Hicks WS (2002). Properties and distribution of South Australian coastal acid sulfate soils and their environmental hazards. 5th International Acid Sulfate Soils Conference, Tweed Heads, NSW

Fitzpatrick R (2005) Atlas of Australian acid sulfate soils. CSIRO Land and Water, Australia

Fitzpatrick RW, Fritsch E, Self PG (1996) Interpretation of soil features produced by ancient and modern process in degraded landscapes; V. Development of saline sulfidic features in non-tidal seepage areas. Geoderma 69:1–29

Fitzpatrick RW, Marvanek SP, Shand P, Merry RH, Thomas M, Raven MD (2008a) Acid sulfate soil maps of the river murray below Blanchetown (Lock 1) and Lakes Alexandrina and Albert when water levels were at pre-drought and current drought conditions. CSIRO Land and Water Science Report 12/08

Fitzpatrick RW, Shand P, Thomas PM, Merry RH, Raven MD, Simpson SL (2008b) Acid sulfate soils in subaqueous, waterlogged and drained soil environments of nine wetlands below Blanchetown (Lock 1), South Australia: properties, genesis, risks and management. CSIRO Land and Water Science Report 42/08, ISSN: 1834-6618

Fitzpatrick RW, Shand P, Thomas B, Marvanek S, Merry RH, Creeper N, Thomas M, Raven MD, Simpson SL, McClure S, Jayalath N (2008e) Acid sulfate soils in the Coorong, Lake Alexandrina and Lake Albert: properties, distribution, genesis, risks and management of subaqueous, waterlogged and drained soil environments. Prepared for Department of the Environment, Water, Heritage and Arts. CSIRO Land and Water Science Report 52/08. CSIRO, Adelaide, pp.177, http://www.clw.csiro.au/publications/science/2008/sr52-08.pdf

Fitzpatrick RW, Shand P, Merry RH (2009) Acid sulphate soils. In: Jennings JT (ed) Natural history of the riverland and murraylands. Royal Society of South Australia Inc., The University of Adelaide, SA 5005, Australia, pp 65–111

Fitzpatrick RW, Grealish G, Chappell A, Marvanek S, Shand P (2010) Spatial variability of subaqueous and terrestrial acid sulfate soils and their properties, for the lower lakes South Australia. Report Number R-00689-01-015

Foy CD (1984) Physiological effects of hydrogen, aluminium, and manganese toxicities in acid soil. In: Adarns F (ed) Soil acidity and liming, 2nd ed. American Society of Agronomy, Madison, Wisconsin, pp 51–97

Georgala D (1980) Paleoenvironment studies of post-glacial black clays in northeastern Sweden. PhD Thesis, Stockh Contrib Geol 36(2):93–151

Glamore WC (2003) Evaluation and analysis of acid sulfate soils impact via tidal restoration. Ph.D. Thesis, University of Wollongong, Wollongong, NSW, Australia, p 469

Goldhaber MB, Kaplan IR (1974) The sulfur cycle. In: Goldberg ED (ed) The sea, vol 5. Marine chemistry. Wiley-Interscience, New York, pp. 627–655

Guo JH, Liu XJ, Zhang Y, Shen JL, Han WX, Zhang WF, Christie P, Goulding KWT, Vitousek PM, Zhang FS (2010) Significant acidification in major Chinese croplands. Science 327:1008–1010

Gutiérrez RA (2012) Systems biology for enhanced plant nitrogen nutrition. Science 336:1673–1675

Hanhart K, Ni DV (1991) The effect of water management on the cultivation of an irrigated dry season rice crop on a severe acid sulphate soil in Hoa An station, Vietnam. Interim report of VH-10 Program, Department of Soil Science and Geology, Wageningen Agricultural University

Harter RD (2007) Acid soils of the tropics, 2007, ECHO Technical Note. 17391 Durrance Road, North Fort Myers, FL 33917, USA

Harrington N (2004) Investigation into the precipitation of Al-hydroxide during pumping of bores at Bookpurnong (Western's Highland site) in relation to groundwater quality in the Loxton Sands aquifer. DWLBC Report 2004/40

Haysom MB, Hurney AP, Nielsen P (1986) The influence of lime on the acid soils of Ingharn. In: Proceedings of Australian society of sugarcane technologists,. 1986 conference, pp 55–62

Helyar KR, Porter WM (1989) Soil acidification, its measurement and the processes involved. In: Robson AD (ed) Soil acidity and plant growth. Academic Press, Sydney, pp 61–101

Hetherington S, Asher CJ, Blarney FPC (1988) Comparative tolerance of sugarcane, navybean, soybean and maize to aluminium toxicity. Austr J Agric Res 38:171–176

Hicks WK, Kuylenstierna JCI, Owen A, Dentener F, Seip HM, Rodhe H (2008) Soil sensitivity to acidification in Asia: status and prospects. Ambio 37:295–303

Howa N, Cuong H (2006) Pollution of Cd, Fe and Ni in canal water in acid sulfate soils as affected by soil acidification and available metal concentrations in soils. In: Fourth international symposium on Southeast Asian water environment, Bangkok, Thailand, 6–8 Dec

Hudd R (2000) Springtime episodic acidification as a regulatory factor of estuary spawing fish recruitment. PhD Thesis, Helsinki University, Finland, 42 pp

Huu TN, Gau TTN, Ngan PN, Van TTB, Khuong NQ (2022) Potential phosphorus solubilizing purple nonsulfur bacteria isolated from acid sulfate soil in improving soil property, nutrient uptake and yield of pineapple (Ananas comosus L. Merrill) under acidic stress. Applied and Environmental Soil Science, https://doi.org/10.1155/2022/8693479

IFA (2012) Global supply and demand outlook for fertilizer and raw materials. IFA. Also available at www.fertilizer.org

Indraratna B, Sullivan J, Netherty A (1995) Effect of groundwater table on the formation of acid sulfate soils. Mine Water Environ 14:71–84

Isbell RF (1996) The Australian soil classification system. CSIRO Publishing, Melbourne

Isbell RF and Edwards DG (1988) Soils and their management in the Australian wet tropics. In National Soils Conference 1988 review papers. Australian Society of Soil Science Incorporated, Canberra, pp.152–180

IUSS Working Group WRB (2015) World reference base for soil resources 2014, update 2015: international soil classification system for naming soils and creating legends for soil maps. World soil resources reports No. 106. FAO, Rome, pp 192

Jackson ML (1963) Aluminum bonding in soils: a unifying principle in soil science. Soil Sci Soc Amer Proc 27:1–10

Jackson WA (1967) Physiological effects of soil acidity. Agron Monogr 12:3–124

Jackson TL and Reisenauer HM (1984) Crop response to lime in Western United States. In Ed. Adams F. Soil Acidity and Liming. ASA, CSSA, SSSA, Madison, WI, USA, pp.333–347

Jewitt TN, Law RD, Virgo KJ (1979). Vertisols of the tropics and subtropics: their management and use. In: Outlook on Agriculture 70, 1 (problem soils issue)

Johnson DW, Turner J, Kelly JM (1982) The effects of acid rain on forest nutrient status. Water Resour Res 18:449–461

Jones RM (1982) White clover (Trifolium repens) in subtropical southeast Queensland. IT. Effects of lime application on nutrient concentrations in soil and white clover. Tropical Grasslands 16:127–130

Ju XT, Xing GX, Chen XP, Zhang SL, Zhang LJ, Liu XJ et al (2009) Reducing environmental risk by improving N management in intensive Chinese agricultural systems. Proc Natl Acad Sci USA 106:3041–3046

Kamprath EJ (1984) Crop response to lime on soils in the tropics. In: Adams F (ed) Soil acidity and liming, 2nd ed. American Society of Agronomy, Madison, WI, pp 349–368

Kang BT, Reynolds L, Atta-Krah AN (1990) Alley farming. Adv Agron 43:315–359. https://doi.org/10.1016/S0065-2113(08)60481-2

Kant S, Bi YM, Rothstein SJ (2011) Understanding plant response to nitrogen limitation for the improvement of crop nitrogen use efficiency. J Exp Bot 62:1499–1509

Kawalec A (1972) World distribution of acid sulfate soils. References and Map. Proceedings of the International Symposium on Add Sulfate SoUs, 13–20 August 1972, Wageningen, Netherlands

Kerridge PC (1978) Fertilization of acid tropical soils in relation to pasture legumes. In: Andrew CS, Karnprath EJ (ed) Mineral nutrition of legumes in tropical and subtropical soils. CSIRO, Melbourne, Australia, pp 395–415

Kerridge PC, Andrew CS, Murtha OO (1972) Plant nutrient status of soils of the Atherton Tableland, North Queensland. Aust J Exp Agric Anim Hus 12:618–627

Kijne JW (2006) Abiotic stress and water scarcity: identifying and resolving conflicts from plant level to global level. Field Crop Res 97:3–18

Kisinyo PO, Opala PA, Gudu SO, Othieno CO, Okalebo JR, Palapala V, Otinga AN (2014) Recent advances towards understanding and managing Kenyan acid soils for improved crop production. Afr J Agric Res 9:2397–2408

Kivinen E (1950) Sulphate soils and their management in Finland. In: Trans 4th international congress soil science, V01.2, pp 259–262

Kochian LV (1995) Cellular mechanisms of aluminum toxicity and resistance in plants. Annu Rev Plant Phys 46:237–260

Konsten CJM, Suping S, Aribawa IB, Widjaja-Adhi IPG (1990) Chemical processes in Acid sulphate soils in Pulau Petak, South and Central kalimantan, Indonesia. In: AARD & LAWOO Papers workshop on acid sulphate soils in the humid tropics, 20–22 November 1990, Bogor, Indonesia, pp 109–136

Kosaka J (1971) Report on acid sulfate soils in the muda irrigation area. (Japanese, English Summary) Trop Agric Res Center (Japan) Bull. 17: 29. Cited in van Breemen [1982]

Lal R (2007) Anthropogenic influences on world soils and implications to global food security. Adv Agronomy 93:69–93

Le TMH, Pham AN, Collins RN, Waite TD (2008) Impacts of compressive pressure and solution composition on hydraulic conductivity and consolidation coefficients of acid sulfate soils. Aust J Soil Res 46:112–121

Lee MT, Wilson GL (1972) The calcium and pH components of lime response in tropical legumes. Aust J Agric Res 23:257–265

Lin C, Long X, Mai S, Xu S, Chu C, Jiang D (2004) Effects of multi-conditioners on minesite acid sulfate soils for vetiver grass growth. Pedosphere 14(2004):371–378

Lin C, Wu Y, Lu W, Chen A, Liu Y (2007) Water chemistry and ecotoxicity of an acid mine drainage-affected stream in subtropical China during amajor flood event. J Hazard Mater 142:199–207

Liua Y, Lina C, Maa Y, Lua W, Wua Y, Huanga S, Zhua L, Lia J, Chena A (2009) Toxic effects of two acid sulfate soils from the Dabaoshan Mine on Corymbia citriodora var. variegata and Daphnia carinata. J Hazard Mater 166:1162–1168

Ljung K, Maley F, Cook A, Weinstein P (2009) Acid sulfate soils and human health—a millennium ecosystem assessment. Environ Int 35:1234–1242

Loc HH, Binh DV, Park E, Shrestha S, Dung TD, Son VH, Tru NHT, Mai NP, Seijger C (2021) Intensifying saline water intrusion and drought in the Mekong Delta: From physical evidence to policy outlooks. Science of the Total Environment 757:143919

Lovett WJ, Johnson AD (1969) Manganese upta1ce by tobacco and bean plants grown on soils of the Mareeba-Dimbulah irrigation area, North Queensland. Australian Joumal of Experimental Agriculture and Animal Husbandry 8:466-469

Macdonald BCT, White I, Åström ME, Keene AF, Melville MD, Reynolds JK (2007) Discharge of weathering products from acid sulfate soils after a rainfall event, Tweed River, eastern Australia. Appl Geochem 22:2695–2705

Maddocks G, Lin C, McConchie D (2004) Effects of BauxsolTM and biosolids on conditions of acid-generating mine soils for plant growth. Environ Pollut 127(2004):157–167

Maji AK, Reddy GPO, Sarkar D (2012) Acid soils of India: their extent and spatial variability. National Bureau of Soil Survey and Land Use Planning (NBSS & LUP), Nagpur. ISBN 978-81-89043-24-7

Mandal SC (1997) Introduction and historical overview. In: Mahapatra IC, Mandal SC, Misra C, Mitra GN, Panda N (eds) Acid soils of India. ICAR, New Delhi, India, pp 3–24

Maneewan N, Sa-nguansubpayakorn C (2007). Effect of organic and chemical fertilizers on acid sulfate soils management for Patumthani rice production. In: 4th INWEPF steering meeting and symposium, paper 2-03, 7P

Marschner H (2011) Marschner's mineral nutrition of higher plants. Academic Press, London

McArthur JM, Turner JV, Lyons WB, Osborn AO, Thirlwall MF (1991) Hydrochemistry on the Yilgarn Block, Western Australia: ferrolysis and mineralisation in acidic brines. Geochim Cosmochim Acta 55:1273–1288

McBride M (1994) Environmental chemistry of soils. Oxford University Press, Inc., USA

Menzies NW (2003) Toxic elements in acid soils: chemistry and measurement. In: Rengel Z (ed) Handbook of soil acidity. Marcel Dekker Inc., New York, pp 267–296

Michon G, Mary F, Bompand J (1986) Multistoried agroforestry garden system in West Sumatra, Indonesia. Agrofor Syst 4:315–338

Millennium Ecosystem Assessment (2000) Ecosystems and human well-being: synthesis. Island Press, Washington, DC

Millennium Ecosystem Assessment (2003) Ecosystems and human well-being: synthesis. Island Press, Washington, DC

Millennium Ecosystem Assessment (2005) Ecosystems and human well-being: synthesis. Island Press, Washington, DC, 800 pp

Minh L, Tuong T, van Mensvoort MEF, Bouma J (1997) Contamination of surface water as affected by land use in acid sulfate soils in the Mekong River Delta, Vietnam. Agri Ecosystem Environ 61:19–27

Moormann FR (1961) Research on acid sulfate soils and their amelioration by liming. Report. Agronomic Library, Ministry of Rural Affairs, Saigon, pp 3–15

Moormann FR (1963) Research on Acid Sulfate Soils and Their Amelioration by Liming. Report. Saigon: Agronomic Library, Ministry of Rural Affairs. pp 3-15

Moraghan JT, Patrick WH Jr (1974) Selected metabolic processes in a submerged soil at controlled pH. In: International congress soil science trans. 10th (Moscow) II, pp 264–268

Mullins CE, Young IM, Bengough AG, Ley GJ (1987). Soil Use Manag. https://doi.org/10.1111/j.1475-2743.1987.tb00715.x

Nair PKR (1984) Soil productivity aspects of agroforestry. ICRAF science and practice of agroforestry I. The International Council for Research in Agroforestry, Nairobi, Kenya

Negene W (2019) Review on the extent of acid soil in Ethiopia, its impact and management methods. J Biol Agric Healthcare 9(15). www.iiste.org. ISSN 2224-3208, https://doi.org/10.7176/JBAH

Nguyen ND (2014) Adequate use of sulphur. Vietnam Agriculture Newspaper, February 10, 2014

Nhung MM, Ponnamperuma FN (1966) Effects of calcium carbonate, manganese dioxide, ferric hydroxide, and prolonged flooding on chemical and electrochemical changes and the growth of rice in flooded acid soil. Soil Sci 102:29–41

NISF (2012) The basic information of main soils unit of Vietnam. National Institute for Soils and Nordmyr L, Österholm P, Åström M (2008) Estuarine behaviour of metal loads leached from coastal lowland acid sulphate soils. Marine Environmental Res 66:378–393

Nordmyr L, Österholm P, Åström M (2008) Estuarine behaviour of metal loads leached from coastal lowland acid sulphate soils. Mar Environ Res 66:378–393

Nordstrom DK, Alpers CN (1999) Geochemistry of acid mine waters. In: Plumlee GS, Logsdon MJ (eds) The environmental geochemistry of mineral deposits, Part A.: Processes, techniques, and health issues. Reviews in Economic Geology, vol. 6A. (Society of Economic Geologists, Inc.), pp 133–160

Nwaga D, The C, Ambassa-Kiki R, Ngonkeu MEL, Tchiegang-Megueni C (2004) Selection of arbuscular mycorrhizal fungi for inoculating maize and sorghum grown in oxisol/ultisol and vertisol in Cameroon. Bationo A, Ed. In Managing Nutrient Cycles to Sustain Soil Fertility in Sub-Saharan Africa. CIAT: London, UK, pp 608

Österholm P, Åström M (2002) Spatial trends and losses of major and trace elements in agricultural acid sulphate soils distributed in the artificially drained Rintala area, W. Finland. Appl Geochem 17:1209–1218

Palko J (1994) Acid sulphate soils and their agricultural and environmental problems in Finland. PhD Thesis, University of Oulu, Finland, 58 pp

Palko J, Räsänen M, Alasaarela E (1985) Happamien sulfaattimaiden esiintyminen ja vaikutus veden laatuun Sirppujoen vesistöalueella (in Finnish). The occurrence and effects of acid sulphate soils on the water quality of River Sirppujoki: National board of waters report, 260, 95 pp

Palm CA, Szott LT (1989) TropSoils technical report 1986–1987, p. 70. North Carolina State University, Raleigh

Parkpian P, Pongsikal P, Sangtong P (1991) Characteristics of acid soils in Thailand: a review. In: Wright RJ et al (eds) Plant-soil interactions at low pH. Kluwer Academic Publishers. PLSO AS44, pp 397–405

Pasricha NS, Ponnamperuma FN (1976) Influence of Salt and Alkali on Ionic Equi. libria in Submerged Soils. Soil Sci Soc Am J 40:374–380

Pearson RW (1975) Soil acidity and liming in the humid tropics. Cornell International Agriculture Bulletin 30. Cornell University, Ithaca, New York

Pham-Huu-Anh, Moonnann FR, Golden JD (1961) Liming Experiments on Acid Sulfate Soils. Researches on acid sulfate soils and their amelioration by liming. Ministry of Rural Affairs

Ponnamperuma FN (1972) The chemistry of submerged soils. Adv Agron 24:29–96

Ponnamperuma FN, Bradfield R, Peech M (1955) Physiological disease of rice attributable to iron toxicity. Nature 175:265

Ponnamperuma FN, Attanandana T, Beye G (1973). Amelioration of three acid sulphate soils for lowland rice. In: Dost H (ed) Acid sulphate soils. Proceedings of the international symposium on acid sulphate soils, 13–20 August 1972, Wageningen, Vol 11. International Institute for Land Reclamation and Improvement Publication No. IS, Wageningen, pp 391–406

Pons LJ (1973) Outline of the genesis, characteristics, classifications and improvement of acid sulphate soils. In: Dost H (ed) International symposium on acid sulphate soils. Introductory papers and bibliography. International Institute for Land Reclaimation and Improvement (ILRI), Wageningen, The Netherlands, pp 3–27

Pons LJ, van der Kevie W (1969) Acid sulfate soils in Thailand. Soil Survey Reports of the Land Development Department. Ministry of National Development

Pons LJ, van Breemen N, Driessen PM (1982) In: Kittrick JA, Fanning DS, Hossner LR (ed) Physiography of coastal sediments and development of potential soil acidity. SSSA special publication. https://doi.org/10.2136/sssaspecpub10.c1

Powell B, Martens M (2005) A review of acid sulfate soil impacts, actions and policies that impact onwater quality in Great Barrier Reef catchments, including a case study on remediation at East Trinity. Marine Poll Bull 51:149–164

Pusat Penelitian (2000) Atlas Sumber Daya Tanah Eksplorasi Indonesia. Skala 1:1.000.000 (Exploration soil resource atlas, at 1:1,000,000 scale). Pusat Penelitian dan Pengembangan Tanah dan Agroklimat. Bogor, Puslitbangtanak

Raintree JB (1987) In: Raintree JB (ed) Land, trees and tenure. Proceeding international workshop on tenure issues in agroforestry, Nairobi, Kenya, pp 35–78. International Council Research Agroforestry and Land Tenure Center, Nairobi, Kenya and Madison, Wisconsin

Rayment GE (1978) Establishment and maintenance fertilizers for grass/legume pastures in coastal southeast Queensland. In: Proceedings of workshop on management of Southeast Queensland Coastal Pastures. Gympie. May 30–31, 1978. Department of Primary Industries, Brisbane, Queensland, pp 105–114

Regassa H, Agegnehu G. (2011). Potentials and limitations of acid soils in the highlands of Ethiopia: a review. In: Mulatu B, Grando S (eds) Barley research and development in Ethiopia. ICARDA, Aleppo, Syria, p 103–112

Rimwanich S, Suebsiri B (1983) Nature and management of problem soils in Thailand. In Proceeding international conference on ecology and management of problem soils in Asia. FFTC Book Series No. 27, pp 13–26

Ritsema CJ, Nugroho K, Obstindie K (1991) Column experiments, The Netherlands. In: AARD & LAWOO simulation model of physical and chemical processes in acid sulphate soils to evaluate water management strategies, Bogor, pp 27–57

Robarge WP (2008) Acidity. In: Chesworth W (ed) Encyclopedia of soil science, encyclopedia of earth sciences series, 860pp

Roberts VG, Smeda Z (2001) The distribution of soil fertility constraints in KwaZulu-Natal, South Africa. In: Proceedings of the 5th international symposium on plant soil interactions at low pH, Bergville, South Africa, 12–16 March 2001; p 12

Roots M, Åström M (2006) Gulf of Bothnia receives high concentrations of potentially toxic metals from acid sulphate soils. Boreal Environ Res 11(5):383–388

Rorison IH (1972) The effect of extreme soil acidity on the nutrient uptake and physiology of plants. AGRIS, pp 223–254. ISSN: 0074-6452

Russo RO, Budowski G (1986) Effect of pollarding frequency on biomass of Erythrina poeppigiana as a coffee shade tree. Agrofor Syst 4:145–162

Saarinen T, Vuori KM, Alasaarela E, Klove B (2010) Long-term trends and variation of acidity, CODMn and colour in coastal rivers of Western Finland in relation to climate and hydrology. Sci Total Environ 408:5019–5027

Sammut J, Lines-Kelley R (2000) Acid sulfate soils, 2nd ed. Environment Australia, ISBN 0-7347-1208-1

Sammut J (2004) An introduction to acid sulfate soils. NSW Department of Primary Industries, Agriculture, Sydney, NSW

Sanchez PA (1976) Properties and Management of Acid Soils in the Tropics. John Wiley and Sons, New York, USA, p 618

Sanchez PA (1979) Soil fertility and conservation considerations for agroforestry systems in the humid tropics of Latin America. In: Mongi HO, Huxley PA (eds) Soils research in agroforestry. ICRAF, Nairobi

Sanchez PA (1987) Management of acid soils in the humid tropics of Latin America. In: Management of acid tropical soils for sustainable agriculture. IBSRAM Proc No. 2. Bangkok, Thailand, pp 63–107

Sanchez PA (1989) Soils. In Eds. Leith H and Werger MJA, Tropical Rainforest Ecosystems: Biogeographical and Ecological Studies, Elsevier, Amsterdam. Vol. 148, pp 73–86

Satawathananont S, Patrick WH Jr, van Breemen N (1991) Changes with time of redox potential, pH and chemical constituents of acid sulphate soils ofThailand. In: Deturck P, Ponnamperuma FN (eds) Rice production on acid soils of the tropics. Proceedings of an international symposium, 26–30 June 1989. Institute of Fundamental Studies, Kandy, pp 59–64

Sen TK, Chamuah GS, Seghal JL (1994) Occurrence and characteristics of some kandi soils in Manipur. J Indian Soc Soil Sci 42:297–300

Shamshuddin J, Azura AE, Shazana MARS, Fauziah CI, Panhwar QA, Naher UA (2014) Properties and measurement of acid sulfate soils in Southeast Asia for sustainable cultivation of rice, oil palm, and cocoa. Adv Agron 124:91–142

Sharma UC, Singh RP (2002). Acid soils of India: their distribution, management and future strategies for higher productivity. Fertilizer News 47:45–52

Sillen LG, Martell AE (1964) Stability constants of metal-ion complexes. Special hbl. No. 17, 754 p. The Chemical Society, Burlington House, London

Sillen LG, Martell AE (1971) Stability constants of metal-ion complexes. Supplement No. 1. Special Publ. No. 25, 865 p. The Chemical Society, Burlington House, London

Siman A, Cradock FW, Nicholls PI, Kirton HC (1971) Effects of calcium carbonate and ammonium sulphate on manganese toxicity in an acid soil Australian Journal of Agricultural Research 22:201–214

Simpson H, Pedini P (1985) Brackish water aquaculture in the tropics: the problem of acid sulfate soil environment. Appl Geochem 19:1837–1853

Sivaguru M, Horst WJ (1998) Differential impacts of aluminum on microtubule organization depend on growth phase in suspension-cultured tobacco cells. Physiol Plant 107:110–119

Smith J, van Oploo P, Marston H, Melville MD, Macdonald BCT (2003) Spatial distribution and management of total actual acidity in an acid sulfate soil environment, McLeods Creek, northeastern NSW, Australia. CATENA 51(1):61–79

Sohlenius G, Öborn I (2004) Geochemistry and partitioning of trace metals in acid sulphate soils in Sweden and Finland before and after sulphide oxidation. Geoderma 122(2–4):167–175

Soil Survey Staff (1975) Soil Taxonomy, a basic system of soil classification for making and interpreting soil surveys. U.S. Department of Agriculture. Washington D.C., U.S. Government Printing Office. pp 754

Soil Survey Staff (1999) Soil taxonomy. A basic system of soil classification for making and interpreting soil surveys, 2nd ed. Agriculture Handbook 436. Natural Resources Conservation Service, United States Department of Agriculture, Washington, DC

Soil Survey Staff (2003) Soil taxonomy: a basic system of soil classification for making and interpreting soil surveys. Agriculture Handbook 436. U.S. Government Printing Office, Washington, DC

Sombatpanit S (1970) Thesis. Uppsala: Royal Agriculture College. Cited in Bloomfield and 179 Coulter, 1973:265–326

Spaargaren Q (1989) Management of acid soils in Africa. In: Craswell ET, Pushparajah E (1989) Management of acid soils in the humid tropics of Asia. ACIAR Monograph No. 13 (IBSRAM Monograph No.1), x + 118 p

Stangel P, von Uexküll HR (1990) Regional food security: demographic and geographic implications. In: IRRI (ed) Phosphorus requirements for sustainable agriculture in Asia and Oceania. IRRI, Los Baños, Philippines, pp 21–43

Steiner C, Teixeira WG, Lehmann J, Nehls T, Macˆedo JLVD, Blum WEH, Zech W (2007) Long-term effects of manure, charcoal, and mineral fertilization on crop production and fertility on a highly weathered central Amazonian upland soil. Plant Soil 291:275–290

Sullivan LA, Bush RT, Fyfe D (2002) Acid sulfate soil drain ooze: distribution, behavior and implications for acidification and deoxygenation of waterways. In Lin C, Melville MD, Sullivan LA (Eds.) Acid sulfate soils in Australia and China. Science Press, Beijing., 91–99

Sullivan LA, Ward NJ, Bush RT, Burton ED (2009) Improved identification of sulfidic soil materials by a modified incubation method. Geoderma 149:33–38

Sullivan LA, Fitzpatrick RW, Bush RT, Burton ED, Shand P, Ward NJ (2010) The classification of acid sulfate soil materials: further modifications. Southern Cross GeoScience Technical Report No. 310. Southern Cross University, Lismore, NSW, Australia, 12 pp

Szott, L. T. (1987). Nutrient stocks in managed and natural humid tropical fallows. Ph.D. Dissertation, North Carolina State University, Raleigh

Szott LT, Davey CB, Palm CA, Sanchez PA (1987) Tropical soils technical report 1985–1986. North Carolina State University, Raleigh, pp 31–35

Szott LT, Palm CA, Sanchez PA (1991) Agroforestry in acid soils of the humid tropics. Adv Agrofor 45:275–301

Tamm CO, Andersson F (1989) Swedish forests in a changing environment. In: Szabolcs I (ed) Environmental threats to forest and other natural ecosystems. Budapest, Hungary, pp 13–24

Tanaka A, Navasero SA (1966) Growth of the rice plant on acid sulfate soils. Soil Sci Plant Nutr 12(3):23–30

Teitzel JK, Bruce RC (1973a) Fertility studies of pasture soils in the wet tropical coast of Queensland. 5. Mixed alluvial soils. Aust J Exp Agric 13:306–311

Teitzel JK, Bruce RC (1973b) Fertility studies of pasture soils in the wet tropical coast of Queensland. 6. Soils derived from beach sand. Aust J Exp Agric 13:312–318

Toivonen J, Toivonen P (2011) Characterization of acid sulfate soils and assessing their impact on a humic boreal lake. J Geochem Explor 110:107–117

Torres F, Raintree J, Davey CB (1983) Research to develop agroforestry systems for the upper basin of the Peruvian Amazon. Report to the The International Development Research Centre (IDRC), Canada, ICRAF, Nairobi, Kenya

Tulau MJ (1999) Acid sulfate soil management priority areas in the lower tweed floodplains. Department of Land and Water Conservation, Sydney

Tully K, Sullivan C, Weil R, Sanchez P (2015) The state of soil degradation in Sub-Saharan Africa: baselines, trajectories, and solutions. Sustainability 7:6523–6552

Tuong TP (1993) An overview of water management of acid sulphate soils. In: Dent D, van Mensvoort MEF (eds) Selected papers on the Ho Chi Minh city symposium on acid sulphate soils. International Institute for Land Reclamation and Improvement Publication 53. Wageningen, 281–287

US Report (1967) The world food problem. A report of the President's Science Advisory Committee, vol II. Washington DC, USA

van Breemen N (1973a) Acid soil forming processes in acid sulphate soils. In: Dost H (ed) Proceedings of the international symposium on acid sulphate soils, 13–20 August 1972. ILRI Publication, vol 18. Wageningen, the Netherlands, pp 66–130

van Breemen N (1975) Acidification and Deacidification asaResult of Flooding. Soc A Mer Proc 39:1153–1157

Van Breemen N (1976) Genesis and solution chemistry of acid sulphate soils. Agricultural report 848. PUDOC, Wageningen

van Breemen N (1982) Genesis, morphology and classification of acid sulfate soils in coastal plains. Add sulfate weathering. SSSA Special Publication Number 10. Soil Science Society of America, Madison, Wisconsin

Van Breemen N (1993) Environmental aspects of acid sulphate soils. In: Dent DL, Van Mensvoort MEF (eds) Selected papers of the Ho Chi Minh City symposium on acid sulphate soils, vol 53. ILRI Publication. Wageningen, The Netherlands, pp 391–401 van Breemen N, Pons LJ (1978) Acid sulfate soils and rice. IRRI, pp 739–761

van der Kevie W, Yenmanas B (1972) Detailed reconnaissance soil survey of Southern Central Plain area. Report SSR-89, Soil Survey Division, Bangkok, 187 pp

van Oploo P (2000) Soil and pore-water relations of a coastal acid sulfate soil in northern New South Wales, Australia. Ph.D. Thesis, The University of New South Wales, Sydney, Australia, p 386

Venter A, Herselman JE, VanderMerwe GME, Steyn C, Beukes DJ (2001) Developing soil acidity maps for South Africa. In: Proceedings of the 5th international symposium on soil plant interactions at low pH, Bergville, South Africa, p 21

von Uexkül HR, Mutert E (1995) Global extent, development and economic impact of acid soils. Global extent, development and economic impact of acid soils. Plant Soil 171:1–15

Vörösmarty C, Leveque C, Revenga C (2005) Chapter 7. Freshwater. In: Hassan R, Scholes R, Ash N, editors. Ecosystems and Human Well-Being: Current State and Trends, vol 1. Washington, DC: Island Press

Wang JP, Harsh R, Guo-Ping Z, Neville M, Zhou MX (2006) Aluminium tolerance in barley (Hordeum vulgare L.): physiological mechanisms, genetics and screening methods. J Zheijang Univ Sci B 7:769–787

Wang X, Yan X, Liao H (2010) Genetic improvement for phosphorus efficiency in soybean: a radical approach. Ann Bot 106:215–222

Ward NJ, Sullivan LA, Fyfe DM, Bush RT, Ferguson AJP (2004) The process of sulfide oxidation in some acid sulfate soil materials. Aust J Soil Res 42(4):449–458

Watts JCD (1969) MaJay Agr J 47:187–202. Cited in Bloomfield and Coulter [1973]

Weppling K (1993) Hydrochemical factors affecting the neutralization demand in acid sulphate waters. Vatten 49:161–170

Williams CH (1980) Soil acidification under clover pastures. Aust J Exp Agric Anim Husb 20:561–567

Williams CH (1990) The management of problem soils in Australia. In: Problem soils of Asia and the Pacific. FAO, Bangkok, Thailand, pp 68–103

Wilson B, White I, Melville M (1999) Floodplain hydrology, acid discharge and change in water quality associated with a drained acid sulfate soil. Marine Freshwater Res 50:149–157

Wood AW (1985) Soil degradation and management under intensive sugar cane cultivation in North Queensland. Soil Use Manag I:120–124

Yli-Halla M, Puustinen M, Koskiaho J (1999) Area of cultivated acid sulphate soils in Finland. Soil Use Manag 15:62–67

Zhao Y, Duan L, Xing J, Larssen T, Nielsen CP, Hao J (2009) Soil acidification in China: is controlling SO_2 emissions enough? Environ Sci Technol 43:8021–8026

Zhao XQ, Chen RF, Shen RF (2014) Coadaptation of plants to multiple stresses in acidic soils. Soil Sci 179:503–513

Chapter 3
Chemistry, Microbiology, and Behaviour of Acid Soils

Abstract Mineral compounds come in a range of particle sizes in soil. Mineral colloids made of hydrated silica, iron hydroxide, aluminium, and aluminosilicates make up the lowest portion. In addition to providing essential nutrients, water, and oxygen absorption, soil colloids also operate as a home for bacteria. Colloids control the water–air interaction in soil. The decomposing remnants of plants, animals, and bacteria are the source of the organic matter found in soil, and the local microbes break them down. Humification is a process that happens when organic materials break down. Humus, or humic substances, are created through a combination of physical–chemical and microbiological processes. A portion of these materials are partially colloidal. It is possible to use the concentrations of aluminium ions (Al^{3+}) and hydrogen ions (H^+) in soils as markers of soil acidity. Reduced plant root growth, which restricts nutrient and water intake, low beneficial microbial activity, and nutrient/element toxicities are some of the effects of acidic soil on agricultural output. Many research have addressed the intricacies of soil acidity in great detail. The organic colloids provide food for the microbes. Additionally, they improve the structural stability of the soil when combined with silty particles. By efficiently absorbing water and promoting the adsorption and exchange of mineral components, humus aids in the growth of higher plants. The liquid fraction of the soil that has dissolved materials is referred to as the soil solution. As the second most significant biological agent in the agricultural environment after plants, soil microorganisms can be used as markers of the quality of the soil. Numerous chemical and biological processes that greatly affect soil fertility, nutrient cycling, and carbon cycling are largely driven by soil bacteria. The rhizosphere and plant roots are home to a wide variety of beneficial microorganisms for plants, including saprophytic microorganisms, biocontrol agents, mycorrhizae, free-living fungi, nitrogen-fixing bacteria that can grow in symbiotic relationships or independently, and rhizobacteria that promote plant growth. The activities of these beneficial bacteria are restricted by the acidity of the soil, with the exception of fungi, which may thrive across a wide range of soil pH levels.

Keywords Chemistry of acid soils · Microbiology of acid soils · Mineral colloids · Development of soil acidity · Behaviour of acid soils

U. C. Sharma et al., *Soil Acidity*, Progress in Soil Science,
https://doi.org/10.1007/978-3-031-76357-1_3

3.1 General Description

A primary focus of current global soil science research is the conservation of soil as a crucial natural resource, ensuring its health and high quality to support the production of economically viable, high-quality crops (Karlen et al. 2003). Due to this focus, the European Commission has chosen to prioritise soil conservation in its policy (COM 2002). Various types of soils exist worldwide, and plants cultivated on them often encounter challenges related to mineral stress. Plants may encounter many types of mineral stress, which can be influenced by various soil characteristics such as moisture, temperature, bulk density, texture, acidity and alkalinity (pH), sodicity and salinity, elemental shortages and toxicities, as well as organic matter content. Clark and Baligar (2000) state that each of these variables has a distinct influence on the type and severity of mineral stresses that plants encounter.

Sumner and Noble (2003) reported that around 4 billion hectares of land, which accounts for roughly 30% of the world's ice-free region, are impacted by acidity. Approximately 13% of Kenya's entire land area is occupied by acid soils, which are extensively distributed across the agricultural regions in the central and western parts of the country. Gudu et al. (2007) reported that a total of over 5 million smallholder farmers grow crops such as maize, legumes, tea, and coffee on a combined area of more than one million hectares of land. In the tropics, the leaching of essential nutrient bases for crops, such as potassium (K), magnesium (Mg), and calcium (Ca), has taken place as a result of extensive soil weathering that has occurred over thousands of years. Subsequently, the cations H, Al, and Mn are substituted, hence contributing to the acid-induced stresses on crop yield (Okalebo et al. 2009). Akinrinade et al. (2006) state that severely weathered soils in humid and subhumid climates exhibit acid infertility traits that restrict crop growth, reduce yield, and hinder soil production. These factors are the result of insufficient nutrients. In regions where acid-forming fertilisers, such as di-ammonium phosphate (DAP), have been consistently applied to already acidic soils for many years, agricultural productivity is poor and decreasing (Nekesa 2007). Due to the presence of various nutrient deficiencies in these soils, the application of mineral fertilisers is necessary to enhance crop productivity. Nevertheless, mineral fertilisers are sometimes limited in availability and costly, and their use in conjunction with an imbalanced diet may exacerbate the problem of soil acidity (Nottidge et al. 2006). The practice of applying lime to acid soils is not commonly done in Sub-Saharan Africa (SSA), likely due to a lack of understanding about the effectiveness of lime, its limited availability, and the high costs associated with transporting lime supplies (Okalebo et al. 2009). The chemical degradation of arable lands has been expedited by continuous cropping and incorrect fertiliser application, resulting in a decrease in the soils' capacity to sustainably support crop production (Mugendi et al. 2007). According to Kisinyo et al. (2005), continuous cropping has led to soil acidity due to excessive levels of aluminium (Al) and subsequent phosphate (P) deficiency, which hinders plant growth. Liming is the most essential and effective approach for alleviating the constraints of soil acidity on crop yield (Haynes 1984). For the purpose of enhancing soil fertility and improving crop productivity in acidic soils, the implementation of well planned and skilfully conducted liming is important. Liming promotes efficient nutrient utilisation, namely phosphorus, which therefore reduces agricultural production risks associated with soil acidity. Applying lime to acidic soils can potentially boost maize yields by reducing exchangeable acidity and elevating pH levels (Onwonga et al. 2010). Lowering soil acidity enhances the proliferation of microorganisms and, as a result, increases their activity in soils. This study was primarily focussed on the potential of soil acidity to enhance food and fibre production, thereby defining the scope of the issue.

Fageria and Baligar (2008) assert that the physical, chemical, and biological characteristics of soil have complex interactions with factors that restrict plant growth. Significant impediments to producing crops on tropical soils include soil erosion and insufficient water retention capacity. The primary chemical factors that limit plant growth in acidic soils are deficiencies or availability of calcium, magnesium, phosphorus, and aluminium. Biologically, soil acidity detrimentally affects the functioning of beneficial microbes. The breakdown of organic matter, mineralization and immobilisation of nutrients, absorption and utilisation by plants, and crop yields are all greatly influenced by this (Fageria and Baligar 2008). A study conducted by Akinrinade et al. (2006) found that acid soils possess the capacity to adsorb crucial components onto their surfaces. This suggests that several essential nutrients are immobilised rather than being made accessible for crop use. Acidic soils have experienced substantial weathering and contain notable quantities of aluminium and iron hydroxides. Acidic soils are formed due to leaching, the absorption of nutrients by plants, and the removal of plants (Wild 1993). This phenomena is especially common in humid tropical regions. Microorganisms play a role in the synthesis of organic acids, which is a contributing component to the development of soil acidity (Wild 1993). Additional factors that influence the formation of soil acidity include the introduction of acidic substances and chemicals, such as ammonium compounds, through the process of nitrification. Root respiration and the decomposition of organic matter in the soil result in the release of carbon dioxide, which combines with water to produce a mild organic acid. This acidification process contributes to the acidity of the soil (Donald 2011). The breakdown of organic waste and the conversion of ammonium and sulphur fertilisers both lead to the creation of potent organic and inorganic acids, specifically nitric and sulphuric acid, respectively.

3.1.1 The Soil Composition

Soil is composed of various components, including air, soil solution, mineral and organic solid particles, as well as living organisms like as edaphon. The proportions of different soil elements generally remain consistent for a particular type of soil (Fig. 1.1). Soil contains mineral compounds in various particle sizes. The lowest section consists of mineral colloids composed of aluminosilicates, hydrated silica, aluminium, and iron hydroxide. Soil colloids have a crucial role in absorbing oxygen, water, and critical nutrients, while also serving as a habitat for bacteria. The water–air interaction in soil is regulated by the colloids. The organic matter in soil originates from the decaying remains of plants, animals, and bacteria, which are decomposed by the resident microorganisms. During the decomposition of organic materials, a process called humification occurs. This process involves a combination of microbiological and physical–chemical processes that produce humic compounds, also known as humus. Some of these substances exist in a partially colloidal state. The microorganisms derive their nourishment from the organic colloids. In addition, when coupled with silty particles, they enhance the soil's structural integrity. Humus facilitates the growth of higher plants by effectively absorbing water and facilitating the adsorption and exchange of mineral components. The soil solution refers to the liquid portion of the soil that contains dissolved substances.

3.1.2 The Soil Solution

The soil solution consists of water that has dissolved gases, organic matter, and minerals. The water is retained in the soil due to the capillary forces present within the soil's particles. The water content and temperature fluctuations have a continuous impact on the chemical composition of the soil solution, either by diluting or concentrating it. Nevertheless, the microorganisms in the area constantly have access to the nitrates, potassium, phosphate, and ammonium salts. The soil solution also contains easily available organic compounds, such as monosaccharides and amino acids. Soil water offers favourable conditions for a range of organisms, such as plants and bacteria.

(i) It transports construction and energy materials via the capillaries, (ii) It influences the pH, osmotic pressure, aeration, amount, and quality of nutrients in the soil solution.

3.1.3 The Soil Environment

The soil environment refers to the conditions and characteristics of the soil.

The soil's air content occupies the empty spaces created by the lack of water between the solid particles. The soil colloids are also fully saturated with air. The soil contains an air component that constitutes between 8 and 35% of its total volume. Nitrogen (N_2), oxygen (O_2), and carbon dioxide (CO_2) are permanent constituents of the atmosphere. The volatile compounds consist of butyric acid, alcohol, esters, NH_3, H_2, CO, NO_x, SO_2, H_2S, CH_4, and C_2H_6. The concentration of CO_2 in soil air is typically ten times higher than in atmospheric air, and soil air is fully saturated with water vapour. When the oxygen concentration in soil drops below 1%, the process of oxygen-based metabolism, which involves the reduction of sulphate and denitrification, transitions to oxygen-free metabolism. Consequently, we can observe

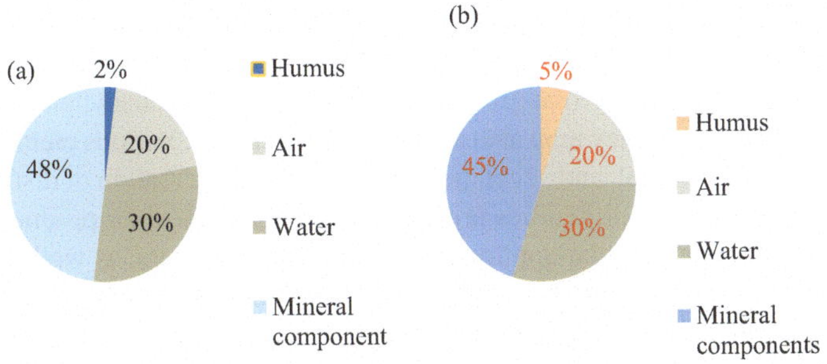

Fig. 3.1 Average fraction of particular phases of the soil: mineral (**a**) and organic (**b**)

the growth and progress of the anaerobic microorganisms. Figure 3.1 displays the mean proportions of minerals and organic matter in soils.

3.1.4 Edaphon

The edaphon refers to a collective of organisms that inhabit the soil. The creatures that inhabit the uppermost layer of soil include bacteria, fungi, unicellular algae, vascular plants, and animals, specifically invertebrates. The metabolic capacities of soil microorganisms contribute to the continuity of element cycles in nature, enhancing their durability. Their effects exert a substantial influence on the formation of organic molecules via mineralization, as well as alterations in mineral compounds. Edaphon constitutes around 1–10% of the total dry mass of organic matter in soil.

3.1.5 Sour Soils

Acidic soils are characterised by a predominance of acidity, which poses challenges for agricultural land use in the context of problematic agricultural soils. Their distinct features include a high level of acidity ranging from severe (pH 5.5–4.5) to extreme (pH 4.5), a limited cation-exchange capacity, and a low base saturation. Acidic soils, which are problematic for agriculture, are mainly classified into the following Reference Soil Groups according to the World Reference Base (WRB) soil classification system: Acrisols, Arenosols, Cambisols, Histosols, Ferralsols, Luvisols, Planosols, Podzols, and Fluvisols. The Reference Soil Groups with the highest prevalence of acid soils are Acrisols, Ferralsols, and Podzols.

In addition to moderate weather, the tropical and subtropical regions are characterised by the presence of acidic soils. Soil formation is influenced by certain climatic, topographical, vegetative, parent material, and temporal characteristics.

Acrisols and ferralsols are more common in older land surfaces found in humid tropical settings. Acid sulphate soils can be found in low-lying coastal areas in the tropics that were once occupied by mangrove swamps. Podzols are characteristic soils of boreal forests, although they may also occur in other regions. Various elements, such as climate, parent material, and vegetation, contribute to the creation of distinct types of acid soils. Despite the difficulties presented by acidic soils, their productivity can be maximised through regular application of lime and fertilisers, as well as diligent soil management. Aluminium poisoning poses a genuine issue with distinct consequences for its management. Aluminium toxicity is likely to occur in soils that have a pH (H_2O) of 5 or lower and an aluminium saturation of more than 60% of the cation exchange capacity (CEC). Elevated concentrations of aluminium (Al) have a significant influence on various physiological processes in plant cells, including cell division, respiration, nitrogen mobilisation, glucose phosphorylation, root growth, nutrient absorption, and translocation (especially the immobilisation of phosphorus in the roots). When considering agricultural issue soils, it is vital to differentiate between acid soils with aluminium toxicity and soils without aluminium toxicity due to the varying management consequences. Thionic Fluvisols, characterised by the presence of a sulphuric horizon in the upper soil layers, are a distinct type of acid soils that exhibit aluminium toxicity. Draining these soils leads to oxidation processes that generate sulphuric acid, causing very acidic soil conditions.

3.1.6 Extremely Acidic Soils

Soils with a pH level below 5, indicate extreme acidity. Soil acidity exerts a substantial influence on the biological, chemical, and physical properties of soils. The presence of low calcium concentrations in the soil solution of acid soils restricts biological activity and compromises the integrity of the soil structure. Nevertheless, the primary obstacle to production in extremely acidic mineral soils is the possibility of high levels of toxic aluminium concentrations in the soil solution, which can be detrimental to plants. As a result of complex soil formation processes, numerous tropical soils, especially those that have been heavily eroded, possess high concentrations of exchangeable aluminium. The reduction in length and increase in diameter of the roots are the initial indications of aluminium poisoning. Diminished ramification and discolouration of the root system. The symptoms of aluminium toxicity are similar to those of calcium or phosphorus deficiency. If aluminium toxicity worsens, it has a significant impact on several plant metabolic functions, such as nutrient absorption and translocation (especially phosphorus immobilisation in the roots), cell division, respiration, nitrogen mobilisation, and glucose phosphorylation. The permissible concentration of aluminium varies depending on the botanical species. Nevertheless, only some crops, such as tea, have the ability to accumulate elevated levels of aluminium without causing any detrimental effects on plant development. Plants' tolerance to aluminium can be affected by several factors, such as differences in root structure, the capacity to increase the pH of the root area, processes that limit the movement of aluminium from roots to shoots, the ability to store aluminium in shoots without harming plant metabolism, and mechanisms that hinder the absorption of nutrients (such as calcium, magnesium, and phosphorus) in the presence of aluminium. The application of lime is a common technique used to increase the pH of soil, which in turn reduces the concentration of aluminium in the soil solution. Lime is, however, limited in abundance in numerous tropical places. Therefore, it is of utmost importance to cultivate crop varieties that possess resistance to aluminium, especially in tropical regions. There are crop varieties that exhibit tolerance to various sorts of crops. These soils have the potential to support the cultivation of different types of upland rice, cassava, mango, cashew, citrus, pineapple, and cowpea. These soils are suitable for cultivating several grasses, including *Brachiaria decumbens*, *Paspalum plicatulum*, *Pueraria lobata*, *Melinis minutiflora*, and *Hyparrhenia rufa*, as well as legumes such as *Stylosanthes* sp., *Desmodium* sp., and *Centrosema* sp.

The following soil Groups may consist of acid soils that provide a concern due to the presence of aluminium.

(i) Ferralsols and Acrisols
(ii) The Podzols

Moderately acidic soils with pH values ranging from 5 to 6.5 do not exhibit aluminium toxicity. When considering the harmful effects of aluminium, there are no notable alterations in the pH needs of crops in moderately acidic soils. Even crops such as sugar beets, barley, and wheat, which typically thrive in neutral soil conditions, can still offer satisfactory harvests in moderately acidic soils as long as they receive the appropriate quantity of nutrients, water, and optimal rooting conditions. An abundance of organic matter can mitigate some of the adverse effects of soil acidity on the physical and chemical characteristics of the soil. As the pH of the soil falls, there should be a corresponding increase in the quantity of organic matter present in the soil. In order to achieve optimal tillage conditions, promote strong aggregate stability, and enhance soil aeration, it is essential to maintain a high concentration of organic matter in the soil. Consequently, ensuring an adequate level of caution is necessary while attempting to enhance the organic matter content in moderately acidic soils. For example, the implementation of mulching or no-tillage methods can achieve this outcome.

3.1.7 Soil Acidification and Acidity

Acidification refers to the process of increasing the acidity of a substance, while soil acidity refers to the level of acidity present in the soil. The words "soil acidity" and "soil acidification" are separate yet interrelated. Low soil pH indicates soil acidity, which can result from either natural processes or human intervention. Soil acidity is characterised by a greater abundance of hydrogen ions dissolved in soil moisture compared to a neutral or alkaline soil. It can result in stunted plant growth and has a substantial influence on the chemical and biological processes occurring in the soil. Conversely, any procedure that raises the acidity of the soil is referred to as "soil acidification." Soil acidification is the act of reducing soil pH by increasing the amount of dissolved hydrogen ions. This can be achieved by adding acid to the soil or by eliminating bases, which are substances that counteract acidity. The cumulative outcome of these activities determines whether the soil becomes increasingly acidic or increasingly alkaline over time. Several processes can either acidify or alkalize the soil. Due to the leaching activity of water seeping through soil, the soils in humid places tend to undergo acidification processes, resulting in increased acidity over time. Naturally occurring acidic soils are abundant, and gradual acidification is a natural phenomenon. According to the estimations of Von Uexküll and Mutert (1995), approximately 3950 million hectares, or 30% of the ice-free land on Earth, is occupied by soils with pH levels below 4.8. Issues arise when human activities such as agriculture or the combustion of fossil fuels, leading to the occurrence of acid rain in the industrialised regions of the northern hemisphere, accelerate soil acidification. When assessing the impacts of rapid soil acidification, it is crucial to take into account the spatial distribution of natural soil acidity. This is because soils that are already mildly acidic are the most vulnerable to significant degradation. Hydrogen ions, dissolved in soil water, are the main catalysts for the chemical weathering of rock and soil minerals. These ions also get neutralised in the process. Minerals undergo weathering or complete dissolution when soils develop and the soil's pH lowers in humid settings. Old, heavily weathered soils contain only the most durable soil minerals, such as quartz, iron oxides, and aluminium hydroxides. Soil acidity can serve as an indicator of weathering in certain instances.

3.1.8 Development of Soil Acidity

The levels of hydrogen ions (H^+) and aluminium ions (Al^{3+}) present in soils have the potential to be utilised as indicators of soil acidity (Fageria and Baligar 2008). Soil acidity has a combination of consequences on agricultural productivity, including nutrient/element toxicities, low beneficial microbial activity, and reduced plant root growth, which limits nutrient and water intake (Fageria and Baligar 2003). The complex aspects of soil acidity have been thoroughly discussed in several studies (Tang and Rengel 2003).

Through a process of hydrolysis, the species of aluminium ions produce hydrogen ions (Lindsay 1979), as given below.

$$Al(OH)_{2+} + H^+ (i) Al^{3+} + H_2O \quad (i)$$
$$Al^{3+} + 2H_2O \text{ results in } Al(OH)_{2+} + 2H^+ \quad (ii)$$
$$Al^{3+} + 3H_2O = Al(OH)_3O + 3H^+ \quad (iii)$$
$$Al^{3+} + 4H_2O \text{ results in } Al(OH)_4^- + 4H^+ \quad (iv)$$
$$Al(OH)_5^{2-} + 5H + (v) Al^{3+} + 5H_2O \quad (v)$$

According to Lindsay (1979), the concentration of exchangeable Al^{3+} decreases by a factor of 1000 for each unit increase in pH. Additionally, when the pH increases, the Al^{3+} precipitates as insoluble Al hydroxyl species. However, as the pH levels above 6.5, aluminates with a negative charge are formed, causing aluminium to gradually dissolve (Haynes 1984).

The Al $(OH)_2^+$ species is only present within a narrow pH range and is of minimal significance. Aluminium ions (Al^{3+}) are the most abundant below a pH of 4.7. From pH 4.7 to pH 6.5, the dominant ions are $Al(OH)_2^+$. From pH 6.5 to pH 8.0, the prevalent ions are $Al(OH)_3$. Beyond a pH of 8.0, the primary ions are $Al(OH)_4^-$. Al $(OH)_5^{2-}$ species can be found at elevated pH levels, above the usual range observed in soils (Bohn et al. 2001).

3.1.9 The Impact of Soil Acidity on Soil Microorganisms and Root Growth

Soil microorganisms, being the second most important biological agents in the agricultural environment after plants (Fageria 2002), can serve as indicators of soil quality. Soil microbes have a crucial role in driving many chemical and biological processes, which significantly influence soil fertility, nutrient cycle, and carbon cycling (He et al. 2003). A multitude of microorganisms that are useful to plants, such as nitrogen-fixing bacteria that can either form symbiotic relationships or exist independently, rhizobacteria that enhance plant growth, saprophytic microorganisms, biocontrol agents, mycorrhizae, and free-living fungi, can be found in the rhizosphere and plant roots. Except for fungi, which may grow in a broad range of soil pH levels, the activities of these helpful microbes are limited by soil acidity (Brady and Weil 2002). Acidic soils exert diverse impacts on plants. Aluminium restricts the growth of plant roots due to its direct influence on metabolism or indirectly by binding soil phosphate to form aluminium phosphates, which reduces agricultural yields in general (Mora et al. 2005). The sensitivity of different plant species and cultivars to the circumstances in acidic soils varies, as stated by Wild (1993). Maize exhibits a moderate tolerance range and is more likely to flourish within the pH range of 5.5 to 6.0. The Githigu maize variety, which is located in Central Kenya, is one of the local cultivars that have been specifically adapted to the lower end of the tolerance range (Kanyanjua et al. 2002). Acidity leads to a complex interaction of factors that limit plant growth, including the physical, chemical, and biological properties of the soil. Biologically, soil acidity detrimentally affects the functioning of beneficial microbes. The decomposition of organic matter, mineralization and immobilisation of nutrients, and their uptake and utilisation by plants are all greatly influenced by this. Ultimately, it has a profound effect on crop yields (Huber et al. 2006). Research has shown that the application of organic amendments and lime treatment can have an impact on soil microorganisms, specifically bacteria and fungi (Magdoff 2001). Forge et al. (2008) demonstrated that organic inputs have a positive impact on the abundance of several components of the soil food web, such as bacterial and fungal populations. Recent studies have demonstrated a strong correlation between variations in soil chemistry and changes in soil microbial populations across different areas (Jenkins et al. 2009; Nilsson et al. 2007). Research has shown that there is a strong correlation between soil pH and the composition, and sometimes the variety, of soil bacterial communities (Hartman et al. 2008; Lauber et al. 2009). Both the overall structure of the bacterial population (Fierer and Jackson 2006; Lauber et al. 2009) and the makeup of particular bacterial groups (Jenkins et al. 2009) adhere to this pattern. Multiple studies have shown that many soil microorganisms produce organic acids such as carbonic, acetic, citric, and others. Sharif et al. (2011) found that these acids provide a beneficial environment that enhances the solubility of phosphorus and its uptake by plants. Kucey et al. (1989) found through experiments in wet environments that the release of organic acids often led to the microbial dissolution of soil phosphate. Hence, the application of mineral acids, organic acids, a blend of organic substances, or biological therapy could potentially enhance the accessibility of phosphorus for plant absorption. The solubility can be enhanced by using organic manures and P materials, as shown in previous studies (Sharif et al. 2011).

3.1.10 Consequences of Soil Acidification

An optimal soil pH for plant growth is slightly acidic as it enhances the activity of soil microorganisms and increases the availability of plant nutrients. The delicate balance is disrupted when the pH levels become either high (alkaline) or low (acidic), causing plant nutrients that were previously adequate to either become inadequate or toxic for plant growth. When the soil pH becomes overly acidic, important nutrients such as phosphorus, calcium, magnesium, and molybdenum become unavailable. Acidic circumstances in farming systems will result in a reduction in plant productivity. Consequently, any form of productive agriculture will become less lucrative and increasingly reliant on fertilisers. Adjusting the soil pH to a more optimal range will enhance the availability of essential nutrients. When the pH of the soil is 4.8 or below, there is a risk of nutrient poisoning in acidic soils (Slattery et al. 1999). Aluminium (Al) and manganese (Mn) toxicity are the primary toxicities seen in acid soils (Slattery et al. 1999). Aluminium and manganese exhibit toxicity to plant roots when they are present in higher concentrations in the soil solution of extremely acidic soils (pHW 4.3). Root stunting is the predominant plant symptom of aluminium toxicity in acidic soils (Slattery et al. 2000). Impaired root growth hinders the absorption of nutrients and water, resulting in decreased production. Certain plants exhibit greater tolerance to elevated levels of aluminium in the soil solution compared to others. However, when the pH lowers, the potential for agricultural utilisation of these tolerant plants also diminishes. Using acid tolerant crops has the potential to expand the usage of soil in very acidic conditions. However, there is a risk in developing more acid tolerant cereal and legume varieties. In the event of this happening, soil deterioration has the potential to advance to a level that cannot be reversed. These practices will lead to a deteriorated soil resource, which cannot be restored to its previous condition and is projected to sustain current crop productivity. It is evident that there is a necessity to preserve this valuable resource to the greatest extent feasible, and one strategy is to prevent the occurrence of highly acidic soils. Productive plants such as lucerne, phalaris, canola, and barley face challenges in establishing and developing in acidic soils. The establishment of lucerne is permanently hindered by both low pH and hazardous levels of aluminium (Yokota and Ojima 1995). In order to counteract the process of acidification, deep-rooted perennial pastures such as lucerne and phalaris are being cultivated (Ridley et al. 2002). If the soil pH is too low, the inability of certain plant species to thrive will result in the continued leaking of nitrates. This will accelerate acidification and enhance aquifer recharge, exacerbating the salinity problems on arid terrain. A study conducted by Slattery et al. (1998) has shown that allowing soils to maintain a pH level of 4.0 or below for an extended period of 10 years has significant implications for the depletion of clay minerals in the soil. By permitting weathered soils to undergo acidification within a single agricultural generation, it seems that there is minimal harm inflicted upon the mineral composition of the soil. If acidity is not controlled over extended periods, it will result in the ongoing depletion of clay and the accumulation of silica in the soil. In order to prevent this permanent soil degradation, it is necessary to utilise liming materials to counteract high soil acidity. As of the current time, just 22% of Australia's entire area of acidic soil has been treated with lime, according to estimations from AACM in 1995. According to Slattery's 2000 study, around 5% of the 90 million hectares of land affected by soil acidity have a high level of acidity, with a pH value of 4.8 (pHCa). In order to ensure long-term production and prevent permanent soil deterioration, it is necessary to lime the remaining areas of this property promptly, even though some parts have already been treated with lime. Acidity is increasing in the top soil layers of cropping soils in all of Australia's major grain-producing regions (Coventry et al. 1997; Slattery et al. 1997). While the deeper soil layers in many of these places are often alkaline, they can also be acidic and inhospitable due to their sodic, saline, or dual nature. The surface soil layers provide a rich and nutritious environment for plant roots, and it is important to protect them from irreversible clay loss caused by soil acidity. If top soil acidity continues without reduction, there is a chance of acidity developing in subsurface that was previously neutral. Remediating subsurface acidity is challenging because neutralising minerals such as lime and base-rich organic matter migrate slowly down the soil profile (Ridley and Windsor 1992; Myers and De Pauw 1995). Granitic soils often encounter significant problems with subsoil acidity in recharge zones, leading to hindered plant growth and intensified dry land salinity. Although agriculture has exacerbated the acidity of these granite subsoils, managing recharge on these soils will require careful attention.

Soil acidity has a negative impact on several duties carried out by natural resources. Certain factors, such as soil degradation and the decline of native biodiversity, can significantly impact the entire community and perhaps harm leisure and tourism activities. Here are a few examples of these areas:

- Soil bacteria
- Forestry
- Plant life
- Soils that contain a high concentration of acid sulphate
- Marine biota
- Infrastructure
- Human wellness
- Animal welfare
- Weed control
- Soil composition.

3.1.11 Limitations of Acidic Soil

Due to the prevalence of acidic soils, plants grown in such conditions frequently encounter a range of mineral nutritional problems. However, variations in soil properties and the degree of acidity in different countries' soils result in varying levels of difficulty in particular parts of the world. According to Eswaran et al. (1997), approximately 26% of the world's ice-free surface area, which is equivalent to 37.8 million square kilometres, consists of acidic soils. Approximately 20% of the world's ice-free region, equivalent to 29.2 million square kilometres, is covered by acidic subterranean soils, which is somewhat less than the amount covered by surface soils. South America has a majority of its soils classified in these classes, with a prevalence of moderate to high surface soil acidity, often with a pH of 4.5. The acidity levels of surface soils in North America and Europe are approximately 31% and 43% respectively. Certain soil orders, such as Oxisols, Ultisols, Alfisols, and Andepts (Van Wambeke 1976), are often associated with acidity and can expose plants to significant mineral deficiencies or toxicities. The relevance of each chemical limitation and/or interaction changes depending on parameters such as soil pII and organic matter, soil type and horizon, concentration and species of nutritional elements, parent material, soil physical qualities, plant species' genotypes, and climatic conditions. The occurrence of these common patterns can be observed in the limitations of mineral deficiency and toxicity that are associated with soil pH.

Soil acidity imposes growth and yield constraints on several crops worldwide. The chemical features of more acidic soils are primarily affected by the presence of hazardous compounds and components such as soluble forms of Al, Fe, and Mn, nitrites, and various dangerous organic acids, in addition to acidity itself. Krstic et al. (2012) reported that over 40% of the world's arable lands have acidic soils, which present a substantial obstacle to crop productivity. Aluminium, which constitutes the third-highest concentration in the earth's crust, poses a risk to plants when it is dissolved in soil solutions with acidic pH levels (Kochian 1995). Von Uexküll and Mutert (1995) reported that there are 3950 million hectares of land with acidic soil, and 15% of this land is used for growing annual and perennial crops. The growth of coniferous forests is facilitated by the acid soils found in the northern belt, which consists mainly of organic acid soils and is located in the rainy northern temperate zone. Within the humid tropics, there exists a region characterised by the presence of mineral acid soils located to the south. Uexküll and Mutert (1995) reported that now only 12% of the area dedicated to agricultural production is characterised by acidity. However, there is a global trend of increasing acidification of soil. Mineral acid soils are formed from parent materials that are naturally acidic and lacking in basic cations (such as Ca, Mg, K, and Na), or by the leaching of these elements from the soil, which reduces the pH and the soil's buffering capacity. Aluminium (Al) is capable of dissolving in soil and becomes increasingly prevalent in the soil solution as phytotoxic aluminium ions. Aluminium is found in most mineral soils in adequate quantities to act as a buffer, maintaining the soil at a pH of approximately 4. Organic acid soils are characterised by high quantities of humic acids and partially degraded plant debris. These soils typically lack aluminium buffering and can have a pH that drops significantly below 4 (Kidd and Proctor 2001). Various factors can alter the toxicity of acid soil, depending on its composition. The primary constraint on plant growth in acidic soils with a high mineral concentration is the toxicity of aluminium. Aluminium (Al) is released from soil minerals as $Al(OH)_2^+$, Al(OH), and $Al(H_2O)_3^+$ when the conditions are acidic. The last form, $Al(H_2O)_3^+$, is also known as Al (Kinraide 1991). Aluminium ions at micromolar concentrations rapidly inhibit root growth in most agriculturally important plants.

Soil properties can vary greatly based on factors such as soil types, topography, climate, vegetation, anthropogenic activities, and soil distribution patterns. Surface modelling is a valuable technique used in the fields of agriculture and environmental science to estimate soil parameters using interpolation. The advancement of this technique is necessary for the implementation of precision agriculture and the control of soil pollution. However, the effectiveness of the application relies on its capacity to effectively capture the geographical variability of soil properties using spatial interpolation. In order to enhance the quality of interpolation for soil properties, it is crucial to do comprehensive study on the interpolating approach (Shi et al. 2009). The root apex is the primary location where Al poisoning occurs. The intricate nature of how aluminium affects cellular processes poses challenges in identifying the primary effect(s) of aluminium poisoning. Exposure to Al inhibits the formation of lateral roots and causes reduced growth in the primary root. The root tips are shortened and stunted as a result of the suppression of cell elongation and cell multiplication. The plant's small root system makes it more susceptible to drought stress as it hampers its capacity to absorb nutrients and water. Plants that are susceptible to aluminium toxicity see a substantial reduction in both the quality and yield of crops (Jovanovic et al. 2006, 2007; Samac and Tesfaye 2003). The physical–chemical characteristics of Serbian vertisols have an impact on the distribution and accessibility of specific elements for plants, as stated by Dugalic et al. (2004), Jelic et al. (2010), and Milivojevic et al. (2011). Potential acidity refers to a situation where there is an abundance of pyrite compared to compounds that can neutralise acid. The interpretation focuses on the physiographic contexts, both current and historical, that are essential for the production, deposition, or sedimentation of pyrite and acid-neutralising components, taking into account the necessary components and environmental conditions. Potential acidity accumulates mainly in sediments rich in kaolinite, which are found in tidal flats below the mean high water level. These sediments are characterised by a dense mangrove vegetation and are regularly flushed by saline or brackish tides. The rate of sedimentation is such that the mangroves can persist well below the mean high water level for several decades, as observed by Pons and Breemen in 1981. This is where the potential acidity is built up. These conditions are favoured by low rates of sedimentation or land subsidence compared to sea level, paired with a tropical humid climate.

Soil acidification is a major obstacle to agricultural productivity and sustainability on a global scale (Kochian et al. 2004). Agricultural residues play a vital role in creating pH gradients in soil profiles, however the quantity and timing of residue application are mostly determined by the farming system and the specific growing season. The pH rises observed after the addition of residues are believed to be caused by the decarboxylation of organic anions (Tang and Yu 1999; Yan et al. 1996). Previous studies (Noble and Randall 1999; Pocknee and Sumner 1997) have proposed that the release of basic cations during decomposition increases the pH level. However, there is a lack of empirical evidence to support this concept. In their study, Yan and Schubert (2000) found that the addition of cations in the form of salts, such as Na_2SO_4, did not provide the same effect of reducing acidity as the addition of sodium malate. The primary mechanism responsible for changes in pH is believed to be nitrogen (N) cycling, wherein organic N is transformed into NH_4^+, resulting in the consumption of H^+. Subsequently, NH_4^+ is turned into NO_3^-, releasing $2H^+$ in the process (Xu et al. 2006a; Xu and Coventry 2003). The plant's uptake of nitrates would otherwise mitigate the overall acidity resulting from nitrate leaching. The pH change will also be affected by the association/dissociation processes of H^+ with residue surfaces, organic molecules, and the soil matrix. The composition of the residues and the soil environment will influence the overall effect of these processes. The liming potential of residues is shown by the surplus cation concentration, which serves as an indicator of ash alkalinity (Noble et al. 1996). The initial pH, cation exchange capacity, organic matter content, and texture are edaphic factors that significantly impact the temporal release of alkalinity due to their influence on association/dissociation reactions and biological activity. The concentration and composition of nitrogen (N) in residues play a crucial role in determining what happens to the N (whether it is mineralized or immobilised) and whether the N cycle leads to acidification. The breakdown of residues by microorganisms is negatively affected by the ratio of carbon to nitrogen (C:N) and is also hindered in soils with low initial pH. Furthermore, the chemical composition of residues is important because a significant portion (50%) of the alkalinity in plant materials can be readily accessible, soluble, and capable of moving through the soil profile.

Acidity and aluminium (Al) toxicity are the primary obstacles that restrict the utilisation of agricultural soil. Lime application can effectively address problems caused by acidic soil. Utilising sulphate-reduced bacteria (SRB) or Acidithiobacillus ferrooxidans (AF) to augment coal can serve as an extra solution for addressing acid soil problems (Firnia and Rohmawati 2022). By incorporating coal enriched with SRB or AF into organic fertilisers, plantations can enhance the value of their organic material and promote sustainable agricultural practices. The objective of this study was to investigate if supplementing organic fertiliser with an ameliorant, microbial agents, or microbial agents enhanced with an ameliorant can enhance its value as a sustainable supply of organic material for plantings in an agricultural system. The study focussed on two Ultisols (Firnia and Rohmawati 2022) with varying levels of exchangeable Al content. The results indicated that the soil's pH, calcium (Ca), and magnesium (Mg) levels were elevated when a dolomite ameliorant or dolomite-enriched AF ICBB 8789 was used. The application of AF ICBB 8789 significantly improved the cation exchange capacity (CEC) in the soils of Jasinga and Lebak. By incorporating sulphate-reducing bacteria (SRB), a coal or lime amendment can improve the accessibility of phosphorus (P) in the acidic soil found in Jasinga, West Java, and Lebak, Banten. Coal, solid rocket boosters (SRB), or air–fuel (AF) enhanced with coal have the capacity to provide organic material for agriculture in a sustainable agricultural system (Firnia and Rohmawati 2022).

Liming is a frequently used remedy for traditional issues with acidic soil. Liming is a technique that can improve the physical, chemical, and biological characteristics of soil. Liming improves chemical characteristics of soil by increasing soil pH, enhancing the availability of vital nutrients, and reducing the activity of toxic elements such as Al, Fe, and Mn, which can be detrimental when present in excessive amounts (Leiwakabessy and Sutandi 2004). Aside from lime ameliorant, lignite has emerged as a noteworthy alternative for treating acid soils. According to Adani et al. (1998), light coal has the ability to produce humic acid containing a carboxyl group, which allows it to adsorb metals. Gerke (1993) asserts that the application of humic chemicals can reduce the phytotoxicity of Al^{3+} and Fe^{3+}. Research has demonstrated that applying coal to the soil can increase soil pH (Yazawa et al. 2000), reduce the availability of heavy metals (Janos et al. 2010), and enhance the absorption of phosphorus from fertilisers (Scheel et al. 2008). Scientists have recently developed an interest in utilising microorganisms due to their environmentally benign nature in reducing the harmful impacts of high concentrations of heavy metals in soil. Research conducted by Kurniawan et al. in 2015 has shown that Acidithiobacillus ferrooxidans is capable of carrying out the bioleaching process, which involves the dissolution of metals. Sulphate-reducing bacteria, along with Acidithiobacillus ferrooxidans, can be employed to improve the chemical properties of acidic soils. These bacteria has the capacity to reduce the amount of sulphate produced during the oxidation of sulfidic minerals. The bacteria's activity leads to an increase in pH and a decrease in metal accumulation. Sulphate-reducing bacteria possess the capability to convert sulphate, sulfite, or thiosulphate ions into sulphide by using these ions as electron acceptors to power their metabolic activities. Due to the strong reactivity of hydrogen sulphide and its rapid interaction with metals to form highly stable metal-sulphides, the presence of significant amounts of dissolved metals will result in the precipitation of these metals (Widyawati 2011). According to Plante (2007), bacteria produce organic acids that will chemically bind with Al and Fe through a process called chelation. This interaction makes the organic phosphorus in the soil more easily broken down by enzymes, resulting in the release of bound phosphorus into the soil. It is necessary to investigate the effects of using ameliorant, microbial, and ameliorant-enriched soil microbes as alternative technologies to enhance soil productivity in sustainable agriculture systems. This research should focus on the potential use of coal, sulphate reduced bacteria (SRB), and Acidithiobacillus ferrooxidans to reduce aluminium toxicity in acidic soils and increase phosphorus availability. In this investigation, dolomite is being utilised for the purpose of comparing it with conventional amelioration methods. This study aimed to enhance the value of organic material for plantations in a sustainable agriculture system by investigating the impact of ameliorant, microbial, and microbial-enriched ameliorant application on changes in chemical properties in acid soils with varying levels of Al saturation.

True peat is created when the production of organic matter exceeds its breakdown because of reasons such as low soil temperatures, water saturation, extreme acidity, nutrient deficiency, or a combination of these conditions. Authentic peat soils possess an organic matter concentration of at least 65% and a minimum depth of 50 cm. Globally, approximately 240 million hectares of land are occupied by peat soils, mainly found in boreal and temperate regions. The coastal lowlands of southeast Asia contain almost 20 million hectares of the anticipated 32 million hectares in the tropics. Approximately 240 million acres of peat soils exist globally. Peat soils exhibit distinct chemical and physical characteristics that set them apart from mineral soils. Consequently, utilising peat for agricultural purposes necessitates the implementation of a whole new reclamation approach. If peat soils are appropriately reclaimed and managed, the bulk of them have the potential to be highly productive in a sustainable manner. The organic fraction consists primarily of lignin, cellulose, hemicellulose, proteins, carbohydrates, and "humus substances" such as humic and fulvic acids. The composition of this fraction is largely determined by the types of plants present. While calcareous peats are somewhat prevalent, most untouched peats are acidic, with a pH ranging from 3.5 to 5.5. Additionally, they have lower plant nutrient levels per unit volume compared to mineral soils. There is often a limited availability of micronutrients. Nevertheless, even compact peats that rely on precipitation have the ability to sustain abundant and mature vegetation. This vegetation assimilates nutrients, transiently retains them, and subsequently releases them back into the soil as litter and other plant debris. The cycle is disrupted when the native vegetation is destroyed as a means of reclaiming the land, and as a result, the natural soil fertility often decreases rapidly thereafter. Plants are unable to easily absorb all of the nutrients included in peat. Plants utilise only a small fraction of the nitrogen that is accessible to them, while the rest is chemically bonded in stable compounds. Phosphorus is also fixed to some extent. While potassium is more easily accessible, its levels are often rather low. Peat and mineral soil materials differ principally in their low bulk densities, composition of colloids, and distinct thermal properties. Reclaiming water-saturated peats for dryland crop production poses challenges due to the material's high porosity. The necessary drainage system prevents the buoyancy of groundwater and typically results in substantial subsidence of the land surface, as well as compaction of the loose peat pile. Moreover, the fibrous peat material exhibits a diminished load-bearing capacity as a result of its substantial pore volume and pliability. Consequently, constructing roads, structures, and water systems becomes challenging, and it also leads to the instability of crops and trees, causing them to tilt or collapse. Peat can undergo substantial desiccation and contraction due to sudden and extensive drainage. In exposed peat soils, the poor heat conductivity of the organic material causes very high temperatures to accumulate in the top few centimetres when they are directly exposed to sunlight. This expedites the process in bare soils. Consequently, colloids experience an irreversible transformation, leading to the disintegration of peat into a dry powder that exhibits unfavourable physical properties and a heightened susceptibility to wind erosion. To protect peats, it is crucial to maintain a shallow watertable and ensure that they are consistently covered by plants. Combining these tactics is rarely effective because most arable crops need a deep root zone for optimum growth. Peat soils that are saturated with water are also colder than most mineral soils, resulting in a slower rate of crop growth. Artificial drainage is typically indispensable. Drain spacings are influenced by various factors such as the necessary drainage depth, the kind of peat, the level of decomposition, the packing density, the heterogeneity, and the presence of wood or mineral admixtures. Over time, drainage systems may require modification because the hydraulic properties of peat are influenced by alterations in the water regime. Reclaimed peat is often stated to possess physical characteristics that make it ideal for agricultural use. Some parts of this are accurate. Most peat soils possess favourable water retention and root penetration characteristics, facilitating the easy gathering of peanut pods, rhizomes, or tubers. In order to mitigate the detrimental effects of shrinkage, settling, compaction, mineralization, burning, and wind erosion, it is necessary to employ advanced soil management and costly approaches due to the perishable nature of peat.

The sorptive quality of soils is attributed to the presence of electrical charges and the extensive surface area of clay minerals and humus. The clay minerals and humus possess a negative charge as a result of their inherent structure and chemical composition. Protons detach from the surfaces of clay minerals (alumiosilicates and hydrated oxides of iron and aluminium) as well as from the acidic groups in humus. This charge remains constant regardless of pH. The detachment of protons leads to a negative charge that varies with pH. Clay minerals consist of iron and aluminium hydrous oxides. When exposed to protons, these oxides acquire a positive charge at low pH levels. The extent of this pH-dependent shift is greater in clay minerals with a more disordered structure, as it primarily occurs at their surfaces and edges. The adsorbed cations effectively neutralise the negative charge in clays. In clays with a 2:1 structure, certain cations become trapped within the gaps between layers and can only diffuse into the surrounding solution at an exceedingly slow rate. Due to the tight binding of the silicate sheets in clays such as illite, the movement of potassium ions in the interlayer is restricted, resulting in the inability to interchange these cations. Conversely, smectites lack a robust connection that binds the silicate sheets together. The interlayer gaps expand, facilitating the permeation of cations into the outer solution. The main discovery is that smectites, due to their larger surface area, are capable of retaining a greater number of exchangeable cations compared to illite, which has a higher charge density.

3.1.12 The Presence of Air, Soil Particles, and Water

Most plant materials require a significant amount of water for production. The transportation of essential plant nutrients from solid soil particles to plant roots and throughout the leaf structure is primarily facilitated by this mechanism. Plants are able to reach water due to gradients created by capillary and gravitational forces. Water is absorbed by plants and then released into the atmosphere through their leaves. The presence of nutritious solutes in water is influenced by concentration gradients. Water that is dispersed over a larger surface area of soil becomes more easily accessible to plants and is more prone to drainage. Water retention is enhanced when it is confined within smaller pores or the interstices between individual clay particles. Colloid clay particles exhibit a substantial surface to volume ratio, enabling them to retain a considerable amount of water. When the soil becomes waterlogged, meaning it is completely saturated with water, it undergoes substantial alterations in its physical, chemical, and biological characteristics. Soil that contains a high concentration of organic matter has the capacity to retain a significant quantity of water in comparison to other types of soil. However, this water is less accessible to plants since the organic matter absorbs and holds onto the water through physical and chemical processes. In this type of soil, microbes rapidly use oxygen as they decompose organic matter through respiration, leading to its depletion. Consequently, due to the absence of adequate air, the presence of excess water in these soils is detrimental to plant growth. Except for rice, most valued crops are unable to thrive in waterlogged soils.

3.1.13 Soil Substances

The essential micronutrients required by plants are boron, chlorine, sodium chloride, copper, iron, manganese, zinc, vanadium, and molybdenum. Plants require both macronutrients and micronutrients to facilitate their growth. They must exist in minimal quantities as their harmful impacts manifest at higher concentrations. Most of them are constituents of essential enzymes. Plants require several important macronutrients, including carbon, hydrogen, oxygen, nitrogen, phosphorus, sulphur, potassium, calcium, and magnesium. The atmosphere and water serve as the primary suppliers of carbon, hydrogen, and oxygen. Certain elements, including chlorine, manganese, iron, zinc, and vanadium, are prone to engage in the process of photosynthesis. Certain plants have the ability to directly extract nitrogen from the atmosphere through the use of nitrogen-fixing bacteria. The soil is the primary source of the essential macronutrients. Nitrogen, phosphorus, and potassium (NPK) are commonly applied to soil as fertilisers. Calcium deficiency in soil is caused by the absorption of calcium by plants, the leaching of carbonic acid in acidic soils, and the competition with elevated levels of sodium, potassium, and magnesium in alkaline soils. Usually, lime is applied to soils that lack calcium (a procedure called liming) in order to provide plants with the required calcium. Plants can obtain magnesium by either absorbing it from organic matter or through the exchange of ions with clays. The presence of large amounts of calcium, sodium, or potassium in the soil reduces the amount of magnesium available, and plants take in sulphur in the form of assimilable SO_4^{2-}. Sulphate ions, in contrast to K^+, are not restricted by ion exchange binding and can be readily absorbed by plant roots. In order to effectively manage nitrogen (N) and phosphorus (P), it is essential to have a thorough grasp of the natural nutrient changes that occur, as these changes are accurately depicted by the nitrogen and phosphorus cycles. The alterations in agricultural systems mostly occur inside the soil and result from complex interplay among the atmosphere, soil microorganisms, flora and fauna, and soil moisture. The fertility of the soil is maintained by nitrogen that is bonded to humus. It serves as a reservoir for the nitrogen required by plants. Another advantage of this feature is its capacity to release nitrogen at a rate that closely aligns with the rate of plant growth. Typically, plants absorb nitrate ions (NO_3^-) from the soil. Nitrifying bacteria convert nitrogen, in the form of NH_4^+ (fertiliser), into NO_3^- in soils, enabling plants to utilise it. The primary modules of certain leguminous plants, such as soy beans, alfalfa, and clover, possess the unique ability to convert atmospheric nitrogen into a usable form. Legumes have the ability to enhance soil by introducing substantial quantities of nitrogen.

Although phosphorus makes only a small proportion of plant matter, it is essential for the growth and development of plants. Plants may access orthophosphate most easily when the pH level is close to neutral. The phosphate species that can be absorbed by plants are $H_2PO_4^-$ and HPO_4^{2-}. These species are present in the soil at a certain pH. In acidic soils, orthophosphate ions can either be precipitated or sorbed by cations like Al^{3+} and Fe^{3+}. Only a little amount of phosphorus given as fertiliser leaches from the soil due to its fixation as hydroxyapatite. Potassium is one of the three essential components of fertilisers that plants require. Typically, soil potassium levels are considered adequate only when an adequate amount of potassium is provided to compensate for the potassium that is depleted by crops. This occurs because any surplus potassium supplied is primarily assimilated by clays and organic substances present in the soil. The crop production gradually declines as the potassium present between the illite layers is gradually depleted in areas where crops have been cultivated for an extended period without sufficient potassium-rich fertilisers. The increase in crop production is not as substantial as expected when potassium fertiliser is subsequently applied. This is because the potassium is not readily accessible for plant growth, but instead it is absorbed back into the illite structure. Consequently, farmers are compelled to pay exorbitant costs for potassium fertilisers without witnessing a commensurate increase in crop yield. In order to sustain high agricultural yields, it is necessary for the crop to replace the nutrients that have been depleted.

3.1.14 Pollutants and Refuse

Significant quantities of industrial, municipal, and agricultural waste are disposed of in the soil without undergoing any form of treatment. Heavy metals such as mercury, lead, cadmium, nickel, and arsenic play a significant role in causing land pollution issues. For example, it has been determined that the presence of high levels of heavy metals in the soil can be attributed to the disposal of waste materials from nearby mines and industrial facilities located in close proximity to agricultural regions. In certain situations, the practice of degrading and disposing of hazardous organic wastes on land is implemented. Pesticides present in soil can disperse throughout the ecosystem by several physical methods, such as being absorbed by the soil, washed away by rainfall, absorbed by plants and animals, and carried by wind. Nevertheless, it is the activities facilitated by bacteria, animals, plants, and sunlight that genuinely make a substantial contribution to the decrease in the overall quantity of residues. Heat and pH are additional factors to take into account. Soil catalysts play a vital role alongside soil enzymes in breaking down pesticides that are rather unstable. Pesticides are broken down by most soil microorganisms, such as actinomycetes, fungi, and bacteria, through processes include oxidation, ether cleavage, ester and acid hydrolysis, oxidation, and epoxidation. The reductive systems are major components of microorganism degradation systems. When fuels containing sulphur are burned, they release SO_2, which eventually results in the deposition of sulphate in the soil. Nitrates are produced in the atmosphere by the conversion of air nitrogen oxides, and subsequently, they are deposited into the soil. In addition to heavily trafficked highways, lead particles from car exhaust also build up in soil. Soils in proximity to lead and zinc mines have a propensity to accumulate elevated concentrations of Pb and Zn. These factors have a detrimental impact on the microfauna, bacteria, fungi, and other organisms, resulting in a decrease in soil quality. The depletion or loss of soil organic matter is directly linked to biological degradation.

Nitrogen

A global environmental concern is the accumulation of atmospheric nitrogen (N). The primary source of nitrogen oxides emissions is the combustion of fossil fuels. From 1980 to 2003, most of Europe had a decrease in oxidised nitrogen emissions, ranging from 20 to 50%, except for Portugal, Spain, and Greece. Fagerli and Aas (2008) found that the decreasing pattern is usually observed in wet deposition (rainfall) but not in dry deposition (as a gas or fine particulate). The pattern of ammonia emissions and deposition exhibits great variation, with certain European countries reporting minor increases while others are witnessing declines. Erisman et al. (2008) found that agricultural sources are responsible for 90% of ammonia emissions. Consequently, local concentrations may exhibit significant variations. The high levels of total nitrogen deposition in many regions of Europe pose a significant threat to biodiversity and ecosystem services. Furthermore, many countries have already surpassed critical loads, which are thresholds indicating a risk of damage to sensitive ecosystem components. Indeed, it is expected that nitrogen deposition would be one of the three main factors influencing changes in global biodiversity by the year 2100, alongside land use and climate change (Sala et al. 2000). High amounts of ammonia, which can cause immediate poisoning, are rare in Europe and usually only found near specific sources. Nevertheless, even at lower quantities, N deposition still has discernible effects on plants, such as heightened susceptibility to secondary stress and the occurrence of eutrophication and acidity in the soil (Bobbink et al. 2010). Nitrogen has the capacity to enhance primary productivity as it serves as the limiting nutrient in numerous semi-natural terrestrial ecosystems. The drop in soil pH, which is directly or indirectly connected to nitrogen deposition, is caused by acidic precipitation, oxidation of dry-deposited compounds, loss of basic cations through ion exchange, plant uptake, and nitrification of ammonium (Bobbink et al. 2010). Stevens et al. (2004, 2009) have found that higher soil acidity can lead to several adverse consequences. These include altering the composition and diversity of plant species and enhancing the solubility of harmful metals and nutrient cycles.

Desorption of phosphorus (P) in acidic soil

The significance of phosphorus desorption from fertilised soils to both agriculture and the environment cannot be overemphasised. The P forms of two limed acid and three calcareous soils in Europe were characterised by chemical extraction, isotopic exchange, and sequential fractionation. Isotopic exchange revealed that the limed acid had a higher amount of surface-adsorbed P compared to the calcareous soils. Calcareous soils were characterised by the prevalence of calcium phosphates, whereas limed acid soils were dominated by Al and Fe phosphates. It is worth noting that one of these soils had a significant proportion of calcium phosphates (Delgado and Torrent 2000). Mixed cation and anion exchange resins (ACER) were shown to be more effective than goethite and anion exchange resin (AER) in desorbing phosphorus (P). The desorption curves exhibited notable variations in the uniformity of the P desorption reactions, with calcareous soils displaying the highest and lowest levels of uniformity. The percentage of ACER-extractable phosphorus that was available to plants in limed acid and calcareous soils was 50% and 20%, respectively. Iron (Fe) and aluminium (Al) phosphates are frequently formed as precipitates in acidic soils, as noted by Lindsay in 1979. Hydroxyapatite (HA) is found as a result of the breakdown of superphosphates in acid and somewhat acid soils (Kumar et al. 1994). However, it is worth noting that some heavily fertilised acid soils do contain significant amounts of calcium phosphates (Lookman et al. 1996) and hydroxyapatite (HA).

Reactive substrates, when used in filter systems, can effectively reduce phosphorus (P) pollution. Once these substrates become saturated with phosphorus, they can be reused in agriculture. These substrates are often calcium carbonate derivatives with a high pH, which can be particularly beneficial for soils with acidity. Three reactive substrates (Filtra P, Polonite, and wollastonite) were introduced as additives to an acidic soil in a pot experiment. When comparing the impact of substrate amendments with control and potassium phosphate treatments, there was a tendency for ryegrass yield and P uptake to be enhanced. Polonite and Filtra P significantly increased the concentrations of phosphorus (P) and calcium (Ca) in the ryegrass. However,

Polonite exhibited the highest yield/amendment ratio among the two. Throughout the experiment, the inclusion of all three substrates resulted in an increase in the pH, Al-extractable P, and cation exchange capacity of the soils. Consequently, the addition of these substrates to acidic soils can facilitate the recycling of phosphorus and improve soil properties (Cucarella et al. 2008).

3.2 Soils Containing Acidic Sulphate

The establishment of a canal network in Central Kalimantan with the purpose of reclaiming a significant piece of a tidal backswamp for the cultivation of wetland rice has raised serious concerns among several stakeholders over its potential negative impacts on soil properties, water quality, and sustainable productivity. The results indicated that following the reclamation of mineral soils, the top layer of peat experienced a reduction or complete disappearance (from 20 to 40 cm to 13 to 21 cm), whereas the fertility of the oxidised and depleted layers of the soil profiles increased. Anda et al. (2009) found that "reclamation" caused a decrease in soil pH values, an elevation in Al^{3+} and Al saturation, and leaching of cations. The content of pyrite in reduced layers of profiles decreased dramatically from 2.6 to 5.2% before "reclamation" to 0.3–1.9% after "reclamation," suggesting that pyrite was undergoing oxidation. Organic soils also had comparable morphological changes, with a surface loss ranging from 8 to 25 cm. Additionally, there was an increase in the decomposition stage, as seen by the decrease in fibre content from 60–73% to 13–27%. The observed trends of decreasing pH values and increasing electrical conductivity (EC) as well as concentrations of cations and anions, from the inlets (used as a control) to the primary canal and secondary canals, indicate a considerable deterioration in water quality due to the leaching of pyrite oxidation products from rice fields. The study's findings suggest that it is important to keep the ground water table above the pyrite location, which is around 60 cm below the soil surface. Alternatively, rainfall or tidal flushing can be used to remove oxidation products and related acidity from the arable land, in order to promote crop growth.

The presence of a sulphuric horizon or sulfidic material is an inherent attribute of acid sulphate soils, which limits their suitability for agricultural cultivation. The sulphuric horizon, as described by Soil Survey Staff (1999), is characterised by a pH value of 3.5 or below due to the presence of sulphuric acid. The sulphide that is capable of undergoing oxidation to form sulphuric acid is also used as an indicator for the existence of materials containing sulphide. When soil is drained or exposed to aerobic conditions, a notable decrease in pH value indicates the presence of sulfidic minerals or pyrite (Dent 1986; Jansen et al. 1992; Tuong 1993; Van Breemen 1993). The process of pyrite oxidation occurs due to the existence of Fe^{3+} and O_2 (Brown and Jurinak 1989). Interpreting and characterising acid sulphate soils is challenging due to their dynamic features (Jansen et al. 1992). Tides, seasons, and drainage systems exhibit dynamics on daily, annual, and longer time scales, respectively. Effective water management is crucial for controlling several dynamic processes, such as soil aeration, ripening, and acidity regulation (Dent 1986; Tuong 1993).

According to Andriesse (1993), the backswamp areas are predominantly composed of acid sulphate soils, which are characterised by problems such as water-logging, tidal flooding, low bearing capacity, and the occurrence of pyrite. The backswamp area in Central Kalimantan is naturally inundated. Consequently, the process of "reclamation," involving the construction of primary, secondary, tertiary, and quaternary canals, was initiated as a necessary step to facilitate the clearance of land for the cultivation of rice. In order to facilitate the provision of water of superior quality and the removal of water of inferior quality, separate canals for supply and drainage were constructed. Analysis of pH field measurements indicated that the pH values of lower strata prior to "reclamation" were consistently 1.0–2.0 units higher than corresponding oxidised layers and 0.5–1.0 units higher than oxidised layers. Any pyritic strata that are not covered may have pH values as low as 3. The low pH value can be explained by the oxidation of pyrite, as shown by the reaction reported by Brown and Jurinak in 1989.

$$Fe^{2+} + 2SO_4^2 + 2H + FeS_2 + 7/2O_2 + H_2O$$

As stated by Konsten et al. (1994), the abundance of protons can be attributed to three factors: (i) the presence of free acidity, (ii) the hydrolysis of exchangeable aluminium, and (iii) the release of adsorbed $SO_4{}^2$, which generates soluble Fe(II)SO_4. This process demonstrates that the oxidation of one mole of pyrite produces sulphuric acid, which dissociates to release two moles of H^+ ions and is accountable for the significant decrease in pH. The oxidation of pyrite resulted in the production of Fe^{2+}, $SO_4{}^{2+}$, and H^+, leading to a reduction in pH. According to Singer and Stumm (1970), the process of pyrite oxidation by H_2O_2 can be described in the following manner:

$$FeS_2 + 7H_2O_2Fe^{2+} + 2H + 6H_2OFe^{2+} + 2SO_4^2$$

The presence of sulphidic material was confirmed by a decrease in pH value to 3.5 over an 8-week incubation period. Additionally, the existence of a sulphuric horizon was qualitatively determined at a pH value of 3.5 (Soil Survey Staff 1999). This method is time-consuming, rendering it unsuitable for analysing several soil samples.

3.2.1 Acid–Base Exchange Reactions

The primary role of soils is to facilitate the exchange of cations, thereby releasing essential trace metals into the environment and making them accessible to plants as nutrients. The exchange of soil cations occurs between the mineral and organic constituents. The value is expressed as 300–400 milliequivalents per 100 grammes. Humus demonstrates an exceptionally elevated cation exchange capability. Soil cation exchange results in the generation of trace metal nutrients. As the exchange takes place, H^+ ions are transferred for metal ions, which are then absorbed by the roots. The soil's buffering capacity is determined by its kind, as it acts as a buffer and withstands changes in pH. Acid sulphate soils, also referred to as "cat clays," form as a result of the oxidation of pyrite in the soil.

3.2.2 Pore Water Chemistry of Acid Sulphate Soils

The process of oxidising iron sulphides, such as pyrite, in the soil results in the creation of acid sulphate soils. These soils can undergo a variety of oxidation, reduction, exchange, desorption, dissolution, and precipitation reactions. Oxidation of the soils commences when sulfidic strata are exposed to air as a result of receding water tables, excavation, or exposure to other oxidants. The equation can be employed to depict the conversion of pyrite into iron (II) and sulphate. However, this process actually encompasses multiple bacterially mediated processes.

$$FeS_2(s) = Fe^{2+} + 2SO_4^{2-} + 2H^+ = 7/2O_2 + H_2O \tag{3.1}$$

By increasing the amount of acid, it is possible for iron to undergo further oxidation, resulting in the formation of iron (III). When there is enough oxygen, if the pH is over 4.0, solid particles of ferric hydroxide will form and settle. Iron (III) remains soluble and can directly oxidise pyrite when the pH is below 4.0. The iron (III)-iron (II) cycle greatly accelerates the oxidation process. Even in the absence of dissolved oxygen, pyrite oxidation by iron (III) can still occur in partially oxidised acid sulphate soils (ASS) that have been inundated after a dry spell, provided that the soils are suitably acidic. The presence of reddish-brown goethite flocs, FeOOH in the surface soils, and increased acidity are all consequences of the oxidation of iron (II) to iron (III), which occurs as follows:

$$Fe^{2+} + 14O_2 + 34H_2O \text{ equals } FeOOH + 2H^+. \tag{3.2}$$

In conclusion, the complete oxidation of pyrite in damp soils can be described as the following (Dent 1986):

$$Fe(OH)_3 + 2SO_4^2 + 4H^+ = FeS_2 + 15/4O_2 + 7/2H_2O \tag{3.3}$$

The characteristic pale yellow mottles of jarosite, $KFe_3(OH)_6\,(SO_4)_2$, which occur when the pH is less than 3.7 under strongly oxidising circumstances, are one example of the frequent partial oxidation products that are observed in surface soils.

$$K + Fe_3(OH)_6(SO_4)_2 + 3H_2O + H^+ = 3Fe(OH)_3 + 2H_2SO_4 + K^+$$

Jarosite hydrolyses gradually and might function as a sizable acidity reserve in oxidised profiles (Lin et al. 2004).

$$KFe_3(OH)_6(SO_4)_2 + 3H_2O = K^+ + 2SO_4^{-2} + 3H^+(5) + 3Fe(OH)_3$$

Furthermore, the soil's exchange sites retain a significant amount of acidic products, such as exchangeable aluminium ions. Acid is liberated from these reserves by reverse hydrolysis reactions in Equation. Tin and Wilander (1994) found that the water chemistry of floods that are rising and retreating varies, and this variation is strongly influenced by the current hydrologic circumstances. In the process of acid hydrolysis of kaolinite, illite, and smectite, which are common clay minerals found in estuarine sediments, the acid solution in oxidised sulfidic sediments reacts with the clay minerals, resulting in the release of silica and metal ions. The metal ions released are generally aluminium, iron, potassium, sodium, and magnesium. In addition to other ions, manganese and small quantities of heavy metals can also be discharged (Willett et al. 1993). Annually, floodplains that have been drained have the capacity to release several metric tonnes of reaction products into coastal streams (White et al. 1997; Johnston et al. 2004).

3.3 Acidic Soils

Approximately 12.5 million hectares out of the total 500 million hectares of fine-textured soils, which are believed to have been created in sediments from rivers and oceans according to the FAO/UNESCO Soil Map of the World from 1971 to 1979, have a high concentration of pyrite. These soils will turn acidic when exposed to air or may have already been acidic. Peat or non-pyritic sediments encompass far larger regions than pyritic sediments that remain consistently saturated with water. The specific process may vary, but the overall reaction that characterises complete pyritization of ferric oxide can be described as follows: The essential constituents for the accumulation of pyrite are:

(i) Sulphate, consistently supplied over a prolonged duration, such as through the use of seawater.
(ii) Sedimentary minerals containing iron.
(iii) Organic substances that can be metabolised (CH_{20}).
(iv) Microorganisms that typically exist and are capable of reducing sulphate.
(v) An environment lacking oxygen.
(vi) Limited exposure to air to enable complete oxidation of sulphide to disulphide.

Acidity can only occur if some of the bicarbonate (HCO^-) produced during sulphate reduction is removed from the solution. It is essential to leach the interstitial solution. The movement of tides greatly supports the process of leaching, which can accelerate the formation of pyrite by disturbing diffusion-controlled rate-limiting mechanisms and by supplying dissolved oxygen and sulphate from seawater.

The elimination of HC⁻ also leads to a decrease in pH, resulting in slightly acidic circumstances that promote the rapid formation of pyrite. Based on the response, the amount of pyrite present is used as an indicator for the potential acidity.

$$Fe(OH)_3 + 2SO_4^{2-} + 4H^+ = FeS_2 + 15/4O_2 + 7/2H_2O$$

One iron molecule, Fe, and two sulphur molecules, S, are combined to form the cubic structure of pyrite.

During the process of neutralisation by silicates, the Fe(III) derived from the oxidation of pyrite can replace the Mg present in smectites, occasionally reaching a level similar to 1% of pyrite-S. Sedimentary rock with a potential for acidity contains a higher concentration of pyrite compared to acid-neutralising chemicals. To create such a substance, one must have chemicals with low concentrations that can neutralise acid, as well as components needed for pyrite production, such as sulphate, sulphate reducers, organic matter, iron, and alternating periods of anaerobic conditions with limited aeration. Additionally, it is important to remove any dissolved alkalinity that is produced during sulphate reduction. Pyritization of ferric iron occurs predominantly in intertidal regions with mangroves or reeds. Pyrite concentrations accumulate most in areas with heavy tidal flushing. Significant amounts of dense silt with high levels of pyrite were deposited due to fast increases in sea level, particularly during the most recent glacier period (Pons et al. 1982). The acidity generated by a 1% concentration of pyrite is counteracted by a 3% concentration of $CaCO_3$.

3.3.1 *Distribution of Acidity Potential*

Potential acidity is most likely to develop in poorly drained inland valleys that receive sulphate-rich water, as well as in salty and brackish lagoon, sea, and lake bottoms, and tidal flats and tidal swamps that have high levels of salinity. Inland valleys with potential acidity are constrained by the limited availability of sulphates. Some examples of specific soil types are the pyritic sandy gley soils found in valleys in the Eastern Netherlands (Poelman 1973), the pyritic papyrus peats in Oeganda (Chenery 1954), and the sulfidic peat soils in Minnesota. Seas, lakes, and brackish lagoon bottoms can exhibit elevated levels of sulphide concentrations. The concentration of primary (synsedimentary) organic matter frequently controls the level of sulphate reduction and the production of sulphides under these conditions (Berner 1971). The presence of subaquatic sediments with potential acidity is often limited to boreal and temperate temperature zones. According to Pons (1965), the amounts of primary organic matter in sediments are quite high, and the breakdown of organic matter in these sediments is slower compared to tropical zones. Subaquatic sediments often contain significant amounts of iron monosulphides, in addition to pyrite, which contributes to their black colouration. This may be owing to inadequate pyritization of sulphide in the underlying sediments caused by insufficient aeration. Instances include the sediment at the base of the ancient Littorina Sea (now the Baltic Sea), which has a concentration of reduced sulphur exceeding 2%, and the sediment at the base of the Black Sea, which has minimal or no carbonates present (Berner 1971; Wiklander et al. 1950). After the retreat of the last glacier, the land experienced isostatic uplift, causing the drainage of extensive areas of Littorina sediments. This process also led to the formation of acid sulphate soils along the Baltic beaches of Sweden and Finland. Despite the presence of about 1% reduced sulphur in the bottom sediments of the ancient Zuyder Sea in the Netherlands, the reclaimed area will not become acidic due to the comparatively high percentage of carbonate.

3.3.2 The Generation of Pyrite in Wetlands and Coastal Areas

Tidal flats and tidal swamps that are saline and brackish are the most important sources of potential acid sulphate soils in terms of quantity. The lower sections of the wetlands and the exposed tidal flats are often inundated by water containing high levels of sulphates, resulting in permanent degradation. Nevertheless, the uppermost areas around and above the average high tide level mostly display aeration and minimal to no reduction in sulphate. Primary organic elements in the sediments of exposed tidal flats and stream bottoms can inhibit the formation of pyrite. Tropical soils often have a low organic matter content, whereas temperate soils, particularly those with a high clay concentration, tend to be rich in organic matter. The reason why temperate regions have higher primary pyrite levels compared to tropical regions is explained in Pons' (1965) study. However, in tidal marshes and swamps, a telmatophytic plant can greatly augment the quantity of primary organic matter. Mangroves, belonging to the Rhizophoraceae family, are telmatophytes found in tropical regions. In temperate tidal marshes, reeds (Phragmites) and rushes (Scirpus) are other examples of telmatophytes that grow roots in reduced muds. The type and distribution of plants are determined by factors like as water temperature, salinity, and duration of flooding, which in turn control the transport of secondary organic elements to the sediment.

While clay contents are often lower in arid and semi-arid regions compared to the wet tropics, rivers from these places primarily transport smectites to their estuaries. This process can help counteract a large amount of acidity that is produced as a result of oxidation (Allen 1964). The clay mineral composition in temperate climates is diverse and includes a significant proportion of exchangeable bases. Rivers in the humid tropics, such as the Amazon and the Mekong, have silt composed of clay that undergoes weathering, resulting in a somewhat neutralising capability and a moderate amount of bases adsorbed onto it. Carbonates are virtually absent in most coastal sediments found in the wet tropics. Most rivers have acidic water and non-calcareous sediments because their catchment areas have old, highly worn soils. Moreover, in tropical regions, the coastal seawater often becomes diluted by acidic river water, which enhances the probability of carbonate dissolution (Brinkman and Pons 1968). In contrast, marine sediments found in temperate and semi-temperate regions generally have a higher concentration of primary carbonates. The timeframe required for the formation of substantial levels of potential acidity, also known as pyrite, is estimated to be between several decades and several millennia. Tidal marsh plants must remain stationary for a minimum duration in order to amass a sufficient quantity of pyrite. Consequently, sedimentation must take place in a slow manner (Moormann and Pons 1974). Tidal areas often contain numerous well-formed creeks with relatively low rates of sediment accumulation. Unlike rapidly developing coastal systems found in Malaysia (Diemont and Van Wijngaarden 1974) and along the coast of Guyana (Brinkman and Pons 1968), estuarine environments generally contain a significant number of creeks.

3.3.3 Formation of Potential Sulphate Acidity in Soil

Acid sulphate soils are formed when the production of sulphuric acid resulting from the oxidation of reduced S-compounds surpasses the capacity of adsorbed bases and easily weatherable minerals to neutralise the acid, causing the pH to drop below 4. Pyrite (cubic FeS_2) is the most prominent sulphur mineral in these sediments, in terms of quantity. The formation of sedimentary pyrite involves three essential steps: (a) the conversion of sulphate to sulphide through reduction, (b) the partial oxidation of sulphide to polysulphides or elementary S, and (c) either the direct precipitation of FeS_2 from Fe and polysulphides or the formation of FeS from Fe-oxides or Fe-silicates. Irrespective of the specific mechanism, the subsequent overall reaction encapsulates the process of ferric oxide pyritization:

The following components are essential for the formation of pyrite:

- Sulphates, consistently supplied over a prolonged period (e.g. with seawater)
- Sedimentary minerals containing iron.
- Metabolizable organic matter (CH_2O)
- The nearly constant existence of microbes that reduce sulphate.
- An environment without oxygen.
- Limited oxygen supply to fully convert sulphide into disulphide.

Acidity cannot occur until a part of the bicarbonate (HCO) generated during sulphate reduction, which is an alkalinity, is removed from the system. It is essential to leach the interstitial solution. The movement of tides substantially promotes leaching, which accelerates the formation of pyrite by interrupting diffusion-controlled processes that restrict the pace of reaction and by supplying dissolved oxygen and sulphate from seawater. The elimination of HCO also results in a decrease in pH, leading to slightly acidic conditions that expedite the formation of pyrite.

The reaction is determined based on the pyrite content, which serves as a signal of possible acidity.

$$Fe(OH)_3 + 2SO_4^{2-} + 4H^+ = FeS_2 + 15/4O_2 + 7/2H_2O$$

The equation for the neutralising capacity is derived when bases are present and are represented by their oxide components.

The soil's capacity to neutralise acid is influenced by the quantity of exchangeable bases, the existence of readily weatherable silicates, and the concentration of carbonates. The presence of an exchange complex in marine clays with a fine texture can neutralise acidity caused by pyrite up to a concentration of 0.5%. Additionally, if these clays are saturated with bases and contain substantial quantities of smectitic clay, they can prevent the pH from dropping below 4.0. An example of neutralisation by silicates is the substitution of Mg in smectites with Fe(III) from oxidising pyrite, which can occur up to the equivalent of 1% pyrite-S (Van Breemen 1980). The acidity generated by a 1% concentration of pyrite is counteracted by a 3% concentration of $CaCO_3$.

3.3.4 Distribution of Acidity Potential

Acidity can be most readily generated in:

- Landlocked valleys with inadequate drainage that are susceptible to an inflow of water containing high levels of sulphates.
- Areas characterised by the presence of saline and brackish water, such as lagoons, seas, and lake bottoms.
- Saline, brackish, and intertidal plains, along with tidal marshes.

Some examples of these types of soils are the pyritic sandy gley soils found in valleys in the Eastern Netherlands (Poelman 1973), the pyritic papyrus peats of Oeganda (Chenery 1954), and the sulfidic peat soils of Minnesota.

Typically, the quantity of substances that function as neutralisers in tidal sediments during the time of deposition, namely. Various climates possess distinct interchangeable foundations, readily erodible silicate minerals, and carbonates. Kaolinite is commonly the predominant clay mineral in tidal sediments found in the humid tropics, particularly in smaller estuaries. These sediments do not have exchangeable bases and are unable to efficiently weather and neutralise acid (Vieillefon 1973). While clay contents are often lower in arid and semi-arid regions compared to the wet tropics, rivers from these places primarily carry smectites to their estuaries. These smectites can help neutralise a large amount of acidity that is produced as a result of oxidation (Allen 1964). The clay mineralogy in temperate climates is diverse and includes a significant quantity of exchangeable bases. Rivers in the humid tropics, including the Amazon and the Mekong, have sediments that are rich in clay. These sediments have the ability to neutralise acidity and contain a moderate amount of bases that are adsorbed onto them. The bulk of volcanic catchment basins produce sediments that are abundant in minerals that can be easily weathered. For example, the existence of these minerals reduces or avoids the occurrence of acidity in the tidal flats of the volcanic island of Java, in contrast to those of Sumatra and Kalimantan (Driessen et al. 1976).

The duration required for the formation of substantial levels of potential acidity, also known as pyrite, is most likely to span from many decades to thousands of years. Consequently, the tidal marsh vegetation needs to persist in a particular location for a minimum duration in order to gather a sufficient amount of pyrite. Consequently, sedimentation must take place in a slow manner (Moormann and Pons 1974). Tidal areas often have numerous, properly formed creeks with relatively low rates of sediment accumulation. Unlike rapidly developing coastal systems found in Malaysia (Diemont and Van Wijngaarden 1974) and along the coast of Guyana (Brinkman and Pons 1968), estuarine environments generally contain a significant number of creeks. Black Fes was found in beach sediments that were accumulating, but not in strata from estuaries, in the local area. Although the accreting coast had minimal changes, the levels of dissolved sulphide in the estuary sediments fell to undetectable amounts during spring tides. These data indicate that tidal marshes with a well-established stream system experience significantly more effective tidal flushing. To maintain a fairly low pH (6.5–7), it is necessary to fully convert ferric iron into pyrite, remove interstitial water through leaching, and eliminate HCO. Tidal flushing would promote intermittent aeration.

Soil acidity in the entire wheatbelt of Western Australia presents a significant risk to both natural resources and the economy. With over 14.25 million hectares of wheatbelt soils currently classified as acidic or at risk of becoming acidic to the extent that it hampers productivity, farmers are greatly concerned about the loss of production and the sustainability of their operations. Approximately 10% of the yearly harvest in the wheatbelt region is estimated to be lost during production due to acidity.

3.3.5 Acidification Rate

The process of soil acidification caused by agricultural activities can be mitigated, but not completely halted. The use of lime will always be necessary to prevent excessive acidity in the soil. The inappropriate application of nitrogen fertilisers is a primary factor contributing to soil acidity. Managing nitrogen fertiliser inputs can effectively reduce nitrogen leaching, particularly in places characterised by high levels of precipitation. Accurately determining the required quantities of nitrogen fertiliser is crucial, and dividing the applications into many parts could be a viable option. It is advisable to utilise nitrogen fertilisers that have a low or no acidifying effect. When considering the appropriate liming, it is important to assess the cost-effectiveness of nonacidifying options in relation to other nitrogen fertilisers. The process of removing crops also leads to soil acidity. The extent of acidification is influenced by the amount of alkalinity that is exported from the field, which varies depending on the type of product being harvested. It is often recommended to maintain a wide range of crops and pastures, while also applying lime to decrease soil acidity. Opting for fewer acidifying options and implementing a recovery liming programme can be advantageous when the soil is already highly acidic.

3.3.6 P Accessibility

The extended process of soil weathering in the humid tropics generally leads to a decrease in the availability of phosphorus (P) over time (Walker and Syers 1976). The production of biomass may be limited in the future due to a scarcity of phosphorus caused by this process (Wardle et al. 2004). However, in Borneo, tropical rainforests may still generate large quantities of biomass and maintain a high level of biodiversity, even in the presence of highly depleted soils (Kitayama et al. 2004; Kitayama 2005). The efficient recycling of phosphorus (P) through ectomycorrhizal and fine root systems, the increased utilisation of P by plants, and the utilisation of non-labile P through the release of organic acids and enzymes (such as phosphatase) are some examples of adaptive mechanisms that have been observed in response to P limitation (Lambers et al. 2008; Kitayama et al. 2004; Liu et al. 2006). Root exudation of organic acids is an important strategy for plants to obtain phosphorus (P) in P-deficient soils seen in tropical rainforests. This is because the di- and tri-carboxylic acids have the ability to dissolve stubbornly bound P that is attached to aluminium or iron oxides, hence increasing P levels in the soil around the roots (known as the rhizosphere). Several plant species have developed the ability to release certain organic acids as a result of their long-term adaptation to soils with insufficient phosphorus content (Grayston et al. 1996; Zhang et al. 1997). Research has shown that when these plants experience a deficit of phosphorus, their roots release higher amounts of organic acids, specifically oxalic, citric, and malic acids (Jones 1998).

Plants in tropical rain forests can obtain phosphorus (P) from soils that lack phosphorus by releasing organic acids through their roots. Nevertheless, the rapid biodegradation of organic acids by rhizosphere bacteria may reduce their efficacy in promoting P mobilisation in the rhizosphere. Podzolization was more common in the older soil, which consisted of tertiary age materials. This soil also had lower levels of inorganic labile P compared to the younger soil, which was composed of quaternary colluvial deposits. The levels of oxalate, citrate, and malate in the soil solution around the roots were significantly higher in the older soil that lacked phosphorus compared to the overall soil. Conversely, these levels were reduced in the younger soil that had sufficient phosphorus (Fujii et al. 2012). The elevated levels of organic acids in the rhizosphere of P-deficient soils can be attributed to two factors: greater root exudation and a diminished capacity for organic acid sorption. The mineralization kinetics data showed that oxalate and citrate in the soil solution were quickly mineralized in both the rhizosphere and bulk fractions of both phosphorus-rich and phosphorus-deficient soils, with short mean residence times of 2.3–13.1 h for oxalate and 0.8–1.6 h for citrate. The rhizosphere section of the soil with phosphorus deficiency had the largest concentration of organic acids, which were rapidly replenished by root exudation. Consequently, the rates of oxalate and citrate mineralization were at their peak in this area. Our research indicates that in soil lacking phosphorus, there is a potential rise in both the generation and consumption of organic acids. Fujii et al. (2012) found that the impact of organic acids on the release of phosphorus in the rhizosphere of tropical montane rain forests is influenced by the level of phosphorus in the soil and the ability to bind anions. Phosphorus (P) is the primary factor that restricts the productivity of natural streams. In order to understand the factors that control the availability of phosphorus (P) in the ecosystem, it is essential to grasp the interrelationships between the movement of P in terrestrial and aquatic environments. Aluminium hydroxide ($Al(OH)_3$) was associated with the primary phosphorus (P) fraction, which also included some iron hydroxide ($Fe(OH)_3$) and organic phosphorus (P). In natural systems, parent materials constitute the sole supply of phosphorus (P), apart from the little amounts obtained via precipitation (Smeck 1973; Walker and Syers 1976). P is predominantly found in the form of apatite ($Ca_5(PO_4)_3(OH, F, Cl)$) in parent materials that have not undergone significant weathering.

A woodland watershed represents a landscape consisting of distinct yet interconnected habitats. The influence of topography and water flow on the process of soil formation and weathering in forest soils has been widely recognised for a considerable period of time (Jenny 1941). Leaching plays a critical role in determining the distribution of aluminium (Al), iron (Fe), and phosphorus (P) within the soil layers of Spodosols. This relationship is highlighted by the significance of leaching in these particular systems, as demonstrated by Wood et al. (1984) and Lundstedt et al. (2006).

3.3.7 *Nutrient Behaviour*

Pérez-Novo et al. (2008) found that the removal of organic matter had a greater impact on the adsorption of copper compared to zinc. The adsorption patterns of the two metals were more similar to each other than to the untreated samples. In the absence of treatment, the inclusion of copper had a substantial impact on diminishing the competitive adsorption of zinc. Conversely, the inclusion of zinc only had a minor effect on reducing the competitive adsorption of copper. While the hydrogen peroxide treatment decreased the competitive adsorption of both metals, copper exhibited greater efficacy than zinc in competing for binding sites on adsorbing surfaces. The desorption of Cu occurred at a significantly slower rate compared to Zn desorption, and the presence of hysteresis is particularly evident for Cu.

Kabata-Pendias and Pendias (2001) state that concentrations of these substances can be elevated due to the impact of agricultural and industrial operations. This rise can be substantial enough to harm plants, decrease agricultural productivity, and degrade the quality of environmental water sources. The concentration of copper and zinc in soils is naturally determined by the composition of the parent material. These concentrations vary depending on the kind of rocks, with larger levels found in basic rocks compared to acid rocks. The movement of metals in soil, which is influenced by their specific interactions with soil colloids, dictates the extent of environmental problems resulting from metal buildup. Research on adsorption has shown that copper has a higher affinity for organic matter than zinc. However, the adsorption behaviour of heavy metals might differ depending on the type of soil and other soil factors (McBride et al. 1997, 1998). The use of insecticides, organic manure, and inorganic fertilisers concurrently might result in the introduction of copper and zinc into the soil. Their movement is primarily influenced by the competition between the two metals for attachment sites in soil colloids. Gomes et al. (2001) found that copper was preferentially adsorbed over zinc in Ultisols, Oxisols, and Alfisols, indicating a competitive process. The soils in NW Spain often have a high acidity level and contain a significant amount of organic matter. The study conducted on soils from this site found that copper (Cu) is more strongly adsorbed and forms stronger links compared to zinc (Zn). In fact, Cu was much less desorbed from the soils than Zn, as reported by Arias et al. in 2005. Furthermore, it was shown that the presence of Cu had a substantial impact on reducing the amount of Zn that could be adsorbed. However, the presence of Zn did not affect the amount of Cu that could be adsorbed (Arias et al. 2006). Studies by Elliott et al. (1986) and Mesquita (1998) have revealed that Cu exhibits a higher affinity for both organic and inorganic soils compared to Zn. Given the significant diversity of soil adsorbing surfaces, there are two methods to examine the impact of soil constituents on the adsorption of certain metals. One aspect of the study involves examining the process of adsorption of different components of soil colloids, such as clays (Spark et al. 1995), organic matter (Shuman 1988), iron and aluminium oxides (Cavallaro and McBride 1984), and organomineral associations (Arias et al. 2002). The second approach involves extracting a specific component from the soil to assess its influence on the absorption of metals (Agbenin and Olojo 2004).

In soil samples that were not treated, copper exhibited a greater degree of adsorption compared to zinc due to the distinct physicochemical properties of the two metals, as well as copper's stronger affinity for organic matter. Nevertheless, the application of H_2O_2 to soil results in the elimination of organic matter and reduces the adsorption of Zn and, specifically, Cu (Pérez-Novo et al. 2008). The reduction in Cu adsorption at the highest concentration examined, 1530 M, was 3.6 mol g^{-1}, while the decrease in the sample with the highest fraction of C was approximately four times larger. The decrease in Cu adsorption can be attributed to the organic C content of the original sample. The adsorption of Copper in the untreated samples was consistently reduced by less than 6%, regardless of the amount of Cu provided, due to competition with Zn. The quantity of metal introduced had an impact on the extent to which Zn adsorption was diminished in the presence of Cu; the greater the amount of metal added, the more significant the reduction. The presence of Zn in the H_2O_2-treated samples led to a considerable decrease in copper adsorption. However, the reduction in Zn adsorption in those samples was less pronounced compared to Cu adsorption. Consequently, at low levels of concentration, both metals were adsorbed in a similar manner. However, at high concentrations in the soils, copper (Cu) exhibited a stronger adsorption than zinc (Zn). The desorption of Cu occurred at a significantly slower rate than the desorption of Zn in both competitive and noncompetitive tests (Pérez-Novo et al. 2008).

The addition of phosphate significantly enhanced the adsorption of copper from copper nitrate solutions by eight acidic soils derived from granite or amphibolite parent materials. The addition of phosphate did not significantly alter the Z-potentials of the clay fractions in these soils. Furthermore, the effect of phosphate was not enhanced when it was added to the soil prior to the copper solution. Based on these findings, the presence of phosphate did not impact the sorption of copper by enhancing the electrostatic attraction between Cu^{2+} ions and soil particles, nor did it cause any interaction between Cu^{2+} ions and sorbed phosphate. MINTEQA2 simulations demonstrated that the introduction of phosphate resulted in an elevation in pH. Consequently, this led to the formation of Cu–P species that exhibited high solubility in the solution (Pérez-Novo et al. 2009).

The presence of elevated copper levels in the uppermost layers of agricultural soils can be attributed to the application of fertilisers, the use of copper-based fungicides, and the close vicinity to metal-processing industries (Kabata-Pendias and Pendias 2001). The regulation of copper levels in plants and soil microorganisms is not solely determined by the overall concentration of copper in the soil. However, the concentration of copper in the soil solution determines its mobility through the soil and subsequently influences the potential for surface and subsurface water pollution. The concentration of copper in the soil solution is determined by two main factors: the physico-chemical characteristics of the soil, such as pH, which affects solubility, and the sorption phenomena, which are influenced by the type and amount of soil colloids. Research conducted by McBride et al. (1998) has shown that in acidic settings, soil organic matter is more efficient at preventing the movement of copper compared to clay or Fe/Mn/Al oxides. Studies have shown that when soils are treated with multiple metals at the same time, their sorption behaviour is different compared to when only one metal is added. This was observed in studies conducted by Serrano et al. (2005) and Arias et al. (2006). However, the presence of other species in the soil solution can affect the sorption processes of copper. There has been a smaller amount of research conducted on the impact of anions compared to other cations on the process of metals being absorbed by soils. However, in the context of phosphate, it holds considerable importance. Phosphate rock has been utilised for the remediation of metal-contaminated soils and streams due to its ability to immobilise metals such as Pb, Cu, and Zn (Cao et al. 2004). The process involves the substitution of calcium with the sorbed metal in hydroxyapatite layers, either by surface complexation or through dissolution and coprecipitation (Ma et al. 1994; Xu et al. 1994). Investigating the influence of phosphate on metal sorption is important due to the widespread use of phosphate fertilisers, especially in acid soils. These fertilisers are often applied in large amounts to overcome the problem of phosphorus becoming immobilised as insoluble ferric phosphates. According to Clark and McBride's (1984) research, the existence of phosphate can have either a positive or negative impact on the movement of metal ions. Phosphate can have either a positive or negative impact on the mobility of zinc in soil acidity, as observed by Barrow in 1987. There is reason to be concerned when acidic soils are treated with high doses of phosphate at the same time as copper-based fungicides, or when soils with already high levels of copper, such as vineyard soils, are treated. In vineyard soils, copper levels can exceed 100 mg kg^{-1} (Arias et al. 2004; Fernández-Calviño et al. 2008). The absorption of phosphate by soil colloids may be the reason for phosphate's impact on Cu sorption. Sequioxides exhibit a remarkably strong attraction to phosphate (Sumner 1992). The adsorption of phosphate onto sorbent particles results in the surfaces of the particles acquiring a negative charge, hence enhancing their attraction to cations.

Pérez-Novo et al. (2011) examined the adsorption–desorption kinetics of Zn in the presence and absence of P using the stirring flow chamber method. In contrast to copper (Cu), zinc (Zn) was solely adsorbed at rapid adsorption sites in the absence of phosphorus (P). Sluggish adsorption sites, where the Zn^{2+} ion acted as a bridge between P and organic matter, were the main reason for the increased adsorption of Zn when P was present. Zinc (Zn) exhibited significantly higher desorption compared to copper (Cu) upon adsorption, regardless of the presence or absence of phosphorus (P) addition. Nevertheless, when P was present, the amount of Zn that was released after being adsorbed was significantly lower compared to when P was not present. Pérez-Novo et al. (2011) demonstrated that the presence of P enhances the binding strength of Zn to adsorbing surfaces compared to its absence.

Zinc is an essential element for numerous crops, making it a vital constituent of soil. However, the use of pesticides, mining, organic and inorganic amendments, agricultural activities, or industrial residues can elevate the amounts of zinc in soil to levels that are harmful to plants. In acidic soils, like other heavy metals, Zn is very mobile and can cause harm to plants (phytotoxic). The phytotoxic potential of Zn in the soil is influenced by various edaphic parameters, including pH, clay type and content, organic matter, cation-exchange capacity, and the presence of Fe and Al oxides. The concentration of Zn in the soil solution is a determining factor in its phytotoxicity (Kiekens 1995; Gaudalix and Pardo 1995; Arias et al. 2005). The presence of other elements also affects the concentration of Zn in the soil solution. Competition tests have shown that the presence of other heavy metals significantly affects the concentration of Zn in the soil solution (Saha et al. 2002; Arias et al. 2006). Nevertheless, the influence of anions such as phosphate on the movement of metals has not been well studied. Although there have been differing results from research regarding the effect of phosphorus (P) on zinc (Zn) adsorption in soil, it is usually observed that phosphates tend to enhance Zn adsorption (Zhao and Selim 2010). Phosphate addition has the effect of increasing the pH of acidic soils, which in turn enhances the process of Zn adsorption. Conversely, phosphate addition decreases the pH of alkaline soils, leading to a reduction in Zn adsorption (Barrow 1987). In order to fully understand the impact of P on Zn adsorption, it is essential to examine the kinetics of adsorption and desorption of Zn. This is because the adsorption–desorption processes in soil do not happen instantaneously, as stated by Harter and Lehmann in 1983. The impact of phosphate on the adsorption–desorption kinetics of Zn has been investigated in batch studies (Dang et al. 1994; Taylor et al. 1995) and stirred flow chamber experiments (Shi et al. 2008; Fernández-Calviño et al. 2010). Two notable exceptions in the field are the studies conducted by Rupa and Tomar (1999) on the influence of phosphorus (P) on zinc (Zn) desorption, and by Zhao and Selim (2010) on the influence of P on Zn adsorption and desorption.

The potential influence of phosphate on the mobility of metals makes it highly intriguing to comprehend the manner in which this element influences the movement of metals in soil. Due to the application of zinc-containing pesticides and soil amendments, as well as the major usage of phosphate to prevent nutritional deficits in crops cultivated in acidic soils, zinc and phosphate can readily interact in agricultural soils. This study investigated the impact of phosphorus (P) on the zinc adsorption capacity of eight surface layers in soils located on top of granite and amphibolites. It was noted that P had a significant impact on the adsorption of Zn, particularly at high concentrations. Due to the larger concentrations of Fe and Al oxides in the soils on amphibolite compared to those on granite, the impact was more significant in the former. Pérez-Novo et al. (2011) propose that the formation of a P–Zn complex on colloid surfaces is the main reason for the increased zinc adsorption resulting from the presence of phosphate.

3.4 Soil Sulphuration

3.4.1 The Activities of S

The diversity of sulphur-containing biological compounds is immense. The sources of S include cofactors such as coenzyme A, molybdenum cofactor (MoCo), and lipoic acid, as well as the chloroplast lipid sufloquinovosyl diacylglycerol and other secondary products (Leustek 2002; Leustek and Saito 1999). Glutathione, along with proteins such as thioredoxin, glutaredoxin, and protein disulphide isomerase, play a vital role as cellular redox buffers. These compounds have essential activities in terms of protein structure, regulation, and catalysis. Disulphide bond formation regulates the functionality of many enzymes. This regulation is employed to coordinate the activity of several carbon dioxide assimilating enzymes with the light-dependent activities of photosynthesis. Thioredoxin is the molecule that regulates the target enzymes by using electrons from ferredoxin. This has been established by Leustek and Saito in 1999, Saito in 2000, Scherer in 2001, and Matsubayashi et al. in 2002.

3.4.2 Organic Soil S

Bloem (1998) states that a significant portion of the soil's sulphur (S) content, perhaps as much as 98%, can be found in the form of organic S compounds. These compounds are associated with various animal remnants, plant residues, and soil microorganisms. Probert (1980) observed that the distribution of organic matter concentration in soils typically corresponds to the distribution of organic S concentration as depth increases. Soil organic S can be classified into two primary groups: the first group consists of S atoms in the oxidised state, while the second group consists of S atoms in the reduced form. Recent research indicates that the soil microbiological biomass S generally makes up 1.5–5% of the total soil organic S (Wu et al. 1993). However, Stevenson (1986) discovered that the component of microbiological biomass can account for just 1–3% of the soil organic S. Microbial species mostly include S in the form of proteins and amino acids (Banerjee and Chapman 1996). The majority of soil microorganisms exhibit sulphur concentrations ranging from 1 to 10 g/g, carbon-to-sulphur ratios ranging from 57:1 to 85:1, and nitrogen-to-sulphur ratios of approximately 10:1 based on dry weight.

3.4.3 Soil Inorganic S

Organically bound S is generally more abundant than inorganic S in the majority of agricultural soils (Bohn et al. 1986). The most common form of inorganic sulphur is sulphate, which can be subdivided into mineral sulphur, adsorbed sulphate ions, and sulphate ions present in the soil (Barber 1995). Sulphur can be deposited as calcium, magnesium, or sodium sulphate in the SO_4^{2-} form. Tidal marshlands accumulate significant quantities of sulphide metals, particularly pyrite (FeS_2). According to Elkins and Ensminger (1971), if the SO_4^{2-} in soil is not easily accessible to plants, any treatment that reduces its retention and increases the amount of SO_4^{2-} in the soil solution should enhance its availability to plants. Upon the removal of liquid from these regions, the S compounds undergo oxidation to form SO_4^{2-} ions, resulting in a simultaneous fall in pH.

3.4.4 Soil Interactions Between Nitrogen and Sulphur

An intense agricultural practice that utilises a high analysis fertilisation strategy, along with improved cultivars, might result in nutrient depletion, leading to an imbalance of nutrients in the soil. Fazli et al. (2008) found that the lack of sulphur (S) affects the efficiency of extra nitrogen (N). Therefore, it is necessary to include sulphur in order to enhance the effectiveness of nitrogenous fertilisers. Kowalenko and Lowe (1975) discovered that the low rate at which minerals were formed in the soil sample during incubation was due to a high ratio of nitrogen to sulphur, which was induced by the addition of nitrogen. Janzen and Bettany (1984) determined that the optimal proportion of easily accessible nitrogen (N) to easily accessible sulphur (S) is 7:1. Reduced ratios led to decreased seed yields. In field settings, a rapeseed and mustard crop showed a 27–31% increase in sulphur recovery without nitrogen, but a 37–38% increase with the addition of 60 kg N ha^{-1} (Sachdev and Deb 1990).

3.4.5 Interactions Between Nitrogen and Sulphur in Plants

The availability of these nutrients in plants is highly interconnected due to the vital role that sulphur (S) and nitrogen (N) play in protein synthesis. Several studies (Zhao et al. 1993; McGrath and Zhao 1996; Ahmad et al. 1998) have shown a correlation between sulphur and nitrogen in terms of the amount of plant material and agricultural productivity in various types of crops. Barney and Bush (1985) found that when researching the tobacco plant, they saw that when one nutrient was restricted, there was an increase in the concentration of the other nutrient. Conversely, when the treatment was reversed, the stored nutrient was used in the synthesis of proteins. Likkineni and Abrol (1994) found that a reduction in the supply of S to crops impacts the utilisation of soil nitrogen, resulting in an increase in nitrate leaching. O'Conner and Vartha (1969) observed that when the soil's nitrogen status was below standard, a significant application of gypsum resulted in a decline in hay yield.

3.5 Buffering for Acidic and Basic Soils

A soil's alkaline, acidic, or neutral reaction depends on a variety of factors, including parent material, climate, vegetation, and management. The chemistry of acidic and alkaline soils as well as the unique traits of each type of soil will be discussed in this section. The production and consumption of protons by various mechanisms in soils and sediments influences pH. Soil pH is determined through a process that involves interactions between parent material, organisms, and the local environment throughout time. Many important procedures that generate or use protons are included in the list below. Protons are produced by these processes and interact with the soil exchange complex and solid phase. A soil's pH is mostly controlled by its primary minerals, clay, organic matter content, and percolation. Proton production and consumption procedures of note.

1. Biological breakdown of proteins
   $\text{Proteins} \rightarrow \text{Proteoses} \rightarrow \text{Peptones} \rightarrow \text{Polypeptides} \rightarrow \text{Amino acids} \rightarrow NH_3$
   $NH_3 + H_2O \rightarrow NH_4^+ + OH^-$
2. Nitrification
   $NH_4^+ + 1½O_2 \rightarrow NO_2^- + 2H^+ + H_2O,$
   $NO_2^- + ½\,O_2 \rightarrow NO_3^-$
3. Denitrification
   $H_2O + NO_3^- + 2e^- \rightarrow NO_2^- + 2OH^-,$
   $2H_2O + NO_2^- + 3e^- \rightarrow ½N_2 + 4OH^-$
   $H_2O + NO_2^- + 3e^- \rightarrow ½N_2 + 4OH^-$
   $1½H_2O + NO_2^- + 2e^- \rightarrow ½N_2O + 3OH^-$
4. Oxidative metabolism
   $\text{Enzymatically available organic matter} + O_2 \rightarrow CO_2 + H_2O$
   $H_2O + CO_2 \rightarrow H_2CO_3,$
   $H_2CO_3 \rightarrow H^+ + HCO_3^-$
5. Other electron acceptors for microbiological respiration.
   $2H_2O + MnO_2 + 2e^- \rightarrow Mn_2^+ + 4OH$
   $Fe(OH)_3 + e^- \rightarrow Fe^{2+} + 3OH^-$
   $6H_2O + SO_4^{2-} + 8e^- \rightarrow H_2S + 10OH^-$

6. Addition of "lime" to an acid soil.
   $H_2O + CO_2 + CaCO_3 \rightarrow Ca^{2+} + 2HCO_3^-$,
   $2HCO_3^- + 2OH^- \rightarrow 2CO_2$
7. Addition of sulfur
   $H_2O + S^0 + 1½O_2 \rightarrow SO_4^{2-} + 2H^+$
8. Protolysis due to dilution
   $Mexch + H_2O \rightarrow Hexch + M^+ + OH^-$
9. Ion uptake
   $Root + K^+ \rightarrow K(Root) + H^+$
   $Root + NO_3^- \rightarrow Root\ (NO_3^-) + OH^-$
10. Fire-rapid mineralization of soil organics
    $Ca\text{–}O.M. \rightarrow CO_2 + CaO$
    $CaO + H_2O \rightarrow Ca^{2+} + 2OH$

Factors affecting soil buffer

(i) clay content and texture
(ii) organic matter content
(iii) oxide components
(iv) carbonate content
(v) initial pH and;
(vi) weatherable mineral content.

The majority of essential nutrients are typically stored in organic matter and primary and secondary minerals. The nutrients that are absorbed onto the colloids clay and humus have a moderate availability. The primary minerals, on the other hand, immediately hold the nutrients but are not easily available. The soil solution fraction consists of ions that are readily available for absorption by plant roots. Humus possesses a far greater potential, per unit of weight, to retain nutrients and water compared to clay. In general, humus has the potential to greatly enhance the soil's capacity to promote plant growth in small quantities (Donahue et al. 1977). Table 3.1 displays the symbol of elements and their respective forms that are absorbed by plants.

3.5.1 Mechanism of Nutrient Uptake

The plant assimilates all essential nutrients via its roots, excluding carbon. With the exception of water and H_3BO_3, nutrients need to be in an ionic state and be located at the root surface in order for plants to absorb them. When parts of a root are in close contact with soil particles, there may be a direct exchange between nutrient ions adsorbed on the surface of soil colloids and hydrogen ions from the surface of root cell walls. Under any circumstances, the nutrients in the soil that were in direct touch with the root would rapidly deplete. Nevertheless, the level of nutritious ions at the surface of the root is preserved by three key mechanisms. There are three primary

Table 3.1 Elements and their ion or molecule uptaken by plants

Element	Symbol	Ion or molecule	Element	Symbol	Ion or molecule
Carbon	C	CO_2	Iron	Fe	Fe^{2+}, Fe^{3+}
Hydrogen	H	H^+, H_2O	Magnesium	Mg	Mg^{2+}
Oxygen	O	O^{2-}, OH^-, $CO_3{}^{2-}$, $SO_4{}^{2-}$, CO_2	Boron	B	H_3BO_3, $H_2BO_3{}^-$, $B(OH)_4{}^-$
Phosphorus	P	$H_2PO_4{}^-$, $HPO_4{}^{2-}$	Manganese	Mn	Mn^{2+}
Potassium	K	K^+	Copper	Cu	Cu^{2+}
Nitrogen	N	$NH_4{}^+$, NO_3	Zinc	Zn	Zn^{2+}
Sulphur	S	$SO_4{}^{2-}$	Molybdenum	Mo	$MoO_4{}^2$
Calcium	Ca	Ca^{2+}	Chlorine	Cl	Cl^-

mechanisms via which nutrient ions, which have been dissolved in the soil solution, interact with plant roots and maintain nutrient levels:

1. Mass flow
2. Diffusion
3. Root interception.

Although all three processes occur simultaneously, one or more of them may be more essential for a certain nutrient. Mass flow alone can transport a sufficient amount of calcium to the surface of the roots, as it is usually found in high quantities in the soil solution. Phosphorus requires diffusion in addition to mass flow since the soil solution is notably lacking in this element compared to the levels needed by plants. Most often, nutrient ions in the soil solution need to travel a considerable distance before they can reach the surface of the root. This movement can happen due to mass flow, for example, when dissolved nutrients are carried along with soil water that is going towards a root that is actively absorbing water from the soil. The nutrient ions exhibit a flow pattern akin to leaves drifting downstream in a stream. In addition, nutritional ions continuously diffuse throughout the root surface, moving from regions of higher concentration to regions of lower concentration when nutrients are scarce. Thus, plants are able to gather nutrients even during the night, when water absorption by the roots is reduced. As the roots extend further into new, unused soil, the process of root interception is initiated.

3.5.2 Benefits of Humus and Organic Matter in Soil

- The conversion of raw organic matter into humus sustains the population of microorganisms and other species in the soil, hence maintaining a robust and thriving soil ecosystem (Vreeken-Buijs et al. 1998; Elo et al. 2006).

- The presence of plants, animals, and microbes in soil is influenced by the speed at which humus is formed. If humus creation is fast, it promotes their coexistence, whereas if it is slow, it restricts their coexistence.
- Stable humus and effective humus serve as supplementary nutrition sources for bacteria, with the latter acting as a reservoir for long-term nutrient storage.
- As deceased plants undergo decomposition, intricate organic compounds undergo either gradual oxidation (resulting in lignin-like humus) or breakdown into simpler forms (such as sugars and amino sugars, aliphatic, and phenolic organic acids). These simpler forms are then converted into microbial biomass (known as microbial humus) or restructured and further oxidised into humic assemblages (fulvic and humic acids). These humic assemblages bind to clay minerals and metal hydroxides. The ability of plants to uptake and metabolise humic substances through their root systems has been a topic of extensive deliberation. Humus in plant physiology is recognised to have both a hormonal and nutritional function (Eyheraguibel et al. 2008).
- Humus is a colloidal substance that enhances the soil's capability to retain nutrients by improving its ability to exchange cations through chelation. While plants have access to these essential nutrients, they are retained in the soil and safeguarded against being carried away by irrigation or rainfall (Szalay 1964).
- Humus enhances the soil's capacity to withstand drought conditions by retaining 80–90% of its weight in moisture, as demonstrated by Olness and Archer (2005).
- Humus possesses the capacity to regulate excessively acidic or alkaline soil conditions due to its biological composition (Kikuchi 2004).
- Mucilages, produced by bacteria during humification, contribute to the maintenance of soil structure by binding soil particles together and enhancing soil aeration (Caesar-Tonthat 2002). Chelated substances, such as heavy metals and excess nutrients, can be prevented from entering the larger ecosystem by forming strong bonds with the complex organic molecules of humus (Huang et al. 2008).
- The dark colour of humus, usually black or dark brown, facilitates the warming of cold soils during springtime.

3.6 Transport and Adsorption of Sulphate

The primary origins of sulphate in the majority of natural soils are the deposition of air S compounds and the oxidation of reduced S, such as pyrite. The sulphur (S) present in the atmosphere can originate from either anthropogenic sources, such as the emission of sulphur dioxide (SO_2) resulting from the combustion of fossil fuels, or natural sources, such as volcanic activity, sea spray aerosols containing sulphur compounds, or gaseous sulphur compounds produced by living organisms. The majority of S inputs ultimately convert into H_2SO_4, which has the capacity to acidify soil and water. Seaspray contains only S, which is accompanied by base cations and does not cause water acidification. The acidifying potential of atmospherically supplied sulphate (often referred to as excess SOi-) is obtained by subtracting seaspray S from the overall S deposition in coastal regions. The adsorption of sulphate greatly diminishes the acidifying effect of H_2SO_4 in soils. Fuller et al. (1987) and Mitchell et al. (1989) conducted research that revealed the significance of Al and Fe oxide surfaces as adsorption sites for soil solution S04. As the pH drops, the process of sulphate adsorption increases, as observed in studies conducted by Nodvin et al. (1986) and Fuller et al. (1987). The pH-dependent adsorption is likely caused by the amphoteric character of the adsorption sites on oxide surfaces. These sites become negatively charged after deprotonation, leading to anion repulsion. The presence of organic matter negatively affects the adsorption of sulphate, even in soils with high amounts of Al and Fe oxides (Johnson and Todd 1983). The decrease in sulphate adsorption is believed to be caused by the obstruction of the sites by organic waste. According to the study conducted by Cosby et al. in 1985, the process of sulphate adsorption is usually described using a Langmuir isotherm model and is often believed to be a reversible phenomenon.

3.6.1 Variability in Chemical Composition Along the Vertical Axis

The creation of soil horizons is a consequence of enduring physical, chemical, and biological processes, as stated by FAO-UNESCO (1988) and the Soil Survey Staff (1975). The development of soil at a certain location is determined by the relative significance of several factors involved in soil formation, including climate, parent material, plant and animal life (including people), terrain, and time. The transportation of specific compounds throughout the process of soil formation often leads to a decrease in certain sections of the soil profile and an increase in others. For example, Podzols display a highly acidified and bleached E horizon at the surface due to the chelating effect of percolating fulvic acids, resulting in the depletion of AI and Fe. Within the depths, chelates of Fe- and AI combine to create a spodic B that contains high concentrations of Fe, Al, and organic C. Podzols often develop an acidic, undecomposed layer of organic matter (0 horizon) on their surface. The presence of different chemical properties, such as pH, CEC, and concentrations of Fe and Al oxides and soil organic matter, in Podzols can result in differing amounts of chemical reactivity in their strata.

3.6.2 Soil Remediation

The soil solution composition of a particular soil horizon can vary greatly during the year due to (i) changes in the composition of infiltration water, (ii) variations in meteorological conditions, and (iii) fluctuations in biological activity. Van Breemen and Van Dijk (1988) observed substantial seasonal fluctuations in the soil solution chemistry of an oak-birch woodland in the Netherlands over a six-year study. High levels of atmospheric NH_3 were found to significantly enhance the leaching of AI, Ca, and Mg, particularly in the growth season. Van Breemen and Van Dijk (1988) found that evapotranspiration led to increased concentration levels over the summer, as reported by CI. The process of mineralization and subsequent nitrification of soil organic nitrogen over the summer can lead to a significant increase in soil solution N03 and Ca in ecosystems that have low levels of atmospheric nitrogen deposition (Foster et al. 1989). Persistent leaching can lead to long-term alterations in the composition of the soil solution. Nevertheless, most recently launched monitoring systems are sometimes too sluggish to detect these alterations.

The humic substances (HSs) in soil's organic matter (SOM) are crucial for enhancing the soil's water retention capacity and facilitating plant growth (Davies and Ghabbour 2001). Arable soils have a maximum of 10% of SOM, whereas HSs account for approximately 80% of SOM (MacCarthy et al. 1990; Stevenson 1994). Dead leaves, which are an important contributor of nutrients to the soil, consist of around 50–60% cellulose, 15–20% lignins, and 15–20% lipids (Wershaw 1999). Humification refers to the gradual process of synthesising and breaking down humic substances (HSs) in the soil. This process begins with the exothermic oxidation of deceased plants and animals (de Bertoldi et al. 1996). Over time, hydrocarbons break down and transform into coal, mostly consisting of aromatic compounds, as well as crude oil and carbon dioxide. Coal can be produced from aliphatic soil components or HSs by an oxidation process known as dehydrogenation.

3.6.3 Soil Microorganism Activity

Acidic soil reduces the growth and reproduction of soil microorganisms, particularly bacteria and fungi, leading to a decrease in their overall functions. Diminished nutrient cycling and decomposition of organic materials. The process of plant uptake can be hindered by the relatively sluggish rate at which soil microbes transform nutrients into forms that are accessible to plants. Significantly, nodulation in legumes can have malfunctions. Nitrogen-fixing rhizobia bacteria work along with pasture and crop legumes in root nodules to generate nitrogen. Rhizobia bacteria may be unable of establishing a successful symbiotic relationship in acidic soils or may have inadequate populations, resulting in a deficiency of nitrogen. Rhizobia bacteria species exhibit varying degrees of tolerance to soil acidity. For example, medic rhizobia are highly susceptible and may not be able to survive. Pasture legumes may fail to flourish on acidic soil, resulting in grass-dominated pastures.

3.6.4 How Plants Are Affected by Soil Acidity

Acidic soils have higher levels of aluminium and manganese, which can be harmful to sensitive plants due to their toxic effects. Phosphorus, calcium, magnesium, and molybdenum are essential elements that are increasingly becoming less accessible for plant growth. These are the potential outcomes of this: inadequate plant establishment and longevity, irregular or uneven growth in pasture or crop, reduced nodulation in legumes, heightened vulnerability of plants to diseases, invasion of weeds in pastures and crops, yellowing or death of leaf tips, and limited or stunted root growth.

The pseudogley soil samples with higher acidity exhibited hazardous levels of mobile aluminium (10–30 mg/100 g), which were identified as elevated. The brown woodland soil samples exhibited significant acidity, with mobile Al values ranging from 12.8 to 90.0 mg/100 g. A correlation was found between a high level of mobile aluminium and a significant suppression of nitrification at low soil acidity. The pseudogley soils exhibited chemical nitrification and denitrification of applied nitrites at pH levels below 4.0 (in 1 M KCl), as observed by Jakovljevi et al. in 2005.

The increase in the extent of acid soils can be attributed, in part, to the utilisation of pure chemical compounds as NPK fertilisers and the contamination caused by acid gases originating from the atmosphere. Consequently, the use of acid soils for plant cultivation is progressively challenging, highlighting the urgent need to enhance their chemical and other characteristics. The chemical features of more acidic soils are primarily affected by the presence of toxic chemicals such as soluble forms of Al, Fe, and Mn, nitrites, and various hazardous organic acids, in addition to acidity itself. Mobile Al is found in elevated amounts in acidic soils with a pH below 5.5 (in water). There is a concurrent rise in the absorption of this element by plants, which negatively affects the roots, and a decline in the absorption of other nutritional elements.

3.7 Sulphur in the Soil and Plant Growth

As soil acidity increases, the availability of molybdenum, phosphorus, magnesium, and calcium to plants decreases, while the availability of aluminium increases to amounts that can be fatal. Manganese toxicity occurs in soils with low weathering capacity when the soil's acidity level rises. The primary factors influencing plant growth in acidic soils are the alterations in the accessibility of aluminium and essential plant nutrients. Robson and Abbott (1989) found that soil acidity has the potential to inhibit microbial activity and negatively impact nodulation and nitrogen fixation in certain legumes.

Aluminium

As the pH level lowers, the amount of accessible aluminium increases. Fenton and Helyar (1999) state that soils vary in the amount of accessible aluminium at a specific pH level, as seen in Fig. 3.2. A well-weathered soil contains a greater amount of aluminium compared to a poorly weathered soil.

Aluminium toxicity mainly hampers root growth by producing disturbances to the structure and functioning of roots. Aluminium poisoning in the leaves hampers a plant's growth in acidic soil by impairing its ability to absorb nutrients and moisture. This condition manifests symptoms that resemble phosphorus deficiency and water scarcity stress. During periods of drought, crops experience rapid wilting and both pastures and crops exhibit reduced water efficiency. Plants have varying levels of tolerance to aluminium. Plants can tolerate high quantities of aluminium in the soil. These plants are also affected because large quantities of aluminium in the soil immobilise phosphorus and can disrupt the absorption and utilisation of calcium and magnesium. Plants' tolerance to soluble aluminium exhibits variability. The tolerance levels for many frequently cultivated agricultural products and pastures in Australia are listed in Table 3.2. The plants are categorised based on four different levels of tolerance, as described by Ring et al. (1993).

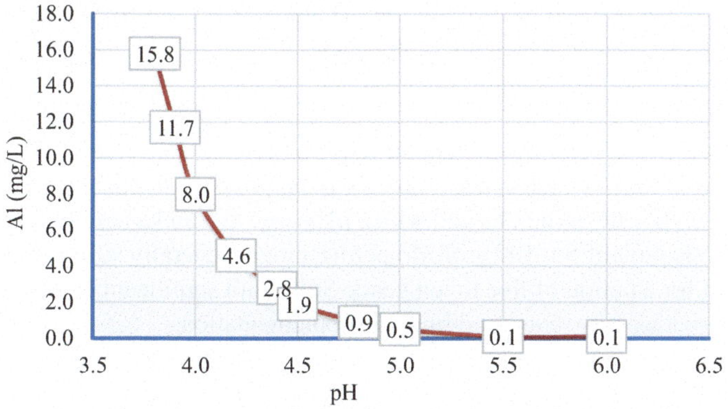

Fig. 3.2 As soil pH falls below 5, aluminium becomes available to plants (Fenton and Helyar 2002)

Table 3.2 Aluminium sensitivity (tolerance) of some crop and pasture plants

Highly sensitive	Durum wheat, faba beans, Chickpeas, Alfalfa, medics, strawberry, Balansa, Berseem and Persian clovers, Buffel grass, Tall Wheatgrass
Sensitive	Canola, some wheats, barley, albus lupins, phalaris grass, red clover, Caucasian, and Kenya white clovers
Tolerant	Some wheats, annual and perennial rye-grass, Tall fescue, Haifa white, and subterranean clovers
Highly tolerant	Lupins, oats, triticale, cereal rye, cocksfoot, kikuyu, paspalum, yellow & slender serradella, Maku lotus, common couch, Consul love grass

Manganese

Plants can be affected by either an excess or a lack of manganese, leading to toxicity or deficiency, respectively. The soil types most susceptible to manganese toxicity are strongly weathered soils with high levels of rapidly reducible manganese oxides, such as red earths and red brown earths. Additionally, highly weathered soils with high levels of iron and aluminium oxides that adsorb and accumulate Mn^{2+} are also prone to acquiring manganese toxicity (Conyers et al. 1997). Not all plants are susceptible to the toxicity of Mn. Sensitive plants exhibit indications of Mn toxicity, such as chlorosis and leaf margin crinkling. Inadequacies should only be expected in circumstances of absolute inadequacies, such as some types of siliceous sands, or in soils with exceptionally high levels of calcium carbonate (i.e. greater than 50%). Soils with high iron content tend to retain manganese rather than seeing a decrease in manganese levels as the soil matures. While the authors acknowledge that there may be concerns in Brazil, their limited familiarity with the literature (Carmargo and Falici 1974) leads them to assume that a deficit of manganese (Mn) could potentially be a more significant problem. To reduce elevated levels of available manganese in the soil, lime can be used to increase the pH above 5.5. Conversely, the process of liming can exacerbate a deficiency of manganese.

Molybdenum

Molybdenum is a crucial trace element required for the conversion of nitrate into the ammonium form, which is necessary for plant proteins and the nitrogen fixation process. To rectify the issue of a soil with a pH below 5.5 and a natural deficiency of Mo, the application of 50–100 g of Mo per hectare is generally sufficient to resolve the problem for a period of five to ten years. Soils with significant levels of iron and aluminium oxides may require additional supplementation.

Calcium

Based on the authors' observations, most agricultural soils have an adequate amount of calcium as a nutrient for cultivated crops and pastures. When there is a slight calcium deficiency, it initially appears in regions of the plant that are located furthest away from the primary water circulation, such as the peanut producing zone. Low soil calcium levels can have a detrimental effect on the nodulation of underground clover. Immature, growing leaves may wilt at the point where they attach to the stem because to temporary, extreme shortages. Instances of highly severe calcium deficiency are more prone to happen in sandy, low-organic matter soils with a pH below 4, or in regions where excessively acidifying fertilisers have been applied. Under these conditions, the exchangeable calcium may decrease to less than 40% of the exchangeable cations other than aluminium, but only in soils with a poor cation exchange capacity. These soils generally include large concentrations of soluble aluminium, which is lethal to all plants except for those that can tolerate highly acidic conditions, such as sugar cane, before the symptoms of calcium deficiency become apparent. The symptoms include stunted, sparsely branched, and discoloured roots, as well as dead shoot tips. Fenton and Helyar (2002) state that differentiating between root symptoms and indicators of aluminium toxicity can provide a significant challenge.

Magnesium

Instances of livestock and crop yield reductions caused by magnesium nutritional shortages are infrequent. The subsurface soil strata in Australia typically contain substantial quantities of magnesium, if not all of them. Excessive amounts of magnesium can occasionally result in the deterioration of soil composition.

Phosphorus

As the acidity of soils increases, phosphorus is bound to iron and aluminium oxides, resulting in reduced availability to plants. In order to avoid phosphorus deficit in plants, it is necessary to increase the amount of phosphorus in acidic soils that are abundant in iron and aluminium oxides.

3.8 Cycle of Carbon

The carbon cycle commences in the plant's leaves, where sugars are synthesised from carbon dioxide (CO_2) and water, facilitated by solar radiation. The sugars undergo partial oxidation, resulting in the production of organic acids. These acids then dissociate into H^+ ions and organic anions. Salts of organic acids form when the positively charged cations Ca^{2+}, Mg^{2+}, K^+, and Na^+ neutralise the negative charge of the anions. The root is accountable for assimilating these fundamental cations. To counterbalance excessive cation intake, the H^+ ions are either used in metabolic activities or excreted at the root surface. The plant does not require a continual acid–base balance in order to absorb water and nutrients. Instead, it is essential for the charge across the root membranes to remain constant. Consequently, the soil's acidity increases while the plant matter's alkalinity also increases. The buffer consisting of malate and pyruvate maintains the plant's internal pH equilibrium. When a plant sheds its leaves or dies, the plant tissue breaks down and turns into organic matter in the soil. This process releases the alkalinity that was previously contained as organic anions. The phenomenon of fire remains consistent, albeit at a significantly accelerated pace. The acidity of the soil will be counteracted by the alkali produced as the organic components decompose.

$$\begin{aligned}&\text{Sunshine}(CO_2 + H_2O) \rightarrow CO_2 + H_2O(\text{in leaf}) \rightarrow RCOO + H^+ \rightarrow RCOO\\&\quad + M(\text{Plant becomes alkaline}) \rightarrow CO_2 + H_2O \text{ and energy}\end{aligned}$$

The addition of acid to the soil is proportional to the accumulation of alkali in the above-ground biomass. As the acid undergoes decomposition, the reintroduction of biomass to the soil counteracts its effects. Otherwise, the ecosystem will sustain an equilibrium between alkaline and acidic conditions that, in the absence of any external influences, would endure indefinitely. Acid will be employed in the process of weathering. Nevertheless, the act of extracting plant matter from the ecosystem, such as during the process of harvesting grain or producing hay, leads to an increase in acidity within the ecosystem.

3.9 The Cycle of Nitrogen

The nitrogen cycle is initially described in relation to the process by which legume plants and rhizobia symbiosis take nitrogen from the atmosphere and integrate it into the plant. Upon the death of a plant, its nitrogen content becomes part of the organic matter, which subsequently decomposes and releases ammonium nitrogen. Ammonium nitrogen undergoes mineralization and nitrification processes, resulting in the conversion of ammonium nitrogen to nitrate nitrogen and the release of acid into the soil. As the plant root absorbs nitrate nitrogen, it excretes an identical amount of anion to neutralise the acidity, ensuring that the charge across the root membrane remains constant. The ecological system's acid-alkaline equilibrium will remain stable if the plant absorbs 100% of the nitrate nitrogen generated. Nevertheless, the soil's acidity increases when a portion of the nitrate is washed away from the area around the plant's roots before it can be taken in by the plant. Afterwards, nitrogen can be released back into the atmosphere through a process called denitrification. The alkaline nature of this step neutralises the acid produced by the mineralization and nitrification processes. It is important to mention that Fenton and Helyar (2002) discovered that soil acidity is caused by the acidification of the soil by acids, rather than the removal of "base" cations such as Ca^{2+}, Mg^{2+}, K^+, and Na^+. This finding contradicts what some older textbooks state.

3.10 Equilibrium State of Soil Acidity

Most ecosystems in nature have reached a state of near equilibrium, where the balance between acidity and alkalinity is steadily tilting towards increased acidity. In this naturally stable condition, the ecosystem is generating a minimal quantity of acid, which in turn releases an adequate number of cations from the parent material to maintain the soil's health. In this book, the term "stable state pH" will be used to describe the alkali/acid balance in an ecosystem that has reached a stage of very gradual variations.

The natural ecosystem is regulated by the following inherent systems that control the addition of acid:

- The age of the soil.
- The extent of leaching and erosion, and therefore the likelihood of alkaline mineral loss, is controlled by the climate, namely the amount of rainfall.

- Acidification is influenced by variations in the vegetation's nitrogen fixation ability, nutrient absorption and recycling capacity from the soil, and its ability to limit nutrient leaching through water usage and stripping of nutrients from leached water. Moreover, the levels of anions in different plants differ, affecting the amount of alkali lost when plant material is extracted and the amount of alkali replenished from the deeper layers of soil.
- The varied capacity of different soil parent materials to buffer or neutralise acids have an impact on the rate of acidification. When comparing weathered minerals such as sandstone, which may have a quartz content of 98% (SiO_2), basalt and alluvium have a significant potential to neutralise acidity, also known as pH buffering capacity.

3.10.1 Distribution of Soil Acidity Throughout the Soil Profile (pH Profile)

To know how to construct a productive stable state pH profile, it is necessary to understand the components that affect the distribution of soil acidity down the profile. It is essential to manage the acid delivered to each layer within the profile in order to create a cropping or pasture programme that will sustain a productive pH profile (Fenton and Helyar 2002). Examples of the impact of pH levels on pastures and crops in acidic soil include the following:

- When the pH drops below 4.7 for a distance more than 20 cm, it will cause damage to plants that are sensitive to acid.
- A plant with a low tolerance for acidity can often grow its roots up to a depth of 10 cm in unfavourable acidic soil, but it is less probable for it to do so in soil that is 20 cm deep.
- Lime's effectiveness decreases gradually as the pH in the layer above exceeds 5.5.
- The pH levels below the layers treated with lime affect the response to the lime application.The pH profile is essential for planning a crop and pasture rotation alongside a liming programme. If there is subsurface acidity, it may take several years before the economic advantages of liming become apparent.

3.11 The Formation of a pH Profile

Diverse agricultural, forestry, and natural ecosystems give rise to the formation of distinct soil profiles. In numerous systems, such as the one mentioned, the pH of the uppermost 5 cm is greater than that of the underlying layers (5–20 cm). At Wagga Wagga, the pH levels in undisturbed soils are 5.5 to 6 in the top 5 cm, 4.5 to 5.5 in the next 5 to 20 cm, and then increase again due to the addition of net alkali in the lower root zone. In contrast, the opposite outcome may occur in different systems. When acid is added to the entire profile, the long-term outcome usually leads to a profoundly acidic profile, with a somewhat less acidic layer in the top 2–3 cm. Hydrogen and hydroxyl ions, along with their chemical counterparts (aluminium, carbonate, and bicarbonate ions), can be redistributed throughout the profile by processes such as leaching, diffusion, soil movement, or mixing.

If it is challenging to prevent the formation of a highly acidic pH profile, an alternative option would be to utilise a deep subsoil with a lower acidity level. If the process of acidification is accelerated, the profiles will rapidly become acidic throughout unless finely ground lime is added to a depth of at least 10 cm to neutralise any increase in acidity. The soil pH profile is influenced by the following factors:

- Leaching of nutrients alters the distribution of H^+ and OH^- (or HCO^{3-} and organic anions) excreted by roots, as well as the positioning of acid and alkaline reactions in the N and C cycles, in response to imbalances in cation and anion absorption.
- If the nitrate generated through the process of nitrification is washed away, it might result in a net increase in acidity occurring near the upper part of the soil profile. Acidity and alkalinity redistribution in the soil profile can be achieved through mass water flow, diffusion, or soil movement. Typically, when the pH falls below 5.0 or rises over 5.5, the flow of water will be predominantly alkaline or acidic, accordingly.
- Unless the surface pH is above 7 and there is a significant level of biological activity, the alkaline effect of lime that is applied travels gradually downwards in the soil (pH > 5.5).
- Plants transport organic anions and nutrients to the surface of the soil.
- According to Baker et al. (1995), several species of worms and ants are capable of excavating soil from deeper layers and transporting lime downwards in the soil profile by both intentional movement and accelerated leaching via the burrows they create.

The addition of acids to the soil from the nitrogen (N) and carbon (C) cycles mostly occurs through nitrification and dissociation of organic acids. Nitrification refers to the conversion of organic acids formed in the soil by microflora, while dissociation involves the breakdown of organic acids introduced to the soil through plant wastes. Ammonium ions in the soil can either be immobilised or volatilized. Additionally, roots release hydrogen ions in response to absorbing more cations than anions. The primary processes involved in the immobilisation of nitrate by microflora include denitrification, conversion of organic nitrogen into ammonium through mineralization, adsorption of hydrogen ions by organic anions present in soil with lower pH, oxidation of organic anions into carbon dioxide and water, and the excretion of OH or HCO_3 by roots in response to a higher uptake of anions compared to cations. The distribution of the principal processes for adding acids and alkalis relies on how they are added to the soil profile.

The oxidation of organic anions in alkaline conditions and the process of ammonification in alkaline conditions are two important processes that frequently occur near the soil surface when organic matter is added. The extent of ammonium leaching prior to nitrification dictates the location of the predominant acid reaction, nitrification. The excretion of acid or alkali by roots, which is another important function, is influenced by the relative abundance of nutritious cations and anions in a specific layer of soil. A useful simplification of this balance is that roots tend to release acid when nitrogen availability is low in a soil layer, as they generally absorb more cations that are not nitrogen and less anions that are not nitrogen. In areas where there is an abundant supply of N, it is expected that there will be large rates of acid excretion, particularly if the majority of the soil's N is in the form of ammonium. Conversely, in cases when most of the nitrogen in the soil exists as nitrate, we can expect a significant amount of alkali excretion. Hence, the biological processes that impact the dispersion of acid addition to the soil profile involve the drainage of ammonium after ammonification and the drainage of nitrate before roots may absorb it. An example of how nutrient leaching might affect the pH profile is when NO_3–N is leached to a deeper level in the profile before being absorbed by the plant. Nitrification will acidify the area where it occurs and increase the pH in the area where it is being absorbed. Although this approach may serve as a method to decrease the acidity of the underlying soil, there is no assurance that the NO_3–N will indeed be assimilated and not be carried away from the area where the roots are present.

3.11.1 Aluminium in Its Mineral Form

Mineral forms of aluminium that can be found in soils include hydrous oxides (such as gibbsite), aluminosilicates (such as feldspars, kaolinite, and imogolite), sulphates (such as jurbanite), and phosphates (such as variscite) (Dixon and Weed 1989; Lindsay and Walthall 1989). The dissolving or precipitating of a chemical is driven by the amount of free energy that is lost during the process. The standard free energy of the reaction, defined by equilibrium thermodynamics, is added to the free energies associated with the formation of precipitates and nucleation during precipitation, or the disruption of the lattice structure of the solid phase during dissolution. Collectively, these free energies constitute the overall quantity of the reaction. Previously, it was believed that the standard free energy of the reaction played a crucial role in causing precipitation or dissolution, and that the principles of equilibrium thermodynamics could be used to understand the extent of the reaction. Considering the inconsistent outcomes of the approach, it has been shown that alternate sources of free energy loss related to the kinetics of the reaction are more relevant. Thermodynamics, as described by Lewis and Randall (1923), encompasses the probability of a reaction occurring and the minimum work required for the reaction. The duration required for a reaction to reach equilibrium is dictated by its kinetics, which are affected by the surface morphology and the transport of reactants and products to and from a surface.

3.11.2 Organic Compounds Containing the Element Al

While aluminium is the most abundant metallic element in soil, it is mainly present in forms that cannot be dissolved when the pH level is over 5. The enhanced solubility and mobility of aluminium in an acidic environment are believed to present a potential hazard to both plants and animals (Rosseland et al. 1990). Various forms of aluminium can be differentiated in soils. Al refers to aluminium, a chemical element. Porous refers to a material or substance that has small holes or spaces. Water or soil solution refers to a mixture of water and soil. Aluminium ions that are capable of being exchanged; When biological molecules are mixed with... Polynuclear hydroxyl-Al is sorbed externally. Aluminium interlayer or a comparable material, Secondary minerals, such as hydrous oxides and aluminosilicates, Primary aluminosilicate minerals. Interlayer Al compounds, such as hydroxyl-Al polymers, hydroxyl-Al-organic matter complexes, or hydroxyl-aluminosilicates, were mentioned by Soon in 1993. Gibbsite and boehmite are predominantly found in heavily eroded and depleted soils in tropical and sub-tropical areas (Jackson 1979), however, other studies (Conyers 1990) have indicated their occurrence in podzolic soils.

Despite being the most abundant element in the Earth's crust, aluminium (Al) is not essential for plant growth. Nitrogen (N) is the most essential mineral for plant growth. It is a non-metal that is often found in low concentrations in soils, leading to nitrogen deficiency in many plants. Plants developing in acidic soils face the challenge of dealing with both nitrogen limitation and aluminium toxicity simultaneously (Zhao and Shen 2018). Significant quantities of nitrogen fertilisers are used on agricultural fields to achieve high crop yields, even when the nitrogen is not well utilised, resulting in soil acidification and potential aluminium toxicity. Aluminium inhibits the growth of roots, leading to a decrease in the uptake of nitrogen by plants and a reduction in their efficiency in using nitrogen. Acidic soils are characterised by a relatively high ratio of NH_4^+ to NO_3^- and elevated amounts of toxic Al. Aluminium-sensitive plants have a preference for NO_3^- as a nitrogen source, whereas aluminium-tolerant plants often choose NH_4^+. NH_4^+ increases the solubilisation of toxic Al in soil solutions compared to NO_3^-. However, in solution culture, NH_4^+ typically decreases the harmful effects of Al on plants because its protons compete with Al^{3+} for adsorption sites on the root surface. Aluminium inhibits both the absorption of nitrate ions (NO_3^-) by plants and the activity of nitrate reductase, although to a smaller degree compared to how nitrate ions (NO_3^-) hinder the uptake of ammonium ions (NH_4^+) by plants. The collective results of many experiments indicate a mutually beneficial relationship between NH_4^+ feeding and the ability of plants to tolerate aluminium. This has profound implications for how plants acclimatise to acidic soils where toxic aluminium and ammonium ions are chemically dominant. In the concluding part of our study, we explore the practical use of this knowledge in enhancing the efficiency of nitrogen utilisation and the ability of plants to tolerate aluminium in acidic soils (Zhao and Shen 2018).

Zama et al. (2022) conducted a study to investigate the effects of nitrogen and phosphorus-enriched soils on the soil concentration correlations, growth kinetics, mineral nutrition, and nitrogen usage efficiencies of Vachellia sieberiana in a greenhouse experiment. The analysis of the soils used as the growth substrate indicated a high level of acidity, with a soil pH of 3.9. Nitrogen-enriched soils exhibited greater acidity when compared to P-enriched soils. The most significant decrease in exchangeable acidity was seen for soil pH values ranging from 3.9 to 4.1 units. The correlation between exchangeable acidity and soil pH was strongly negative. Plant saplings exhibited higher root biomass, shoot biomass, total biomass, and concentrations of nitrogen and phosphorus when growing in soils with high phosphorus concentrations. Extreme soil acidity in N-enriched soil was a significant factor contributing to the unavailability of P, resulting in impaired seedling growth. Applying lime to highly acidic soils can effectively decrease the levels of toxic heavy metals, such as aluminium, that have been caused by severe soil acidification (Zama et al. 2022).

The insufficient nutritional composition of soils in Africa poses a challenge to the sustainable growth of plants and agricultural practices (Goldman 1995). Mafongoya et al. (2006) classify South African soils as having low nitrogen levels and acidic pH levels. The acidic soil pH and restricted cation exchange capacity have a detrimental effect on the availability of nutrients such as potassium (K^+) and calcium (Ca^{2+}) (Aprile and Lorandi 2012). In acidic conditions, certain nutrients, such phosphorus (P), can become insoluble and sequestered due to cation binding (Sharma et al. 2013). In addition, the process of P sequestration and the reduction of P availability for plant uptake have been associated with soils that have elevated levels of cations, including aluminium (Al), iron (Fe), and manganese (Mn) (Afonso et al. 2020). Plant growth requires both organic and inorganic forms of nitrogen (N) and phosphorus (P) (Pueyo et al. 2021). Adenosine triphosphate (ATP), an important energy component for plant metabolism, is particularly utilised by legume plants during biological nitrogen fixation (BNF). It is worth noting that ATP also contains phosphate (Mitran et al. 2018). The presence of acidic soil conditions is anticipated to have an effect on the process of nodulation and nitrogen fixation due to the fixation of phosphorus by cations in acidic soils, which makes it unattainable for plants (Balemi and Negisho 2012). On the other hand, a plentiful supply of phosphorus (P) is essential for legume plants to carry out processes such as production, nodulation, growth, and nitrogen fixation (Zaidi et al. 2009).

The pH range of 6.0 to 6.5 exhibited the most pronounced association between pHBC and the content of soil organic matter (Nelson and Su 2010). Soil acidity is a significant concern for agriculture in many regions globally, while alkalization is a problem in drier areas with irrigation (Brady and Weil 2008). The pH of the soil is influenced by the initial pH, the amount of acid or alkali added, and the pH buffering capacity (pHBC) of the soil. The primary factors that control soil pHBC are protonation and deprotonation of acidic groups on organic matter, oxides, and hydroxides, dissolution and precipitation of carbonates, complexation and decomplexation of aluminium by organic matter, and ion exchange. Clay mineral and primary mineral dissolution also play a significant role in soils with high levels of weatherable minerals or over longer time scales (Bloom 2000). Within the pH range of 4.0–6.5, there is a strong correlation between soil pHBC (pH buffering capacity) and the quantity of organic matter present. The relationship with clay content is also significant, although to a lower extent (Dolling et al. 1994; Noble et al. 1997). The significance of these processes changes in relation to pH.

3.11.3 Acid Soil Characteristics

The pH scale, with a range of 0 to 14, quantifies the hydrogen ion concentration. Neutrality is rated as seven on the scale. An acidic pH is defined as a pH value between 0 and 7, whereas a basic or alkaline pH falls between 7 and 14, with 14 being the highest level of basicity. Carbon dioxide is produced through the process of respiration by soil microbes and plant roots. It is formed through the combination of soil organic matter with water, resulting in the production of carbonic acid. This acid then interacts with the insoluble primary minerals, clay, and humic micelle found in the soil.

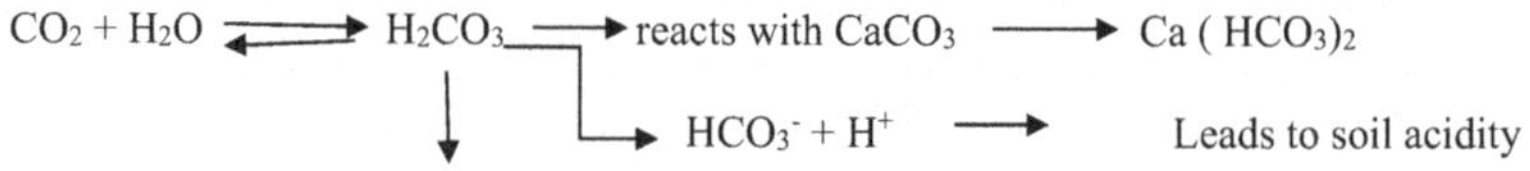

H-ions released from weak carbonic acid also lead to soil acidity. Carbonic acid reacts with clay micelle forming soluble bicarbonates.

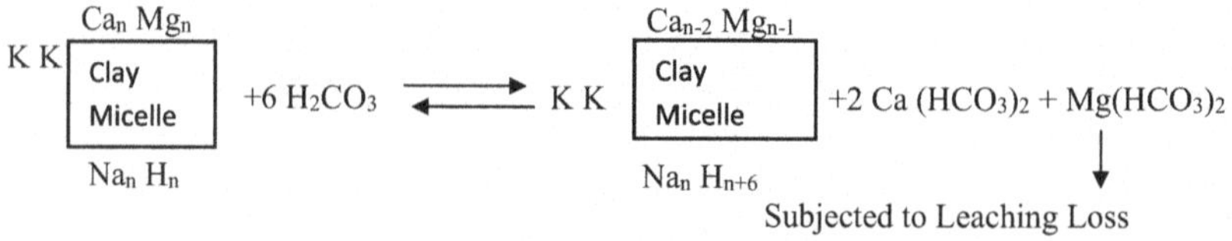

Thus, in the moist climate of the highlands, abundant rainfall causes the erosion of soluble bicarbonates of alkaline metal ions, including Ca^{2+}, Mg^{2+}, Na^{+}, and K^{+}. This leads to the reversible reaction progressing in the forward direction and increasing the number of H-ions from n to $n + 6$, as depicted in the reaction, at the expense of calcium and magnesium ions in the clay lattice. Aluminium exhibits solubility as the pH of the soil drops, leading to an elevation in the concentration of aluminium in the soil solution. The roots of sensitive plant species exhibit toxicity when exposed to soil aluminium concentrations ranging from 2 to 5 parts per million (ppm), whereas tolerant plant species demonstrate toxicity at concentrations exceeding 5 ppm. Aluminium is the most abundant metal in the Earth's crust. It constitutes approximately 7% of the total mass or weight of the earth's crust. The pH of the soil is perhaps the most critical management factor that influences the concentration of Al^{3+} in the soil solution. When the pH drops below 6.0, soluble aluminium is present in the soil. However, unless the pH drops below a pH level of 5.5, it is of little importance in the vast majority of soils. However, unless the soil pH drops below a pH of 5.0, it is seldom a concern. Nevertheless, when the soil pH drops below pH 5.0, there is a significant increase in the concentration of soluble aluminium in almost all types of soil. Prior studies conducted by Mukherjee and Mitra (1944) and Mukherjee and Chatterjee (1945) suggested that Al^{3+} ions play a significant role in soil acidity. This was demonstrated through experiments involving the interaction between hydrogen clays and neutral salts, resulting in the formation of H^+ and Al^{3+} ions, and to a lesser extent, Fe^{3+} ions in the salt extracts. Mukherjee et al. (1943) gathered valuable information about the properties of hydrogen clays by analysing titration curves, buffer capacities, base exchange capacities, and identifying the mineral constituents using physical methods such as X-ray, optical, and thermal analysis.

Uchida and Hue (2000) state that during the process of soil weathering, clay minerals such as Kaolinite decompose due to the acidic nature of the soil. This breakdown results in the formation of soluble Al-ions, which are listed below. Al^{3+} undergoes hydration in aqueous environments, resulting in the formation of a complex containing 6 water molecules, known as Al.

$$(H_2O)_6^{3+}.\underset{\text{Kaolinite}}{Al_2Si_2O_5(OH)_4} + 6H^+ = 2Al^{3+} + 2Si(OH)_4 + 2H_2O$$

Chernov's (1947) hypotheses state that the hydrolysis of Al results in the production of hydrogen ions, which play a crucial role in maintaining the active acidity (hydrogen ion concentration) of soil solutions. The mechanism of hydrogen release undergoes changes with pH. Most of the aluminium in soil can dissolve when the pH is below 4.7, forming Al^{3+} or $Al(OH)_2{}^+$ cations. These cations are strongly adsorbed and are preferred over hydrogen. When Al^{3+} and water combine, they form $Al(OH)_2{}^+$ and release one hydrogen ion. This reaction occurs in a solution when Al^{3+} is in equilibrium with exchangeable Al. Adsorbed hydrogen ions, which are in balance with the hydrogen ions in the soil solution, serve as a secondary (minor) source of hydrogen. Datta and Sharma (2006) state that the process of hydrolysis of aluminium leads to the formation of both monomeric and polymeric Al complexes, which in turn release hydrogen ions. Additional hydrolysis gradually releases hydronium ions by extracting protons from the water in hexaaquoaluminium.

$$Al(H_2O)_6^{3+} + H_2O = Al(OH)\,(H_2O)_5^{2+} + H_3O^+\ Al(H_2O)_6^{3+} + H_2O$$
$$\downarrow_{H_2O}$$
$$= Al(OH)(H_2O)_5^{2+} + H_3O^+$$
$$\rightarrow Al(OH)_2(H_2O)_4^{+} + H_3O^+ \text{ and so on}$$

Aluminium ions (Al^{3+}) may not be present in moderately acidic soils due to the abundance of bases in that environment. Instead, Al^{3+} is converted into $Al(OH)_2{}^+$. Due to the higher concentration of hydroxyl ions in this acidic condition of soil, more complex $Al(OH)_2{}^+$ ions are formed, resulting in a reduced probability of hydrolysing to produce hydrogen ions. However, the amount of substances that undergo hydrolysis to produce hydrogen ions in the soil solution is smaller compared to the very acidic soil. Aluminium's solubility is highly influenced by pH, and an increase in soil pH decreases the toxicity of aluminium. The element can be found in the soil solution either as soluble ionic species or as a basic structural component in secondary clay minerals such as kaolinite and vermiculite. Aluminium undergoes polymerization on clay surfaces and between clay interlayers, it gets adsorbed on both organic and mineral colloidal surfaces, and it exists in amorphous, crystalline, and hydroxy-Al forms. Aluminium ions contribute significantly to the cation exchange capacity in acidic soils, resulting in a high level of aluminium saturation in these soils. The Al^{3+} ions, commonly known as exchangeable Al^{3+}, are actually a combination of monomeric Al ions (Al^{3+}, $AlOH_2{}^+$, and $Al(OH)_2{}^+$), with an average charge per Al ranging from 2 to 3. This charge lowers as the pH level increases. Aluminium strongly adheres to exchange sites, resulting in soil solution concentrations that are usually moderate, often ranging from 10 to 250 M.

Consequently, soils can be classified into three distinct categories based on their acidity levels: The acidity in soil can be classified into three types: active acidity, which is caused by the presence of H^+ ions in the soil solution; exchangeable acidity, which consists of both H and Al- ions located in the exchangeable positions of silicate clays; and non-exchangeable acidity, also known as residual or reserve acidity, which is caused by organic matter and bound Fe/Al components. Variable charge arises from the process of protonation and deprotonation of functional groups that include oxygen on the surfaces of organic colloids, as well as on Fe and Al oxides, and broken edges of silicate minerals (Reeve and Sumner 1971; Sparks 1995). The pH-dependent negative charge is generated through the deprotonation of functional groups such as carboxylic or phenolic groups in organic substances.

M–O–H in equilibrium with soil water:

$$\begin{array}{ll}\text{At higher pH values} & MOH + OH^- \rightarrow MO^- + H_2O \\ \text{At lower pH} & MOH + H^+ \rightarrow MOH_2^+\end{array}$$

Deprotonation of functional groups in organic matter is also a phenomenon having pH dependency and negative charge increases with rise in soil pH.

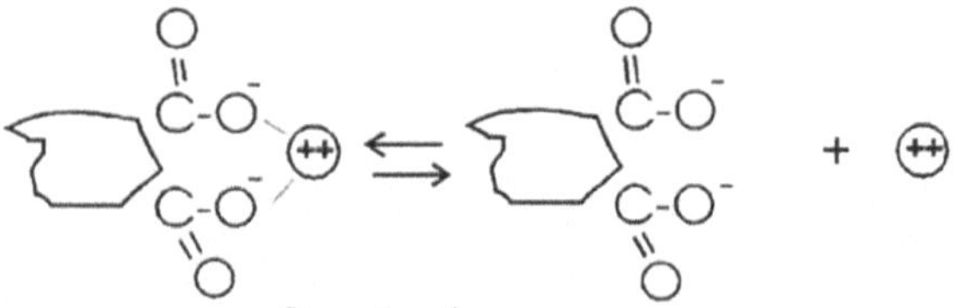

Deprotonation

The pKa values of the functional groups govern the impact of organic matter on charge fluctuation with pH. For instance, carboxylic acids have pKa values between 3 and 9, while phenolic groups have pKa values starting at 9 (Talibudeen 1981; Hayes and Swift 1985). At pH 2, silanol (SiOH), an inorganic functional group, begins to dissociate (Talibudeen 1981). Because of the existence of sesquioxides and organic colloids, soils can produce pH-dependent negative or positive charges depending on pH and pOH (Mott 1988). According to Parfitt's (1980) report on variable charge soil components, the amphoteric character of sesquioxides—primarily those of Fe and AI, and to a lesser extent Mn—is the primary source of the pH-dependent positive charge, especially at very low pH levels. As the soil solution gets increasingly acidic below the zero point of charge (zpc), the oxygen and hydroxyl groups that are present at the surface of oxides have a significant increase in their ability to take protons. A higher pH is reached at the point of zero charge (Parks and de Bruyn 1962) due to the prevalence of positive charges in soils. Calculated (Gillman and Uehara 1980; Stoop 1981) and presented in Figs. 3.5 and 3.6, the Point of Zero Charge (pH^0) of soil samples from two soil profiles, Pedon 1 (typic dystrophchrults) and Pedon 2 (typic haplaquent), collected from Soil Series, Lalchara and Vidyanagar in Upland (100–200 m above msl) and Lowland (< 100 m) in Tripura in North East India (Datta et al. 1991). The pH differential (pH 0.05 M–pH 0.002 M) is displayed graphically versus pH 0.002 M values. The ZPC, or pH0, measured for two soil series' worth of soils ranges from 1.90 to 3.42, suggesting a predominance of net negative charge in the soils. Researchers have also identified similar variations in ZPC (Dolui et al. 1987; Mishra et al. 1989). In kaolinitic soils, ZPC is approximately 3.0 (Gallez et al. 1976); research have also recorded values as high as 6.3 in soils with high sesquioxide content (van Raij and Peech 1972, Ilton et al. 1976). Juo (1981) reported that pure iron and aluminium oxide minerals have a zpc over pH 7, while kaolinite and quartz have a zpc below pH 3. The zpc is often shifted by the interaction of soil constituents like kaolinite with iron or aluminium oxide. Figure 3.3 shows ZPC (pH^0) in Pedon 1 (Typic Dystrochrults) in Tripura under upland conditions.

ZPC (pH^0) which is a special condition for which the presence of equal amounts of positive and negative charge exists, in Pedon 1 was found to vary from 1.90 to 2.60 and in Pedon 2 (Fig. 3.4), from 2.27 to 3.42. As indicated by Shahoo et al. (2020) in soils of Manipur, pH in KCl solution is always less than in water and negative pH indicates the occurrence of amorphous sesquioxides with variable charges (Bhattacharayya et al. 1994; Chandran et al. 2005). The soils are also not at their point of zero charge and contain considerable amount of reserve acidity in soils of North East India (Sen

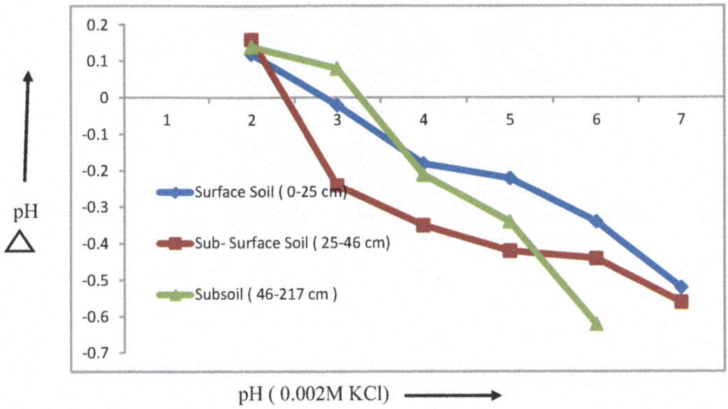

Fig. 3.3 ZPC (pH^0) in Pedon 1 (Typic Dystrochrults) in Tripura under upland situation

et al. 1997). Surface electric potential of two soils as studied varied from − 185.9 to − 258.4 mV at 25 °C. ZPC is close to pH 3.0 in Kaolinitic soils (Gallez et al. 1976) but in soils containing high amount of sesquioxides, ZPC values are around 6.3 as reported by van Raij and Peech (1972). Variable charge mainly arises from the protonation /deprotonation processes of the surface functional groups in the soils and are present mainly in clay minerals and metal oxides (Dultz et al. 2019; Ponto and Berg 2018).

As reported by Datta et al. (1991), water soluble (2.90–3.50 mg/kg) and NH_4OAc exchangeable (16.6–18.25 mg/kg) Fe are appreciably high in Pedon 2 (Lowland) as compared with $Fe_{water\ soluble}$ (0.40–0.42 mg/kg) and $Fe_{Am.acetate}$ (1.66–3.33 mg/kg) in upland soils (Pedon 1). Aluminium (KCl extractable) however showed reverse trend

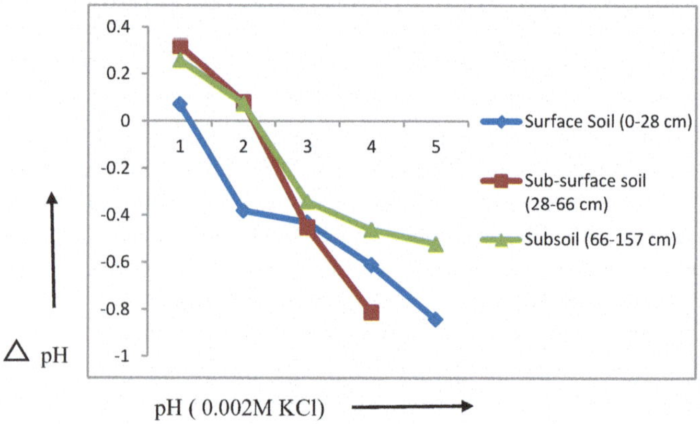

Fig. 3.4 ZPC (pH^0) in Pedon 2 (Typic Haplaquent) in Tripura under lowland situation

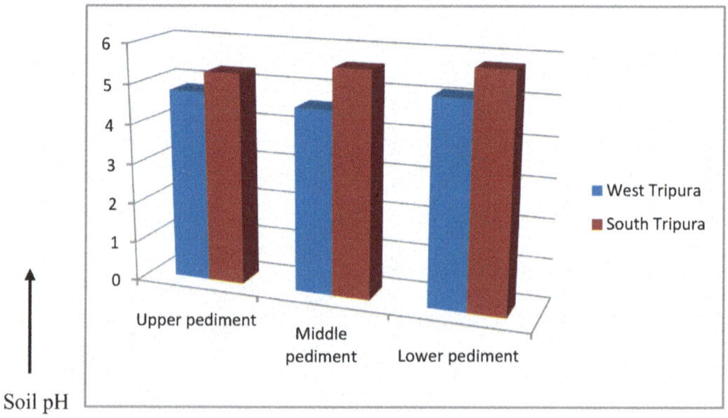

Fig. 3.5 Soil pH in hill toposequence in shifting cultivation sites

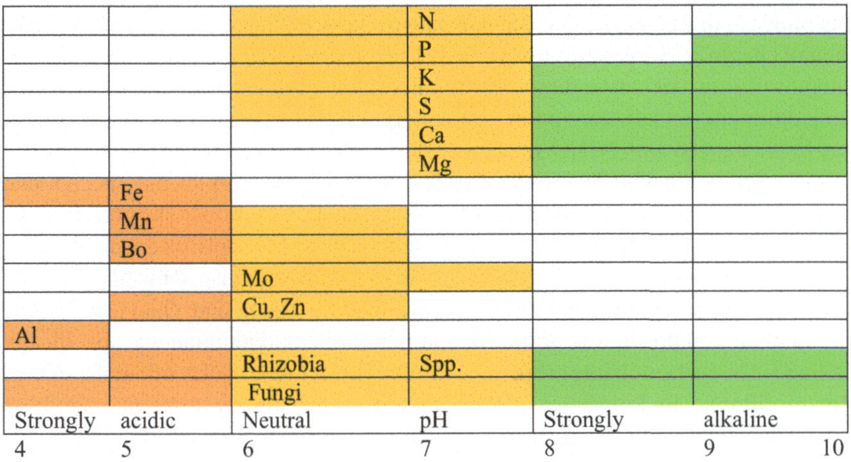

Fig. 3.6 Showing pH zones where maximum nutrients availability occures

possibly due to low pH in Pedon 1 (0.50–2.50 mg/kg). Oxalate extractable iron (151–572 mg/kg), as expected, is markedly less than that of dithionite extractable one (2.6–5.6%) being higher in lowlands as compared to soils in upland toposequence. Oxalate extractable Al (170–215 mg/kg) is somewhat higher than the dithionite extractable Al (140–195 mg/kg).Correlation values as worked out between ZPC (pH^0) and soil properties are presented in Table 3.3.

The presence of organic carbon is strongly and inversely associated with ZPC (pH^0), suggesting that organic components tend to shift towards lower pH values. The existence of oxyhydroxides of Fe and Al in the silt fraction of soils is suggested by the positive and significant association between the silt/clay ratio and ZPC. Conversely,

Table 3.3 Correlation values between ZPC (pH^0) and soil properties in Tripura in North East India

Soil properties	Correlation coefficient (*r*)
pH (water)	0.776*
Organic carbon	– 0.730*
Sand (%)	– 0.445
Silt (%)	0.602
Silt/clay ratio	0.715*
Fe (Water soluble)	0.682
Fe (NH_4OAc)	0.607
Fe (Dithionite)	0.803*
Al ((NH_4OAc)	– 0.826*
Al (Oxalate)	0.715*
Al (Dithionite)	0.645

*Significant at 5% level

clay minerals that possess a persistent negative charge exhibit a movement of the zero point of charge (ZPC) towards a lower pH. There is a positive correlation between the amounts of iron that can be dissolved in water and the amounts that can be extracted using dithionite, and the zero point charge (ZPC) at pH^0. Except for Al in a form that may be exchanged with NH_4OAc, its oxalate and dithionite extractable forms show a positive correlation with ZPC. Sesquioxides, unlike organic carbon, have been found to experience a rise in pH at the point of zero charge in acidic soils. The presence of a persistent positive charge was observed in Spodosols and Oxisols, as documented by Gillman and Uehara in 1980. Variable charge soils are distinguished by their elevated levels of iron and aluminium oxides, as described by Zhang and Zhao (1997). The clay mineral composition is primarily composed of 1:1 type minerals, specifically kaolinite. The surface charge properties of soils with 2:1 type of clay minerals, such as Montmorillonite, hydrous mica, and vermiculite, are distinctly different from those of constant charge soils due to these two factors. Due to the presence of both variable charge components and permanent charge surface elements in varying proportions, the surface charge properties of soils become increasingly complex. The presence of the zpc at low pH levels can be linked to the interaction between organic matter and iron and aluminium oxide minerals, as well as the presence of kaolinite. Juo (1980) documented that soils containing mostly kaolinite exhibited zero point of charge (zpc) values ranging from 3.5 to 5.5. The surface charge characteristics of soils have a significant impact on various phenomena such as the adsorption of cations and anions, the formation of organo-mineral complexes, dispersion, flocculation, swelling, and shrinking. This was discussed in a review by Zhang and Zhao (1997).

3.11.4 *The Occurrence of Aluminium Toxicity*

Aqueous aluminium (Al) also forms inorganic complexes with fluoride ions (F–) and sulphate ions (SO_4^{2-}). The formation of these complexes is influenced by factors such as pH, concentration of the inorganic ligands, ionic strength, and temperature. The ease of transitioning from a solid to a liquid phase and its high solubility in an acidic environment are crucial elements that determine its significant role in the environment (May and Nordstrom 1991). Previously, aluminium was believed to be innocuous for the human body since it is easily eliminated through urine. Recent research in environmental toxicity have suggested that aluminium may be a contributing factor to several diseases in humans, animals, and plants. The occurrence of acid rain and the subsequent acidification of the environment cause changes in the pH of water and soil. These changes lead to the release of poisonous aluminium ions, which in turn cause harmful effects such as plant poisoning and disruptions in the functioning of both human and animal organisms. The solubility and mobility of inactive forms of aluminium are enhanced by alterations in the soil environment. The pH is the most significant factor among many that directly and indirectly affect the mobility of aluminium. The acidity primarily causes the degradation of the soil structure and the displacement of Al^{3+} ions by means of cation exchange, rendering them mobile.

The threshold soil pH at which aluminium becomes available in harmful proportions is influenced by various parameters, such as the major clay minerals, organic matter content, levels of other cations and anions, overall salt concentrations, and the specific soil type. Elevated concentrations of sodium ions have the ability to disintegrate the structure of soil and displace cations that are attached to the surfaces of the soil through cation exchange, particularly when they are present in significant quantities. This could potentially enhance soil acidity and directly facilitate the movement of aluminium ions in the soil. The World Health Organisation recommends that the maximum quantity of Aluminium in ground water should be below 0.1 mg/L. The level of Aluminium in natural bodies of water can vary considerably due to a variety of physical, chemical, and mineral causes. The concentration of dissolved Aluminium in fluids with pH values close to neutral typically varies from 0.001 to 0.05 mg/L. However, it increases to 0.5–1 mg/L in more acidic waters or waters that have a high content of organic matter. Water polluted by acid mine drainage can reach severe levels of acidity. In such cases, measurements have shown dissolved Aluminium concentrations of up to 90 mg/L (WHO 1997). The human body contains an approximate total content of 9 mg/kg of Aluminium. Concentrations of up to 100 mg/kg may be seen in certain organs, such as the spleen, kidneys, and lungs. The daily intake of Aluminium is roughly 5 mg, with just a minute portion being absorbed. As a result, the acute toxicity is quite modest. The rate of absorption is around 10 μg per day. These quantities are deemed non-toxic to humans. However, once the element is absorbed by the body, it is difficult to eliminate, leading to various health issues such as nerve damage, kidney illness, allergies, and lung disorders. Aluminium is potentially mutagenic and carcinogenic. Elevated consumption of aluminium can also lead to osteomalacia, which is characterised by deficiencies in vitamin D and calcium. Aluminium intake primarily occurs via the consumption of food and drinking water. The latest standards ranged from 50 to 200 μg/L.

3.11.5 Soil Mineralogy

The soil qualities in soils formed by weathering processes are influenced by the chemical composition and mineralogy of the parent material. The chemical and physical characteristics of soils resemble those of the rocks and minerals that make up the parent materials. The North East region of India, characterised by a humid tropical climate, has the largest area affected by soil acidity. This region experienced significant land development due to geodynamic forces during the Cretaceous period, which occurred approximately 60–110 million years ago. As a result, it is believed to have generated one of the most diverse collections of plant and animal life during the longest period of the Mesozoic era, known as the phanerozoic. The extensive deposits of sediment on the Earth's surface underwent deformation, resulting in the formation of ridges and basins. This had a significant impact on the overall shape of the Earth's surface during later periods. Geological mapping and related investigations have shown that the rock formations in the North East have a diverse range of geomorphological and geological characteristics, spanning from the Precambrian to recent times. Rocks encompass a diverse range of sediments, which are generated through external processes, as well as metamorphic and magmatic rocks, which are formed through internal processes occurring deep within the Earth. According to Datta's compilation in 2013, the rock composition in the NE region can be found in Table 3.4.

The soils in North East India consist of 5 different soil orders: Alfisols, Entisols, Inceptisols, Mollisols, and Ultisols. These soils are formed on different types of rocks that range in geological age from Paleozoic to modern formations. Additionally, there are 22 main groups and 45 subgroups of these soils, as documented by Patiram and Datta in 2006. The ancient rocks are stratified with tertiary and quaternary layers. The North East India region is predominantly characterised by Proterozoic gneissic rocks, which consist of gneisses, magmatites, metasedimentary bands, amphibolites, metadolerite, and metapyroxenite. The shales in the vicinity of Myanmar have undergone lithification and have a slaty texture, often in association with quartz veins and serpentine. The Tipam series, found extensively in Tripura, Nagaland, and Arunachal Pradesh, are sedimentary formations mostly composed of sandstones. These sandstones also contain ferruginous components and mottled clay/

Table 3.4 Rock composition in North East India

Series of rock formation	North East India	Rock composition
Disang	Nagaland, Manipur, Mizoram and Tripura	Dark grey shales associated with sandstones, quartz and serpentines
Tipam	Tripura, Nagaland, Arunachal Pradesh	Grits, sandstones, conglomerates and clay
Jaintia	Meghalaya	Sandstones, calcareous sandstones and limestones
Barail	Nagaland, Manipur, Mizoram and Tripura	Shales, weathered mica, feldspathic sandstones
Shillong	Meghalaya	Gneissic complex consisting of a thick series of quartzites, slates and schists
Surma	Naga Hills, North Cachar Hills and Assam	Sandstones, sandy shales, mudstones, conglomerates, ferruginious sandstones
Darjeeling	Sikkim, Darjeeling, Teesta valley	Metamorphosed slates, magmatics, granites and schists
Deptila	Nagaland, Manipur, Mizoram and Tripura	Ferruginous in nature
Potin, Bomdila, Tenga	Arunachal Pradesh	Low grade Schists, gneisses, phyllites, amphiboles
Lohit	Arunachal Pradesh	Meta sediments, chloritic quartz, mica schists, carbonate rocks and lime rocks

sand in Assam. The Jaintia series in North Cachar, Assam, and the Jaintia Hills of Meghalaya are characterised by the presence of abundant coal deposits and a dense layer of sandstones, calcareous sandstones, and limestones. The Shillong series is a gneissic complex composed of metamorphosed igneous and sedimentary rocks, including quartzites, slates, schists, and masses of granitic intrusions. Meghalaya possesses abundant reserves of precious minerals such as coal, limestone, uranium, granite, kaolin, and glass sand. The Barail series in Mizoram consists of sedimentary rocks that include shales, micaceous sandstone, and feldspathic sandstone, which are frequently found in layers with siltstone.

The geological formations in Manipur consist of shales, slates, siltstone, mudstone, limestone, and varying amounts of sandstone and quartzite. The meta-sedimentary sequence in Arunachal Pradesh consists of quartzite ranging in colour from grey to pale brown, magnetite-quartzite, carbonaceous/graphitic phyllite, dark grey phyllite, black to grey slate, and dolomite. In North and West Sikkim, there are hard and large gneiss rocks that are resistant to erosion. However, in South Sikkim, the rocks are predominantly composed of mica. Limestone sedimentary deposits have been documented in Sikkim. The majority of limestone rock is found in the southern margin of the Meghalaya plateau. It belongs to the Cretaceous-Tertiary sedimentary rock formation groups and has a reserve of 149.59 million tonnes, which accounts for 12% of the country's total reserve (Lamare and Singh 2016). In addition to Meghalaya, limestone formations can also be found in the states of Assam, Nagaland, and Manipur.

The chemical breakdown of rocks and minerals in soils results in the formation of clay minerals. These clay minerals are secondary minerals that are generated through the re-crystallisation or transformation of primary minerals. They take the form of various types of phyllosilicates, which can have either 1:1 or 2:1 type layers.The predominant secondary minerals are kaolinite (resulting from the chemical weathering of feldspar), chlorite (formed by the weathering of biotite, pyroxene, and amphibole), sericite (arising from the weathering of feldspar), and serpentine (resulting from hydrothermal alteration of olivine). Gibbsite in the Ultisols of North East India can be generated through either the weathering of primary Al–silicate minerals or the alteration of clay mineral intermediates, as documented by Bhattacharyya et al (2000). The soils in North East India have a low activity of kaolinitic clay due to their clay cation exchange capacity (CEC) and effective cation exchange change (ECEC), which are both less than 16 and 12 cmole (p+)/kg, respectively (Smith 1986; Bhattacharyya et al. 2009). Kandic horizons are characterised by low activity clay that is predominantly composed of kaolinite and sesquioxides. According to Sen et al (1994), four pedons developed on sedimentary parent material in hilly terrain of Manipur exhibited kandic features and were designated as Typic Kandihaplohumult and Typic Kandihapludult.

It is important to note that ECEC in acid soils refers to the total quantity of exchangeable bases, including Ca, Mg, K, Na, and exchangeable Al. The $BaCl_2$-TEA method indicates that the total acidity is 3–5 times greater than when evaluated by IN KCl. This suggests that acid weathering in humid tropics leads to the formation of hydroxyl interlayering in vermiculite (HIV). According to Chandran et al

(2004) and Pal et al (2014), acid soils are believed to include a combination of minerals including kaolinite, vermiculite with hydroxy interlayering, chlorite, mica, and small amounts of feldspar. Vermiculites have absorbed Al^{6+} ions in their interlayer to generate hydrated aluminium species (HIV), which then react with kaolinite to form HIV-chlorite. The presence of high levels of Al^{3+} in North East India results in the formation of ferruginous acid soils. These soils have interstratifications of both 2:1 and 1:1 minerals, together with hydroxyl-Al interlayering (Pal 2017, 2019; Bhattacharyya et al 1993). The soils found in the humid tropics of North East India are characterised by the presence of kaolinite and other hydroxy-interlayered minerals. These soils can be classified as extremely acidic Ultisols, as well as slightly acidic to neutral Alfisols and Mollisols.

3.11.6 Soil Acidity in Northeast India

In the humid tropical climate (HT) of North East India, the soils generated are diverse and can be classified into three main soil orders: Ultisols, Alfisols, and Mollisols. Mollisols are predominantly found in regions characterised by extensive forest vegetation, while Alfisols are primarily found in regions with little forest vegetation and/or in agricultural soils (Bhattacharyya et al. 2005). Ultisols are typically present in agricultural regions (Bhattacharyya et al. 2000; Sen et al. 1997), while Mollisols are more prevalent in places with extensive forest cover. Conversely, Ultisols have a markedly acidic pH, while Alfisols and Mollisols span from mildly acidic to neutral. Based on a thorough study conducted by ICAR-NAAS (2010) on the degraded and wastelands of India, it has been determined that water erosion is the predominant form of soil degradation observed in the soils of North East India. The highest proportion of land classified as degraded is located in Nagaland (87%), followed by Meghalaya (78%), Arunachal Pradesh (73%), Assam (66%), Manipur (53%), Tripura (38%), and Sikkim (37%). Assam has the smallest proportion of land that is classified as being degraded. High temperatures in climatic conditions can lead to the formation of acidic soils due to a decrease in soil fertility. The increase in soil acidity is a clear sign of the chemical degradation of the soil. Only soils with a pH level below 4.5 (indicating high acidity) and between 4.5 and 5.5 (indicating moderate acidity) were considered in the calculation of the area affected by acidity degradation. According to this assumption, it has been estimated that over 6.98 million hectares, which is around 9.4% of the total land area of the country, is affected by soil acidity (ICAR-NAAS 2010). The soils found in the regions of Northeast India under the HT (high temperature) climate are predominantly strong to moderately acidic Alfisols, Ultisols, and Mollisols. Over time, the ongoing weathering process in the HT climate would lead to the formation of Ultisols that include a significant amount of layer silicate minerals. Sehgal (1998) states that acid soils rich in organic carbon (OC) can be effectively managed and can support abundant forest vegetation, as well as various crops such as horticulture, grain, tea, coffee, and spice crops. Therefore, it would be unwise to classify them as soils that have undergone chemical degradation.

The main cause of soil acidity in North East India is the high amount of rainfall, which results in the depletion of basic cations from the soil matrix. The occurrence of organic acids, specifically humic, hymetamalonic, and fulvic acids, in humus contributes to soil acidity in temperate, sub-temperate, and alpine regions of North East India through several mechanisms. The North East India region can be categorised into three distinct climatic zones. The first zone is characterised by a cool humid monsoon climate in the hilly regions, the second zone is a wet subtropical monsoon climate and third zone experiences a humid mesothermal monsoon climate (Satapathy and Datta 2006).

The soils in a wet environment beneath forest in Sikkim have a high organic carbon (OC) concentration, which is significant enough to be classified as Mollisols. These Mollisols make up around 14.64% of the soils in Sikkim. The presence of dense, mixed, semi-evergreen and moist, deciduous forests in undulating terrain in North East India led to the development of soil acidity, which was influenced by high rainfall exceeding 2000 mm. The addition of OC to Ultisols and acidic Alfisols has proved feasible as a result of the favourable soil temperature and moisture conditions. In addition to heavy rainfall, soil acidity is caused by the continuous depletion of lime and other basic elements by crops, the accumulation of acids found in manure, the use of acid-forming fertilisers, microbial activity, and the formation of soil on acidic rocks containing at least 10% quartz (Turner and Verhoogen 1960). Rhyolite, obsidian, and acidic tuff are the main types of volcanic rocks, while granite and granodiorite are the most prevalent acidic rocks among deep-seated formations. In the Himalayan region, there are three types of rocks that are linked to soil acidity. These include metamorphic rocks found in the upper Himalayas, sedimentary and metamorphic rocks in the middle and lower Himalayas, and Shiwalik sediments from the Tertiary period in the sub-Himalayan zone (Dhir 1967). Red ferruginous soils mostly form on granite, granite-gneiss, ferruginous sandstone, and schist from the Achaean Period. They can also be found in certain areas on Deccan basalt, shale, limestone, and marine deposits (Bhattacharyya et al. 2009; Sehgal 1998).

Laterization is a pedogenic process that commonly occurs in Alfisols and Ultisols, which are types of soils prevalent in tropical and subtropical regions. Elevated temperatures and intense precipitation lead to the swift erosion of rocks and minerals. The transportation of substantial quantities of water through the soil results in the phenomena of eluviation and leaching. The majority of the by-products of weathering, which consist of uncomplicated tiny molecules or nutritional ions, are transported out of the soil profile by leaching if they are not absorbed by plants for nourishment. Iron and aluminium compounds are the only two exceptions to this procedure. Tropical soils acquire their distinct reddish hue from iron oxides. The process of heavy leaching results in an acidic pH in these soils due to the overall depletion of base cations. In regions with a humid and tropical climate, the soils found in Sikkim are classified as Entisols and Inceptisols. In Arunachal Pradesh, the soils are classified as Entisols, Inceptisols, Alfisols, and Ultisols. Similarly, in Assam, Manipur, Mizoram, Meghalaya, Nagaland, and Tripura, the soils are classified as Entisols, Inceptisols, Alfisols, and Ultisols. The soils (Typic Dystrochrepts) obtained from shifting agriculture sites in West and South Tripura were found to be relatively more acidic, with a pH range from 4.5 to 5.8. The soil pH gradually increased from the upper to the lower regions of the hills, as described by Datta et al. (2001). The soil pH in the hill toposequence of shifting agriculture sites is illustrated in Fig. 3.5.

The rise in pH levels in the lower areas of hills in the hilly toposequence is caused by soil erosion from the slopes of the hills, namely from areas where 1st, 2nd, and abandoned shifting cultivation practices occur. The estimated amount of soil erosion per hectare per year is around 147, 170, and 30 tonnes, as reported by Singh and Singh in 1981. Approximately half of the land area in North East India is characterised by moderate slopes, with a gradient ranging from 9 to 15%. Strong slopes, with a gradient of 15 to 30%, cover 17.5% of the land area, while severe steep slopes, with a gradient of 30 to 70%, cover 16% of the land area (Velayutham and Reddy 1990). Aside from variations in topography, changes in soil pH can also be influenced by land use. A study conducted in Mon District, Nagaland, indicated that the pH levels ranged from 4.62 to 4.91 on cultivated land and from 4.56 to 5.09 in forest area, with average values of 4.70 and 4.79, respectively (Konyak et al. 2020). The pH value of soils under cultivated land use was somewhat lower compared to soils under forest land use, possibly due to the presence of a higher concentration of exchangeable aluminium (Al^{3+}) ions in the cultivated soils. The prevalence of soil acidity in North East Indian soils (Sharma and Prasad 1995; Singh and Sharma 2002) affects the availability and movement of essential nutrients such as nitrogen, phosphorus, sulphur, and basic cations. Ling et al. (2007) found that, after a period of 21 days, simulated acid rain with a pH of 2.5 removed approximately 20% of the available soil potassium (K^+).

It is worth noting that acid soils in Eastern India have a deficiency of B and Mo. Various crops have shown positive reactions when these elements are applied, either through the soil or as a spray substance (Panda 1987). When 15 kg of $ZnSO_4$/ha, 5 kg of Boric acid/ha, and 1.5 kg of Ammonium Molybdate/ha were applied, there was a significant increase in grain yield in direct seeded rice in upland soils (Typic Paleochrults). The increase was 15.7%, 30.8%, and 37.2% compared to the control. In contrast, in transplanted rice in lowland soils (Typic Haplaquents) in Tripura, the increase was 39.3%, 22.8%, and 19.8% (Datta and Dhiman 2001). The high response of upland rice to B and Mo can be attributed to their limited availability in upland soils (0.4 mg B/kg and 0.13 mg Mo/kg), which is lower than in lowland soils (0.7 mg B/kg and 0.16 mg Mo/kg). Conversely, lowland soils in waterlogged conditions have an abundance of Fe, which leads to an antagonistic effect with Zn. As a result, the application of Zn fertiliser produces a high response in lowland soils. There is evidence that high levels of DTPA extractable iron (23.8 mg/kg) and manganese (33.6 mg/kg) may have caused a decrease in the availability of zinc to crops. This antagonistic effect was observed in certain acidic soils in Manipur. In these soils, the availability of iron was negatively correlated with the availability of zinc, confirming the occurrence of antagonism. The study conducted by Singh et al. (1990) found similar results regarding the antagonistic effects of Zn $\times$ Mn and Zn $\times$ Fe. This is likely due to the adsorption or fixation of these elements with Al and Fe-oxides, as well as leaching during heavy rainfall in high-altitude soils in Meghalaya. These findings demonstrate the response of rice to micronutrients in acidic soils that are abundant in sesquioxides.

3.12 Deposition of N

The deposition of anthropogenic nitrogen has the potential to change the biogeochemistry of ecosystems. This can lead to acidification of soils and surface waters, reduced availability of some nutrient cations, and increased concentrations of hazardous metals. Ecological changes caused by nitrogen deposition are been observed in remote areas of western North America. The soil's ability to resist changes in acidity was assessed through laboratory acid titration. Additionally, the levels of extractable cations and soil pH were measured in soils collected from field plots. These plots were subjected to different levels of nitrogen deposition: ambient deposition (8 kg N ha^{-1} yr^{-1}), and deposition at 20, 40, and 60 kg per hectare per year above the ambient level. The soil's ability to buffer acid reduced as nitrogen inputs increased, with a 40% decrease observed at the maximum input level. This decrease was accompanied by a fall in pH, a loss of extractable magnesium, and an increase in aluminium and manganese levels (Lieb et al. 2011). The acidity barrier was around 28 kg N ha^{-1} yr^{-1}. Considerable variability in soil acid buffering capacity, concentrations of base cations, and pH was observed across the experimental blocks, regardless of treatment effects. This variability may be attributed to differences in the extent of winter snow cover and its impact on the introduction of anthropogenic nitrogen, as well as variations in local dust deposition. Soil acidification is an inherent phenomenon that has substantial impacts on plant growth. Cultivating food crops gets increasingly difficult when soil acidity levels rise, particularly when the pH drops below 4.5. As soil pH decreases, most plant nutrients are lost, while aluminium and a few micronutrients become more soluble and detrimental to plants (Harter 2007). These problems are especially intense in hot and humid tropical regions with heavily weathered conditions. According to Sanchez and Logan (1992), approximately 1.7 billion hectares, which is equivalent to one-third of the tropics, have acidic soil with levels of soluble aluminium that are detrimental to most crop plants.

3.12.1 What Causes Soil Acidity?

The primary factors contributing to soil acidification include precipitation and leaching, acidic geological material, decomposition of organic matter, and the cultivation of high-yield crops. Humid places are more prone to having acidic soils. Excessive rainfall eventually washes away the essential components (calcium, magnesium, sodium, and potassium) in the soil profile that act as buffers against soil acidity. Soils derived from weathered granite are likely to be more acidic when compared to soils created from shale or limestone. Acidity arises from the presence of hydrogen ions

(H^+), which are electrically charged elements and are generated during the decomposition of organic matter. The short-term consequences of acidic soil formation resulting from the decomposition of organic waste are insignificant, much like those caused by rainfall. The primary factor responsible for the increase in soil acidity is the cultivation of high-yielding crops. Crops acquire vital nutrients such as calcium, magnesium, and potassium during their growth to fulfil their nutritional requirements. As crop yields increase, a greater quantity of lime-like minerals are extracted from the field. The grain has a significantly lower amount of these crucial components in comparison to the leaf and stem of the plant. The acidity of soil is more affected by the harvesting of high-yielding forages such as bermudagrass and lucerne, compared to the harvesting of grains. Nitrogen fertiliser use has been identified as the cause of the increasing soil acidity problems in the region. Indeed, the application of ammoniacal fertiliser elements leads to the soil becoming acidic. Nevertheless, the amount of nitrogen absorbed by the crop is comparable to the nitrogen present in the fertiliser. Nitrogen fertiliser, in reality, acidifies soil by increasing crop productivity and the subsequent removal of basic elements.

Soil acidification is a gradual and inherent process that occurs as a result of natural weathering. Farming practices often lead to accelerated soil acidity. The time scale for measuring changes in soil pHCa under agricultural usage is typically in the range of tens or hundreds of years, rather than thousands of years as observed in the natural environment. Agriculture contributes to accelerated soil acidification through four main mechanisms: (i) the use of fertilisers containing ammonium or urea, (ii) the leaching of nitrate nitrogen derived from legume fixation or ammonium fertilisers, (iii) the removal of harvested crops, and (iv) the accumulation of soil organic matter.

The concept of acidity is one of the initial chemical principles introduced in primary school science classrooms. Children are educated about the concept of pH, where a pH value of 7.0 is considered neutral. pH values above 7.0 suggest alkaline conditions, while values below 7.0 indicate acidic ones. Plant life is generally adversely affected by highly acidic and highly alkaline situations. While it is accurate, there is a scarcity of chemically neutral reactions in nature. While it is true that the majority of rain is acidic, acid rain has generated specific issues. When carbon dioxide (CO_2) combines with water in the atmosphere, pure rainwater is essentially distilled water. This water has a pH of approximately 5.6 (Harter 2007).

$$H_2O + CO_2 = H_2CO_3 = H^+ + HCO_3^- = 2H^+ + CO_3$$

The pH of pure water in equilibrium with ambient carbon dioxide is raised to 5.6 due to the concentration of hydrogen ions (H^+). Plants produce carbon dioxide as a result of their respiration mechanisms. Additionally, while plants are actively growing, their roots can significantly increase the concentration of carbon dioxide in the soil. Consequently, there is an increase in the concentration of carbon dioxide dissolved in soil water, resulting in a decrease in pH and an increase in acidity (Harter 2007). Nevertheless, hydrogen ions present in the soil originate from sources other than carbon dioxide. Fertiliser can serve as a substantial contributor of hydrogen ions in cultivated soils. Ammonium (NH_4^+), a commonly used nitrogen source in contemporary fertilisers, undergoes a conversion into nitrate (NO_3^-), which is subsequently accompanied by the liberation of hydrogen ions.

$$NH_4^+ + 2O_2 = NO_3^- + 2H_+ + H_2O$$

Put simply, the conversion of each ammonium ion to nitrate results in the creation of two hydrogen ions. Urea, a widely used nitrogen fertiliser, undergoes decomposition in soil with a pH below 6.3 according to the following reaction:

$$CO(NH_2)_2 + 2H^+ + 2H_2O = 2NH_4^+ + H_2CO_3$$

One urea molecule undergoes a breakdown process that leads to the removal of two hydrogen ions. This leads to an increase in the pH of the soil in the nearby region. However, the ammonium ion is subsequently converted into nitrate as explained earlier, and the presence of two ammonium ions leads to the release of four hydrogen ions. Thus, the administration of urea may lead to a transient rise in pH, but the overall effect will still be to acidify. If the soil pH exceeds 6.3, a similar reaction occurs, and the temporary increase in pH might result in the loss of ammonium through volatilization. Although the monocalcium phosphate [$Ca(H_2PO_4)_2$] commonly employed as a fertiliser component is less acidifying than ammonium, it can nevertheless contribute to acidity. Upon contact with water, it will undergo a chemical reaction to produce dicalcium phosphate ($CaHPO_4$) and phosphoric acid (H_3PO_4) (Harter 2007).

$$Ca(H_2PO_4)_2 + H_2O = CaHPO_4 + H_3PO_4$$

The phosphoric acid progressively releases hydrogen ions to the soil as pH is raised from about 3.0 to above 7.0.

$$H_3PO_4 = H^+ + H_2PO_4^- = 2H^+ + HPO_4^{2-} = 3H^+ + PO_4^{3-}$$

The presence of the third hydrogen is usually inconsequential as it is only relinquished at pH values that exceed neutrality. Nevertheless, the presence of two hydrogen ions is worrisome due to their susceptibility to being lost within the pH range of acidic soil. Phosphorus fertiliser is usually administered in a specific area or band that is aligned with a row of plants. The band can experience very low pH values due to the tendency of H_3PO_4 to discharge hydrogen ions. The acidity will be progressively absorbed by the soil surrounding the band. Lindsay and Stephenson (1959) reported that the pH values in the vicinity of a fertiliser belt can go as low as 1.5.

As hydrogen ions are a result of similar reactions in soil, all of these mechanisms lead to acidification. Sulphur oxidation is a significant inorganic acidification process. Indeed, when a soil's pH exceeds the desired level and a decrease in pH is necessary, sulphur is commonly employed. Landscapes that have a naturally high amount of sulphur, such as areas with mine waste and reclaimed mangrove zones, are especially susceptible to acidification (Harter 2007). For example, the reaction will take place in the presence of pyrite.

$$2FeS_2 + 6H_2O + 7O_2 = 4SO_4^{2-} + 8H^+ + 2Fe(OH)_2$$

The recorded concentrations of free H_2SO_4 can reach up to 300 cmol (H^+) kg^{-1}. It is important to note that one cmol is equivalent to 0.01 times the molecular weight, as stated by Thomas and Hargrove in 1984. For better understanding, a negative charge of 300 cmol (H^+) kg^{-1} is ten to thirty times greater than the usual negative charge present in soils. Consequently, the majority of soil particles have an insufficient amount of negative charge to attract all hydrogen ions to their surfaces. In the absence of a sufficient amount of a substance that increases soil pH, the acidity of the soil can get extremely low, as observed in drained mangrove swamps in Africa. This is caused by the presence of unbound hydrogen ions remaining in the soil solution. Various forms of organic matter can have acidifying properties, depending on the plant they are derived from. Certain plants contain significant quantities of organic acids. The organic acids inherently modify the soil's pH as they decompose (Harter 2007). Other plants are acidifying due to their low quantities of bases. Leguminous plants are more acidifying because they have a higher capacity to absorb cations compared to non-leguminous plants, specifically in terms of anions. Due to the occurrence of microbial nitrogen fixation within the plant structure, legumes have a limited uptake of nitrate from the soil. The uptake of nitrates partially offsets the uptake of base cations (Ca^{2+}, Mg^{2+}, K^+) in non-leguminous plants, leading to a decrease in the quantity of hydrogen exchanged from the root to acquire these nutrients (Harter 2007).

While acid rain does contribute to acidification, its impact on soils is often limited to those that are highly sensitive, such as acid sands with very poor cation exchange capacities. Typically, the small amount of precipitation input will have no effect due to the existing very acidic nature and abundance of sesquioxides in tropical soils. Soil acidity imposes growth and yield constraints on several crops worldwide. The chemical features of more acidic soils are primarily affected by the presence of hazardous compounds and components such as soluble forms of Al, Fe, and Mn, nitrites, and various dangerous organic acids, in addition to acidity itself. Aluminium (Al) toxicity is a significant barrier to achieving high agricultural yields on acid soils, which are present on around 40% of arable lands globally (Krstic et al. 2012). Aluminium, which constitutes the third-highest concentration in the earth's crust, poses a risk to plants when it is dissolved in soil solutions with acidic pH levels (Kochian 1995). Von Uexküll and Mutert (1995) reported that there are 3950 million hectares of land with acidic soil, out of which 15% are utilised for cultivating both annual and perennial crops. The growth of coniferous forests is sustained by the acid soils found in the northern region. These soils consist mainly of organic acids and are prevalent in the rainy northern temperate zone. Within the humid tropics, there exists a region characterised by a band of mineral acid soils located to the south. Uexküll and Mutert (1995) reported that now only 12% of the area dedicated to agricultural production is characterised by acidity. However, there is a global trend of increasing acidification of soil. Mineral acid soils are formed from parent materials that are inherently acidic and lacking in essential basic cations (such as Ca, Mg, K, and Na), or from the leaching of these elements from the soil, resulting in a decrease in pH and the soil's capacity to function as a buffer. Aluminium (Al) is capable of dissolving in soil and its concentration in the soil solution increases, leading to the presence of phytotoxic aluminium ions. Aluminium is commonly found in mineral soils in adequate quantities to maintain the soil's pH at around 4. Organic acid soils consist of significant amounts of humic acids and partially degraded plant debris. These soils typically lack aluminium buffering and can have a pH level that falls below 4 (Kidd and Proctor 2001).

3.12.2 Base Cation Depletion and Soil Acidification

The solution chemistry of aluminium is intricately linked to the chemical characteristics of acidic soils. Trivalent aluminium undergoes hydrolysis to produce both monomeric and polymeric hydroxyaluminium compounds. As an example:

$$Al^{3+} + 1H_2O\, Al(OH)_2^+ + H^+ \text{ low pH} \rightarrow Al^{3+} + 2H_2O\, Al(OH)_2^+ + 2H^+$$

$$Al^{3+} + 3H_2O\, Al(OH)_3 + 3H^+ \rightarrow Al^{3+} + 4H_2O\, Al(OH)_4^- + 4H + \text{high pH}$$

All the above products: $Al(OH)_2^+$, $Al(OH)_2^+$, $Al(OH)_3$, and $Al(OH)_4^-$ are called monomeric hydroxyaluminium species because only one Al atom is involved in those complexes.

The process of liming an acidic soil typically decreases the accessibility of essential micronutrients for plants, such as Zn, Fe, and B. Additionally, liming decreases the levels of harmful metals like Cd, Ni, and Cr. Acid soils are frequently linked to deficits in Ca and Mg. Nevertheless, the limited availability of Ca or Mg is not caused directly by soil acidity, as Ca remains the primary exchangeable cation in the majority of soils (with pH levels ranging from 4.0 to 4.5) in the mainland USA. Nevertheless, shortages of Ca and Zn frequently occur in the acidic soils of Hawaii.

3.12.3 Changing Soil pH

The soil's pH level is among several environmental factors that impact the quality of plant growth. Nutrient availability is directly influenced by the soil pH value. Plants exhibit optimal growth in specific soil pH ranges. Azaleas, rhododendrons, blueberries, and conifers exhibit optimal growth in acidic soils with a pH range of 5.0 to 5.5. Vegetables, grasses, and the majority of ornamental plants thrive on mildly acidic soils with a pH range of 5.8 to 6.5. There are three categories, namely primary, secondary, and micronutrients, that are used to define nutrients for promoting healthy plant growth. When compared to other nutrients for plants, nitrogen (N), phosphorus (P), and potassium (K) are the essential elements that are needed in comparatively large amounts. Secondary nutrients, such as calcium, magnesium, and sulphur, are as crucial for promoting healthy plant growth, although they are required in smaller quantities compared to critical nutrients. Manganese (Mn) and zinc (Zn) are micronutrients that are required by plants in very small quantities. Ensuring the soil maintains the optimal pH level facilitates the resolution of most secondary and micronutrient deficiencies. The presence of plant nutrients or the concentration of plant-toxic substances in the soil are the primary factors that significantly impact plant growth during pH extremes. In highly acidic soils, the availability and toxicity of aluminium and manganese to plants may increase. At low pH levels, the plant's availability to calcium, phosphorus, and magnesium is reduced. Phosphorus and most micronutrients become less available at pH levels of 6.5 and above.

Acidic soils hinder plant growth by decreasing the accessibility of important plant nutrients and increasing the presence of toxic ions in the soil solution, such as manganese and aluminium. These components are the primary contributors to the subpar crop performance and failure observed in acidic soils. Soil pH below 5.5 may have elevated aluminium content, which might hinder or impede root growth. Plants exhibit stunted growth and display symptoms of nutritional deficiency due to their incapacity to uptake water and nutrients. Excessive amounts of manganese hinder the regular growth of the above-ground sections of the plant, resulting in stunted growth, discolouration, and reduced productivity.

3.12.4 How Much Lime is Needed?

Although harvested crops deplete a significant amount of components such as lime each year, the soil's pH remains rather stable. This indicates that the soil possesses buffering capacity, meaning it is capable of resisting changes. The primary mechanism responsible for buffering acidic soil is the exchange of lime-like components, particularly calcium, which attach to the surface of soil particles. Calcium is the main element engaged in this exchange. When the crop removes elements from the soil solution, they are substituted by connected elements that transfer from the soil particles to the solution and restore it. Eventually, the reserve components will be depleted, leading to an increase in acidity. When applying lime, it is crucial to consider the reservoir or buffer's capacity. Clay soils, which often have a greater capacity for holding water compared to sandy soils, necessitate a larger amount of lime to attain the desired pH level within the permissible range.

The optimal pH level for a certain crop affects the need for liming. In order to manage acidity without compromising crop yields, liming to a pH of 5.5 can be employed, as grasses generally have a higher tolerance for acidic soils compared to legumes. Contrarily, legumes have a higher calcium requirement and perform optimally within a pH range of 6.5 to 7.5. The recommended pH range for nutritional availability is between 6.0 and 7.0. Table 3.2 displays the pH ranges that common field crops favour. To maintain continuous wheat growth without any legume component, it is necessary to apply a minimum of 0.5 tonnes of ECCE lime, which is equivalent to 25% of the soil test recommendation to raise the pH to 6.8. The purity of lime material is determined by the quantity of calcium carbonate it contains, as measured by its ECCE value. This value is influenced by two factors: the quantity of ground lime material and the degree of fineness of the ground lime material. As the ECCE increases, the amount of calcium carbonate also increases, while the size of the material decreases.

3.12.5 What is the Time Frame for Lime to Become Effective?

Due to the requirement of water for the lime reaction to initiate, its functionality is limited in arid soil, resulting in a delayed operation. Despite adequate soil moisture, it may take up to one year for a noticeable change in pH to occur. Mixing lime with soil enhances the efficacy of acidity neutralisation by facilitating the interaction between soil and lime particles, which is necessary for neutralisation to occur. For optimal growth of high-yielding perennial forages, it is important to regularly test the soil to identify any lime deficit. By detecting this shortfall early, the pH can be adjusted with unincorporated spread treatments. Ensuring optimal soil fertility necessitates the careful management of pH levels. Periodic soil testing provides recommendations for applying lime and indicates the pH levels of the soil. Farmers often experience reduced fodder production due to their neglect of soil acidity issues caused by lime deficiency.

3.12.6 The Effects of Soil Acidity

Farmers observe the immediate consequences of soil acidity in the form of:

(i) Decreased output and diminished income, resulting in financial losses.
(ii) Reduced harvests from acid-vulnerable crops and pastures.
(iii) Inadequate development of long-lasting grazing fields.
(iv) Lack of perennial pastures' persistence.

Acidic soils also have an impact on the community.

- The soil undergoes irreversible deterioration when the acidity seeps down to a level that cannot be practically or economically remedied. This is a gradual process that is expected to have a greater impact on future generations than on current land managers.
- The recharge of aquifers is a result of reduced water consumption by plants that are impacted by soil acidity. This can result in dryland salinity and infrastructural degradation, such as the deterioration of roadways.

- Acidic soils with annual vegetation contribute to increased soil erosion and the deposition of silt and organic matter into streams. This occurs because the predominance of annual vegetation leaves the soils exposed to erosion for a considerable portion of the year.

3.12.7 Treating Acidic Soils with Lime

The level of soil acidity directly affects both the amount and quality of a crop. An acidic soil can impede the growth of plant roots and tops, limit the availability of essential plant nutrients, diminish beneficial biological processes, and elevate the presence of harmful materials in the soil. In order to fully benefit from other costly and time-consuming soil management measures, it is essential to effectively manage soil acidity. An acid soil is a type of soil that has a low pH level, typically below 7. It is characterised by a high concentration of hydrogen ions, which makes it acidic in nature. Soil acidity is caused by the presence of soil particles that contain significant amounts of hydrogen or aluminium, or both. The soil and organic matter particles possess a negative charge that attracts and retains positively charged substances such as hydrogen, calcium, magnesium, potassium, sodium, and aluminium.

3.12.8 What is the Purpose of Limestone?

Limestone is effective in neutralising soil acidity, providing calcium and magnesium, enhancing the accessibility of certain plant nutrients, stimulating beneficial biological processes, and enhancing the composition of specific types of soil. Effective use of lime, together with other favourable soil management techniques, typically leads to higher crop yields and improved crop quality. Excessive application of lime, particularly on sandy soils with low organic matter content, can significantly decrease crop yields by limiting the accessibility of certain essential nutrients. Acidifying a soil that has been excessively limed poses a challenging task. If a soil test reveals an excessively high soil pH, recommendations will be provided on methods to decrease the pH level. What substances can be utilised as liming agents? The selection of liming material is typically based on factors such as the requirement for magnesium, its availability, cost, speed of reaction with soil, and convenience in handling and storage.

Soil acidity, which is defined by a low pH and elevated levels of aluminium (Al), is a significant limitation for crop cultivation in numerous global regions. In Hawaii, a significant number of agricultural fields have acidic soil, especially those classified as Oxisol, Ultisol, and Andisol. In order to rectify soil acidity, the application of lime (traditional approach) or bio-residuals, such as compost or crop residues (modern approach), has shown effective in reducing aluminium toxicity and enhancing crop productivity. Biochar is a stable type of charcoal that is made from natural organic materials including wood waste, corn stalks, and macadamia shells. It is generated by a high temperature, low oxygen process called controlled pyrolysis or even through natural forest fires (Berek et al. 2011). Owing to its molecular composition, biochar has greater chemical and biological stability compared to its initial carbon source, rendering it resistant to conversion back into CO_2. Consequently, biochar has the capacity to effectively store carbon for an extended period, thereby facilitating carbon sequestration. Conversely, the surface of biochar may possess numerous chemically reactive groups, including COOH, OH, and ketone, which enable biochar to effectively adsorb harmful compounds like aluminium (Al), manganese (Mn) in acidic soils, and arsenic (As) and cadmium (Cd) in soils loaded with heavy metals. Therefore, biochar has the potential to be utilised in the restoration of habitats that are unfavourable for plant growth, such as acidic soils, or pose a risk to human health due to contamination by heavy metals. The new focus on adding biochar to agricultural soils is owing to its perceived advantages in improving soil quality, increasing crop yields, and the possibility of earning carbon credits through carbon sequestration (Berek et al. 2011).

Berek et al. (2011) investigated the potential of biochar to mitigate aluminium toxicity in an acidic soil in Hawaii. The soil, namely the Leilehua series which belongs to the Ultisol order, was obtained from Waiawa, Oahu. The soil, without any changes, has a pH of 4.6, 1.9 cmolc/kg acidity, and 1.3 cmolc/kg exchangeable Al. Due to these characteristics, it is difficult for Desmodium intortum, a tropical forage legume that is particularly sensitive to soil acidity, to thrive in this soil. The presence of biochar had a positive impact on the levels of aluminium in the shoot. At all three lime levels (0, 3, and 6 tons/ha), the average concentration of shoot Al was 129 ppm (or μg/g) without biochar. In comparison, the concentrations were 85 ppm and 70 ppm at 2.5% and 5.0% biochar, respectively.

3.12.9 *Impact of Land Usage on the Labile Carbon Pool*

Sharma et al. (2014) undertook a study to evaluate the effects of various land use systems on labile carbon pools and soil organic carbon stocks in the foothill Himalayas. Soil samples were taken from four distinct land use systems, namely. The study collected soil samples from seven areas, specifically focussing on agriculture, forest, horticulture, and degraded lands. The samples were taken up to a depth of 0.5 m. The levels of potassium permanganate oxidizable (KOC) and microbial biomass carbon (MBC) followed the following pattern: forest N > horticulture N > agricultural N > degraded lands, at each of the depths. Regarding soil organic carbon (SOC), both farmed and degraded fields exhibited nearly equivalent levels. The largest soil organic carbon stocks in the top 0.5 m of soil were found in forest soils (47.5 Mg ha^{-1}), followed by horticulture systems (42.4 Mg ha^{-1}), degraded lands (36.3 Mg ha^{-1}), and agricultural lands (35.1 Mg ha^{-1}). Agricultural and degraded lands exhibited a reduction of up to 25% in soil organic carbon (SOC) stocks compared to forest soils. This suggests that deforestation or the conversion of forest land for agricultural purposes is causing a decline of up to 12.4 Mg ha^{-1} SOC over time in the top half-metre layer of soil. The carbon content in agricultural soils exhibited the most significant decline with increasing depth. In the lowest level (30–50 cm), the values for MBC and SOC were approximately 50% of those found at the surface, while KOC values were almost 50% lower. According to Sharma et al. (2014), the carbon pool known as KOC showed a strong correlation with soil textural qualities, specifically CEC. However, the primary determining factor for carbon levels in soils appeared to be land use.

Post and Kwon (2000) state that the buildup and turnover of soil organic matter (SOM) significantly impact soil fertility and ecosystem health. Additionally, this process determines whether soils contribute to or absorb carbon in the global carbon cycle. The viability of industrial systems relies on carbon dynamics, which also plays a substantial role in the global carbon cycle (Chen et al. 2004). The dynamics of terrestrial carbon reservoirs are directly controlled by the nature and kind of land use regimes. Anthropogenic activities have significantly altered the global carbon stocks and flows by modifying the land cover and land use (Bolstad and Vose 2005). The carbon cycle has been affected by the transformation of forest lands into various production systems and land uses (Ciais et al. 1995; Sellers et al. 1997). Farming or land use modifications can significantly influence soil organic matter (Howard et al. 1998). The agricultural production system has an impact on the rate of carbon loss from the soil (Buyanovsky and Wagner 1998). The Siwaliks, often known as the foothills of the Himalayas, possess a fragile ecological system. Sharma et al. (2009) found that the soils are predominantly coarse to medium in texture, with low levels of organic carbon and accessible nutrients, and a very limited ability to retain water. Approximately 80% of the precipitation occurs during the months of July and August, when powerful storms result in significant soil erosion. The erratic fluctuations in the weather result in both low and very unpredictable soil production (Sharma and Sharma 2003).

3.13 Significance of Soil pH

Soil pH is defined as the negative logarithm of the proton (H^+) activity, and it is used to measure the acidity or basicity of soil. The pH scale ranges from 0 to 14. A pH level of 7.0 is considered neutral, whereas values below 7.0 are considered acidic and those above 7.0 are considered basic. According to the USDA Natural Resources Conservation (1993), soil pH typically varies from highly alkaline (pH > 9.0) to acidic (pH 3.5). Soil pH is a crucial aspect of soil chemical properties because it regulates several chemical reactions. Specifically, pH affects the availability of nutrients by controlling their chemical forms. For example, most crop species cannot readily absorb ferric iron, making it an unavailable type of iron for them. When the pH level is high, crop plants can experience iron deficiency because ferric iron is the primary form of the nutrient at that pH (Liu and Hanlon 2012). The soil's pH level regulates the availability of nutrients, the lack or excess of nutrients, the presence of harmful substances, and the need for adding lime or sulphur. These two soil chemical properties, lime and sulphur addition, are essential for achieving the highest possible crop yield. Ronen (2007) states that maintaining a pH within this range ensures optimal absorption of essential nutrients for crop growth and development. Iron and/or manganese have limited bioavailability for most crops when the soil pH levels are 8.0 or above. Nevertheless, when the pH of the soil solution drops to 5.0 or below, the presence of aluminium, iron, manganese, and/or zinc becomes detrimental to the growth of most crops (Osakia et al. 1997).

3.13.1 Methods for Reducing Soil pH

Organic matter

Adding any type of organic material to soil can increase its acidity, but sphagnum peat and pine bark mulch have a particularly strong influence in this regard. Nevertheless, a substantial quantity of organic matter must be incorporated into the soil in order to exert a noteworthy impact on pH levels. Due to the expenses and labour involved, this approach is typically restricted to small-scale garden areas.

Ammonium fertilisers

Ammonium fertilisers consist of urea, ammonium nitrate, and ammonium sulphate. The bacteria present in the soil transform ammonium into acidic chemicals. Ammonium fertilisers offer the advantage of both fertilising plants and acidifying the soil simultaneously. Nevertheless, in order to prevent excessive application of nitrogen, it is advisable to utilise these products just for the purpose of making incremental adjustments to the soil pH. However, the use of ammonium fertilisers carries a lesser risk of excessively reducing pH compared to the use of elemental sulphur. Ammonium fertilisers can also be utilised to preserve a pre-existing low soil pH in acid-loving plants. Ammonium sulphate is an ammonium-based fertiliser that has a higher acidity per unit of nitrogen compared to other regularly used ammonium fertilisers, with a ratio of two to three times.

3.13.2 Types of Lime

The following section provides descriptions of various forms of lime that can be utilised for soil pH management and/or the adjustment of calcium and magnesium levels. The percentage of CaO and MgO in a typical batch of lime is provided after each topic. Calcitic lime, commonly known as high-calcium lime, has 50–56% CaO and 1–4% MgO. It is very soluble and is the recommended choice when the soil has low calcium and high magnesium levels. Typically, it has the highest reactivity and is the predominant type found in certain regions. Magnesian lime, commonly known as hi-mag lime, has 32–42% CaO and 5–15% MgO. It has moderate solubility and is the recommended choice when the levels of pH, calcium (Ca), and magnesium (Mg) are low. Using high-Mg liming materials for an extended length of time increases the amount of Mg in the soil and decreases the amount of Ca. This might lead to a deficiency of Ca during periods of stress. The preferred variety of lime when magnesium levels are extremely low is dolomitic lime, which contains 30% calcium oxide (CaO) and 20% magnesium oxide (MgO). Dolomitic lime exhibits the lowest solubility among the other materials. Hydrated lime, which contains 60% CaO and 12% MgO, exhibits the fastest reaction with the soil. However, unlike the aforementioned ground limestones, it does not sustain its liming activity for an extended period of time. Hydrated lime possesses caustic properties that can harm both persons and plants, therefore, those applying it must exercise caution to prevent plant burn. Reserve the use of hydrated lime for urgent situations that require swift alterations in soil pH. Gypsum does not function as a liming agent and does not have any impact on the pH of the soil. The substance is a raw calcium sulphate product mostly composed of calcium sulphate with bound water ($CaSO_4$ $2H_2O$). While gypsum does not possess the ability to neutralise soil acidity, it does serve as a valuable supply of calcium and sulphur. Fluid lime is a colloidal mixture of finely pulverised limestone in water, maybe including other dispersants. Pulverised limestone has a higher reactivity with soil compared to regular limestone. For fluid lime, all of the liming material must be able to go through a 100-mesh filter, and around 80 to 90% must be able to go through an even smaller 200-mesh screen. The principles governing the effectiveness of ground agricultural lime are likewise applicable to fine or fluid lime. Lime suspensions do not have any unique advantages compared to normal agricultural lime that has a high concentration of particles measuring 60-mesh or smaller. Pelletized lime, also known as pell-lime, refers to finely ground lime that has been compressed into pellet form for convenient use. Due of its thin texture, it will rapidly undergo chemical reactions in the soil. Unlike conventional agricultural lime, it does not have long-lasting liming effects over a span of several years.

3.13.3 Utilisation of Organic Matter and Cover Crops

Organic matter has an impact on plant growth and is often described as the substance that binds soil particles together. Additionally, it stimulates the formation of soil aggregates, hence enhancing drainage, soil tilth, and soil structure. Organic matter enhances the water retention ability of sandy and sandy loam soils. Organic matter can be incorporated into the soil by a variety of means, such as utilising green manure crops, cover crops, agricultural leftovers, animal manures, mulches, and composts. Green manure crops consist of sweet clover, alfalfa, densely planted field corn and summer plantings of soybean. Typically, these crops are ploughed under before reaching maturity. During this phase, the plants typically possess the highest concentration of nitrogen and other essential nutrients, along with a sufficient level of moisture to facilitate rapid decomposition. Nevertheless, green manure crops can also be incorporated into the soil when they have reached the mature dry stage. At that point, they do not break down easily and more nitrogen may be required to assist in the decomposition process. Growers typically cultivate cover crops after the harvest to safeguard the soil from erosion, and they normally plough the cover crops under the soil in the subsequent spring. Supplemental nitrogen may be required to accelerate the breakdown of the cover crop. It is particularly crucial to plough under rye before it reaches a height of 18 inches. Various cover crops often necessitate certain soil conditions to achieve optimal development. Alfalfa thrives in well-drained soils, whereas Ladino clover can tolerate poorly drained soils. Certain crops, like rye, possess fibrous root systems, while others, such as sweet clover, have substantial taproots that can penetrate the soil to significant depths. When it is feasible to utilise a blend of these crops, the amalgamation yields a greater amount of organic matter to be incorporated into the soil by ploughing. The impact of soil pH on the development and growth of crops is examined in section.

3.13.4 Effects of pH on Crop Growth

Cations and anions nutritional effects

Soil pH determines the solubility and bioavailability of nutrients essential for crop production. Seventeen components are necessary for the regular growth and development of vegetable crops. The source indicates that the seventeen nutrient elements can be categorised into two main groups. The first group consists of fourteen soil nutrients, namely nitrogen, phosphorus, potassium, calcium, magnesium, sulphur, iron, zinc, copper, manganese, boron, chlorine, molybdenum, and nickel. The second group includes three nutrients obtained from air and water, which are carbon, hydrogen, and oxygen. All soil nutrients exist in ionic form, with some being anionic (carrying a negative charge, such as nitrate ions), some being cationic (carrying a positive charge, such as ammonium ions), and some having both charges. Common cation nutrients are K, Ca, Mg, Fe, Zn, Cu, Mn, and Ni, while common anion nutrients are P, S, Cl, and Mo. Nevertheless, N can function as either an anion or a cation. At pH 7.5 or lower, less than 2% of element B exists as the anion $B(OH)_4^-$, whereas the rest of the element remains in its undissociated forms, boric acid (H_3BO_3) or $B(OH)_3$. The soil's pH directly affects the solubility, or bioavailability, of each of these fourteen essential components. Iron (Fe), copper (Cu), manganese (Mn), and zinc (Zn) exhibit high solubility when the pH is below 5.0. At a pH that is below the normal range, these small amounts of essential nutrients can combine with phosphate, causing the phosphorus to become inaccessible. Nevertheless, Ca and Mg exhibit a significant solubility and are capable of sequestering P at pH levels over 7.0. In general, vegetable plants efficiently utilise all nutrients within the pH range of 5.5 to 7.0.

3.13.5 Modifications to Nutrient Uptake in the Vicinity of the Root Zone

The alteration of soil particle characteristics by soil pH affects the nutritional intake of vegetable plants. If the soil pH is unfavourably low, the positive charges on the surfaces of soil particles can strongly bind nutrients such as phosphorus (P), potentially causing a deficiency of P in vegetable plants. Vegetable plants may face difficulties in utilising Fe, Mn, and Zn when the soil pH is excessively high. Based on a study conducted by Thomson et al. in 1993, it was shown that bean plants (*Phaseolus vulgaris* L.) absorbed 93.3% more phosphorus (P), 53.8% more iron (Fe), and 44.1% more zinc (Zn) at a pH level of 5.4 compared to a pH level of 7.3. The absorption of P, Fe, and Zn is more efficient at pH 5.4 compared to pH 7.3, resulting in increased bioavailability of these elements.

3.13.6 Alterations to the Toxicity of Metals

Metal poisoning occurs when the soil pH is below 5.0 and metals such as Al, Fe, Mn, and Cu are more soluble than what plants need. The addition of lime is necessary to increase the pH of the soil and minimise the risk of toxicity, so preventing this problem.

3.13.7 Implications for Plant Pathogens

Soil-borne diseases have a direct correlation with the pH of the soil. For instance, the clubroot disease caused by Plasmodiophora brassicae affects mustard, cabbage, and other crucifers. This disease becomes a severe epidemic when the soil pH is below 5.7, but its occurrence is greatly reduced within a pH range of 5.7 to 6.2. When the soil pH exceeds 7.3, this disease is nearly eliminated. Similarly, the occurrence of common potato scab is promoted when the pH levels exceed 5.2, but it is significantly inhibited when the pH levels drop below 5.2 (Kioke et al. 2003). Table 3.5 displays the optimal pH range for the growth of various crops.

3.13.8 Bioavailability of Nutrients

The limited availability of nutrients in commercial crops is often caused by solubility limits or the immobilisation of plant nutrients by soil colloids. Bioavailability refers to the proportion of a nutrient that can be dissolved or made mobile by root exudates, such as protons (which are linked to soil pH), chelates, mucilage and mucigel, or

Table 3.5 pH range for horticultural and vegetable crops

Crop	pH	Crop	pH	Crop	pH	Crop	pH
Apple	5.6–7.5	Egg plant	5.5–6.5	Pepper	5.5–6.5	Sweet potato	5.0–5.9
Asparagus	6.0–8.0	Endive	6.0–7.0	Potato	4.5–6.5	Tomato	5.5–6.8
Beet	6.0–7.5	Kale	5.3–7.0	Pumpkin	5.5–7.5	Water melon	5.5–6.5
Blueberry	4.5–5.5	Lettuce	6.0–7.0	Radish	5.5–6.5	Alfalfa	6.5–8.4
Cabbage	5.8–7.0	Lima bean	5.5–6.8	Snap bean	5.5–6.5	Corn	5.5–8.4
Carrot	5.8–7.0	Muskmelon	6.0–7.5	Spinach	6.0–7.0	Garlic	6.5–8.4
Cauliflower	6.0–7.0	Mustard	5.5–6.5	Squash	5.5–6.5	Small grains	5.5–8.4
Celery	6.0–7.0	Onion	6.0–6.7	Strawberry	5.2–6.5	Pastures	5.5–8.2
Cucumber	5.5–6.8	Pea	6.0–7.0	Sweet corn	5.5–7.5	Turf grass	5.5–8.0

Source Liu and Hanlon (2012)

microbial products (Neumann and Romheld 2012). As an example, the Everglades Agricultural Area has soil that has been cultivated and has a total phosphorus (P) level of up to 1227 parts per million (ppm). However, only 1.3 ppm of that phosphorus is available for biological processes (Wright et al. 2009). The bioavailability of phosphorus in that specific soil is only 0.1% of the overall amount. P deficit, in this context, specifically denotes an insufficiency of absorbable or utilisable phosphorus for crop plants, rather than a deficiency of phosphorus in the soil itself. Soil pH effectively controls the availability of most nutrients. The bioavailability of P, Fe, Mn, B, Zn, and Cu decreases when soil pH increases. The bioavailability of Ca, Mg, and Mo decreases when the soil pH decreases. Vegetable plants necessitate substantial quantities of macronutrients such as nitrogen (N), phosphorus (P), potassium (K), calcium (Ca), magnesium (Mg), and sulphur (S). Additionally, they require small amounts of micronutrients or trace elements including iron (Fe), manganese (Mn), boron (B), zinc (Zn), copper (Cu), chlorine (Cl), molybdenum (Mo), and nickel (Ni). The soil pH (Fig. 3.3) affects both the solubility and bioavailability of macronutrients and micronutrients. Ferrous iron undergoes rapid oxidation and is limited in dry soil, particularly in soil with a pH of 7.3 or higher. As a result, ferric hydroxide ($Fe(OH)_3$) predominates as the primary form of iron in dry soil. The solubility of ferric hydroxide in water is limited to only 6.3×10^{-20} mol/L, which is equivalent to 1.34×10^{-11} mg of iron per 1000 L of water at pH 7.3. However, when the pH is 5.3, the solubility of the substance is 1.34×10^{-5} mg of Fe per 1000 L of water. A reduction in soil pH by just three pH units results in a million-fold increase in solubility. The sudden shift in solubility could potentially explain the frequent manifestation of iron shortage symptoms in soils with a pH of 7.3 or above. When the soil is adequately moist and has a neutral or slightly acidic pH, there is usually an adequate amount of ferrous iron available to fulfil the iron needs of crops.

3.13.9 Soil pH and Absorption of Cation and Anion Nutrients

When the pH of the soil is below 3.0, the hydrogen ion in low-pH soils is present as a hydrated proton and can be dangerous (Liu et al. 2007). Nevertheless, the increasing solubility of toxic elements such as aluminium (Al) in the soil results in mostly indirect effects on nutrient intake. The solubility of aluminium is also influenced by the pH of the soil. As the soil's pH decreases, the solubility of Al increases. The solubility of Al is significantly higher at pH 5.5 or below compared to any other pH level, with a difference of 1000-fold. The solubility of Al, for example, is only 0.05 parts per million (ppm) at pH 5.0, but it increases significantly to a lethal 51 ppm at pH 4.0. Elevated concentrations of aluminium can have detrimental effects on root structure and lead to a deficiency of phosphorus in the soil. According to Lidon and Barreiro (1998), maize roots can be harmed or experience slowed growth when the content of aluminium exceeds 9 parts per million (ppm). The detrimental effect on plants is evidence of the toxicity of Al. Aluminium and phosphate form a solid substance in soil with low acidity. Al and P have a mutually supportive relationship. As previously stated, the solubility of aluminium (Al) is 1000 times higher at a pH of 4.0 compared to a pH of 5.0. The bioavailability of phosphorus (P) at pH 4.0 reduces to one thousandth of the concentration present at pH 5.0 due to an increase in aluminium (Al) concentration, which was caused by the rise in aluminium. Figure 3.3 demonstrates that other factors have similar impacts. The absorption of cation nutrients by soil particles can be replaced by protons in the soil solution, which is why low pH levels worsen the problem of nutrient leaching. Nutrient leaching reduces the efficiency of nutrient utilisation in many crops.

Modifications to the process of nutrient absorption in the vicinity of the root zone

Root development and absorption of water are hindered when there are high levels of harmful elements such as aluminium at low pH. This can cause plants to show symptoms of phosphorus deficiency and drought stress. Plants experiencing aluminium-induced stress exhibit impaired nitrogen absorption from the soil. There are two reasons why the absorption of cation nutrients might be inhibited and nutrient scarcity can arise. First, there may be a decrease in the excretion of protons. Second, the amounts of cation nutrients such as calcium, magnesium, zinc, and manganese in the soil solution may diminish, making them less accessible.

The impact of pH on microbial activity in soil

The pH level has a direct influence on the bioavailability of both macronutrients and micronutrients, as well as on microbial activity. Soil microorganisms thrive in a slightly acidic pH range (6–7) because most nutrients are highly available in that range (Sylvia et al. 2005). Under such conditions, vegetable crops can thrive due to the ability of microbes to enhance the availability of nutrients and promote the absorption of nutrients by plants (Das et al. 2010).

Nutrient sources affect the pH of the soil in the root zone

Acid-forming fertilisers are substances that, when absorbed by plants, lower the pH of the rhizosphere. Fertilisers containing cation nutrients, such as ammonium-N, potassium, calcium, and magnesium, have an acidifying effect. On the other hand, fertilisers containing anion nutrients, such as nitrate N, P, and sulphur, have an alkalizing effect. Ammonium chloride, potassium chloride, calcium chloride, and magnesium chloride are examples of fertilisers that have an acid-forming effect. Some examples of fertilisers that have alkaline properties are sodium nitrate, sodium dihydrogen phosphate, and sodium sulphate. The acidity or basicity of applied fertilisers is not influenced by whether the fertiliser itself is acidic or basic. The crop plants' preferential absorption of nutrients results in either acidity or alkalinity. For example, potassium chloride possesses a chemically neutral composition. Both potassium and chlorine (Cl) are essential for the growth and development of vegetable crops. Nevertheless, plants require a higher proportion of potassium (K) to chlorine (Cl) than 80. Based on this proportion, plants must absorb more than 80K^+ ions for every one Cl^- ion they absorb. These two nutrients can be either positively or negatively charged. If plant cells were to accumulate these two types of cation and anion ions without undergoing electrical neutralisation, they would accumulate significant positive charges. The presence of these disproportionate charges can rapidly cause cell death. Plant cells have developed two mechanisms to prevent this. In the initial approach, upon receiving K, they release charges of the same nature, such as protons (H^+), in a stoichiometric manner. The cells can neutralise imbalanced charges by absorbing an equivalent amount of ions with countercharges, such as OH^- or HCO_3^-. This occurs when they absorb K^+ ions. Regardless of the approach taken, the ultimate outcome remains consistent: the pH of the growth media is reduced, particularly in the area surrounding the roots. Like potassium nitrate, sodium nitrate is chemically inert. However, since nitrate (N) is negatively charged and serves as the primary nutrient for crop production, while sodium is not essential for crop growth and development, the pH in the root zone rises when plants take up nitrogen from sodium nitrate. Thus, if needed, the pH of the soil in the root zone can be effectively modified by deliberately selecting fertilisers, such as potassium chloride or sodium nitrate.

Soil pH and nutrient depletion

Ammonia volatilization is a process that occurs when ammonium-N, which is one of the two primary forms of commercial nitrogen fertilisers, is released into the atmosphere. Ammonium and ammonia can reach a state of dynamic chemical equilibrium in a soil solution. The pH of the soil solution dictates the direction in which the chemical equilibrium between ammonium and ammonia is altered. Ammonium and ammonia have comparable concentrations at a pH of 9.2. Ammonium exists in a state of being dissolved in water, while ammonia in solution can exist in both a dissolved state and a gaseous state. Ammonia exhibits a solubility of 31% in water at a temperature of 77 °F (25 °C). Converting this dissolved ammonia into gaseous ammonia is a straightforward process, after which it is released into the atmosphere. Ammonia volatilization refers to the release of this gas. Soil pH is the primary factor that affects the extent of ammonia volatilization. Ammonia volatilization occurs when fertilised soils are treated with ammoniacal-N sources such as ammonium sulphate or urea, particularly when the soil pH is elevated (above 7.2). The rate of ammonia volatilization in Homestead, Florida's Krome Very Gravelly Loam soil, during potato cultivation, was recorded to be as high as 26% (Liu et al. 2007). Leaching of anionic nutrients occurs when the pH of the soil exceeds 7.0, causing hydroxide ions to displace anionic nutrients that carry positive charges. This displacement reduces the capacity of soil particles to retain anionic nutrients. Costa and Seidel (2010) found that there is a direct correlation between soil pH and the increase in nitrate leaching. Consequently, increased soil pH exacerbates the leaching of anionic nutrients and reduces the efficiency of fertiliser utilisation. In order to enhance the profitability of vegetable crops and minimise leaching problems, it is crucial to maintain the soil pH at an appropriate level. Aside from soil pH, the ionic charge of micronutrients, including those with multiple charges such as Mn and Fe, is also significant. The ionic charge of micronutrients is often influenced by oxidation–reduction potential and microsite conditions. Iron (Fe) and Manganese (Mn), for example, exhibit higher bioavailability in adequately moist soil (between field capacity and wilting point) compared to dry soil due to the decreased oxidation–reduction potential of wet soil. The oxidation–reduction potential increases as the pH of the soil increases. The occurrence of iron (Fe) or manganese (Mn) deficiency in high pH soils can be attributed to the influence of pH levels over 7.0 and dry soil moisture conditions, which promote a deficiency.

Nutritional efficiency

Nutrient utilisation efficiency is defined as the amount of vegetable production obtained per unit of nutrient intake. Due to the decrease in profit margins and the rise in fertiliser prices, it is now more imperative than ever. By determining the productivity of individual units of a certain nutrient, one can compute the efficiency of nutrient utilisation. In 2012, two research studies on snap beans were carried out in Belle Glade and Lake Harbour, both located in Palm Beach County. The findings from the two tests indicated that the optimal phosphorus rate was 120 pounds of phosphorus pentoxide (P_2O_5) per acre. The experiment locations had an impact on the efficacy of P in snap bean production. To increase the pH of the soil when it is severely low, it is typically essential to add agricultural lime. Calcium (Ca), magnesium (Mg), and molybdenum (Mo) exhibit restricted bioavailability, which may adversely affect the yield of vegetables. The previous examination of the effects of toxicity may also play a part. An elevated soil pH can enhance nutrient availability and mitigate toxicity.

3.13.10 Lime Demand

Agricultural lime is the most commonly used soil amendment to increase soil pH. It is usually finely pulverised. The size of the limestone particles used and the soil's buffering capacity determine the amount of lime required to increase the soil's pH. The soil's buffering capacity mitigates the extent to which the acidity of a solution changes when an acid or base is added. The rate of the neutralisation reaction increases as the particle size of the ground lime decreases. The soil's buffering capability is influenced by the quantity of organic matter and clay present. Soils that have a higher proportion of clay have a greater buffering capability compared to soils with a lower clay content. Soils that contain higher amounts of organic matter have a better capacity to act as a buffer compared to soils with smaller amounts of organic matter. Soils with high buffering capabilities necessitate a greater amount of agricultural lime compared to soils with low buffering capacities, even for the same incremental change in soil pH. In contrast to clay soils, sandy soils have a lower buffering capacity and need less lime to achieve the same pH change. Calcium silicate can be used to neutralise active acidity in soil. All local organic sources, such as sphagnum moss peat and garden waste compost, have an acidic pH. It is possible to have pH levels as low as 3.6–4.2. These sources can be utilised to neutralise free hydroxide and/or bicarbonate ions.

3.13.11 Utilise Nitrogen Fertilisers Containing Nitrates

Liming has the ability to modify the pH of the entire soil profile. Vegetable crops require electrical equilibrium after assimilating negatively charged nitrate ions. Consequently, the utilisation of nitrate nitrogen fertilisers can elevate the pH of the root zone without incurring additional expenses. This process operates at a sluggish pace throughout the entire soil profile, as nitrogen (N) needs to be supplied in accordance with recommended fertiliser levels. However, it rapidly increases the pH in the area surrounding the plant roots. The table provides a comprehensive summary of the symptoms associated with nutritional deficiency. Table 3.6 gives the deficiency symptoms due to the shortage of different nutrients in the soil.

3.13.12 Effect of Soil Acidity on Plant Nutrient Availability

Soil acidity affects nutrient availability and microbial activity as well as plant growth. Figure 3.6 shows the relationships between soil pH and plant nutrient availability. Each nutrient is represented by a band indicating its availability across the normal pH range of soils. When the band is narrow, a nutrient is relatively unavailable, while a wide band indicates a high availability. Plant nutrients such as nitrogen (N), phosphorus (P), potassium (K), sulfur (S), calcium (Ca), magnesium (Mg) and molybdenum (Mo) have low availability at strongly acidic pH values. Other nutrients such as manganese (Mn), copper (Cu) and zinc (Zn) tend to be more available until the soil gets very acid (less than pH 5). Iron (Fe) and aluminium (Al) availability increases as soil acidity increases and Al becomes toxic to plants at pH values less

Table 3.6 Nutrient deficiency symptoms

Element	Symptoms
Nitrogen	Thin, erect, and stiff stems. Leaves are tiny and golden, with crimson undersides on some crops (tomatoes). Lower leaves are the first to be impacted on sandy soils, particularly after heavy rain or overirrigation. During cool growing seasons, it can also be found on organic soils
Phosphorus	The stems are slender and shortened. The leaves turn purple. The older leaves were the first to be impacted. Plants become stunted and mature later. On acidic or highly basic soils. Also when the earth is chilly and damp
Potassium	On the leaf margins of older leaves, grey or tan patches appear. A scorch eventually forms along the entire margin. Following leaching rains or overirrigation on sandy soils
Magnesium	Magnesium (Mg) is a mineral. Yellowing between veins appears first on older leaves, then on new leaves. Older leaves fall off quickly. On highly acidic soils or leached sandy soils
Sulphur	Yellowing of younger leaves and growth in general. On very sandy soils with limited organic matter, especially with continued use of sulfur-free fertilisers and in places with little atmospheric sulphur
Zn	Tiny rust-coloured dots on the cotyledon leaves of the bean plant and lighter areas (white bud) on the leaves of the maize plant. When the soil is cold and moist in the early spring, or when there is an overabundance of phosphorus present
Mn	A mottled yellow appearance between the veins of the youngest leaves; not as severe as iron insufficiency in soils with a pH greater than 6.4
Cu	New leaves turning yellow and plants becoming stunted as a result. Onion bulbs are tender and have scales that are very thin and pallid
Fe	Clearly visible patches of yellow or white between the veins on the youngest leaves. where the pH of the soil is more than 6.8
Mo	Cauliflowers infected with whiptail disease have leaves that are pale, deformed, and narrow, and the older leaves have some interveinal yellowing. Rare. on soils that are extremely acidic
B	The growing tips of the plant perish, and the leaves become misshapen. Boron deficiency can lead to a number of specific diseases, the most common of which are brown curd and hollow stem of cauliflower, cracked stem of celery, blackheart of beetroot and internal browning of turnip. on soils with a pH more than 6.8, as well as on sandy, leached soils, or on crops with an extremely high demand, such as cole crops
Ca	The growth of growing points was restricted on the shoots and the roots. Specific deficiencies include blossom-end rot, which can affect tomatoes, peppers, and watermelons; brownheart, which affects escarole; blackheart, which affects celery; and tipburn, which affects cauliflower or cabbage. On soils that are highly acidic, or when there is a severe lack of moisture
Chlorine	Deficiencies extremely rare. In most cases, only in the conditions of a laboratory

than 5. Activity of N-fixing organisms such as rhizobia related to legumes decreases as soils become more acid but fungi are tolerant of soil acidity. Legume crops (alfalfa, clovers, soybeans and dry beans) are generally more sensitive to low soil pH. Activity of some herbicides is increased and decreased in others as soils become more acid, creating problems with effectiveness or carryover to following crops.

3.13.13 Aluminium Effects on Plants

Al^{3+} bonds with P to form less available P, causing P deficiency symptoms. Therefore, plants exposed to toxic Al levels show poor growth, water stress and nutrient deficiencies. A visual examination of the plant roots will show a poorly developed root system as well as poor root nodulation in legumes. Most crop species are unaffected by free Al^{3+} down to pH 5, but below that pH, plants may be affected either directly by the Al^{3+} or by its interaction with availability of plant nutrients such as P resulting in decreased growth and productivity. Often this occurs without obvious visual plant symptoms.

Soil acidity is a common characteristic of the humid and subhumid tropics. Acid soil infertility encompasses several soil chemical and biological functions: aluminium toxicity, calcium and magnesium deficiencies and low populations of symbiotic nitrogen-fixing bacteria and mycorrhizae (Sanchez 2019). To a lesser extent, acid soil infertility also involves manganese toxicities and molybdenum deficiency as well as hydrogen toxicity in Histosols. Soil acidity often occurs together with phosphorus deficiency. The chemistry of soil acidity is complex (van Raij 1991; Sumner 1997). Ultisols of the southeastern United States, similar to the tropical soils of this order, showed that additional H^+ produced by acid rain had no significant effect on soil pH or base saturation (Richter 1986). Acid rain produced by oil refineries and coal-burning plants that use materials rich in sulfur are still abundant in the tropics and can result in acid rain downwind. Soils that are most susceptible are sandy, low in SOM and effective cation exchange capacity (ECEC), and high in base saturation (Rechcigl and Sparks 1985).

- Root absorption of cations in excess of anions.
- Secondary acidity from long-term applications of ammonium-containing fertilisers.
- Oxidation of added sulphur, also causing secondary acidity. The resulting H^+ ions react with aluminium-containing soil minerals and produce exchangeable aluminium ions (Al^{3+}). Exchangeable Al^{3+} ions displace basic cations such as calcium (Ca^{2+}), magnesium (Mg^{2+}), potassium (K^+) and sodium (Na^+) ions from the cation exchange sites in clay and SOM surfaces into the soil solution. These basic cations can leach out of the soil. In variable-charge soils, this process has taken place through millennia in high-rainfall areas (also under previous climates in currently semi-arid and arid areas), and often the soils are naturally depleted of basic cations. In most Andisols, it is a relatively recent process. Soil acidity

is different from the acidity of a liquid, where it is defined as acid when the pH is below 7.0, where H^+ concentrations are larger than those of hydroxide (OH^-) ions. The work of Coleman and Thomas (1967) and colleagues in North Carolina proved that exchangeable Al^{3+} is the dominant cation associated with soil acidity. Hydrogen ions are unstable in exchangeable form because they attack aluminium octahedral layers of 1:1 or 2:1 clay minerals as well as aluminium oxides and hydroxide coatings, releasing exchangeable Al^{3+} which is held by the ECEC, eventually going into the soil solution by cation exchange. Likewise, OH^- ions generated by liming are also unstable because they react with other soil constituents. Some processes and equations involved are shown in Table 3.7.

Table 3.7 Some processes and general equations on acid soils

Process	Equation
Generation of primary	**Acidity**
CO_2 Hydrolysis	$CO_2 + H_2O \rightarrow H_2CO_3 \leftrightarrow H^+ + HCO_3$
Decomposition of litter and SOM (respiration)	$[RCH_2OH] + O_2 + H_2O \leftrightarrow RCOOH \leftrightarrow RCOO + H^+$ SOM organic matter generalised strong organic acid
Nitrification	$NH_4{}^+ + 2O_2 \rightarrow NO_3{}^- + H_2O + 2H^+$
Atmospheric decomposition	$H_2SO_4 \rightarrow 2H^+ + SO_4{}^{2-}$(Same with HNO_3)
Generation of Al ions	
Exchangeable Al^{3+} formation and Al^{3+}in soil solution	$3H^+ + \underset{\text{Al in octahedral layer}}{[3O{-}Al{-}3OH]} \rightarrow \underset{\text{Exchangeable Al}}{Al^{3+}[^{3-}] + 3H_2O} \leftrightarrow \underset{\text{Soil solution}}{Al^{3+}}$
	$Al^{3+} + H_2O \rightarrow Al(OH)^{2+} + H^+$
Al^{3+} hydrolysis	$Al(OH)^{++} + H_2O \rightarrow Al(OH)_2{}^{2+} + H^+$
	$Al(OH)_2{}^+ + H_2O + (Al\ OH)_3\downarrow + H^+$
Liming	
Liming 1	$CaCO_3 + H_2CO_3{}^- \rightarrow Ca(HCO_3)_2$
Liming 2	$3Ca(HCO_3)_2 + 2Al^{6+}[^{6-}] \rightarrow 3Ca\ 1^{6+}[^{6-}] + 2Al(OH)_3\downarrow +$ $6CO_2$
Generation of secondary	**Acidity (Oxidation)**
Anhydrous ammonia	$NH_3 + 2O_2 \rightarrow H^+ + NO_3{}^- + H_2O$
Urea	$CO(NH_2)_2 + 2H_2O \rightarrow (NH_4)_2CO + 4O_2 \rightarrow 2H^+ +$ $2NO_3{}^- + CO_2 + H_2O$
Ammonium sulphate	$(NH4)_2SO_4 + 4O_2 \rightarrow 4H^+ + 2NO_3 + SO_4{}^{2-} + 2H_2O$
Diammonium phosphate	$(NH4)_2HPO_4 + O_2 \rightarrow 3H^+ + 2NO_3{}^- + H_2PO_4 + H_2O$
Sulphur oxidation	$S + 1.5O_2 + H_2O \rightarrow H_2SO_4 \rightarrow 2H^+ + SO_4{}^{2-}$

Source Sanchez (2019)

3.13.14 Exchangeable H^+

Exchangeable H^+ is found in small amounts in acid mineral soils. But in Histosols, exchangeable H^+ can be stable because it is held by carboxyl groups of SOM. Also, H^+ causes direct injuries to plant roots, which decreases the uptake of cations below pH 4.2 (Kamprath 1984), but they are rapidly converted into exchangeable Al^{3+}.

3.13.15 Exchangeable Al^{3+}

Aluminium, the third most common element in soils as well as in the Earth's crust, is not an essential element for plants and animals, and at high concentrations can be toxic to plants and animals, including humans. In fact, most natural organic compounds, such as leaf litter and manures, contain essentially no aluminium in their pure form. The exceptions are those few plant species that accumulate aluminium in their tops, such as ferns and some trees. Exchangeable Al^{3+} is determined by extracting soils with unbuffered 1 M KCl and titrating the extract with a base (Lin and Coleman 1960). Any exchangeable H^+ present is included in this measure. Some call this exchangeable acidity (Al^{3++} H^+), or KCl^- extractable acidity, but some prefer the term exchangeable Al^{3+} with the understanding that it includes whatever H^+ is present, which is small, except in Histosols (Sanchez 2019). Exchangeable Al^{3+} is precipitated at pH values of about 5.5. Thus, little or no exchangeable Al^{3+} is found at higher soil pH values, as demonstrated in Table, which occur at progressively higher pH values. SOM can decrease the concentration of exchangeable Al^{3+} and aluminium in the soil solution by forming chelates with organic acids (Kamprath 1984; Keltjens 1997). This is why humic subgroups of Oxisols, Ultisols and Alfisols have less exchangeable Al^{3+} than typic subgroups. Humic subgroups are usually found in tropical highlands and subtropics. Lower soil temperatures favour the accumulation of SOM.

3.13.16 Aluminium Saturation

In addition to actual exchangeable Al^{3+} values, a very useful metric of soil acidity is the percent aluminium saturation of the effective cation echange capacity (ECEC).

$$\mathrm{ECEC} = \mathrm{Ca}^{2+} + \mathrm{Mg}^{2+} + \mathrm{K}^{+} + \mathrm{Na}^{+} + \mathrm{Al}^{3+} + \left(\mathrm{H}^{+}, \text{if present}\right)$$

All are expressed as cmolc/kg of soil. Sodium ions are not worth measuring in acid soils. Aluminium saturation is calculated by dividing the exchangeable Al^{3+} by the ECEC:

$$\text{Percent aluminum saturation} = \left(\text{Exchangeable Al}^{3+}/\text{ECEC}\right) \times 100$$

Since aluminium is precipitated at a pH of about 5.5, soils are essentially base-saturated above such pH levels. Base saturation is defined as follows:

$$\text{Percent base saturation} = \left(Ca^{2+} + Mg^{2+} + K^{+} + Na^{+} + Al^{3+}\right)/\text{ECEC} * 100$$

All units are also in $cmol_c$/kg of soil.

3.13.17 *Aluminium in the Soil Solution*

Exchangeable Al^{3+} is held tightly to the negative charges on clay surfaces. Work in North Carolina (Nye et al. 1961; Evans and Kamprath 1970), Guyana (Cate and Sukhai 1964) and Brazil (Gonzalez-Erico et al. 1979) showed that there is less than 3 ppm (0.3 cmolc/L) aluminium in the soil solution when the aluminium saturation is below 40–60%. Aluminium in the soil solution rises sharply above 60% aluminium saturation. In highly oxidic soils (with gibbsitic, ferritic, sesquic, ferruginous and allitic mineralogy), the relationship between aluminium saturation and aluminium in the soil solution is a linear one, resembling the solubility of gibbsite (Ayers et al. 1965). The levels of aluminium in the soil solution also depend on the soil-organic-matter content and the salt content (Kamprath 1972). Aluminium in the soil solution decreases as SOM increases because organic matter radicals (mainly carboxyls) form strong complexes with aluminium, as mentioned before. Aluminium in the soil solution increases with increasing salt content (after soluble fertiliser applications or during dry spells) because other cations then displace exchangeable Al^{3+} by mass action (Brenes and Pearson 1973). Aluminium is toxic to plant growth at concentrations of about 4 ppm (mg/kg) in maize (Kamprath 1984). Since there is a close relationship between soil solution aluminum and percent aluminium saturation, the latter metric is widely used to express plant tolerance levels and calculate the amounts of liming to correct it. The tipping point in general is 60% aluminium saturation, and that is used to define aluminium-toxic soils.

3.13.18 *Buffering Capacity*

Buffering capacity is the soil's resistance to changes in pH when an acid or a base is added. Buffering capacity increases with increasing clay, SOM and ECEC (Thomas and Hargrove 1984; Ngatunga et al. 2001). In the FCC system, the C soils are the most buffered while the Se soils are the least buffered. Soils are more buffered than aquatic systems. Buffered solutions at pH 7 or 8.2, are used to estimate the lime requirement (Kamprath 1984; Ngachie and Smyth 1989). Most buffer methods seek to obtain a soil pH of 6 to 7, the rationale being that this pH range is suitable for optimal plant growth in mineral soils.

3.13.19 Non-exchangeable Acidity

Residual, total, titratable, buffer-extractable or nonexchangeable acidity is the amount of H^+ ions extracted with buffered solutions at pH 7 and above. Kamprath (1984) indicates that one reason why the ECEC is low in acid soils is the large proportion of negative charges in SOM that are complexed with Al^{3+}. This complexed Al^{3+} is not exchangeable with other cations and represents an important buffer. In Brazil, non-exchangeable acidity is considered a component of lime recommendations (van Raij et al. 1997). Non-exchangeable acidity has no detrimental effect on plant growth (Kamprath 1972, 1997).

3.13.20 Distribution of Acid Soils

Aluminium-toxic soils—defined as having 60% or more aluminium saturation in the top 50 cm in the FCC system, cover about 15% of the world, most of it in the tropics. Over one-quarter of tropical land (about 1.4 billion hectares) is aluminium-toxic to most crops. Aluminium-toxic soils are also found in about 8% (500 million hectares) of the temperate region, mostly in the southeastern United States, southeastern China and subtropical southern Brazil. Tropical America has a much higher proportion (43%) of aluminium-toxic soils than tropical Africa and Asia. In Africa, these soils are concentrated in Rwanda, Burundi, the Congo rainforest, KwaZulu-Natal Province of South Africa and small areas elsewhere. The vast majority of African smallholder farmers luckily do not have to cope with this problem. Indonesia, Malaysia and the hill country of mainland Southeast Asia harbour the majority of aluminium-toxic soils in tropical Asia. Tropics-wise, the vast majority of them are in udic and ustic soil moisture regimes. Aluminium toxicity is highly prevalent in the tropical– subtropical moist broadleaved forest biomes and also prevalent in tropical–subtropical savannas. It is usually the overwhelming constraint to crop production in these soils, but there is a plethora of trees and grasses that thrive in aluminium-toxic natural systems. Hydrogen toxicity can exist in organic soils, those that have more than 12 percent organic carbon in the topsoil. This does not happen when the subsoil is calcareous, like many Histosols of Florida, but it happens in other organic soils, particularly when drained. Histosols occupy only 400 million hectares in the world, mostly in temperate and boreal regions, and only 45 million hectares (1%) in the tropics. They are most prevalent in tropical Asia, particularly along the coasts of Indonesia and Malaysia.

3.14 Acid Soil Infertility

Aluminium toxicity

Hardy (1926) was probably the first tropical soil scientist who warned about the toxic effects of aluminium in acid soils. Early studies by Abruña et al. (1970) on tobacco, Villagarcía (1973) on potatoes and Gonzalez-Erico et al. (1979) on maize show that the primary effect of aluminium toxicity is direct injury to the root system. Root development is restricted and the roots become thicker, stubby and show dead spots. These studies, as well as many similar ones conducted in the temperate region, also indicate that aluminium tends to accumulate in the roots and impede the uptake and translocation of calcium and phosphorus to the plant tops (Foy 1974). Thus, aluminium toxicity may produce or accentuate calcium and phosphorus deficiencies.

Calcium deficiency

Although aluminium toxicity is the primary culprit for soil acidity, calcium and magnesium deficiencies are probably the second most common causes. When aluminium was precipitated by liming to pH 5.6 with $MgCO_3$, maintaining calcium at the low levels, root growth stopped within 60 hours. When aluminium was precipitated and the calcium level increased to 4.4 cmolc Ca/kg by liming with $CaCO_3$, root growth progressed normally (Sanchez 2019). Symptoms of calcium deficiency in plants are difficult to see. In maize, it shows as the failure to develop leaves that fully separate from the stem. Critical levels for most crops are about 1.0 cmolcCa/kg or 25–30% Ca saturation of the ECEC in the topsoil (Kamprath 1984). In Oxisol subsoils, Ritchey et al. (1982) found that 0.10–0.15 cmolc Ca/kg was needed for normal root growth. Calcium must be present near the root tips because it is not translocated from the plant tops. Since many subsoils in Oxisols have little or no exchangeable calcium, a small amount of calcium leached from the topsoil greatly increases root growth. Small amounts of lime (0.5 t/ha) are used for calcium fertilisation of aluminium-tolerant plants in typic subgroups of Oxisols and Ultisols (Spain et al. 1975; Sanchez and Salinas 1981). Peanuts and tomatoes, because of their high calcium requirements, often respond to direct calcium fertilisation.

Magnesium deficiency

Many acid soils in the tropics show magnesium deficiencies without having calcium deficiency problems. Oxisols in the Cerrado of Brazil are low in exchangeable magnesium (< 0.5 cmolc Mg/kg), and respond directly to magnesium fertilisation (Lopes and Cox 1977, Gonzalez-Erico et al. 1979). Small doses of dolomitic limestone can take care of these deficiencies. Soybean, cotton, potato and tobacco are commonly deficient in magnesium (Adams 1984).

Manganese toxicity

Manganese (Mn) is very soluble at pH values lower than 5.5. If this element is present in high amounts, manganese toxicity can occur along with aluminium toxicity. Certain acid soils may be low in aluminium but high in exchangeable manganese.

Common beans are susceptible to manganese toxicity. This situation is aggravated by the fact that common beans prefer tropical highlands where the soils are often high in SOM and are manganese-toxic to beans, where they are grown in Hawaii (Hue et al. 2001), Central America and East Africa. Coffee is also generally susceptible to manganese toxicity, and is grown preferably in similar climates and soils as the common bean.

Molybdenum deficiency

Molybdenum (Mo) availability decreases linearly with decreasing pH, so soils with pH below 5 are often deficient in molybdenum. Deficiency is usually more common with nitrogenfixing legume species in acid soils, because of the higher molybdenum requirements of nitrogen fixation. Although liming can correct it, small amounts, 30 g Mo/ha, can also be used, with NaMoO4 mixed with the seed. Acid, sandy soils are where molybdenum deficiencies most commonly occur (Brady and Weil 2008). When soybeans or peanuts respond to liming to > pH 5.5 the effect can be due to increased solubility of molybdenum (Pearson 1975). The mixing of molybdenum with legume pasture seeds has made an enormous impact on pasture production in Australia.

Effects of soil biota

Soil biota is affected by soil pH, and in some cases biota affects soil pH as well. Of the two main groups of microbes, bacterial abundance and diversity peaks at about pH 5.5–6.0, while fungi prefer lower pH values. The nitrifying bacteria are generally limited to soils above pH 5.5 (Siqueira et al. 1997). However, many very acid soils with pH values of 4.0–4.5 contain plenty of nitrate ions that must have resulted from nitrification, but the processes are not well understood. Rhizobial bacteria are more sensitive to low pH than their legume partner in the nitrogen-fixation symbiosis. Several genes associated with aluminium tolerance of the rhizobia species and strains have been identified (Glenn et al. 1997). Termites and earthworms thrive in a wide range of pH values, and exist in high abundance at pH 3.8–4.0 in the tropics. Liming is not expected to make much of a difference since macrofaunal communities are largely acid tolerant (Lavelle et al. 1995).

3.14.1 Secondary Acidity

Secondary acidity is that induced by the nitrification of ammonium-containing mineral fertilisers; ammonium ions produced by biological nitrogen fixation; the decomposition of SOM, manures and other organic fertilisers; and to a much lesser extent, elemental sulfur applications for pest-control purposes. The H^+ ions generated by nitrification attack clay minerals and produce exchangeable Al^{3+}, as described before. They can be precipitated out in soils with high pH, or when acid soils are limed. Acidification is evident usually after several years of continuous high nitrogen-fertiliser applications, such as those practiced in intensively fertilised pastures and in tea plantations. The secondary acidity assumes all the NO_3—produced stays in the soil. What counts is the NO_3—that stays in the soil beyond that taken up by plants; or that which is leached down, accompanied by a basic cation like K^+; or that which is denitrified. If all the NO_3—produced is taken up by plants, they also take up one H^+ from the nitrification equation, leaving the other H^+ in the soil (Binkley and Richter 1987). If the NH_4^+ ions are taken directly, there is no acidifying affect. The degree of acidification is ameliorated in soils that have a high buffering capacity (those high in clay, SOM and ECEC). In natural forest systems, however, secondary acidity seldom occurs because nitrification is usually synchronised with plant uptake (Helyar 1991). Furthermore, there are plenty of NO_3—anions that remain in the soil upon inorganic nitrogen fertilisation to allow the accompanying H^+ ions to acidify the soil because the efficiency of most nitrogen fertilisers (the proportion taken up by the crop) is on the order of 30–70%. However much of these NO_3—anions are often taken up by nitrogen-deprived bacteria. The residual effects of nitrogen fertilisers on soil acidity are most common in heavily fertilised pastures. Villamízar and Lotero (1967) evaluated the effects of annual applications of urea, ammonium sulphate and sodium nitrate at rates up to 1000 kg N/ha per year for 5 years on an Andisol from Colombia. Optimum yields of 30 t/ha per year of pangola grass were obtained at an annual rate of 500 kg N/ha. At this rate, the topsoil pH decreased from 5.8 to 4.2 with ammonium sulphate, remained the same with urea and increased to about 6.8 with sodium nitrate. Secondary acidity produced by nitrogen fertilisers incorporated in the topsoil and lime needed to neutralise it after complete nitrification. Ngatunga et al. (2001) evaluated soils of the Makonde Plateau in Tanzania where sulphur was successfully added by foliar application for controlling powdery mildew in cashew at an annual rate of 90 kg S/ha. When the sulphur is oxidised by Thiobacillus, two H^+ ions are released. They reported that pH decreased by 0.3 units in 4 years depending on the soil's buffering capacity (increasing with clay content and SOM) in Oxisols, Ultisols and Alfisols. The solution to this problem was an annual application of 0.2 t/ha of lime.

3.14.2 Advantages of Acid Soils

Well-drained aluminium-toxic Oxisols and Ultisols provide substantial advantages when well managed:

- They can help solubilise some phosphate rocks that would be unreactive in non-acid soils.
- Oxisols and similar soils classified as Ci in the FCC system usually have less exchangeable Al^{3+} throughout the profile (< 1 to 2 cmolc Al/kg) than many other aluminium-toxic soils including Ultisols –requiring less lime to reach the desirable aluminium-saturation level.

- Subsoil anion exchange capacity will hold nitrate and sulphate anions and provide a safety net against further leaching, allowing deep roots to recycle them and accompanying cations back to the topsoil if plant residues are returned.
- Weed control is usually less problematic, particularly in ustic soil moisture regimes.
- The subsoils can become major reservoirs of water and nutrients that can be recycled by plant roots. However, the greatest advantage of such soils is their excellent physical properties, allowing earlier planting time and offering less susceptibility to erosion.

Today's agriculture needs to follow the principles of sustainability that include building up and maintaining long-term soil productivity. On soils where acidity limits crop yields, soil and subsoil acidity amelioration constitutes an important part of best management practices (BMPs) to achieve sustainability (Prochnow 2014). A major aspect of soil acidity management is the application of lime, but other practices may also be needed to correctly address the problem. Proper soil acidity management, among other benefits, increases the efficiency of applied fertilisers, improves the effectiveness of some herbicides, protects the environment, and enhances the profit potential for the farmer.

3.14.3 Principles of Soil Acidity

In aqueous systems, an acid is a substance that donates hydrogen ions or protons (H^+) to some other substance. Conversely, a base is any substance that accepts H^+. The H^+ ions, or active acidity, increase with the strength of the acid. The undissociated H^+ contribute to a soil's potential acidity. Buffer systems can maintain the pH of a solution within a narrow range when small amounts of an acid or a base are added. Buffering defines the resistance to a change in pH (Prochnow 2014). Generally, buffer solution systems are composed of a weak acid (HA) and one of its salts (BA) or a weak base and one of its salts. Soils differ in terms of active and potential acidity. Also, soils behave like buffered weak acids, with the H^+ in the cation exchange complex (CEC) of humus and clay minerals providing the buffer for soil solution pH.

3.15 Soils of Humid Regions

Many of the soils in humid regions are very acid (< pH 5) and crop growth is hindered by toxicities of aluminium or manganese or both. The major exchangeable cation in these acid soils is aluminium. Crop production is drastically reduced when aluminium saturation of the active cation exchange sites is greater than 60% and tends to be optimum when aluminium saturation is zero. Although base saturation is a criterion in soil taxonomy, it does not identify soils with acidity problems Kamprath 1980). Soils with aluminium toxicity problems can be identified by extracting aluminium with potassium chloride and determining the proportion of the active cation exchange sites, sum of aluminium + calcium + magnesium + potassium, occupied by aluminium. Many well-drained soils in the humid tropics are acid and contain low amounts of basic cations. Early studies in which soils were limed to near neutrality often gave variable results (Pearson 1975). Many of the soils had high buffer capacities, and the amounts of lime required to raise the pH to neutrality were not economical. Liming practices that worked in temperate areas where 2:1 clays dominated were not applicable to highly weathered soils with 1:1 clays and 2:1 Al-interlayered clays (Kamprath 1980).

3.15.1 Soil pH, Exchangeable Aluminium and Aluminium Saturation of Soils

Major constraints to crop production in acid soils are toxicities of aluminium, manganese, and hydrogen and deficiencies of calcium (Jackson 1967). Research in the 1950s with Ultisols showed that potassium chloride-exchangeable aluminium is the predominant cation in acid soils and that more meaningful values of cation saturation are obtained when CEC is based on the sum of exchangeable bases plus potassium chloride-exchangeable aluminium. Exchangeable aluminium is a major cation in many of the acid soils, particularly at pH 4.5 and less. It occupies a large proportion of the effective CEC in these acid soils. Aluminium toxicity is definitely a problem in many soils in the tropics. Another factor affecting the exchangeable aluminium content of acid soils is the kind of clay minerals present (Juo and Kamprath 1979). Acid topsoils with kaolinite and 2:1 clays with aluminium-interlayers had, on the average, 2½ times more exchangeable aluminium than those with only kaolinite even though the clay content was considerably less. Subsoils with aluminium-interlayered clays also had much higher exchangeable aluminium contents even though their pH's were higher than subsoils with only kaolinite. Highest concentrations were in coarse-textured Ultisols from North Carolina, which contained 2:1 clays with Al-interlayers. The lowest concentration was in a clay-textured Oxisol from Brazil in which the clay mineral was kaolinite (Kamprath 1980).

3.15.2 Relationship Between Soil pH and A1 Saturation

Soil pH is related to aluminium saturation. At pH 5.4 the aluminium saturation of the effective CEC (sum of exchangeable Al, Ca, Mg, and K) was essentially zero; at pH 4.2 it was approximately 60%. Juo (1977) found a high correlation ($r = -0.81$) between soil pH and aluminium saturation of subsoils from West Africa. It must be kept in mind, however, that electrolyte concentration influences soil pH and amount of aluminium in the soil solution (Pearson 1975). Therefore, high rates of fertilisation may intensify aluminium problems because the higher electrolyte concentration increases aluminium in the soil solution.

3.15.3 Crop Growth as Related to Exchangeable Aluminium Saturation and Soil pH

Toxic concentrations of aluminium in the soil solution limit crop growth on acid soils. Methods for determining soil solution aluminium are not suitable for rapid analysis of large numbers of samples. There is, however, a fairly good relationship between the percentage of aluminium saturation of the effective CEC and soil solution aluminium (Nye et al 1961; Evans and Kamprath 1970). Aluminium concentration in the soil solution increases rapidly when the percentage aluminium saturation is greater than 60%. An example of the relationship between crop growth and percentage of aluminium saturation is given in Table 3.8. For eight Ultisols and Oxisols the relative yield of corn on nonlimed plots as compared to that in limed plots was highly correlated with percentage of aluminium saturation. Only on the acid soil with a relatively high aluminium saturation was there a severe depression in yield. Responses to liming on the other soils, which had relatively low aluminium saturations, ranged from 10 to 22%. This suggests that the detrimental effect of pH per se or manganese toxicity on growth of corn was not as great as that of high aluminium saturation.

Table 3.8 Relationship between soil acidity parameters of Puerto Rican soils and relative yield of corn on nonlimed treatments as compared to that on limed treatments (Abruna et al 1975)

Soil order	Initial pH	Exchangeable Al (meq/100 g)	Al saturation (%)	Relative yield (%)
Ultisol	4.3	9.9	68	10
Ultisol	3.9	6.3	66	22
Ultisol	4.2	6.9	63	18
Ultisol	4.3	4.5	47	47
Oxisol	4.6	0.9	40	61
Ultisol	4.6	5.5	35	62
Oxisol	4.2	1.1	33	85
Oxisol	4.8	0.4	18	85

3.15.4 Critical Values of Aluminium Saturation and pH for Crop Production

To use soil acidity parameters for predicting crop response to lime application, one must have information about critical values for a given parameter. The levels of aluminium saturation and soil pH at which yields are less than 50% and more than 90% of the maximum (Table 3.9) have been summarised with data from a number of liming studies with various crops. The critical levels at the 90% yield level are less than 20 to 10% aluminium saturation for maize and generally less than 10% aluminium saturation for soybean, green bean, sugarcane, and cotton; and soil pH values were in the range of 5–5.3 for corn, 5.6 for soybean, and 5.3–6 for cotton. Critical levels at yields less than 50% of the maximum are:

(a) Aluminum saturations of more than 40 to 60% in most instances, and,
(b) Soil pH values of < 4.5 for maize, and < 5.0 for soybean and cotton.

When establishing critical levels, the wide differences in varietal tolerance for aluminium within a given plant species must be considered. Another factor is the sensitivity to soil acidity of the crops in the rotation. The liming practice used must provide a favourable environment to the most sensitive crop. For most crops a liming programme should reduce aluminium saturation to near zero. This will generally provide a favourable pH and adequate calcium supply.

Table 3.9 Critical levels of aluminium saturation and soil pH for a number of crops

Region	Crop	Critical level			
		< 50% max	Yield	90% max	Yield
		pH	Al saturation (%)	pH	Al saturation (%)
Brazil[a]	Maize	–	–	5.3–5.6	< 10
Puerto Rico[b]	Maize	4.4	50	5.0	< 15
USA[c]	Maize	4.5	60	5.1	< 40
Philippines[d]	Maize	4.4	46	5.0	< 20
Brazil[a]	Soybean	–	–	5.6	< 10
Puerto Rico[e]	Soybean	5.0	20	5.6	< 10
USA[c]	Soybean	4.9	40	5.6	< 20
Puerto Rico[f]	Sugarcane	4.4	60	5.0	< 10
USA[c]	Cotton	5.0	45	6.0	< 7
Uganda[g]	Cotton	–	–	5.3	0
Puerto Rico[b]	Green bean	4.3	50	5.2	< 10

[a]Soares et al. (1975), [b]Abruna et al. (1975), [c]Kamprath (1970), [d]Samonte and Ocampo (1977), [e]Abruna et al. (1978), [f]Abruna and Vicente-Chandler (1967), [g]Foster (1970)

3.15.5 *Effect of the Soil's Acidification*

Crops and pastures on acid soils (pH 5.5) experience lower microbiological activity, decreased nutrient availability, and higher phytotoxicity (Al, Fe, Mn, etc.) as a result (Cronan and Grigal 1995; Whitfield et al. 2010). By speeding the leaching of minerals including manganese, calcium, magnesium, and potassium, onsite soil acidification lowers net primary productivity and carbon sequestration and causes nutritional deficits in plants (Haynes 1986). The growth of subsurface acidity (Tang 2004), the breakdown and consequent loss of clay minerals from the soil (Chen 2007), and the erosion brought on by a decline in groundcover (Slattery and Hollier 2002) are all caused by on-site soil acidification. Additionally, soil acidification has off-site impacts such as groundwater enrichment in soluble metals and acidification of surface water due to sediment losses. In turn, these processes disperse heavy metals into the food chain and water supplies (Slattery and Hollier 2002; Driscoll et al. 2003).

Acidification of the soil is a sneaky process. It starts out gradually and, if not stopped, can worsen until the soil is permanently harmed (Edmeades and Ridley 2003; Slattery and Hollier 2002). An rise in pH and acid-neutralising capacity (ANC) may help with biological recovery (Marschner and Noble 2000). The primary cause for worry is subsoil acidity, which is extremely challenging to address using traditional techniques (Liu and Hue 2001; Hue and Licudine 1999). Pollutants that contribute to soil acidification, such as sulphur dioxide (SO_2), can be reduced by measures to decrease global warming (NADP 2014; Vestreng et al. 2007). However, the recovery of acid-affected locations may take several decades due to the sluggish response of the soil to reductions in acid deposition (Zhao et al. 2009).

3.15.6 *Acidification of Soil Quality and Trends Worldwide*

Food production is severely hampered by acidic soil all over the world. The topsoil has traditionally been treated with lime to combat it, but little could be done to raise the pH of the subsoil. Largely in industrialised nations, where it is possible to carry out soil management programmes to conserve soil characteristics and afford the cost of lime to buffer soil acidity, have programmes to enhance soil pH been implemented. However, there have been instances even in industrialised nations, like Australia, where subsoil acidity increased as a result of failures to address topsoil acidity. The situation is worse in developing nations where access to lime is limited by abject poverty. As a result, the amount of farmland being acidified is growing (Sumner and Noble 2003). In addition to agricultural areas, woodlands and grasslands are also impacted by soil acidity. Sumner and Noble (2003) estimate that topsoil acidity (pH 5.5) impacts about 30% of the world's ice-free land area and subsurface acidity affects up to 75%. Figure 6.8 depicts the pH situation around the world. The world's most acidic topsoils (pH 3.5) are found in South America in regions where intense agriculture and deforestation are practised, as well as in river deltas home to mangroves, such as the Amazon and Orinoco Deltas (Marchand et al. 2006).

3.15.7 *Acidification Filtration*

According to Guo et al. (2010), they currently inhabit about one-third of the ice-free territory on Earth, mostly in temperate zone forests and the wet tropics. Experiments at local level prove that hydrolytic acidity is the main primary liming requirement indicator on moraine sandy loam Albeluvisols and glacio-fluvial sand Luvisols and Arenosols. The most effective means of neutralisation is primary liming (single application) at one rate according to soil hydrolytic acidity. The main acidity indexes in the Albeluvisols topsoil reveal: pH changes from 4.4 to 6.6 and mobile Al changes from toxic to plants (78 mg kg^{-1}) to non-toxic (4 mg kg^{-1}). One rate (6.6–7.8 t ha^{-1} $CaCO_3$) of lime remains effective (i.e. until soil reaches pH 5.5) in moraine loam for about 10 years, while in glacio-fluvial sandy soil one rate of lime (6.5 t ha^{-1} $CaCO_3$) is effective (until soil reaches pH 5.3) for only 5 years. Soil pH returns quicker to initial levels in sandy soils than in moraine loams, after 20–22 and 23–28 years, respectively.

3.15.8 *Management of Soil Acidity in Limed Soils*

Soil acidification proceeds at different rates: in the initial 7– 14 years after liming pH decreases faster (0.1–0.3 pH units per year), but later the process slows down. Acidification and consequent Al release are suppressed for > 30 years in sandy and ~50 in loamy soils. To control soil reaction, repeated and periodic liming cycles and rates on the background of primary liming with one rate should be applied to prevent pH decreased and increased concentrations of toxic Al. In moraine loamy soil, the recommended pH level of 5.6–6.2 for plants in the topsoil is achieved on the periodic third or fourth liming using a 50-year liming system. Periodic liming of glacio-fluvial sandy soil with one rate of dust limestone every 10 years maintains topsoil pH at 5.2–5.4. During periodic liming, mobile Al changes into immobile forms in loams or its concentrations are non-toxic (1.6–2.2 mg kg^{-1} in sand). Periodic liming at one rate, followed by different intensities (every 3–4 years and every 10 years) repeated over two decades changes acid soil characteristics both in topsoil and the whole profile to $\leq$ 100 cm depth (Eidukeviciene et al. 2001). The relationship between soil pH and distribution of different forms of Al is shown in Fig. 3.7.

Spatial patterns of topsoil pH in Lithuania are fundamentally controlled by soil parent material. The pH spatial structure of topsoils, limed once or twice, basically echoes initial soil reaction conditions. Topsoil, subsoil and parent material collectively influence soil reaction and should not be treated as separate components. The effect of economic activities on agricultural topsoil in space and time is evident,

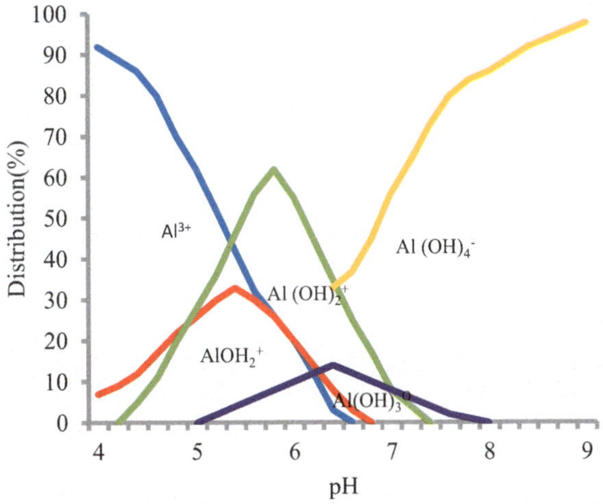

Fig. 3.7 Relationship of pH with distribution of different forms of Aluminium

although the consequences are not as important as natural background effects. Natural and economic processes affect soil in specific areas and the effectiveness of management activities is strongly influenced by pedogenic processes.

3.15.9 Problems with Decreasing Soil pH

- Root growth issues (acute).
- Aluminium toxicity.
- Nutrient deficiencies (P, Mo).
- Limits nitrogen fixation by legumes.
- Disease interactions.
- Herbicide interactions.
- Reduced nutrient use efficiency.
- Reduced biological activity.
- Reduced cation exchange capacity (CEC) …and the list goes on.

3.16 Soil Acidity and Influencing Factors

3.16.1 Types of Acidity

Active acidity

Measured in a soil and water mixture, active acidity is the amount of hydrogen ions in the soil solution. The pH of the soil is a general indicator of the existence of free lime (calcium carbonate), the availability of nutrients, and the excessive availability of particular ions, such as sodium, hydrogen, aluminium, and manganese.

- $[H^+]$ in solution, soil water solution.
- Measure directly with pH metre, e.g. 1:1 soil:water suspension.
- pH controls chemical properties.
- pH affects both biological and physical properties.

Reserve acidity

Reserve acidity is defined as the amount of hydrogen ions bound to organic matter and clay, expressed as buffer pH in a buffer solution. Reserve and active acidity have a link, however it varies according on the soil. It is affected by the kind and quantity of clay in the soil as well as by the amount of organic matter and free lime. The reserve to active acidity ratio indicates the soil's buffer capacity, or its ability to withstand pH changes. The ratio of reserve to active acidity and the cation exchange capacity of soil both rise with an increase in clay and organic matter content. Thus, the buffer capacity, or reserve acidity, of a sandy soil is much less than that of a soil containing more clay, such as a silt loam. When the soil pH is 6.3 or lower, the buffer pH is measured to determine the amount of lime required to neutralise a major portion of the reserve acidity.

Exchangeable acidity. Buffer pH

- Amount of aluminium $[Al^{+++}]$, Hydrogen $[H^+]$ and (some) Iron $[Fe^{+++}]$ that occupies exchange sites on clays, soil aggregates, and organic matter.

Residual acidity

- Not readily available
- Bound aluminium and hydrogen in clay minerals and soil aggregates.

3.16.2 Soil Factors Influencing Acidity

- pH
- Organic Matter
- Mineralogy
- Age and Climate
- Texture.

3.16.3 Processes that Cause Soil Acidity

- Carbonic (CO_2!) and Organic Acids.
- Decomposition of Organic Matter.

- Oxidation of Nitrogen (Nitrification).
- Oxidation of Sulphur.
- Plant Uptake/Leaching of Cations.
- Acidic Phosphate Fertilisers.

3.16.4 Sources of Soil Acidity

1. *CO_2*

   - Conversion of carbon dioxide to carbonic acid.
   - Respiration = uptake up O_2 and subsequent release of CO_2.
   - Plant roots, microbial activity, soil fauna, etc. are primary source of CO_2 in soil.

$$CO_2(g) + H_2O(l) \leftrightarrow H^+(aq) + HCO_3^-(aq)$$

2. *N Fertilisers*

$$NH_3 + 2O_2 = H^+ + NO_3^- + H_2O(\text{nitrification of ammonia})$$

3. *Humus*

   - Soil organic matter contains compounds that behave like weak acids.
   - Hydrogen ions are released as crop residues decompose.
   - Organic acids contribute to soil acidity.
   - Humic and fulvic acids.
   - SOM serves as pH buffer.
   - Acts as both [H^+] donor and [H^+] acceptor.
   - Buffering capacity of soil increases with increasing soil organic matter content.
   - Humus has higher CEC than most clay minerals.
   - SOM binds Al^{3+}.

4. *Straw removal* (Huggins et al. 1998)

   - Cation removal by crop uptake will either increase or decrease soil acidification.
   - Removal of base cations when baling straw.
   - Burning retains cations.
   - N, C, and S are oxidised—and float away in the breeze.
   - Fertiliser Cost to balance nutrient losses.
   - Legumes tend to acidify soils more than cereals because the removal rate of base cations by legumes is significantly higher.

3.16.5 Managing Acid Soils

Amendments

- Add organic matter
- Chelates aluminium, iron and manganese,
- Amend with calcium carbonate.
- Calibrated application rates raise soil pH.

Types of Soil Acidity
Soil acidity is estimated by pH measurement. Soil pH is an index of active acidity.

3.16.6 Nitrogen and Phosphorus Fertilisers

Nitrogen fertilisers used in Nebraska are primarily ammonium based. Following application, ammonium-nitrogen is transformed to nitrate-nitrogen in a process called nitrification. Nitrification is a major cause of soil acidification as two hydrogen ions are released to the soil solution for each nitrate ion produced. This is shown as:

$$NH_4^+ + 2O_2 \rightarrow 2H^+ + NO_3^- + H_2O$$

When anhydrous ammonia is applied, however, it reacts with water to form ammonium and a hydroxide ion (OH^-) with the effect of neutralising some acidity:

$$NH_3 + H_2O \rightarrow NH_4^+ + OH^-$$

(OH^- reacts with H^+ to produce water while neutralising acidity). This reaction, followed by nitrification, results in a net increase in one hydrogen ion for each nitrate ion produced from anhydrous ammonia. Acidity produced by applying nitrogen fertiliser is similar for anhydrous ammonia, ammonium nitrate, and urea, but is higher with ammonium sulphate, diammonium phosphate (DAP), and monoammonium phosphate (MAP). More acidity is produced per unit of nitrogen applied with the phosphate fertiliser. This may not be of concern as the rates of application are typically much less than for the major nitrogen fertilisers. Nonetheless, producers may want to consider other phosphate sources such as triple superphosphate to minimise soil acidification.

3.16.7 Sulphur Fertilisers

Sulphur is a soil-acidifying element. Approximately 3 kg of calcium carbonate are needed to neutralise the effect of 1 kg of sulphur applied as elemental sulphur or ammonium polysulphide.

Acidification is negligible with sulphate salt (calcium sulphate) fertilisers, intermediate for ammonium thiosulphate, and high for ammonium polysulphide and elemental sulphur. Plant uptake of sulphate ions has the effect of neutralising acidity, as explained below. Acidification is expected to be greatest, therefore, when sulphur application does not result in increased plant uptake of sulphate. With the exception of sandy soils, and possibly some low organic matter soils under no-till, major crops may not be responsive to applied sulphur. Matching sulphur application with crop needs may be an opportunity to reduce costs while reducing soil acidification.

3.16.8 Organic Fertilisers

Applying animal dung may cause the soil's acidity to rise or fall. Manure's organic matter decomposition, which produces organic and inorganic acids, contributes to some of the soil acidity. However, manure frequently includes sufficient amounts of basic cations and carbonate ions to counteract both the effect of acidification and some already-present soil acidity. The liming value of manure can be determined by analysing sample material.

3.16.9 Legumes and Nitrogen Fixation

Legumes can derive much of their nitrogen from the atmosphere through biological nitrogen fixation, thereby reducing fertiliser needs. This is, however, a soil-acidifying process as hydrogen ions are products of nitrogen fixation. The acidity resulting from the harvest of one ton of alfalfa hay could require more than 80 kg of aglime for neutralisation. However, the acidification effect is spread throughout the root zone, resulting in less effect on surface soil pH than may be implied.

3.16.10 Leaching of Nitrate

Acidity generated by nitrification of ammonium-based fertilisers is largely due to poor recovery of nitrate ions by plants. Nitrate leached from the root zone does not neutralise acidity as occurs when nitrate is taken up by plants. Matching the fertiliser application rate to the crop's nitrogen uptake minimises leaching losses and the net effect of nitrification on soil acidity.

3.16.11 Desirable pH

The adverse effect of these toxic elements is most easily (and economically) eliminated by liming the soil. Liming raises the soil pH and causes the aluminium and manganese to go from the soil solution back into solid (non-toxic) chemical forms. For grasses, raising the pH to 5.5 will generally restore normal yields. Legumes, on the other hand, do best in a calcium rich environment and often need the pH in a range of 6.5 to 7.0 for maximum yields. A soil pH in the range of 6.0 to 7.0 is also desirable from the standpoint of optimum nutrient availability. However, the most common nutrient deficiencies in Oklahoma are for nitrogen, phosphorus, and potassium, and availability of these elements will not be greatly changed by liming. Nutrients most affected by soil pH are iron and molybdenum. Molybdenum is critical for nitrogen fixation by legume crops.

3.16.12 Innate Elements Influencing Soil pH

When soil is first produced, its pH is determined by minerals in the parent material. Rainfall and temperature affect how much leaching occurs as well as how quickly soil minerals deteriorate. Due to leaching from heavy rains, soil pH falls over time in warm, humid areas, a process known as soil acidification. However, soil weathering and leaching are less severe in dry regions, and the pH might be neutral or alkaline. Sandier soils have a lower capacity for buffering pH changes than soils with higher levels of clay and organic matter. Clay composition cannot be adjusted, however management can alter the amount of organic matter. Sandy soils frequently have low levels of organic matter, which results in a low buffering capacity, high rates of water infiltration, and a greater susceptibility to acidification.

3.16.13 Soil pH Control

Land usage and management have an impact on soil pH. Type of vegetation affects soil pH. For instance, sections of forest land typically have higher levels of acidity than areas of grassland. After a few years, converting grassland or forestland to farmland can cause significant pH shifts. The loss of organic matter, the removal of soil minerals after crop harvest, the eroding of the top layer, and the effects of nitrogen and sulphur fertilisers are the main contributors to these changes. Fertilisers containing sulphur and nitrogen may gradually reduce the pH of the soil.

3.16.14 Limiting or Reversing the Acidity

- Adding lime to an acidic soil to improve its pH and applying nitrogen and sulphur to crops when needed
- Diversified crop rotations to interrupt the acidifying effects of N fertiliser application. Application of irrigation water, manure, and other organic materials that have a high content of calcium or magnesium bicarbonates. Applying N fertiliser in the right amounts, on time (relative to crop uptake), and with good irrigation management.
- Use of continuous no-till, cover crops, applications of solid manure, and various crop rotations with high-residue crops to boost the amount of organic matter in the soil and enhance its ability to act as a buffer against pH shifts.

How pH affects soil function

The pH of the soil is a great predictor of how well a soil will support plant growth. As demonstrated in Table 3.10, a range of 6 to 7.5 is ideal for the majority of crops. A lack of numerous nutrients, a loss in microbial activity, a reduction in crop output, and a degradation in soil health are all effects of too high or low soil pH values. For instance, soil pH levels between 7.5 and 8.5 and below 5.5 limit the amount of phosphate that plants can absorb. There are four main management techniques to make more phosphate available to plants:

1. Liming acidic soils to raise pH levels to 6.5 to 7.0.
2. Using tiny, numerous applications of phosphorus fertiliser rather than a single, massive application.
3. By banding or injecting, phosphorus source (manure and/or fertiliser) and soil contact can be reduced.
4. Putting P-fertilisers where the roots are most active, close to the crop row.

Table 3.10 Crop yield relative to soil pH (%)

Crop	pH				
	4.7	5.0	5.7	6.8	7.5
Corn	34	73	83	100	85
Wheat	68	78	69	100	99
Soybeans	65	79	80	100	93
Oats	77	93	99	98	100
Barley	0	23	80	95	100
Alfalfa	2	9	42	100	100
Timothy grass	31	47	67	100	95

Source Methods for Assessing Soil Quality (SSSA 1996)

Low pH inhibits nitrogen cycling. The pH affects the breakdown and efficacy of insecticides and herbicides as well as the solubility of heavy metals. When the soil is acidic or alkaline, several diseases thrive. The pH of the soil also affects the efficacy and potential carryover of several herbicides applied to the soil.

3.16.15 Quantifying Soil Acidification

The amount of acid (i.e. hydrogen ions) added to or produced in the soil per acre per year can be used to measure soil acidification. When soil is acidified, the pH of the soil decreases. The attribute known as pH buffer capacity describes the differences between soils' resistance to pH drop. Calculating the rate of pH change in soil involves dividing the pH buffer capacity by the rate of acid input. Therefore, the pH buffer capacity has a significant impact on how vulnerable the soil is to acidification-induced degradation. When it is high, the soil is more resistant to the effects of acid inputs, therefore it takes longer for the pH to drop and start affecting plant growth. The soils with the most organic matter and clay have the highest pH buffer capability. Australian research demonstrate that a variety of farming practises cause the soil profile to become more acidic. 17 papers addressing the rates of acidification in 35 agrarian and forestry ecosystems were summarised by Slattery et al. (1999). Out of these, only one production method—tobacco production—alkalized the soil, with the other systems acidifying it to varying degrees. The trait known as pH buffer capacity allows all soils to some extent fend off the tendency of soil acidification processes to lower soil pH. Because low buffer capacities are so common, it is normal to expect significant pH reductions when farming introduces acid to the soil. On the plus side, lime is proportionately more successful at bringing an acidic soil's pH up.

Numerous studies investigating the impact of lime on acidic soils have been conducted since the 1940s (Cregan et al. 1989). According to the mechanisms described by Helyar and Porter (1989), soil acidity in southern Australia has significantly increased since the 1940s. The first soil degradation assessments that placed acidity and acidification into perspective just started to appear in the middle of the 1980s (AACM International 1995; Chartres et al. 1992; Coventry 1985), in spite of the significant amount of early research into agriculturally generated acidification.

According to estimates made by Von Uexküll and Mutert (1995), soils with pHw 5.5 make up 30% or 3950 Mha of the world's ice-free land area. Over 49 Mha of Australia's soils fall into the highly and moderately acidic category (NLWRA 2001). The majority of these soils are inherently acidic, and their low fertility has a bearing on the type of ecosystems that have grown there. Because they can only withstand a slight pH decrease before it reaches a critical value, soils that are already somewhat acidic are typically the most susceptible to the negative consequences of soil acidification. Soil acidity can only be seen by poor crop and pasture growth, changes in species composition, and soil chemical analysis. By pelleting the seed with lime or banding it with superphosphate and small amounts of lime at sowing, it is sometimes possible to improve poor legume nodulation during pasture establishment

(Cregan et al. 1989). Even when lime is applied in large quantities, it merely addresses the immediate issue and not the underlying cause (such as excessive water and nitrate salt leaking below the root zone).

The acidity and alkalinity of the soil solution are gauged by the soil pH. The negative logarithm of the hydrogen ion concentration, or H^+, is what pH is defined as, or $pH = -\log [H^+]$. Depending on the pH values of the soil, which range from around 0 to 14 on a scale, soils are classified as acidic, neutral, alkaline (or basic) or basic (Fig. 3.8). Less than 7 is acidic, more than 7 is alkaline, and 7 is neutral (clean water). Each unit on the pH scale is ten times more alkaline (less acidic) than the unit below it because pH is a logarithmic function. For instance, a solution with a pH of 6 has a concentration of H^+ ions that is 10 times higher than a solution with a pH of 7, and 100 times higher than a solution with a pH of 8. Both the base- and acid-forming ions in the soil have an impact on soil pH. Iron (Fe^{2+} or Fe^{3+}), aluminium (Al^{3+}), and hydrogen (H^+) are frequent cations (positively charged dissolved ions) that form acids, whereas calcium (Ca^{2+}), magnesium (Mg^{2+}), potassium (K^+), and sodium (Na^+) are common cations that form bases. This is mainly because soils and irrigation waters naturally contain base cations linked to carbonates and bicarbonates. There is little base cation leaching because of the comparatively low precipitation levels, which leads to a relatively high level of base saturation and pH values higher than 7. On the other hand, acid conditions develop in soils with parent materials rich in silica and other elements (such as rhyolite and granite), as well as in areas with a lot of sand and low buffering capabilities (ability to resist pH shift). As precipitation increases, base cations are more likely to be washed away, which lowers the pH of the soil.

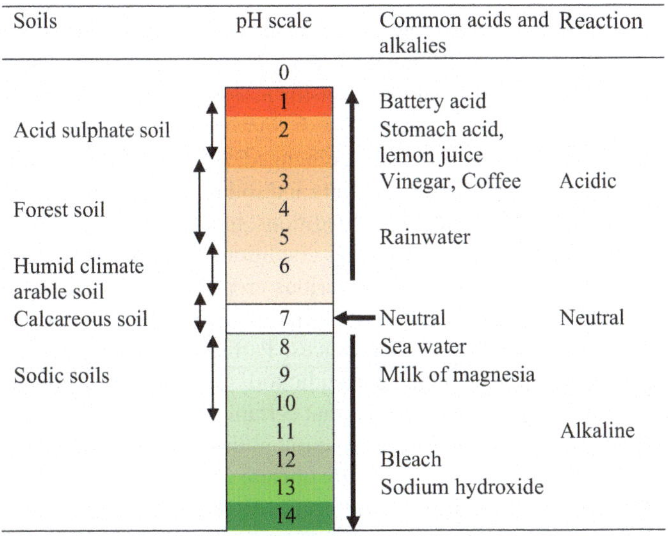

Fig. 3.8 Showing pH range with examples of soils and commodities. *Source* McCauley (2009)

3.17 Nutrient Availability

3.17.1 Exchange Capacity

Exchange capacity, in a nutshell, is the soil's capacity to store and give nutrients to a crop. The majority of soils in the Western US region of Montana and Wyoming have a net negative charge, thus their cation exchange capacity (CEC) is greater than their anion exchange capacity (AEC). More cations, such as Ca^{2+} or K^{+}, can bind to the surfaces of clay and organic matter particles, which are known as exchange sites, in soils with high CEC values. A soil with a high CEC will also have a larger capacity for buffering, which will improve the soil's ability to withstand pH shifts. In contrast to more silty or sandy soils, soils with high clay and/or organic matter content often have higher CEC and buffering capabilities. H^{+} will compete with other cations for exchange sites because it is a cation. More base cations will be on the particle exchange sites and be less prone to leaching when the soil pH is high (i.e. more basic, low concentration of H^{+}). Contrarily, more H^{+} ions are available to "exchange" base cations, removing them from exchange sites and releasing them to the soil solution (soil water), when the soil pH is lower (i.e. less basic, higher concentration of H^{+}). Nutrients that are exchanged as a result are either absorbed by the plant or lost due to leaching or erosion.

3.17.2 Nutritional Status

Soil pH has a significant influence on the availability of nutrients to plants. P is is mostly available within a pH range of 6 to 7. Less nutrients are accessible to plants outside of these ideal levels. Except for molybdenum (Mo), the availability of micronutrients declines when soil pH levels approach 8 because cations are more tightly bonded to the soil and less easily exchanged. Acid soil crops may be exposed to metal toxicity as a result. Individual plants and soil organisms differ in their ability to withstand alkaline and/or acidic soil conditions, in addition to how pH affects the availability of nutrients. The best growing conditions for crops seem to be neutral ones. However, the ideal pH ranges for various crops will differ. However, the pH ranges for each type of soil microorganism differ, with the activity of these organisms being greatest close to neutral circumstances. Particularly, microbial activity and population are significantly lower in very acidic soils (less than 5) than in more neutral soils. Additionally, research has revealed that certain "specialised" microorganisms, such as nitrifying bacteria (which convert ammonium to nitrate) and nitrogen-fixing bacteria linked to certain legumes, typically operate poorly when soil pH falls below 6 (Haby 1993; Sylvia et al. 1998). The optimal pH range shown for potatoes is also reflective of a microorganism relationship; the bacteria causing common scab infection are more prevalent as pH rises (Walworth 1998). Therefore, the optimal pH range for growing potatoes is 5.0 to 5.5 because the risk of common scab infection is minimised at lower pH (due to lower microbial activity). When soil pH is extreme, either too acidic or alkaline, pH modifications may be needed to obtain optimal growing conditions for specific crops.

In addition to the acidity itself, the primary issues that arise when it comes to the chemical features of highly acidic soils are the existence of hazardous chemicals and elements. Some examples of these include soluble forms of aluminium, iron, and manganese as well as nitrites and a variety of harmful organic acids. According to research carried out by Krstic et al. (2012) and others, one of the primary factors limiting agricultural output on acid soils is the toxicity of aluminium (Al). This factor can be found on up to forty percent of the world's arable land.

The humid tropics are home to a mineral-rich belt of acidic soils that run south to north. According to Von Uexküll and Mutert (1995), approximately 12% of the land that is used for crop cultivation is acidic. Mineral acid soils can be caused by parent materials that are naturally acidic and deficient in the basic cations (Ca, Mg, K, and Na), or they can be caused by the leaching of these elements from the soil, which lowers the pH of the soil and the buffering capacity of the soil. When the pH of the soil drops, aluminium (Al) becomes more soluble, and the fraction of phytotoxic aluminium ions that are present in the soil solution rises. In almost all mineral soils, the amount of aluminium present is adequate to buffer the soil to a pH level of about 4. According to Kidd and Proctor (2001), organic acid soils, which are characterised by high concentrations of humic acids and partially decomposed plant material, often have a low level of Al buffering, which allows the pH of these soils to range quite a bit lower than 4.

The toxicity of acidic soil is caused by a number of variables, which vary depending on the makeup of the soil. Al toxicity is the principal component that inhibits plant growth in soils that are acidic and have a high concentration of minerals. According to Kinraide (1991), the aluminium that is liberated from soil minerals in acidic conditions appears in the forms of $Al(OH)_2^+$, Al(OH), and $Al(H_2O)_3^+$, the last of which is simply known as Al. At micromolar concentrations, aluminium ions quickly impede root growth in the vast majority of agriculturally significant plant species.

3.17.3 Acidifying Agents in the Soil

The production of acid soil is the end consequence of a complicated series of chemical reactions known as soil acidification. It can be thought of as the sum of all natural and manmade activities that work to bring the pH of soil solution down (Krug and Frink 1983). This definition gives it the broadest possible meaning. Rainfall and leaching, the presence of acidic parent material, the decomposition of organic matter, the harvesting of high-yielding crops, the nitrification of ammonium, and a low buffer capacity resulting from a lack of clay and organic matter (Zhang 2013) are the primary factors that cause soils to become acidic. When we consider that a soil is considered acid when there is a greater abundance of acidic cations, such as hydrogen (H^+)

and aluminium (Al^{3+}), in the soil than there is of alkaline cations, such as calcium (Ca^{2+}), magnesium (Mg^{2+}), potassium (K^+), and sodium (Na^+), then understanding soil acidity becomes very simple.

3.17.4 Leaching Due to Precipitation

Rainfall that is excessive is a powerful agent that can be used to remove basic cations over a protracted period of time. The amount of precipitation that falls causes the concentration of soluble salts to drop to a low level, and it also removes any calcium carbonate and gypsum that may have been present. Runoff has the greatest potential to turn soils acidic when a significant amount of water moves through the soil in a short amount of time. Sandy soils are frequently the first to become acidic because water percolates quickly, and sandy soils only contain a tiny base reservoir (buffer capacity) due to low amounts of clay and organic matter (Wolf and Snyder 2003). This causes sandy soils to be the first to become acidic. Excessive precipitation causes the basic components of the soil profile (Ca, Mg, Na, and K) to be washed away, which leads to an increase in the acidity of the soil (Foy 1984). According to Brady and Weil (2016), excessive precipitation causes the leaching of soluble nutrients like Ca and Mg, which are then specifically supplied by Al from the exchange sites.

3.17.5 Acidic Parent Material

Due to the varied chemical make-up of parent materials, the rate at which soils undergo acidification depends on the amount of time that has passed since they were created. According to Guo et al. (2010) and Hue (1992), soil that was formed from granite rock is likely to have a higher acidity level than soil that was formed from calcareous shale or calcareous stone. Granite and rhyolite are two examples of acid rocks. Acid rocks are distinguished from basic rocks by the presence of a greater proportion of quartz or silica relative to their overall composition of basic materials or basic elements. The accumulation of acidic material in soil occurs when rocks that are lacking in bases are dissolved or decomposed during the process of soil formation. This occurs despite the fact that there is no loss of base during the process of soil creation. According to Yilmaz (1989), there are extensive regions that were formed from acidic parent rocks of silica and sandy soils, and these regions have always needed lime. However, leaching losses and the removal of bases by crops are responsible for the development of the majority of acid soils (Brady and Weil 2016). If the pH value is less than 4, the soil has a very strong acidic quality, while the pH values of the vast majority of soils fall anywhere between 4.5 and 6.5. According to Bekele and Hofner (1993), the majority of the time, the soils that may be found in high altitude regions of the country have an acidic reaction, are deficient in exchangeable cations, and have a low base saturation.

3.17.6 Organic Matter Breakdown

The breakdown of organic matter produces hydrogen ion, which is the primary contributor to acidity. When decomposing organic matter releases carbon dioxide (CO2) into the atmosphere, this gas then combines with the water in the soil to produce carbonic acid, which is a weak acid. This particular acid is the same one that is produced when CO2 undergoes a natural reaction with precipitation in the atmosphere to produce acid rain. Many organic acids are produced by the decomposition of organic materials; nonetheless, many organic acids are also weak acids. As is the case with rainfall, the contribution of decomposing organic matter to the creation of acid soil is typically very minimal, and the only effects that could ever be evaluated in a field would be the cumulative effects of several years' worth of time. (Prochnow 2014). Organic matter in the soil, also known as humus, often contains reactive carboxylic, enolic, and phenolic groups that function in a manner similar to that of weak acids. The process of their dissociation results in the release of H^+ ions. In addition, during the process of decomposition, CO_2 and organic acids are formed, both of which result in the replacement of bases on exchange complexes with H^+ ions (Somani et al. 1996).

3.17.7 The Repercussions of Having Acidic Soil

Poor plant growth and reduced productivity of plants that are sensitive to acidity will reduce in growth and productivity, allowing weeds to develop or reducing oil cover, both of which can lead to soil erosion (Mcfarland et al. 2001). Poor plant growth and reduced productivity of plants that are sensitive to acidity will lower in growth and productivity. According to Gebregziabher et al. (2016), the most significant contributor to soil acidity is the presence of aluminium toxicity in the subsurface soil. A low pH in the top soils largely impacts the availability of nutrients, which in turn inhibits the legumes' ability to nodulate and pastures' capacity to fix nitrogen. According to Chris and Stephen research from 2009, these issues can be mitigated to some degree if the pH of the topsoil is kept at or above 7.0.

Soils are considered acidic when their pH levels remain persistently lower than 5.5 for long periods of time. Agricultural practises such as flooding and the application of agrochemicals can contribute to the acidification of soil. These soils are exceedingly difficult to cultivate because they are connected with metallic toxicities, including aluminium (Al) and iron (Fe^{2+}), which are responsible for a significant drop in crop production. Cultivation in these soils is highly difficult. Higher levels of soil Al inhibit the growth of plant roots, and an excess of Fe^{2+} is responsible for nutritional problems. The acidic situation is being addressed with a variety of tried-and-true approaches, including the installation of an adequate water management system, the utilisation of lime, and others. However, these approaches only provide momentary gratification and can have an impact on the fertility of the land. As a more sustainable and environmentally acceptable option, the use of microorganisms to reduce the toxicity of metals is rising in popularity (Varghese et al. 2022).

3.17.8 Initiation of the Formation of Acid Sulphate Soils

In most cases, acidic soils can be found in regions that receive a lot of rainfall. As a result of excessive rainfall, soluble basic salts of potassium, calcium, magnesium, and sodium are washed away from soils, leaving behind significant amounts of insoluble oxides and silicates of aluminium, iron, silicon, and manganese. These compounds are very acidic. In addition, the acid rain has a greater impact since it adds surplus quantities of hydrogen ions (H^+) to the soil, which helps to contribute to the acidity of the soil by displacing all of the soil's cations. Other key processes that contribute to the creation of acid soil are the intensive use of acid-forming ammonium-based nitrogenous fertilisers, the uptake of base elements and lime by crops, and the degradation of organic matter (Bian et al. 2013; Rahman et al. 2018).

In the coastal plains, acid soils that contain pyrite (FeS_2) are referred to as "acid sulphate soils" because of the submerged circumstances. The majority of rice is cultivated in these acid sulphate soils; the submerged circumstances and intermittent draining processes that occur during rice cultivation help oxidise pyrites in these soils to produce sulphuric acid (Fig. 3.9). According to research from Indraratna et al. (2006), acid production causes aluminium and iron ions to get mobilised, which further lowers the pH of the soil. Both Acidithiobacillus ferrooxidans and Acidithiobacillus thiooxidans, which were once known as Thiobacillus thiooxidans, are the most active oxidizers in soils with a pH lower than 4 and are responsible for the development of acid sulphate soil. Both of these strains were originally designated Thiobacillus ferrooxidans. These bacteria transform dissolved sulphates into sulphides, which, when combined with metals, especially iron, make metal sulphide.

Oxidation of pyrite (FeS_2) and formation of sulphuric acid.

$$2FeS_2 + 2H_2O \rightarrow FeSO_4 + H_2SO_4$$

Thiobacillus ferrooxidans

$$4FeSO_4 + O_2 + 2H_2SO_4 \rightarrow 2Fe_2(SO_4) + 2H_2O$$

$$Fe_2(SO_4)_3 + FeS_2 \rightarrow 3FeSO_4 + 2S$$

Thiobacillus thiooxidans

$$2S + 3O_2 + H_2O \rightarrow 2H_2SO_4$$

Acidic soils have pH values that are persistently lower than 5.5 over a long period of time. Acidity in the soil can be brought on by agricultural practises such flooding and the use of agrochemicals. These soils are exceedingly difficult to cultivate because they are contaminated with the harmful metals aluminium (Al) and iron (Fe^{2+}), which significantly reduce crop production. Increased soil Al levels impede plant root growth, while too much Fe^{2+} disrupts the balance of nutrients. In order to lessen the acidic condition, a number of traditional approaches are being used,

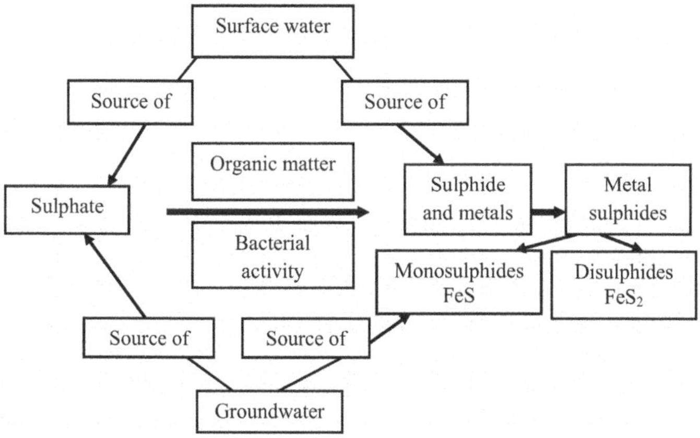

Fig. 3.9 Formation and accumulation of acid sulphate soil. *Source* Varghese et al. (2022)

including a good water management system and the application of lime. However, these techniques only provide short-term satisfaction and may reduce soil fertility. As an environmentally beneficial solution, using microorganisms to reduce such metal toxicity is gaining appeal. The main methods by which agriculturally significant microorganisms confer Al and Fe stress tolerance on plants are covered in this chapter, along with the difficulties of growing plants in acidic soils and the effects of Al and Fe toxicity on crop plants.

Most soils cultivated for crop production fall within the pH range of pH 6–8, where nutrient availability to the plant is optimal. Acid soils (pH 8), however, fall outside this optimal pH range and pose challenges for the plant such as low nutrient availability, ion toxicities and nutrient imbalances (Läuchli and Grattan 2012). Most nutrients are not equally available to plants across the pH spectrum. Several mineral nutrients are severely affected in these non-optimal pH soils, particularly Ca, K, P and Fe.

The pH of the soil solution has a profound influence on nutrient availability, nutrient uptake and ion toxicity to plants. The vast majority of soils that are cultivated for crop production around the world fall within the neutral, slightly acid and slightly basic pH range (i.e. pH 6–8). This is the general range where nutrient availability is optimal. However, there are those soils where the pH falls far from this normal range and these, if not corrected to an adequate range, can pose adverse health effects on plants. Soils that are highly acidic ($pH < 5.5$) present a spectrum of challenges for the plant, including nutrient availability, ion toxicities, and nutrient imbalances influencing the ion relations and nutrition within the plant itself. The pH of the soil solution is an indication of the activity of the hydrogen ion. The pH numerical value is expressed on a negative log scale such that a one unit increase or decrease corresponds to a ten-fold increase or decrease in the hydrogen ion activity. Therefore a change of soil pH from 6 to 8 corresponds to a 100-fold decrease in the hydrogen ion activity. Plants often differ in their response when grown on soils with extreme pH values. Native plants that have adapted on acid soils are referred to as calcifuges while those that have adapted on alkaline soils are called calcicoles (Lee 1998). Most crop plants are neither and grow optimally when grown in soils near neutrality. Therefore these extreme soils can provide challenges for the farmer as the plant often suffers nutritional disorders and toxicities should the soil not be adjusted to a more optimal pH range.

Acid soils are characterised by having a soil solution pH below neutrality (pH < 7) but strongly acidic soils, those with a pH < 5.5, can be detrimental to the plant either by imposing nutrient deficiencies or ion toxicities (Marschner 1995). In highly acidic mineral soils, the activity of the H + ion is very high, which in itself can be toxic to plants and affect ion transport mechanisms across cell membranes such as membrane bound ATPases. Acid soils also increase the solubility of Mn^{2+} and Al^{3+}, leading to plant toxicity. Aluminium and manganese toxicity harm the plant differently. Al^{3+} toxicity damages roots, adversely affecting their development, which in turn limits nutrient and water uptake (Rout et al. 2001). Although Al^{3+} toxicity varies among plants, its toxicity to roots results in damage to the apical meristems, similar to the effect of Ca^{2+} deficiency, giving the root system a stubby appearance (Delhaize and Ryan 1995). As pH of the soil solution decreases, Al-hydroxide species are deceased with a corresponding increase in the free Al^{3+} species. The mechanism of Al^{3+} toxicity in plants is unclear but it may involve some interaction with Ca^{2+} since apoplastic Ca^{2+} is displaced by Al^{3+} and inhibits its uptake (Delhaize and Ryan 1995). Al^{3+} has over a 500-fold affinity for certain phospholipids over Ca^{2+}, thereby potentially affecting membrane function (Marschner 1995). The actual mechanisms regarding Al^{3+} toxicity to plants are poorly understood but it involves exclusionary mechanisms as well as internal tolerance mechanisms. Rout et al. (2001) also point out that differential tolerance probably involves differences in the structure and function of roots. Mn^{2+} toxicity, on the other hand, is controlled by the scion's ability to control this trace metal from accumulating in leaves (Marschner 1995). Interestingly, the presence of Mg^{2+} has been shown to increase the tolerance to high manganese in shoot tissue of wheat and increased the plant's ability to discriminate against Mn^{2+} ions (Goss and Carvalho 1992). Therefore both Al^{3+} and Mn^{2+} toxicity can be alleviated by increasing the soil solution pH and perhaps by addition of calcium and magnesium, respectively. Some plants, "the calcifuges," have evolved in these extreme acid environments. These plants tolerate low levels of Ca^{2+} and high acidity. Clearly they have adapted a mechanism to Al^{3+} tolerance, but adaptation can occur either to "tolerance of stress" or "avoidance of stress" (Marschner 1995). At the same time, these tolerant species must have a mechanism allowing for highly efficient uptake and utilisation of Ca^{2+}, Mg^{2+} and P, which occur in very low concentration in these soils. However, the vast majority of plants cannot tolerate these acidic conditions and would require soil amendments to increase soil pH and nutrient concentration in order to perform well. Acid soils are also often deficient in Ca^{2+}, Mg^{2+}, P and Mo (Marschner 1995). With the high activity of H^+ in the soil solution, the loading of the divalent cations into the apoplast of root cortical cells decreases. Moreover, acid soils are characteristic of highly leached soils where these bases have been naturally depleted from the soils thereby aggravating the deficiency.

The rapid decline of soil pH was attributed to long-term high atmospheric acid deposition (nitrogen and sulphur) therein. The forest at earlier succession stage with originally higher soil pH appeared to be more vulnerable to acid deposition than that at later succession stage with originally low soil pH (Liu et al. 2010). Soil acidification can be a natural process of proton-generating reactions, including the protolysis of carboxylic acid and natural organic acids and the uptake of cations. However, this process can be accelerated, particularly by deposition of sulphur (S) and nitrogen (N) compounds (Richter 1986; Kazda and Zvacek 1989). Soil acidification likely results in loss of base cations and mobilisation of aluminium (Al) and other potentially toxic elements (Pichtel et al. 1997; Gilliam et al. 2005). The decrease of soil pH occurred throughout the soil profiles and was accompanied by considerable losses of exchangeable of base cations, which decreased by about 50% of the pool in (Falkengren-Grerup 1987). Similarly, the calcium pools in forests in Norway and the northern US have been reported to decline about 50% in the last 45 years (Likens et al. 1996). Soil pH has reached as low as 3.2–3.6 in areas of southern Germany (Z¨ottl et al. 1989). Southern China has become the third-largest producer of acid rain in the world, after the northeastern US and central Europe (Larssen et al. 1999). The total annual SO_2 emissions in China increased constantly from 1982 to 1997 and slightly decreased in 1998–2004, but then increased and reached the highest point of 25.5 million tons in 2005 (MEPC 1996–2006).

3.17.9 pH Indicator

A solution's pH (acidity or basicity) can be visually assessed by adding a small amount of a halochromic chemical substance known as a pH indicator to the solution. In the Arrhenius model, a pH indicator is therefore a chemical detector for hydronium ions (H_3O^+) or hydrogen ions (H^+). According to the pH, the indicator typically causes the colour of the solution to change. Olfactory indicators, for instance, display changes in their odour as indications of other physical qualities (Huang et al. 2008). The pH of a neutral solution is 7.0 when the temperature is 25 °C, which is regarded as the standard. Acidic solutions are those with a pH value less than 7.0, and basic (alkaline) solutions are those with a pH value more than 7.0. As carboxylic acids and amines, which are weak protolytes, make up the majority of naturally occurring organic molecules, pH indicators have various uses in analytical chemistry and biology. Additionally, one of the three primary categories of indicator chemicals utilised in chemical analysis is pH indicators. Complexometric indicators are favoured for the quantitative measurement of metal cations [1, 2], whereas titrations using a redox reaction as the analysis's foundation employ redox indicators from the third chemical class. pH sensors are usually weak acids or bases on their own. A pH indicator's general reaction pattern can be expressed as follows:

$$H_2O + +\text{Ind-Here} = \text{Hind} + H_2O$$

The abbreviations HInd and Ind- denote the indicator's conjugate base and acid, respectively. The ratio of these defines the colour of the solution and establishes a link between the colour and pH level. We may express pH indicators as weak protolytes using the Henderson-Hasselbalch equation as follows:

$$\text{pH is equal to pka} + \log([\text{Ind-}])/([\text{HIInd}]).$$

The equation, derived from the acidity constant, asserts that both species are present in a 1:1 ratio when pH equals the pKa value of the indicator. When the pH is higher than the pKa value, the conjugate base concentration is greater than the acid concentration, and the conjugate base colour predominates. The opposite is true if pH is lower than the pKa value. At the pKa value, the colour usually does not change immediately, but there is a pH range where a variety of colours are present. As a general rule, this pH range lies between the pKa value plus or minus one, albeit it varies amongst indicators. This is based on the idea that solutions keep their colour for as long as at least 10% of the other species are still around. For instance, if the ratio of the concentration of the conjugate base to the concentration of the acid is 10:1, the pH will be pKa + 1. In contrast, the ratio is 1:10 and the pH is pKa − 1 if the acid is ten times more abundant than the base.

An indicator may be used to obtain quite precise measurements of pH by measuring absorbance quantitatively at two or more wavelengths. The principle can be illustrated by taking the indicator to be a simple acid, HA, which dissociates into H^+ and A^-.

$$HA \rightarrow H^+ + A^-$$

3.18 Acid, Base, and Ion Exchange in the Soil

If pH is too high, elemental sulphur, concentrated sulphuric acid or aluminium sulphate can be added to the soil to decrease alkalinity. Caution should be taken when adding these acidifying materials. If they are over applied the pH can drop into the 2 to 3 range in small areas of the soil and potentially kill plants in these areas. Iron chlorosis, a deficiency of iron in the plant, can be corrected by applying chelated iron or iron sulphate to the soil or foliage. The term chelate comes from the Greek word claw. Chelates are chemical bonds (claws) that help hold metal ions, such as iron, in solution so that plants can absorb them. Different chemicals, from relatively simple natural chelates like citrate, to more complex manufactured chemicals, can act as chelates. When a chelated metal is added to the soil, the nutrient held by the chelate will remain available to the plant longer than if the sulphate salt is applied to the soil. Most nutrients do not require the addition of a chelate to help absorption. Only a few of the metals, such as iron and zinc, benefit from the addition of chelates. The types of chelate used will depend on the nutrient needed and the soil pH. Iron is normally applied foliarly, just wetting the leaf, using a 2% iron solution. This is because the soil binds with the iron so rapidly that as much as 10–15 lbs of iron may not meet the plant iron requirement. Zinc can be applied foliarly using a 0.5% zinc solution or applied to the soil. The chelated form works best for soil application. The sulphate or chelated form can be used for foliar application. Acidification also occurs when base cations such as calcium, magnesium, potassium and sodium are lost from the soil. Losses occur when these bases are leached from the soil. This leaching increases with increasing precipitation. Acid rain accelerates the leaching of bases. Plants take bases from the soil as they grow, donating a proton in exchange for each base cation. Where plant material is removed, as when a forest is logged or crops are harvested, the bases they have taken up are permanently lost from the soil. Many plants produce organic acids. Where plant litter accumulates on or is incorporated to the soil, these acids (including acetic acid, humic acid, oxalic acid, and tannic acid) are liberated. This is especially acute in soils under coniferous trees such as pine, spruce and fir, which return fewer base cations to the soil than do most deciduous trees. Certain parent materials also contribute to soil acidification. Granites and their allied igneous rocks are called "acidic" because they have a lot of free quartz, which produces silicic acid on weathering. Also, they have relatively low amounts of calcium and magnesium. Some sedimentary rocks such as shale and coal are rich in sulphides, which, when hydrated and oxidised, produce sulphuric acid which is much stronger than silicic acid.

3.19 pH and Soil Fertility of Sub-merged Soils

The pH values of acid soils increase on submergence whereas the opposite occurs in calcareous and sodic soils. For most soils the fairly stable pH attained after several weeks of submergence is between 6.5 and 7.0. The pH of a submerged soil exerts a marked influence on the capacity of the soil to supply nutrients through

1. Direct effects on nutrient absorption.
2. Direct effects on the concentrations of nutrients or toxic substances in the soil solution.
3. Indirect effects on chemical equilibria, and sorption and desorption.
4. Influence on microbial processes connected with the release or loss of plant nutrients and the generation of toxic substances.

Direct effects on nutrient absorption

Plants can absorb nutrients and grow in the pH range from 4 to 8 if the secondary effects of pH on nutrient deficiencies and toxicities are eliminated (Arnon and Johnson 1942). Rorison (1972), however, suggests that at low concentrations of other ions, hydrogen ions per se could be more toxic than Arnon and Johnson (1942) suggested. Cation absorption by roots reaches its maximum between pH 5 and 7, and anion absorption decreases above pH 6 (Moore 1972). The increase in pH of acid soils and the decrease in pH of alkaline soils brought about by flooding (Ponnamperuma et al. 1966a) favour nutrient uptake by rice.

Effects on concentration of nutrients or toxins

A soil's pH has marked, direct influence on the concentrations of iron, manganese, and aluminium in submerged soils (Table 3.11). For most oxisols and ultisols that undergo reversible oxidation–reduction seasonally, the activity of Fe^{2+} in the soil solution (the concentration will be higher) is given by the equation pH–1/2 pFe = 5.4. The iron (II) hydroxide potential (pH–1/2 pFe) appears to be a characteristic for a given soil and remains constant during long periods of submergence (IRRI 1968; Ponnamperuma 1972). If its value is determined for a soil, it can be used to calculate the activity of Fe. The only additional parameter required for the calculation of concentration is the specific conductance from which the ionic strength is derived (Ponnamperuma et al. 1966b). The peak values for more than 100 wetland rice soils ranged from 6600 ppm at pH 5.67 for an acid sulphate soil to 0.07 ppm at pH 7.25 for a calcareous soil low in organic matter.

The practical implications of the pH–Fe^{2+} relationship for the fertility of submerged soils are that at high pH values rice suffers from iron deficiency, and that at low pH values rice is affected by iron toxicity (Ponnamperuma 1955; Tanaka and Yoshida 1970) and by phosphorus and potassium deficiencies induced by excess Fe^{2+} (Yamada 1959). High concentrations of aluminium and manganese are common causes of crop failure in acid aerobic soils. Except in certain acid sulphate soils, aluminium toxicity disappears within a few weeks of soil submergence because of the increase in pH.

Effects on chemical equilibria and sorption and desorption

Mineral phosphorus is present in soils mainly as aluminium and iron phosphates at low pH, and as calcium phosphates at high pH, and is found sorbed on the surfaces of clay and oxides of aluminium and iron (Russell 1973). An increase in pH values favours the solubility of aluminium and iron phosphates and desorption of phosphorus, and a decrease in pH favours the solubility of Thus the solubility of calcium phosphates (Stumm and Morgan 1970). Phosphorus in soils tends to be maximum in the pH range from 6 to 7. The increase in pH of acid soils and the decrease in pH

Table 3.11 Influence of soil properties on the Fe(II) hydroxide potential $pFe(OH)_2$ and the concentration of water-soluble Fe^{2+} at pH 6.5 and ionic strength of 0.03 mol/litre

pH	OM (%)	Fe (%)	Mn (%)	$pFe(OH)_2$	Fe^{2+} (ppm)
4.6	4.1	2.8	0.02	5.41	629
5.9	3.3	1.7	0.33	5.09	195
6.6	2.0	1.6	0.31	4.90	64
7.6	1.5	0.3	0.06	4.83	43

Source Ponnamperuma (1977)

of calcareous and sodic soils increase the availability of phosphorus in submerged soils. This is a benefit of flooding rice soils (Ponnamperuma 1965).

The soils of southern Karnataka are, in general, acidic with poor base saturation and low exchange capacity. Free oxides of Fe and Al and kaolinite clays dominate in these soils. The productivity potential of these soils is limited (Dhananjaya and Ananthanarayana 2009). Further, there is a concern that the severity of soil acidity may increase because of poor management and heavy rainfall (Krishnamurthy Rao and Ramaiah 1993; Ananthanarayana and Ravindra 1998). Soil acidity is usually defined in terms of KCI extractable (permanent) and pH-dependent acidity (Coleman and Thomas 1967). The permanent acidity is also ascribed to isomorphous substitution whereas the second type is ascribed to the polymers of Fe and AI, the kaolinite clays and the organic matter. The proportional amount of these two types of acidities determines the production potential, phosphate fixation and lime requirement of soils.

Dhananjaya and Ananthanarayana (2009) analysed two surface soil samples representing acidic U1tisol and Entisol to study the nature of soil acidity and exchange chemistry. The soils under study were sandy loam to sandy clay loam in texture with low pH, CEC and base saturation. The negative values of ΔpH in both the soils indicate that the soils carry net negative charge under natural soil pH conditions. The low ZPC (zero point charge) values recorded for these acid soils were low which might be due to high organic matter in the surface soils. The negative pH (KCI)-ZPC values indicate highly weathered nature of these acid soils. The variation in the values of the ZPC, ΔpH, surface potential and pH (KCI)-ZPC values is attributed to the varying contents of organic matter, clay, Fe and AI oxides and hydroxides and dominance of kaolinite clay in these soils. The values of total potential acidity, pH-dependent acidity and total acidity were very high, exchange and extractable acidity values were very low thus contributing little towards total potential and total acidity. The pH-dependent acidity contributed more than 94 per cent towards total potential acidity. The extractable acidity value was higher than exchange acidity. The values of hydrolytic acidity were higher than exchange and extractable acidity. The lime requirement (LR) values for the soils revealed that LR as estimated by Adams–Evans buffer method, Incubation method, SMP buffer method and exchangeable Ca method were higher than those determined by other methods in Sirsi soil whereas LR as estimated by SMP-buffer method was higher than that determined by other methods in case of Mudigere soil (Dhananjaya and Ananthanarayana 2009).

3.19.1 Production Constrains of Acid Soils

A. *Related to soil physical properties*

- Leaching of bases, fine earth and organic matter
- Coarse texture soils (high macro pore)
- Low organic matter, weak soil structure

- High infiltration and permeability rate,
- Low moisture holding capacity
- Formation of soil crust, particularly in red and laterite soils (seed germination is affected)
- More susceptible to drought, dry spells
- Poor drainage of peat, marshy and acid sulphate soils

B. *Related to soil chemical properties*

- Low P-availability and high P-fixation capacity
- Toxicity of Fe and Mn (Al in some cases)
- Low organic matter status of lateritic and red soils
- Deficiency of secondary nutrients (Ca, Mg), and micronutrients (Mo, B and Zn)
- Leaching losses of nutrients are high
- Nutrient imbalance, low cation exchange capacity
- Dominance of kaolinite and sesquioxides clay fraction (low water holding capacity)
- Low level of base saturation percentage.

3.19.2 Soil Amendments

Due to the serious issues with soil acidification, it is essential to create new soil additives to reduce soil acidity. In three acidic soils, Dai et al. (2014) investigated the impact of biochars made from various feedstocks on soil pH, N transformation, and pH buffering as well as the mechanisms underlying changes in these variables. By Day 100, the pH of the soil had stabilised. By Day 180 of incubation, the addition of biochar raised the pH of the soil by 0.5 to 1 units at a 1% incorporation rate and by 1 to 2 units at a 3% rate. The alkalinity (excess cations) of biochar and N nitrification in soils both had a role in determining the impact of biochar additions on pH alterations. In the Plinthudult (clayey loamy, kaolinitic, thermic Typic Plinthudult) and Paleudalf (clayey loamy, illitic, thermic Typic Paleudalf) soils, both the alkalinity of the biochars and nitrification in the soils contributed to the soil pH changes (Dai et al. 2014). Additionally, the pH increase was significantly influenced by biochar alkalinity while the pH reduction was primarily caused by N nitrification. During the incubation, biochar undoubtedly had a favourable priming impact on soil organic N. The pH buffering of soil was generally increased by biochar, while the increases in pH buffering varied between soils and biochar.

Acidification of soils in intensive agricultural systems has been documented (Barak et al. 1997; Zhang et al. 2008); it is linked to overuse of NH_4^+-based fertilisers, plant harvest (Bolan et al. 1991), and acid deposition (Zhang et al. 2007). Finding more efficient soil additives to address soil acidity and other qualities is urgently needed due to the nutrient deficits and rising availability of hazardous metals in soils brought on by acidification. In an effort to reduce soil acidity, lime, minerals, industrial wastes (such as alkaline slag, coal fly ash, and red mud), and integration of plant residues are frequently employed (Xu et al. 2006b; Li et al. 1990). However, given the potential for long-term soil improvements as well as additional advantages including effective waste management, reducing the effects of climate change, and energy production, biochar is seen as an efficient and all-encompassing soil supplement to reduce soil acidity. The pyrolysis temperature and the kind of feedstock affect the fundamental characteristics of biochar (Enders et al. 2012; Meng et al. 2013). According to Tsai et al. (2012), when the pyrolysis temperature for the formation of biochar rose, the pH of the resulting materials rose to a maximum of 11.64. According to Yuan and Xu (2011), biochar made from legumes increased soil pH more than biochar made from nonlegumes. When Singh et al. (2010) characterised a variety of biochars made from various feedstocks, they found that poultry litter biochar had the highest pH. The processes behind the dynamics of pH rises and pH buffering changes in soils brought on by biochar are currently unclear. Even though research on biochar's ability to improve acid soils is growing (Yuan and Xu 2011; Yuan et al. 2011a; Hass et al. 2012), the majority of studies have only focussed on one type, and there have been very few comparisons of the effects of biochars on a variety of soil types (Yuan and Xu 2012).

3.19.3 *Transformations of NO_3 and NH_4^+*

The three soils treated with biochar at 1% and 3% all experienced a rise in NO_3 concentrations over time (Dai et al. 2014). The Psammaquent soil had the lowest final NO_3^- concentration, followed by the Plinthudult, and the Paleudalf soil had the highest final NO_3^- concentration. For instance, NO_3 concentrations increased by 21.4 mg kg^{-1} with swine dung biochar, 10.0 mg kg^{-1} with reed straw biochar, 9.5 mg kg^{-1} with pineapple peel biochar, and 4.1 mg kg^{-1} with rape straw biochar in the Psammaquent soil from Day 2 to Day 180 when these biochars were applied. In contrast to the changes in NO_3 concentrations, there was a minor increase in NH_4^+ concentration following the addition of biochar from Day 2 to 9 and subsequently a significant decrease from Day 9 to 180. From Day 2 to Day 180, the Plinthudult and Paleudalf soils experienced substantially faster declines in NH_4^+ concentrations than did the Psammaquent soil. In the Plinthudult and Paleudalf soils, all biochar additions resulted in larger drops in NH_4^+ concentrations than their controls, however in the Psammaquent soil, only the addition of swine manure-derived biochar resulted in a decrease that was greater than the control (Dai et al. 2014).

3.19.4 *PH Buffering Alterations*

Correlation coefficients (R^2) for the pH buffering of the three soils treated with the four types of biochar at 1% and 3% were all > 0.92 and frequently > 0.95.

Swine manure biochar significantly increased the pH buffering capacity (b) at 3% in comparison to the control, resulting in pH values of 20.96 for Psammaquent soil (compared to 13.33 for the control), 26.81 for Plinthudult soil (compared to 11.34), and 32.26 for Paleudalf soil (compared to 14.66). Rape straw biochar and pineapple peel biochar were next, with reed straw biochar being the lowest. In contrast to the 3% rate, however, the inclusion at 1% only resulted in a minor improvement or perhaps a decrease in buffering capacity. Reed straw biochar had little to no effect on the three soils' ability to act as a buffer at either the 1% or the 3% concentrations (Dai et al. 2014).

3.19.5 Integrated Natural Resource Management (INRM)

We believe that in order to align with current soil quality notions, soil quality needs to be seen in a larger context and as a part of an integrated natural resource management framework (INRM). This INRM technique includes the soil component as a crucial component (Dumanski et al. 1998). After years of experimenting with methods to conduct inter-disciplinary research, the international agricultural research centres have established an INRM research framework, depicted in Fig. 3.10. INRM is described as "the sustainable use of the resource base of agriculture." (Izac and Sanchez 2001; CIFOR 2000).

3.19.6 Acid Soil Aluminium Chemistry

The amount of hydrogen (H^+) ion activity in the soil solution, as well as edaphic, environmental, and biological conditions, determine how acidic the soil is. Compared to soils formed from calcareous parent materials, those formed from granite parent materials acidify more quickly. Sandy soils with less clay tend to become acidic more quickly because of their lower alkaline cation reserves and increased leaching potential. High rainfall has an impact on how quickly water percolates through the soil profile, which in turn impacts how quickly soil becomes acidic. Acid precipitation from contaminated air and the loss of cations through crop harvesting both intensify soil acidification (Ulrich et al. 1980). Acidification is also a result of organic materials decomposing and producing weak acids such as carbonic acid. Al is extremely pH-dependent in terms of solubility, and a rise in soil pH reduces Al's toxicity. It can be found in soil solution as soluble ionic species or as a fundamental structural element in secondary clay minerals like kaolinite and vermiculite. Aluminium is polymerized on clay surfaces and between clay interlayers, adsorbed on organic and mineral colloidal surfaces, and exists in amorphous, crystalline, and hydroxy-Al forms. Al ions satisfy a sizeable amount of the cation exchange capacity in acidic soils, meaning that these soils have a high Al saturation level. Despite being referred to as exchangeable Al^{3+}, these Al ions are actually a combination of monomeric Al ions [Al^{3+}, $AlOH_2{}^+$, and

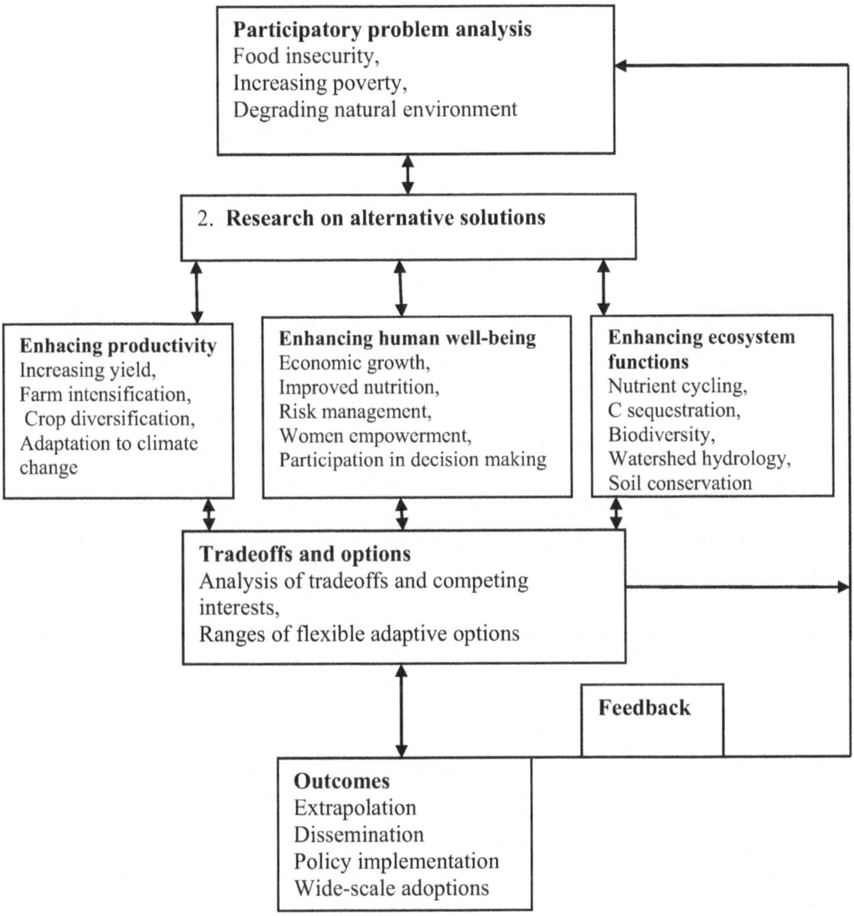

Fig. 3.10 A Worldwide agricultural research community-use model of the integrated natural resource management research process. *Source* CIFOR (2000), modified by Sanchez et al. (2003) and authors of this book

$Al(OH)_2^+$] has a pH-dependent average charge per Al between 2 and 3, declining. Aluminium is firmly bound to exchange sites, therefore soil solution concentrations are typically modest, frequently falling between 10 and 250 μM.

3.20 Subsoil Acidity

Reducing aluminium toxicity and calcium deficiency in the subsoil is a major but difficult management imperative in many areas of the tropics, in order for roots to fully tap the entire soil profile for moisture and nutrients. Deep lime incorporation by machinery is usually not feasible much beyond 30 cm, but other ways are possible. The use of aluminium-tolerant cultivars, together with moving calcium to the subsoil, became an important breakthrough. Downward calcium and magnesium movement has been observed after liming the topsoil in Oxisols, Andisols and other well-aggregated soils within a few years of application. Early reports by Pearson et al. (1962) and Amaral et al. (1965) obtained field evidence of downward calcium and magnesium movement in Oxisols in Puerto Rico and Brazil, while Mahilum et al. (1970) and Morelli et al. (1971) did so in Andisols in Hawaii and Costa Rica. The downward movement of calcium and magnesium occurs in wellaggregated soils when they are limed at the rates needed to neutralise exchangeable Al^{3+} or at higher rates. Once the topsoil's exchange sites are base-saturated, exchangeable Ca^{2+} and Mg^{2+} held on these sites moves down. However, much depends on the accompanying anion. Therefore, the calcium and magnesium movement depends on anions already present in the topsoil, which are free to accompany the cations because topsoils seldom have anion exchange capacity. The nitrate (NO_3^-) anion (a product of nitrification) and the sulphate anion (SO_4^{2-}) in gypsum are the principal accompanying anions. In an early laboratory study with a South African Oxisol, Reeve and Sumner (1972) showed much faster movement of calcium and magnesium to the subsoil when gypsum ($CaSO_4$) was used instead of $Ca(OH)_2$. Sanchez (2019) summarised his experiences as below:

- Unlike lime, gypsum does not neutralise exchangeable Al^{3+} or increase pH.
- Gypsum reduces aluminium saturation, by adding exchangeable Ca^{2+} to the ECEC.
- Gypsum eliminates calcium deficiency in the subsoil, which is probably more important than subsoil aluminium toxicity in many Oxisols and Ultisols. Calcium must be present in the soil around the root tips since it is not translocated from the plant tops.
- Gypsum increases the uptake of soil moisture from the subsoil, preventing drought stress during dry spells in the rainy season (veranicos in Brazil).
- The sulphate anions in gypsum form non-toxic insoluble aluminium complexes, decreasing aluminium in the soil solution and aluminium saturation in the subsoil as well.
- Nitrate ions not taken up by plants can rapidly leach to the subsoil where they are captured by the anion exchange capacity present in oxidic subsoils. Deep roots can capture this nitrate.
- Gypsum applications can decrease exchangeable Mg^{2+} in the topsoil and move it downwards, necessitating dolomitic lime on $MgSO_4$ applications if the topsoil levels are low.

- Decreases in exchangeable K^+ have happened occasionally in fruit crops, but seldom in cereal crops, because crops recycle leached potassium with their deep roots, promoted by gypsum, and add it to the soil surface as stover.
- Gypsum increases the sulphur content of forage, or cereal stover that is used as forage, to the required level of 4 g S/kg of dry mass.
- The residual effects of gypsum surface applications can last for at least 5 years in the subsoil.
- Crop yields often double, and the effects of more frequent dry spells and rainfall variability, often attributed to climate change, are attenuated. Gypsum is now recommended for Oxisols, oxidic Ultisols and acid, sandy Entisols and Inceptisols.

3.20.1 Forms of Acidity in Soil

1. Strong acidity—This is due to presence of soluble Al and absorbed Al. Aluminium being trivalent preferably adsorbed as compared to H^+ ions. This Al undergoes hydrolysis forming Al hydroxide. If pH is < 5.0, than Al may be present in the form of hexa hydrated Al ions and it may occupy more than 40% of exchange capacity.

$$[Al(H_2O)_6] + H_2O \rightarrow [Al(H_2O)_5(OH)]^{++} + H^+$$
$$[(Al(H_2O)_5)OH]^{++} + H_2O \rightarrow [Al(H_2O)_4(OH)_2]^{+} + H^+$$
$$[Al(H_2O)_4(OH)_2]^{+} + H_2O \rightarrow [Al(H_2O)_3(OH)_3] + H^+$$

Due to this formation effective cation exchange capacity will decreases, but soil pH increases.

2. Moderate acidity: In this case per cent base saturation is higher than strong acidity. Aluminium present not as on Al ion but as hydroxy ions. This hydroxy ion move in between the crystal unit and are tightly adsorbed thereby presenting expansion and blocking the exchange site.
3. Weak acidity: It is due to the presence of hydrated Al^{3+} ion and H^+ ions. pH of the weak acidity is 5.2.
4. Very weak acidity (pH 5.2–6.5): This acidity is due to the presence of COOH group of organic origin and hydrogen group of hydroxy Al polymer in the interlayer position and due to carbonic acid and basic Al sulphate.
5. Very–Very weak acidity: This acidity due to phenol group of organic acids, hydrogen group of hydroxy Al polymers. Bicarbonate of Ca and Mg.
6. Extremely weak acidity: Acidity is due to presence of alcoholic group of organic matter, salicylic acid, gypsic acid etc.

3.20.2 Effect of Soil Acidity on Plants Growth

Soil acidity influence plant growth by the production of organic acid which are produces due to decomposition of organic matter or due to root secretion. Effect of soil acidity may be separated into two group.

Direct

1. Toxicity effects of H^+ ions are observed in root tissues.
2. Permeability of the plant membranes for cations is affected.
3. Balance between basic and acidic constituents through roots is disturbed.
4. Enzyme changes occur in plants due to pH changes.

Indirect

1. Availability of some nutrients like phosphorous.
2. Higher availability of Al, Mn, Fe, Zn, Cu etc.
3. Adverse effects on the beneficial activities of micro-organism.
4. Increase of plant diseases.
5. Nutrients like Ca, K, Mo, etc. become deficient.

Nutrient availability

Acid soils are robust because of this loss of nutrients like potassium, calcium, and magnesium that can be taken in hay and grain or depleted by leaching. A soil's structural issues may also be caused by a calcium deficiency. As soil acidity rises, nutrients like manganese and phosphorus may become unavailable (McFarland et al. 2001). Soil pH has an impact on the nutrients that are available to plants. The major plant nutrients nitrogen, phosphorous, potassium, sulphur, calcium, manganese, and the trace element molybdenum are less available and may even be insufficient in acidic soils (Fig. 3.11). The availability of these nutrients will rise when the pH of acidic soil is raised through liming. In acidic soils, there is a greater abundance of iron, manganese, copper, zinc, and aluminium (Ngoune et al. 2018). Nitrate, other nutrients, and heavy metals from the soil profile are being leached, which is leading to increased groundwater and surface water pollution (Mcfarland and colleagues 2001).

3.20.3 Acidity of the Soil Microbes

Reduced microbial activity In moderately and strongly acidic soils, microbial activity that fixes nitrogen or breaks down organic matter is reduced. Reduced land values can result from acidity-related loss of productive capacity (McFarland et al. 2001). Because the growth and reproduction of soil microbes, principally bacteria and fungi, as well as the breakdown of organic matter and the cycling of nutrients, are all reduced in acidic soil, the majority of microbial processes are also affected. Plant uptake may

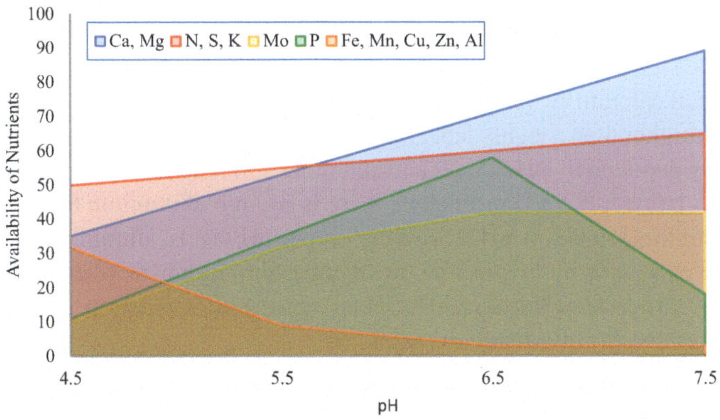

Fig. 3.11 Relationship between soil pH and nutrient availability, in acidic soils. *Source* Nelson and Su (2010)

be hampered by the slower rate at which soil bacteria convert nutrients into forms that plants can use. In root nodules, nitrogen-fixing rhizobia bacteria collaborate with crop and pasture legumes to produce nitrogen. According to Quilliam et al. (2013), low pH can have an impact on the microbial activity in top soils, particularly by reducing the nodulation of legumes. In acidic soils, rhizobia bacteria are significantly decreased. Some pasture legumes may not survive because diminished rhizobia populations are unable to adequately nodulate roots and form a functional symbiosis (Unkovich 2013).

3.21 Reactions Involving Aluminium

Although Al is one of the most abundant elements in soils, it is not very mobile at the pH values encountered in most soils (between 6 and 8). The solubility of Al increases in acidic soils (pH < 6) and in the presence of complexing ligands (Driscoll 1989). In naturally acidic soils dissolved Al is predominantly in an organically complexed form (David and Driscoll 1984; Driscoll et al. 1985). Introduction of high loads of mineral acidity results in a significant mobilisation of inorganic AI, so that it may become the dominant cation in solution (Mulder et al. 1989). Dissolved Al is potentially toxic to terrestrial biota and has been hypothesised to be one of the major causes of the occurrence of widespread forest decline in central Europe (e.g. Ulrich et al. 1980). Major sources of dissolved Al include a number of operationally defined pools, such as exchangeable AI, organically bound AI, amorphous Al oxide and crystalline Al oxide (e.g. gibbsite). Unfortunately, fluxes between these (physicochemically poorly defined) pools and solution are unknown.

3.21.1 Aluminium and pH: A Connection

The amount of calcium available determines how well the soil can withstand pH shifts at pH levels higher than roughly 5.5. However, the calcium content has typically been substantially depleted by pH 5.5. As a result, other soil reactions must assume the job of the buffer below pH 5.5. Fortunately, there is enough aluminium in most soils for this ion to fill this job below pH 5.5. At the proper pH levels, aluminium compounds [$Al(OH)_3$] react with hydrogen ions to liberate aluminium in what are known as "hydrolysis" processes (Harter 2007). The overall buffer reaction is, even though there are a number of intermediary reactions:

$$Al(OH)_3 + 3H + \text{equals } Al^{3+} + 3H_2O.$$

Particle surfaces in soils often have a negative charge, which attracts and holds positively charged base cations including calcium, manganese, potassium, sodium, and others. With very few exceptions, soil particles preferably retain hydrogen ions over these other ions. This means that some of the hydrogen ions will displace these base ions anytime one of the acidification reactions raises the hydrogen ion concentration in the soil, making the base ions more likely to leach out of the soil. If the bases aren't supplied by the natural world or by humans, well-drained, humid region soils eventually become quite acidic and devoid of base ions. As this process goes on, the concentration of hydrogen (H^+) on the surface eventually rises to the point where it attacks the mineral structures directly, liberating silica ($Si_4{}^+$), iron ($Fe_3{}^+$), and aluminium (Al^{3+}) as well as other ions from the minerals. The released iron and aluminium ions frequently emerge from the soil solution and transform into hydrous oxides (sesquioxides), which remain in the soil, whilst the silica ions drain to lower levels (Harter 2007). In time, Kaolinite and the insoluble amorphous hydrous oxides of aluminium and iron take over as the main solids in the soil. Due to their propensity for being heavily worn, tropical soils typically have limited mineral nutrient retention capacities. Through the weathering process described above, the majority or all of the minerals with considerable negative charge have vanished, leaving very low charged Kaolinite as the predominant crystalline mineral in the soils. In this case, the non-crystalline amorphous materials take on the role of the main negative charge sources. Sesquioxides, which are created as weathering progresses, make up the majority of these materials. As mentioned before, the aluminium oxides start to dissolve at a pH of around 5.5. The amount of free aluminium ions in the soil solution will rise to levels that are hazardous to many plants if the pH falls to below 4.0. There are also fewer bonding sites for the nutritional cations due to the loss of negative charges that goes along with it. Additionally, as the concentration of aluminium ions in the solution rises, this ion successfully competes with other cations for the remaining negative sites (Harter 2007).

3.21.2 What Transpires After Liming?

When lime is applied to an acid soil, a variety of things happen, most of them simultaneously, as documented by McLean (1971). Whatever form calcium is delivered to the soil in, it will be transformed to calcium carbonate ($CaCO_3$) in the presence of atmospheric carbon dioxide. For example,

$$CaO + CO_2 = CaCO_3$$

or

$$Ca(OH)_2 + CO_2 = CaCO_3 + H_2O$$

In an acid soil, the calcium carbonate then dissociates:

$$CaCO_3 + H_2O = Ca^{2+} + HCO_3^- + OH,$$

and,

(1) The hydrogen ions in the soil are neutralised

$$OH^- + H^+ = H_2O$$

(2) The base saturation 5 increases as calcium (and magnesium, if present) replaces the hydrogen ions on the soil surface

$$2H\text{-soil} + Ca^{2+} = Ca\text{-soil} + 2H^+$$

The neutralisation of the hydrogen ions as described above causes an increase in pH and a change in the ratios of adsorbed (on soil particles) and in solution basic cations. Both outcomes often increase the soil's ability to support plant development. Sesquioxides, such as $Al(OH)_3$, are generated when aluminium ions in solution combine with those already present in the soil and are displaced from their positions as negative sites on soil particle surfaces.

$$Al(OH)_3 = Al^{3+} + 3OH^-$$

Lowering the level of aluminium below what would be harmful to plants. The soil's ability to store and exchange cationic plant nutrients will rise along with an increase in the amount of solid aluminium sesquioxide. A decrease in the pH-dependent ability of the soil to store and exchange anions, on the other hand, forces anions like sulphate into solution where they are more readily absorbed by plants or leached from the soil. As the aforementioned processes take place, the soil goes through a number of related modifications. The solubility of plant nutrients is impacted by soil pH changes, with most becoming more and others becoming less accessible to the plant. Aluminium, manganese, and perhaps other chemicals with toxic quantities are neutralised. When conditions are more favourable for both nitrogen fixation and nitrogen mineralization from plant leftovers, nitrogen normally becomes more readily available. Last but not least, the acid weathering of primary and secondary minerals is reduced, possibly resulting in a reduction in the nutrients accessible to the plant. Liming a soil can also change its microbiological and physical characteristics. Depending on the type of soil to which lime is added, permeability of the soil may increase or decrease. As previously mentioned, increased calcium content and pH of the soil will generally encourage microbial growth, which will alter the availability of nitrogen as well as certain other nutrients. On the other hand, the accelerated loss of soil organic matter could arise from the increased microbial development, which is typically considered unfavourably.

3.21.3 Overliming Risks

The risk of applying too much lime is often low in temperate zone soils, but that is not the case for tropical soils. The goal pH should probably not exceed about 6.0, with optimal being more in the range of 5.0 to 5.5. In reality, liming of most tropical soils is better considered as calcium fertilisation than pH correction. The quantities of aluminium and manganese in the soil solution are significantly reduced in this pH range, and these ions are no longer hazardous to plants. However, a sustained rise in pH can make molybdenum hazardous. Additionally, nutrients including manganese, copper, zinc, boron, and zinc can become insufficient in plants. This is due to the fact that these nutrients are less soluble at higher pH levels as well as a decrease in the acid weathering of the few minerals remaining present in the soil that carry nutrients (Harter 2007). Physical rather than chemical issues with overliming tropical soils may be among the most significant ones. As mentioned, liming can have an impact on soil permeability. Many tropical soils exhibit high infiltration rates and, as a result, rapid base leaching due to their very stable structure, which is caused by the propensity of iron and aluminium oxides to bind soil particles together forming what are known as aggregates. Overliming can weaken this structure, which in turn causes soil aggregates to crumble and lose some of their permeability and capacity for drainage. A soil gains more tiny aggregates at the price of larger ones when calcium and magnesium are added—from lime—to it. This is probably a case of Ca^{2+} dissolving sesquioxide (Fe and Al)-stabilised clumps. Due to aggregate destabilisation's reduced permeability, wet soils and a total change in the environment may follow since sesquioxide stabilised soils are often located in the humid tropics.

3.21.4 Climate Change and Soil Productivity

To mitigate global climate change and simultaneously increase soil productivity, the use of biochar in agriculture can be a modern agro-technology that can help in reducing greenhouse gas emissions, enhancing soil carbon sequestration, and ultimately increasing crop yield. Mosharrof et al. (2021) conducted a 60-day incubation study with eleven treatments (T) in which two different biochar produced from rice husk (RHB) and oil palm empty fruit bunches (EFBB) at two rates (10 and 15 t ha^{-1}) and on three rates of dolomitic limestone (100%, 75%, and 50%), recommended rate of NPK and a control (no amendment). The result showed that biochar and lime significantly increased soil pH, available P, and decreased exchangeable Al compared to the control. The pH increase was 44.02% compared to the control treatment on day 15, and the available P was found to be 22.44 mg kg^{-1} on day 30 when 75% lime + 15 t ha^{-1} RHB) was appied. The cumulative CO_2 emission was 207.40 μmol CO_2 m^{-2} that decreased 139.41% compared to the control.

Liming is the most prominent and efficient practice to neutralise soil acidity of agricultural soil (Cho et al. 2019). Besides, lime can increase Ca and Mg accessibility in soils (Kunhikrishnan et al. 2016). To alleviate soil acidity, lime is broadcasted onto the soil surface or incorporated into the soil via tillage operation. However, under no-tillage conditions, surface application of lime is the most appropriate way because lime applied on the surface does not alter the quality of soil which is proved in many previous types of research works (Yagi et al. 2014). Application of lime increases soil pH, availability of P, cation exchange capacity (CEC), and base saturation and decreases the concentration of Al (Jafer and Hailu 2017). Adding biochar converted from organic residues is an environmentally sound agronomic practice and acts as an ecological risk-free soil conditioner globally (Gonzaga et al. 2018). Biochar is a carbon-rich organic material that had gone through the pyrolysis cycle of thermal decomposition of biomass when heated to temperatures, usually in the range of 300 and 1000 °C under low oxygen concentration (Tomczyk et al. 2020). Pyrolysis is a thermochemical strategy utilised for changing biomass into biochar at a temperature going from 350 to 700 °C in an anaerobic condition. Thermochemical methods include pyrolysis and gasification to produce biochar in the solid phase (Bhutto et al. 2016). Depending on the pyrolysis process and the variety of feedstock used, biochar's physico-chemical properties vary (Uslu et al. 2020). Biochar is utilised as an amendment that can recoup soil fertility and richness, plant development, improve carbon sequestration just as waste administration, and immobilisation of contamination (Sistani et al. 2019). It can be used to enhance plants' growth and increase the yield of crops by increasing a significant amount of soil carbon (Mosharrof et al. 2021). Many previous researchers reported that soil infertility and low crop productivity are common phenomena in tropical soil. These are influenced by the low soil pH and low cation exchange capacity (CEC), insufficient nutrients, Al and Fe toxicity, structural imbalance, etc. (Wei et al. 2019; Syuhada et al. 2016). Biochar increases the organic carbon in soil and serves as a C sequester to suppress soil organic decomposition (Schofield et al. 2019). The application of biochar usually raises soil pH and enhances the availability of nutrients for plant uptake (Hossain et al. 2020). The ash content of biochar helped to accelerate the pH of the soil towards neutral (Rabileh et al. 2015). According to Chang et al. (2019), phosphorus availability significantly increased by applying biochar, which was fixed by the high amount of Al and Fe. The bioavailability of macronutrients like K, Ca, Mg, and P increased by using biochar in the soil (Syuhada et al. 2016).

In many parts of Ethiopia, soil acidity is a serious environmental and economic concern, as it results in large losses in food production and poses a threat to human health. According to Fageria and Baligar (2008), soil acidity is the result of a complex interaction among various factors, including nutritional deficiencies and toxicities in plants, low activity levels in beneficial microbes, and reduced plant root growth that inhibits the plant's ability to absorb nutrients and water. Poor crop production on acidic soil has long been associated with aluminium (Al^{3+}) toxicity, reduced nodulation or mycorrhizal infections and low phosphorus (P) availability (Kochian et al. 2004; Wang et al. 2006). Soil acidity and Al^{3+} toxicity in surface soil can be ameliorated through liming (Fageria and Baligar 2008; Rebecca et al. 2010). Changes in soil pH brought about by liming may have profound effects on the availability of many elements absorbed by crops. Liming increases soil pH and thus decreases Al^{3+} and Mn toxicities which also increase NO_3–N, Ca and P availability (Arshad and Gill 1996; Caires et al. 2005). In Ethiopia, large areas of highlands with altitude of > 1500 m located in almost all regional states of the country are affected by soil acidity.

Barley (*Hordeum vulgare* L.) is the dominant cereal crop grown in the high lands of Ethiopia where soil acidity is rampant. Barley production covers a total area of 1.02 million ha with a national average productivity of 1.87 t/ha (CSA 2014). Farmers practice barley– bare fallow–oats rotation system reduces the negative effects of soil acidity and improve soil fertility (Hailu and Getachew 2011). However, this rotation system is not sustainable and in the long run will degrade the soil resource due to severe soil erosion on the bare fallow efforts to ameliorate the deleterious effects of soil acidity must therefore be accompanied by measures to increase available P in soils.

3.22 Crop Yields

Management practices to sustain crop yields are necessary to conserve or enhance soil quality (Coulter et al. 2009). A difference in management practices often results differences in biological, chemical and physical soils properties which in turn results changes in functional quality of the soil (Islam and Weil 2000). Soil quality has become a focal point for attempts to quantify modification in soil due to various soil management systems (Islam and Weil 2000; Islam 2008). To develop sustainable agro ecosystem a knowledge about soil quality is essential which gives present and future trends and early warning signs, provide information about the problems of soils and gives a precious base which can be utilised to assess successive and future capacity of soil (McGrath and Zhang 2003). Roberson et al. (1991) reported that cover crops have rapid and significant effects on the stability of soil macroaggregates, even when the total amount of C is apparently unaffected. On fine-textured soils, cover crops have improved physical properties for example aggregation, porosity, bulk density and permeability (Karlen and Colvin 1992). In recent years, it was found that biological properties of soil act as early and sensitive indicators in response to alteration in management systems (Kennedy and Papendick 1995). Biological parameters together with soil chemical and physical properties are recognised to be necessary to assess SQ as affected by changes in management operations (Parkin et al. 1996). The particulate organic matter, active C, total N, Cmic, biological activities, enzymes, soil pH, cation exchange capacity, salinity, bulk density and soil aggregation are important indicators of dynamic soil quality because they are responsive to changes in soil management (Wander 2004). As a dynamic component of the SOM, the microbial life of the soil is often considered as a key indicator of soil quality (Islam and Weil 2000). It is the main mediator of organic matter turnover and responds more quickly than does total organic C to changes in management practices (Powlson et al. 1987).

3.22.1 Utilisation of Crop Leftovers and Animal Manure

Farmers claim that there has been considerable deforestation, a reduction in per capita land ownership, and a lack of firewood as a result of the growing human population and the urgent need for food and fuel. As a result, farmers were compelled to use crop stalks like maize and livestock dung as fuel instead of replenishing the soil. According to reports, farmers with little resources gather dried cow dung from shared grazing lands for local markets. Farmers said that they heavily rely on crop residue for animal feed because collective grazing pastures are typically unprofitable. Small cereal stalks like wheat, barley, and triticale that would typically be returned to the soil were also used for thatching dwellings. For roofing, stalks from crops like wheat and triticale were favoured over feed. 54% of residential buildings in the Amhara region have thatch roofs, according to the CSA (2007). Due to their high presence of basic minerals, grain, pasture, and crop wastes typically have an alkaline pH (Upiohn et al. 2005). Therefore, the entire removal of nutrients from the system in the form of grain and crop residue can be linked to a drop in pH of the soils in the studied locations. Since the pH of the homestead soil samples across the study sites was greater than that of the other land-uses, this link seems legitimate.

3.22.2 Abandoning Conventional Methods of Fertility Management

According to the farmers, the abandonment of traditional fertility management practises like fallowing, "Chichet" or "hura" (night parking or corralling of cattle on out fields during the rainy season as a method of in situ manure application), and crop rotation, manure application, was caused by high population pressure and the ensuing shortage of arable land. Double cropping has been possible in the Banja region thanks to adequate rainfall that has been distributed quite evenly. The farmers claimed that a rising degree of soil acidity in their region made outfields less suitable for producing pulses. Legumes could only be produced in homestead areas as a result, and planting cereal after cereal on the outfields became customary. Farmers were inclined to produce the staple crops required for basic food security even when the soil was not an issue. On homesteads, pulses were grown alongside crops like maize, potato, and Brassica spp. for domestic use. Due to the substantial amount of lime that is required to neutralise soil acidity, its high initial cost, and the challenges connected with its transport and application, farmers in the majority of the study locations were hesitant to use lime. Compared to homestead soils, which benefited more from manure and family waste, outfields often had more acidic soils and were fertilised more frequently with mineral fertiliser. Other studies have demonstrated that one of the factors contributing to the depletion of soil nutrients in Ethiopian agriculture is the low level of fertiliser consumption (IFPRI 2010; ATA 2013).

3.22.3 Farmer Perceptions of Soil Acidity Indicators

According to the farmers, flattop and bottomlands are thought to be more fruitful and less acidic than sloped lands. According to the farmers, the flattop lands experience little nutrient outflow through erosion while the bottomlands gain from nutrient and organic matter inflow from uplands through sedimentation. The farmers cited several key signs of acid soils, including poor establishment, stunted development, pale green young seedlings, low stands and poor tillering capacity of cereals, and consequently poor crop and straw production. Other signs of soil acidity were the declining suitability of the soil for once-common crops like barley, faba beans, and field peas as well as the varying suitability of such soils for acid-tolerant plants like triticale, oats and white lupin. "Yemote" or "dead" soils were described as being extremely acidic and were frequently planted with eucalyptus. These soils rarely allowed the development of oat, triticale, and lupin. Farmers connected an increase in acidity with a poor crop response or rising mineral fertiliser need.

Incidence of zinc (Zn) deficiency in soils and plants has been reported worldwide including India. Results revealed that concentrations of total as well as extractable Zn varied widely among the acids soils and the amount of Zn extracted by different extractants also differed. Zn deficiency was observed in 7 to 82% soil and 2 to 57% soils based on the diethylenetriaminepentaacaeticacid (DTPA) extractable Zn and 0.1 M hydrochloric acid (HCl) extractable Zn, respectively. The contribution of soil organic carbon (OC) content towards total and DTPA-Zn was higher as compared to soil pH. Geostatistical analysis revealed that total and DTPA-Zn concentrations in acid soils of India were influenced by different sets of environmental factors (Behera et al. 2011).

3.22.4 *Factors Affecting Nutrient Availability*

Nutrient availability can be impacted by soil chemical and physical properties, including parent material and naturally occurring minerals; amount of organic matter; depth to bedrock, sand, or gravel; and permeability, water holding capacity, and drainage. In addition, environmental conditions and crop characteristics have an important impact on nutrient availability. It is not unusual for crops in fields or portions of fields to show nutrient deficiencies during periods of the growing season, even where an adequate nutrient management plan is followed. The fact that nutrients are applied does not necessarily mean they are available. Plants obtain most of their nutrients and water from the soil through their root system. Any factor that restricts root growth and activity has the potential to restrict nutrient availability. This is not because nutrients are not plant-available in the soil, but because the ability of the crop to take up those nutrients is restricted. Understanding how these factors can cause nutrient deficiency in crops is important to avoiding excessive concern about the need for additional fertilisation when a sound nutrient programme is already in place. Soil compaction can limit or completely restrict root penetration and effectively reduce the volume of soil, including nutrients and water, which can be accessed by the plant. To limit soil compaction, avoid entering fields that are too wet, and minimise the weight per axle by decreasing load weight and/or increasing tire surface area in contact with the soil. Planting when soils are wet can create a compacted wall next to the seed that will prevent the seedling from developing an adequate root system. Tilling wet soils will result in clods that become hard and dry out quickly on the surface, preventing roots from accessing resources inside the clod.

Soil water content is critical not only to supply the water needs of the crop but also to dissolve nutrients and make them available to the plant. Excess water in the soil, however, depletes oxygen (O_2) and builds up carbon dioxide (CO_2) levels. While O_2 is needed by roots to grow and take up nutrients, high CO_2 levels are toxic. Temperature is important in regulating the speed of soil chemical processes that make nutrients available. Under cool soil temperatures, chemical reactions and root activity decrease, rendering nutrients less available to the crop. Portions of the plant nutrients are taken up as roots extract soil water to replenish water lost through the leaves. Cool air temperatures can lower evapotranspiration and reduce the convective flow of water and nutrients from the soil to the root.

3.23 Microbiology of Acid Soils

The soil microorganisms promote the durability (continuity) of element cycles in nature due to the diversity of their metabolic capacities. Their actions have a significant impact on the creation of organic compounds through mineralization as well as changes in mineral compounds. About 1–10% of the dry mass of the soil's organic matter is made up of edaphon.

Viruses

The only way for viruses to proliferate is inside of host cells, including those of humans, plants, animals, and bacteria. The bacteriophages (phages), or viruses that live inside bacterial cells, are the most significant type of viruses in the soil environment. The ability of phages to eradicate specific bacterial populations and to select microorganisms both negatively and favourably determines their function in the soil environment. Phage attacks on the Rhizobium bacteria, which cause the fall of papilionaceous plants crops, are an illustration of their detrimental impact.

Bacteria

The fundamental unit of all soil microorganisms is bacteria. They have a high rate of metabolic activity. The majority of soil bacteria are distinguished by their propensity to cling to soil colloids and mineral molecule surfaces. Particularly large concentrations of bacteria congregate around the remains of animal and plant tissues, as well as in animal waste that makes its way into the soil. Plant roots and other underground areas provide an environment that is particularly conducive to the development of bacteria. The two main types of soil bacteria are those that always exist in every type of soil (autochthonous) and those that only appear when a significant amount of organic matter has been released into the soil (zymogenous). The actinomycetes and rod-coccus bacteria of the Arthrobacter genus make up the biggest group of soil bacteria.

Actinomycetes

The bacteria known as Actinomycetes are (chemo) organotrophic. They develop long, branching threads that resemble mycelium and are packed with bacterial cells. The threads' width ranges from 1 to 5 cm. They primarily inhabit dirt or decaying plants. The majority of them live a saprophytic lifestyle, while some of them are harmful to both plants and animals. Their capacity for growing at 40–50 °C gives them a wide range of potential for decomposing different compounds. The Actinomycetes break down substances that are difficult for other bacteria to break down, such as steroids, lignin, chitin, hydrocarbons, fatty acids, and humic acids. The aforementioned materials break down and release aromatic chemicals. The actinomycetes bacteria are responsible for the distinctive scent of recently tilled soil, especially in the spring.

Fungi

The eucaryotic organisms known as fungi are the only true heterotrophs. The majority of them are fermenting fungus or aerobes. They use the energy and carbon released during the breakdown of organic substances to create their own cells. The fungi lack chlorophyll. In contrast to bacteria, the fungal cell wall contains polysaccharides such as chitin, glucans, and others. Although they can be found as deep as 1 m into the soil, they are more frequently found in the topmost layers. They form symbiotic relationships with higher plants, insects, and algae. Numerous fungal species can cause disease in people, plants, and animals. Their vegetative phases produce thread-like shreds that are typically multicellular and somewhat branching. Their dense weaving produce thallus or mycelium. The size of a single cell is around 10 cm. The genera Penicillium, Aspergillus, Trichoderma, Verticillium, Fusarium, Rhizopus, Mucor, Zygorhynchus, and Chaetomium are the most prevalent soil fungi. In acidic soils, fungi thrive and have a significant impact on the pH-changing process.

3.23.1 The Function of Fungus and Bacteria

They both contribute to the formation of humus, which is the most crucial element of soil and has a significant impact on its composition, sorption abilities, and richness in organic compounds.

By creating mucous capsules, they, like filamentous bacteria and fungi, have a significant impact on how soil is created with a crumb texture and a spongy structure.

3.23.2 Soil Fauna

The protozoans, which mostly consume bacteria, are an example of the soil microfauna. They have the responsibility for selection and population renewal of soil microorganisms. Among them, amoebae and flagellates predominate. Nematodes (eelworms), snails, insects, myriapods, mites, and other animals make up the mesofauna. They consume decaying organic material that helps humus form. Rodents, moles, and earthworms are examples of macrofauna. The creatures disintegrate soil components and transport them to a considerable depth. The earthworms are the most significant invertebrate because they consume dead organic matter and absorb it along with the soil's mineral components. Non-digested residue combined with mineral soil and metabolites are excreted in the form of lumps (coprolites), which helps the soil develop a crumb-like texture and loosen. On a surface of one hectare, earthworms have the capacity to transfer 7000 kg of soil via their digestive systems in a single year. It subsequently enables improved water movement, oxygenation, and aeration.

3.23.3 Quantity of Soil Microbes

The kind of soil and its content affect the quantity and make-up of soil microorganisms. Structure, humidity, and the amount of biological material present.

Viruses

It is unknown how many viruses are present in soil. They are thought to weigh less than 0.01 tonnes per hectare.

Bacteria

Per 1 g of soil, there are anything between a few million and a few billion bacteria cells. Up to roughly 30 cm of cultivable soil contains the maximum concentration of microorganisms. Their numbers rapidly decrease in deeper layers. For every hectare of land, there may be several hundred kg to several tonnes of bacterial mass in the top 30 cm of cultivable soil. The bacteria discover higher concentrations of organic chemicals emitted by plants, such as organic acids, amino acids, and vitamins, near and on the surface of the roots. Because of this, there are several times more bacteria in the soil close to the roots than in the layer around them, known as the rhizosphere. More bacteria can be found in soils high in organic compounds; typically, 1 g of cultivable soil contains between 0.5 and 5.0 billion bacteria (1.5 to 15 tons/ha). Acidic soils have a high concentration of fungi but a low concentration of bacteria.

Fungi

Except for yeast, fungi only exist as mycelium or spores. Depending on the soil's humidity and the availability of organic materials, the number of these units per 1 gramme of soil may reach several tens of thousands, with spores making up anywhere from several to a few dozen percent of those units. The majority of fungi are found in acidic peat and woodland soils. They might outnumber the bacteria in that location. The upper 20–30 cm stratum contains the majority of the fungus. In the upper layers, the total mass of fungi is roughly identical to that of bacteria, and in soils from forests, it may even be higher. It contains, on average, 1.5 tons/ha (0.001–1.0 billion fungus).

Algae

Algae often exist in soil levels between 0 and 10 cm deep where sunshine can reach them (rarely below 50 cm). Per 1 g of soil (0.2 tons/ha), their numbers can range from 100 k to 3 m. The populations may rise in a favourable environment, like heavily irrigated tropical soils.

3.23.4 Soil Life

Protozoa can grow widely in soils with sufficient moisture. Per 1 g of dry soil, there can be anywhere from a few hundred and several million of them. Protozoa make up between 0.1 and 0.5 tons/ha of soil, nematodes between 0.0 and 0.2 tons/ha, earthworms between 0 and 2.5 tons/ha, and other soil creatures between 0 and 0.5 tons/ha. Both macroorganisms and different microorganisms can live in soil naturally. Their edaphic elements play a role in how they develop in soil. The totality of soil classification criteria, including humidity, fertility (the availability of food ingredients in usable forms), pH reaction, and other physical parameters, govern how living things evolve in ecosystems. Some important requirements of active soil life are:

Water

Water-containing habitats are necessary for the correct development of all microorganisms. Water enables the movement of microorganisms inside soil, the diffusion of nutrients and substrates into cells, and the elimination of waste products from metabolism. It also affects a cell's ability to maintain a healthy osmotic pressure and response. The growth of the bacteria is either slowed down or prevented altogether by an excessive concentration of nutrients and by an excessive amount of moisture. A soil with too much water slows down the transport of oxygen and nitrogen and encourages the growth of predators that eat the microorganisms. The predatory protozoans may be prevented from moving by a lack of water, which encourages the growth of the bacteria. The bacteria might not have access to water that is tightly bound to the solid soil particles. Moulds and yeast are generally significantly more tolerant of environmental dryness than are bacteria. The soil humidity between 50 and 70% of its maximum water capacity (WHC) is ideal for the aerobic breakdown of organic compounds in soil. Higher values restrict the amount of soil oxygenation, whilst lower values reduce water activity.

Osmotic force

The osmotic pressure, which is linked to soil humidity and gradually rises as the soil dries out, has a significant impact on the development of microorganisms. The solution pressure in soils with medium humidity varies between 0.5 and 5 atm. Salty soils may allow it to rise to 100 atm. The pressure fluctuates from 3 to 6 atm inside the cells of the microorganisms. The mechanism by which the microorganisms' cells absorb water is hampered by the osmotic pressure, which is higher in soil solution than within the cells and consequently slows down the growth of the bacteria.

The Redox capacity

The tendency of a substance to receive or lose electrons is reflected in its redox potential. It plays a big part in how physical and chemical processes work in the soil. Salts of elements with varying valence are dissolved in soil solutions. These include, for example, Fe, Mn, and S. The interaction between the oxidised and reduced molecules affects both oxygen-based and oxygen-free processes. The direction and nature of metabolic transformation depend on the Redox patterns that the elements in their oxidised or reduced forms produce. The water content of the soil impacts the redox potential value through the dissociation processes, which in turn affects the composition and development of soil microorganisms. The processes of mineralization and oxidation are accelerated by overdrying the soil and the improved aeration that results. Contrarily, excessive soil moisture leads to oxygen evaporation, which is regulated by microaerophiles and anaerobes. These organisms lower the soil's redox potential, which promotes the reduction and fermentation processes.

Influence of soil pH

The pH of the reaction, which controls the activity of the enzymes and the transportation mechanism, greatly influences how microbiological processes in soil proceed.

The ability to endure reaction changes to some amount is one of the characteristics of soil's solution, which is characterised by buffer properties. The pH of the soil affects the availability and solubility of nutrients. The pH value of soil depends on its chemical composition; however, during the biological processes of decomposition of organic matter, changes in pH may occur as a result of metabolism and the physiology of the microorganisms. Iron and manganese are only available under conditions when the pH level is low, whereas molybdenum is only available in high pH. Acid rains, fertilisation, the presence of sulphur oxidising bacteria, and other factors can cause soil acidity to rise, which affects a variety of metabolic processes. The nitrification process is one of the most pH-sensitive processes in soil. The conversion of NH_4^+ to NO_3^- occurs in this manner. The uptake of ammonium ions (NH_4^+) by microbes from the environment contributes to soil's acidity, whereas the uptake of nitrates (NO_3^-) contributes to soil's alkalinity. Other salts' solubility and accessibility to microbes are then impacted by these changes. Numerous types of known bacteria may grow in a pH range of 4 to 9. However, the pH range of 6.5 to 8.0 provides the best circumstances for bacterial development. The extreme acidophiles may thrive in pH values as low as 1–3, while many acid-tolerant microbes can tolerate a pH range of 1–6. Some of them produce sulphuric acids by oxidising mineral sulphides, such as Thiobacillus, Thermophilus, and Sulfolobus species. The majority of fungus enjoy an acidic atmosphere. The majority of fungi are strongly acidophilic; their ideal pH range for growth is between 4 and 6. The bacteria that make up the Nitrosomonas group are moderately alkaliphilous microorganisms that thrive in a pH range of 7.3 to 9.6.

Temperature

Because of their heat tolerance and ideal development temperatures, soil microorganisms differ from one another. According to how sensitive they are to temperature, microorganisms can be divided into three groups: thermophilic, mesophilic, and psychrophilic. The growth temperature for thermophilic organisms is between 40 and 70 °C, while that for mesophilic organisms is between 15 and 45 °C. Although some species may be more or less tolerant of temperature changes than others, the majority are mesophilic, meaning they can survive temperatures of up to 30 °C. The cell membranes of species that can survive at temperatures as low as 0 °C include unique lipids that keep the membranes semi-fluid. Lipids in thermophilous kinds have high melting points. Due to increased energy consumption for respiration, decreased output, and the emergence of side effects, an excessive increase in temperature significantly reduces the process of biosynthesis. Microorganisms inhibit their life processes at temperatures below 6 °C by entering anabiosis or resting forms. In soils, the temperature can rise to 70 °C at noon and fluctuate by roughly 50 °C during the day. The temperatures in the deeper levels of the soil profile are unaffected by variations in surface temperature over the course of a 24-h period.

Oxidation

One of the most important components influencing how microorganisms develop is oxygen. Oxygen controls how well they can grow, how quickly they can grow, how much mass they can add, and how productively they can produce certain metabolites. The kind of metabolism, concentration, and energy source, as well as the stages of population expansion and the physiological state of the cells, all affect the microbes' need for oxygen. Only bacteria can endure anaerobic conditions over extended periods of time. Typical soil pores contain 50% water and 50% air. When the oxygen concentration in the soil's air is more than 0.2 mg O_2/l, the biodegradation process happens more quickly. Microorganisms use 70% of the oxygen in soil and plant roots use 30% of it, whereas chemical reactions only use trace amounts of oxygen. When the oxygen content of a soil is less than 1%, oxygen-free circumstances exist. Anaerobic bacteria carry out the biological process of decomposing organic matter when oxygen supply is not possible. They use substances like sulphate or nitrate as an oxygen source. Organic matter decomposition processes are slower and less energy-efficient in oxygen-free environments.

Even though isolated ammonia oxidising bacteria in laboratory cultures cannot thrive below neutral pH, autotrophic ammonia oxidation occurs in acidic soils. Because there were distinct populations living in acidic and neutral soils, the organisational structures of both communities responded differently to changes in pH. Phylogenetic reconstructions of crenarchaeal 16S rRNA and amoA genes confirmed selection of separate lineages within the pH gradient. High similarity in phylogenies suggested a high degree of congruence between 16S rRNA and amoA genes, and phylogenetic reconstructions of crenarchaeal 16S rRNA and amoA genes. Different amounts of mRNA transcripts, as well as archaeal and bacterial amoA gene copies, were found at various points along the pH gradient. According to Nicol et al. (2008), both the amount of archaeal amoA gene abundance and the amount of its transcript reduced as the pH of the soil rose. This was in contrast to the situation for bacterial amoA gene abundance, which was typically lower and saw an increase in its transcript abundance.

This difference in the rates of oxidation in the two profiles may be brought about by one or both of two factors, the chemical and physical nature of the vegetation, or the activity of the microbial population. It is improbable tbat very substantial chemical differences exist between the two grasses other than in their content of nitrogen, and it is clear from the values which are recorded in this paper that H. mollis, particularly during its season of active growth, is much richer in nitrogen, either on a dry-weight basis or in terms of carbon content, than is *D. flexuosa* (Scurfield and Boswell 1952). The physical nature of the vegetation may well be an important factor in determining the compactness of the surface layers and in consequence the degree of aeration. Observation suggests that H. mollis forms a looser, more friable and better aerated litter than does *D. flexuosa.* In this connexion it may be noted that the profiles under *D. flexuosa* in localities such as the Lady Canning Plantation, which lie at higher altitudes than the other localities studied and in which therefore lower temperatures and higher rainfalls would tend to reduce surface aeration and rates of oxidation, are much more strongly developed, with deeper layers of humus than elsewhere. The activity of the microbial population will depend upon available nutrient and in particular nitrogen; the supply will determine the activity and maybe even the constitution of the microflora and the nature of the end-products. Before consideration is given to the nitrogen contents of the profiles, two particular points may be mentioned with regard to the constitution of the microflora.

This is despite the fact that this is not always observed (Webster et al. 2002). These investigations, on the other hand, quantify the "standing crops" of ammonia oxidizers rather than the activities themselves, and as a result, they may indicate the consequences of long-term differences in development and colonisation. They demonstrate the ability of certain evolutionary lineages to select for and adapt to a low pH environment. Recent studies reveal that crenarchaea that do not require heat to survive play an important part in the oxidation of ammonia found in soil. There is evidence that archaeal ammonia oxidizers in soil have higher transcriptional activity than bacterial ammonia oxidizers. This data comes from a number of studies (Treusch et al. 2005; Leininger et al. 2006; Tourna et al. 2008).

The nutrients present

In addition to carbon, other nutrients like nitrogen, phosphorus, sulphur, calcium, magnesium, and potassium are needed to increase the biomass of microorganisms. Nitrogen and phosphorus are particularly significant elements since they are required for the creation of proteins and nucleic acids. All the necessary elements are present in fertile soils in suitable amounts, however the ratios of the various elements are thrown off in contaminated soils. It is recommended that soils have a weight ratio of roughly 10:1:0.1 for carbon, nitrogen, and phosphorus. The physical, chemical, and structural characteristics of the soil are all improved by calcium.

Toxic substances

Toxic substances in soil can slow down or stop microbiological activity entirely. Pesticides, aliphatic and aromatic hydrocarbons, formaldehydes, chloroorganic compounds, heavy metals, and salts found in high quantities are among the substances that are particularly dangerous.

Light

Only the first few cm of earth are illuminated. The kind and density of plants that are growing there will determine how much light is produced. Only algae that do photosynthesis require light to function. Since earthworms migrate to the surface at night in search of food and in order to reproduce, light exposure has an impact on their activities.

Decomposition of humus and synthesis

Amorphous and typically dark in colour, humus is an organic colloidal structure with a vast surface area that can adsorb gas and water ions. It contains portions of organic materials with a low C:N ratio (between 10 and 15), compared to the leftovers from dead plants, where the ratio is roughly 40:1. The components of humus are fulvic, humic acids, and humins. These are aggregates of chemical compounds that have undergone some degree of carbonization and are characterised by the presence of carboxyl, phenyl, and methoxyl groups that contain the letters C, O, N, P, and S, as well as an aromatic skeleton with multiple side chains. The main humus-forming process is carried out by fungi and bacteria in the soil, particularly actinomycetes.

3.23.5 Humification

Humification is the name given to the process of humus production. Lignin, hydrocarbons, and nitrogen compounds are the primary building blocks from which humus compounds are created. But the sort of humus depends on the soil type and the weather. Following the decomposition of fresh organic matter, microorganisms produce metabolites, which are the precursors of humus compounds. They also produce biomass, which, after atrophy and autolysis, provides the additional initial substrates required for the formation of humus. The primary method of producing humus compounds is through the synthesis of nitrogen components from protein-derived fragments, such as polyphenols. The breakdown of lignin, the conversion of hydrocarbons, and numerous microbial synthesis processes could all be sources of polyphenols. Many polyphenols are produced as byproducts of certain bacteria. The oxidation of polyphenols, which produces chinoid chemicals, is the next stage in humification. The phenol oxidases generated by several microbes, including the fungus Serpula lacrymans, catalyse these conversions. The polymerization of oxidised phenols is the procedure's last step. In environments with oxygen present, the transformation of organic matter results in the production of humus, whereas environments without oxygen result in the formation of peat deposits.

Since the 1960s, soil microbiology underwent major changes in methods and approaches and this review focuses on the developments in some selected aspects of soil microbiology. Research in cell numbers of specific bacterial and fungal groups was replaced by a focus on biochemical processes including soil enzyme activities, and flux measurements of carbon and nutrients (Insam 2001). Ecologists focussed on soil microbial pools whereas soil microbial biomass as an important source and sink of nutrients were recognised in agriculture. Soil microbiologists started to use structural components like phospholipid fatty acids for quantification of specific microbial groups without the need to cultivate them. In the last decade, molecular approaches allowed new insights through the analysis of soil extract DNA showing an unexpected diversity of genomes in soil. At the end of the review a brief outlook is given on the future of soil microbiology which ranges from in situ identification of bacteria, to routine assays of microbial communities by microarray technology (Insam 2001).

3.23.6 Microbiology of Soil and Nutrients Cycling

A huge variety of life forms can be found in soil, which is a dynamic ecosystem. It provides plants with a mechanical base from which they can collect nutrients.

From invertebrates like worms and insects to mammals like rabbits, moles, foxes, and badgers, it provides shelter for a wide variety of animals. Additionally, it offers environments that are home to a dizzying array of microbes. These various life forms interact with one another and the soil to produce ever-changing conditions. Soil ecosystems can continue to evolve as a result. The activity of living things in soil influences its depth, structure, and other characteristics. The soil's composition varies depending on the region's bedrock, slope, and climate. The variety of soil types is caused by the interactions between these many components. As a result, it may be discovered that the same basic soil structure supports widely varied biological communities depending on the location. These intricate communities make a significant contribution to the ongoing cycle of nutrients around the world.

3.23.7 Types of Ecosystems in Soil

An very acidic environment is produced by thermoacidophilic bacterial activity, such as that found in coal slag heaps. Acid from slag heaps may leach and modify the chemical composition of bedrock. Uncovered rocks offer an extremely hostile environment. However, even these could become colonised. Everywhere we look, colonisation is evident. Check the gravestones for lichens the next time you go there. Rocks are colonised by microbes called lichens. Even if the rock is not in its natural setting, this is still true. The dates on gravestones are convenient. You can estimate how long it takes to colonise native rocks by comparing the ages of various headstones and the level of colonisation. Cyanobacteria are among the first organisms to colonise rocks. Parent rocks don't contain nitrogen in a form that biological processes can easily access. The ability of bacteria to fix atmospheric nitrogen for use by other organisms sets them apart from other life forms. Because they are nitrogen-fixing photolithotrophs, cyanobacteria are perfectly suited to colonise rock surfaces. To grow, all they need is light and inorganic fertilisers. Other species can employ the fixed nitrogen and carbon molecules that cyanobacteria can produce as nutrition. They are in charge of how organic matter first accumulates on exposed rocks. This starts the biological processes that result in the development of soil and the cycling of nutrients. The initial step in converting bare rock into soil fit for the sustenance of plant and animal life is the colonisation of rocks by cyanobacteria. The soil's microbes play a significant role in the soil's dynamic renewal by recycling both organic and inorganic materials. The smallest particles are located closest to the ground's surface as soils grow and change, and they get increasingly bigger all the way down to the bedrock.

The organic composition of a soil ecosystem may eventually take the stage as it develops. Bogs contain the highest quality organic soil. Bogs form an anaerobic environment because they are soggy. The facultative microbes swiftly exhaust any dissolved oxygen. This creates a setting that is particularly hostile to fungi and aerobic bacteria. Bogs make great locations for the preservation of organic matter since these organisms frequently cause the degradation of organic structures.

The soil's microbial population is not evenly distributed. Higher plants' roots frequently harbour large concentrations of microbes. The term "rhizosphere effect" refers to this occurrence. Rhizo means "root" in Greek, therefore the rhizosphere is the area around a plant's roots. Although fungi and protozoa are also present in the rhizosphere, bacteria make up the majority of the microorganisms there. The rhizosphere has been found to contain auxotrophic mutants that require different amino acids, supporting the theory that microorganisms obtain nutrition from plants. Additionally, this layout might be advantageous to plants. Nitrogen may be fixed by bacteria in a way that plants can absorb and utilise. The relationship between microbes and higher plants can become exceedingly intimate under certain conditions. Roots develop mycorrhizas when they become closely entwined with fungus. Another significant illustration of the strong relationship between nitrogen-fixing bacteria and leguminous plants is found in root nodules. In this case, the interaction with plants is caused by bacteria rather than fungi.

3.23.8 How Do Bacteria Participate in the Cycling of Nutrients?

The basis of life on Earth is carbon, which gets elaborated into complex organic structures from water and simple chemical compounds like carbon dioxide. Other elements besides carbon, oxygen, and hydrogen are present in these compounds. Proteins, amino acids, and nucleic acids all include nitrogen. Nucleic acids, lipids, molecules that store energy, and other organic phosphates all include phosphorus. Sulphur is mostly present in several proteins and amino acids. The ecology continually cycles through each of these components. It is necessary for elements to be in various chemical states at different stages of many natural biological cycling processes. The exception is phosphorus. It is constantly assimilated as inorganic phosphates. After entering living things, phosphorous undergoes metabolic reactions that change it into more complicated forms. Although they are commonly found in nature as insoluble salts, inorganic phosphates are relatively broadly distributed. In natural ecosystems, phosphates frequently serve as a limiting nutrient despite an apparent abundance of phosphorous. This means that unchecked proliferation of organisms is inhibited as phosphate supplies run out. Phosphates can be changed from insoluble to soluble. The activity of the acid byproducts of bacterial fermentations may help to accomplish this. These might then be absorbed by microorganisms. It is also possible to artificially introduce soluble phosphates to the soil in the form of organophosphate insecticides or plant fertilisers. Numerous detergents are also made with the help of phosphoric acids. These substances may wind up in rivers and lakes, artificially raising the amount of phosphates that are available to life. Algal blooms are the outcome of this, which allows algae to overgrow in the impacted waters. These may obstruct other plants' access to light, killing them and altering the water's natural ecosystem. Animals may potentially be poisoned by some algal blooms. Microorganisms are involved in the metabolic modification of metal ions in addition to the cycling of non-metal elements. Iron bacteria of the genus Gallionella and Thiobacillus ferrooxidans are both capable of oxidising ferrous (Fe^{2+}) iron into ferric (Fe^{3+}) iron. Small amounts of ferric iron can be converted by many bacteria to ferrous iron. Additionally, there is a class of bacteria that breathes iron and gets its energy from respiration. Instead of using oxygen, they use ferric iron as an electron acceptor. Iron can be converted into its magnetic salt magnetite by magnetotactic bacteria, such as Aquaspirillum magnetotacticum. They serve as biological magnets, these germs. In the transformation of manganese ions, where similar interactions to those seen with iron are present, bacteria are also crucial. The continuation of life on Earth would be difficult without the cycling of elements because vital nutrients would be quickly absorbed by organisms and locked in a form that other organisms could not use. Although the events involved in the cycling of elements are frequently chemical in nature, biological reactions also play a significant role. The role of microbes in this process is crucial.

In a fully functioning ecosystem, photolithotrophs or chemolithotrophs are present alongside chemoorganotrophs or photoorganotrophs, and nutrients are constantly exchanged between these various organismal kinds. While organotrophs require a supply of complex organic molecules to function, lithotrophs get their energy from the metabolism of inorganic substances like carbon dioxide. Chemotrophs can develop in the dark and get their energy from chemical molecules, in contrast to phototrophs, which need light as a source of energy. Different compounds cycle at highly different rates, according to estimates of how quickly inorganic compounds change over time. Every water molecule on the earth is thought to split as a result of photosynthesis and then be renewed by various life forms over the course of 2 million years. Plants or photosynthetic microorganisms can both mediate photosynthesis. The oxygen in the atmosphere is released during photosynthesis. All of the oxygen in the atmosphere is likely biologically produced, and its cycle is expected to take roughly 2000 years. Carbon dioxide is also taken up by photosynthesis and converted into organic molecules. In the course of some fermentations and respiration, these emit carbon dioxide. The carbon dioxide cycle in the atmosphere only lasts 300 years.

It may be difficult to fathom living without photosynthesis because we are accustomed to seeing green vegetation. After all, it is this process that gives us the oxygen we require to survive. However, it is important to keep in mind that photosynthesis is what creates molecular oxygen. Numerous biological types are highly poisonous to this element. When life first appeared on Earth, the atmosphere contained little or no oxygen. The only reason that aerobic species can exist is because they have developed sophisticated defence mechanisms to reduce the toxicity of oxygen. Not all life is dependent on sunlight, either. Thermal vents in the crust of the Earth can be found at the gloomy depths of both the Atlantic and Pacific oceans. These offer chemolithotrophic bacteria a source of heat and chemical energy. A variety of invertebrates use these bacteria as a food source in return. All of these vibrant and diversified groups live in complete darkness surrounding the "black smokers."

3.23.9 What is the Carbon Cycle?

The aerobic carbon cycle is well known to the majority of people. Organic molecules are produced during photosynthesis as a result of the fixation of carbon dioxide. The main producers of organic carbon compounds, which serve as nourishment for other creatures, are photosynthetic plants and microorganisms. Through the processes of fermentation and respiration, these organisms consume organic carbon and decompose organic matter. Chemoorganotrophic microorganisms emit carbon dioxide by decomposing organic carbon molecules. In the dark, chemolithotrophic bacteria can transform inorganic carbon into organic matter. Anaerobic carbon cycling is also a capability of some microorganisms. The conversion of organic molecules into carbon dioxide or methane occurs as a result of fermentation reactions, which are frequently seen in bacteria found in water and anaerobic soils. Some fermentations have the potential to produce hydrogen gas as a byproduct. Methane-oxidising bacteria can use methane as a source of carbon and energy. These microorganisms can produce sugars and amino acids from the methane in their surroundings, contributing once again to the cycle of carbon compounds.

3.23.10 What is the Cycle of Nitrogen?

The development of agriculture was one of the key phases in the evolution of human civilization. In order to increase the productivity of livestock and food crops, the natural environment is being artificially modified. The necessity to increase soil fertility arose with the growth of agriculture. The primary factor affecting soil fertility is the availability of fixed nitrogen in a usable form for crop plants. As a result, both agriculture and natural ecosystems depend critically on the biological nitrogen cycle. Nitrogen assimilation is the process by which inorganic nitrogen substances like nitrates, nitrites, and ammonia are transformed into organic nitrogen substances like proteins and nucleic acids. Numerous bacteria convert nitrates to nitrites, whereas a small number of bacteria convert nitrites to ammonia. Assimilatory nitrate reduction can then incorporate ammonium salts into organic polymers. Amino acids like glutamate and glutamine are the main amino acids used to repair ammonia into organic materials. These can be used to produce other nitrogen compounds. Organic nitrogen compounds need to be broken down to release ammonia in order for nitrogen to cycle continuously. Large amounts of ammonia are produced by putrefactive metabolism from nitrogen-containing biopolymers. The enzyme urease, which breaks down urea to release carbon dioxide, water, and ammonia, may also be produced by bacteria. The amount of ammonia generated by Helicobacter pylori's urease is adequate to shield this bacteria from the stomach's acidic pH.

The inorganic cycling of nitrogen compounds also involves bacteria. The biological oxidation of ammonia is carried out by nitrifying bacteria. These bacteria are chemolithotrophs, which means they get their chemical energy from oxidation. Utilising this energy, carbon dioxide is converted into organic molecules. When ammonia is oxidised, nitrifying bacteria, such as those of the genus Nitrosomonas, create nitrite ions. Nitrites can be converted to nitrates by bacteria of the genus Nitrobacter and a few other genera. Nitrifying bacteria may engage in additional evil activities aside from their function in the nitrogen cycle. It has been demonstrated that they colonise the sandstone used to construct these churches. Up to five centimetres of the stone's matrix can be reached by the bacteria that water carries from the surface. Here, they generate sufficient levels of nitrous and nitric acid to erode the stone. As a result, it's possible that industrial pollution-induced acid rain is not the only factor contributing to the deterioration of grand public structures. Some bacteria may use nitrates instead of oxygen for a process known as dissimilatory nitrate reduction. Nitrate is converted throughout this process to nitrite, and finally to ammonia. This could then be incorporated into organic molecules as previously mentioned. However, not all bacteria go along this route. Pseudomonas bacteria, micrococci, and Thiobacillus species have the ability to decrease nitrates and release nitrogen gas into the environment. Common places to find bacteria that can produce nitrogen gas from the reduction of nitrates are compost piles, sewage treatment facilities, and soils that are rich in organic matter.

Nitrogen gas needs to be fixed in a way that allows it to be utilised by living things in order for the inorganic nitrogen cycle to be complete. If it weren't for this, life on Earth would have only persisted until all the nitrogen compounds were broken down into nitrogen gas. In the process of nitrogen fixation, nitrogen is transformed into ammonia. The only living things that can biologically fix nitrogen are bacteria. If life is to continue on this planet, they are crucial. The primary creators of organic matter in the biosphere are green plants, and they need a source of fixed nitrogen to do so. Free-living nitrogen-fixing bacteria can be killed and lysed to produce fixed nitrogen. However, nitrogen-fixing bacteria usually develop intimate relationships with plants. In certain instances, the bond between the two organisms grows so close that bacteria reside inside plant tissues as endosymbionts. The plant receives all of the fixed nitrogen requirements from bacteria. They get a supply of organic carbon molecules in exchange. Nitrogen fixing is not always brought about by biological processes. The agrochemical business produces enormous amounts of nitrogenous fertilisers. Additionally, environmental natural and man-made processes produce nitrates of nitrogen. Nitrogen oxidation is aided, especially in the high atmosphere, by ultraviolet radiation and lightning. These processes are accelerated at ground level by electrical discharges, particularly the operation of automobile engines. Nitrogenase is the enzyme that catalyses biological nitrogen fixing. In the presence of oxygen, this enzyme's activity is quickly eliminated. Numerous mechanisms have been developed by nitrogen-fixing bacteria to shield nitrogenase from the negative effects of oxygen. Growing in anaerobic environments is the most straightforward solution. Obligate anaerobes include nitrogen fixers like Desulfovibrio desulfuricans and Clostridium pasteurianum. They can therefore only grow in environments that safeguard nitrogenase activity. Facultative anaerobes are prevalent in microorganisms. Normally, facultative nitrogen-fixing bacteria can only fix nitrogen when they are growing in anaerobic conditions. Some of the Enterobacteriaceae, like Enterobacter species, and facultative members of the genus Bacillus are examples of these bacteria. The Gram-negative bacterium Klebsiella pneumoniae is capable of nitrogen fixation in both anaerobic and microaerophilic environments. This is due to the fact that bacterial respiration can successfully lower the local oxygen tension to zero in a microaerophilic atmosphere. This enables nitrogenase to work. This serves as an illustration of how breathing can satisfy two needs.

When it comes to plant health and soil quality, microbial populations in the soil are crucial. To better understand soil biology and plant–microbe interactions, researchers have developed a wide range of techniques for examining the structure, diversity, and activity of microbes (Rincon-Florez et al. 2013). The rhizosphere's functional microbiological investigations have revealed fresh information about the function of microbial communities in plant nutrition and disease defence. The most popular conventional and innovative culture-independent molecular techniques to evaluate the diversity and functionality of soil microbial communities are presented in this study. Additionally, we go over the benefits and drawbacks of these methods and offer an outlook on recently developed technology for microbial community characterisation in soil (Rincon-Florez 2013).

3.23.11 Soil as a Habitat

Survival capacity of rhizobia was higher in soil than in nutrient medium at low pH 3.5–6.5 levels (Appunu et al 2009). The cowpea symbionts are classified into slow-growers, Bradyrhizobium spp. and fast-growers, *Sinorhizobium* spp. (Zhang et al. 2007). Many biotic and abiotic factors affect the persistence of symbiotically effective introduced rhizobial strain in soil. Soil acidity is the one of the factors which restricts production of cowpea through its impact on nitrogen fixation. The failure of nodulation under acid soil conditions is common, especially in soils of pH less than 5. Soil acidity limits symbiotic nitrogen fixation by limiting Rhizobium survival and persistence in soils, as well as reducing nodulation (Ibekwe et al. 1997). High symbiotic effective rhizobial inoculation is a common practice in agricultural legume production (Catroux et al. 2001) which requires survival and establishment of inoculated rhizobia in the soil environment (Da and Deng 2003). A wide variation in the tolerance to acid-soil conditions have been reported among Bradyrhizobium strains of many agriculturally important legumes from various countries (Graham et al. 1994; Raza et al. 2001) and in India (Appunu et al. 2005). Differential symbiotic performance of Bradyrhizobium isolates has already been reported (Zhang et al. 2002). Only one of the symbionts needed to be acid tolerant for good nodulation to be achieved at pH 4.5 (Vargas and Graham 1988).

The absorption of P by plants and crop production are both increased by using microorganisms that solubilise phosphate as inoculants. It is generally recognised that many bacterial species, especially those found in the plant rhizosphere, have the ability to positively influence plant growth (Glick 1995). Pseudomonas, Azospirillum, Burkholderia, Bacillus, Enterobacter, Rhizobium, Erwinia, Serratia, Alcaligenes, Arthrobacter, Acinetobacter, and Flavobacterium strains are among the "plant growth promoting rhizobacteria" (PGPR) group of bacteria (Kloepper and Schroth 1978). Both in the lab and in field tests, PGPR has been shown to stimulate many crops. In field trials utilising a combination of Bacillus megaterium and Azotobacter chroococcum, a yield increase of 10–20% in the same crop was recorded (Brown 1974). Wheat yield increased by up to 30% with Azotobacter inoculation and up to 43% with Bacillus inoculants. Maize, sorghum, and wheat yields have increased due to *Azospirillum* spp. (Baldani et al. 1987; Sarig et al. 1990); meanwhile, peanut, potato, sorghum, and wheat yields have increased due to *Bacillus* spp. (Capper and Campbell 1986). There are two different kinds of processes that PGPR might use to influence plant growth: direct mechanisms and indirect mechanisms (Glick 1995). The reduction or avoidance of pathogenic microorganisms' harmful effects is known as indirect growth promotion, and it is primarily caused by bacteria that produce siderophores (Leong 1986) or antibiotics (Sivan and Chet 1992). By producing phytohormones (Xie et al. 1996), fixing N_2 (Christiansen-Weneger 1992), and lowering the membrane potential of the roots (Bashan and Levanony 1991), direct growth stimulation can be achieved.

3.23.12 Availability of Phosphate in Soil

One of the most important macronutrients for biological growth and development is phosphorus (P). It is prevalent in soil in concentrations of 400–1200 mg/kg. According to Goldstein (1994), the content of soluble P in soil is typically quite low, at 1 ppm or less. The cell may absorb P in several forms, although the majority of it is absorbed as HPO_4 or H_2PO_4 (Beever and Burns 1980). Rocks and other deposits, such as primary apatites and other primary minerals created during the geological epoch, are the largest sources of phosphorus (Odum 1986). For instance, it is predicted that India has about 40 million tonnes of deposits of phosphatic rock, which should serve as a cheap supply of phosphate fertiliser for agricultural production (Roychoudhury and Kaushik 1989). Primary minerals including apatite, hydroxyapatite, and oxyapatite are examples of phosphorus mineral types found in soil. They are found as a component of stratum rock, and their main quality is insolubility. Despite this, because they may be solubilised and made available to plants and microbes under the right circumstances, they make up the largest reservoirs of this element in soil. Additionally, the surface of the poorly soluble and assimilable hydrated oxides of Fe, Al, and Mn can be discovered to have mineral phosphate attached to it.

3.23.13 Bacteria that Dissolve Phosphate

In soil and plant rhizospheres, phosphate-solubilizing bacteria are abundant (Sperberg 1958; Raghu and MacRae 1966; Alexander 1977). These comprise both aerobic and anaerobic strains, with submerged soils being more likely to contain aerobic bacteria (Raghu and MacRae 1966). In comparison to nonrhizosphere soil, the rhizosphere often contains a significantly larger concentration of phosphate-solubilizing bacteria (Raghu and MacRae 1966). After multiple cycles of inoculation, some strains' ability to dissolve phosphate has been observed to be unstable (Halder et al. 1990; Illmer and Schinner 1992; Kucey 1983). The characteristic appears to be persistent in the majority of isolates, but (Arora and Gaur 1979).

3.23.14 Solubilisation of Organic Phosphate

There are several different organic substrates in soil that can provide P for plant growth. This type of P needs to be hydrolysed into inorganic P in order to be used for plant nutrition. Phosphatase enzymes are used to mineralize the majority of organic phosphorus molecules. It has been noted that soil contains a sizable quantity of phosphatase activity (Lynch 1990; El-Sawah et al. 1993; Bishop et al. 1994; Feller et al. 1994). Most soils have pH values that vary from acidic to neutral. Acid phosphatases should therefore be a key player in this process. In two forested Spodosoils, there was detectable acid phosphatase activity in the rhizosphere of slash pine (Fox and Comerford 1992). When phosphatase activity was examined in the rhizospheres of maize, barley, and wheat by Burns (1983), he found that it was significant at both acidic and neutral soil pH levels in the inner rhizosphere. A significant number of acid phosphatases is expressed by soil bacterial strains belonging to the following genera: *Rhizobium* (Abd-Alla 1994), *Enterobacter, Serratia, Citrobacter, Proteus, Klebsiella* (Thaller et al. 1995), *Pseudomonas* (Gügi et al. 1991), and *Bacillus* (Skrary and Cameron 1998).

3.23.15 Phosphate-Solubilizing Bacteria as Growth Stimulators for Plants

Although there are several phosphate-solubilizing bacteria in soil, their populations are typically too low to effectively compete with other bacteria that are frequently present in the rhizosphere. As a result, the P they release typically isn't enough to significantly stimulate in situ plant growth. To take advantage of the property of phosphate solubilisation for the enhancement of plant yield, it is therefore necessary to inoculate plants with a specific microorganism at a much higher concentration than that typically found in soil. Numerous studies on the encouragement of plant growth by bacteria that can solubilise inorganic and/or organic P from soil after being introduced into soil or plant seeds have been published (Gerretsen 1948; Cooper 1959; Datta et al. 1982; Kucey et al. 1989).

3.23.16 Future Potential

By increasing P absorption by the plant, phosphate-solubilizing bacteria play a significant role in plant nutrition, and their usage as PGPR makes a significant contribution to the biofertilisation of agricultural crops. For better outcomes, more focus should be placed on research and application of novel phosphate solubilizing bacterial and other PGPR combinations. Further study on the processes underlying the synergistic interaction is necessary to understand the molecular underpinnings of these interactions. However, it appears to be conceivable to genetically modify phosphate-solubilizing bacteria to increase their capacity for phosphate-solubilisation and/or to introduce this characteristic into strains that already have other traits that promote plant growth.

To determine the importance of Al-availability for soil micro-organisms 95 forest soils from Tyrol/Austria with comparable topography, vegetation, climatic conditions, soil type and with low soil pH (median = 3.9) were investigated for their physical (percentage of stable aggregates, water holding capacity), chemical (pH, electrical conductivity, contents of organic matter, concentrations of easily extractable aluminium, calcium, potassium, magnesium and phosphorus) and microbiological characteristics (microbial biomass and respiration, metabolic quotient, content of ATP, activities of protease and CM-cellulase, cfu-values of total and Al-tolerant bacteria and fungi). A highly significant negative correlation was detected between concentrations of KCl-extractable aluminium and soil pH. By the application of multivariate statistical methods, the effect of the concentration of KCl-extractable aluminium on abundance and activities of soil micro-organisms could be revealed. Al turned out to be of great importance for micro-organisms and often outmatched the significance of other well-known soil properties like organic matter, pH or water holding capacity. However, due to very healthy trees at the sites under investigation no effect of Al or soil acidification on forest decline could be detected.

3.23.17 Soil Microbiology

The study of organisms found in soil, their functions, and the ways in which these organisms alter the qualities of the soil is known as soil microbiology. It is thought that the first ancient bacteria and microbes came into being in the seas that were on Earth in their rudimentary state between two and four billion years ago. These bacteria had the ability to fix nitrogen, and as a result of their reproduction, they released oxygen into the environment. This release of oxygen resulted in bacteria that were more evolved. The structure and fertility of various soils are both affected by the microorganisms in the soil, making these creatures very significant. Bacteria, actinomycetes, fungus, algae, and protozoa are the several kinds of microorganisms that can be found in soil. According to Subba Rao (1999), each of these groupings possesses a unique set of features that characterise the creatures it inhabits and the unique functions it performs in the soil.

3.23.18 Bacteria

Other than viruses, the smallest creatures that can be found in soil are bacterial and archaeal cells. Prokaryotic organisms include both bacteria and archaea. Every other type of microorganism is classified as eukaryotic, which indicates that it possesses a more complex cell structure than prokaryotic organisms, including organelles located within the cell, as well as the ability to reproduce sexually. According to Subba Rao (1999), the cell structure of a prokaryote is quite straightforward and does not contain any organelles. According to Wood (1989), bacteria and archaea make up the majority of the microorganisms found in soil. These organisms provide a variety of vital functions, one of which is the fixation of nitrogen, in addition to other biochemical activities.

Biochemical processes

The ability of bacteria to adapt to different biochemical environments is one of the characteristics that sets them apart from other organisms. A kind of bacteria known as Pseudomonas is capable of metabolising a large variety of different chemicals and fertilisers. In contrast, a different genera known as Nitrobacter is the only one that is able to obtain its source of energy by converting nitrite into nitrate, a process that also produces oxygen and is known by its other name, oxidation. In addition, the genus Clostridium offers an additional illustration of the adaptability of bacteria because, in contrast to the majority of species, it is able to grow and reproduce even in the absence of oxygen by engaging in anaerobic respiration. There are a few different species of Pseudomonas, including Pseudomonas aeruginosa, that are able to perform respiration in both an aerobic and anaerobic environment, with nitrate serving as the terminal electron acceptor (Wood 1989).

Nitrogen fixation

Bacteria are responsible for the process of nitrogen fixation, which is the conversion of atmospheric nitrogen into nitrogen-containing compounds (like ammonia) which can be used by plants to uptake. Autotrophic bacteria, or bacteria that derives its energy making its own food by oxidation, like the Nitrobacters species, rather than feeding on plants or other organisms. The bacteria that are autotrophic are responsible for nitrogen fixation, and the amount of autotrophic bacteria is small compared to heterotrophic bacteria (the opposite of autotrophic bacteria, heterotrophic bacteria acquires energy by consuming plants or other microorganisms), but are very important because almost every plant and organism require nitrogen in some way, and would have no way of obtaining it if not for nitrogen-fixing bacteria (Subba Rao 1999).

3.23.19 Actinomycetes

Actinomycetes are soil microorganisms. They are a type of bacteria. They are similar to both bacteria and fungi, and have characteristics linking them to both groups. Actinomycetes are often believed to be the missing evolutionary link between bacteria and fungi, but they have many more characteristics in common with bacteria than they do fungi (Sylvia et al. 1998). Actinomycetes are similar to bacteria because they, like bacteria, are prokaryotic, are sensitive to antibacterial and affected in the same way that bacteria is by them. Actinomycetes can hardly be distinguished from bacteria at its early stages because of how much they resemble bacteria in size, shape and gram-staining properties. Gram staining is a common technique used to classify organisms into two main groups: Gram-positive and Gram-negative, by staining organisms to distinguish its cell wall properties. Gram-positive means that the cell has a thick cell wall and gram-negative means the opposite, that the cell wall is thin. Cell wall properties can help distinguish different types of microorganisms from each other (Subba Rao 1999).

Ability to produce antibiotics

The ability of actinomycetes to create antibiotics is one of the most noteworthy aspects of this group of microorganisms. Antibiotics generated from actinomycetes include, but are not limited to, streptomycin, neomycin, erythromycin, and tetracycline, to name just a few examples. Neomycin is utilised to lower the danger of bacterial infection during surgical procedures, while streptomycin is utilised in the treatment of tuberculosis as well as infections brought on by specific bacteria. Erythromycin is a very essential antibiotic that is used to treat a variety of bacterial diseases, including bronchitis, pertussis (whooping cough), pneumonia, and infections of the ear, intestine, lung, urinary tract, and skin. Erythromycin is also used to prevent some types of bacterial infections. This capacity to create these helpful antibiotics is the foundation of our entire pharmaceutical business and has resulted in the preservation of human life (Sylvia et al. 1998).

3.23.20 Fungis

The population of fungi in soil is second only to that of bacteria when compared to that of other microorganisms. As food sources for other, larger species, diseases, and beneficial symbiotic connections with plants or other organisms, fungi play an essential role in the soil. Fungi also contribute to the reduction of crop wastes and the biochemical processing of minerals, both of which help to enhance the soil that they occupy. According to Sylvia et al. (1998), different species of fungi can be distinguished from one another primarily by the size, shape, and colour of the spores that are utilised in the process of reproduction.

Fungi are affected by the majority of the environmental conditions that also have an effect on bacteria and actinomycetes in terms of their growth and spread. The growth of fungi has been shown to have a direct association with the quality as well as the quantity of organic matter present in the soil. This is due to the fact that the majority of fungi rely on organic matter as a source of sustenance (Wood 1989). Acidic conditions are ideal for the growth of fungus, while bacteria and actinomycetes cannot tolerate their presence. As a consequence, acidic surroundings are home to a preponderance of fungi. Because fungi are aerobic, or dependent on oxygen, and the higher the moisture level in the soil, the less oxygen is present for fungus (Subba Rao 1999), fungi grow well in dry, desert soils. Fungi can grow well in soils that are extremely acidic.

3.23.21 Algae

Algae can make its own nutrients through a process known as photosynthesis. Photosynthesis is when light energy is converted to chemical energy that can be stored as nutrients. For algae to grow, it must be exposed to areas of light because photosynthesis requires light, so algae is typically distributed evenly wherever sunlight and moderate moisture is available. Algae, however, do not have to be on the soil surface or directly exposed to sun rays, but it can live below the soil surface as long as the algae has uniform temperature and moisture conditions. Bacteria are not the only organism that can fix nitrogen, because algae are capable of performing nitrogen fixation as well (Subba Rao 1999).

Types of algae

Algae can be split up into three main groups: the Cyanophycease, the Chlorophycease, and the Bacillariacease. The Cyanophycease contain chlorophyll, which is the molecule that absorbs sunlight and uses that energy to make carbohydrates from carbon dioxide and water, and also pigments that make it blue-green to violet in colour. The Chlorophycease usually only has chlorophyll in it which makes it green, and the Bacillariacease contains chlorophyll as well as pigments that make the algae brown in colour (Subba Rao 1999).

Nitrogen fixation

Blue-green algae, or Cyanophycease, are the algae that are responsible for nitrogen fixation. The amount of nitrogen fixed by these algae depends more on physiological and environmental factors rather than the organism's abilities. Some of these factors include intensity of sunlight, concentration of inorganic and organic nitrogen sources, and temperature and stability of the environment (Sylvia et al. 1998).

3.23.22 Protozoa

Protozoa are eukaryotic organisms which are some of the first microorganisms to develop a means of sexual reproduction, which is a huge evolutionary step from duplication of spores, like many of the other soil microorganisms depend on. Protozoa can be split up into three categories: flagellates, amoebae, and ciliates (Sylvia et al. 1998).

3.23.23 Flagellates

Flagellates are the smallest members of the protozoa group, and can be divided further based on whether they can participate in photosynthesis. Nonchlorophyll-containing flagellates are not capable of photosynthesis because chlorophyll is the green pigment that absorbs sunlight in the process. These flagellates are found mostly in soil and flagellates that contain chlorophyll typically occur in aquatic conditions. Flagellates can be distinguished by their flagella, which is their means of movement. Some have several flagella, and other species only have one, but it resembles a long branch or appendage that helps the flagella move (Sylvia et al. 1998).

3.23.24 Amoebae

Amoebae are larger than flagellates and move in a different way. Amoebae can be distinguished from other protozoa by their slug-like properties and pseudopodia. A pseudopodia or "false foot" is a temporary obtrusion from the body of the Amoebae that helps pull it along surfaces for movement or helps to pull in food. The amoebae does not have permanent appendages and the pseudopodia is more of a slime-like consistency than a flagella (Sylvia et al. 1998).

3.23.25 Ciliates

Ciliates are the largest of the protozoa group, and move by means of short, numerous cilia that produce beating movements. Cilia resembles small, short hairs, and they can move in different directions to propel the organism in different direction, giving it more mobility than flagellates or amoebae are capable of (Sylvia et al. 1998) It is important to understand the many different groups and species of microorganisms in different soils because they affect so much of the soil. Microorganisms contribute to nutrient availability in soil, manage soil stability by means of different biochemical processes such as nitrogen fixation, and they contribute to the growth and success of the plants and overall ecosystem of a soil environment.

3.23.26 Influence of Soil pH

The availability of nutrients and metal toxicity are just two examples of how soil acidity affects numerous chemical and biological aspects of soil (McBride 1994). Microbial populations can also be impacted by soil acidity in a variety of ways (Sylvia et al. 2005). The relative importance of the two main decomposer groups in soil, fungus and bacteria, was examined by Rousk et al. (2009) over a continuous soil pH gradient at Rothamsted Research in the United Kingdom's Hoosfield acid strip. In a silty loam soil that has been continuously used to grow barley for more than 100 years, this experimental site offers a consistent pH gradient, ranging from pH 8.3 to 4.0, within 180 m. By evaluating the incorporation of acetate into ergosterol to evaluate fungal growth and the incorporation of leucine and thymidine to test bacterial growth, they were able to directly quantify the importance of fungus and bacteria. Lower pH caused fungal growth to increase fivefold and bacterial growth to decrease fivefold, according to measurements based on growth. As seen by the fungal growth/bacterial growth ratio, this led to a rise in fungal importance of about 30 times, from pH 8.3 to pH 4.5. In contrast, equivalent impacts on the biomass markers of bacteria (bacterial PLFAs) and fungus (ergosterol and phospholipid fatty acids (PLFAs)) revealed only a two- to threefold difference in the relevance of fungi in the same pH range. Only roughly one-third of the total carbon mineralization, assessed as baseline respiration, decreased with the shift in the relative importance of fungi and bacteria throughout the pH gradient, presumably indicating functional redundancy. All microbiological variables were universally inhibited below pH 4.5, which was probably caused by enhanced inhibitory effects brought on by the release of free aluminium or declining plant production. Growth measures offered much higher sensitivity than measurements based on biomass for analysing decomposer group relevance (Rousk et al. 2009).

The majority of nutrient transformations in soil, which regenerate minerals that restrict plant production, are carried out by the soil microbial population. The microbial decomposer community is dominated by fungi and bacteria, and as their shared task of rotting organic matter in soil can be roughly specified, there is a good chance that they will interact. Tolerance to metals (Rajapaksha et al. 2004), temperature dependence (Pietikäinen et al. 2005), turnover rate (Rousk and Bååth 2007a, b), nutrient demand (De Vries et al. 2006; Rousk and Bååth 2007a, b; Van Groenigen et al. 2007; Wallenstein et al. 2006), and food web linkage (Hedlund et al. 2004; Hunt et al. 1987). Soil management, nutrient input changes, climate change, and other anthropogenic activities have the ability to either directly or indirectly affect the bacterial and fungus composition, which will have an effect on soil performance.

Biochemical markers such as phospholipid fatty acids (PLFAs) (Zelles 1999) and glucose amines (Appuhn et al. 2004), as well as DNA-based molecular techniques (Boyle et al. 2008; Lauber et al. 2008) have almost exclusively been used in attempts to distinguish between these two components of the microbial community. However, measures based on biomass are only tangentially related to process contribution, and determining process contribution from biomass involves understanding its condition (which can range from dormant to highly active), which is frequently challenging to distinguish. Gross activity (respiration) and the biomass makeup of bacteria and fungus have been attempted to be correlated (Beare et al. 1992; Holland and Coleman 1987). The decoupling caused by variable growth efficiency (Six et al. 2006; Thiet et al. 2006), or variation in the amount of substrate C used for biomass production compared with the amount of energy used in respiration, makes this causal link weak (Rousk and Bååth 2007a, b; Rousk et al. 2008). This may explain why basal respiration does not correlate with direct estimates of fungal and bacterial growth.

The pH of the soil is one of the most important elements impacting the microbial community there. Abiotic factors like the availability of carbon, nutrients, and metals are all highly influenced by pH (Andersson et al. 2000; Kemmitt et al. 2006; Firestone et al. 1983; Flis et al. 1993). Additionally, in both agricultural (Arao 1999; Bardgett et al. 2001) and forest (Bååth and Anderson 2003; Blagodatskaya and Anderson 1998; Frostegård et al. 1993) soils, soil pH may regulate biotic variables, such as the biomass composition of fungus and bacteria (Fierer and Jackson 2006). The fact that soil pH has a wide range of effects on a number of parameters makes it difficult to analyse those effects. It may be difficult to distinguish changes in numerous parameters while experimenting with soil pH. In contrast, comparing the pHs of various natural soils includes complicating elements that are usually impossible to pinpoint since they result from variations in soil type and management practises that also exist among soils.

There were opposing pH impacts, with fungal and bacterial development being the most affected by pHs over pH 4.5. The growth ratio showed a 30-fold increase in the relative importance of fungi as a result; the pH range with the highest growth ratio was 4.5. It has been studied how pH affects bacterial development in the past. According to Bååth and Arnebrant (1995), adding lime and ash to forest soils enhanced bacterial growth by around a factor of five, as determined by TdR incorporation, and caused pH alterations in the range of pH 4 to 7. Similar to this, a study that examined 19 distinct soils from diverse land uses and covered a pH range of 4 to 8 revealed that higher pHs resulted in an increase in bacterial growth as determined by Leu incorporation (Bååth 1998). The growth of bacteria quadrupled between pH 4 and pH 8. The activities of microbial communities and the biogeochemical processes that they mediate are known to be significantly influenced by soil pH. According to Kemmitt et al. (2006), soil pH will have an impact on substrate chemical form, concentration, and availability, as well as cell activity and growth. Furthermore, there is compelling evidence that worldwide bacterial diversity and community structure are significantly influenced by soil pH (Fierer and Jackson 2006).

Beneficial soil microbes and plants prefer a near-neutral pH range of 6 to 7, so increased soil acidity is often accompanied by shifts in the types of microbes in soils and their activities. This means significant changes in the rate of decomposition which can lead to immobilisation of basic nutrients and decreased nutrient availability to plants. The structure and function of the soil microbial community can be both directly and indirectly affected by soil pH. The direct interaction of hydrogen ions (H^+),which are at high concentrations when pH is low, with microbial cells may influence microbial communities in a number of ways, including disruption of cell membranes, altered enzyme production, and limited reproduction. This equates to reduced overall microbial function towards the health and productivity of soils (Birgander et al. 2014). Also, soil fungi are favoured by a low soil pH, meaning that with a lower soil pH the community shifts from a balance between bacteria and fungi to a much more fungi-dominated soil (Rousk et al. 2010). This may allow the invasion of fungal root pathogens, but also change the way organic residues are decomposed. Because soil fungi and bacteria play different roles in the decomposer community and interact to release nutrients to plants, a lower soil pH will change those relationships and soil carbon and plant nutrients may become immobilized, slowing turnover and nutrient release (Rousk et al. 2009).

In a recent review by Slattery et al. (2001), the impact of various soil environmental variables in pastures and broadacre crops on the Rhizobium legume symbiosis is extensively discussed. Elevated amounts of aluminium have been demonstrated to be directly harmful to the growth of Rhizobia, impacting nodule formation and the nitrogen fixation process (Slattery et al. 2001). Soil acidity affects Rhizobium survival and persistence. The productivity of pasture and subsequent crop systems that depend on the biological input of nitrogen is significantly impacted by the fact that many Victorian soils are already at a pH where clover nodulation is diminished. If the nitrogen produced is not fully absorbed by plant roots, it may seep out and contribute to additional acidification. It has been demonstrated that aluminium can have a metagenetic effect on rhizobia at very low soil pH and under conditions of high soluble aluminium (Wood 1995). The consequences that altering soil pH may have on plant growth through root symbiosis, microclimate pH, and nutrient bioavailability are unknown, however these data show that the microbial biodiversity may change over time. The effects of soil acidity on the symbiosis of the indigenous Rhizobia are still being studied. Nutrient toxicity and nutrient scarcity in heavily weathered acidic soil prevent bacterial development. Acidity alters the microbial population of the soil and lessens the effects on the rhizosphere and roots. It improves mycorrhizal occurrence and establishment, both crucial for nitrogen uptake under extremely acidic nutrient-poor environments (Siqueira et al. 1997), while decreasing breakdown and nutrient transformation for root growth. Acidity reduces nitrifying bacteria's ability to break down organic matter into ammonium and nitrate for later absorption by plants. However, if the soil pH is extremely acidic, then biologically driven processes like stubble breakdown are inhibited due to low fungal activity. Fungi often have a pH optimum that is lower than that of bacteria. Additionally, as hazardous acetic acid is created during the fungi's degradation of plant remains in wet conditions, the pH of the soil would further decline (Lynch 1995a, b).

Although pH has an effect on microbes directly, it's also important to take into account how pH may affect (i) rhizodeposition and crop residue utilisation, which serve as the main substrate for microscopic growth, and (ii) the physico-chemical environment of soil, which serves as a source of nutrients for both microorganisms and plants as a water-stabilised structure (Lynch 1995a, b). Since different microbial groups have distinct pH optimums, acid soils differ from neutral to alkaline soils in terms of population balance or biodiversity. Low pH inhibits nitrification, however humic acids can override this inhibition. The negative impact of acidity on the nodulation of legumes is the element of microbial activity in acid soils that has received the greatest attention from researchers. Acidity, which makes aluminium accessible at amounts poisonous to Rhizobium, is largely responsible for this (Lynch and Wood 1988).

The negative impact of acidity on legume nodulation is the element of microbial activity in acid soil that has received the greatest attention from researchers. Acidity, which makes aluminium accessible at amounts poisonous to Rhizobium, is largely responsible for this (Lynch and Wood 1988). All microbial cells will be impacted by acidity, as will the metabolic processes that they mediate. It is essential that any assessments of an agricultural system take the role of the soil biota into account.

3.23.27 Autotrophic Nitrogen Fixation

The two-step process known as autotrophic ammonium oxidation converts ammonium to nitrite and back to nitrate. These two processes are carried out by ammonia-oxidising bacteria (AOB) and nitrite-oxidising bacteria (NOB), in that order. For growth and carbon dioxide fixation, these organisms can only use the energy provided by the oxidation of inorganic N molecules (Hooper et al. 1997). Nitrosococcus is a solitary species belonging to the g-subclass of Proteobacteria that exhibits ammonia-oxidising capabilities. However, due to its apparent limitation to marine environments (Ward and Carlucci 1985; Teske et al. 1994), it is not the subject of this study. Representatives of the NOB can be found in several phylogenetic groups. It is frequently believed that the rate-limiting phase in autotrophic nitrification is ammonium oxidation. Ammonia mono-oxygenase may oxidise a variety of organic, non-polar molecules in addition to ammonia (Hooper et al. 1997). Many of these substances act as ammonia oxidation competition inhibitors.

3.23.28 Heterotrophs

It can be challenging to count different heterotrophic nitrifiers since they may relate different metabolic processes to nitrification. As a result, heterotrophic nitrifier enumeration has typically only been determined for particular metabolic categories, like oxime degraders. To determine whether these substances may significantly contribute to nitrate formation in acid soils, oxime degraders were counted (Lang and Jagnow 1986). They were few in number, which is consistent with the widespread belief that oximes and other metabolites containing N-oxides are only synthesised sparingly (Wood 1987).

3.23.29 Heterotrophic Nitrification

Islam et al. (2007) conducted laboratory incubation experiments, using 15N-labelling techniques and simple analytical models, were conducted to measure heterotrophic and autotrophic nitrification rates in two acid soils (pH 4.8–5.3; 1/5 in H_2O) with high organic carbon contents (6.2–6.8% in top 5 cm soil). The soils were from pastures located near Maindample and Ruffy in the Northeast Victoria, Australia. Gross rates of N mineralization, nitrification and immobilisation were measured. Heterotrophic nitrification accounted for 19% and 7% of the total nitrification in the Maindample and Ruffy soils, respectively. Active nitrification under acid conditions has been suggested as evidence of heterotrophic nitrification (Focht and Verstraete 1977; Papen and von Berg 1998). However, it has been suggested that acid tolerant autotrophs could also be responsible for nitrification in such soils (De Boer and Kowalchuk 2001). Honda et al. (1998) and De Boer and Kowalchuk (2001) found that heterotrophic nitrifiers can use both inorganic and organic substrates for nitrification. In addition, because some heterotrophic nitrifiers denitrify simultaneously (Matheson et al. 2003), and therefore, accumulate little or no nitrite, actual heterotrophic nitrification rates may have been underestimated (Robertson et al. 1988). Despite the fact that heterotrophic nitrification was identified more than 100 years ago, the biochemistry of heterotrophic nitrifiers is poorly known and their contribution to nitrification in soil is still speculative. Heterotrophic nitrifiers need organic compounds as their energy source in contrast to the chemolithotrophic nitrifiers. Most of the potential pathways for nitrite/nitrate production by heterotrophs can be considered as secondary metabolism. Only nitrification and simultaneous denitrification by some heterotrophic bacteria is known to have connection to the energy metabolism. Evidently, the nitrification pathways of bacteria and fungi differ. Some heterotrophic bacteria oxidising ammonia have ammonia monooxygenase (AMO), but there are also heterotrophic bacteria oxidising ammonia without AMO (Martikainen 2022). The structure of AMO of chemolithotrophic ammonia oxidizers and heterotrophs differs. AMO has not been found in nitrifying fungi. The conditions for heterotrophic nitrification in soil highly differ from those in waste waters where heterotrophic nitrification activity can be high. Heterotrophic nitrification in soil is limited by the low availability of easily decomposable organic substrates. Possible nitrification by heterotrophs in the rhizosphere and endophytic root microbes gaining a good supply of substrates from plants is not known. Fungi are of special interest in heterotrophic nitrification in soil because fungi can evidently nitrify not only easily decomposable substrates but also when decomposing recalcitrant organic compounds (like lignin) which are abundant in soil (Martikainen 2022). Nitrite/nitrate production from easily decomposable nitrogenous organic compounds, such as amino acids, is common among heterotrophic bacteria and fungi. The general conclusion that heterotrophs use solely an "organic pathway" in their nitrification is, however, not valid because also ammonia can be oxidised. Owing to the diverse, poorly known biochemistry of heterotrophic nitrification and methodological difficulties to differentiate heterotrophic and chemolithotrophic nitrification in soil, the role of heterotrophic nitrification in soil nitrogen cycle remains uncertain (Martikainen 2022).

3.23.30 Influence of Electrochemical Properties on Nitrifying Microorganism

Soil surface electrochemical properties may have a strong influence on nitrifying microorganisms, H^+ and NH_4^+ activities, and therefore on the nitrification process. A gradient of surface electrochemical parameters was obtained by amendment of a subtropical acid pine soil (Oxisol) with 0% (control), 3%, 5%, 8%, 10%, and 12% pure Ca-Montmorillonite by weight (Jiang et al. 2011). The H^+ and NH_4^+ activities, the abundance of the ammonia-oxidising bacterial (AOB) and archaeal (AOA) amoA gene copies, and time-dependent kinetics of net nitrification were investigated. Soil particle surface specific area ranged from 53 to 103 $m^2\ g^{-1}$ and increased with increasing montmorillonite application rate. Similar to specific area, surface charge quantity, surface charge density, electric field strength and surface potential increased after montmorillonite amendment. The H^+ and NH_4^+ activities decreased linearly after montmorillonite addition (Jiang et al. 2011). AOB amoA gene copy number was 1.82×105 copies g^{-1} for unamended soil, and the highest AOB amoA gene copy numbers were found for the 10% montmorillonite amendment ($3.11 \times 107\ g^{-1}$ soil), which was more than 150 times higher than unamended soil. AOA amoA gene copy numbers were 9.19×103 copies g^{-1} dry unamended soil, and the highest AOA amoA gene copy numbers were found in the 8% montmorillonite amendment ($1.22 \times 105\ g^{-1}$ soil). Although pH significantly decreased during the first three weeks of incubation, no significant difference was observed between the unamended control and different rates of montmorillonite addition treatments during the whole incubation. The largest net nitrification (103 mg N kg^{-1}) was observed in the 10% montmorillonite amendment and the lowest in unamended soil (62 mg N kg^{-1}). While montmorillonite did not change the kinetic patterns of net nitrification, the highest nitrification potential (275 mg N kg^{-1}) for the 10% montmorillonite treatment was more than 3 times higher than unamended soil from simulation of time-dependent kinetics. Nitrification was significantly stimulated after montmorillonite amendment in acid soil mainly due to an increase in the quantity and activity of AOB and AOA. We concluded that soil particle surface parameters can significantly influence nitrification, especially in acid soils (Jiang et al. 2011).

Differences in nitrification could also be driven by biotic factors such as the composition of the soil bacterial community (Balser and Firestone 2005; De Boer and Kester 1996; De Boer et al. 1996; Wheatley et al. 2003). Community members can have specific positive or negative effects on the growth and activity of nitrifying bacteria. For example, chitinolytic soil bacteria can produce antibiotics against nitrifying bacteria (De Boer et al. 1996), and ammonia-oxidising bacteria (AOB) are poor competitors for NH_4 þ relative to NH_4-assimilating heterotrophs when NH_4 is limiting (van Niel et al. 1993; Verhagen and Laanbroek 1991; Verhagen et al. 1992). Recently, Nugroho et al. (2009) examined the bacterial community structure and interactions between members of the bacterial community in acid forest soils to see if they prohibited growth of ammonia-oxidising microorganisms and reduced their nitrifying activity. They concluded that low net nitrification rates in acid Scots pine forest soils cannot be solely explained by unfavourable abiotic soil conditions, but that still unknown biotic factors contributed to suppression of nitrification (Nugroho et al. 2009). However, the most important biotic factor in nitrification is often the nitrifying microorganisms themselves. Historically, microbial ammonia oxidation was thought to have been performed by NH_4^- oxidising bacteria (AOB) and only bacteria that possess the amoA gene for the ammonia monooxygenase (AMO), the key enzyme of nitrification (Arp et al. 2002; Koops et al. 2007; Di et al. 2009; Jia and Conrad 2009). However, ammonia-oxidising archaea (AOA) have been reported to possess the amoA gene (Francis et al. 2005; Leininger et al. 2006; Wuchter et al. 2006; Lam et al. 2007), and the AOA amoA gene copies can be much higher than AOB in soil (Schleper et al. 2005; Leininger et al. 2006; Roesch et al. 2007; Nugroho et al. 2009), which indicate that AOA plays an important role in nitrification.

Although the importance of soil surface electrochemical parameters in nitrification is acknowledged, no quantitative information related to soil particle surface electrochemical parameters, and their effects on nitrifying microorganisms and nitrification, is available. The nitrification process in this soil was significantly affected by soil particle surface parameters. Although pH was significantly altered by Ca-montmorillonite amendment, pH was adjusted to the same as the unamended control before incubation. One may argue that Ca addition with montmorillonite amendment may have some side-effects on nitrification in acid soils because Ca will increase base saturation and change pH buffering capacity.

An interesting feature of this acid soil is the presence of micro-organisms which in pure culture do not multiply and are not metabolically active at ^H values near to that of an aqueous suspension of the soil. The most important examples are a bacterium which decomposes cellulose and the nitrogen-fixing bacteria. While the view that cellulose decomposition in acid soils is chieffy brought about by fungal attack is supported by observations made on this soil, it is of interest to isolate a bacterium capable of decomposing cellulose in aqueous media only at acidities much less than that of the whole soil. The degree of metabolic activity of the organism in the soil is as yet unknown, but it must be sufficient to maintain at least a small number of active colonies. Since chemical analysis has shown that oxidation, molecular degradation and increased acidity march together, it may well be that the *pH* at the surface of the cellulose fibre is higher than that in the general body of the soil. These organisms, therefore, whose substrate is a solid fibre may exist in isolated pockets and be little affected by the acidity of the whole soil. Such cellulose-decomposing micro-organisms may be widespread in acid soils, since a similar organism has been isolated from the *Eriophorum* peat located on the higher parts of the moor. Nitrogen-fixing bacteria have been isolated from this soil, although the ^H is much below that at which they can be grown in synthetic media lacking fixed nitrogen. There is, however, no evidence that they do fix nitrogen in the soil, and in view of the known ^H limits of their capacity for nitrogen fixation in pure culture it would seem probable that they grow in acid soil by absorbing fixed nitrogen from the medium and by the fermentation of the carbohydrate present.

3.23.31 Significance of Soil Microorganisms

- Involved in nutrient transformation process
- Decomposition of resistant components of plant and animal tissue
- Role in microbial antagonism
- Participate in humus formation
- Predator to nematodes
- Surface blooming reduces erosion losses
- Improves soil structure
- Maintenance of biological equilibrium.

3.24 Impact of Agricultural Management

Agricultural management alters physical and chemical soil properties, which directly affects microbial life strategies and community composition. The microbial community drives important nutrient cycling processes that can influence soil quality, cropping productivity and environmental sustainability. In this research, a long-term agricultural experiment in a subtropical Acrisol was studied in south Brazil. The plots at this site represent two tillage systems, two nitrogen fertilisation regimes and three crop rotation systems. Using Illumina high-throughput sequencing of the 16S rRNA gene, the archaeal and bacterial composition was determined from phylum to species level in the different plot treatments. The relative abundance of these taxes was correlated with measured soil properties (de Quadros et al. 2012). The P, Mg, total organic carbon, total N and mineral N were significantly higher in the no-tillage system. The microbial diversity was higher in the no-tillage system at order, family, genus and species level. In addition, overall microbial composition changed significantly between conventional tillage and no-tillage systems. Anaerobic bacteria, such as clostridia, dominate in no-tilled soil as well as anaerobic methanogenic archaea, which were detected only in the no-tillage system. Microbial diversity was higher in plots in which only cereals (oat and maize) were grown. Soil management influenced soil biodiversity on Acrisol by change of composition and abundance of individual species.

Agricultural management affects soil chemical, physical and biological features. Tillage typically involves complete removal of vegetation followed by plantings often designed to completely switch the crop cover from the previous year and this process increases soil erosion and degradation, which triggers many changes in the soil community composition (Lee 1985; Chan 2001; Fierer et al. 2012). Tillage also results in nitrous oxide (N_2O) and carbon dioxide (CO_2) emission to the atmosphere (Bayer et al. 2006; Omonode et al. 2011). When the soil is not tilled, these gases remain stored within soil pores, which significantly reduce field emissions of these greenhouse gases (Lal 1997).

No-tillage is a sustainable cropping management system that protects soil, water, air, and biodiversity (Campbell et al. 1996; Calegari et al. 2008), and authors have shown that even a reduction in tillage leads to increased microbial activity and biomass in contrast to surface soil under conventional tillage (Von Lützow et al. 2002; Feng et al. 2003). Agricultural areas under no-tillage systems are increasing worldwide largely as result of improved soil conservation efforts. In recent years, the United States, Brazil, and Argentina have converted about 26, 25 and 20 million hectares, respectively, into no-tillage cropping systems (Derpsch et al. 2010). The residues left on soil after harvest increase soil organic matter (SOM), conserve soil moisture, and reduce soil temperature in summer (Wang et al. 2011). Feng (2003) found that a no-till management system led to significantly higher soil organic carbon (SOC) and microbial biomass in the soil surface compared to the conventional tillage treatment. Wang (2011) found significant increases in soil N, organic C and SOM fraction with no-tillage system, while conventional tillage had deleterious impact on soil microbial biomass and also reduced SOC. In conventional tillage and no-tillage systems, nitrogen (N) fertilisation and crop rotations are also factors that can modify soil quality and fertility thereby affecting microbial community composition. Recent studies have demonstrated that microbial community composition changes with N fertiliser. Fierer et al. (2012) showed that N addition can result in a shift of microbial composition from oligotrophic to copiotrophic bacteria. Lu et al. (2011) showed that all nitrogen cycle processes increased with N addition. The crops rotation where plants predominate that are succulent and rich in proteins and sugars such as legumes release nutrients more quickly, whereas plants that are more fibrous, such as grasses and cereals, release nutrients more slowly, but promote more stable organic matter (Sarrantonio and Gallandt 2003). After harvest, the cover crop residues that are left on the soil improve soil fertility, since they release nutrients to soil. Conventional tillage releases the nutrients more quickly because the residues are crushed and mechanically incorporated to soil, which stimulates microorganisms that degrade organic matter. In no-tillage conditions, the nutrient releasing is progressive, since the residues remain on the soil surface and therefore decomposition is slower. The crop rotation composition and the tillage management, together, determine the soil C:N ratio, which has significant effect on microbial composition (Fierer et al. 2009), crop residue decomposition, and crop nutrient cycling.

Soils harbour more diverse microbial populations than any other habitat on earth. In fact, the vast majority of these microscopic soil organisms are highly beneficial in terms of nutrient cycling, soil tilth, and soil health. Because of their important roles in these crucial soil properties and their direct interactions with plants, beneficial soil microorganisms are also absolutely critical to soil fertility and plant nutrition. Unfortunately, the rapid acidification of soils in the inland Pacific Northwest is having detrimental impacts on the populations and effectiveness of beneficial soil microorganisms (Sullivan et al. 2017).

Every teaspoonful of soil typically contains hundreds of millions of microorganisms, the majority of which are absolutely essential to healthy, productive soils (Chaparro et al. 2012). Microbes in soil are important to healthy soil processes and good soil quality. Many aspects of most nutrient cycles are controlled by soil microbes. For example, without microbes, organic matter decomposition simply wouldn't occur, legumes would not be able to fix nitrogen, and ammonia would not be converted to plant-available nitrate. Without important beneficial soil fungi, most plants would be much more limited in their ability to acquire nutrients and water from the soil, resist drought, and produce economically viable yields. Additionally, soil microbes also play a key role in the breakdown and degradation of a huge number of widely used herbicides (Forlani et al. 1999).

3.24.1 Rhizobia and Soil

Rhizobia are the soil bacteria that form nodules on plant roots, fixing nitrogen (N) from the atmosphere to provide available N to the plant. This relationship is critical to plant success in both agricultural and natural systems due to the steady supply of plant-available N that the bacteria provide to the plant. Like most other soil bacteria, rhizobia flourish and function best in neutral to basic soil pH. When soil pH is around 6 or 7, legume roots naturally form an association with rhizobia in the soil and symbiotically fix N. However, application of ammonia-based fertilisers reduces the efficiency of this symbiotic relationship and how effectively rhizobia can provide N to their host plant (Weese et al. 2015), while also further acidifying the soil solution. While some rhizobia can survive in acidic soils, low soil pH can significantly reduce nodule numbers, nodule function, and N-fixing capabilities within the roots of legumes, such as lentils and chickpeas (Tang and Thomson 1996). This can result in reduced plant vigour and productivity, as well as significant crop loss where soil pH drops particularly low. In soils where the pH drops below 5, when compared to soil with a pH over 6, the number of nodules per soybean plant can decrease by 40–60% (Lin et al. 2012). Low pH can inhibit nodulation by limiting the legume's ability to secrete the required signals into the rhizosphere that attract the rhizobia and cause the formation of the root nodules (Hungria and Stacey 1997). Also, calcium (Ca^{2+}) and molybdenum (Mo), both known to be essential to root nodule formation and N fixation, become basically unavailable below pH 5, further limiting rhizobial Nfixation. Furthermore, metals like Al and manganese (Mn), which become increasingly available in soil solution as pH drops, are toxic to the legume-rhizobia symbiosis (Bordeleau and Prévost 1994). Beneficial Fungi and Mycorrhizal Symbiosis As stated above, acid soils favour a higher proportion of fungi in soil communities because many soil bacteria do not tolerate acidic conditions well (Sylvia et al. 2005). Fungi typically account for as much as 75% of the soil microbial biomass, and hyphal length can be as great as almost 176 miles per ounce of soil in agroecosystems (Ritz et al. 2009). Many of these soil fungi function primarily in decomposition processes and nutrient cycling, but they may also help with remediation of metals, including Al, in acidic conditions. Fungal-driven binding of Al at low soil pH is one of the many ways fungi can help improve soil and plant health (Gadd 2007). According to research by Seguel et al. (2013), plant-symbiotic fungi known as mycorrhizas have been shown to shield plant root systems from a variety of challenges, including metal toxicity, drought, and nutrient depletion. These fungi increase access to limiting nutrients such as phosphorus (P), which is particularly important in acid soils due to the reduced P availability at lower pH (Seguel et al. 2013). The hyphal networks surrounding the rooting system constitute the "body" of fungi sometimes called a mycorrhizosphere. This hyphal network is often complex with many plant growth-promoting characteristics.

Pyrite, or iron disulphide, is the most common sulphide mineral on the Earth's surface and is widespread through the geological record. Because sulphides are mainly produced by sulphate-reducing bacteria (SRB) in modern sedimentary environments, microorganisms are assumed to drive the formation of iron sulphides, in particular, pyrite. However, the exact role played by microorganisms in pyrite formation remains unclear and, to date, the precipitation of pyrite in microbial cultures has only rarely been achieved (Duverger et al. 2020). Excess sulphide delivered by the bacteria and oxidants (such as polysulphides) promoted the evolution of mackinawite into greigite (Fe_3S_4) and the nucleation of pyrite spherules. These spherules were several hundreds of nanometers wide and occurred within the extracellular polymeric substance (EPS) of the biofilm after only 1 month. Altogether, the present study demonstrates that the mineral assemblage induced by the metabolic activity of sulphate-reducing bacteria strongly depends on the source of iron, which has strong implications for the interpretation of the presence of pyrite and vivianite in natural environments.

Iron is the fourth most abundant element on the Earth's surface, but due to modern oxic conditions, dissolved Fe^{2+} concentration in modern ocean is low (around 20 nM). Pyrite (FeS2) is the most thermodynamically stable iron sulphide mineral in anoxic low-temperature conditions (Schoonen 2004) and is ubiquitous in both modern environments and the sedimentary record (Rickard et al. 2017). Although surface sulphide minerals can have diverse origins, sedimentary processes predominate over magmatic, hydrothermal, and volcanic ones (Rickard et al. 2017). In modern environments, sulphide formation is mainly driven by the metabolism of sulphate reducing microorganisms (SRM), which are prevalent in anoxic environments, e.g., marine sediments or water column of permanently stratified euxinic waters such as the Black Sea (Vetriani et al. 2003) or Lake Cadagno (Tonolla et al. 2004) or even in ferruginous environments such as Lake Pavin (Lehours et al. 2005; Berg et al. 2016, 2019). Understanding the mechanisms of a possible biogenic pyrite formation pathway is crucial as pyrites are used as paleoenvironmental proxies and might also be good candidates for the search of biosignatures of early life (Shen and Buick 2004). Sedimentary sulphides are generally enriched in light sulphur isotopes, suggesting their microbial origin (Thode et al. 1953). However, many reports of pyrite framboids formed under strictly abiotic conditions undermined their possible use as biosignatures (Berner 1969; Farrand 1970; Butler and Rickard 2000; Ohfuji and Rickard 2005). Later studies of sedimentary framboids using scanning transmission X-ray microscopy (Maclean et al. 2008) or nano-SIMS (Wacey et al. 2015) and revealing the presence of organic matter associated with framboids revived the biogenic interpretation.

3.24.2 Pseudomonas Toxicity to Aluminium

At pH 6.7 and 5.5, both Pseudomonas bacteria grew to > 107 cfu mL-I but perished at pH 4.5. Strain IE grew at pH 5.0, but strain 1G perished. As the pH value dropped, the rate of growth decreased. Up until a cell density of around 108 cfu mL-I, the pH of the prescribed medium stayed constant. At that point, the buffering capacity was exceeded, and the pH rose. The data showed that the bacterium could reproduce at that pH level rather than only creating more advantageous acidic circumstances before reproduction took place. The pH of soil invertebrates' guts is frequently higher than that of the soil as a whole; suggesting that acidity-inhibited chemical processes which may flourish in this milieu. The character and quantity of soil fauna communities frequently change dramatically as a result of liming acidic soils (Lavelle et al. 1995). As huge endogeic and anecic invertebrates establish major populations and successfully compete with the less active arthropods of the litter system, the overall biological activity is significantly raised. In order to maintain high levels of soil faunal activity in acid soils and hence boost crop productivity and sustainability, emphasis is placed on managing organic matter (Lavelle et al. 1995). The pH has a big impact on the faunal communities' composition. Earthworm populations grow with pH, and they have a major impact on how litter gets buried. Since litter is quickly buried in the soil and decomposition occurs there rather than in the litter system, the drilosphere system triumphs over it, and the majority of faunal activity is found there. It may be noted that these natural, humus-rich, acid soils present a recognisable microbiological pattern with a clearly defined group of fungi, a uniformly distributed nitrogen-fixing bacterium, Clostridium butyricum, a vigorous protein decomposition by a group of bacteria, cellulose decomposition which, while it must occur slowly from the very nature of the soils, is primarily a fungal process, and nitrifying bacteria present in general in the more highly oxidising soils.

References

AACM International (1995) Social and economic feasibility of ameliorating soil acidification—a national review. Land and Water Resources Research and Development Corporation, Canberra

Abd-Alla MH (1994) Use of organic phosphorus by Rhizobium leguminosarum biovar. viceae phosphatases. Biol Fertil Soils 18:216–218

Abruña F, Vicente-Chandler J, Pearson RW (1970) Crop response to soil acidity factors in Ultisols and Oxisols. Soil Sci Soc Am Proc 34:629–635

Abruna F, Vicente-Chandler J (1967) Sugarcane yields as related to acidity of a humid tropic Ultisol. Agron J 59:330–331

Abruna F, Pearson RW, Perez-Escolar R (1975) Lime response of corn and beans grown on typical Oxisols and Ultisols of Puerto Rico. In: Bornemisza E, Alvarado A (eds) Soil management in tropical America. North Carolina State University, Raleigh, North Carolina, pp 261–282

Abruna F, Rodriguez J, Badillo-Feliciano J, Silva S, Vicente-Chandler J (1978) Crop response to soil acidity factors in Ultisols and Oxisols in Puerto Rico; soybeans. J Agric Univ P.R. 62:90–112

Adams F (ed) (1984) Soil acidity and liming, 2nd edn. American Society of Agronomy, Madison, WI

Adani F, Genevini P, Zaccheo P, Zocchi G (1998) The effect of commercial humic acid on tomato plant growth and mineral nutrition. J Plant Nutr 21(3):561–575

Afonso S, Arrobas M, Rodrigues M (2020) Soil and plant analyses to diagnose hop fields irregular growth. J Soil Sci Plant Nutr 20:1999–2013

Agbenin JO, Olojo LA (2004) Competitive adsorption of copper and zinc by a Bt horizon of a Savanna Alfisol as affected by pH and selective removal of hydrous oxides and organic matter. Geoderma 119:85–95. https://doi.org/10.1016/S0016-7061(03)00242-8

Ahmad A, Abraham G, Gandotra N, Abrol YP, Abdin MZ (1998) Interactive effect of nitrogen and sulphur on growth and yield of rapeseed-mustard (Brassica juncea [L.] Czern and Coss. and Brassica campestris L.). J Agron Crop Sci 181:193–199

Akinrinade EA, Iroh I, Obigbesan GO, Hilger T, Romheld V, Neumann G (2006) Response of cowpea varieties to phosphorus supply on an acidic alumi-haplic-Acrisol from Brazil. Nigerian J Soil Sci 16:115–120

Alexander M (1977) Introduction to soil microbiology. John Wiley, New York, USA

Allen JRL (1964) Sedimentation in the modern delta of the river Niger, West Africa. In: Developments in sedimentology, vol I. Elseviers Publication Company. Amsterdam, London, New York, pp 26–35

Allison VJ, Yermakov Zm Miller RM, Jastrow JD, Matamala R (2007) Using landscape and depth gradients to decouple the impact of correlated environmental variables on soil microbial community composition. Soil Biology and Biochemistry, 39:505–516

Amaral AZ, Verdade FC, Schmidt NC, Wutke ACP, Igue K (1965) Parcelamento e intervalo de aplicação de calcário. Bragantia 24:83–96

Ananthanarayana R, Ravindra MR (1998) Soil acidity and liming in Kamataka. Technical Bulletin, Department of Soil Science and Agricultural Chemistry, UAS, Bangalore

Anda M, Siswanto AB, Subandiono RE (2009) Properties of organic and acid sulfate soils and water of a 'reclaimed' tidal backswamp in Central Kalimantan, Indonesia. Geoderma 149:54–65. https://doi.org/10.1016/j.geoderma.2008.11.021

Andersson S, Nilsson I, Saetre P (2000) Leaching of dissolved organic carbon (DOC) and dissolved organic nitrogen (DON) in moor humus as affected by temperature and pH. Soil Biol Biochem 32:1–10

Andriesse W (1993) Acid sulphate soils: diagnosing the illness. In: Dent DL, Van Mensvoort MEF (eds) Selected papers of the Ho Chi Minh City symposium on acid sulphate soils, vol 53. ILRI Publication. Wageningen, The Netherlands, pp 11–29

Appuhn A, Joergensen RG, Raubuch M, Scheller E, Wilke B (2004) The automated determination of glucosamine, galactosamine, muramic acid, and mannosamine in soil and root hydrolysates by HPLC. J Plant Nutr Soil Sci 167:17–21

Appunu C, Sen D, Dhar B (2005) Acid and aluminium tolerance of Bradyrhizobium isolates from traditional soybean growing areas of India. Indian J Agric Sci 75(12):727–728

Appunu C, Reddy LML, Reddy CVCM, Sen D, Dhar B (2009) Symbiotic diversity among acid-tolerant bradyrhizobial isolates with cowpea. J Agric Sci 4:126–131

Arao T (1999) In situ detection of changes in soil bacterial and fungal activities by measuring 13C incorporation into soil phospholipid fatty acids from 13C acetate. Soil Biol Biochem 31:1015–1020

Arias M, Barral MT, Mejuto JC (2002) Enhancement of copper and cadmium adsorption on kaolin by the presence of humic acids. Chemosphere 48(10):1081–1088

Arias M, López E, Fernández D, Soto B (2004) Copper distribution and dynamics in acid vineyard soils treated with copper-based fungicides. Soil Sci 169:796–805

Arias M, Pérez-Novo C, Osorio F, López E, Soto B (2005) Adsorption and desorption of copper and zinc in the surface layer of acid soils. J Colloid Interface Sci 288(1):21–29. https://doi.org/10.1016/j.jcis.2005.02.053. PMID: 15927557

Arias M, Pérez-Novo C, López E, Soto B (2006) Competitive adsorption and desorption of copper and zinc in acid soils. Geoderma 133:151–159

Arnon DI, Johnson CM (1942) Influence of hydrogen ion concentration on the growth of higher plants under controlled conditions. Plant Physiol 17:525–539

Arora D, Gaur C (1979) Microbial solubilization of different inorganic phosphates. Indian J Exp Biol 1979(17):1258–1261

Arp DJ, Sayavedra-Soto LA, Hommes NG (2002) Molecular biology and biochemistry of ammonia oxidation by Nitrosomonas europaea. Arch Microbiol 178:250–255

Arshad MA, Gill KS (1996) Field pea response to liming of an acid soil under two tillage systems. Can J Soil Sci 76:549–555

Asghar M, Kanehiro Y (1980) Effects of sugarcane trash and pineapple residue on soil pH, redox potential, extractable Al, Fe and Mn. Trop Agric 57:245–258

ATA (Agricultural transformation Agency of Ethiopia) (2013) Status of soil resources in Ethiopia and priorities for sustainable management. Paper presented at: The Launch of the Global Soil Partnership in Eastern and Southern Africa; Nairobi, Kenya

Ayres AS, Hagihara HH, Stanford G (1965) Significance of extractable aluminum in Hawaiian sugar cane soils. Soil Sci Soc Am Proc 29:387–392

Bååth E (1998) Growth rates of bacterial communities in soils at varying pH: a comparison of the thymidine and leucine incorporation techniques. Microb Ecol 36:316–327

Bååth E, Anderson TH (2003) Comparison of soil fungal/bacterial ratios in a pH gradient using physiological and PLFA-based techniques. Soil Biol Biochem 35:955–963

Bååth E, Arnebrant K (1995) Growth rate and response of bacterial communities to pH in limed and ash-treated forest soils. Soil Biol Biochem 26:995–1001

Baker et al (1995)

Baker AJM, McGrath SP, Reeves RD, Smith JAC (2000) Metal hyperaccumulator plants: a review of the ecology and physiology of a biological resource for phytoremediation of metal–polluted soils. In: Terry N, Banuelos G (eds) Phytoremediation of contaminated soil and water. Lewis Publisher, Boca Raton, pp 85–107

Baldani VLD, Baldani JI, Döbereiner J (1987) Inoculation on field-grown wheat (Triticum aestivum) with Azospirillum spp. in Brazil. Biol Fert Soils 4:37–40

Balemi T, Negisho K (2012) Management of soil phosphorus and plant adaptation mechanisms to phosphorus stress for sustainable crop production: A review. J Soil Sci Plant Nutr 12:547–562

Balser TC, Firestone MK (2005) Linking microbial community composition and soil processes in a California annual grassland and mixed-conifer forest. Biogeochemistry 73:395–415

Bambara S, Ndakidemi PA (2010) The potential roles of lime and molybdenum on the growth, nitrogen fixation and assimilation of metabolites in a nodulatedlegume. A special reference to Phaseolusvulgaris L. Afri J Biotechnol 8:2482–2489

Banerjee MR, Chapman SJ (1996) The significance of microbial biomass sulphur in soil. Biol Fertil Soils 22:116–123

Barak P, Jobe BO, Krueger AR, Peterson LA, Laird DA (1997) Effect of long-term soil acidification due to nitrogen fertilizer inputs in Wisconsin. Plant Soil 197:61–69. https://doi.org/10.1023/A:1004297607070

Barber SA (1995) Soil nutrient bioavailability. John Wiley & Sons, New York

Bardgett RD, Jones AC, Jones DL, Kemmitt SJ, Cook R, Hobbs P (2001) Soil microbial community patterns related to the history and intensity of grazing in sub-montane ecosystems. Soil Biol Biochem 33:1653–1664

Barney PE Jr, Bush LP (1985) Interaction of nitrate and sulphate reduction in tobacco. I. Influence of availibility of nitrate and sulphate. J Plant Nutr 8:507–515

Barrow NJ (1987) The effects of phosphate on zinc sorption by a soil. J Soil Sci 38:453–459

Bartoli F, Poulenard AJ, Schouller BE (2007) Influence of allophane and organic matter contents on surface properties of Andosols. Eur J Soil Sci 58:450–464

Bashan Y, Levanony H (1991) Alterations in membrane potential and in proton efflux in plant roots induced by Azospirillum brasilense. Plant Soil 137:99–103

Batjes NH (1996) Total carbon and nitrogen in the soils of the world. Eur J Soil Sci 47:151–163

Bayer C, Martin-Neto L, Mielniczuk J, Pavinato A, Dieckow J (2006) Carbon sequestration in two Brazilian Cerrado soils under no-till. Soil till Res 86:237–245

Beare MH, Parmelee RW, Hendrix PF, Cheng W, Coleman DC (1992) Microbial and faunal interactions and effects on litter nitrogen and decomposition in agroecosystems. Ecol Monogr 62:569–591

Behera SK, Singh MV, Singh KN, Todwal S (2011) Distribution variability of total and extractable zinc in cultivated acid soils of India and their relationship with some selected soil properties. Geoderma 162:242–250

Bekele T, Hofner W (1993) Effects of different phosphate fertilizers on yield of barleyand rapeseed on reddish brown soils of theEthiopian highlands

Berek AK, Hue N, Ahmed A (2011) Beneficial use of biochar to correct soil acidity. Hānai'Ai/The Food Provider

Berendsen RL, Pieterse CMJ, Bakker PAHM (2012) The rhizosphere microbiome and plant health. Trends Plant Sci 17:478–486

Berg G (2009) Plant–microbe interactions promoting plant growth and health: Perspectives for controlled use of microorganisms in agriculture. Appl Microbiol Biotechnol 84:11–18

Berg G, Smalla K (2009) Plant species and soil type cooperatively shape the structure and function of microbial communities in the rhizosphere. FEMS Microbiol Ecol 68(1):1–13. https://doi.org/10.1111/j.1574-6941.2009.00654.x

Berg et al (2016)

Berg et al (2019)

Berner RA (1969) Migration of iron and sulfur within anaerobic sediments during early diagnosis. Amer J Sci 267:19–42

Berner RA (1971) Principles of chemical sedimentology. McGraw-Hill Book Co., 240 pp

Berthrong ST, Jobbagy EG, Jackson ARB (2009) A global meta-analysis of soil exchangeable cations, pH, carbon, and nitrogen with afforestation. Ecol Appl 19:2228–2241

Bhattacharyya T, Pal DK, Deshpande SB (1993) Genesis and transformation of minerals in the formation of red (Alfisols) and black (Inceptisols and Vertisols) soils on Deccan

Bhattacharyya T, Sen TK, Singh RS, Nayak DC, Sehgal JL (1994) Morphology and classification of Ultisols with kandic horizon in north-eastern region. J Indian Soc Soil Sci 42:301–306

Bhattacharyya T, Pal DK, Srivastava P (2000) Formation of gibbsite in the presence of 2:1 minerals: an example from Ultisols of northeast India. Clay Miner 35(5):827–840. https://doi.org/10.1180/000985500547269

Bhattacharyya T, Pal DK, Chandran P, Ray SK (2005) Land-use, clay mineral type and organic carbon content in two Mollisols–Alfisols–Vertisols catenary sequences of tropical India. Clay Res 24:105–122

Bhattacharyya T, Sarkar D, Sehgal JL,Velayutham M, Gajbhiye KS, Nagar AP, Nimkhedkar SS (2009) Soil taxonomic database of India and states (1: 250,000) NBSSLUP Bulletin 143, NBSS & LUP, Nagpur, India, pp 266

Bhutto AW, Qureshi K, Abro R, Harijan K, Zhao Z, Bazmi AA, Abbas T, Yu G (2016) Progress in the production of biomass-toliquid biofuels to decarbonize the transport sector—prospects and challenges. RSC Adv 6:32140–32170

Bian M, Zhou M, Sun D, Li C (2013) Molecular approaches unravel the mechanism of acid soil tolerance in plants. Crop J 1(2):91–104

Binkley D, Richter D (1987) Nutrient cycles and H^+ budgets of forest ecosystems. Adv Ecol Res 16:1–51

Birgander J, Rousk J, Olsson PA (2014) Comparison of fertility and seasonal effects on grassland microbial communities. Soil Biol Biochem 76:80–89. https://doi.org/10.1016/j.soilbio.2014.05.007

Bishop ML, Chang AC, Lee RWK (1994) Enzymatic mineralization of organic phosphorus in a volcanic soil in Chile. Soil Sci 1994(157):238–243

Blagodatskaya EV, Anderson T-H (1998) Interactive effects of pH and substrate quality on the fungal-to-bacterial ratio and qCO_2 of microbial communities in forest soils. Soil Biol Biochem 30:1269–1274

Blancaflor EB, Jones DL, Gilroy S (1998) Alterations in the cytoskeleton accompany aluminum-induced growth inhibition and morphological changes in primary roots of maize. Plant Physiol 118:159–172

Bloem EMS-Bv (1998) Agrarokosystemen unter besonderer Beruksichtigung hydrologischer and bodenphysikalischer Standorteigenschaften, vol 192, p 156

Bloom PR (2000) Soil pH and pH buffering. In: Sumner ME et al (eds) Handbook of soil science. CRC Press, Boca Raton, FL, pp B333–B352

Bobbink R, Hicks K, Galloway J, Spranger T, Alkemade R, Ashmore M, Bustamante M, Cinderby S, Davidson E, Dentener F, Emmett B, Erisman JW, Fenn M, Gilliam F, Nordin A, Pardo L, De Vries W (2010) Global assessment of nitrogen deposition effects on terrestrial plant diversity: a synthesis. Ecol Appl 20:30–59

Bohn HL, Barrow NJ, Rajan SS, Parfitt RL (1986) Reactions of inorganic sulfur in soils. In: Tabatabai MA (ed) Sulfur in agriculture. Agron Monogr 27. ASA, CSSA, and ISSSA, Madison, WI, pp 233–249

Bohn H, McNeal BL, O'Connor GA (2001) Soil chemistry, 3rd edn. John Wiley & Sons, New York

Bolstad PV, Vose JM (2005) Forest and pasture carbon pools and soil respiration in the southern Appalachian mountains. For Sci 51:372–383

Bordeleau LM, Prévost D (1994) Nodulation and nitrogen fixation in extreme environments. Plant Soil 161:115–125. https://doi.org/10.1007/BF02183092

Boyle SA, Yarwood RR, Bottomley PJ, Myrold DD (2008) Bacterial and fungal contributions to soil nitrogen cycling under Douglas fir and red alder at two sites in Oregon. Soil Biol Biochem 40:443–451

Brady NC, Weil RR (2002) The nature and properties of soils, 13th edn. Prentice Hall, Upper Saddle River, New Jersey

Brady NC, Weil RR (2008) The nature and properties of soils, 14th edn. Pearson Prentice Hall, Upper Saddle River, NJ

Brady NC, Weil RR (2016) The nature and properties of soils, 15th edn. Pearson Prentice Hall, Upper Saddle River, NJ

Brallier S, Harison RB, Henry CL, Dongsen X (1996) Liming Effects on availability of Cd, Cu, Ni, and Zn in a soil amended with sewage sludge 16 years previously. Water Air Soil Pollut 86:195–206

Brenes E, Pearson RW (1973) Root responses of three Gramineae species to soil acidity in an Oxisol and an Ultisol. Soil Sci 116:295–302

Brinkman R, Pons LJ (1968) A pedogeomorphological classification and map of the Holocene sediments in the coastal plain of the three Guiana's. Soil Survey Papers No. 4. Soil Survey Inst. Wageningen, 40 p

Brockwell J, Pilka A, Holliday RA (1991) Soil pH is a major determinant of the numbers of naturally-occurring Rhizobium meliloti in non-cultivated soils of New South Wales. Aust J Exp Agric 31:211–219

Bromfield SM, Jones OL (1972) The initial leaching of hayed-off pasture plants in relation to the recycling of phosphorus. Aust J Soil Res 23:811–824

Brown ME (1974) Seed and root bacterization. Annu Rev Phytopatol 12:181–197

Brown AD, Jurinak JJ (1989) Mechanism of pyrite oxidation in aqueous mixtures. J Environ Qual 18:545–550

Bulgarelli D, Rott M, Schlaeppi K, van Themaat EVL, Ahmadinejad N, Assenza F, Rauf P, Huettel B, Reinhardt R, Schmelzer E et al (2012) Revealing structure and assembly cues for Arabidopsis root-inhabiting bacterial microbiota. Nature 488:91–95

Burns RG (1983) Extracellular enzyme-substrate interactions in soil. In: Slater JH, Whittenbury R, Wimpenny JWT (eds) Microbes in their natural environment. Cambridge University Press, Cambridge, pp 249–298

Buyanovsky GA, Wagner GH (1998) Carbon cycling in cultivated land and its global significance. Glob Change Biol 4:131–141

Caesar-Tonthat TC (2002) Soil binding properties of mucilage produced by a basidiomycete fungus in a model system. Mycol Res 106:930–937. https://doi.org/10.1017/S0953756202006330

Caires EF, Feldhaus IC, Barth G, Garbuio EJ (2002) Lime and gypsum application on the wheat crop. Sci Agric 59:357–364

Caires EF, Alleoni LRF, Cambri MA, Barth G (2005) Surface application of lime for crop grain production under a no-till system. Agron J 97:791–798

Campbell CA, McConkey BG, Zentner RP, Dyck FB, Selles F, Curtin D (1996) Tillage and crop rotation effects on soil organic C and N in a coarse-textured Typic Haploboroll in south-western Saskatchewan. Soil till Res 37:3–14

Cao X, Ma LQ, Rhue DR, Appel CS (2004) Mechanisms of lead, copper, and zinc retention by phosphate rock. Enviorn Poll 131:435–444

Capper AL, Campbell R (1986) The effect of artificially inoculated antagonistic bacteria on the prevalence of take-all disease of wheat in field experiment. J Appl Bacteriol 60:155–160

Cardon ZG, Gage DJ (2006) Resource exchange in the rhizosphere: molecular tools and the microbial perspective. Annu Rev Ecol Evol Syst 37(1):459–488

Carmargo MN, Falesi IC (1974) Soils of central plateau and Transamazonic highway of Brazil. In: Bornemisza E, Alvarado A (eds) Soil management of tropical America. Pub. Nth Carolina State University N.C. USA

Carter JM, Gardner WK, Gibson AH (1994) Improved growth and yield of faba beans (Vicia faba cv. Fiord) by inoculation with strains of Rhizobium leguminosarum biovar viciae in acid soils in south-west Victoria. Aust J Agric Res 45:613–623

Carter CM, Vander Sloot HA, Cooling D (2009) PH-dependent extraction of soil and soil amendments to understand the factors controlling element mobility. Eur J Soil Sci 60:622–637

Cate RB, Sukhai AP (1964) A study of aluminum in rice soils. Soil Sci 98:85–93

Catroux G, Hartmann A, Revellin C (2001) Trends in rhizobial inoculant production and use. Plant Soil 230:21–30

Cavallaro N, McBride MB (1984) Zinc and copper sorption and fixation by an acid soil clay: effect of selective dissolutions. Soil Sci Soc Am J 48(5):1050–1054

Chan KY (2001) An overview of some tillage impacts on earthworm population abundance and diversity implications for functioning in soils. Soil. till. Res. 57:179–191

Chandran P, Ray SK, Bhattacharyya T, Dubey PN, Pal DK, Krishnan P (2004) Chemical and mineralogical characteristics of ferruginous soils of Goa. Clay Res 23:51–64

Chandran P, Ray SK, Bhattacharyya T, Srivastava P, Krishnan P, Pal DK (2005) Lateritic soils of Kerala, India: their mineralogy, genesis and taxonomy. Aust J Soil Res 43:839–852

Chang HY, Haruna AO, Majid NMNA, Jalloh MB (2019) Improving soil phosphorus availability and yield of Zea mays L. using biochar and compost derived from agro-industrial wastes. Ital J Agron 14:34–42

Chaparro JM, Sheflin A, Manter D, Vivanco JM (2012) Manipulating the soil microbiome to increase soil health and plant fertility. Biol Fertil Soils 48:489–499

Chen J (2007) Rapid urbanization in China: a real challenge to soil protection and food security. CATENA 69:1–15

Chen CR, Xu ZH, Mathers NJ (2004) Soil carbon pools in adjacent natural and plantation forests of subtropical Australia. Soil Sci Soc Am J 68:282–291

Chenery EM (1954) Acid sulphate soils in Central Africa. In: Trans. 5th International Congress Soil Science, Lepoldville, 4, pp 195–198

Chernov V (1947) Historical Aspects of Soil Chemistry. https://www.Udel.edu/soilchem/Historical%20Aspects%20of%20Soil%20Chemistry.PDF

Cho SR, Jeong ST, Kim GY, Lee JG, Kim GW (2019) Evaluation of the carbon dioxide (CO2) emission factor from lime applied in temperate upland soil. Geoderma 337:742–748

Christiansen-Weneger C (1992) N_2-fixation by ammonium-excreting Azospirillum brasilense in auxin-induced tumours of wheat (Triticum aestivum L.). Biol Fertil Soils 12:85–100

Ciais P, Tans PP, Trolier M, White JW, Francey RJ (1995) A large Northern Hemisphere terrestrial CO_2 sink indicated by the 13C/12C ratio of atmospheric CO_2. Science 269:1098–1102

CIFOR (2000) Integrated natural resource management research in the CGIAR. Center for International Forestry Research, Bogor, Indonesia, 50 pp

Clark RB, Baligar VC (2000) Acidic and alkaline soil constraints on plant mineral nutrition. Plant-Environment Interactions, 133–177

Clark CJ, McBride MB (1984) Adsorption of Cu(II) by allophane as affected by phosphate. Soil Sci 38:412–421

Coleman NT, Thomas GW (1967) The basic chemistry of soil acidity. Agron Monogr 12:1–41

COM (2002) Communication from the Commission to the Council, the European Parliament, the Economic and Social Committee and the Committee of the regions towards a thematic strategy for soil protection. Commission of the European Communities, Brussels, p 2002

Conyers M (1990) The control of aluminium solubility in some acidic Australian soils. J Soil Sci 41:147–156

Conyers MK, Uren NC, Helyar KR, Poile GJ, Cullis BR (1997) Temporal variation in soil acidity. Aust J Soil Res 35:1115–1129

Cooper R (1959) Bacterial fertilizers in the Soviet Union. Soils Fertilizers 22:327–330

Costa ACSD, Seidel EP (2010) Nitrate leaching from a latosol after application of liquid swine manures in different pH values and organic matter contents. ActaSci Agron 32(4):743–748

Coulter JA, Nafziger ED, Wander MM (2009) Soil organic matter response to cropping system and nitrogen fertilization. Agron J 101:592–599

Coventry DR (1985) Changes in agricultural systems on acid soils in southern Australia. In: Proceedings 3rd Australian agronomy conference, pp 126–145

Coventry DR, Slattery WJ, Burnett VF, Ganning GW (1997) Liming and re-liming needs of acidic soils used for crop-pasture rotations. Aust J Exp Agric 37:577–581

Cregan PD, Hirth JR, Conyers MK (1989) Amelioration of soil acidity by liming and other amendments. In: Robson AD (ed) Soil acidity and plant growth. Academic Press, Sydney, Australia, pp 205–264

Crews TE, Kitayama K, Fownes JH, Riley RH, Herbert DA, Mueller-Dombois D, Vitousek PM (1995) Changes in soil phosphorus fractions and ecosystem dynamics across a long chronosequence in Hawaii. Ecol Lett 76:1407–1424

Cronan CS, Grigal DF (1995) Use of calcium/aluminum ratios as indicators of stress in forest ecosystems. J Environ Qual 24:209–226

CSA News (2014) Annual meeting. 59(10):31. https://doi.org/10.2134/csa2014-59-10-15

Cucarella V, Zaleski T, Mazurek R, Renman G (2008) Effect of reactive substrates used for the removal of phosphorusfrom wastewater on the fertility of acid soils. Biores Technol 99:4308–4314. https://doi.org/10.1016/j.biortech.2007.08.037

Da HN, Deng SP (2003) Survival and persistence of genetically modified Sinorhizobium meliloti in soil. Appl Soil Ecol 22:1–14

Dai Z, Wang Y, Muhammad N, Yu X, Xiao K, Meng J, Liu X, Xu J and Brookes PC (2014) The effects and mechanisms of soil acidity changes, following incorporation of biochars in three soils differing in initial pH. Soil Sci Soc Am J 78:1606–1614.·https://www.researchgate.net/publication/273937080

Dang YP, Dalal RC, Edwards DG, Tiller KG (1994) Kinetics of zinc desorption from vertisols. Soil Sci Soc Am J 58:1392–1399

Das A, Prasad R, Srivastava A, Giang PH, Bhatnagar K, Varma A (2010) Fungal siderophores: structure, functions and regulation. In: Varma A, Chincholkar SB (eds) Microbial soderophores (soil biology). Springer-Verlag, Heidelberg, Germany, pp 1–42

Datta M, Dhiman KR (2001) Yield and response of rice (Oryza sativa) as influenced by methods and sources of micronutrient application. Indian J Agri Sci 71(3):190–192

Datta M, Sharma UC (2006) Soil acidity and amelioration. In: Sharma UC, Datta M, Samra JS (eds) Soils and their management in North East India. ICAR research Complex for NEH Region, Umroi, Meghalaya, India, pp 220–241

Datta M, Banik S, Gupta RK (1982) Studies on the efficacy of phytohormone producing phosphate solubilizing Bacillus firmus in augmenting paddy yield in acid soils of Nagaland. Plant Soil 69:365–373

Datta M, Laskar S, Prasad RN (1991) Appraisal of charge characteristics in some soils of Tripura. J Indian Soc Soil Sci 39:767–769

Datta M, Bhattacharya BK, Saikh H (2001) Soil fertility—a case study of shifting cultivation sites in Tripura. J Indian Soc Soil Sci 49(1):104–109

David MB, Driscoll CT (1984) Aluminum speciation and equilibria in soil solutions of a Haplorthod in the Adirondack mountains (New York, USA). Geoderma 33:297–318

Davies PJ (2010) The plant hormones: their nature, occurrence, and functions. In: Davies PJ (ed) Plant hormones—biosynthesis, signal transduction, action, 3rd ed. Kluwer Academic Publishers, Dordrecht, The Netherlands, pp 1–15

Davies G, Ghabbour EA (2001) Humic acids: marvelous products of soil chemistry. J Chem Educ 78:1609–1614

de Bertoldi M, Sequi P, Lemmes B, Papi T (eds) (1996) The science of composting: parts 1 and 2. Blackie Academic and Professional, Glasgow

de Boer W, Kowalchuk GA (2001) Nitrification in acid soils: micro-organisms and mechanisms. Soil Biol Biochem 33:853–866

De Boer W, Klein Gunnewiek PJA, Veenhuis M, Bock E, Laanbroek HJ (1991) Nitrification at low pH by aggregated autotrophic bacteria. Appl Environ Microbiol 57:3600–3604

De Boer W, Gunnewiek PJAK, Parkinson D (1996) Variability of N mineralization and nitrification in a simple, simulated microbial forest soil community. Soil Biol Biochem 28:203–211

de Quadros PD, Zhalnina K, Davis-Richardson A, Fagen JR, Drew J, Bayer C, Camargo FAO, Triplett EW (2012) The effect of tillage system and crop rotation on soil microbial diversity and composition in a subtropical acrisol. Diversity 4:375–395. https://doi.org/10.3390/d4040375

De Vries FT, Hoffland E, van Erkeren N, Brussaard L, Bloem J (2006) Fungal/bacterial ratios in grasslands with contrasting nitrogen management. Soil Biol Biochem 38:2092–2103

Delgado A, Torrent J (2000) Phosphorus forms and desorption patterns in heavily fertilized calcareous and limed acid soils. Soil Sci Soc Am J 64:2031–2037

Delhaize E, Ryan PR (1995) Update: aluminium toxicity and tolerance in plants. Plant Physiol 107:315–321

Dent D (1986) Acid sulphate soils: a baseline for research and development. ILRI, Wageningen, The Netherlands

Derpsch R, Friedrich T, Kassam A, Hongwen L (2010) Current status of adoption of no-till farming in the world and some of its main benefits. Int J Agric Biol Eng 3:1–25

Dhananjaya BC, Ananthanarayana R (2009) Soil acidity and exchange chemistry in soils of southern Karnataka Introduction. Agropedology 19(2):99–105

Dhir RP (1967) Pedological characteristics of some soils of north western Himalayas. J Indian Soc Soil Sci 15:61–69

Di HJ, Cameron KC, Shen JP, Winefield CS, O'Callaghan M, Bowatte S, He JZ (2009) Nitrification driven by bacteria and not archaea in nitrogen-rich grassland soils. Nat Geosci 2:621–624

Diemont WH, van Wijng'aarden W (1974) Sedimentation patterns, soils, mangrove vegetation and land use in tidal areas of West Malaysia. In: Walsh G et al (ed) Proceedings of international symposium biology and management of mangroves, vol 2:513–528

Dixon JB, Weed SB (1989) Minerals in the soil environment. Soil Science Society of America, Madison, WI, USA

Dolling PJ, Porter WM, Rowland IC (1994) Acidification rates in the central wheatbelt of Western Australia. 2. On a sandy duplex soil. Aust J Exp Agric 34:1165–1172. https://doi.org/10.1071/EA9941165

Dolui AK, Dey S, Nayek AK (1987) Charge Characteristics in two Soil profiles. J Indian Soc Soil Sci 35:353–357

Donahue RL, Miller RW, Shickluna JC (1977) Soils: an introduction to soils and plant growth. Prentice-Hall, Upper Saddle River

Donald B (2011) Soil pH: what it means. State University of New York College of Environmental Science and Forestry. Retrieved from http://www.esf.edu on 21st Dec 2011

Driessen PM, Buurman P, Permadhy (1976) The influence of shifting cultivation on a 'podzolic' soil from Central Kalimantan. Bull. 3 Peat and podzolic soils and their potential for agriculture in Indonesia. Soil Research Institute, Bogor

Driscoll CT (1985) Aluminum in acidic surface waters: chemistry, transport, and effects. Environ Health Perspect 63:93–104

Driscoll CT (1989) The chemistry of aluminum in surface waters. In: Sposito G (ed) The environmental chemistry of aluminum. CRC Press, Boca Raton, FL, USA, pp 241–277

Driscoll CT, van Breemen N, Mulder J (1985) Aluminum chemistry in a forested Spodosol. Soil Sci Soc Am J 49:437–444

Driscoll CT, Driscoll KM, Mitchell MJ, Raynal DJ (2003) Effects of acidic deposition on forest and aquatic ecosystems in New York State. Environ Pollut 123:327–336

Duff RB, Webley DM (1959) 2-Ketogluconic acid as a natural chelator produced by soil bacteria. Chem Ind 1376–1377

Dugalić G, Gajić B, Bokan N, Jelić M, Tomić Z, Dragović R (2012) Liming increases alfalfa yield and crude protein content in an acidic silty loam soil. Afr J Biotechnol 11:11552–11558

Dultz S, Woche SK, Mikutta R, Schrapel M, Guggenberger G (2019) Size and Charge constraints in microaggregation: model experiments withmineral particle size fractions. Appl Clay Sci 170:29–40

Edmeades DC, Ridley AM (2003) Using lime to ameliorate topsoil and subsoil acidity. In: Rengel Z (ed) Handbook of soil acidity. Eastern Hemisphere Distribution, Switzerland, Basel. ISBN 0-8247-0890-3

Eidukeviciene M, Ozheraitiene DJ, Tripolskaja LN, Marcinkonis SI (2001) The effect of long-term liming on the chemical properties of Lithuanian soils. Eurasian Soil Sci 34(9):999–1005

Elberling B, Matthiesen H (2007) Methodologically controlled variations in laboratory and field pH measurements in waterlogged soils. Eur J Soil Sci 58:207–214

Elkins DM, Ensminger LE (1971) Effect of soil pH on the availability of adsorbed sulfate. Soil Sci Soc Am Proc 35:931–943

Elliott HA, Liberati MR, Huang CP (1986) Competitive adsorption of heavy metals by soils. American Society of Agronomy, Crop Sci Soc Am Soil Sci Soc Am 15(3):214–219

Elo S, Maunuksela L, Salkinoja-Salonen M, Smolander A, Haahtela K (2006) Humus bacteria of Norway spruce stands: plant growth promoting properties and birch, red fescue and alder

colonizing capacity. FEMS Microbiol Ecol 31:143–152. https://doi.org/10.1111/j.1574-6941.2000.tb00679.x
El-Sawah MMA, Hauka FIA, El-Rafey HH (1993) Study on some enzymes cleaving phosphorus from organic substrates in soil. J Agric Sci 1993(18):2775–2785
Enders A, Hanley K, Whitman T, Joseph S, Lehmann J (2012) Characterization of biochars to evaluate recalcitrance and agronomic performance. Bioresour Technol 114:644–653. https://doi.org/10.1016/j.biortech.2012.03.022
Ente PJ (1964) Initial decalcification due to oxidation of sulphides in young marine soils in The Netherlands. In: Transactions 8th international congress of soil science, Bucharest 1964, vol 2, pp 779–783
Erisman JW, Bleeker A, Hensen A, Vermeulen A (2008) Agricultural air quality in Europe and the future perspectives. Atmos Environ 42:3209–3217
Evans CE, Kamprath EJ (1970) Lime response as related to percent aluminum saturation, soil solution aluminum and organic matter content. Soil Sci Soc Am Proc 34:893–896
Eyheraguibel B, Silvestrea J, Morard P (2008) Effects of humic substances derived from organic waste enhancement on the growth and mineral nutrition of maize. Biores Technol 99:4206–4212. https://doi.org/10.1016/j.biortech.2007.08.082
Fageria NK (2002) Soil quality vs. environmentally based agricultural management practices. Commun Soil Sci Plant Anal 33:2301–2329
Fagerli H, Aas W (2008) Trends of nitrogen in air and precipitation: model results and observations at EMEP sites in Europe 1980–2003. Environ Pollut 154:448–461
Fageria NK, Baligar VC (2003) Fertility management of tropical acid soils for sustainable crop production. In: Rengel Z (ed) Handbook of soil acidity. Marcel Dekker, New York, pp 359–385
Fageria NK, Baligar VC (2008) Ameliorating soil acidity of tropical Oxisols by liming for sustainable crop production. Adv Agron 99:345–399. https://doi.org/10.1016/S0065-2113(08)00407-0
Falkengren-Grerup U (1987) Long-term changes in pH of forest soils in southern Sweden. Environ Pollut 43(2):79–90
FAO-UNESCO (1988) Soil map of the world. Revised Legend. World Soil Resources Rep 60, FAO Rome, 119pp
Fazli IS, Jamal A, Ahmad S, Masoodi M, Khan JS, Abdin MZ (2008) Interective effect of sulphur and nitrogen on nitrogen accumulation and harvest in oilseed crops differing in nitrogen assimilation potential. J Plant Nutri 31:1203–1220
Feller C, Frossard E, Brossard M (1994) Phosphatase activity in low activity tropical clay soils. Distribution in the various particle size fractions. Can J Soil Sci 74:121–129
Feng Y, Motta AC, Reeves DW, Burmester CH, van Santen E, Osborne JA (2003) Soil microbial communities under conventional-till and no-till continuous cotton systems. Soil Biol Biochem 35:1693–1703
Fenton G, Helyar KR (1999) The causes and management of acid soils. Invited paper for the soils workshop 'describing, analysing and managing our soil' (25 pp), 22–26th November. University of Sydney and NSW Branch Australian Soil Science Society
Fenton G, Helyar K (2002) The role of the nitrogen and carbon cycles in soil acidification. Acid Soil Action Res Report 2002:183–199
Fernández-Calviño D, Pateiro-Moure M, López-Periago E, Arias-Estévez M, Nóvoa- Muñoz JC (2008) Copper distribution and acid–base mobilization in vineyard soils and sediments fromGalicia (NWSpain). Eur J Soil Sci 59:315–326
Fernández-Calviño D, Pérez-Novo C, Bermúdez-Couso A, López-Periago E, Arias- Estévez M (2010) Batch and stirred flow reactor experiments on Zn sorption in acid soils. Cu Competition. Geoderma 159:417–424
Fierer N, Jackson RB (2006) The diversity and biogeography of soil bacterial communities. Proc Natl Acad Sci USA 103:626–631
Fierer N, Strickland MS, Liptzin D, Bradford MA, Cleveland CC (2009) Global patterns in belowground communities. Ecol Let 12:1–2

Fierer N, Lauber CL, Ramirez KS, Zaneveld J, Bradford MA, Knight R (2012) Comparative metagenomic, phylogenetic, and physiological analyses of soil microbial communities across nitrogen gradients. ISME J 61:1007–1017

Firestone MK, Killham K, McColl JG (1983) Fungal toxicity of mobilized soil aluminum and manganese. Appl Environ Microbiol 46:758–761

Firnia D, Rohmawati I (2022) Chemical properties of acid soil after amelioration of dolomite, coal, and microbes. In: The 3rd international conference on agriculture and rural development, IOP Conf Ser: Earth Environ Sci 978:012010. IOP Publishing. https://doi.org/10.1088/1755-1315/978/1/012010

Flis SE, Glenn AR, Dilworth MJ (1993) The interaction between aluminium and root nodule bacteria. Soil Biol Biochem 25:403–417

Focht DD, Verstraete W (1977) Biochemical ecology of nitrification and denitrification. Adv Microb Ecol 1:135–214

Forge TA, Hogue EJ, Neilsen G, Neilsen D (2008) Organic mulches alter nematode communities, root growth and fluxes of phosphorus in the root zone of apple. Appl Soil Ecol 39:15–22

Forhne Y, Rinklebe J, Diaz- Bone RA, Du L (2011) Controlled variation of redoxconditions in a flood plain soil: impact on metal mobilization and biomethylation of arsenic and antimony. Geoderma 160:414–424

Forlani G, Mangiagalli A, Nielsen E, Suardi CM (1999) Degradation of the phosphonate herbicide glyphosate in soil: evidence for a possible involvement of unculturable microorganisms. Soil Biol Biochem 31(7):991–997. https://doi.org/10.1016/S0038-0717(99)00010-3

Foster HL (1970) Liming continuously cultivated soils in Uganda. East Afr Agric for J 36:58–69

Foster NW, Nicholsen JA, Hazlett PW (1989) Temporal variation in nitrate and nutrient cations in drainage waters from a deciduous forest. J Environ Qual 18:238–244

Fox TR, Comerford NB (1992) Rhizosphere phosphatase activity and phosphatase hydrolysable organic phosphorus in two forested spodosols. Soil Biol Biochem 24:579–583

Foy CD (1974) Effects of aluminum on plant growth. In: Carson EW (ed) The plant root and its environment. University Press of Virginia, Charlottesville, VA, pp 601–642

Foy CD (1984) Physiological effects of hydrogen, Al and manganese toxicities in acid soil. In: Adams F (ed) Soil acidity and liming. American Society of Agronomy, Madison, Wisconsin, pp 57–97

Francis T, Whittaker K, Shandas V, Mills AV, Graybill JK (2005) Incorporating science into the environmental policy process: a case study from Washington State. Ecol Soc 10(1):35. http://www.ecologyandsociety.org/vol10/iss1/art35/

Fujii K, Aoki M, Kitayama K (2012) Biodegradation of low molecular weight organic acids in rhizosphere soils from a tropical montane rain forest. Soil Biol Biochem 47:142–148

Fuller RD, Driscoll CT, Lawrence GB, Nodvin SC (1987) Processes regulating sulphate flux after whole-tree harvesting. Nature 325:707–709

Furrer G, Zysset M, Charlet L, Schindler PW (1991) Mobilisation and fixation of aluminium in soils. Metal Compounds Environ. Life 4:89–97

Gadd GM (2007) Geomycology: biogeochemical transformations of rocks, minerals, metals and radionuclides by fungi, bioweathering and bioremediation. Mycol Res 111:3–49

Gallez A, Juo ASR, Herbillon AJ (1976) Surface and charge characteristics of selected soils in the tropics. Soil Soc Am J 40(4):601–608

Gaudalix ME, Pardo MT (1995) Zinc sorption by acid tropical soils as affected by cultivation. J Soil Sci 46:317–322

Gebregziabher G, Abera DA, Gebresamuel G, Giordano M, Langan S (2016) An assessment of integrated watershed management in Ethiopia. In: IWMI Working Papers, vol 170. https://doi.org/10.5337/2016.214

Gerke J (1993) Aluminum complexation by humic substances and aluminum species in the soil solution. Geoderma 63:165–175

Gerretsen FC (1948) The influence of microorganisms on the phosphate intake by the plant. Plant Soil 1:51–81

Gillman GP, Uehara G (1980) Charge characteristics of soils with variable and permanent charge minerals. II. Experimental, Soil Sci Soc Am J 44(2):252–255. Available from https://doi.org/10.2136/sssaj1980.03615995004400020009x

Gilliam FS, Lyttle NL, Thomas A, Adams MB (2005) Soil variability along a nitrogen mineralization and nitrification gradient in a nitrogen-saturated hardwood forest. Soil Sci Soc Am J 69:247–256

Glenn AR, Tiwari RP, Reeve WG, Dilworth MJ (1997) The response of root nodule bacteria to acid stress. In: Moniz AC, Fulani AMC, Schaffert RE, Fageria NK, Rosolem CA, Cantarella H (eds) Plant–soil interactions at low pH: sustainable agriculture and forestry production. Brazilian Soil Science Society, Campinas, pp 123–138

Glick BR (1995) The enhancement of plant growth by free-living bacteria. Can J Microbiol 41:109–117

Goldman A (1995) Threats to sustainability in African agriculture: Searching for appropriate paradigms. Hum Ecol 23:291–334

Gomes PC, Fontes MPF, Silva AG, Mendonça ED, Netto AR (2001) Selectivity sequence and competitive adsorption of heavy metals by Brazilian soils. Soil Sci Soc Am J 65:1115–1121

Gonzaga MIS, Mackowiak C, De Almeida AQ, Junior JITDC, Andrade KR (2018) Positive and negative effects of biochar from coconut husks, orange bagasse and pine wood chips on maize (Zea mays L.) growth and nutrition. Catena 162:414–420

Gonzalez-Erico E, Kamprath EJ, Naderman GC, Soares WV (1979) Effect of depth of lime incorporation on the growth of corn in an Oxisol of central Brazil. Soil Sci Soc Am J 43:1155–1158

Goss MJ, Carvalho MJGPR (1992) Manganese toxicity: the significance of magnesium for the sensitivity of wheat plants. Plant Soil 139:91–98

Graham PH, Draeger K, Ferrey ML, Conroy MJ, Hammer BE, Martinez- Romero E, Naarons SR, Quinto C (1994) Acid pH tolerance in strains of Rhizobium and Bradyrhizobium, and initial studies on the basis for acid tolerance of Rhizobium tropici UMR1899. Can J Microbiol 40:198–207

Grayston SJ, Vaughan D, Jones D (1996) Rhizosphere carbon flow in trees, in comparison with annual plants: the importance of root exudation and impact on microbial activity and nutrient availability. Appl Soil Ecol 5:29–56

Gudu SO, Okalebo JR, Othieno CO, Kisinyo PO, Obura PA, Ligeyo DO (2007) Improving P acquisition and Al tolerance of plants. Presented at the France workshop, October, 2007, 69

Gügi B, Orange N, Hellio F, Burini JF, Guillou C, Leriche F, Guespin-Michel JF (1991) Effect of growth temperature on several exported enzyme activities in the psychrotropic bacterium Pseudomonas fluorescens. J Bacteriol 1991(173):3814–3820

Guo JH, Liu XJ, Zhang Y, Shen JL, Han WX, Zhang WF, Christie P, Goulding KWT, Vitousek PM, Zhang FS (2010) Significant acidification in major Chinese croplands. Science 327:1008–1010

Haby VA (1993) Soil pH and plant nutrient availability. FertiGram X(2)

Halder AK, Mishra AK, Bhattacharyya P, Chakrabartty PK (1990) (1990) Solubilization of rock phosphate by Rhizobium and Bradyrhizobium. J Gen Appl Microbiol 36:81–92

Hardy F (1926) The role of aluminum in soil infertility and toxicity. J Agric Sci 16:616–631

Harter RD (2007) Acid soils of the tropics. Echo Technical Note, University of New Hampshire, North Fort Myers

Harter RD, Lehmann RG (1983) Use of kinetics for the study of exchange reactions in soils. Soil Sci Soc Am J 47:666–669

Hartman WH, Richardson CJ, Vilgalys R, Bruland GL (2008) Environmental and anthropogenic control of bacterial communities in wetland soils. Proc Natl Acad Sci USA 105:17842–17847

Hass A, Gonzalez JM, Lima IM, Godwin HW, Halvorson JJ, Boyer DG (2012) Chicken manure biochar as liming and nutrient source for acid Appalachian soil. J Environ Qual 41:1096–1106. https://doi.org/10.2134/jeq2011.0124

Haynes RJ (1984) Lime and phosphate in the soil–plant system. Adv Agron 37:249–315

Haynes RJ (1986) The decomposition process: mineralization, immobilization, humus formation and degradation. In: Haynes RJ (ed) Mineral nitrogen in the plant-soil system. Academic Press, Orlando, pp 53–126

Hayes MHB, Swift RS (1985) The chemistry of soil organic colloids. In: Greenland DJ, Hayes MBH (eds) The chemistry of soil constituents. John Wiley & Son, New York, pp 179–320

He Z, Yang X, Baligar VC, Calvert DV (2003) Microbiological and biochemical indexing systems for assessing quality of acid soils. Adv Agron 78:89–138

Hedlund K, Griffiths B, Christensen S, Scheu S, Setälä H, Tscharntke T, Verhoef H (2004) Trophic interaction in changing landscapes: responses of food webs. Basic Appl Ecol 5:495–503

Helyar KR (1991) The management of acid soils. In: Wright RJ, Baligar VC, Murrmann RP (eds) Plant–soil interactions at low pH. Proceedings of the second international symposium on plant–soil interactions at Low pH, 24–29 June 1990, Beckley,West Virginia, USA, Kluwer, Dordrecht, pp 365–382

Helyar KR, Porter WM (1989) Soil acidification, its measurement and the processes involved. In: Robson AD (ed) Soil acidity and plant growth. Academic Press, Sydney, pp 61–101

Herrick JE (2000) Soil quality: an indicator of sustainable land management? Appl Soil Ecol 15:75–83

Holland EA, Coleman DC (1987) Litter placement effects on microbial and organic matter dynamics in an agroecosystem. Ecology 68:425–433

Honda N, Hirai M, Ano T, Shoda M (1998) Antifungal effect of a heterotrophic nitri®er Alcaligenes faecalis. Biotech Lett 20:703–705

Hooper AB, Vannelli T, Bergmann DJ, Arciero D (1997) Enzymology of the oxidation of ammonia to nitrite by bacteria. Antonie Van Leeuwenhoek 71:59–67

Hossain Z, Bahar M, Sarkar B, Donne SW, Ok YS, Palansooriya KN, Kirkham MB, Chowdhury S, Bolan N (2020) Biochar and its importance on nutrient dynamics in soil and plant. Biochar 2:379–420

Houlden A, Timms-Wilson TM, Day MJ, Bailey MJ (2008) Influence of plant developmental stage on microbial community structure and activity in the rhizosphere of three field crops. FEMS Microbiol Ecol 65:193–201

Howard PJA, Howard DM, Lowe LE (1998) Effects of tree species and soil physicochemical conditions on the nature of soil organic matter. Soil Biol Biochem 30:285–297

Howeler RH, Sieverding E, Saif S (1987) Practical aspects of mycorrhizal technology in some tropical crops and pastures. Plant Soil 100:249–283

Huang DL, Zeng GM, Feng CL, Hu S, Jiang XY, Tang L, Su FF, Zhang Y, Zeng W, Liu HL (2008) Degradation of lead-contaminated lignocellulosic waste by Phanerochaete chrysosporium and the reduction of lead toxicity. Environ Sci Technol 42:4946–4951. https://doi.org/10.1021/es800072c

Huber C, Baier R, Gottlein A, Weis W (2006) Changes in soil, seepage water and needle chemistry between 1984 and 2004 after liming an N-saturated Norway spruce stand at the Hoglwald, Germany. For Ecol Manage 233:11–20

Hue NV (1992) Correcting soil acidity of a highly weathered Ultisol with chicken manure and sewage sludge. Commun Soil Sci Plant Anal 23:241–264

Hue NV, Licudine DL (1999) Amelioration of subsoil acidity through surface application of organic manures. J Environ Qual 28:623–632

Hue NV, Ikawa H, Silva JA (1994) Increasing plant-available phosphorus in an Ultisol with a yard-waste compost. Commun Soil Sci Plant Anal 25:3291–3303

Hue NV, Vega S, Silva JA (2001) Manganese toxicity in a Hawaiian Oxisol as affected by soil pH and organic amendments. Soil Sci Soc Am J 65:153–160

Huggins DR, Guyanovsky GA, Wagner GH et al (1998) Soil organic C in the tallgrass prairie-derived region of the corn belt: effects of long-term crop management. Soil till Res 47(3–4):219–234

Hungria M, Stacey G (1997) Molecular signals exchanged between host plants and rhizobia: basic aspects and potential application in agriculture. Soil Biol Biochem 29:819–830

Ibekwe AM, Angle JS, Chaney RL, Van Berkum P (1997) Enumeration and nitrogen fixation potential of Rhizobium leguminosarum biovar trifolii grown in soil with varying pH values and heavy metal concentrations. Agric Ecosyst Environ 61:103–111

ICAR-NAAS (Indian Council of Agricultural Research-National Academy of Agricultural Sciences) (2010) Degraded and wastelands of India—status and spatial distribution. Indian Council of Agricultural Research and National Academy of Agricultural Sciences. Published by the Indian Council of Agricultural Research, New Delhi, 56 pp

IFPRI (2010) Fertilizer and soil fertility potential in Ethiopia: constraints and opportunities for enhancing the system. IFPRI, Washington, DC

Illmer P, Schinner F (1992) Solubilization of inorganic phosphates by microorganisms isolated from forest soil. Soil Biol Biochem 24:389–395

Ilton MF, Page AL, Lund LJ (1976) The effect of pH, salt concentration, and nature of electrolytes on the charge characteristics of Brazilian tropical soils. Soil Sci Soc Am J 40:521–527

Insam H (2001) Developments in soil microbiology since the mid 1960s. Geoderma 100(3–4):389–402

IRRI (International Rice Research Institute) (1968) Annual report 1967. Los Bafios, Philippines, 308 p

Islam KR, Weil RR (2000) Land use effects on soil quality in a tropic forest ecosystem of Bangladesh. Agric Ecosyst Environ 79:9–16

Islam KR (2006) Test of active organic matter as a measure of soil quality. 18th World Soil Science Congress, Philadelphia, Pennsylvania, USA, July 9-15, 2006

Islam A, Chen D, White RE (2007) Heterotrophic and autotrophic nitrification in two acid pasture soils. Soil Biol Biochem 39:972–975

Izac A-MN, Sanchez PA (2001) Towards a natural resource management paradigm for international agriculture: the example of agroforestry research. Agric Syst 69:5–25

Jackson WA (1967) Physiological effects of soil acidity. In: Pearson RW, Adams F (eds) Soil acidity and liming. American Society of Agronomy, Madison, Wisconsin

Jackson ML (1979) Soil mineralogy. In: Fairbridge RW, Fink CW Jr (eds) The encyclopedia of soil science, part I. Dowden, Hutchiinson and Ross Inc., Stroudsburg, Pennsylvania, USA, pp 475–482

Jafer DG, Hailu G (2017) Application of lime for acid soil amelioration and better soybean performance in South Western Ethiopia. J Biol Agric Healthc 7:95–100

Jakovljevi M, Kresovi M, Blagojevi S, Antic-Mladenovi S (2005) Some negative chemical properties of acid soils. J Serb Chem Soc 70(5):765–774

Janos P, Vavrova J, Herzogova L, Pilarova V (2010) Effects of inorganic and organic amendments on the mobility (leachability) of heavy metals in contaminated soil: a sequential extraction study. Geoderma 159:335–341

Jansen JAM, Andriesse W, Prasetyo H, Bregt AK (1992) Guidelines for soil survey in acid sulphate soils in the humid tropics; the main problems considered. AARD and LAWOO. ILRI, Wageningen, The Netherlands

Janzen HH, Bettany JR (1984) Sulfur nutrition of rapeseed. Influence of fertilizer nitrogen and sulfur rates. Soil Sci Soc Am J 48:100–107

Jelic M, Djalovic I, Milivojevic J, Krstic D (2010) Mobile aluminium content of vertisols as dependent upon fertilization system and small grains genotypes. In: Proceedings of 3rd international scientific/professional conference agriculture in nature and environment protection, Vukovar, Croatia, May 31–June 2, 2010, pp 137–142. ISBN 978-953-7693-008

Jenkins SN, Waite IS, Blackburn A, Husband R, Rushton SP, Manning DC, O'Donnell AG (2009) Actinobacterial community dynamics in long term managed grasslands. Antonie Van Leeu-Wenhoek 95:319–334

Jenny H (1941) Factors of soil formation. Dover Publications, Inc., New York, NY

Jia ZJ, Conrad R (2009) Bacteria rather than Archaea dominate microbial ammonia oxidation in an agricultural soil. Environ Microbiol 11:1658–1671

Jiang X, Maa Y, Yuan J, Wright AL, Li H (2011) Soil particle surface electrochemical property effects on abundance of ammonia-oxidizing bacteria and ammonia-oxidizing archaea, NH_4 þ activity, and net nitrification in an acid soil. Soil Biol Biochem 43:2215–2221

Johnson DW, Todd DE (1983) Relationship among iron, aluminum, carbon and sulfate in a variety of forest soils. Soil Sci Soc Am J 47:792–800

Johnson D, Leake JR, Ostle N, Ineson P, Read DJ (2002) In situ $13CO_2$ pulse labelling of upland grassland demonstrates a rapid pathway of carbon flux from arbuscular mycorrhizal mycelia to the soil. New Phytol 153:327–334

Johnston SG, Slavich P, Hirst P (2004) The acid flux dynamics of two artificial drains in acid sulphate soil backswamps on the Clarence River floodplain, Australia. Aust J Soil Res 42:623–637

Jones DL (1998) Organic acids in the rhizosphere a critical review. Plant Soil 205:25–44

Juo ASR (1977) Soluble and exchangeable aluminum in Ultisols and Alfisols in West Africa. Commun Soil Sci Plant Anal 8:17–35

Juo ASR (1980) Mineralogical characteristics of alfisols and ultisols. In: Theng BKG (ed) Soil with variable charge. New Zealand Society of Soil Science, Lower Hutt, New Zealand, pp 69–84

Juo ASR (1981) Chemical characteristics. In: Greenland DJ (ed) 81 Characterisation of soils. Clarendon Press, Oxford, pp 51–79

Juo ASR, Kamprath EJ (1979) Copper chloride as an extractant for estimating the potentially reactive aluminum pool in acid soils. Soil Sci Soc Am J 43:35–38

Kabata-Pendias A, Pendias H (2001) Trace elements in soils and plants. CRC Press, Boca Raton, FL, p 2001

Kamprath EJ (1970) Exchangeable aluminum as a criterion for liming leached mineral soils. Soil Sci Soc Am Proc 34:252–254

Kamprath EJ (1972) Soil acidity and liming. Soils of the Humid Tropics. National Academy of Sciences, Washington, DC, pp 136–149

Kamprath E (1980) Soil acidity in well-drained soils of the tropics as a constraint to food production. In: Priorities for alleviating soil-related constraints to food production in the tropics, Proceedings of a symposium, International Rice Research Institute and the New York State College of Agriculture and Life Sciences, and Cornell University in Cooperation with The University Consortium of Soils for the Tropics

Kamprath E (1984) Crop response to lime on soils in the tropics. In: Adam F (ed) Soil acidity and liming, 2nd ed. Agronomy 12. Soil Science Society of America, USA, Madison, WI, pp 349–368

Kamprath EJ (1997) Introductory comments. In: Moniz AC, Fulani AMC, Schaffert RE, Fageria NK, Rosolem CA, Cantarella H (eds) Plant–soil interactions at low pH: sustainable agriculture and forestry production. Brazilian Soil Science Society, Campinas, pp 193–194

Kanyanjua SM, Ireri L, Wambua S, Nandwa SM (2002) Acidic soils in Kenya. KARI Technical Paper No 11, pp 28

Karlen DL, Colvin TS (1992) Alternative farming system effects on profile nitrogen concentration on 2 Iowa farms. Soil Sci Soc Am J 56:1249–1256

Karlen DL, Ditzler CA, Andrews SS (2003) Soil quality: why and how? Geoderma 114:145–156

Kazda M, Zvacek L (1989) Aluminium and manganese and their relation to calcium in soil solution and needles in three Norvay spruce (Picea abies, L. Krast.) stands of Upper Austria. Plant Soil 114:257–267

Keltjens WG (1997) Plant adaptation and tolerance to acid soils: Its possible Al avoidance—a review. In: Moniz AC, Fulani AMC, Schaffert RE, Fageria NK, Rosolem CA, Cantarella H (eds) Plant–soil interactions at low pH: sustainable agriculture and forestry production. Brazilian Soil Science Society, Campinas, pp 109–117

Kemmitt SJ, Wright D, Goulding KWT, Jones DL (2006) PH regulation of carbon and nitrogen dynamics in two agricultural soils. Soil Biol Biochem 38:898–911

Kennedy AC, Papendick RI (1995) Microbial characteristics of soil quality. Soil Water Conserv 50:243–248

Keyser, Munns (1979) Effects of calcium, manganese, and aluminum on growth of rhizobia in acid media 1 H. Soil Sci Soc Am J 43:719–722

Kidd PS, Proctor J (2001) Why plants grow poorly on very acid soils: are ecologists missing the obvious? J Exp Bot 52:791–799

Kiekens L (1995) Zinc. In: Alloway BJ (ed) Heavy metals in soils. Blackie Academic and Professional Press, London, pp 284–305

Kikuchi R (2004) Deacidification effect of the litter layer on forest soil during snowmelt runoff: laboratory experiment and its basic formularization for simulation modeling. Chemosphere 54:1163–1169. https://doi.org/10.1016/j.chemosphere.2003.10.025

Kinraide TB (1991) Identity of the rhizotoxic aluminium species. Plant Soil 134:167–178

Kioke S, Subbarao RV, David RM, Turini TA (2003) Vegetable diseases caused by soilborne pathogens. Publication 8099. University of California Division of Agriculture and Natural Resources, Davis. http://ucanr.org/freepubs/docs/8099.pdf

Kisinyo PO, Othieno CO, Okalebo JR, Kilpsat MJ, Serema AK, Obiero DO (2005) Effects of lime and phosphorus application on early growth of Leucaena in acid soils. Conf Proc 7:1233–1236

Kitayama K (2005) Comments on "Ecosystem properties and forest decline in contrasting long-term chronosequences." Science 308:633

Kitayama K, Aiba S, Takyu M, Majalap N, Wagai R (2004) Soil phosphorus fractionation and phosphorus-use efficiency of a Bornean tropical montane rain forest during soil aging with podozolization. Ecosystems 7:259–274

Kloepper JW, Schroth MN (1978) Plant growth-promoting rhizobacteria on radishes. In: Station de Pathologie vegetale et Phyto-bacteriologie (ed) Proceedings of the 4th international conference on plant pathogenic bacteria, vol II. Gilbert-Clary, Tours, pp 879–882

Kochian LV (1995) Cellular mechanisms of aluminium toxicity and resistance in plants. Ann Rev Plant Physiol Plant Mol Biol 46:237–260

Kochian LV, Hoekenga OA, Piñeros MA (2004) How do crop plants tolerate acid soils? Mechanisms of aluminum tolerance and phosphorous efficiency. Annu Rev Plant Biol 55:459–493

Kochian LV, Pineros MA, Hoekenga OA (2005) The physiology, genetics and molecular biology of plant aluminum resistance and toxicity. Plant Soil 274:175–195. https://doi.org/10.1007/s11104-004-1158-7

Konsten CJM, Van Breemen N, Suping S, Aribawa IB, Groenberg JE (1994) Effects of flooding on pH of rice-producing, acid sulfate soils in Indonesia. Soil Sci Soc Am J 58:871–883

Konyak L, Sharma YK, Sharma SK, Bordoloi J (2020) Fertility status, potassium fractions and acidity nature of the soils of Mon District, Nagaland in relation to land uses. J Indian Soc Soil Sci 68(2):201–209. https://doi.org/10.5958/0974-0228.2020.00023.7

Kopáček J, Marešová M, Hejzlar J (2007) Natural inactivation of phosphorus by aluminum in preindustrial lake sediments. Limnol Oceanogr 52:1147–1155

Kowalenko CG, Lowe LE (1975) Mineralization of sulfur from four soils and its relationship to soil carbon, nitrogen and phosphorus. Can J Soil Sci 55:9–14

Kreitinger JP, Klein TM, Novick NJ, Alexander M (1985) Nitrification and characteristics of nitrifying micro-organisms in an acid forest soil. Soil Sci Soc Am J 49:1407–1410

Krishnamoorthy Rao W, Ramaiah PK (1993) Soil and weather profiles of Indian coffee tracts. Indian Coffee 57:3–5

Krstic D, Djalovic I, Nikezic D, Bjelic D (2012) Aluminium in acid soils: chemistry, toxicity and impact on maize plants. https://www.researchgate.net/publication/221922795

Krug EC, Frink CR (1983) Acid rain on acid soil: a new perspective. Science 221:520–525. https://doi.org/10.1126/science.221.4610.520

Kucey RMN (1983) Phosphate solubilizing bacteria and fungi in various cultivated and virgin Alberta soils. Can J Soil Sci 63:671–678

Kucey RMN, Janzen HH, Leggett ME (1989) Microbially mediated increases in plant-available phosphorus. Adv Agron 42:199–228

Kumar V, Gilkes RJ, Armitage TM, Bolland MDA (1994) Identification of residual P compounds in fertilized soils using density fractionation, X-ray diffraction, scanning and transmission electron microscopy. Fert Res 37:133–149

Kunhikrishnan A, Thangarajan R, Bolan N, Xu Y, Mandal S, Gleeson D, Seshadri B, Zaman M, Barton L, Tang C et al (2016) Functional relationships of soil acidification, liming, and greenhouse gas flux. Adv Agron 139:1–71

Kurniawan R, Juhanda S, Banimulyanty V, Marita (2015) Application of bioleaching in metal separation from pyrite mineral rocks using Thiobacillus ferrooxidans Bacteria and Aspergillus niger fungi. Chemical Engineering Study Program, Itenas Bandung

Lal R (1997) Residue management, conservation tillage and soil restoration for mitigating greenhouse effect by CO_2. Enrichment Soil Tillage Res 43:81–107

Lam P, Jensen MM, Lavik G, McGinnis BM, Schubert CJ, Amann R, Thamdrup B, Kuypers MMM (2007) Linking crenarchaeal and bacterial nitrification to anammox in the Black Sea. Proceedings of the National Academy of Sciences of the United States of America 104:7104–7109

Lamare RE, Singh OP (2016) Application of CCME water quality index in evaluating the water quality status in limestone mining area of Meghalaya, India. The Ecoscan 10:149–154

Lambers H, Raven JA, Shaver GR, Smith SE (2008) Plant nutrient-acquisition strategies change with soil age. Trends Ecol Evol 23:95–103

Lang E, Jagnow G (1986) Fungi of a forest soil nitrifying at low pH values. FEMS Microbiol Ecol 38:257–265

Larssen T, Seip HM, Semb A, Mulder J, Muniz IP, Vogt RD, Lydersen E, Angell V, Tang DG, Eilertsen O (1999) Acid deposition and its effects in China: an overview. Environ Sci Policy 2:9–24

Lauber CL, Strickland MS, Bradford MA, Fierer N (2008) The influence of soil properties on the structure of bacterial and fungal communities across land-use types. Soil Biol Biochem 40:2407–2415

Lauber CL, Hamady M, Knight R, Fierer N (2009) Pyrosequencingbased assessment of soil pH as a predictor of soil bacterial community composition at the continental scale. Appl Environ Microbiol 75(5111–5120):71

Läuchli A, Grattan SR (2012) Soil pH extremes. In: Shabala S (ed) Plant stress physiology

Lavelle P, Chauvel A, Fragoso C (1995) Faunal activity in acid soils. In: Date RA, Grundon NJ, Rayment GE, Probert ME (eds) Plant-soil interactions at low pH: principles and management. Kluwer, Dordrecht, pp 201–211

Lee KE (1985) Earthworms: their ecology and relationships with soil and land use. Academic Press, New York, NY, USA

Lee JA (1998) The calcicole–calcifuge problem revisited. Adv Bot Res 29:1–30

Lehours A-C, Bardot C, Thenot A, Debroas D, Fonty G (2005) Anaerobic microbial communities in Lake Pavin, a unique meromictic lake in France. Appl Environ Microbiol 71(11):7389–7400

Leininger S, Urich T, Schloter M, Schwark L, Qi J, Nicol GW et al (2006) Archaea predominate among ammonia-oxidizing prokaryotes in soils. Nature 442:806–809

Leiwakabessy FM, Sutandi A (2004) Fertilizer and soil fertilization. Department of Soil, Faculty of Agriculture, IPB, Bogor

Leong J (1986) Siderophores: their biochemistry and possible role in the biocontrol of plant pathogens. Annu Rev Phytopathol 24:187–208

Leustek T (2002) Sulfate metabolism. In: Somerville CR, Meyerowitz EM (eds) The Arabidopsis book. American Society of Plant Biologists, Rockville, MD

Leustek T, Saito K (1999) Sulfate transport and assimilation in plants. Plant Physiol 120:637–643

Lewis GN, Randall M (1923) Thermodynamics. Mc Graw-Hill, New York, USA, 653 p

Leyval C, Berthelin J (1989) Interaction between Laccaria laccata, Agrobacterium radiobacter and beech roots: influence on P, K, Mg and Fe movilization from minerals and plant growth. Plant Soil 117:103–110

Li GC, Mahler RL, Everson DO (1990) Effects of plant residues and environmental factors on phosphorus availability in soils. Commun Soil Sci Plant Anal 21:471–491

Lidon FC, Barreiro MG (1998) Threshold aluminum toxicity in maize. J Plant Nutr 21(3):413–419
Lieb AM, Darrouzet-Nardi A, Bowman WD (2011) Nitrogen deposition decreases acid buffering capacity of alpine soils in the southern Rocky Mountains. Geoderma 164:220–224
Likens GE, Driscoll CT, Buso DC (1996) Long-term effects of acid rain: response and recovery of a forest ecosystem. Science 272:244–246
Likkineni KC, Abrol YP (1994) Sulfur requirement in crop plants: physiological analysis. Fert News 39:11–18
Lin CF, Coleman NT (1960) The measurement of exchangeable aluminum in soil and clays. Soil Sci Soc Am Proc 23:12–18
Lin MH, Gresshoff PM, Ferguson BJ (2012) Systemic regulation of soybean nodulation by acidic growth conditions. Plant Physiol 160:2028–2039
Lindsay WL (1979) Chemical equilibria in soils. John Wiley & Sons, New York
Lindsay WL, Stephenson HF (1959) Nature of the reactions of monocalcium phosphate monohydrate in soils: I. The solution that reacts with the soil. Soil Sci Soc Am Proc 23:12–18
Lindsay WL, Walthall PM (1989) The solubility of aluminum in soils. In: Sposito G (ed) The environmental chemistry of aluminum. CRC Press, Boca Raton, FL, USA, pp 221–240
Ling D, Zhang J, Ouyang Y, Huang Q (2007) Role of simulated acid rain on cations, phosphorus and organic matter dynamics in Latosol. Arch Environ Contam Toxicol 52:16–21
Liu W, Ten TY, Wu Y, Huang HW (1991) A47, 553-559 Phase Determination for Membrane Diffraction by Anomalous Dispersion Acta Cryst, 47: 553–559
Liu M (2005) Biological properties and soil organic matter physical protection of paddy soil under different organic management practices. Ph.D. Thesis, Nanjing Agricultural University, Nanjing, China, pp 1–219
Liu G, Hanlon E (2012) Soil pH range for optimum commercial vegetable production. EDIS 2012(11), Department of Horticulture, University of Florida
Liu J, Hue NV (2001) Amending subsoil acidity by surface applications of gypsum, lime, and composts. Commun Soil Sci Plant Anal 32:2117–2132
Liu GD, Dunlop J, Phung T, Li YC (2007) Physiological responses of wheat phosphorus-efficient and -inefficient genotypes in field and effects of mixing other nutrients on mobilization of insoluble phosphates in hydroponics. Comm Soil Sci Plant Anal 38(15–16):2239–2256. https://doi.org/10.1080/00103620701549249
Liu Q, Loganathan P, Hedley MJ, Skinner MF (2006) Root processes influencing phosphorus availability in volcanic soils under young Pinus radiate plantations. Can J for Res 36:1913–1920
Liu KH, Fang YT, Yu FM, Liu Q, Li FR, Peng SL (2010) Soil acidification in response to acid deposition in three subtropical forests of subtropical China. Pedosphere 20(3):399–408
Lookman R, Geerts H, Grobert P, Merckx R, Vlassak K (1996) Phosphate speciation in excessively fertilized soils: a 31P and 27Al MAS NMR spectroscopic study. Eur J Soil Sci 47:125–130
Lopes AS, Cox FR (1977) A survey of the fertility status of surface soils under "Cerrado" vegetation in Brazil. Soil Sci Soc Am J 41:742–747
Lu M, Yang Y, Luo Y, Fang C, Zhou X, Chen J, Yang X, Li B (2011) Responses of ecosystem nitrogen cycle to nitrogen addition: a meta-analysis. New Phytol 189:1040–1050
Lukkari K, Hartikainen H, Leivuori M (2007) Fractionation of sediment phosphorus revisited. I: Fractionation steps and their biogeochemical basis. Limnol Oceanogr: Methods 5:433–444
Lundberg DS, Lebeis SL, Paredes SH, Yourstone S, Gehring J, Malfatti S, Tremblay J, Engelbrektson A, Kunin V, del Rio TG et al (2012) Defining the core Arabidopsis thaliana root microbiome. Nature 488:86–90
Lundstedt H, Liszka L, Lundin R, Muscheler R (2006) Long-term solar activity explored with wavelet methods. Ann Geophys 24:769–778
Lungu OI, Temba J, Chirwa B, Lungu C (1993) Effects of lime and farmyard manure on soil acidity and maize growth on an acid Alfisol from Zambia. Trop Agric 70:309–314
Lynch JM (1980) Effects of organic acids on the germination of seeds and growth of seedlings. Plant Cell Environ 3:255–259

Lynch JM (1990) Microbial metabolites. In: Lynch JM (ed) The rhizosphere. John Wiley and Sons Ltd, Baffins Lane, Interscience, Chichester, England, pp 177–206

Lynch JM (1995a) Microbial activity in acid soils. In: Date RA, Grundon NJ, Rayment GE, Probert ME (eds) Plant-soil interactions at low pH: principles and management. Proceedings of the third international symposium in plant-soil interactions at low pH, Brisbane, Queensland, Australia, 12–16 September 1993, pp 167–172

Lynch JM, Wood M (1988) Interactions between plant roots and micro-organisms. In: Wild A (ed) Russell's soil conditions and plant growth, 11th ed. Longman Scientific & Technical, Harlow, pp 526–563

Ma LQ, Traina SJ, Logan TJ, Ryan JA (1994) Effects of aqueous Al, Cd, Cu, Fe(II), Ni, and Z on Pb immobilization by hydroxyapatite. Environ Sci Technol 28:1219–1228

MacCarthy P, Clapp CE, Malcolm RL, Bloom PR (1990) Humic substances in soil and crop sciences. Selected readings. American Society of Agronomy, Madison, USA

MacLean EL, Merritt DJ, Brannon EM (2008) Social organization predicts transitive reasoning in prosimian primates. Anim Behav 76(2):479–486

Mafongoya PL, Bationo A, Kihara J, Waswa BS (2006) Appropriate technologies to replenish soil fertility in southern Africa. Nutr Cycl Agroecosyst 76:137–151

Magdoff F (2001) Concept, components, and strategies of soil health in agro-ecosystems. J Nematol 33:169–172

Mahilum BC, Fox RL, Silva JA (1970) Residual effects of liming volcanic ash soils in the tropics. Soil Sci 109:102–109

Marchand C, Lallier-Vergès E, Baltzer F, Albéric P, Cossa D, Baillif P (2006) Heavy metals distribution in mangrove sediments along the mobile coastline of French Guiana. Mar Chem 98:1–17

Marine Environment Protection Committee, MEPC (1996–2006) Marine Maritime Organization

Marschner H (1995) Mineral nutrition of higher plants, 2nd edn. Academic Press, San Diego, CA

Marschner B, Noble AD (2000) Chemical and biological processes leading to the neutralisation of acidity in soil incubated with litter materials. Soil Biol Biochem 32:805–813

Martikainen PJ (2022) Heterotrophic nitrification—an eternal mystery in the nitrogen cycle. Soil Biol Biochem 168:108611

Matheson FL, Nguyen ML, Cooper AB, Burt TP (2003) Shortterm nitrogen transformation rates in riparian wetland soil determined with nitrogen-15. Biol Fertil Soils 38:129–136

Matsubayashi Y, Ogawa M, Morita A, Sakagami Y (2002) An LRR receptor kinase involved in perception of a peptide plant hormone, phytosulfokine. Science 296:1470–1472

May HM, Nordstrom DK (1991) Assessing the solubilities and reactions kinetics of aluminuous mineral in soils. 40(2):125–148

McBride M (1994) Environmental chemistry of soils. Oxford University Press, Inc., USA

McBride MB, Martínez CE, Sauvé S (1998) Copper(II) activity in aged suspensions of goethite and organic matter. Soil Sci Soc Am J 62:1542–1548

McCauley A (2009) Soil pH and organic matter. Nutrient management module No. 8. Montana State University, Extension, 4449-8

Mcfarland ML, Haby VA, Redmon L, Bade D (2001) Managing soil acidity. http://overton.tamu.edu/soils. Bulletin No. B-1720

McGrath D, Zhang CS (2003) Spatial distribution of soil organic carbon concentrations in grassland of Ireland. Appl Geochem 18:1629–1639

McGrath SP, Zhao FJ (1996) Sulphur uptake, yield response and the interactions between N and S in winter oilseed rape (Brassica napus L.). J Agric Sci 126:53–62

McLean EO (1971) Potentially beneficial effects from liming: chemical and physical. Soil Crop Sci Soc Fla Proc 31:189–196

Meng J, Wang L, Liu X, Wu J, Brookes PC, Xu J (2013) Physicochemical properties of biochar produced from aerobically composted swine manure and its potential use as an environmental amendment. Bioresour Technol 142:641–646. https://doi.org/10.1016/j.biortech.2013.05.086

Mesquita ME (1998) Copper and zinc competitive adsorption in schistic and granitic acid soils. Agrochimica (Italy) 42:235–245

Milivojevic J, Nikezic D, Krstic D, Jelic M, Djalovic I (2011) Influence of physical chemical characteristics of soil on zinc distribution and availability for plants in vertisols of Serbia. Pol J Environ Stud 20:993–1000

Mishra UK, Satapathy S, Panda N (1989) Characterization of some acid soils of Orissa, I—nature of soil acidity. J Indian Soc Soil Sci 37:22–28

Mitchell MJ, Driscoll CT, Fuller RD, David MB, Likens GE (1989) Effect of whole tree harvesting on the sulfur dynamics of a forest soil. Soil Sci Soc Am J 53:933–940

Mitran T, Meena RS, Lal R, Layek J, Kumar S, Datta R (2018) Role of soil phosphorus on legume production. In: Legumes for soil health and sustainable management. Springer, Singapore, pp 487–510

Moore DP (1972) Mechanism of micronutrient uptake by plants. In: Mortvedt JJ, Giordono PM, Lindsay WL (eds) Micronutrients in agriculture. Soil Science Society of America, Madison, Wisconsin, USA, pp 171–198

Moormann FR, Pons LJ (1974) Characteristics of mangrove soils in relation to their agricultural land use potential. In: Walsh G et al (eds) Proceedings international symposium on biology and management of mangroves, vol 2, pp 529–547

Mora ML, Demanet R, Vistoso E, Gallardo F (2005) Influence of sulphate concentration in mineral solution on ryegrass grown at different pH and aluminium levels. J Plant Nutr 28:1117–1132

Morelli M, Igue K, Fuentes R (1971) Effect of liming on the exchange complex and on the movement of calcium and magnesium. Turrialba 21:317–322

Mosharrof M, Uddin MK, Jusop S, Sulaiman MF, Shamsuzzaman SM, Haque ANA (2021) Changes in acidic soil chemical properties and carbon dioxide emission due to biochar and lime treatments. Agriculture 11:219. https://doi.org/10.3390/agriculture11030219

Mott CJB (1988) Surface chemistry of soil particles. In: Wild A (ed) Russell's soil conditions cmd plcmt growth, 11th ed. Longman Scientific & Technical, New York, pp 239–281

Mugendi DN, Mucheru-Muna M, Mugwe J, Kung'u JB, Bationo A (2007) Improving food production using 'best bet' soil fertility technologies in the Central Highlands of Kenya. In: Bationo et al (eds) Advances in integrated soil fertility management in Sub-Saharan Africa: challenges and opportunities, pp 345–351

Mukherjee JN, Chatterjee B (1945) Liberation of H^+, Al^{3+} and Fe^{3+} ions from hydrogen clays by neutral salts. Nature 155:268–269. https://doi.org/10.1038/155268aO

Mukherjee JN, Mitra RP (1944) Differentiation of Hydrogen Clays and identification of their mineral constitutents by electrochemical and viscous methods. Nature 154(3922):824–824. https://doi.org/10.1038/154824aO

Mukherjee JN, Mitra RP, Mitra DK (1943) Electrochemical properties of Clay minerals and the differentiation of hydrogen clays and Bentonites by Electrochemical methods. I. Kaolinite and Kaolinitis clays. J Phys Chem 47(8):543–549. https://doi.org/10.1021/j150431a002

Mulder J, van Breemen N, Eijck HC (1989b) Depletion of soil aluminium by acid deposition and implications for acid neutralization. Nature 337:247–249

Myers RJK, De Pauw E (1995) Strategies for the management of soil acidity. In: Date RA, Grundon NJ, Rayment GE, Probert ME (eds) Plant-soil interactions at low pH: principles and management. Kluwer Academic Publishers, Dordrecht, The Netherlands, pp 729–741

NADP (National Atmospheric Deposition Program) (2014) National trends network. NADP. Also available at http://nadp.sws.uiuc.edu/ntn/

Nannipieri P, Ascher J, Ceccherini MT, Lan Dl L, Pietramellar G, Renell G (2003) Microbial diversity and soil functions. European Journal of Soil Science 54: 655–60, https://doi.org/10.1046/j.1351-0754.2003.0556.x

Nekesa AO, Okalebo JR, Othieno CO, Thuita MN, Kipsat M, Bationo A, Sanginga N, Kimetu J, Vanlauwe B (2005) Afr Crop Sci Conf Proc 7:1121–1128

Nekesa AO (2007) Effect of Minjingu phosphate rock and agricultural lime in relation to maize, groundnut and soybean yields on acid soils of western Kenya. M. Phil Thesis. Moi University Eldoret, Kenya pp 79

Nelson PN, Su N (2010) Soil pH buffering capacity: a descriptive function and its application to some acidic tropical soils. Aust J Soil Res 2010(48):201–207

Neumann G, Romheld V (2012) Rhizosphere chemistry in relation to plant nutrition. In: Marschnar P (ed) Marschnar's mineral nutrition of higher plants, 3rd edn. Elservier, New York, pp 347–368

Ngachie V, Smyth TJ (1989) Relationships between buffer acidity and exchangeable acidity in lime trials with Ultisols and Histosols. Commun Soil Sci Plant Anal 20:1427–1438

Ngatunga EL, Cools N, Dondeyne S, Deckers JA, Merckx R (2001) Buffering capacity of cashew soils in southeastern Tanzania. Soil Use Manag 17:155–162

Ngoune Tandzi L, Mutengw CS, Ngonkeu ELM, Gracen V (2018) Breeding maize for tolerance to acidic soils: Areview. Agronomy 8(6), p.84

Nicol GW, Leininger S, Schleper C, Prosser JI (2008). The influence of soil pH on the diversity, abundance and transcriptional activity of ammonia oxidizing archaea and bacteria. Environ Microbiol 10(11):2966–2978

Nilsson LO, Baath E, Falkengren-Grerup U, Wallander H (2007) Growth of ectomycorrhizal mycelia and composition of soil microbial communities in oak forest soils along a nitrogen deposition gradient. Oecologia 153:375–384

NLWRA (2001) Australian agriculture assessment 2001, vol 1. National Land and Water Resources Audit, Canberra

Noble AD, Randall PJ (1999) Alkalinity effects of different tree litters incubated in an acid soil of NSW, Australia. Agrofor Syst 46:147–160

Noble AD, Zenneck I, Randall PJ (1996) Leaf litter ash alkalinity and neutralisation of soil acidity. Plant Soil 179:293–302

Noble AD, Cannon M, Muller D (1997) Evidence of accelerated soil acidification under Stylosanthes-dominated pasture. Aust J Soil Res 35:1309–1322. https://doi.org/10.1071/S97053

Nodvin SC, Driscoll CT, Likens GE (1986) Simple partitioning of anions and dissolved organic carbon in a forest soil. Soil Sci 142:27–35

Nottidge DO, Ojeniyi SO, Asawalam DO (2006) Effect of levels of wood ash on soil chemical properties in an acid Utisol of Southeastst Nigeria. Niger J Soil Sci 16:109–114

Nugroho RA, Röling WFM, Laverman AM, Verhoef HA (2007) Low nitrification rates in acid Scots pine forest soils are due to pH-related factors. Microbial Ecol 53:89–97

Nugroho RA, Röling WFM, van Straalen NM, Verhoef HA (2009) Changes in nitrification and bacterial community structure upon crossinoculation of Scots pine forest soils with different initial nitrification rates. Soil Biol Biochem 41:243–250

Nye P, Craig D, Coleman NT, Ragland JL (1961) Ion exchange equilibrium involving aluminum. Soil Sci Soc Am Proc 25:14–17

O'Conner KF, Vartha EW (1969) Responses of grasses to sulfur fertilizers. Plant Soil 60(3):451–459

Odum EP (1986) Fundamentos de Ecologìa. Interamericana, Mexico

Ohfuji H, Rickard D (2005) Experimental syntheses of framboids—a review. Earth Sci Rev 71:147–170

Okalebo JR, Othieno CO, Nekesa AO, Ndungu-Magiroi KW, KifukoKoech MN (2009) Potential for agricultural lime on improved soil health and agricultural production in Kenya. African Crop Sci Conf Proc 9:339–341

Olness A, Archer D (2005) Effect of organic carbon on available water in soil. Soil Sci 170:90–101

Omonode RA, Smith DR, Gál A, Vyn TJ (2011) Soil nitrous oxide emissions in corn following three decades of tillage and rotation treatments. Soil Sci Soc Am J 75:152–163

Onwonga RN, Joyce JL, Mochoge BB (2010) Mineral nitrogen and microbial biomass dynamics under different acid soil management practices for maize production. J Agric Sci 2:16–30

Osakia M, Watanabe T, Tadano T (1997) Beneficial effect of aluminum on growth of plants adapted to low pH soils. Soil Sci Plant Nutr 43(3):551–563

Pal DK (2017) A treatise of Indian and tropical soils. Springer International Publishing AG

Pal DK (2019) Ecosystem services and tropical soils of India. SpringerNature, Switzerland, p 113
Pal DK, Wani SP, Sahrawat KL, Srivastava P (2014) Red ferruginous soils of tropical Indian environments: a review of the pedogenic processes and its implications for edaphology. CATENA 121:260–278. https://doi.org/10.1016/j.catena2014.05.023
Panda N (1987) Acid soils of Eastern India: their Chemistry and management. J Indian Soc Soil Sci 35(4):568–581
Papen H, von Berg R (1998) A most probable number method (MPN) for the estimation of cell numbers of heterotrophic nitrifying bacteria in soil. Plant Soil 199:123–130
Parfitt R A (1980) Chemical properties of variable charge soils. In: Theng BKG (ed) Soils with variable charge. New Zealand Society of Soil Science, Lower Hutt, New Zealand, pp 167–188
Parkin TB, Doran JW, Franco-Vizcaino E (1996) Field and laboratory tests of soil respiration. In: Doran JW, Jones AJ (eds) Methods for assessing soil quality. SSSA Special Publication 49; Soil Science Society of America, Madison, WI, pp 231–246
Parks GA, de Bruyn PL (1962) The zero point of charge of oxides. J Phys Chern 66:967–973
Patiram and Datta M (2006) Geology, mineralogy and genesis of soils. In: Sharma UC, Datta M, Samra JS (eds) Soils and their management in North East India. ICAR research Complex for NEH Region, Umroi, Meghalaya, India, pp 63–120
Pearson RW (1975) Soil acidity and liming in the humid tropics. Cornell International Agriculture Bulletin 30. New York State College of Agriculture and Life Sciences, Cornell University, Ithaca, NY
Pearson RW, Abruña F, Vicente-Chandler J (1962) Effects of lime and nitrogen applications on the downward movement of Ca and Mg in two humid tropical soils of Puerto Rico. Soil Sci 93:77–82
Pennington PI, Ellis RC (1993) Autotrophic and heterotrophic nitrification in acidic forest and native grassland soils. Soil Biol Biochem 25(10):1399–1408
Pérez-Novo C, Pateiro-Moure M, Osorio F, Nóvoa-Muñoz JC, López-Periago E and Arias-Estévez M (2008) Influence of organic matter removal on competitive and noncompetitive adsorption of copper and zinc in acid soils. J Coll Interface Sci 322:33–40. www.elsevier.com/locate/jcis
Pérez-Novo C, Bermúdez-Couso A, López-Periago E, Fernández-Calviño D, Arias-Estévez M (2009) The effect of phosphate on the sorption of copper by acid soils. Geoderma 150:166–170
Pérez-Novo C, Bermúdez-Couso A, López-Periago E, Fernández-Calviño D, Arias-Estévez M (2011) Zinc adsorption in acid soils Influence of phosphate. Geoderma 162:358–364
Pichtel J, Sawyerr HT, Czarnowska K (1997) Spatial and temporal distribution of metals in soils in Warsaw, Poland. Environ Pollut 98:169–174
Pietikäinen J, Pettersson M, Bååth E (2005) Comparison of temperature effects on soil respiration and bacterial and fungal growth rates. FEMS Microbiol Ecol 52:49–58
Pinstrup-Andersen P, Pandhya-Lorch R, Rosegrant MW (1999) World food prospects: critical issues for the early twenty-first century. International Food Policy Research Institute, Washington
Pinto M, Rodriguez M, Besga G, Balcazar N, Lopez FA (1995) Effects of Linz-Donawitz (LD) slag on soil properties and pasture production in the Basque country (Northern Spain). N Z J Agric Res 38:143–155
Pirt SJ (1975) Principles of microbe and cell cultivation. John Wiley and Sons, New York
Plante AF (2007) Soil biogeochemical cycling of inorganic nutrients and metals. In: Soil microbiology, ecology and biochemistry, 3rd ed, pp 389–432
Poelman JNB (1973) Soil material rich in pyrite in non-coastal areas. In: Dost H (ed) Acid sulphate soils. Proceedings international symposium Wageningen. ILRI Publ. no. 18, vol 2, pp 197–207
Ponnamperuma FN (1972) The chemistry of submerged soils. Adv Agron 24:29–96
Ponnamperuma FN (1977) Physicochemical propert1es of subkerged soils in relatin to fertility. The International Rice Research Institute P.O. BaK 933, Manila, Philippines
Ponnamperuma FN (1955) The chemistry of submerged soils in relation to growth and yield of rice. Ph.D. thesis, Cornell University, Ithaca, New York, 208 p

Ponnamperuma FN (1965) Dynamic aspects of flooded soils. In Proceedings of a symposium in International Rice Research Institute, Proceedings of mineral nutrition of the rice plant, February 1964, Los Baiios, Philippines. Johns Hopkins Press, Baltimore, Maryland, pp 295–328

Ponnamperuma FN, Martinez E, Loy T (1966a) Influence of redox potential and the partial pressure of carbon dioxide on the pH values and suspension effect of flooded soils. Soil Sci 101:421–431

Ponnamperuma FN, Tianco EM, Loy TA (1966b) Ionic strengths of the solutions of flooded soils and other natural aqueous solutions from specific conductance. Soil Sci 102:408–413

Pons LJ (1965) Pyrites as a factor controlling chemical ripening and formation of cat clay, with special reference to the coastal plain of Suriname. Agric Exp Stat Paramaribo Bull 82:141–162

Pons LJ, van Breemen N (1981). Factors influencing the formation of potential acidity in tidal swamps. In: Proceedings of the Bangkok symposium on acid sulphate soils: second international symposium on acid sulphate soils, Bangkok, Thailand, 18–24 Jan 1981, pp 37–51

Pons LJ, van Breemen N, Driessen PM (1982) In: Kittrick JA, Fanning DS, Hossner LR (eds) Physiography of coastal sediments and development of potential soil acidity. SSSA Special Publication. https://doi.org/10.2136/sssaspecpub10.c1

Ponto BS, Berg JC (2018) Clay particle charging in a polar media. 161:76–81

Poschenrieder C, Gunsé B, Corrales I, Barceló J (2008) A glance into aluminum toxicity and resistance in plants. Sci Total Environ 400:356–368. https://doi.org/10.1016/j.scitotenv.2008.06.003

Post WM, Kwon KC (2000) Soil carbon sequestration and land-use change: processes and potential. Glob Change Biol 6:317–327

Powlson DS, Brookes PC, Christensen BT (1987) Measurement of soil microbial biomass provides an early indication of changes in tot soil organic matter due to straw incorporation. Soil Biol Biochem 19:159–164

Probert ME (1980) Sulphur in Australia. In: Nicholson AJ (ed) Freney JR. Australian Academy of Science, Canberra, pp 158–169

Prochnow BL (2014) Soil acidity evaluation and management. Better Crops 98(1):22–25

Pueyo JJ, Quiñones MA, de la Peña TC, Fedorova EE, Lucas MM (2021) Nitrogen and phosphorus interplay in lupin root nodules and cluster roots. Front Plant Sci 12:644218

Quilliam RS, DeLuca TH, Jones DL (2013) Biochar application reduces nodulation but increases nitrogenase activity in clover. Plant Soil 366(1–2):83–92

Rabileh MA, Shamshuddin J, Panhwar QA, Rosenani AB, Anuar AR (2015) Effects of biochar and/or dolomitic limestone application on the properties of Ultisol cropped to maize under glasshouse conditions. Can J Soil Sci 95:37–47

Raghu K, MacRae IC (1966) Occurrence of phosphate-dissolving microorganisms in the rhizosphere of rice plants and in submerged soils. J Appl Bacteriol 196629:582–586

Rahman MA, Lee SH, Ji HC, Kabir AH, Jones CS, Lee KW (2018) Importance of mineral nutrition for mitigating aluminum toxicity in plants on acidic soils: current status and opportunities. Int J Mol Sci 19(10):3073

Rajapaksha RMCP, Tobor-Kapłon MA, Bååth E (2004) Metal toxicity affects fungal and bacterial activities in soil differently. Appl Environ Microbiol 70:2966–2973

Raza S, Jornsgard B, Abou-Taleb H, Christiansen JL (2001) Tolerance of Bradyrhizobium sp. (Lupini) strains to salinity, pH, $CaCO_3$ and antibiotics. Lett Appl Microbiol 32:379–383

Rebecca EH, Richard JS, Emmanuel D, Peter JH, Alan ER (2010) Effect of lime on root growth, morphology and the rhizosheath of cereal seedlings growing in an acid soil. Plant Soil 327:199–212

Rechcigl JE, Sparks DL (1985) Effect of acid rain on the soil environment: a review. Commun Soil Sci Plant Anal 16:653–680

Reeve NG, Sumner ME (1971) Cation exchange capacity and exchangeable aluminium in Natal Oxisols. Soil Sci Soc Am Proc 35:38–42

Reeve NG, ME Sumner (1972) Amelioration of subsoil acidity in Natal Oxisols by leaching of surface applied amendments. Agrochemophysica 4:1–5

Rengel Z (2011) Soil pH, soil health and climate change, soil health and climate change. Springer, Berlin, pp 69–85

Rich CI, Black RW (1964) Potassium exchange as affected by cation size, pH and mineral structure. Soil Sci 97:384–390

Richardson SJ, Peltzer DA, Allen RB, McGlone MS, Parfitt RL (2004) Rapid development of phosphorus limitation in temperate rainforest along the Franz Josef soil chronosequence. Oecologia 139:267–276

Richter DD (1986) Sources of acidity in some forested Udults. Soil Sci Soc Am J 50:1584–1589

Ridley AM, Windsor SM (1992) Persistence and tolerance to soil acidity of phalaris and cocksfoot in north-eastern Victoria. Aust J Exp Agric 32:1069–1075

Ridley AM, Avery AL, Oram RN, Hunter J, Shovelton JB, Mahoney GP, Muller WJ (2002) Long-term persistence of aluminium-tolerant and sensitive Phalaris lines on acidic soils and associated changes in soil acidity. Aust J Exp Agric 42:1033–1042

Rincon-Florez VA, Carvalhais LC, Schenk PM (2013) Culture-independent molecular tools for soil and rhizosphere microbiology. Diversity 5(3):581–612

Ring SM, Fisher RP, Poile GJ, Poile GJ, Conyers MK, Morris SG (1993) Screening species and cultivars for their tolerance to acidic soil conditions. Plant Soil 155(156):521–524

Ritchey KD, Silva JE, Costa UF (1982) Calcium deficiency in clayey B horizons of savannah Oxisols. Soil Sci 133:378–382

Ritchie GSP (1995) Soluble aluminium in acidic soils: Principles and practicalities. Plant Soil 171:17–27

Ritz K, Black HIJ, Campbell CD et al (2009) Selecting biological indicators for monitoring soils: a framework for balancing scientific and technical opinion to assist policy development. Ecol Indic 9:1212–1221. https://doi.org/10.1016/j.ecolind.2009.02.009

Robarge WP, Johnson DW (1992) The effects of acidic deposition on forested soils. Adv Agron 47:1–83

Roberson E, Sarig S, Firestone MK (1991) Cover crop management of polysaccharide mediated aggregation in an orchard soil. Soil Sci Soc Am J 55:734–739

Robertson GP (1982) Nitrification in forested ecosystems. Philos Trans R Soc Lond 296B:445 ± 457

Robertson LA, Van Niel EW, Torremans RA, Kuenen JG (1988) Simultaneous nitrification and denitrification in aerobic chemostat cultures of Thiosphaera pantotropha. Appl Environ Microbiol 54(11):2812–2818

Robson AD, Abbott LK (1989) The effect of soil acidity on microbial activity in soils. In: Robson AD (ed) Soil acidity and plant growth. Academic Press

Rodríguez H, Fraga R (1999) Phosphate solubilizing bacteria and their role in plant growth promotion. Biotechnol Adv 17:319–339

Roesch LFW, Fulthorpe RR, Riva A, Casella G, Hadwin AKM, Kent AD, Daroub SH, Camargo FAO, Farmerie WG, Triplett EW (2007) Pyrosequencing enumerates and contrasts soil microbial diversity. ISME J 1:283–290

Ronen E (2007) Micro-elements in agriculture. Practical Hydroponics & Greenhouses July/August:39–48

Rorison IH (1972) The effect of extreme soil acidity on the nutrient uptake and Proceedings physiology of plants. In Dost H (ed) Acid sulfate soils. Proceedings of the International Symposium, vol 1, August 13–20, 1972. Wageningen, pp 223–254

Rousk J, Bååth E (2007a) Fungal and bacterial growth in soil with plant materials of different C/N ratios. FEMS Microbiol Ecol 62:258–267

Rousk J, Bååth E (2007b) Fungal biomass production and turnover in soil estimated using the acetate-in-ergosterol technique. Soil Biol Biochem 39:2173–2177

Rousk J, Demoling LA, Bahr A, Bååth E (2008) Examining the fungal and bacterial niche overlap using selective inhibitors in soil. FEMS Microbiol Ecol 63:350–358

Rousk J, Brooks PC, Baath E (2009) Contrasting soil pH effects on fungal and bacterial growth suggest functional redundancy in carbon mineralization. Appl Environ Microbiol 75(6):1589–1596. https://doi.org/10.1128/AEM.02775

Rousk et al (2010)
Rout GR, Samantaray S, Das P (2001) Aluminium toxicity in plants: a review. Agronomie 21:3–21
Roychoudhury P, Kaushik BD (1989) Solubilization of Mussorie rock phosphate by cyanobacteria. Curr Sci 58:569–570
Rupa TR, Tomar KP (1999) Zinc desorption kinetics as influenced by pH and phosphorus in soils. Commun Soil Sci Plan 30:1951–1962
Russell EW (1973) Soil conditions and plant growth, 10th edn. Longman, London
Sachdev MS, Deb DL (1990) Nitrogen and S uptake and efficiency in the mustard-moong-maize cropping systems. Fertilizer News 35:49–55
Saha UK, Taniguchi S, Sakurai K (2002) Simultaneous adsorption of cadmium, zinc, and lead on hydroxyaluminium- and hydroxyaluminosilicate–montmorillonite complexes. Soil Sci Soc Am J 66:117–128
Sahoo S, Vasu D, Paul R, Sen TK, Ray SK, Chandran P (2020) Acid soils of Manipur in the North Eastern Region of India: Their Mineralogy, Pedology, Taxonomy and Edaphalogy, Clay Research, 39(1):31–43
Saito K (2000) Regulation of sulfate transport and synthesis of sulfur-containing amino acids. Curr Opin Plant Biol 3:188–195
Sala OE, Chapin FS, Armesto JJ, Berlow E, Bloomfield J, Dirzo R, Huber- Sanwald E, Huenneke LF, Jackson RB, Kinzig A, Leemans R, Lodge DM, Mooney HA, Oesterheld M, Poff NL, Sykes MT, Walker BH, Walker M, Wall DH (2000) Biodiversity e global biodiversity scenarios for the year 2100. Science 287:1770–1774
Samac DA, Tesfaye M (2003) Plant improvement for tolerance to aluminium in acid soils. Plant Cell, Tissue Organ Cult 75:189–207
Samonte HP, Ocampo AM (1977) Liming and its residual effect on corn-planted soils. Philipp Agric 60:420–430
Sanchez PA (2019) Properties and management of soils in the tropics, 2nd ed. Cambridge University Press, University Printing House, Cambridge CB2 8BS, United Kingdom
Sanchez PA, Logan TJ (1992) Myths and Science about Chemistry and fertility of soils in the tropics. In: Lal R, Sanchez PA (eds) Muths and science of soils in the tropics. Soil Science Society of American. Special Publication, 29, ASA and SSSA, Madison, pp 35–46
Sanchez PA, Salinas JG (1981) Low-input technology for managing Oxisols and Ultisols in tropical America. Adv Agron 34(279–406):1982
Sanchez PA, Palm CA, Buol SW (2003) Fertility capability soil classification: a tool to help assess soil quality in the tropics. Geoderma 114:157–185
Sarig S, Okon Y, Blum A (1990) Promotion of leaf area development and field in Sorghum bicolor inoculated with Azospirillum brasilense. Symbiosis 9:235–245
Sarrantonio M, Gallandt E (2003) The role of cover crops in North American cropping systems. J Crop Prod 8:53–74
Satapathy KK, Datta M (2006) Soil conservation and management strategies. In: Sharma UC, Datta M, Samra JA (eds) Soils and their management in North East India. ICAR research Complex for NEH Region, Umroi, Meghalaya, India, pp 372–388
Scheel T, Jansen B, Van Wijk AJ, Verstraten JM, Kalbits K (2008) Stabilization of dissolved organic matter by aluminum: a toxic effect or stabilization through precipitation. Eur J Soil Sci 59:1122–1132
Scherer HW (2001) Sulphur in crop production—invited paper. Eur J Agron 14:81–111
Schimel JP, Weintraub MN (2003) The implications of exoenzyme activity on microbial carbon and nitrogen limitation in soil: a theoretical model. Soil Biol Biochem 35:549–563
Schleper C, Jurgens G, Jonuscheit M (2005) Genomic studies of uncultivated archaea. Nature Rev Microbiol 3:479–488
Schoonen MAA (2004) Mechanisms of sedimentary pyrite formation. In: Amend JP, Edwards KJ, Lyons TW (eds) Sulfur biogeochemistry—past and present. Geological Society of America, Boulder, pp 117–134

Schofield HK, Pettitt TR, Tappin AD, Rollinson GK, Fitzsimons MF (2019) Biochar incorporation increased nitrogen and carbon retention in a waste-derived soil. Sci Total Environ 690:1228–1236

Scurfield G, Boswell JG (1952). The microbiology of acid soils, iii. the analysis of soil profiles under Deschampsia Flexuosa (l.) trin. and Holcus Mollis. New Phytologist 52(2), Article first published online: 4 May 2006

Seguel A, Cumming JR, Klugh-Stewart K, Cornejo P, Borie F (2013) The role of arbuscular mycorrhizas in decreasing aluminium phytoxicity in acidic soils: a review. Mycorrhiza 23:167–183. https://doi.org/10.1007/s00572-013-0479-x

Sehgal JL (1998) Red and lateritic soils: an overview. In: Sehgal J, Blum WE, Gajbhiye KS (eds) Red and lateritic soils. Managing red and lateritic soils for sustainable agriculture, 1. Oxford and IBH Publishing Co. Pvt. Ltd., New Delhi, pp 3–10

Sellers PJ, Dickinson RE, Randall DA, Betts AK, Hall FG, Berry JA, Collatz GJ, Denning AS, Mooney HA, Nobre CA, Sato N, Field CB, Henderson-Sellers A (1997) Modeling the exchanges of energy, water, and carbon between continents and the atmosphere. Science 275:502–509

Sen TK, Chamuah GS, Seghal JL (1994) Occurrence and characteristics of some kandi soils in Manipur. J Indian Soc Soil Sci 42:297–300

Sen TK, Dubey PN, Maji AK, Chamuah GS (1997) Status of micronutrients in some dominant soils of Manipur. J Indian Soc Soil Sci 45(2):388–390

Serrano S, Garrido F, Campbell CG, García-González MT (2005) Competitive sorption of cadmium and lead in acid soils of central Spain. Geoderma 124:91–104

Sharif AR, Ghazi-Tabatabaei M, Hejazi E, Askarabad MH, Dehshiri GR (2011) Confirmatory factor analysis of the university student depression inventory (USDI). Procedia Soc Behav Sci 30:4–9

Sharma UC, Sharma V (2003) Socio-economic aspects of tillage and its effect on soil erosion and agricultural productivity in Kandi (drought prone) region of Jammu. In: Proceedings 16th international soil tillage research organisation conference on soil management for sustainability, Brisbane, Australia, pp 1108–1113

Sharma UC, Singh RP (2002) Acid soils of India: their distribution, management and future strategies for higher productivity. Fertil News 47(3):45–48 & 51–52

Sharma UC, Prasad RN (1995) Socio- economic aspects of acid soils management and alternative land use system for north eastern states of India. In: Plant and soil interactions at low pH: principles and management. Kluwer Academic Publisher, The Netherlands, pp 689–696

Sharma V, Mir SH, Arora S (2009) Assessment of fertility status of erosion prone soils of Jammu Siwaliks. J Soil Water Conserv 8:37–41

Sharma SB, Sayyed RZ, Trivedi MH, Gobi TA (2013) Phosphate solubilizing microbes: sustainable approach for managing phosphorus deficiency in agricultural soils. Springer plus 2:587

Sharma V, Hussain S, Sharma KR, Arya V (2014) Labile carbon pools and soil organic carbon stocks in the foothill Himalayas under different land use systems. Geoderma 232–234(2014):81–87

Shi Z, di Toro DM, Allen HE, Sparks DL (2008) A WHAM-based kinetics model for Zn adsorption and desorption to soils. Environ Sci Technol 42:5630–5636

Shi W, Liu J, Du Z, Song Y, Chen C, Yue T (2009) Surface modelling of soil pH. Geoderma 150:113–119

Shuman LM (1988) Effect of organic matter on the distribution of manganese, copper, iron, and zinc in soil fractions. Soil Sci 146(3):192–198

Singer PC, Stumm W (1970) Acid mine drainage: rate determining step. Science 167:1121–1123

Singh and Sharma (2002)

Singh A, Singh MD (1981) Soil erosion hazards in NEH region, Research Bulletin No 10, ICAR research Complex for NEH region, Meghalaya, India, pp 20–22

Singh BP, Singh A, Singh BN (1990) Response of rice (Oryza sativa) to Zinc and Boron application in acid Alfisols under mid altitude conditions of Meghalaya. Indian J Agri Sci 60(1):70–71

Singh B, Singh BP, Cowie AL (2010) Characterisation and evaluation of biochars for their application as a soil amendment. Soil Res 48:516–525. https://doi.org/10.1071/SR10058

Siqueira JO, Moreira FMS (1997) Microbial populations and activities in highly weathered acidic soils: highlights of the Brazilian research. In: Moniz AC, Fulani AMC, Schaffert RE, Fageria

NK, Rosolem CA, Cantarella H (eds) Plant–soil interactions at low pH: sustainable agriculture and forestry production. Brazilian Soil Science Society, Campinas, pp 139–156
Sistani KR, Simmons JR, Jn-Baptiste M, Novak JM (2019) Poultry litter, biochar, and fertilizer effect on corn yield, nutrient uptake, N_2O and CO_2 emissions. Environment 6:55
Sivaguru M, Horst WJ (1998) Differential impacts of aluminum on microtubule organization depend on growth phase in suspension-cultured tobacco cells. Physiol Plant 107:110–119
Sivan A, Chet I (1992) Microbial control of plant diseases. In: Mitchell R (ed) Environmental microbiology. Wiley-Liss, New York, pp 335–354
Six J, Frey SD, Thiet RK, Batten KM (2006) Bacterial and fungal contributions to carbon sequestration in agroecosystems. Soil Sci Soc Am J 70(2):555–569
Skrary FA, Cameron DC (1998) Purification and characterization of a Bacillus licheniformis phosphatase specific for D-alpha-glycerphosphate. Arch Biochem Biophys 1998(349):27–35
Slattery B, Hollier C (2002) The impact of acid soils in Victoria. State of Victoria. Department of Natural Resources and Environment
Slattery ML, Benson J, Berry TD, Duncan D, Edwards SL, Caan BJ, Potter JD (1997) Dietary sugar and colon cancer. Cancer Epidemiol Biomark Prev 6:677–685
Slattery WJ, Edwards DG, Bell LC, Coventry DR, Helyar KR (1998) Soil acidification and the carbon cycle in a cropping soil of north-eastern Victoria. Soil Res 36(2):273–290
Slattery WJ, Conyers MK, Aitken RL (1999) Soil pH, aluminium, manganese and lime requirement. In: Peverill KI, Sparrow LA, Reuter DJ (eds) Soil analysis: an interpretation manual, pp 103–128
Slattery C, Kavakli I, Okita T (2000) Engineering starch for increased quantity and quality. Trends Plant Sci 5:291–298. https://doi.org/10.1016/S1360-1385(00)01657-5
Slattery JF, Coventry DR, Slattery WJ (2001) Rhizobial ecology as affected by the soil environment. Austr J Exp Agric 41:289–298
Smeck N (1973) Phosphorus: an indicator of pedogenic weathering processes. Soil Sci 115:199–206
Smiley RW, Cook RJ (1973) Relationship between take-all of wheat and rhizosphere pH in soils fertilized with ammonium vs. nitrate nitrogen. Phytopathology 63:882–890
Smith GD (1986) The Guy Smith interviews: rationale for concept in soil taxonomy. SMSS Technical Monograph 11, SMSS, SCS,USDA, USA
Soares WV, Lobato E, Gonzalez E, Naderman GC (1975) Liming soils of the Brazilian Cerrado. In: Bornemisza E, Alvarado A (eds) Soil management in tropical America. North Carolina State University, Raleigh, North Carolina, pp 283–299
Soil Survey Staff (1975) Soil taxonomy. Agriculture Handbook 436, Washington DC, 754 pp
Soil Survey Staff (1999) Soil taxonomy a basic system of soil classification for making and interpreting soil surveys. USDA Natural Resources Conservation Services. Agriculture Handbook, vol 436. Washington, DC
Somani L, Totawat K, Sharma RA (1996) Liming technology for acid soils. Agrotech Publishing Academy, Udaipur, p 240
Spain JM, Francis CA, Howeler RH, Calvo F (1975) Differential species and varietal tolerance to soil acidity in tropical crops and pastures. In: Bornemisza E, Alvarado A (eds) Soil management in tropical America. North Carolina State University, Raleigh, NC, pp 308–324
Sparks DL (1995) Environmental soil chemistry. Academic Press, New York
Spark KM, Wells JD, Johnson BB (1995) Characterizing trace metal adsorption on kaolinite. Eur J Soil Sci 46:633–640
Sperberg JI (1958) The incidence of apatite-solubilizing organisms in the rhizosphere and soil. Aust J Agric Res 9:778
Sposito G (1989) The chemistry of soils. Oxford University Press, New York, Oxford, xiii + 277 pp. ISBN 0-19-504615-3
Sposito G (2008) The chemistry of soils, 2nd ed. Chap. 11. Oxford University Press, US, New York
SSSA (1996) Methods for assessing soil quality. Special Publication Number 49 Soil Science Society of America, Inc. Madison, Wisconsin, USA 1996

Stams AJM, Flameling EM, Marnette ECL (1990) The importance of autotrophic versus heterotrophic oxidation of atmospheric ammonium in forest ecosystems with acid soil. FEMS Microbiol Ecol 74:337–344

Steenwerth K, Jackson L, Carlisle E, Scow K (2006) Microbial communities of a native perennial bunchgrass do not respond consistently across a gradient of land-use intensification. Soil Biol Biochem 38(7):1797–1811

Stevens CJ, Dise NB, Mountford JO, Gowing DJ (2004) Impact of nitrogen deposition on the species richness of grasslands. Science 303:1876–1879

Stevens CJ, Dise NB, Gowing DJ (2009) Regional trends in soil acidification and metal mobilisation related to acid deposition. Environ Pollut 157:313–319

Stevenson FJ (1986) Cycles of soil: carbon, nitrogen, phosphorus, sulfur, micronutrients. John Wiley & Sons, New York

Stevenson FJ (1994) Humus chemistry: genesis, composition, reactions. Wiley, New York

Stoop WA (1981) Evaluation of the cation status of oxidic soils using adsorption—release curves, Geoderma, 26 (1–2):107–119. https://doi.org/10.1016/0016-7061(81)/90079-3

Strong DT, Sale PWG, Helyar KR (1997) Initial soil pHaffects the pH at which nitrification ceases due to self-induced acidification of microbial microsites. Aust J Soil Res 35:565–570

Stumm W, Morgan JJ (1970) Aquatic chemistry. Wiley Interscience, New York, p 583

Stumm E, Wieland E (1990) Dissolution of oxide and silicate minerals: rates depend on surface speciation. In: Stumm W (ed) Aquatic chemical kinetics. John Wiley, New York, USA, pp 367–400

Subba Rao NS (1999) Soil microbiology, 4th ed. Enfield: Science Publishers, New Delhi, India. ISBN 9781578080700, 1578080703

Suhadolc M, Schroll R, Hagn A, Dörfler U, Schloter M, Lobnik F (2010) Single application of sewage sludge-impact on the quality of an alluvial agricultural soil. Chemosphere 81:1536–1543

Sullivan TS, Barth VP, Lewis RW (2017) Soil acidity impacts beneficial soil microorganisms. Soil Acidification Series, Washington State University Extension

Sumner ME (1992) The electrical double layer and clay dispersion. In: Sumner ME, Stewart BA (eds) Soil crusting: chemical and physical process. Lewis Publishers, Boca Raton, FL, pp 1–32

Sumner ME (1997) Procedures used for diagnosis and correction of soil acidity: a critical review. In: Moniz AC, Fulani AMC, Schaffert RE, Fageria NK, Rosolem CA, Cantarella H (eds) Plant-soil interactions at low pH: sustainable agriculture and forestry production. Brazilian Soil Science Society, Campinas, pp 195–204

Sumner ME, Noble AD (2003) Soil acidification: the world story. In: Rengel Z (eds) Handbook of soil acidity. Marcel Dekker, New York, pp 1–28

Sustr V, Chalupsky J, Kristufek V (1997) Effects of artificial acidification and liming on the activity of digestive enzymes in Cognettia sphagnetorum (Vejdovsky, 1878) (Annelida, Enchytraeidae). Biol Fertil Soils 24:227–230

Sylvia DM, Fuhrmann JJ, Hartel PG, Zuberer DA (eds) (1998) Principles and applications of soil microbiology. Prentice Hall, Upper Saddle River, NJ, 550p

Sylvia DM, Fuhrmann JF, Hartel PG, Zuberer DA (2005) Principles and applications of soil microbiology. Pearson Education, Upper Saddle River, NJ

Syuhada A, Shamshuddin J, Fauziah C, Rosenani A, Arifin A (2016) Biochar as soil amendment: impact on chemical properties and corn nutrient uptake in a Podzol. Can J Soil Sci 96:400–412

Szalay A (1964) Cation exchange properties of humic acids and their importance in the geochemical enrichment of UO2++ and other cations. Geochim Cosmochim Acta 28:1605–1614. https://doi.org/10.1016/0016-7037(64)90009-2

Talibudeen O (1981) Cation exchange in soils. In: Greenland DJ, Hayes MBH (eds) The chemistry of soil processes. John Wiley & Sons, New York, pp 115–177

Tanaka A, Yoshida S (1970) Nutritional disorders of rice in Asia. International Rice Research Institute, Technical Bulletin, 10, 51 p

Tang C (2004) Causes and management of subsoil acidity. In: Super Soil 2004, 3rd Australian New Zealand soils conference, December 5–9, 2004. Australia, University of Sydney, 6 pp

Tang C, Rengel Z (2003) Role of plant cation/anion uptake ratio in soil acidification. In: Rengel Z (eds) Handbook of soil acidity. Marcel Dekker, New York, pp 57–81

Tang C, Thomson B (1996) Effects of solution pH and bicarbonate on the growth and nodulation of a range of grain legume species. Plant Soil 186:321–330

Tang C, Yu Q (1999) Impact of chemical composition of legume residues and initial soil pH on pH change of a soil after residue incorporation. Plant Soil 215:29–38

Taylor RW, Hassan K, Mehadi AA, Shuford JW (1995) Kinetics of Zn sorption by soils. Commun Soil Sci Plan 26:1761–1771

Teske A, Elm E, Regan JM, Toze S, Rittmann BE, Stahl DA (1994) Evolutionary relationships among ammonia- and nitriteoxidizing bacteria. J Bacteriol 176:6623–6630

Thaller MC, Berlutti F, Schippa S, Iori P, Passariello C (1995) Rossolini GM (1995) Heterogeneous patterns of acid phosphatases containing low-molecular-mass Polipeptides in members of the family Enterobacteriaceae. Int J Syst Bacteriol 4:255–261

Thiet RK, Frey SD, Six J (2006) Do growth yield efficiencies differ between soil microbial communities differing in fungal:bacterial ratios? Reality check and methodological issues. Soil Biol Biochem 38:837–844

Thomas GW, Hargrove WL (1984) The chemistry of soil acidity. In: Adams F (ed) Soil acidity and liming. ASA, CSSA, SSSA, Madison, cap. 1, p 3–56

Thomson CJ, Marschner H, Romheld V (1993) Effect of nitrogen fertilizer form on pH of the bulk soil and rhizosphere, and on growth, phosphorus, and micronutrient uptake of bean. J Plant Nutr 16:493–506. http://hdl.handle.net/102.100.100/244499?index=1

Tin NT, Wilander A (1994) Chemical condition in acidic water in the Plain of Reeds, Vietnam. Water Res 29:1401–1408

Tomczyk A, Sokołowska Z, Boguta P (2020) Biochar physicochemical properties: pyrolysis temperature and feedstock kind effects. Rev Environ Sci Biotechnol 19:191–215

Tonolla M, Peduzzi S, Demarta A, Peduzzi R, Dittmar H (2004) Phototropic sulfur and sulfate-reducing bacteria in the chemocline of meromictic Lake Cadagno, Switzerland. J Limnol 63:161–170. https://doi.org/10.4081/jlimnol.2004.161

Tourna M, Freitag TE, Nicol GW, Prosser JI (2008) Growth, activity and temperature responses of ammonia oxidising archaea and bacteria in soil microcosms. Environ Microbiol 10:1357–1364

Treusch AH, Leininger S, Kietzin A, Schuster SC, Klenk H-P, Schleper C (2005) Novel genes for nitrite reductase and Amo-related proteins indicate a role of uncultivated mesophilic crenarchaeota in nitrogen cycling. Environ Microbiol 7:1985–1995

Tsai WT, Liu SC, Chen HR, Chang YM, Tsai YL (2012) Textural and chemical properties of swine-manure-derived biochar pertinent to its potential use as a soil amendment. Chemosphere 89:198–203. https://doi.org/10.1016/j.chemosphere.2012.05.085

Tuong TP (1993) An overview of water management of acid sulphate soils. In: Dent D, van Mensvoort MEF (eds) Selected papers on the Ho Chi Minh city symposium on acid sulphate soils. International Institute for Land Reclamation and Improvement Publication 53. Wageningen, 281–287

Turner FJ, Verhoogen J (1960) Igneous and metamorphic petrology. McGraw-Hill, New York

Uchida R, Hue NV (2000) Soil acidity and liming. In: Silva JA, Uchida R (eds) Plant nutrition management in Howaii's soils, approaches for tropical and subtropical agriculture, college of tropical agriculture and human resources, University of Howaii, Manoa, USA, pp 101–111

Uexkull HR, Bosshart RP (1989) Management of acid upland soils in Asia. In: Craswell ET, Pushparajah E (eds) Management of acid soils in the humid tropics of Asia. ACIAR Monograph No. 13 (m SRAM Monograph No. 1), Australian Centre for International Agricultural Research, Canberra, pp 2–19

Ulrich B, Mayer R, Khanna PK (1980) Chemical changes due to acid precipitation in a loess derived soil in central Europe. Soil Sci 130:193–199

Unkovich M (2013) Nitrogen fixation in Australian dairy systems: review and prospect. Crop Pasture Sci 63(9):787–804

Upiohn B, Fenton G, Conyers M (2005) Soil acidity and liming, Agfact AC. 19. New South Wales (NSW): NSW Department of Primary Industries
USDA Natural Resources Conservation (1993) Historical changes in soil erosion, 1930–1992: The Northern Mississippi Valley Loess Hills
Uslu OS, Babur E, Alma MH, Solaiman ZM (2020) Walnut shell biochar increases seed germination and early growth of seedlings of fodder crops. Agriculture 10:427
Van Breemen N (1980) Acidity of wetland soils, including Histosols, as a constraint to food production. In: IRRI (ed) Priorities for alleviating soil-related constraints to food production in the tropics. IRRI, Los Baños, Philippines, pp 189–202
Van Breemen N (1993) Environmental aspects of acid sulphate soils. In: Dent DL, Van Mensvoort MEF (eds) Selected papers of the Ho Chi Minh city symposium on acid sulphate soils, vol 53. ILRI Publication, Wageningen, The Netherlands, pp 391–401
Van Breemen N, Van Dijk HFG (1988) Ecosystem effects of atmospheric deposition of nitrogen in the Netherlands. Environ Pollut 54:249–274
Van Groenigen KJ, Six J, Harris D, Van Kessel C (2007) Elevated CO_2 does not favor a fungal decomposition pathway. Soil Biol Biochem 39:2168–2172
van Niel EWJ, Arts PAM, Wesselink BJ, Robertson LA, Kuenen JG (1993) Competition between heterotrophic and autotrophic nitrifiers for ammonia in chemostat cultures. FEMS Microbiol Ecol 102:109–118
van Raij B (1991) Fertilidade do Solo e Adubação. Ceres, São Paulo. 2008. Gesso na Agricultura. Instituto Agronômico de Campinas, São Paulo (cited in Sanchez, 2019)
Van Raij B, Peech M (1972) Electrochemical properties of some Oxisols and Alfisols of the tropics. Soil Sci Soc Am Proc 36:587–593
van Raij B, Cantarella H, Quaggio JA, Furlani AMC (1997) Methods for diagnosis and correction of soil acidity in Brazil: an overview. In: Moniz AC, Fulani AMC, Schaffert RE, Fageria NK, Rosolem CA, Cantarella H (eds) Plant–soil interactions at low pH: sustainable agriculture and forestry production. Brazilian Soil Science Society, Campinas, pp 205–214
Van Wambeke A (1976) Formation, distribution and consequences of acid soils in agricultural development. In: Wright MJ (ed) Plant adaptation to mineral stress in prohlem soils. NY Cornell University Agricultural Experiment Station, Ithaca, pp 15–24
Vargas AAT, Graham PH (1988) Phaseolus vulgaris cultivar and Rhizobium strain variation in acid-pH tolerance and nodulation under acid conditions. Field Crop Res 19:91–101
Varghese EM, Kour B, Ramya S, Kumar NS, Jisha MS, Ramakrishnan B (2022). Rhizosphere microbe-mediated alleviation of aluminum and iron toxicity in acid soils. Academic Press, Rhizosphere Engineering, pp 499–526
Velayutham M, Reddy KCK (1990). Nutrient management based on soil test crop response correlation research. Soil fertility and fertilizer use. Vol IV. Nutrient management and supply
Verhagen FJM, Laanbroek HJ (1991) Competition for ammonium between nitrifying and heterotrophic bacteria in dual energy-limited chemostats. Appl Environ Microbiol 57:3255–3263
Verhagen FJM, Duyts H, Laanbroek HJ (1992) Competition for ammonium between nitrifying and heterotrophic bacteria in continuously percolated soil columns. Appl Environ Microbiol 58:3303–3311
Vestreng V, Myhre G, Fagerli H, Reis S, Tarrasón L (2007) Twenty-five years of continuous sulphur dioxide emission reduction in Europe. Atmos Chem Phys 7:3663–3681
Vetriani C, Tran HV, Kerkhof LJ (2003) Fingerprinting microbial assemblages from the oxic/anoxic chemocline of the Black Sea. Appl Environ Microbiol 69:6481–6488
Vieillefon J (1973) Sur quelques transformations sedimentologiques et mineralogiques dans les sols sulfates acides du Senegal. In: Dost H (ed) Acid sulphate soils. Proceedings international symposium on acid sulfate soils Wageningen. ILRI Publ. No. 18; vol 2, pp 99–113
Villagarcía S (1973) Aluminum tolerance in the Irish potato and the influence of substrate aluminum on growth and mineral nutrition of potatoes. PhD Thesis, North Carolina State University

Villamízar F, Lotero J (1967) Respuesta del pasto pangola a diferentes fuentes y dosis de nitrógeno. Revista Instituto Colombiano Agropecuario 2:57–70
Von Lützow M, Leifeld J, Kainz M, Kögel-Knabner I, Munch JC (2002) Indications for soil organic matter quality in soils under different management. Geoderma 105:243–258
Von Uexküll HR, Mutert E (1995) Global extent, development and economic impact of acid soils. In: Date RA, Grundon NJ, Rayment GE, Probert ME (eds) Plant-soil interactions at low pH: principles and management. Kluwer Academic Publishers, Dordrecht, The Netherlands, pp 5–19
Vreeken-Buijs MJ, Hassink J, Brussaard L (1998) Relationships of soil microarthropod biomass with organic matter and pore size distribution in soils under different land use. Soil Biol Biochem 30:97–106. https://doi.org/10.1016/S0038-0717(97)00064-3
Wacey D, Saunders M, Kong C, Brasier A, Brasier M (2015) 3.46 Ga Apex chert "microfossils" reinterpreted as mineral artefacts produced during phyllosilicate exfoliation. Gondwana Res 36:296–313. https://doi.org/10.1016/j.gr.2015.07.010
Walker TW, Syers JK (1976) The fate of phosphorus during pedogenesis. Geoderma 15:1–19
Wallenstein MD, McNulty S, Fernandez IJ, Boggs J, Schlesinger WH (2006) Nitrogen fertilization decreases forest soil fungal and bacterial biomass in three long-term experiments. For Ecol Manag 222:459–468
Walworth JL (1998) Potatoes-field crop fertilizer recommendations for Alaska. Crop Production and Soil Management Series. FGV-00246A. [Available online] http://www.uaf.edu/coop-ext/. Accessed 17 Feb 2003
Wander MM (2004) Soil organic matter fractions and their relevance to soil function. In: Magdoff F, Weil R (eds) Advances in agroecology. CRC Press, Boca Raton, FL, pp 67–102
Wang JP, Harsh R, Guo-Ping Z, Neville M, Zhou MX (2006) Aluminium tolerance in barley (Hordeum vulgare L.): physiological mechanisms, genetics and screening methods. J Zheijang Univ Sci B 7:769–787
Wang Y, Tu C, Cheng L, Li C, Gentry LF, Hoyt GD, Zhang X, Hu S (2011) Long-term impact of farming practices on soil organic carbon and nitrogen pools and microbial biomass and activity. Soil till Res 117:8–16
Warby RAF, Johnson CE, Driscoll CT (2009) Continuing acidification of organic soils across the northeastern USA: 1984–2001. Soil Sci Am J 73:273–284
Wardle DA, Bardgett RD, Klironomos JN, Setala H, van der Putten WH, Wall DH (2004) Ecological linkages between aboveground and belowground biota. Science 304(5677):1629–1633
Watmough SA, Aherne J, Alewell C, Arp P, Bailey S, Clair T, Dillon P, Duchnese L, Eimers C, Fernandez I, Foster N, Larssen T, Miller E, Mitchell M, Page S (2005) Sulphate, nitrogen and base cation budgets at 21 forested catchments in Canada, the United States and Europe. Environ Monit Assess 109(1–36):146
Webster G, Embley TM, Prosser JI (2002) Grassland management regimens reduce small-scale heterogeneity and species diversity of b-proteobacterial ammonia oxidizer populations. Appl Environ Microbiol 68:20–30
Weese DJ, Heath KD, Dentinger BTM, Lau JA (2015). Long-term nitrogen addition causes the evolution of less-cooperative mutualists. Int J Org Evol. https://doi.org/10.1111/evo.12594
Wershaw RL (1999) Molecular aggregation of humic substances. Soil Sci 1999(164):803–813
Wheatley RE, Caul S, Crabb D, Daniell TJ, Griffiths BS, Ritz K (2003) Microbial population dynamics related to temporal variations in nitrification in three arable fields. Eur J Soil Sci 54:707–714
White I, Melville MD, Wilson BP, Sammut J (1997) Reducing acidic discharges from coastal wetlands in eastern Australia. Wetlands Ecol Manag 5:55–72
Whitfield CJ, Aherne J, Watmough SA, McDonald M (2010) Estimating the sensitivity of forest soils to acid deposition in the Athabasca Oil Sands Region, Alberta, Canada. J Limnol 69(1):201–208
WHO (2011) Zinc for better health: The World Health Report, Geneva, Switzerland
Widyawati E (2011) Formulation of sulfate-reducing bacterial inoculum isolated from paper industry sludge to treat acid mine drainage. J Tekno Hut Tan 4(3):119–125

Wiklander L, Hallgren G, Brink N, Jonsson E (1950) Studies on gyttja soils. II. Some characteristics of northern Sweden. Ann Royal Agric College Sweden 17:24–36

Wild A (1993) Soils and the environment-an introduction. Cambridge University Press Great Britain

Williams CH (1980) Soil acidification under clover pasture. Aust J Exp Agric Anim Hus 20:561–567

Willett IR, Melville MD, White I (1993) Acid drain waters from potential acid sulphate soils and their impact on estuarine ecosystems. In: Dent DL, van Mensvoort MEF (eds) Selected papers of the Ho Chi Minh City symposium on acid sulphate soils. ILRI, Wageningen, The Netherlands, pp 419–425

Wood AW (1985) Soil degradation and management under intensive sugar cane cultivation in North Queensland. Soil Use and Management I, 120–124

Wood PM (1987) Mono-oxygenase and free radical mechanisms for biological ammonia oxidation. In: Cole JA, Ferguson SJ (eds) The nitrogen and sulphur cycles. Proceeding symposium 42 of the Society for General MicrobiologyCambridge University Press, Cambridge, pp 219–243

Wolf B, Snyder G (2003) Sustainable soils: the place of organic matter in sustaining soils and their productivity. CRC Press. Woomer, and M.J.swift(eds.). The biological management of tropical soil fertility.195204. Woomer P, Nair PKR, Mugwe JN, O'neill MK, Mugendi DN 1999. Alley cropping of maize with calliandra and leucaena in the subhumid highlands of Kenya: Part 2. Soilfertility changes and maize yield. Agroforestry systems 46(1), pp.39–50

Wood M (1989) Soil biology. Chapman and Hall, New York

Wood (1995)

Wood T, Bormann FH, Voigt GK (1984) Phosphorus cycling in an northern hardwood forest: biological and chemical control. Science 223:391–393

Wright AL, Hanlon EA McCray JM (2009) Fate of phosphorus in everglades agricultural soils after fertilizer application. SL290. University of Florida Institute of Food and Agricultural Sciences, Gainesville. http://edis.ifas.ufl.edu/ss503

Wu J, O'Donnell AG, Syers JK (1993) Influences of glucose, nitrogen and plant residues on the immbolisation of sulphate-S in soil. Soil Biol Biochem 27:1363–1370

Wuchter C, Abbas B, Coolen MJL, Herfort L, van Bleijswijk J, Timmers P, Strous M, Teira E, Herndl GJ, Middelburg JJ, Schouten S, SinninDamsté JS (2006) Archaeal nitrification in the ocean. Proc Natl Acad Sci USA 103:12317–12322

Xie H, Pasternak JJ, Glick BR (1996) Isolation and characterization of mutants of the plant growth-promoting rhizobacterium Pseudomonas putida GR12-2 that overproduce indoleacetic acid. Curr Microbiol 32:67–71

Xu Y, Schwartz FW, Traina SJ (1994) Sorption of Zn^{2+} and Cd^{2+} on hydroxyapatite surfaces. Environ Sci Technol 28:1472–1480

Xu JM, Tang C, Chen ZL (2006a) Chemical composition controls residue decomposition in soils differing in initial pH. Soil Biol Biochem 38:544–552

Xu J, Tang C, Chen ZL (2006b) The role of plant residues in pH change of acid soils differing in initial pH. Soil Biol Biochem 38:709–719. https://doi.org/10.1016/j.soilbio.2005.06.022

Yagi R, Fidalski J, Tormena CA (2014) The incorporation of limestone in consolidated no-tillage system reduces carbon stock in soil macroaggregates. Ciência Rural 44:1962–1965

Yamada N (1959) Some aspects of the physiology of bronzing. IRC Newsl. 8(3):11–15

Yan F, Schubert S (2000) Soil pH changes after application of plant shoot materials of faba bean and wheat. Plant Soil 220:279–287

Yan F, Schubert S, Mengel K (1996) Soil pH increase due to biological decarboxylation of organic anions. Soil Biol Biochem 28:617–624

Yazawa Y, Wong MTF, Gilkes RJ, Yamaguchi T (2000) Effect of additions of brown coal and peat on soil solution composition and root growth in acid soil from wheatbelt of Western Australia. comm. Soil Science Plant Anal. 31:743–758

Yilmaz Y (1989) An approach to the origin of young volcanic rocks of western Turkey. In: Tectonic evolution of the Tethyan region. Springer, Dordrecht, pp 159–189

Yokota S, Ojima K (1995) Physiological response of root tip of alfalfa to low pH and aluminium stress in water culture. In: Plant-soil interactions at low pH: principles and management. Springer, Dordrecht, pp 329–331

Yuan JH, Xu RK, Qian W, Wang RH (2011a) Comparison of the ameliorating effect on an acidic Ultisol between four crop straws and their biochars. J Soils Sediments 11:741–750. https://doi.org/10.1007/s11368-011-0365-0

Yuan JH, Xu RK (2011) The amelioration effect of low temperature biochar generated from nine crop residues on an acidic Ultisol. Soil Use Manage. 27:110–115. https://doi.org/10.1111/j.1475-2743.2010.00317.x

Yuan JH, Xu RK (2012) Effects of biochars generated from crop residues on chemical properties of acid soils from tropical and subtropical China. Soil Res 50:570–578. https://doi.org/10.1071/SR12118

Zaidi A, Khan M, Ahemad M, Oves M (2009) Plant growth promotion by phosphate solubilizing bacteria. Acta Microbiol Immunol Hung 56:263–284

Zama N, Kirkman K, Mkhize N, Tedder M, Magadlela A (2022) Soil acidification in nutrient-enriched soils reduces the growth, nutrient concentrations, and nitrogen-use efficiencies of Vachellia sieberiana (DC.) Kyal. & Boatwr Saplings. Plants 11:3564. https://doi.org/10.3390/plants11243564

Zelles L (1999) Fatty acid patterns of phospholipids and lipopolysaccharides in the characterisation of microbial communities in soil: a review. Biol Fertil Soils 29:111–129

Zhang H (2013) Cause and effects of soil acidity. Oklahoma cooperative extension fact sheets. Division of Agricultural Sciences and Natural Resources, Oklahoma State University. Available at: http://osufacts.okstate.edu

Zhang XN, Zhao AZ (1997) Surface charge. In: Yu TR (ed) Chemistry of variable charge soils. Oxford (Academic), pp 17–63. https://doi.org/10.1093/oso/9780195097450.03.0005

Zhang FS, Ma J, Cao YP (1997) Phosphorus deficiency enhances root exudation of low-molecular weight organic acids and utilization of sparingly soluble inorganic phosphates by radish (Raghanus satiuvs L.) and rape (Brassica napus L.) plants. Plant and Soil 196:261e264

Zhang H, Daoust F, Charles TC, Driscoll BT, Prithiviraj B, Smith DL (2002) Bradyrhizobium japonicum mutants allowing improved nodulation and nitrogen fixation of field grown soybean in a short season area. J Agric Sci 138:293–300

Zhang B, Liu P, Yang YS, Xu GD (2007) Effect of Al in soil on photosynthesis and related morphological and physiological characteristics of two soybean genotypes. Bot Stud 48:435–444

Zhang H, Wang B, Xu M (2008) Effect of inorganic fertilizer inputs on grain yields and soil properties in a long-term wheat–corn cropping system in South China. Commun Soil Sci Plant Anal 39:1583–1599. https://doi.org/10.1080/00103620802071721

Zhao K, Selim HM (2010) Adsorption–desorption kinetics of Zn in soils: influence of phosphate. Soil Sci 175:145–153

Zhao XQ, Shen RF (2018) Aluminum-nitrogen interactions in the soil-plant system. Front Plant Sci 9:807. https://doi.org/10.3389/fpls.2018.00807

Zhao FJ, Evans EJ, Bilsborrow PE, Syers JK (1993) Influence of S and N on seed yield and quality of low glucosinolate oilseed rape (Brassica nepus L.). J Sci Food Agric 63:29–37

Zhao Y, Duan L, Xing J, Larssen T, Nielsen CP, Hao JM (2009) Soil acidification in China: is controlling SO_2 emissions enough? Environ Sci Tech 43:8021–8026

Chapter 4
Toxicity and Tolerance

Abstract The main causes of acid toxicity are a deficiency of vital nutrients in the soil and an oversupply of harmful metals in the root zone of plants. Among the harmful and most common hazardous elements is aluminium (Al). Different plant species have evolved to tolerate aluminium at varying degrees, which has made it possible to create cultivars with high levels of aluminium tolerance. Numerous pathways of aluminium toxicity in higher plants have been identified through the use of physiological and molecular techniques. Plants can withstand Al stress in two ways: by absorbing Al and by keeping Al out of their root tips, but yet tolerant of it in their root cells. Many higher plants have the ability to exude organic acids that can chelate to form complexes with aluminium. The identification of novel mechanisms of tolerance and the fusion of various processes to attain greater tolerance will be the next big challenges for Al tolerance research. When it comes to marker-assisted selection in breeding programmes, gene-specific molecular markers are preferable over linked marker techniques. Although adding calcium, magnesium, or lime to the soil can lessen negative impacts on plant growth, doing so is expensive and unsustainable from an ecological standpoint. When it comes to marker-assisted selection in breeding programmes, gene-specific molecular markers are preferable over linked marker techniques. A combination of limited water absorption, lack of vital nutrients, and harmful minerals prevents acid soil from developing. Plant defence mechanisms are divided into two groups: (i) internal defence, which is Al-tolerant and involves chelating Al inside the cell before it is stored and compartmentalised in membrane-bound organelles like the vacuole, a structure found in the cytoplasmic matrix of a cell; and (ii) external defence, which involves the release of organic acids from the root tips. A number of external barriers, such as root structures like the cell wall and membrane and chemical exudates like phosphates, organic acids, and phenolic compounds, may prevent Al from entering and building up in cells. Internal processes are used to chelate aluminium and produce non-toxic compounds, including Al-chelating in the cytosol, compartmentalisation in the vacuole, Al-binding proteins, and Al-tolerant isoenzymes.

Keywords Toxicity · Tolerance · Gene-specific molecular marker · Chelation of Al · Organic acids

U. C. Sharma et al., *Soil Acidity*, Progress in Soil Science,
https://doi.org/10.1007/978-3-031-76357-1_4

4.1 General Description

The presence of acid soil poses a worldwide challenge for the growth of plants. Acid poisoning primarily occurs due to an abundance of toxic metals in the root zone of plants and a lack of essential nutrients in the soil. Aluminium (Al) is the most widespread and damaging among the toxic metals. Various plant species have evolved different levels of tolerance to aluminium, enabling the development of cultivars that exhibit high levels of aluminium tolerance. Physiological and molecular approaches have been employed to uncover several routes of aluminium toxicity in higher plants (Biana et al. 2013). Two methods of plant tolerance to Al stress are the exclusion of Al from the root tips and the absorption of Al, but tolerance of Al in root cells. Numerous higher plants possess the capacity to release organic acids that can form complexes with aluminium through chelation. The future challenge for Al tolerance studies will be the finding of new tolerance mechanisms and the integration of different mechanisms to achieve more tolerance (Biana et al. 2013). Considerable advancements have been achieved in the identification of the genes accountable for aluminium tolerance, as well as in comprehending the fundamental systems involved. Gene-specific molecular markers are more advantageous than linked marker approaches for marker-assisted selection in breeding programmes. The development of acid soil is hindered by a combination of mineral toxicity, insufficient essential nutrients, and inadequate water absorption. The initial indication of plant growth is a decrease in root length (Kid and Proctor 2001). While the application of lime, magnesium, or calcium to the soil can mitigate adverse effects on plant growth, it is both costly and ecologically unsustainable. The most effective approach to address soil acidity is to selectively develop cultivars that are tolerant to it. Plants exhibit significant variation in their capacity to withstand acidic soil conditions. The development of acid soil-tolerant cultivars is facilitated by the presence of genetic heterogeneity in tolerance levels. Comprehending the physiological aspects, genetic makeup, and gene regulatory data related to acid soil tolerance is crucial for the effectiveness of breeding schemes. Extensive research has demonstrated that both internal and external mechanisms have a role in the development of tolerance. Higher plants often release organic acid as an external method. Various species regulate different tolerance mechanisms through distinct genes and QTL. Molecular markers have been produced to aid in gene cloning and provide valuable resources for marker-assisted selection in the breeding of tolerant cultivars. Merino-Gergichevich et al. (2010) conducted study that found that soil pH has a notable adverse effect on various aspects of soil health. These include the availability of essential nutrients for plants, the solubility of harmful heavy metals, the activity of microorganisms in the soil, the degradation of root cells, and the cation exchange capacity of soils. The effects of toxicity can be differentiated based on their physical and physiological characteristics. Both lead to insufficient plant growth, resulting in decreased output (Samac and Tesfaye 2003).

Soil acidity imposes growth and yield constraints on several crops worldwide. The chemical properties of very acidic soils are primarily characterised by the presence of hazardous compounds and components such as soluble forms of Al, Fe, and Mn, nitrites, and various harmful organic acids. These factors, along with acidity itself, are the main concerns (Krstic et al. 2012). Aluminium (Al) toxicity is a significant barrier to achieving high agricultural yields on acid soils, which are present on around 40% of arable areas globally. Aluminium, which has the third-highest concentration in the earth's crust, poses a risk to plants when it is dissolved in soil solutions with acidic pH levels (Kochian 1995). Von Uexküll and Mutert (1995) reported that around 3950 million hectares of land has acidic soil, with approximately 15% of this land being utilised for the cultivation of annual and perennial crops.

Although low pH levels can restrict plant growth, it is mostly the presence of toxic metals, particularly aluminium (Al) that hinders plant development. Aluminium constitutes around 7% of the mass of the earth's crust, making it the most abundant metal in that region. The dissolution of even a small fraction of the aluminium compounds present in soils can lead to severe aluminium poisoning in susceptible species. Fortunately, not all types of aluminium are dangerous. As previously mentioned, it is believed that acid soil toxicity is produced by the soluble forms of aluminium. Al^{3+} is considered the most harmful form of trivalent cations in terms of its toxicity to plants. Additionally, certain studies have also associated toxicity with the divalent and monovalent forms. Trivalent cations typically exhibit phytotoxicity towards plants. The presence of Al^{3+} is more pronounced in acidic conditions, as aluminium undergoes hydrolysis in solution. On the other hand, $Al(OH)_2{}^+$ and $Al(OH)_2{}^+$ species are frequently observed at pH levels ranging from 5 to 7. Aluminium hydroxide ($Al(OH)_3$) can form as the pH increases, and the dominant species shifts to aluminium hydroxide ion ($Al(OH)_4{}^-$) when the pH becomes alkaline. Due to the strong affinity of Al^{3+} for many ligands, understanding and predicting the chemistry of Al in soils is a difficult task (Ritchie 1995). Studying the impact of various types of aluminium on roots can be difficult, even in solutions with a known composition. In acidic soils, aluminium is widely recognised as the primary obstacle to plant growth. However, it is important to note that other metals, such as manganese, can also be present in lethal concentrations. Many acidic soils exhibit reduced fertility due to deficiencies in calcium, magnesium, phosphorus, and molybdenum. Recent reviews have extensively discussed the mechanisms of aluminium tolerance and toxicity in plants. Among the review articles, those authored by Kochian (1995), Matsumoto (2000), Ma et al. (2001), and Ryan et al. (2001) stand out.

Plants have successfully overcome the limitations caused by Al toxicity using various strategies. Plants' tolerance to aluminium also relies on symplastic mechanisms, in addition to the common occurrence of systems that include the excretion of organic acids. Various creatures possess different tolerance systems, which are genetically complex traits. Arabidopsis, which has a moderate sensitivity to aluminium (Al), and rice, which has a high tolerance to Al, both show the involvement of many mechanisms. This indicates that the intrinsic level of tolerance in each species is not a determining factor. Molecular and genomic research has greatly facilitated the identification of genes associated with aluminium tolerance. These genes consist of membrane transporters belonging to the ALMT, MATE, and ABC families. Various animals possess functional counterparts of these transporter genes. The expression of Al tolerance genes is predominantly regulated by cis-elements and transcription factors. A comprehensive comprehension of Al tolerance in various plant species can be achieved by integrating information of the QTL, genes, and underlying mechanisms. Utilising the process of pyramiding these genes and implementing tolerance mechanisms through genetic transformation or marker-assisted introgression of better alleles can significantly contribute to the development of cultivars with high aluminium tolerance through molecular breeding. This, in turn, will enhance crop yield on acidic soils. Aluminium (Al) accounts for the largest proportion of metals found in the earth's crust, comprising around 7% of its total mass. Soils can be classified as A1 hazardous due to the sensitivity of many plant species to micromolar concentrations of aluminium (Al). Fortunately, most of the A1 is complexed with ligands or occurs in non-phytotoxic forms such as aluminosilicates and precipitates. Nevertheless, acidic conditions enhance the solubility of A1, and the toxicity of A1 is a major constraint on plant growth in acidic soils (Delhaize and Ryan 1995). Acid rain and different farming practices can accelerate soil acidification, which can happen naturally when basic cations are washed out from soils (Kennedy 1992). Two techniques for maintaining productivity on acidic soils are using plants that are well-suited to such conditions and applying lime to increase the soil's pH. Although A1 toxicity in acid soils has been acknowledged as a problem for over 70 years, our comprehension of the specific sites of toxicity and the sequence of events that hinder plant growth remains mostly speculative.

Acidic soil containing aluminium (Al^{3+}) is toxic to plants and hinders their growth, causing stunted development. Aluminium ions (Al^{3+}) can impact several biological processes, including the elongation of root hair, the absorption of nutrients such as calcium (Ca) and potassium (K), the occurrence of oxidative stress, the disruption of the cytoskeleton and apoplastic processes, and the transport within cells. The references cited are Kochian (1995), Horst et al. (2010), Yamamoto et al. (2002), and Kochian et al. (2004). Al^{3+} can affect these mechanisms. The main indication of the toxicity caused by Al^{3+} is a decrease in the rate of root growth. Due to the presence of acid soils on 35% of arable land, the toxicity of Al^{3+} serves as a substantial selective force that promotes adaptation. Many diverse species possess adaptive defence systems that enhance their survival in acidic soils. Prior to a complete understanding of these processes, they could be categorised into two groups: mechanisms that would probably hinder the entry of Al^{3+} into the root (referred to as exclusion or resistance mechanisms), and mechanisms that would allow plants to tolerate Al^{3+} after it had already entered the symplast. These classes are commonly known as tolerance mechanisms. Previously, it was thought that the two main factors contributing to exclusion mechanisms were the release of ligands that bind to Al^{3+} and block its entry into the cytosol, or the presence of transport systems that remove Al^{3+} from the symplast. Nevertheless, subsequent study has revealed that these two processes are not the exclusive ones implicated. Tolerance mechanisms possess the capacity to either chelate aluminium prior to its entry into the cytosol, forming stable complexes, or sequester aluminium in subcellular compartments, shielding it from harmful effects. We now have a good understanding that these two processes are genuine and that they can occur simultaneously. These topics have been thoroughly examined in numerous comprehensive analyses (Kochian et al. 2004; Ma

2007; Poschenrieder et al. 2008), available for access on the internet. The toxicity of aluminium (Al) is increasingly becoming a significant issue that restricts plant growth on acidic soils. During periods of Al stress, certain plant species, cultivars, or genotypes demonstrate heritable tolerance through the secretion of organic acids. This response is specific to Al stress and occurs mostly at the root apex. Citric acid, oxalic acid, and malic acid are considered the most potent organic acids for detoxifying aluminium. Organic acid neutralises the toxic effects of aluminium both outside and inside the cells.

Approximately 30–40% of the land that is suitable for growing crops and up to 70% of the area that has the potential to be suitable for growing crops is characterised by acidity (Haug and Foy 1984). The primary constraint on plant growth in acidic soils is the presence of active aluminium ions (Al^{3+}), however there are additional limiting variables such as phosphorus (P), calcium (Ca), magnesium (Mg) deficiencies, and hazardous levels of manganese (Mn). The presence of active Al ion has a significant harmful effect on plants, particularly on the root tip meristem, since it hinders the growth of roots. Aluminium poisoning can also disrupt various plant functions, including food absorption, elongation of root hairs, apoplastic processes, cytoskeleton, and inter- and intra-cellular transport (Kochian 1995; Kochian et al. 2004; Kopittke et al. 2016). The impact of aluminium (Al) on the growth and development of plants has primarily been studied using hydroponic methods and focusing on the hazardous concentration of Al. Furthermore, the positive impact of aluminium on plant growth has also been discovered. While Al is not classified as a necessary nutrient, little amounts of Al can enhance plant development (Foy et al. 1973). Over the past twenty years, there has been increasing evidence supporting the potential impact of aluminium (Al) in enhancing phosphorus (P) efficiency, reducing proton (H^+), iron (Fe), and manganese (Mn) toxicities, and promoting rhizobacteria. However, the available reports on these effects are sporadic and lack systematic investigation compared to the extensive research on Al toxicity in plants.

Soil minerals consist of aluminium (Al) and silica (Si) oxides (Al_2O_3, SiO_2), which mix to form clay minerals (aluminosilicate minerals) in different quantities as the soil undergoes weathering. These minerals exhibit variable concentrations of aluminium (Al) and silicon (Si). The structure of clay minerals can vary, but simpler clay minerals like kaolinite (with a 1:1 ratio of aluminium to silicon) are less reactive compared to clay minerals like montmorillonite (with a 2:1 ratio of silicon to aluminium), which are more reactive. Cation exchange capacity (CEC) refers to the reactivity of soils or minerals in attracting and retaining positively charged basic cations, such as Ca^{2+} and Mg^{2+}. It is a measure of the soil's negative charge. Soils with a high cation exchange capacity (CEC) possess a stronger capacity to retain basic cations, hence exhibiting resistance to acidification. Nevertheless, soils with poor cation exchange capacity (CEC) exhibit reduced resistance to acidification and undergo rapid acidification, leading to increased acidity. Soil organic matter possesses a high cation exchange capacity (CEC) and aids in mitigating soil acidification. Hydrated aluminium species, when coupled with hydroxyl ions (OH^-), are generally non-toxic to plants due to their weak charge, which is insufficient to displace basic cations (Ca^{2+}, Mg^{2+}) from soil exchange sites. As the soil pH decreases, the H^+ ion activity increases, leading to a reaction with the OH^- ions that are bound to the Al^{3+} ion. This reaction removes the OH^- ions from the Al^{3+} ion, resulting in an increase in the charge of the Al species to either +2 or +3. These species will replace Ca^{2+} and Mg^{2+} cation exchange sites, leading to an increase in the concentration of Al^{3+} in the plant root zone. Aluminium ion (Al^{3+}) is not a necessary element for plant growth. However, when its concentration in the soil solution rises, it leads to a fall in the levels of calcium ion (Ca^{2+}) and magnesium ion (Mg^{2+}), which in turn negatively impacts plant growth.

4.1.1 Solid-Phase Control of Aluminium Ion Activity

The dissolution of Al-bearing minerals in the solid phase can significantly influence the amounts of AlT in soil solution. The solubility of these minerals is mainly determined by pH, as stated by Lindsay (1979) and Sumner et al. (1991). The correlation between the concentration of Al^{3+} in a soil solution and the equilibrium solubility of known solid phases containing Al^{3+} can be used to determine the dominant solid phases that influence the concentration of Al^{3+} in a specific soil system (Sumner et al. 1991; Xu 1991). When the pH is greater than 5.5, the data points in this study are closely aligned with the solubility line for amorphous $Al(OH)_3(s)$. This indicates that a phase of Al hydroxide is responsible for controlling the concentration of Al^{3+} in this pH range. The concentration of Al^{3+} at a pH below 5.5 was lower than what was expected for amorphous $Al(OH)_3$. This indicates that there is another solid phase or process, other from the dissolution of $Al(OH)_3(s)$, that regulates the concentration of Al^{3+} at a pH below 5.5 in these soils. It is anticipated that other solid phases containing aluminium would also lead to a rise in the concentration of Al^{3+} ions as the pH decreases during dissolution. However, the relationship between the concentration of Al^{3+} ions and pH becomes almost flat below a pH of 5.5. The behaviour seen under acidic conditions is indicative of the expected response of aluminium when it is influenced by the formation of complexes with organic materials in the soil (Zysset et al. 1999). Another potential explanation is that the presence of $Al(OH)_3$ and other minerals caused non-equilibrium conditions due to kinetic limitations during the relatively brief (24-h) equilibration period. However, it is improbable that the dissolution of Al would be slower at pH levels below 5.5 compared to pH levels above 5.5, as the dissolution rates of Al hydroxide minerals typically rise as pH decreases (Stumm and Wollast 1990). Curiously, the concentration of Al^{3+} is regulated to around 10–6 at pH < 5.5 (Fig. 4), which is close to the concentration at which phytotoxicity has been recorded (Delhaize and Ryan 1995; Reddy et al. 1995). The level of $Al(OH)_3(s)$ is approximately 1–2 log (Al^{3+}) units lower than the saturation point between pH 4 and 5. Additionally, a significant amount of AlT is expected to be complexed by FA in the solution. This indicates that Al-organic matter complexes play a significant role in controlling the concentration of (Al^{3+}) at pH < 5.5 in the surface layers of this DS system. Wong and Swift (1995) found that the addition of humic acid to soil with a pH of 3.2–4.6 resulted in an undersaturation of 0.6–0.7 log (Al^{3+}) units compared to gibbsite. The interaction between Al and humic and low molecular weight organic acids was shown to decrease the concentration of Al^{3+} in the solution and the amount of extractable Al. This, in turn, resulted in an increase in root growth in acid soils that were treated with organic materials, as demonstrated by Muhrizal et al. (2003). The presence of a significant amount of organic matter in the top layer of DS soils allows for the formation of solid-phase and dissolved organic matter that can interact with aluminium (Al). Our interaction helps explain why the concentration of Al^{3+} deviates from the expected solubility line of $Al(OH)_3(s)$ at pH levels below 5.5 in our study.

4.1.2 Wheat's Tolerance to Soil Acidity and Aluminium

Significant differences in how different varieties respond to soil acidity (low soil pHCa and high Al), liming, and P fertiliser on acidic loamy sand soils were observed in terms of anthesis biomass, grain production, and quality. Soil acidity significantly hampers wheat productivity in Western Australia. Combining the technique of liming with the selection of aluminium (Al)-tolerant wheat types can greatly diminish the effects of soil acidity. In their study, Tang et al. (2003) conducted glasshouse tests to assess the tolerance of soil acidity in 115 genotypes by measuring root length. Significant variations in genotype were discovered regarding the ability to tolerate acidity at a pH level of 3.9. In 2012, a hydroponic method was employed to assess the resistance of 24 wheat cultivars to Al (Amjad 2013), using the Al-tolerant cv. Cotipora and Al are delicate cultivars. Egrets are being used for checks. The criterion used to determine Al tolerance was the decrease in root length observed in plants cultivated in a nutrient solution with a pH of 4.2 and an Al concentration of 12 ppm (12 mg/kg), in comparison with a solution with a pH of 6.5.

4.2 The Toxicity of Aluminium

The presence of aluminium (Al) in acidic soils hampers both root development and agricultural productivity. Aluminium is toxic and constitutes almost forty percent of the theoretically cultivable area on Earth. Simes et al. (2012) identified two main processes by which plants develop resistance to aluminium. The procedures referred to involve internal detoxification of aluminium and the exclusion of aluminium from root cells. The main emphasis of any molecular breeding programmes aimed at enhancing the Al tolerance of crop species should be on investigating these genes and the other variables. Genes responsible for controlling aluminium tolerance in plants are those that encode membrane transporters, auxiliary transcription factors, and cis-elements that promote gene expression. A comprehensive comprehension of the genetic, molecular, and physiological components of Al tolerance in plants is crucial for developing cultivars that can grow in acidic soils. This will ultimately improve the safety of the global food supply.

Ma et al. (2001) found that aluminium (Al) toxicity in acidic soils is a significant abiotic stress factor that reduces crop productivity. Agricultural practices can exacerbate soil acidity, which is already prevalent in tropical and subtropical climates. This is especially true in locations where food supplies are more uncertain (Ciotta et al. 2002). Aluminium is released from soil minerals in the form of ions in acidic conditions. These ions include $Al(OH)_2^+$, $Al(OH)_2^+$, and $Al(H_2O)_6^{3+}$, which is also known as Al^{3+} (Kinraide et al. 1992). Aluminium ions with a charge of +3 can dissolve in acidic soil and be released into the rhizosphere, causing significant harm to plants that are sensitive to it (Delhaize and Ryan 1995).

The metabolic systems of plants can be influenced by the impacts of aluminium within a few minutes of the onset of stress. These initial effects are then followed by secondary effects, as shown by Kochian (1995). Al has a strong attraction towards substances that donate electrons, such as oxygen (Dale and Sutcliffe 1986). Consequently, it has the ability to bind to phosphate and carboxyl groups in the cell wall, so altering its structure and impeding the growth of the wall (Ma et al. 2004a). Due to the detrimental effects of Al toxicity on root systems, their capacity to absorb water and nutrients is significantly reduced, resulting in a decrease in agricultural output (Ma et al. 2001; Kochian et al. 2004). Aluminium toxicity is believed to be the primary component that limits the growth of plants in acidic soils (Foy et al. 1978; Foy 1984; Carver and Ownby 1995; Jayasundara et al. 1998). The roots are the primary site of the plant's response to aluminium stress, as documented by Foy et al. (1978), Foy (1984), Taylor (1988), and Jayasundra et al. (1998). The roots that have been affected by aluminium have a stunted and delicate nature. The lateral roots and tips of the root thicken and undergo a change in colour to brown. The root system is damaged due to the absence of fine branching and the presence of numerous stubby lateral roots. According to the study conducted by Foy et al. (1978), these roots are not efficient in the process of absorbing nutrients and water. The main indication of aluminium toxicity is the rapid inhibition of root development. The interactions occurring inside the cell wall, plasma membrane, or root symplasm are potential mechanisms that may be responsible for the observed effects (Taylor 1988; Marschner 1991; Horst 1995; Kochian 1995). According to Ryan et al. (1993), the root apex is the primary location for Al poisoning. In maize (Zea mays L.), it has been demonstrated that the terminal 2–3 mm of the root, which includes the root cap and meristem, has to be exposed to aluminium (Al) in order to limit the growth of the root. When administered to the elongation zone of the root, Al did not have a substantial impact on root growth. In their study, Ryan et al. (1993) investigated the potential protective role of the root cap against the growth-inhibiting effects of Al. They found that both undamaged roots and roots with the cap removed were equally susceptible to Al's effects. They concluded that the root cap did not provide protection against aluminium damage. In contrast to previous studies, this indicates that the root cap provides protection against Al by facilitating hormone transfer and signal perception (Bennet and Breen 1991).

Stunting of root growth is the most easily recognisable indication of A1 toxicity and is currently employed as a standard marker for A1-induced stress in plants. Within a span of 60 min, the introduction of micromolar quantities of A1 might hinder the growth of roots in basic nutrient solutions. Nevertheless, the overall suppression of growth does not provide significant insights into the stressors that may precede or occur simultaneously with changes in growth. It is essential to identify the primary anatomical and metabolic locations implicated in order to comprehend the mechanisms of A1 toxicity. It is important to note that A1 can have diverse effects and exhibit varying behaviour depending on the species. Multiple studies have been conducted on the harmful effects of Al poisoning, as evidenced by the reviews conducted by Haug (1984), Taylor (1988), and Rengel (1992a). Therefore, our focus will be on examining the specific locations in higher plants where A1 toxicity occurs, as well as any potential role that calcium may play in the primary mechanism of A1 toxicity.

The growth of coniferous forests is facilitated by the acidic soils found in the northern region. These soils are predominantly composed of organic acids and are prevalent in the rainy temperate zone of the north. Within the humid tropics, there exists a region characterised by mineral acid soils located to the south. According to von Uexküll and Mutert (1995), currently only 12% of the land used for agricultural production is characterised by acidity. However, there is a global increase in the extent of acidic soil. Mineral acid soils are formed from parent materials that have an inherent acidity and lack essential basic cations (such as Ca, Mg, K, and Na), or from the leaching of these elements from the soil, resulting in a decrease in pH and the soil's buffering capacity. Aluminium (Al) is capable of dissolving in soil and becomes increasingly prevalent in the soil solution as phytotoxic aluminium ions. Aluminium is found in the majority of mineral soils in adequate quantities to maintain the soil's pH at around 4. Organic acid soils are characterised by high quantities of humic acids and partially degraded plant matter. These soils typically lack aluminium buffering and can have a pH that falls below 4 (Kidd 2001).

The toxicity of acid soil is influenced by various factors, which depend on the makeup of the soil. The primary constraint on plant growth in acidic soils with a high mineral concentration is the toxicity of aluminium. Aluminium (Al) is released from soil minerals as $Al(OH)_2^+$, Al(OH), and $Al(H_2O)_3^+$ when the conditions are acidic. The last form, $Al(H_2O)_3^+$, is also known as Al (Kinraide 1991). Aluminium ions at micromolar concentrations rapidly inhibit root growth in most agriculturally important plants. The root apex is the primary location where Al poisoning occurs. The intricate nature of how aluminium affects cellular processes poses challenges in identifying the primary effect(s) of aluminium poisoning. Exposure to Al inhibits the formation of lateral roots and causes stunting of the primary root. The root tips are shortened and stunted as a result of the suppression of cell elongation and cell multiplication. The plant's small root system makes it more susceptible to drought stress as it hampers its capacity to absorb nutrients and water. Plants that are susceptible to aluminium toxicity see a substantial reduction in both the quality and production of their crops (Jovanovic et al. 2006, 2007; Samac and Tesfaye 2003). The physical–chemical characteristics of soil in Serbia's vertisols have been found to influence the distribution and availability of various elements to plants, according to several researches (Krstic et al. 2007; Dugalic et al. 2010a, b; Jelic et al. 2010; Milivojevic et al. 2011). This study aimed to examine the pH, exchangeable acidity, and mobile aluminium (Al) levels of pseudogley soil profiles in the aak-Kraljevo basin. Aluminium toxicity is the primary factor that restricts the growth of plants in acidic soils, as stated by Foy (1992). The impact of organic matter on the levels of exchangeable aluminium in soil has been highlighted by numerous studies (Thomas 1976; Lundstrom 1993). Nevertheless, the levels of aluminium in natural waterways remain exceedingly low as a result of the extremely poor solubility of minerals containing aluminium. The concentrations of substances in groundwater are significantly influenced by the pH level (Nordstrom and May 1996). Aluminium toxicity is widely recognised as a major component that restricts plant growth in acidic soils (Adams and Moore 1983; Kochian 1995). Soil aluminium appears in different chemical forms, ranging from soluble ionic species in the soil solution to being a fundamental structural element of secondary clay minerals like kaolinite and vermiculite. It is important to note that acidic soil conditions increase the concentration of trivalent Al^{3+} ions, which are the most harmful form of aluminium for plants (Kochian et al. 2005). Aluminium is strongly bound to exchange sites, leading to low concentrations in the soil solution, typically ranging from 10 to 250 μM. The likely mechanisms of aluminium toxicity involve the interaction of aluminium with the cell wall of the root, resulting in a decrease in root growth (Blamey et al. 1983). Aluminium also interferes with the plasma membrane and the processes of membrane transport, and it suppresses the uptake and metabolism of minerals, particularly calcium and phosphorus (Bengtsson et al. 1988). At the cellular level in plants, the presence of toxic Al can cause an excessive formation of reactive oxygen species (ROS) within cells, resulting in a disruption of the functionality of biomembranes and promoting oxidative damage in plants.

Various strategies are available for plants to endure or tolerate the detrimental impact of aluminium in soil.

The defence mechanisms in plants have been categorised into two groups: (i) external defence, which involves the release of organic acids from the root tips, and (ii) internal defence, which is Al-tolerant and involves the chelation of Al inside the cell, followed by its storage and compartmentalisation in membrane-bound organelles such as the vacuole, which is a structure found in the cytoplasmic matrix of a cell. Aluminium ions exhibit a remarkably slow translocation rate to the upper sections of plants (Ma et al. 1997a, b, c). The majority of plants possess a mere 0.2 milligrams of aluminium per gram of dry mass. Certain plants, known as Al accumulators, have the ability to withstand roughly 10 times the normal amount of aluminium without experiencing any negative effects. Tea plants (Camellia sinensis), commonly cultivated in North East India, particularly in Assam, have the characteristic of being efficient accumulators of aluminium (Al). This is because tea plants thrive in acidic soils with a pH range of 4.5–5.0. The plants contain old tea leaves with an aluminium concentration of up to 30 mg/g of dry mass, as reported by Matsumotao et al. (1976). Baker et al. (2000) discovered that there are 400 species of land plants belonging to 45 groups that have the ability to accumulate multiple dangerous metals.

Aluminium toxicity in plants is evident in soils with a pH below 5.0, and can be identified by observing morphological and physiological signs. Variations in the susceptibility to aluminium poisoning are likely due to variations in the root structure and function. Aluminium disrupts the process of cell division in roots, reduces root respiration, and impairs the absorption and utilisation of water and nutrients, specifically calcium and phosphorus, as well as metabolic pathways. Another effective method for studying metal toxicity in tolerant and sensitive plant genotypes is to analyse the uptake and transportation of metals in different plant organs. Additionally, investigating the mechanism of interaction between metals and mineral nutrients, identifying specific genes responsible for tolerance, assessing the levels and types of organic and amino acids that function as metal chelators and detoxifiers, examining the levels and forms of enzymes, and studying changes in.

4.2.1 The Chemistry of Aluminium in Soil

Aluminium is third in terms of abundance among the elements found in the Earth's crust. It is primarily found bound as oxides and complex aluminosilicates. Despite decades of research, the precise mechanism behind the phytotoxicity of aluminium remains incompletely understood. This concept was first postulated by Hartwell and Pember about 90 years ago, suggesting that soluble aluminium plays a crucial role in inhibiting plant growth in acidic soils. Aluminium, when present in micromolar concentrations, can hinder the growth of roots at the organ, tissue, and cellular levels (Ciamporová 2002). Acidic soils are prevalent in humid tropical and subtropical regions of the world due to an excess of H^+, Mn^{2+}, and Al^{3+} ions, as well as a shortage of Ca^{2+}, Mg^{2+}, and PO_4^{3-} ions. Sulphur dioxide and other air pollutants also contribute to the occurrence of acid soil stress in places outside the tropics (Foy 1984). Aluminium compounds that include a high amount of hydroxyl groups have the ability to dissolve to a certain degree in the liquid portion of the soil in acidic conditions. As a result of the dissolution of the widely present aluminium, approximately 40% of the world's cultivable soil has acidic conditions, leading to a major decrease in agricultural productivity. The presence of the mononuclear Al^{3+} species in an octahedral hexahydrate form at low pH values appears to be the most hazardous. $Al(H_20)_6^{3+}$ undergoes successive deprotonation reactions, resulting in the formation of insoluble $Al(OH)_3$ as the pH increases to 7.0. The aluminate ion, $Al(OH)_4^-$, is formed at a pH of 7.4 in the cytosol. Aluminium can exist in polynuclear forms, meaning it can have multiple aluminium atoms, under nearly neutral circumstances. $AlO_4Al_{12}(OH)_24(H_2O)_{12}^{7+}$, sometimes referred to as Al13 (Parker and Bertsch 1992), is a very important polymer triskaideka aluminium compound and is considered to be the most dangerous form of aluminium.

4.2.2 Aluminium Specifically Affects the Root Apices

The root apex (including the root cap, meristem, and elongation zone) attracts and accumulates a greater amount of aluminium compared to the mature root tissues. Indeed, Ryan et al. (1993) discovered that just the uppermost 2–3 mm of maize roots (consisting of the root cap and meristem) required exposure to Al in order to impede growth. The growth is unaltered when aluminium is applied specifically to the elongation zone or to the entire root, save for the apex (Ryan et al. 1993). Following a 2-h period of being exposed to aluminium, Al, Bennet, and Breen (1991) observed several changes in the microscopic composition of cap cells in maize roots. The researchers proposed a hypothesis that aluminium (Al) would indirectly inhibit the growth of roots by activating a signal-response circuit that involves the root cap, hormones, and secondary messengers. This explanation seemed convincing as it aligned with the widely recognised function of the root cap in perceiving signals and distributing hormones. Subsequent studies demonstrated that both undamaged and decapped maize roots were equally affected by the reduction of root development caused by aluminium (Ryan et al. 1993). These findings highlight the importance of the meristem but contradict the idea that the root cap plays a large role in Al toxicity or tolerance.

4.2.3 Does the Entry of Al into the Symplasm Result in Toxicity?

The straightforward response to this query is unknown. The plasma membrane serves as a barrier to the entry of aluminium (Al) due to the fact that ions, especially polyvalent ions such as Al^{3+}, have very low solubility in lipid bilayers. However, apart from some aluminium (Al) entering the plasma membrane, maybe as a neutral Al compound, by endocytosis, membrane-bound proteins, or stress-related damage, it is also quite plausible that up to half of the total Al at the root apex is located in the symplasm (Tice et al. 1992). It is established that wheat genotypes (*Triticum aestivum* L.) that are tolerant to aluminium (Al) accumulate lower levels of Al in their root apices compared to genotypes that are sensitive to Al. Aluminium exposure for a relatively short duration of 60 min hinders the growth of roots. However, it remains uncertain if the aluminium enters the symplasm rapidly and in sufficient amounts to initiate this reaction. The lack of ability to differentiate between the symplasmic and apoplasmic portions of Al has posed challenges in obtaining precise short-term measurements of Al influx. In a recent study, Lazof et al. (1994) discovered the presence of aluminium (Al) in the symplasm of soybean (Glycine max) roots using secondary-ion MS. This was observed after only 30 min of exposure to Al. This demonstrates that Al has the ability to penetrate cells prior to impeding root growth, which suggests that Al poisoning may occur in the symplasm. The pH range of 6.5–7.5 and the abundance of potential ligands will effectively maintain a low concentration of Al^{3+} ions as they enter the symplasm. Al's high affinity for several metabolically important chemicals indicates that it can cause severe damage in the symplasm, even at low doses (Haug 1984; Martin 1988; Haug et al. 1994). Consequently, we can deduce that the primary sources of toxicity stem from the formation of an Al-ligand combination when Al needs to reach the symplasm to become dangerous. Aluminium has the ability to hinder the critical function of the ligand it binds to, such as enzymes, calmodulin, tubulin, ATP, GTP, and DNA. Alternatively, the complex formed between aluminium and the ligand might negatively affect other metabolic processes.

Undoubtedly, Al has convenient and uncomplicated entry to the apoplasm, however the quantification of Al's entry into the symplasm is now being accurately measured. For any substance to enter the symplasm, it must first interact with the cell wall and cell membranes, which might potentially cause harm. Aluminium has the ability to displace ions from crucial locations on the cell wall or membranes, attach to the lipid bilayer or proteins within the membrane to hinder the transport of nutrients, or disrupt secondary messenger routes (Haug 1984; Taylor 1988; Haug et al. 1994). Aluminium can also attach to the pectic residues or proteins in the cell wall, resulting in a decrease in extensibility or hydraulic conductivity. Once again, it remains ambiguous if any of these interactions are harmful, and there is contradictory data that supports the notion that the apoplasm may be a site of toxicity.

4.2.4 Impact of Aluminium Toxicity on Plant Growth

Aluminium ions (Al^{3+}) cause substantial damage to plants. Exposure to aluminium can cause both structural and functional changes in the body. As per Rout et al. (2001), the term "physiological effects" describes the powerful binding action of soluble Al^{3+} in acidic soils. This binding action can occur in several parts of the cell, such as the cell wall, cell membrane, and cell cytosol. Morphological impacts encompass the observable symptoms that occur on different plant structures. The early and most significant physical manifestations of aluminium toxicity are the inhibition and reduction of root growth. Following the introduction of aluminium, its presence can be detected within a few minutes (Kochian et al. 2004). Aluminium diminishes the extent of penetration and hampers the roots' capacity to search for minerals, resulting in an inadequately developed root system, insufficiencies in nutrients, and ultimately a decline in grain yield (Foy 1988). Hecht-Buchholz and Foy (1981) observed typical symptoms of aluminium toxicity on the recently formed lateral roots of barley. Barley varieties with different levels of tolerance exhibited stunted and inhibited root tips. However, the tolerant genotype displayed these indications many days after the sensitive genotype. In their study, Tamas et al. (2003) discovered that when barley roots were exposed to aluminium during germination, it resulted in the reduction of root growth and the loss of cell viability. The application of an Al solution to the root of white clover resulted in a decrease in the quantity of root hairs. An increased concentration of Al^{3+} caused a decrease in root development and the disappearance of root hairs (Care 1995).

Identifying harmful indications on leaves is more challenging compared to roots. One of the signs is nutritional shortage, which is probably caused by inadequate nutrient transport from damaged roots (Rout et al. 2001). Phosphorus deficiency is indicated by many symptoms such as overall stunting, small and dark green leaves, delayed maturity, purplish coloration of stems, leaves, and leaf veins, as well as yellowing and death of leaf tips (Foy et al. 1978). When there is a lack of calcium, young leaves may contort or curl, and the growing tips or stalks may become weak and collapse (Roy et al. 1988). Aluminium-induced stress in the roots may be a contributing factor to the inhibition of leaf development caused by aluminium (Larsen et al. 1997). Thornton et al. (1986) found that honey locust seedlings had reduced leaf size and slower development rates compared to the control group. According to Konarska (2010), when red pepper seedlings were exposed to Al stress, the leaf blades became smaller and thinner, and the leaf cells also decreased in size.

Several plants exhibit toxicity towards the aluminium cation Al^{3+} at micromolar concentrations. Many plant species have evolved defences that enable them to flourish in acidic soils, where the presence of dangerous quantities of Al^{3+} could otherwise impede plant growth. The aluminium tolerance mechanisms rely significantly on organic acids (Ma et al. 2001). Certain plants excrete organic acids that bond with aluminium in order to neutralise its toxicity in the rhizosphere. Aluminium-activated anion channels in the plasma membrane of wheat and maize facilitate the release of organic acid anions from the root cells. Through the process of creating compounds with organic acids, certain plant species, such as those that accumulate aluminium in their leaves, internally neutralise the toxic effects of aluminium (Ma et al. 2001).

Aluminium (Al), comprising 7% of the earth's crust, is the third most abundant element after oxygen and silicon. Consequently, Al is commonly supplied to plant roots in some manner. Fortunately, most of this element Al resides in the form of inactive oxides and aluminosilicates. Nevertheless, aluminium gets transformed into the perilous trivalent cation, Al^{3+}, when soils undergo acidification due to either natural or human activities. Aluminium toxicity has been recognised as a significant component that restricts plant productivity on around 40% of the world's arable land, which is characterised by acidic soils. Micromolar concentrations of Al^{3+} can rapidly inhibit root growth in certain agriculturally important plant species (Kochian 1995). The ultimate result of the subsequent impacts on nutrient and water acquisition is reduced growth and output. Aluminium exhibits great affinities for oxygen-donor compounds (Martin 1986), enabling it to engage with many locations in the apoplasm and symplasm of root cells. Nevertheless, the specific chemical processes responsible for the toxicity of Al have yet to be determined. Al's strong attraction to these compounds is likely a major factor in its toxicity. Certain plant species have evolved mechanisms to tolerate aluminium stress, allowing them to thrive in acidic soils. The objective of ongoing research in the field of stress physiology is to understand the nature of these tolerance mechanisms. Most of the existing data indicates that specific organic acids, which neutralise Al^{3+} by forming complexes with these ions either inside the cytosol or at the interface between the roots and soil, have a vital function.

4.2.5 The Relationship Between Root Growth and the Harmful Effects of Aluminium

Undoubtedly, soil acidity and aluminium toxicity are among the most detrimental soil variables that significantly affect the growth of most crops. The main signs of aluminium poisoning are the suppression of root elongation and the simultaneous promotion of glucan(-1,3-callose) production. These effects can be observed even after a short period of exposure. Aluminium-induced root damage leads to insufficient absorption of ions and water. Aluminium has been observed to decrease the activity of the mitotic process in numerous plants, resulting in a decrease in the number of cells undergoing proliferation. Potentially, it is feasible to enhance crop output in tropical regions while minimising environmental harm and expenses by genetically adapting plants to tolerate aluminium toxicity. This approach is especially fascinating for maize because there is germplasm available that exhibits tolerance to aluminium, which can be utilised for the purposes of selection and genetic investigation. High quantities of aluminium pose a particularly difficult challenge when it comes to assessing physiological responses. A significant portion of aluminium in the nutritional growth medium can become inactive by precipitation, polymerisation, or complexation. The concentration of free Al that induces toxicity in plant metabolism can be considerably lower when compared with the concentration found in the growth media.

Ca additions are commonly employed in agriculture to reduce acidity and Al toxicity in acidic soils, thus addressing the issue of Al phytotoxicity (Toma and Saigusa 1997; Mora et al. 2002). Several commercially feasible options, such as lime and gypsum or phosphogypsum (PG), are utilised to administer calcium to various crops (Campbell et al. 2006; Takahashi et al. 2006a, b). Carvalho and van Raij (1997) state that the regular addition of these substances to soils is done in order to reduce acidity and improve the accessibility of calcium and magnesium for plants. Although extensive research has been conducted on the impact of different calcium amendments on the physical and chemical properties of acidic soils, there is a lack of information regarding the effectiveness of calcareous amendments on the physiological and biochemical processes of plants. These processes include the absorption of water and nutrients, photosynthesis, and antioxidant systems. This paper aims to present a concise overview of the impact of aluminium toxicity on plants, the interactions between toxic aluminium and different calcium supplements, and how these interactions affect the physiological and biochemical responses of cultivated plants in acidic soils.

One approach to mitigate the harmful effects of acid soil is through the selective breeding of plants that are tolerant to soil acidity. Given that low pH is the main cause of toxicity in acid soils, the most common agricultural approach to address this problem has been the application of lime to the soil. Reports indicate that the utilisation of lime in Western Australia had a significant growth of 57,143 tonnes per year from 2004 to 2010. The addition of lime serves to enhance the formation of root cells, decrease the absorption of aluminium, and enhance cellular protection (Guo et al. 2006; Kinraide 1998). Nevertheless, there are disadvantages associated with this practice (Ma et al. 2001), such as a deficiency in zinc (Zn) and manganese (Mn) (Rao et al. 1993). According to Bose et al. (2011), the inclusion of magnesium has been proposed as a more effective method than lime for lowering aluminium toxicity, as it enhances the release of organic acids. Magnesium, on the other hand, becomes toxic when it exceeds a certain amount (Venkatesan and Jayaganesh 2010).

4.2.6 Strategies for Mitigating the Harmful Effects of Aluminium on Plants

Peripheral devices

External mechanisms that might impede the entry and buildup of Al in cells include root structures including the cell wall, cell membrane, and chemical exudates such as organic acids (Ma et al. 2001), phenolic compounds (Tolra et al. 2009), and phosphates (Pellet et al. 1996). According to Ma (2000), organic acids are the subject of extensive research among the many chemicals generated by cells. Malate exudation (Delhaize et al. 1993a, b, c) and citrate exudation (Ryan et al. 1997) have been found to be linked to tolerance in wheat.

The presence of oxalate secretion in both buckwheat (*Fagopyrum esculentum* M.) and taro (*Colocasia esculenta* [L.] Schott) was identified by Ma and Miyasaka (1998). Esculent buckwheat is an alternative name for buckwheat. The organic acids in question form chelates with aluminium, resulting in the formation of non-toxic complexes known as aluminium organic acids (Matsumoto 2000). This inhibits the interaction between aluminium and the root apices. Prijambada and Proklamasiningsih (2010) conducted a study where they added organic acids externally to investigate their effect on plant growth under aluminium stress. Oxalic acid exhibits a greater ability to form complexes with aluminium compared to citric acid, malic acid, or succinic acid, mostly because of its higher number of carboxyl groups. Succinic acid exhibits the lowest affinity for chelating aluminium. Wheat (Ryan et al. 1995a, b) and tobacco (Nicotiana tabacum) are two plant species that exhibit rapid release of organic acids upon exposure to aluminium (Al) (Delhaize et al. 2001). Both triticale (Triticosecale Wittmack) and soybean showed a delay between the application of the Al treatment and the release of the organic acid, as seen in studies conducted by Ma et al. (2000) and Yang et al. (2001). The exudation of organic acids triggered by Al occurs regardless of protein synthesis, organic acid production, and the total metabolic activity of the cell. There have been reports of other extrinsic factors, such as the impact of cell wall composition and cell membrane. Cultivars of buckwheat that exhibited resistance to aluminium demonstrated a significantly reduced amount of cell wall pectin in comparison with cultivars that were susceptible to aluminium. When subjected to aluminium (Al), a cultivar that was susceptible to the element exhibited an increased level of low-methyl-ester pectins and a decreased level of high-methyl-ester pectins. Yang et al. (2011) found that Al was primarily found in the hemicellulose 1 fraction of cell walls, and the absorption of Arabidopsis decreased when the hemicellulose 1 component was eliminated. Yang et al. (2008) found that Al-tolerant cultivars had higher levels of cell wall polysaccharide compared to Al-sensitive cultivars. This indicates that the Al-tolerant cultivars were able to bind more aluminium in their cell walls. Kim et al. (2010) conducted study that found a correlation between the activity of plasma membrane H^+-ATPase and the inhibition of root development caused by exposure to aluminium.

Internal processes

Internal mechanisms such as Al-chelating in the cytosol, compartmentalisation in the vacuole, Al-binding proteins, and Al-tolerant isoenzymes are employed to chelate aluminium and generate non-toxic substances (Kochian et al. 2004). The internal process that lowers Al toxicity is not well understood due to its complexity and the involvement of numerous chemicals and targets that react to aluminium toxicity (Rounds and Larsen 2008). Watanabe and Osaki (2001) discovered that the melastoma plant has the ability to gather substantial quantities of aluminium in its leaves. Aluminium migrated from the roots to the leaves, where it formed different compounds such as Al-citrate and Al-oxalate complexes. Phenolic chemicals of the flavonoid class can detoxify aluminium (Al) within plant cells. Kidd et al. (2001) found that when maize was treated with Al and Si, phenolic compounds such catechol and quercetin were shown to be produced. The release of these compounds was found to be influenced by the concentration of Al. Our understanding of the internal mechanisms responsible for Al tolerance in plants is still restricted due to the lack of adequate approaches.

4.2.7 Different Forms of Aluminium

Multiple studies have demonstrated the presence of many forms of aluminium in soils, including monomers, polymers, and solid phases. The concentration of these forms is influenced by the extent and duration of hydrolysis of aluminium compounds (Delhaize and Ryan 1995). Mora et al. (2006) reported that Rout et al. (2001) found a significant correlation between low pH and elevated levels of phytotoxic aluminium species. This correlation is linked to the decrease in exchangeable bases in the soil solution. Mora et al. (2006) state that soil acidity is influenced by the application of ammoniacal fertiliser (urea), high winter precipitation leading to the leaching of bases (Na^+, K^+, Ca^{2+}, and Mg^{2+}), and inappropriate agricultural practices. Hede et al. (2001) found that the process of OM mineralisation can enhance the toxicity of Al by releasing hydrogen ions (H^+) into the soil, resulting in a decrease in pH and an increase in the release of Al^{3+}.

4.2.8 Al's Phytotoxic Substance

The intricate chemistry of A1 can be partially attributed to the challenges associated with studying Al-related processes in plants (Martin 1988; Kinraide 1991). Under acidic conditions with a pH of 5, A1 undergoes hydrolysis in a solution, resulting in the formation of the trivalent $A1^{3+}$ species. However, as the pH increases, $A1(OH)_2{}^+$ and $Al(OH)_2{}^+$ species start to evolve. Gibbsite, or solid-phase Al(OH), is present in near-neutral pH conditions, while aluminate, or $Al(OH)^-$, is dominant in alkaline pH conditions. When A1 solutions are partially neutralised with a strong base, a hazardous polynuclear A1 species known as Al, can also be produced (Parker and Bertsch 1992). However, the prevalence and impact of this species on soil toxicity remain uncertain. Due to its preference for acidic surroundings and the toxicity of several trivalent cations to plants, it is widely acknowledged that A13' is the most poisonous species. Despite the difficulty in proving this, around 11 of the previously described monomeric species have been found to be dangerous, according to a study conducted by Kinraide (1991). This connection ultimately determines the level of stress, irrespective of the surrounding circumstances.

4.2.9 General Function of Ca in Plants

The general function of calcium (Ca) in plants is to play a crucial role in several physiological processes. Calcium is a vital nutrient for plants, as stated by White and Broadley (2003). The significance of it lies in its role as a regulator of growth and development (Hepler 2005), and its importance in many metabolic processes/pathways (Plieth 2005). Ca^{2+} is a divalent cation that plays a crucial role in the structure of cell membranes and walls. Furthermore, it aids in the elongation of the roots and stems (White and Brodley 2003). Ryan et al. (1997) reported that a substantial quantity of this mineral is located in the apoplast, where it maintains the vitality of cells and tissues. Researchers have been investigating the vital importance of the cell wall's rigidity and structure since the eighteenth century (Silva et al. 2005; Shukry et al. 2007). Calcium forms ionic and covalent bonds with carboxylates of pectins in the polysaccharide matrix of the cell wall. These bonds are crucial for the growth and texture of plants, fruits, and mature vegetables (Gilroy et al. 1993). Hepler (2005) discusses the role of indol acetic acid (IAA) as a chelator of calcium (Ca) and magnesium (Mg), and its antagonistic interaction with calcium ions (Ca^{2+}). This contact stimulates the growth and elongation of cells. Nevertheless, a plethora of papers provide compelling evidence that the inhibition of cell elongation by Ca does not impede the ability of IAA to enhance cell wall formation. At the cellular level, the interactions between phospholipids and membrane proteins play a crucial role in regulating membrane permeability. These interactions stabilise the lipid bilayer and maintain the structural integrity of cell membranes (Hepler 2005). On the contrary, Ca^{2+} plays a crucial role in both extracellular and intra-cellular signalling (Yocum 2008). It regulates the balance of the chloroplast- and mitochondrial-activating ATP kinases, regulates enzymatic function (Cheng et al. 2002), and produces electrochemical energy (Ryan et al. 1997). Calcium (Ca) serves as a second messenger in several signal transduction pathways within eukaryotic cells (Silva et al. 2005). Calcium (Ca) is modified within cells in response to several stimuli, such as hormones, light, mechanical stress, abiotic and biotic stress, and mechanical disruption (Cheng et al. 2002). The liberation of O_2 from photosystem II (PSII) during water photolysis is the consequence of the Mn redox chemistry, which is triggered by Ca^{2+} (Yocum 2008). The inhibition of this mechanism may occur if Ca^{2+} were substituted with other cations such as K^+, rubidium (Rb^+), or cesium (Cs^+) (Ono et al. 2001).

4.2.9.1 The Role of Ca–Al Interactions in Plant Physiology

Based on common belief, the first reactions of plants to Al^{3+} damage are observed in the roots and result in a decrease in nutrient absorption (Wang et al. 2006a, b). The effects of aluminium on roots have been thoroughly studied, however, there is limited knowledge on the impact of aluminium on leaf structure and the functioning of the photosynthetic apparatus. Multiple studies have shown that exposure to aluminium (Al) can cause various symptoms in leaves, such as necrosis (Zhang et al. 2007), yellowing (Foy 1984), decreased growth (Wang et al. 2006a, b), and delayed maturity (Rout et al. 2001). The alterations were also linked to a decrease in photosynthetic rate (Reich et al. 1994), a reduction in chlorophyll concentration (Wang et al. 2006a, b), and abnormal chloroplast structure (Akaya and Takenaka 2001; Peixoto et al. 2002). The photosynthetic ability of Scots pine (Pinus sylvestris) was reduced and respiration increased due to high Al/Ca ratios, as shown by Reich et al. (1994).

The relevance of Ca in the cell wall is due to its interaction with Ca^{2+}-pectate, which functions as a growth regulator, and its antagonistic association with IAA (Hepler 2005). The cell wall components and intercellular gaps of the apoplast play a critical role as the initial point of interaction between the roots and potentially hazardous aluminium species in the soil solution (Ryan et al. 1997; Rengel and Zhang 2003). Aluminium ions (Al^{3+}) interact with the components of the cell wall in the roots of plants that are growing in acidic soils. Specifically, Al^{3+} binds to Ca-pectate, a substance found in the cell wall. This binding occurs by the attachment of Al^{3+} to carboxyl groups, as seen by Blamey et al. (1997) and Poschenrieder et al. (2008). According to Ma et al. (2007), Al^{3+} has a higher affinity for pectin, a component of cell walls, than Ca^{2+} does. As a result, Al^{3+} displaces Ca^{2+} from negative binding sites in the apoplasm, as stated by Kinraide (1998). Pectin is a major constituent of cellular membranes. Plants undergo biochemical interactions between calcium and aluminium. In higher plants, physiological processes gradually convert oxygen to water, leading to the generation of minimal amounts of reactive oxygen species (ROS) under normal conditions (Scandalios 2002; Khan et al. 2007). The term "ROS" encompasses free radicals such as superoxide anion (O_2^-) and hydroxyl (OH), as well as molecules like H_2O_2, singlet oxygen ($1O_2$), and ozone (O_3). These substances, which are constantly produced by photosynthetic organisms, pose a significant risk (Guo et al. 2004; Khan et al. 2007). Yamamoto et al. (2002) and Mittler et al. (2004) have identified that the main sources of reactive oxygen species (ROS) in plants are organelles that have a high rate of electron flow or oxidising metabolic activity, such as chloroplasts, mitochondria, and peroxisomes. Under different biotic or abiotic environmental conditions, plants generate an increased amount of reactive oxygen species (ROS) (Foyer and Noctor 2005). This can result in oxidative harm to diverse biomolecules such as lipids, proteins, and nucleic acids (Guo et al. 2004, 2006). Tamás et al. (2005) have established a correlation between heavy metals and the augmentation of reactive oxygen species (ROS) production and oxidative stress in plants. In Ma (2005) study, it was discovered that increased toxicity of Al^{3+} in acidic soils, specifically Andisols, leads to elevated generation of reactive oxygen species (ROS) in plant cells. The occurrence of this phenomena is strongly associated with oxidative stress in plants (Ma et al. 2007). Ca^{2+} may enhance plants' ability to withstand various environmental challenges by regulating antioxidant metabolism (Jiang and Huang 2001). ROS sensing signalling pathways encompass the involvement of Ca^{2+} and Ca^{2+}-binding proteins like calmodulin, along with the activation of G proteins and the signalling of phospholipids. This ultimately leads to the accumulation of phosphatidic acid (Foyer and Noctor 2005).

4.2.9.2 Mitigation of Aluminium Toxicity by Calcium Ions

Several studies have extensively established the ability of calcium to mitigate aluminium toxicity in several crops cultivated in acidic soils (Mora et al. 1999, 2002). Based on Akaya and Takenaka's research in 2001, the presence of certain basic cations, specifically Ca^{2+} and Mg^{2+}, can mitigate the negative impacts of aluminium toxicity on the photosynthetic oxygen synthesis of Quercus glauca. Based on previous studies (Alva and Sumner 1988; Mora et al. 1999), the increase in soil pH after the addition of calcium supplements is attributed to the replacement of Al^{3+} and H^+ ions by Ca^{2+} ions from the exchange sites, which then enter the solution.

Soils that have a pH level below 4.8 often contain high concentrations of aluminium and manganese, which can be poisonous. Aluminium toxicity hinders the movement of phosphorus, harms plant roots, and limits their ability to explore.

Additional essential elements and hydration: This phenomenon is most pronounced in soils that have undergone extensive weathering, such as podzolic soils. Plants exposed to aluminium typically exhibit stunted growth, whereas clovers display a pale colouration and have little leaves. Highly delicate plants, particularly lucerne and to a lesser degree certain types of phalaris, exhibit poor health and an extremely slender growth pattern. They may perish within two years of seeding on highly acidic soils. Manganese toxicity is prevalent in soils that have undergone minimal weathering, such as red, black clay loam, and self-mulching soils. Manganese toxicity is exacerbated by arid, high-temperature circumstances or by damp soil with elevated temperatures. Aluminium toxicity is more prevalent than manganese toxicity. It induces chlorosis of leaf edges and overall dwarfing of vulnerable plants. Its prevalence is irregular and dependent on soil moisture and temperature. Lucerne leaves exhibit diverse levels of damage as a result of acidity.

4.2.9.3 Nutrient Deficiencies

When the pH of the soil drops below 4.8, the accessibility of some nutrients is significantly diminished. Aluminium has the ability to immobilise phosphorus in both plants and soil, leading to symptoms that resemble phosphorus shortage. In acid soils with too high quantities of soluble aluminium, the absorption of calcium and magnesium is similarly limited. Molybdenum, a crucial trace element, becomes very insoluble in acid soils, rendering it unavailable for plant use. Molybdenum is essential for the legume nitrogen fixation process, as well as for protein synthesis in grasses. Microbial activity is inhibited in soils with a pH below 4.6. The process of mineralising organic matter and releasing other nutrients is negatively impacted until the soil acidity is rectified by applying significant quantities of lime.

4.2.9.4 The Impact of Soil Acidity on Pastures

Soil acidity is a frequent and immediate cause of low pasture yield. Another factor related to soil acidity is:

- A decline in species that contribute to productivity.
- Proliferation of weeds and invasion of insects.
- Soil erosion refers to the process of the gradual loss of soil due to factors such as water, wind, or human activities.
- Heightened susceptibility to dryland salinity.
- An overall decrease in plant life.
- The discharge of nutrients into water reservoirs.
- Animal production has decreased.
- Decreased microbial activity.
- Decreased agricultural revenue.

Soil acidity can lead to one or, in most situations, a combination of the following issues. The legume component of the pasture is essential as it turns atmospheric nitrogen into a form that plants can use, resulting in high-quality feed for grazing animals. The conversion of atmospheric nitrogen into a form that plants can use relies on the presence of living bacteria, specifically rhizobia, which inhabit and create nodules on the root systems of legumes. These bacteria exhibit host plant specificity and vary in their tolerance to soil acidity. All rhizobia are limited in their growth at low pH values, with the rhizobia that colonise acid-sensitive plants being the most susceptible to acidity. Utilising lime pelleting will provide immediate benefits. Nevertheless, if acid soil conditions are not rectified, the conversion of nitrogen will be significantly diminished after a span of 12 months (Duncan 1999). The resultant pasture will exhibit low productivity, with the occurrence of weed invasion and a subsequent drop in animal output.

4.2.9.5 Factors Influencing Soil pH

Parent material

Factors that affect soil pH include: parent material refers to the underlying geological substance from which soil is formed. The geological components (soil parent material) from which a soil has developed are the main factor that determines its natural pH. After the glaciers receded from Ontario some 10,000–15,000 years ago, the remaining materials exhibited consistent pH levels from the top layer to the underlying bedrock. These sediments were either directly deposited by the glaciers, deposited by meltwaters flowing from the retreating ice front, or settled from the waters of postglacial lakes. The unique characteristics of these glacial habitats led to variations in the composition of the soil, namely in terms of the textures of sand, silt, and clay. In regions such as Tillsonburg, where there was a rapid flow of water resulting from the melting of glaciers, only the larger particles remained while the finer materials were carried away by the water. In regions where the meltwater accumulated and created lakes, such as a substantial portion of Essex, Lambton, and Kent counties, the clay and silt particles settled at the bottom, leading to the formation of soils that are highly enriched with clay. Across extensive regions of Wellington and Waterloo counties, the glacial materials were deposited directly onto the ground with minimal water sifting, leading to the formation of predominantly loam grained soils. The soils of southern Ontario, the Ottawa valley, and the northern clay belt have predominantly developed on limestone-based parent material, as stated in the Soil Fertility Handbook. This substance will lead to a soil pH level of around 8.2. Nevertheless, both natural and human-induced processes have caused acidification, leading to a gradual decrease in soil pH over time. This has resulted in the occurrence of acid soils in certain regions. The mechanisms that acidify soils mostly affect the uppermost layers of soil. Over millennia following the departure of glaciers from Ontario, the surface layers became increasingly acidic as the additional acids interacted with the limestone present in the soil. The soil horizons at deep remain unaltered from the glacial deposits, retaining significant amounts of limestone and exhibiting a somewhat alkaline soil pH. Consequently, the soil profile often displays minor acidity to neutrality near the surface and becomes more alkaline at greater depths. The predominant parent material in most of Northern Ontario consists of acidic igneous geological formations, specifically the Canadian Shield, which leads to the prevalence of acidic soils.

The soil's cation exchange capacity, along with the geological components it originated from, contributes to its ability to resist changes in pH, also known as

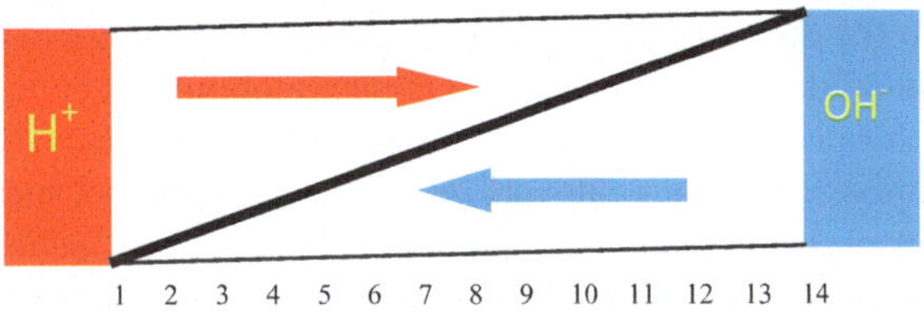

Fig. 4.1 Showing pH range from 1 to 14 with change in H^+ and OH^- ions

buffering capacity. The quantity and composition of clay, as well as the quantity of organic matter in the soil, will influence the rate at which the soil's pH can alter. Therefore, a soil with a rough texture and low organic matter content will have a limited capacity to withstand changes in pH, in contrast to clay soils that contain a significant amount of organic matter. Figure 4.1 illustrates the pH scale ranging from 1 to 14.

4.3 Plant Response to Acidity

4.3.1 Physiological Responses of Root Hairs

Plants adapt to acid soils through various physiological responses, including alterations in root architecture, an increased ratio of roots to shoots, enhanced growth of root hairs, increased interaction with mycorrhizal fungi, and forming mutually beneficial relationships with bacteria that promote plant growth. There are three prominent theories that explain the variations across plant species in their capacity to absorb phosphates, namely: the theories mentioned are the root character theory, the P-requirement theory, and the ionic equilibrium theory (Black 1968). The uptake of phosphorus by plants in soil is affected by the root's uptake properties and the soil's characteristics in supplying phosphorus. The soil's capacity to sustain an adequate phosphorus supply to the roots is contingent upon the phosphorus level in the soil solution. The uptake of P is regulated by the rate at which P is supplied from the soil and the features of root absorption for P (Barber 1978). The absorption of phosphorus in soil is also influenced by the physiological characteristics of roots that determine the kinetics of phosphorus uptake. The deep rooting characteristic of maize hybrids is of greater significance than other root absorption traits for increased phosphorus uptake in acidic soils (Baker et al. 1970). Additionally, variations in phosphorus accumulation properties among maize genotypes are also influenced by root size, the capacity to effectively compete for phosphorus, and the ability to withstand detrimental effects of other elements such as aluminium. The genotypes with high P efficiency exhibited reduced shoot:root ratios and increased dry matter output per unit of P. The genotype that exhibited efficient phosphorus (P) uptake demonstrated lower rates of P uptake, larger root systems, and a stronger ability to transfer P from dormant to active tissue. Similarly, the productive tomato lines generated over

100% more dry matter compared to the unproductive lines, despite the fact that the unproductive lines exhibited greater mobility for phosphorus within the plant under phosphorus stress (Clarkson and Scattergood 1982). The effectiveness of phosphorus uptake in tomato cultivars was affected by factors such as the root surface area per plant, the pace of phosphorus uptake, and the kinetics of the roots, including root hairs (Itah and Barber 1983). The length of the root hairs was strongly correlated with the phosphorus (P) content in both the root and shoot. However, the development of root hairs was not directly influenced by the concentration of P at the surface of the root, but rather by the overall P content of the plant. Additionally, it has been shown that root hairs contribute up to 90% of the total uptake of phosphorus in tomato plants. Genotypic heterogeneity in pigeon pea cultivars has been observed in terms of root growth, shoot growth, and P uptake. In a situation when there is a high supply of phosphorus (P), the presence of root hairs did not have any impact on the uptake of P by the plant or the amount of P present in the plant itself. This was observed in a study conducted on three-week-old Arabidopsis plants by Bates and Lynch (2000). When there was a limited amount of phosphorus available, the normal Arabidopsis plants had a larger overall surface area of roots and a higher concentration of phosphorus per unit length. Therefore, the reaction of root hairs to limited phosphorus availability is a highly effective approach for phosphorus uptake. The strong correlation between root phosphatase activity and pH levels in the rhizosphere suggests that rhizosphere pH plays a crucial role not only in determining the availability of inorganic phosphorus, but also in influencing the activity of root enzymes and the breakdown of organically bound nutrients. In their study, Tadano and Sakai (1991) found that the addition of secretary acid-phosphatases from lupin roots to the rhizosphere of tomato and beetroot plants resulted in an increase in dry matter production. Plants possess many acid-phosphatase isoenzymes in their tissues, including axis, cotyledon, and roots. Certain factors contribute to the recycling of bound phosphorus (P) inside the cytoplasm, while others are released from roots into the rhizosphere to facilitate the conversion of organic-P into a form that can be utilised by plants. The acid-phosphatases that are released can catalyse the hydrolysis of organic-P compounds in soils, such as nucleotides, phospholipids, and sugar phosphates, into inorganic phosphate, which plants can readily take up. It is noteworthy that during phosphorus (P) deficiency, there was an observed rise in the enzymatic activity of phosphatases and phytase in various parts of the plant and in the rhizosphere (Richardson et al. 2009).

4.3.2 Plant Biochemical Reactions

Plants respond to Al toxicity by adapting their metabolism to increase the levels of endogenous and soil inorganic P (Pi). This is achieved through processes such as Pi mobilisation by exudation of organic acids, alternate glycolytic pathways, and Pi transporters. The release of organic molecules from roots is a widespread occurrence. Organic acid metabolism plays a crucial role at the cellular level in several metabolic processes, such as energy production and the creation of building blocks for amino acid biosynthesis. It also has a significant impact on the rhizosphere environment at the whole plant level. The organic acids that build up in the rhizosphere are determined by the specific plant species, the age of the plant, and the kind of tissue. Organic acids can serve as metabolically active solutes to regulate osmotic adjustment and maintain the equilibrium of excess cations. Lately, numerous research have specifically examined the correlation between root exudation and P nutrition. There have been numerous reports indicating that organic acids, which are a significant part of the exudates, have the ability to dissolve organic-P. The study by Lipton et al. (1987) documented the induction of citrate exudation in lucerne under P-starvation conditions utilising a specially developed device for the sterile collection of root exudates. The excretion of organic acids, believed to be a part of the P-starvation rescue system in higher plants, has been found to enhance the acquisition of phosphorus from aluminium phosphate (Lipton et al. 1987). The root secretions are believed to have a role in the development of a crucial adaptive mechanism to cope with phosphate deprivation. This mechanism allows the plant to modify its immediate surroundings and consequently influence the availability of phosphate in the rhizosphere. The proteoid roots of lupin plants that are deficient in phosphorus exude a substantial amount of citric acid, causing the release of a significant quantity of inorganic phosphate from insoluble phosphate ferric hydroxide through the production of variable levels of citrate (Gardner and Parbery 1983).

4.3.3 Plant Molecular Reactions

In recent years, there has been a specific focus on studying how to manage nutrients in acidic soils. This research has primarily been around manipulating plant genotypes to generate varieties that are efficient in phosphorus uptake and tolerant to aluminium toxicity. These manipulations are based on understanding the molecular mechanisms involved in nutrient uptake by plants. Significant progress has been made in comprehending the mechanisms of phosphate transporters, Ca-ATPases, vegetative storage proteins, α-glucosidase, PEPCase, MDH, and PEP phosphatase in relation to improving P uptake efficiency and tolerance to Al toxicity in different crop plants. However, the practical implementation of this knowledge in real-world agricultural settings still requires further refinement. Researchers have shown that roots secrete acid-phosphatases to break down organic phosphate molecules and release orthophosphate in the rhizosphere. This process helps prevent phosphorus shortage stress in acidic soils that lack phosphorus (Richardson et al. 2009). Both plant root and microorganism-derived acid-phosphatases in soil have the ability to catalyse the mineralizstion of organic-P. Tarafdar and Jungk (1987) discovered that the acid-phosphatase activity at the root surface of rapeseed plants was up to 8 times more than in the surrounding soil. They also noticed a significant reduction in organic phosphorus levels in the soil near the root surface. The presence of a significant acid-phosphatase activity was seen in the root compartment of plants with a deficiency in phosphorus, as reported by Li et al. (1997).

Certain plant species exhibit a higher ability than others to flourish in soils with a low pH. Given that a substantial number of soils in the humid tropics possess a low pH (Kamprath and Foy 1985; Maranville et al. 1994), the cultivation of crops that can tolerate low pH in the soil has become a highly significant component of agricultural advancement in those areas. One possible explanation for the success of sugarcane in Hawaii could be its significant ability to withstand low levels of calcium and high levels of aluminium (Yost 2000). Pineapple is known for its ability to tolerate and, in fact, need low soil pH. Hawaii serves as an illustration of a location where crops have been chosen to suit soil conditions instead of modifying soils to suit crops, which is the conventional approach to soil management in numerous temperate climates. Initial research on identifying and measuring the ability of crops to tolerate low pH soils has been conducted in tropical regions (Spain et al. 1975; Silva 1976). Multiple writers have observed that the ability to withstand harmful levels of aluminium (Al) or manganese (Mn) in acidic soils involves both tolerance (the ability to tolerate the toxic element within the plant) and avoidance (preventing the poisonous element from entering the plant). For example, Marschner (1991) discussed this phenomenon.

Maize (*Zea mays* L.) cultivation is hindered by acidic soils, resulting in potential yield decreases of up to 69%. Soils with a low pH, known as acidic soils, can result in the production of toxins due to the presence of aluminium (Al), manganese (Mn), or iron (Fe). Tandzi et al. (2018) found that there is genetic diversity across different types of maize that affects their ability to tolerate low soil pH. This diversity can be used to create maize varieties that are capable of thriving in acidic soil and providing high yields. Although there has been considerable advancement in developing resistance to the toxic effects of aluminium through breeding, there has been no effort in addressing the toxicity of manganese and iron. Consequently, further investigation is necessary to achieve the following goals:

(1) Devise screening techniques for assessing tolerance to manganese and iron toxicities.
(2) Clarify the mechanisms underlying maize tolerance to manganese and iron toxicities.
(3) Create screening methods for evaluating tolerance to other forms of toxicities.
(4) Identify the quantitative trait loci (QTL) in maize cultivars that are responsible for manganese (Mn) and iron (Fe) tolerance.

In order to boost the adoption rate of acid-tolerant maize genotypes, it is necessary to enhance the understanding of farmers and other stakeholders regarding the issue of soil toxicities caused by aluminium (Al), manganese (Mn), and iron (Fe) (Tandzi et al. 2018). This will provide a greater level of acceptance and implementation of the acid-tolerant maize cultivars.

4.3.4 The Influence of Ca on the Toxicity of Al

The association between A1 toxicity and A1–Ca interaction is expected. Preliminary studies discovered that the application of calcium in the form of gypsum (CaSO) or lime (CaCO) decreased the stress caused by A1 and alleviated the symptoms of severe A1 poisoning, which resembled calcium insufficiency, in field conditions. Additional interactions between A1 and Ca can result in stress, despite the fact that the decrease in root growth caused by A1 occurs too rapidly to be attributed to a systemic deficiency of Ca. The vital role of Ca in metabolism is gaining recognition, leading to a surge of research in this area that has resulted in the revival of old notions and the introduction of new ones.

4.3.5 AI's Inhibition of Ca Uptake

Calcium (Ca) is carried into cells through passive mechanisms, and it is probable that membrane-spanning channels play a role in facilitating this process. Studies have demonstrated that A1 reduces the absorption and movement of Ca in plants, which aligns with the observation that certain polyvalent cations including La^{3+}, Ga^{3+}, and Gd^{3+} restrict the transport of Ca (Huang et al. 1992; Rengel 1992a). Given the crucial role that calcium (Ca) plays in metabolism, it is reasonable to argue that AI toxicity is directly linked to this antagonistic interaction. The discovery that A1 significantly reduced calcium absorption in wheat lines sensitive to aluminium more than in aluminium-tolerant lines (Huang et al. 1992) further supported the notion. These studies have established a correlation between A1 toxicity and the inhibition of calcium uptake, which meets the criteria for being the primary cause of toxicity. This correlation is supported by the fact that the effect is measurable within minutes, it affects the root apex (which is the critical site for toxicity), and it is consistent with the long-term symptoms of calcium deficiency. Although first promising, it was ultimately proven that this idea was inadequate in explaining all cases of A1 toxicity. In a recent study conducted by Ryan et al. (1994), it was shown that the application of A1 to wheat seedlings enhanced root growth but also restricted the absorption of calcium. Additionally, it was observed that low doses of A1 could potentially reduce root growth without affecting the uptake of calcium. If the inhibition of Ca transport is caused by A1 toxicity, one would expect the opposite outcome. The inhibition of root growth by low levels of A1 appears to be caused by additional interactions, while certain amounts of A1 hinder the absorption of Ca and may contribute to A1-induced stress.

4.3.6 Apoplasmic Ca Displacement by Al

The apoplasm contains the largest portion of calcium in root tissue, which is essential for promoting cell proliferation and maintaining the stability of cell membranes (Kauss 1987). A1 can displace apoplasmic Ca by either competing for ligands (Rengel 1992a) or by reducing the negative potential difference on the membrane surface (Kinraide et al. 1992). Thus, a counterargument to the notion outlined above is that the toxicity of A1 is caused by the displacement of Ca from crucial binding sites in the apoplasm. This idea proposes a rapid interaction between A1 and Ca, which elucidates the phytotoxicity of numerous different cations. However, the Ca-displacement hypothesis does have several theoretical concerns. According to Kinraide et al. (1992), short-term growth studies have demonstrated that Mgzi, Sr^{2+}, and Ca^{2+} are similarly effective in lowering A1 toxicity. Various cations, including as protons and trivalent cations, actually alleviate A1 stress through a process that is not connected to changes in ionic strength. These data contradict the Ca-displacement hypothesis since adding additional cations to an A1 solution is more likely to decrease the calcium content of the apoplasm rather than raise it. In addition, in contrast to the expected outcome based on this idea, the growth of roots does not align with the anticipated level of Ca activity at the surface of the membrane (Kinraide et al. 1994). These points underscore the necessity for further investigation, despite not disproving the Ca-displacement notion.

4.3.7 Impact of Al

The impacts of aluminium: Aluminium can affect the calmodulin in the symplast, the pectin matrix in the apoplast, and the DNA in plant root cells (Sade et al. 2016). Multiple research indicate that the majority of plants are toxic to aluminium (Sade et al. 2016). Aluminium commonly forms bonds with phosphorus (P) in soils and plant roots, resulting in reduced accessibility and solubility of P, leading to a deficiency of P for plant growth.

Simoes et al. (2012) found that Al toxicity symptoms in plants may be categorised into two types of disorders: long-term reactions that take hours to develop and short-term responses that can be observed within minutes or even seconds after exposure to Al. After a duration of one hour, the initial manifestations of these aluminium-related responses became apparent (Ownby and Popham 1989). Furthermore, as stated by Panda and Matsumoto (2007), the primary response to the injection of Al is the temporary inhibition of Ca influx through the plasma membrane. The perilous ramifications are, essentially:

- Root growth suppression.
- Limitations on the assimilation of nutrients and water.
- Lipid peroxidation.
- The alteration of the cytoskeleton and the suppression of cell division.

Aluminium and other metals present a significant danger to humans, animals, and plants due to their non-biodegradable nature, long-lasting presence in the environment, and ability to accumulate in the food chain (Exley 2003; Ashenef 2014). Tea plants contain a substantial quantity of aluminium in their leaves, which is why it is present in tea leaf infusions. While the gastrointestinal tract can only absorb a small quantity of aluminium and the kidneys are effective at eliminating it, aluminium nevertheless presents significant risks to human health (Exley 2003; Mehra and Baker 2007).

Aluminium (Al) is a highly abundant element in the earth's crust and can be found everywhere in the soil. However, it is mostly present in forms that are not easily accessible to plants when the soil pH is neutral or close to neutral. In acidic soil conditions, with a pH below 5, aluminium becomes soluble and takes on forms that are harmful to plants. This negatively impacts the growth, strength, and productivity of plant roots, leading to a significant reduction in crop yield ranging from 25 to 80%. Legumes are crucial for both human nutrition and the overall functioning of the ecosystem. Legumes are significantly impacted by aluminium toxicity due to their high sensitivity to soil acidity. This poses a significant challenge for legume production on acidic soils (Mahanta et al. 2022).

Tropical and subtropical regions of the world are mostly characterised by the prevalence of acidic soils. The initial and most significant sign of aluminium toxicity is the slowing of root growth. This is then followed by other alterations, such as changes in cell division and DNA replication at the tip of the root, as well as the occurrence of severe oxidative stress, lipid peroxidation, and the buildup of callose at the tips of the roots. Changes in rhizosphere pH and chelation of aluminium through the exudation of organic acids are important strategies for tolerating aluminium toxicity. The most accepted and reliable ways of screening plants in hydroponics culture are evaluation based on root staining and root-regrowth criteria. Seedling root architectural (SRA) features are becoming increasingly important in technology-based phenotyping because of their enhanced precision and accuracy. Multiple significant genes and pathways have been discovered through diverse gene expression investigations. The genes phosphoenolpyruvate carboxylase (PEPC), citrate synthase (CS), isocitrate dehydrogenase (ICDH), malate dehydrogenase (MDH), and ion-transporters such as voltage-dependent anion channel (VDAC), Al-activated malate transporter (ALMT), and their homologues are the primary genes linked to malate/citrate-mediated aluminium detoxification in legumes (Mahanta et al. 2022).

Legumes, especially pulses, are a crucial part of the human diet as they provide substantial amounts of protein, carbs, and many vitamins and minerals. Legumes enhance soil fertility and ecosystem functioning by utilising their ability to fix atmospheric nitrogen through biological nitrogen fixation in symbiotic association with rhizobium bacteria. As a result, they play a crucial role in various cropping systems. Legumes cover around 78.3 million hectares of land and contribute to more than 35% of the global protein consumption (Werner and Newton 2005). Acid soils, referred to as Ultisols or Oxisols, are prevalent in tropical and subtropical locations worldwide, characterised by a pH of 5.5 or below (Sade et al. 2016). Aluminium (Al), as one of the most plentiful metals in the earth's crust, is recognised as the primary abiotic stress factor responsible for yield losses of 25–80% in crops grown in soils with high aluminium content (Singh et al. 2012). When the pH of acidic soils falls below 5, aluminium dissolves into harmful forms that are poisonous to plants, leading to inhibited root growth and lower vigour and production in agricultural plants. Aluminium stress is a major constraint on crop productivity in acidic soils. In fact, 67% of the world's acidic soils are affected by aluminium toxicity (Lin et al. 2012). The presence of elevated levels of heavy metals, such as aluminium, in soil hampers the functioning of soil microbes, resulting in a decline in soil organic matter and subsequent degradation in soil fertility (Yaman and Kumar 2021). Aluminium (Al) is classified as a non-essential plant nutrient and its beneficial effects are limited to only a few plant species (Bojórquez-Quintal et al. 2017). The majority of aluminium (Al) remains inaccessible in near-neutral soil pH as insoluble alumina silicates or oxides, which are typically not exposed to plants (Kochian et al. 2004). As the soil pH decreases and becomes more acidic, aluminium undergoes a conversion into harmful forms in the soil solution, which negatively impacts the growth and development of plant systems. Aluminium exhibits distinct behaviour in soil due to its reduced ionic radius and increased ionic charge, in contrast to other soluble metals. Research on root growth parameters, namely cell division and elongation, has shown a decrease in mitotic activity in the root tips when exposed to hazardous levels of aluminium. This effect has been reported in several species such as wheat, maize, and faba beans (Li et al. 2008; Frantzios et al. 2001). Reactive oxygen species (ROS) formation can result in the oxidation of many biological components, ultimately causing cell death. Moreover, the attachment of aluminium to cellular membranes induces stiffening (Jones and Kochian 1997), triggering a cascade of events involving metal ions, particularly iron, that hastens the process of lipid peroxidation (Yamamoto et al. 2003). Aluminium-induced lipid peroxidation has been observed in various legumes, including pea (Yamamoto et al. 2001), soybean (Cakmak and Horst 1991), greengram (Panda et al. 2003), lentil (Singh et al. 2016), and chickpea (Choudhary and Sharma 2014). Yamamoto et al. (2001) found that callose production, lipid peroxidation, and aluminium accumulation were evenly distributed on the surface of the root apex in pea seedlings treated with aluminium, leading to inhibition of root growth. Plants possess a diverse range of enzymatic antioxidants in their defence system, which function to detoxify reactive oxygen species (ROS). The presence of reactive oxygen species (ROS) in the root apex of pea plants was observed two hours after being subjected to aluminium stress. The levels of ROS further intensified as the exposure period rose (Yamamato et al. 2003). The apoplast of the peripheral cells of roots in plants is the primary and main location for the accumulation of aluminium (Al) (Guo et al. 2004; Liu et al. 2008), from where it is transported to other locations. The presence of elevated levels of heavy metals, such as aluminium, in soil has a detrimental effect on the activity of soil microorganisms. This, in turn, results in a loss in soil organic matter and ultimately leads to a decline in soil fertility (Yaman and Kumar 2021).

Aluminium toxicity is the primary factor that restricts plant growth in acidic soils worldwide (Wright et al. 1989). It is mainly seen in soils with a pH below 5.0 (Woolhouse 1983), however there have been reports of its occurrence in soils with a pH as high as 5.5 (Foy 1974). This metal constitutes roughly 7.5% of the Earth's crust by weight (Haug 1984). The acidic soil has mostly evolved through the process of nitrification of ammonium ions, which has been caused by the continuous application of fertilisers that produce acid, as stated by McNeilly (1994). In non-acidic soils with a pH greater than 5, aluminium is mainly present as insoluble oxide and complex aluminosilicates. As the soil pH lowers, there is a release of Al^{3+} ions and a decrease in the availability of exchangeable cations such as Ca^{2+}, Mg^{2+}, and K^{+}. At micromolar concentrations, ionic aluminium becomes toxic to plants (Parker et al. 1989).

The reduction in soil acidity hampers cell elongation, negatively impacting the growth of meristematic tissues in the root tip where water and nutrient absorption is most active.

The toxicity of aluminium is influenced by the characteristics and amount of surrounding substances. When muriate of potash, which is the salt of muriatic acid, is applied to acidic soils, the chloride ions from potassium chloride (KCl) will intensify the poisonous impact of aluminium (Al). Aluminium accumulates on the surface of plant roots and impairs their permeability. The permeability of the root will be negatively impacted, leading to the accumulation of the substance in the cortex, especially in the nuclei and cell protoplasm. This substance has a colloidal form. If the accumulation occurs, the protoplasm exhibits colloidal characteristics, which in turn hinders cell division or cell development, resulting in the formation of binucleate cells. Under acidic conditions, manganese (Mn) exhibits behaviour similar to aluminium (Al). As the concentration of the solution increases, the pH lowers. Plants can tolerate manganese (Mn) concentrations ranging from 1 to 4 μg per ml. However, if the Mn concentration exceeds this range, the plant will experience toxicity. In field conditions, Mn toxicity is not typically observed. However, in a pot experiment, Mn toxicity occurs due to the steaming effect, which results in an increased amount of Mn in the soil solution. Consequently, plants can suffer from Mn toxicity. The tolerance of manganese toxicity varies among different crops, with paddy rice being able to tolerate higher concentrations of manganese.

4.4 Nutrient Economy in Acidic Soils

4.4.1 Nitrogen Economy

Plants primarily acquire nitrogen by the uptake of NO_3^- ions, which is contingent upon the action of nitrifying bacteria. The nitrification process is most efficient when the pH level ranges from 6.5 to 7.5. Nitrogen is found in an inorganic form at a concentration of 2%, while the remaining 98% is present in an organic form. Mineralisation process is hindered and nitrogen reduction is prevented under acidic conditions. Azotobacter, a type of nitrogen-fixing bacteria, is unable to operate effectively when the pH level drops below 6.0. The availability reduces as a result of denitrification loss.

4.4.2 *Phosphorus Economy*

The availability of P is maximised when the response is between 6.5 and 7.5. Under highly acidic conditions (with a pH of 5.0 or lower), iron (Fe), aluminium (Al), and manganese (Mn) exist in a soluble form. These phosphates (Fe, Al, and Mn) react with water to generate insoluble phosphates.

$$Al^{3+} + H_2PO_4^- + 2H_2O \leftrightarrow 2H^+ + Al(OH)_2 * H_2PO_4 \text{ (Insoluble)}$$

The phosphate undergoes a chemical reaction with hydrated oxides of iron and aluminium, resulting in the formation of insoluble hydroxy phosphate compounds of iron and aluminium. The phenomenon of phosphorous being unavailable is referred to as P-fixation. The monovalent $H_2PO_4^-$ ions are the most abundant in extremely acidic solutions with a pH range of 4.0–5.0.

4.4.3 *Potassium Economy*

Potassium (K) is depleted from acidic soil due to the process of leaching. The lack of potassium (K) is caused by the transformation of exchangeable potassium into non-exchangeable potassium. The plant does not convert it into any organic form, resulting in a decrease in the availability of potassium (K). In acidic soils, the availability of calcium (Ca), magnesium (Mg), and sulphur (S) also declines. The presence of Fe, Mn, and Al components in acidic soil can have a harmful impact on plants. The abundance of Mo increases as soil pH increases.

4.5 Toxification of Acid Soils

Bian et al. (2013) found that acid soil toxicity is caused by low soil pH, deficiencies in essential nutrients (phosphorus, magnesium, calcium, potassium, and sodium), and the presence of toxic heavy metals (iron, copper, manganese, zinc, and aluminium). Consequently, the presence of elevated amounts of aluminium, iron, and manganese can be attributed to a low soil pH, which impedes the growth of crops (Zeigler et al. 1995). High levels of aluminium and iron oxides, hydroxide, and low soil pH immobilise phosphorus, making it unavailable to plants (Kisinyo et al. 2013). When breeding maize for global low soil pH tolerance, it is important to include the elements Al, Mn, and Fe.

4.5.1 Harmful Effects of Aluminium

The restriction of crop growth to their maximum potential leads to a decrease in agricultural output due to the presence of hazardous substances (Kochian et al. 2015). Barabasz et al. (2002) found that the presence of aluminium toxicity, at concentrations ranging from 60 to 300 grams per litre of water in soil, can lead to a significant decrease in agricultural yields. This reduction can range from 25 to 80% in various crop plants. The presence of aluminium (Al) inhibits the absorption of nitrogen (N), phosphorus (P), and potassium (K), essential nutrients required for stimulating root development (Osaki et al. 1997). The presence of high levels of aluminium toxicity in the subsoil reduces the depth of plant roots, increases their vulnerability to drought, and hinders the absorption of nutrients from the subsurface (Ulrich et al. 1980). The adverse effects of Al result in root damage, which hinders a plant's capacity to take up nutrients, resulting in nutritional inadequacy (Steiner et al. 2012). The deficiency of phosphorus due to aluminium toxicity results in inhibited plant growth, slender and weak stems, and leaves with a purplish hue, leading to a decrease in grain production (Ouma et al. 2012). In order to evaluate the risk of plant production in acidic soils, it is necessary to determine the quantity of available aluminium (both exchangeable and in the soil solution). Although acidic soils have mostly been focused on in regard to Al toxicity, there has been comparatively little emphasis given to Fe and Mn toxicities. Iron is the fourth most abundant mineral in the Earth's crust, following oxygen (O_2), silicon (Si), and aluminium (Al). The presence of high concentrations of ferrous iron (Fe^{2+}) in the soil solution, particularly in flooded soils, is associated with the condition known as iron toxicity (Becker and Asch 2005). The hydrolysis of Fe is more acidic than the hydrolysis of Al. Al hydrolysis mechanisms usually function as a means of mitigating acidity caused by Fe toxicity. Once most of the aluminium ions in the soil have undergone a reaction, the process of iron hydrolysis becomes dominant. This leads to a significant reduction in the soil's pH (Havlin et al. 2005). Under conditions of low soil pH, anaerobic bacteria generate excessive amounts of ferrous ions, posing a threat to plants. Inadequately oxygenated or densely packed acidic soils can lead to an increase in iron concentration to hazardous levels.

Soils can be classified as A1 hazardous because many plant species are sensitive to even small amounts (micromolar levels) of aluminium (Al). Fortunately, most of the A1 is either bound by ligands or found in non-phytotoxic forms such as aluminosilicates and precipitates (Delhaize and Ryan 1995). Nevertheless, a low pH enhances the solubility of A1, and the toxicity of A1 is a substantial constraint on plant growth in acidic soils. Acid rain and different farming practices can accelerate soil acidification, which can naturally happen when basic cations are washed off from soils (Kennedy 1992). Two strategies for maintaining productivity on acidic soils are using plants that are well-suited to these conditions and applying lime to increase the soil's pH. Although A1 toxicity in acid soils has been acknowledged as a concern for over 70 years, our comprehension of the fundamental locations of toxicity and the sequence of processes that eventually hinder plant growth remains mostly speculative.

4.5.2 Indicators of Aluminium Toxicity

Root growth inhibition is the most easily recognisable indication of A1 toxicity and is currently employed as a standard marker for A1-induced stress in plants. Within a span of 60 min, the introduction of micromolar quantities of A1 might hinder the growth of roots in basic nutrient solutions. However, the overall suppression of growth does not provide significant information about the stressors that will occur either prior to or concurrently with changes in growth. It is essential to identify the primary anatomical and metabolic locations implicated in order to comprehend the mechanisms of A1 toxicity. It is important to note that A1 can have various outcomes and exhibit varied behaviour depending on the species. This occurs due to the hydrolysis of A1 in the solution. The presence of gibbsite, or solid-phase $Al(OH)_3$, indicates a near-neutral pH, while an alkaline pH is characterised by the predominance of $Al(OH)_4^-$, or aluminate. These monomeric A1 cations interact with various chemical and inorganic ligands, such as PO_4^{3-}, SO_4^{2-}, F^-, organic acids, proteins, and lipids. The concentrations of the monomeric A1 species and different A1 molecules in solution can be predicted using available equilibrium constants for these processes. When A1 solutions are partially neutralised with a strong base, a highly toxic polynuclear A1 species called Al13 can also form (Parker and Bertsch 1992), although its natural prevalence and contribution to soil toxicity remain uncertain.

4.5.3 Aluminium Toxicity Specifically Affects the Root Apices

Following a 2-h period of being exposed to aluminium, Al, Bennet and Breen (1991) observe several changes in the microscopic composition of cap cells in maize roots. The researchers postulated that the presence of aluminium (Al) would indirectly inhibit the growth of roots by activating a signal-response circuit that involves the root cap, hormones, and secondary messengers. Aluminium toxicity hampers root growth, whereas nitrogen is a crucial nutrient for plant growth and development. The findings demonstrated that the presence of N intensified the suppression of root extension and the buildup of aluminium in the root. The genotype that is susceptible to Al experienced more severe Al toxicity compared to the genotype that is tolerant to Al (Ma et al. 2022). The histochemical observation clearly demonstrated that aluminium has a preference for binding to the root apex in the 7–10 mm zones. Additionally, the genotype that is susceptible to aluminium accumulated a higher amount of aluminium in these specific zones.

Regrettably, a majority of over 60% of acidic soils are situated in developing nations, and these soils play a crucial role in food production (Liu et al. 2014). Aluminium (Al) is the most abundant metallic element found in the Earth's crust. It often occurs in the form of non-toxic aluminosilicates and oxides in neutral soils. Nevertheless, in soils with a pH lower than 5, poisonous aluminium ions (Al^{3+}) are dissolved into the soil solution and directly harm root systems, leading to a substantial decrease in global crop production (Kochian et al. 2015; Liu et al. 2014). The primary manifestation of aluminium toxicity is the suppression of root development, as aluminium predominantly accumulates in the root system (Sun et al. 2020). The scientific literature has extensively recorded that the root apex serves as both the primary location for perceiving and responding to aluminium, as well as the destination for aluminium accumulation (Kollmeier et al. 2001). The strong attraction to the cell wall of the root apex leads to various negative effects, including the separation of the plasma membrane, disruption of cellular signalling, and excessive formation of reactive oxygen species (ROS) (Ma 2007). These drawbacks alter the composition of the cell wall and disrupt its organisation, thereby diminishing its flexibility and malleability. This is the underlying cause for impeding the elongation of root cells (Horst et al. 2010). In addition, Liu et al. (2018) found that the changes in reactive oxygen species (ROS) caused by aluminium (Al) are localised to certain locations, with a noticeable decrease observed from the tip to the base of the root. Plants have developed many methods to deal with aluminium stress in order to sustain satisfactory growth and productivity (Kochian et al. 2015). One way that plants tolerate aluminium is by producing organic acid anions from the root apex, which then form a stable complex with aluminium. This complex helps to reduce the toxicity of aluminium (Ryan et al. 2011). Another mechanism that provides aluminium tolerance is the augmentation of antioxidative defence capacities (Inostroza-Blancheteau et al. 2012). There is increasing evidence that Al stress can modify the functioning of enzymes involved in the removal of reactive oxygen species (ROS) (Sun et al. 2020; Ma et al. 2007). Nitrogen is a vital nutrient for the growth and development of plants; the impact of nitrate on root development has been extensively researched. The presence of nitrate has been found to hinder the growth of primary roots, while having the reverse impact on lateral roots (Celis-Arámburo et al. 2011). Various crucial genes associated with nitrate signalling pathways have been discovered, such as the nitrate sensor, transcription factors, protein kinases, and molecular components. In a recent study, Chu et al. (2021) discovered a new transcription factor called HBI1. This transcription factor plays a crucial role in regulating the transmission of nitrate signals by maintaining the balance of reactive oxygen species (ROS). Additionally, it was discovered that the application of nitrate reduces the formation of H_2O_2, which in turn hinders nitrate signalling. This creates a feedback regulatory loop that controls the development of plant roots. A prior investigation has also documented that nitrate has the ability to impede the growth of the main root by controlling the generation of reactive oxygen species (ROS) in the tips of the root (Zang et al. 2020). Aluminium (Al) and nitrogen (N) are crucial elements that significantly influence root growth, ultimately impacting the yield of crops. The release of OH^- from roots leads to an increase in the number of negatively charged sites on the root surface. This increase in negative charge allows for the binding of Al^{3+} ions. Additionally, the release of OH^- also leads to an increase in the presence of pectin and hemicellulose, which are both rich in negatively charged functional groups such as COO^- and –OH. These functional groups have a strong ability to bind positively charged Al^{3+} ions. Typically, root tips of 0–10 mm in length are utilised to study the negative effects of aluminium on the growth of plant roots (Sun et al. 2016). Nevertheless, the impact of N on the toxicity and accumulation of Al in root tips remains uncertain. This study aimed to examine the impact of nitrogen (N) on aluminium (Al) toxicity in the root growth and Al accumulation in root tips of two wheat genotypes with varying levels of Al tolerance. This was achieved by analysing the root phenotype, histochemical staining, reactive oxygen species (ROS), antioxidant enzyme activity, and cell wall fractions in root tips.

4.5.4 The Solubility of Al in Soils

Aluminium can be found in several mineral forms in soils, such as hydrous oxides, aluminosilicates, sulphates, and phosphates. In acidic soils, a significant amount of the cation exchange capacity is fulfilled by aluminium ions, indicating that the soils have a high aluminium saturation. The Al ions, known as exchangeable Al_3C, consist of a combination of monomeric Al ions [Al_3C, $AlOH_2C$, $Al(OH)_2C$]. On average, each Al ion has a charge between 2 and 3, which decreases as the pH level increases (Wild 1988). Aluminium strongly binds to exchange sites, leading to low concentrations in soil solution, typically ranging from 10 to 250_M (Adams and Moore 1983; Kamprath 1978). Due to the pH-dependency of Al solubility, the addition of lime to an acidic soil causes the formation of hydroxy-Al species, resulting in the precipitation of exchangeable and soluble Al. The monomeric species $AlOH_2C$ and $Al(OH)_2C$, which have a positive charge, can undergo polymerisation to create both large and tiny polynuclear complexes with a positive charge. These complexes then become attached to the surfaces of clay minerals and organic matter (Stole et al. 1976). Therefore, raising the pH of an acidic soil to a level above approximately 5.5 significantly reduces the concentrations of soluble and exchangeable aluminium to negligible levels, thereby eliminating the limitation of aluminium toxicity on crop development. This review explores the use of organic residues as an alternative to liming for lowering the concentration of phytotoxic aluminium in soil solution. An alternate method has the potential to decrease the quantity of lime needed to achieve sufficient agricultural yields in acidic soils.

4.5.5 Soil Aluminium Phytotoxicity

Aluminium toxicity is typically the primary factor that limits crop growth in acidic soils (Foy 1988). The primary and easily noticeable indication is the suppression of root development, resulting in stunted and underdeveloped roots with limited growth of the main root and hindered production of side roots (Foy 1988). Root growth is inhibited by the obstruction of both the elongation and division of cells

(Kochian 1995). Due to limited root growth, the plant's capacity to search for nutrients and water in the soil is significantly diminished. Consequently, plants experiencing aluminium toxicity often encounter nutritional and/or water deficiencies. Aluminium toxicity hinders the active ion uptake mechanisms that occur across the plasma membrane of root cells (Kochian 1995; Wright 1989). Specifically, there have been numerous reports of Al-induced decreases in the absorption of calcium (Ca) and phosphorus (P) (Huang et al. 1992; Mugwira 1980). The Al/P interaction mechanism is believed to involve an adsorption-precipitation process between aluminium (Al) and phosphorus (P) at the surface of the root or within the cell walls of the root free space (Naidoo et al. 1978). The P can either be absorbed by hydroxy-Al that has already precipitated in the root free space or it can be precipitated as insoluble aluminium phosphates. Both of these possibilities are highly probable. Plants experiencing aluminium poisoning often exhibit symptoms of phosphorus deficiency (Foy 1988; Haynes 1984). Therefore, the improvement of aluminium toxicity through liming typically leads to a significant increase in phosphorus uptake by plants, despite the fact that the supply of soil phosphorus may remain the same or even decrease (Haynes 1982). Although both exchangeable Al and Al saturation have been utilised as indicators of possible Al toxicity (McCray and Sumner 1990), the concentration of Al in soil solution provides a more direct assessment of the conditions experienced by plant roots. It has been discovered that the concentration of aluminium in the soil solution is strongly correlated with the inhibition of plant growth in acidic soils (Adams and Moore 1983; Evans and Kamprath 1970). However, the inhibition of growth is not solely determined by the amount of aluminium available in the soil solution, but also by the specific types of soluble aluminium that are present. The current consensus is that the presence of inorganic and organic complexing ions in soil solution leads to a decrease in Al toxicity (Hue et al. 1986). The activity of Al_3C and/or monomeric hydroxy-Al species is strongly associated with reduced yields caused by Al toxicity (Wright 1989). Blamey et al. (1983) proposed that either Al_3C or $Al(OH)_2C$ could be the main ions accountable for the reduction in root elongation of soybean [Glycine max (L.) Merr]. Nevertheless, both Alva et al. (1986) and Kinraide and Parker (1990) discovered that plant development exhibited greater sensitivity to $Al(OH)_2C$ and $Al(OH)_2C$ compared to Al_3C. It is believed that the presence of HC at low pH levels, where Al_3C is dominant, helps reduce the harmful effects of Al_3C on plant growth by competing with Al_3C at the plasma membrane of root cells (Kochian 1995). When the pH is higher, the presence of monomeric hydroxy-Al species is more dominant, resulting in a significant decrease in the activity of HC. As a result, the competitive effects of HC are nullified. Aluminium has the ability to create stable complexes in the soil solution with soluble humic molecules and simple aliphatic organic acids (Ritchie 1989). The subsequent sections provide a detailed discussion on the characteristics of these complexes. According to Kinraide (1991), aluminium in a form that is complexed to soluble organic matter (OM) does not have a harmful effect on plants. Additionally, the process of complexing aluminium with OM seems to be a crucial mechanism for detoxifying aluminium in the soil solution, as stated by Kochian (1995) and Ritchie (1989). Organic matter plays a crucial role in binding with aluminium (Al) during the process of breaking down plant and animal garbage. As the debris decomposes, a variety of organic compounds are released or created by the microbes responsible for decomposition. Aluminium has a high affinity for many of these compounds. Soil organic matter forms complexes with aluminium and other polyvalent cations, which can be classified into two primary groups (Stevenson and Vance 1989). The compounds can be categorised into two groups: (i) distinct biological substances, such as basic organic acids, phenols, phenolic acids, hydroxamate siderophores, sugar acids, and polymeric phenols, and (ii) intricate humic substances. The precise significance of individual biochemical components compared to humic substances in the process of binding aluminium is uncertain and can differ depending on the characteristics of the soil and the surrounding environment. Humic compounds are usually believed to play a significant role in lowering the phytotoxicity of aluminium in acidic soils, according to Harper et al. (1995) and Wong et al. (1995).

The issue of aluminium phytotoxicity is a significant concern in agricultural regions characterised by low soil pH (Foy 1984; Brown et al. 2007). These issues are worsened by the susceptibility of numerous crops to micromolar levels of Al in solution and the variability in Al tolerance among different plant species (Delhaize and Ryan 1995; Reddy et al. 1995). The presence of Al phytotoxicity in acidic soils is determined by the level of toxic forms of Al in the soil solution. This level is influenced by the complexation of Al, the control of Al solubility by solid-phase factors, and the characteristics of the soil environment (Kinraide 1997). The majority of aluminium (Al) in the soil solution is bound to both inorganic and organic ligands, which modifies the ratio and effectiveness of Al forms that are harmful to plants (Berggren and Mulder 1995). In acid soils, the activity of Al^{3+} (Aluminium ion) is a more accurate indicator of potential harm to plants than measurements of total soluble or exchangeable forms of aluminium. However, it should be noted that Al^{3+} may not be the only poisonous form of aluminium (Shann and Bertsch 1993). Lime raises the pH of the soil, enhances the growth of crops (Arshad and Gill 1996), and reduces the amount of extractable Al^{3+} (Wildey 2003) as well as Al^{3+} in DS systems. In the Inland Pacific Northwest (IPNW), lime is infrequently utilised due to its relatively expensive price and research indicating minimal or non-existent increase in crop productivity (Bezdicek et al. 2003; Wildey 2003). Implementing alternative application strategies, such as placing lime in a concentrated area below the row during seeding, could potentially reduce the amount of lime needed and address economic limitations caused by high application rates. However, the impact of this practice on soil pH and crop yield has not been thoroughly investigated. According to Delhaize and Ryan (1995), the level of aluminium activity is a more dependable measure of the possibility of aluminium phytotoxicity compared to the quantity of aluminium.

Aluminium phytotoxicity is a significant issue in agricultural regions characterised by low soil pH (Foy 1984; Brown et al. 2007). These problems are worsened by the susceptibility of numerous crops to micromolar levels of aluminium in a liquid solution and the differences in aluminium tolerance among various plant species (Delhaize and Ryan 1995; Reddy et al. 1995). The presence of Al phytotoxicity in acid soils is determined by the level of toxic forms of Al in the solution, which is influenced by the complexation of Al, the control of Al solubility by solid-phase, and the characteristics of the soil environment (Kinraide 1997). The majority of aluminium (Al) in the soil solution is bound to inorganic and organic ligands, which modifies the ratio and effectiveness of phytotoxic forms of Al (Berggren and Mulder 1995). In acid soils, the activity of Al^{3+} (aluminium ion) is a more accurate indicator of potential harm to plants than measures of total soluble or exchangeable forms of aluminium. However, it should be noted that Al^{3+} may not be the only poisonous form of aluminium (Shann and Bertsch 1993). Lime raises the pH of the soil, enhances the growth of crops (Arshad and Gill 1996), and reduces the amount of extractable Al^{3+} (Wildey 2003) as well as Al^{3+} in DS systems. In the Inland Pacific Northwest (IPNW), lime is infrequently utilised due to its comparatively expensive price and research indicating minimal or non-existent increase in crop productivity (Bezdicek et al. 2003; Wildey 2003). Utilising alternative application strategies, such as placing lime in a concentrated area below the row during seeding, may enable the use of lower lime rates. This can help overcome economic limitations caused by high application rates. However, the impact of this practice on soil pH and crop yield has not been thoroughly investigated. According to Delhaize and Ryan (1995), aluminium activity is a more dependable measure of potential aluminium phytotoxicity compared to aluminium concentration.

4.5.6 Interactions with Humic Substances

Characteristics of the products

Humic substances constitute the majority (70–80%) of the organic matter in soil and are composed of large organic molecules with high molecular weight. These molecules are primarily derived from the decomposition of plant and animal residues, as well as the metabolic processes of microorganisms (Stevenson 1994). They are present as diverse, intricate, three-dimensional formless formations. Humic compounds have the ability to create complexes with polyvalent metal ions, such as Al, since they possess a remarkably high quantity of O-containing functional groups, including COOH, phenolic, enolic, alcoholic OH, and C=O. Because humic substances are made up of different types of molecules, complexation can be thought of as happening at many different places where the strength of the binding can vary from mild ionic forces to the development of stable coordinate linkages (Stevenson and Vance 1989). The interactions of aluminium with organic matter are intricate and likely involve simultaneous processes such as chelation, complex formation, adsorption, and coprecipitation (Haynes 1984). Due to the lack of differentiation among the several methods mentioned in the literature, aluminium is commonly referred to as being "bound" to organic matter (Ritchie 1989). Aluminium can create both soluble and insoluble connections with humic substances.

Commonly found low molecular weight organic acids in soils include formic, acetic, propionic, butyric, %-crotonic, lactic, oxalic, succinic, fumaric, tartaric, and citric acids (Stevenson 1967). Plant leaves frequently contain significant levels of organic acids, including malic and citric acids, and to a lesser amount succinic, fumaric, and oxalic acids (Stevenson and Vance 1989). These acids are introduced into the soil through plant remains. Furthermore, the soil microbial biomass is responsible for the production of a diverse array of organic acids (Rovira and McDougall 1967). Organic acids have the ability to create stable chelate complexes with Al_3C and other polyvalent cations. According to Stevenson and Vance (1989), hydroxy acids, like citric acid, have a higher affinity for forming complexes compared to compounds with only one COOH group. The concentration of organic acids in the soil solution is typically modest, ranging from 1 to 5 mM (mM). However, significantly greater concentrations are frequently observed in the rhizosphere (Stevenson and Vance 1989). Concentrations of these substances are notably elevated in soils that have been enriched with organic manures, as demonstrated by Iyamuremye et al. (1996). The contents of organic acids in cultivated soils can be notably lower compared to paired locations with native vegetation (Hue et al. 1986).

Soluble aluminium detoxification

Aluminium in the soil solution can establish connections with soluble humic substances. The outcome is commonly known as organically "complexed soluble Al," although it should be highlighted that many binding methods can occur. Multiple researchers have demonstrated that the toxicity of monomeric Al to plants is alleviated when it becomes associated with soluble humic substances. For instance, Tan and Binger (1986) shown in controlled greenhouse studies using sand culture that the development of maize plants declined in a linear fashion as the amount of aluminium added increased, when humic acid (HA) was not present. Nevertheless, the use of humic acid at a dosage of 100–350 mg kg^{-1} significantly enhanced plant growth and mitigated the adverse impact of rising aluminium concentrations. In solution culture tests, Suthipradit et al. (1990) discovered that the inclusion of fulvic acid (FA) significantly decreased the concentration of monomeric aluminium in the solution. This reduction in aluminium levels helped mitigate the harmful impact of aluminium on the growth of soybean, cowpea, and green gram. The Al-fulvate complex, which is not harmful, remained dissolved in the solution. The precise mechanism by which organic acids detoxify monomeric aluminium in the soil solution remains uncertain. Hue et al. (1986) showed that soluble organic acids (5–50_M) in nutrient solutions effectively detoxified Al and enhanced the growth of cotton seedlings' roots. The organic acids that show the most efficacy, in ascending order, were citric, oxalic, and tartaric acid, whereas malic, malonic, and salicylic acids demonstrated comparatively lower effectiveness, in descending order. However, Suthipradit et al. (1990) discovered that in solution culture, the presence of malic and oxalic acids at a concentration of 50_M did not provide any relief from aluminium toxicity in soybean, cowpea, and green gram seedlings. In contrast, FA at a concentration of 65 mg L^{-1} shows remarkable efficacy in both complexing and detoxifying aluminium. These observations support the conclusions made by Ritchie et al. (1982) that HA is superior to simple carboxylic acids in forming complexes with Al. It is important to note that some organic acids that might form complexes with aluminium may exist in the soil solution at high concentrations for brief durations, as they are easily broken down by microorganisms (Ritchie 1994; Wong et al. 1995). One could argue that a consistent supply of organic acids is necessary to continue the process of Al detoxification in the soil solution (Ritchie 1994). However, the formation of complexes between organic acid anions and Al likely provides some defence against breakdown. Prior to making conclusive claims, it is necessary to ascertain the relative biodegradability of organic acid anions when they are present alone versus when they are complexed with aluminium. It is worth mentioning that organic acids play a role in the detoxification of aluminium in the rhizosphere, as stated by Kochian and Jones (1996). The release of malic acid from root tips seems to be a mechanism by which wheat genotypes exhibit resistance to aluminium, whereas the release of citric acid appears to be associated with aluminium resistance in maize and snapbean genotypes (Delhaize and Ryan 1995; Pellet et al. 1994). Organic acids are released to protect the root apex by binding with aluminium in the rhizosphere, as demonstrated by Kochian (1975). This process effectively prevents the aluminium from being poisonous to plants.

4.6 Plant Aluminium Tolerance

4.6.1 External Tolerance

- Alterations in the pH levels of the rhizosphere.
- Characteristics of the plasma membrane and composition of the cell wall.
- Excretion of chelating molecules and release of mucilage.
- Alleviating toxicity of aluminium by using elements, auxins, and other substances.
- There are cap-like structures located at the tip of the root.

4.6.2 Intrinsic Tolerance

- Chelation of aluminium occurs in the cytosol through the binding of organic acids, proteins, and other organic ligands.
- All transporters.
- The process of vacuolar compartmentalisation and vesicular trafficking.
- Altering plant metabolism and DNA checkpoints.
- Toxicity reduction.

Among certain species, there exists considerable diversity in the ability to tolerate A1, a trait that has facilitated the development of Al-tolerant varieties in various crops by breeders, as well as the study of the physiological and biochemical aspects of A1 tolerance by researchers. Wheat has been found to be particularly beneficial in this area because to significant variations in A1 tolerance between different genotypes, which can be as high as tenfold. Near-isogenic lines, which are genetically modified to have a single difference in the Al-tolerant locus, offer efficient platforms for studying A1 tolerance processes. This is despite the presence of multiple major and minor genes in certain wheat cultivars that contribute to A1 tolerance (Berzonsky 1992). The intentional removal of additional genes during the creation of these lines avoids the possible complication of multiple pathways that enable tolerance. Given that wheat has been the primary focus of A1 tolerance research, the subsequent discussion will primarily focus on summarising recent progress in our understanding of A1 tolerance in this particular species.

Various strategies are available for plants to endure or withstand the detrimental impacts of aluminium in soil. The defence mechanisms in plants are categorised into two groups: (i) external defence, which involves the exudation of organic acids from the root tips, and (ii) internal defence, known as Al tolerance, where Al chelation occurs within the cell and is subsequently stored and compartmentalised in membrane-bound organelles such as the vacuole, which is a structure found in the cytoplasmic matrix of a cell. The translocation of Al ions to the upper parts of plants occurs at an exceptionally sluggish pace (Ma et al. 1997a, b, c). The majority of plants contain a mere 0.2 milligrams of aluminium per gram of dry weight. Certain plants, known as Al accumulators, can withstand roughly 10 times the normal amount of Al without experiencing any detrimental effects. Tea plants (Camellia sinensis), commonly cultivated in North East India, particularly in Assam, are considered typical accumulators of aluminium (Al) due to their ability to thrive on acidic soils with a pH range of 4.5–5.0. The plants contain old tea leaves with an aluminium concentration of up to 30 mg/g of dry mass, as reported by Matsumotao et al. (1976). Baker et al. (2000) discovered that there are 400 species of land plants belonging to 45 different groups that have the ability to accumulate multiple dangerous metals.

To understand the techniques for plant adaptation to Al toxicity (Thakuria et al. 2016), it is necessary to alter root architecture and establish potential root associations with mycorrhizal fungi, as well as symbiotic and mutualistic relationships with plant growth boosting bacteria. Research has been conducted to identify plant genotypes that are efficient in phosphorus uptake and tolerant to aluminium toxicity. This research has focused on understanding the molecular mechanisms involved in phosphate transporters, Ca-ATPases, vegetative storage proteins, β-glucosidase, PEPCase, MDH, and PEP phosphatase. The goal is to improve the efficiency of phosphorus uptake and enhance tolerance to aluminium toxicity in different crop plants (Bhalerao and Prabhu 2013).

Aluminium toxicity in plants is evident in soils with a pH below 5.0, as indicated by morphological and physiological symptoms. Variations in root structure and function are likely responsible for the differential tolerance to aluminium toxicity. Aluminium hinders the process of cell division in roots, reduces root respiration, and impairs the absorption and utilisation of water and nutrients, specifically calcium and phosphorus, as well as metabolic pathways. Additional viable methods for investigating metal toxicity in tolerant and sensitive plant genotypes include assessing metal absorption and distribution in different plant organs, elucidating the underlying mechanism of interaction with mineral nutrients, identifying specific genes accountable for tolerance, quantifying the levels and types of organic and amino acids that function as metal chelators and detoxifiers, analysing the quantity and forms of enzymes involved, and examining alterations in root permeability to ions and molecules and the corresponding mechanisms.

Tea, azalea, rye, cranberry, weeping love grass, bermuda grass, star grass, buckwheat, peanut, and plants belonging to the Proteaceae family contain significant amounts of aluminium (Al). Additional plant species that have been identified as having the capacity to withstand soils with high levels of aluminium include pangola grass, rubber, blueberry, and Norway spruce (Kamprath and Foy 1985). A notable discovery has been made on the variation in tolerance to acidity variables among different variants within plant species. Foy (in Kamprath and Foy 1985) noted that varietal variations in aluminium tolerance have been observed in a range of plants including rice, alfalfa, tomato, soybean, ryegrass, snap bean, cotton, corn, sunflower, pea, sweet potato, green algae, and even among viruses. Research indicates that there are variations in taro cultivars' tolerance to aluminium. The discovery of varietal tolerance to aluminium toxicity has resulted in the development of aluminium-tolerant soybean and maize cultivars in Brazil (Shaffert 1994). CIAT (Colombia) and IRRI (Philippines), two international agricultural research centres, have worked together to create new types of rice that can withstand high levels of Al saturation (50–75%) without significantly reducing crop yield. The achievement of identifying and implementing tolerance in plant varieties has exceeded expectations. The CIAT pasture and savannas programme has achieved significant success in the selection and breeding of acid-tolerant forage and pasture species. Indeed, they have successfully identified numerous species that exhibit a significant capacity to withstand high levels of acidity, so eliminating it as a limiting factor. CIAT's forage species exhibit a high level of tolerance against aluminium toxicity.

4.7 Mechanisms of Aluminium Tolerance

The global issue of acidification of soils is progressively hindering crop productivity. Several variables contribute to the phytotoxicity of these soils. However, in acidic soils with a high mineral concentration, aluminium (Al) is the primary cause of toxicity. Aluminium toxicity primarily affects the root tip, resulting in the suppression of cell elongation and cell division. This leads to the slowing of root growth and a decrease in the absorption of water and nutrients. Aluminium tolerance has been established in several crop species through natural variation. In certain crops, the ability to tolerate aluminium has been incorporated into high-yielding and well-suited variants. The resistance to aluminium seems to be a multifactorial trait with a complex genetic basis. The selection approach continues to be a constraining factor in variety development due to the specific limitations associated with all strategies. Molecular markers have been linked to Al tolerance genes or quantitative trait loci in Arabidopsis and several crops, which will aid in the creation of more tolerant cultivars. Several genes have been discovered that are activated or suppressed when exposed to aluminium (Samac and Tesfaye 2003). The majority of identified induced genes are not exclusively responsive to aluminium exposure, but are also activated by other stressful situations. Overexpression of certain genes has led to increased tolerance to aluminium. Furthermore, the activation of genes responsible for the synthesis of organic acids has led to an improved production of these acids and a corresponding increase in tolerance to aluminium (Samac and Tesfaye 2003).

The phytotoxicity of aluminium (Al) has emerged as a significant issue in impeding plant growth in acidic soils. The effects of Al stress on various aspects of wheat root physiology, including rhizosphere pH, root elongation, Al absorption by roots, exudation of organic acids from roots, and factors related to Al-tolerant mechanisms, have been investigated using hydroponics, fluorescence spectrophotometry, and high performance liquid chromatography (HPLC) (Wang et al. 2000). The study selected two wheat cultivars, Brevor and Atlas66, which have different levels of tolerance to aluminium (Al). The pH of the rhizosphere has a beneficial impact on aluminium tolerance. Atlas66, which is tolerant to aluminium (Al), has a greater ability to maintain a high pH in the rhizosphere compared to Brevor, which is susceptible to aluminium. Elevated pH levels can diminish the activity and toxicity of Al^{3+} ions, while simultaneously enhancing the efficacy of excreting organic acids from the roots. Brevor exhibits greater inhibition of root elongation due to the exposure of roots to a low pH Al^{3+} solution. Brevor also accumulates more Al in its roots compared to Atlas66, even at higher pH levels. In Atlas66 roots, the exudation of malic and citric acids induced by Al has been observed, whereas no Al-induced organic acids have been detected in Brevor. The results suggest that the release of organic acids from Atlas66 roots triggered by Al has a favourable association with Al tolerance (Wang et al. 2000). The thorough examination of Al^{3+} and H^+ treatment reveals that wheat is negatively affected by an excessive amount of Al^{3+} rather than a low pH.

Based on the research by Taylor (1988), Carver and Ownby (1995), and Kochian (1995), aluminium tolerance can be classified into two categories: mechanisms that enable plants to tolerate aluminium in their cells and systems that prevent aluminium from entering the root apex. The former are referred to as external tolerance mechanisms, whereas the latter are referred to as internal tolerance mechanisms. Due to the prevailing belief that most of the aluminium in the root exists in the apoplasm and that the entry of aluminium into the symplasm is typically minimal, there is limited research on internal tolerance mechanisms compared to the extensive research on external mechanisms. One reason for the disparity in research is the relative scarcity of studies on internal mechanisms compared to those on external mechanisms. Studies have shown that the symplasm can include around 50–70 percent of the overall aluminium content (Tice et al. 1992). Furthermore, it has been observed that the symplasm can still retain aluminium even after being exposed to an aluminium-containing solution for just 30 min (Lazof et al. 1994).

The most major internal tolerance mechanisms for aluminium (Al) include Al-binding proteins, chelation in the cytosol, compartmentation in the vacuole, the formation of Al-tolerant enzymes, and enhanced enzyme activity (Taylor 1991). The experimental results from many studies (Picton et al. 1991; Delhaize et al. 1991; Basu et al. 1994; Somers et al. 1996; Basu et al. 1997) clearly support the synthesis of Al-binding proteins.

The global population is projected to increase by an additional two billion individuals by the year 2050. To meet the growing need for food, fibre, and animal feed, crop production must increase by a minimum of 50%. To achieve most of these advantages, it would be necessary to enhance crop productivity and intensify agricultural practices. Additionally, it will be crucial to cultivate new terrain that was previously considered unfit for farming. Acidic soils are prevalent in sub-Saharan Africa, Asia, and other regions that are projected to see significant population expansion. While the application of lime (calcium carbonate) is the most efficient method for reducing soil acidity and enhancing its agricultural appropriateness, it is not feasible or widespread in developing countries, where food production mostly occurs on small subsistence farms. Whenever feasible, a common strategy in management is to merge Al^{3+}-resistant germplasm with the use of lime or another ameliorant. Certain crops can flourish in acidic soils, while others can be improved by the application of conventional breeding methods. Nevertheless, other methodologies are required for cultivating crops that do not possess sufficient inherent diversity to be selectively bred for resistance against Al^{3+}. In order to address the challenges posed by population growth and climate change, it is essential to consider all potential options, including transgenic approaches, in our agricultural systems (Hoisington 2002; Bhalla 2006). After recognising the importance of organic anion efflux in Al resistance, many researchers attempted to incorporate this characteristic into model organisms using genetic engineering. The two main strategies were to enhance the transport of organic anions across the plasma membrane and to stimulate the synthesis of organic anions in order to increase their outflow. Based on Kochian's research in 1995 and subsequent studies by Kochian et al. (2004), Al tolerance systems can either prevent Al ions from accessing the root apical cells through apoplastic processes or neutralise internal Al toxicity. Aluminium enters the cytoplasm through symplastic processes, where it undergoes detoxification by forming complexes with organic molecules once within the cell (Kochian 1995). When Al is present, certain plant species respond by releasing organic acids through membrane transporters. This process results in the formation of metal complexes that are not harmful. This process effectively blocks the entry of Al into the symplast by preventing it from flowing past the plasma membrane.

The detrimental impact of aluminium on plant growth has been ascribed to many physiological processes, while the exact mechanism remains elusive. The suggested mechanisms of aluminium toxicity involve the interaction of aluminium with the cell wall of the root, the disturbance of the plasma membrane and its transport processes by aluminium, and the suppression of mineral absorption and metabolism, particularly calcium and phosphorus (Bengtsson et al. 1988). The primary and most apparent indication is the suppression of root development, resulting in stunted and underdeveloped roots with limited expansion of the primary root and hindered production of lateral roots. Root growth is inhibited by the obstruction of both the elongation and division of cells (Kochian 1995). Due to limited root growth, the plant's capacity to search for nutrients and water in the soil is significantly diminished. Consequently, plants experiencing aluminium toxicity often encounter nutritional and/or water deficiencies. Aluminium toxicity hampers the proper functioning of active ion uptake activities that occur across the plasma membrane of root cells (Kochian 1995). Specifically, the decrease in calcium (Ca) and phosphorus (P) absorption caused by aluminium (Al) has been well shown (Huang et al. 1992). The Al/P interaction process involves an adsorption-precipitation reaction between aluminium (Al) and phosphorus (P) at the surface of the root or within the empty space of the root, such as cell walls (Naidoo et al. 1978). The hydroxy-Al that has already precipitated in the root free space may absorb the P, or the P may be precipitated as insoluble aluminium phosphates. Both of these possibilities are quite probable. Consequently, plants experiencing aluminium poisoning frequently exhibit symptoms of phosphorus deficiency. Therefore, the improvement of aluminium toxicity through liming typically leads to a significant increase in phosphorus uptake by plants, despite the fact that the availability of soil phosphorus may remain the same or even decrease (Havlin et al. 2009). Although both exchangeable Al and Al saturation have been utilised as indicators of possible Al toxicity, the concentration of Al in soil solution provides a more direct assessment of the conditions experienced by plant roots. It has been discovered that the content of aluminium in the soil solution is directly linked to the suppression of plant growth in acidic soils (Adams and Moore 1983). However, the suppression of growth is not solely dependent on the amount of aluminium available in the soil solution, but also on the specific types of soluble aluminium that are present. Indeed, the presence of Al^{3+} ions and/or individual hydroxy-Al molecules is strongly associated with reduced crop yields caused by Al toxicity. The decreased root elongation of soybean [Glycine max (L.) Merr] may be caused by either Al^{3+} or $Al(OH)_2{}^+$ ions, according to Blamey et al. (1983). Nevertheless, the plants' development exhibited greater sensitivity to $Al(OH)_2{}^+$ and $Al(OH)_2{}^+$ compared to Al^{3+}. At low pH levels, it is believed that the presence of H^+ ions helps reduce the harmful effect of Al^{3+} ions by competing with them at the plasma membrane of root cells (Kochian 1995). At elevated pH levels, when monomeric hydroxy-Al species are more prevalent, the activity of H^+ is significantly diminished, rendering its competitive effects insignificant. Based on the research conducted by Kochian et al. (2005), it was found that acidic conditions enhance the presence of trivalent Al^{3+} in plants. Trivalent Al^{3+} is the most toxic form of aluminium available to plants. Due to its poisonous nature, aluminium induces alterations in the physiological and biochemical processes undergone by plants, thereby diminishing their overall productivity. At the cellular level, the presence of toxic aluminium leads to an excessive formation of reactive oxygen species (ROS) within cells (Ma 2005). Guo et al. (2006) discovered that this alters the functioning of the biomembranes, leading to a preference for oxidative damage in plants. Due to its poisonous nature, aluminium induces alterations in the physiological and biochemical processes undergone by plants, thereby diminishing their overall productivity. At the cellular level, the presence of toxic aluminium leads to an excessive formation of reactive oxygen species (ROS) within cells (Ma 2005). According to Guo et al. (2006), this alters the way biomembranes work and promotes oxidative damage in plants.

4.8 Genes that Allow Tolerance of Aluminium

4.8.1 Other Species and Mechanisms

There are various potential methods for detoxifying A1, and it is probable that plants have developed multiple remedies to counteract the issue of A1 toxicity. It seems that species other than wheat have also developed mechanisms to tolerate A1 based on the release of organic acids. A snapbean cultivar that is tolerant to AI excreted almost ten times more citric acid from its roots when exposed to AI treatment compared to an AI-sensitive cultivar (Miyasaka et al. 1991). Pellet et al. (1995) demonstrated that a genotype of maize (Zea mays L.) that is tolerant to AI excretes a significantly higher amount of citric acid from its root apices in response to AI compared to a genotype that is sensitive to AI. The formation of a robust combination between citric acid and $A1^{3+}$ is more potent in counteracting AI toxicity compared to succinate or malate, as demonstrated by Ownby and Popham (1989). Al-induced exudation of organic acids can be observed in many species, indicating the presence of Al tolerance systems that operate through distinct processes. For instance, several animals have the ability to store significant amounts of Al and must have efficient internal mechanisms for detoxification. Taylor (1991) provides a comprehensive analysis of several potential pathways that are influenced by cell wall characteristics, pH effects, AI^{3+} concentration reduction, AI compartmentation, compound exudation, and the formation of Al-resistant proteins.

4.8.2 Genes Responsible for Aluminium Tolerance and Their Genetic Mechanism

Applying agricultural lime can effectively neutralise the soil's surface acidity. Enhancing the upper layer alone will not facilitate the infiltration of plant roots through the acidic subsoil layers to access the crucial water and nutrient reservoirs beneath, regardless of any improvements made to the surface layer. The most logical solution is to choose and cultivate genotypes that possess enhanced tolerance to acidic soils and excessive levels of aluminium. This is the sole method that is logical. Extensive research has been carried out on the genetics and chromosomal location of genes that confer tolerance to aluminium in cereal crops, particularly in wheat. According to Aniol and Gustafson (1984), possessing information about the specific chromosomes in wheat and rye that contain the genes responsible for aluminium tolerance is crucial for conducting chromosome manipulation in wheat and triticale breeding.

4.8.3 Genetic Techniques for Improving Aluminium Tolerance

Plant genetic resources are abundant reservoirs of diverse and valuable traits that can be harnessed for enhancing crop species. To enhance aluminium tolerance in wheat, it is beneficial to introduce tolerance genes that naturally exist in the primary, secondary, and tertiary genepools of the Triticeae family. Triticeae is a botanical family of plants. The core genepool for wheat consists predominantly of hexaploid, tetraploid, and diploid wheats. Transferring genes between species within the same primary gene pool is a rather manageable task. Triticale is classified as a hybrid of the primary and secondary gene pools, whereas Aegilops species and rye are classified as members of the secondary gene pool. Wheat may readily be crossed with most species in the secondary genepool, although there may be challenges in expressing foreign genes when they are introduced into a wheat background. The tertiary genepool consists of annual and perennial fodder grasses, which are known for being challenging to access without the use of sophisticated breeding techniques. Wheat has a considerable level of variability in its response to aluminium. An effective approach to identify aluminium-resistant accessions in a genebank collection is to evaluate the accessions originating from places with highly acidic soils and aluminium-related issues. Brazil is home to wheats and ryes, such as BH1146 and Blanco, that exhibit exceptional tolerance to high levels of aluminium. This is due to the country's extensive areas of acidic soil. The research has been primarily motivated by the efforts of the plant breeding community, as their main goal is to enhance production on acid soils by creating resistance to aluminium toxicity (Garvin and Carver 2003; Hede et al. 2001). This research has been conducted for many purposes. The majority of research has focused on cereals, namely those classified within the Triticeae family, such as wheat and rye. This can be attributed to a combination of variables, such as national focus and geography, as well as the inherent tolerance inherited within that tribe.

Research on the physiological and genetic aspects of Al tolerance provides a valuable foundation for genomics-based investigations. The physiological characteristics of Al tolerance and toxicity clearly indicate that OA release and internal detoxification play significant roles in the processes that lead to Al tolerance. Having a more comprehensive understanding of these pathways from a mechanistic perspective is beneficial for improving the identification of candidate genes during the map-based cloning process. By examining the dynamic patterns of gene or protein expression, it may be able to discover key components that play a critical role in aluminium (Al) tolerance, as Al tolerance can potentially be produced in several systems (Li et al. 2000; Magalhaes 2002). In this section, we will assess the progress made in identifying the genes and proteins essential for Al tolerance. Additionally, we will explore the potential of using transgenic plants or yeast to validate the significance of these candidate genes. Experiments conducted on wheat have also prompted endeavours to monitor fluctuations in gene expression throughout the entirety of the stress response process. Several research teams have investigated the genetic expression in wheat roots that have been exposed to aluminium, as documented by Milla et al. (2002),

Richards et al. (1994), and Snowden and Gardner (1993). Milla et al. (2002) found that most of the genes identified through these experiments demonstrated responsiveness to both abiotic and biotic stimuli. Moreover, the majority of these genes were discovered to be expressed equally in both sensitive and tolerant genotypes. With just a few noteworthy deviations, it remains uncertain whether genes that are activated by aluminium are pertinent to the mechanisms that contribute to aluminium tolerance.

By employing the process of screening for S. Pombe, Delhaize, and their colleagues have identified wheat genes in a more functional manner. The yeast Saccharomyces cerevisiae has been genetically modified by introducing plant genes in order to enhance its resistance to aluminium (Delhaize et al. 1999). The utilisation of this approach resulted in the identification of six unique clones, with one of them containing a gene that encodes a phosphatidylserine synthase. This enzyme is involved in the metabolic pathway of phospholipids. In yeast, the gene's overexpression caused resistance to Al. However, in Arabidopsis, it resulted in the formation of spontaneous necrotic lesions and dwarfism (Delhaize et al. 1999). The probable cause of this was an imbalance in the equilibrium between phosphatidylserine and phosphatidylglyercol. The latter is a crucial phospholipid present in leaves and is an essential part of chloroplast lamellae. Aluminium toxicity is known to be associated with lipid peroxidation, as established by Yamamoto et al. in their studies conducted in 2002 and 2001. However, there is a possibility that phosphatidylserine is less susceptible to peroxidation compared to other phospholipids. This suggests that a higher concentration of phosphatidylserine could potentially safeguard yeast membranes.

4.8.4 Genetic Variations in Aluminium Tolerance

The growth of various agricultural, pasture, and tree species, as well as their cultivars, in acidic soils mostly relies on their varying levels of tolerance to aluminium and their varying needs for calcium. Significant variations occur among different farmed species in terms of their ability to tolerate soil acidity variables, both between and within species.

Some crops in humid tropical woods and tropical savanna biomes thrive at pH levels that would be lethal to maize or soybeans due to aluminium toxicity. Pineapple is widely recognised as the most prominent example, but, coffee, tea, rubber, sugar cane, and cassava also have a notable ability to withstand elevated levels of aluminium saturation. Cowpeas and pigeon peas exhibit greater acid tolerance compared to common beans or soybeans among the grain legumes. In 1978, Duke created a comprehensive list of more than 1000 plant species that have economic significance and are known to tolerate unfavourable environmental conditions. Included were the ability to tolerate "acid soils," "lateritic soils," and "aluminium toxicity." Duke's list, while preliminary and not fully comprehensive, demonstrates the wide range of aluminium-tolerant genetic resources. Out of the total number of species, 397

Table 4.1 Some important food crops and other crops generally tolerant to acid soil conditions in the tropics

Acid-tolerant species	Susceptible species with acid-tolerant cultivars	Acid susceptible species
Cassava	Common bean[a]	Peanut[b]
Cowpea	Maize	Cotton
Pigeon pea	Sorghum	Alfalfa
Plantain/ banana	Soybean	Perennial soybeans
Potato	Sweet potato[c]	Cacao
Rice	Wheat	–
Sugar cane[d]	–	–

Updated from Sanchez and Salinas (1981)
[a]Sensitive to manganese toxicity
[b]Sensitive to calcium deficiency, but generally aluminium-tolerant
[c]Tolerant to manganese toxicity (Ray Weil, personal communication 2006)
[d]Seldom responds to liming

(which accounts for 38%) were identified as having the ability to withstand acid soils, lateritic soils, or aluminium toxicity. Table 4.1 provides a comprehensive list of the primary tropical food crops, including fruit and tree crop species that have the ability to withstand high levels of acidity.

4.9 Breeding for Aluminium Tolerance

The tolerance of aluminium (Al) exhibits significant variation both within and among plant species. This heterogeneity has enabled breeders to develop genotypes that can flourish in acidic soil conditions. Aluminium-tolerant germplasm is employed in combination with liming procedures to effectively neutralise acidity. Nevertheless, relying solely on lime as a means of addressing acid soils in many agricultural systems proves to be either cost-prohibitive or time-consuming, particularly when dealing with deep-seated acidity. On order to sustain production on acidic soils, it may be beneficial to replace aluminium-sensitive species with aluminium-tolerant ones. For example, alfalfa in pastures can be substituted with more alfalfa-tolerant species of pasture. However, this approach faces the challenge that the nutritional quality of the alternative pastures may not be as good as that of alfalfa. Breeders may enhance the tolerance to aluminium in elite genotypes by utilising a diverse range of species (Berzonsky 1992; Johnson et al. 1997; Kerridge and Kronstad 1968). Several prominent dominant genes regulate the aluminium tolerance of wheat (Triticum aestivum). These genes have been utilised in breeding endeavours to augment the Al tolerance of plants. The presence of highly acidic soil in Brazil has led to the development of wheat genotypes that are tolerant to aluminium (Al). As a result, a considerable portion of the wheat germplasm with Al tolerance may be attributed to Brazil. The manifestation of aluminium toxicity is observed during the growth of roots, and the utilisation of particular amounts of aluminium in hydroponic cultivation has been demonstrated as a reliable measure of aluminium tolerance for various species. The utilisation of hematoxylin staining has demonstrated potential in evaluating plant resistance to aluminium (Polle et al. 1978). In this case, the roots are dyed after being cultivated in an Al solution for a day or more to determine which ones have purple root tips, indicating sensitivity to the particular concentration of Al used in the growth solution. The lack of colour in the root tips of the Al-tolerant genotypes indicates that these genotypes either prevent the entry of Al (as explained below) or form Al complexes that cannot be stained by hematoxylin. Hematoxylin forms a purple complex by binding with aluminium. Wheat, maize, soybean, phalaris, and barley are species that have been successfully developed with increased tolerance to acid soils through the screening of aluminium under specified conditions in laboratory or glasshouse settings. There is much evidence indicating that organic acids play a role in the tolerance of different plant species to aluminium. Certain organic acids possess the capacity to form complexes with Al^{3+} ions, transforming them into forms that are safe for plants. Hue et al. (1986) assessed the effectiveness of several organic acids in protecting plant roots from aluminium toxicity in a hydroponic culture. The researchers found that organic acids containing hydroxyl and carboxyl groups, which are capable of creating stable ring structures with Al^{3+} through 5- or 6-bonds, provided the most effective protection against Al toxicity (Fig. 3). Citric, oxalic, and malic acids are common organic acids present in plants that fulfil this requirement. Certain plant species exhibit the utilisation of both internal and exterior detoxification processes for aluminium tolerance, which are believed to include the detoxification of organic acids.

4.9.1 Avoiding Aluminium from Roots by Using Organic Acids

Some plant genotypes are able to resist aluminium because it is not found in the root apices. For example, several methods used in wheat research have shown that wheat genotypes that are tolerant to aluminium (Al) accumulate lower levels of Al in the tips of their roots compared to genotypes that are susceptible to Al. As a response to Al, certain species with genotypes that can tolerate Al release organic acids (Ryan et al. 2001; Ma et al. 2001). The organic acids form chelates with aluminium, which protect the roots by preventing its harmful effects. For example, when exposed to aluminium, soybeans, snap beans, maize, and Cassia toru release citrate, while buckwheat releases oxalate. Al, triticale, rapeseed, oats, radish, and rye release both malate and citrate. Given that the root apices are the most susceptible part of the root system to Al toxicity, it is reasonable to expect that in many cases, the release of organic acids predominantly occurs from the root apices. Moreover, the finding that certain genotypes of these species that are tolerant to aluminium release a considerably higher amount of organic acid compared to their corresponding genotypes that are susceptible to aluminium does provide evidence for the hypothesis that organic acid efflux is a mechanism for aluminium tolerance. An investigation of almost identical wheat lines that vary in aluminium tolerance at a specific genetic location (the Alt1 locus) offers convincing evidence for the correlation between the release of organic acids and aluminium tolerance. Due to their essentially identical nature, save for their tolerance to Al, these lines are especially valuable for studying the physiology of Al tolerance. This avoids comparisons between genotypes that, despite belonging to the same species, exhibit genetic differences, and excludes the potential of multiple tolerance mechanisms influencing the observed phenotype.

4.9.2 Al Is Internally Cleansed with Organic Acids

Certain plant species possess the remarkable ability to accumulate and retain aluminium in their roots and branches. These Al-tolerant species have undoubtedly developed defence systems to maintain the metal in non-toxic forms within the plant, as well as mechanisms to allow the metal to move within the plant and across different membranes. Research indicates that organic acids form complexes with internal aluminium in buckwheat, hydrangea, and melastoma. Aluminium complexed with citrate has been found in hydrangea, while aluminium complexed with oxalate has been found in buckwheat. These findings were made using 27Al-nuclear magnetic resonance (Watanabe et al. 1998a, b; Ma et al. 1998; Ma et al. 1997a, b, c). Citrate and oxalate, which are powerful chelators of aluminium, protect cellular components against the phytotoxic effects of aluminium. The organic acid that forms compounds with aluminium in buckwheat differs across various plant tissues. Buckwheat is anticipated to absorb aluminium in the form of Al^{3+}, although the mechanism behind this process remains unclear. Aluminium is bound to oxalate within the roots to form a complex with a ratio of 1 part aluminium to 3 parts oxalate. The predominant form of aluminium in the xylem is Al-citrate, which exists as a 1:1 complex. This complex is then transported to shoot cells, where it is transformed into the 1:3 Al:oxalate complex. Buckwheat, a kind of wheat, produces oxalate in response to aluminium. The plant's ability to tolerate high levels of aluminium may be due to both internal and external detoxification mechanisms. Organic acid complexes are anticipated to include certain transporters in their process of crossing the membrane, while the precise mechanism remains unknown.

Although there is much evidence establishing a correlation between organic acids and the aluminium tolerance mechanisms in numerous species, it seems that certain species employ other approaches. Brachiaria decumbans, a plant that is very tolerant to aluminium (Al), must employ alternate mechanisms to deal with toxic levels of Al in the soil solution. This is because it does not produce organic acids as a response to Al, as stated by Wenzl et al. (2001). Given that pH strongly influences the phytotoxicity of aluminium, a potential approach to mitigate its harmful effects could involve increasing the pH in the vicinity of the root tips. An investigation of an Arabidopsis mutant (alr1) that is resistant to aluminium provides supporting evidence for this mechanism. The mutant exhibited an increase in pH in the solution directly surrounding the root apex due to the presence of aluminium, resulting in a decrease in the activity of Al^{3+} ions (Dengenhardt et al. 1998). Previous studies have established a correlation between phosphate outflow and aluminium tolerance. Aluminium forms phosphate complexes and research has found that these complexes are released from the root apices of Atlas, a highly resistant wheat cultivar, along with malate (Pellet et al. 1996). Phosphate efflux was shown to be constitutive, indicating that the presence of Al was not required to initiate the efflux process, unlike the efflux of malate.

Furthermore, studies conducted on yeast have shown that the overexpression of genes that are not related to the outflow of organic acids can provide tolerance to aluminium. MacDiarmid and Gardner (1998) examined a yeast genome library to identify genes that, when overexpressed in yeast, provided resistance to Al toxicity. They identified two genes, ALR1 and ALR2, that exhibited similarities and subsequently assigned names to them. These genes encode membrane-bound proteins that possess capabilities of transporting Mg ions. Indeed, the yeast necessitates substantial amounts of external Mg^{2+} in order to thrive when the genes are deactivated. The mechanism by which these genes are overexpressed to confer aluminium (Al) tolerance is not yet fully understood. However, one hypothesis suggests that the proteins produced by these genes are the primary targets of Al toxicity in yeast. Cell death occurs when the uptake of Mg^{2+} by Al is interrupted. However, when the ALR genes are expressed at high levels, Mg transport can still occur even in the presence of normally fatal amounts of Al. One of these genes encodes phosphatidyl serine synthase, an enzyme that plays a role in the synthesis of phospholipids (Delhaize et al. 1999). The improved Al tolerance may be attributed to a shift in the lipid makeup of the plasma membrane, as phospholipids are crucial components of membrane lipid bilayers.

4.9.3 Genetic Engineering to Increase Al Tolerance

There is a limited availability of germplasm that exhibits Al tolerance in several species. Alfalfa and barley possess certain germplasm that exhibits resistance to aluminium, albeit their tolerance range is more limited compared to certain other species. An effective approach is to modify the synthesis and release of organic acids, which have been strongly linked to aluminium tolerance. The genes responsible for encoding organic anion transporters have not been cloned as of yet. The capacity to modify these pathways is possible due to the cloning of many genes that encode enzymes involved in the creation or breakdown of organic acids. The first publication utilising this approach revealed the manifestation of a Pseudomonas aeruginosa citrate synthase gene in tobacco (de la Fuente et al. 1997). The initial decisive action in the tricarboxylic acid cycle is the merging of oxaloacetate and acetylCoA, leading to the formation of citrate.

4.9.4 Future Challenges

Despite major advancements in understanding Al tolerance processes based on organic acid outflow, there is still much to discover about the molecular mechanisms that activate anion channels by aluminium. Specifically, we require a more comprehensive understanding of the cellular processes involved in the first detection of aluminium, subsequent opening of channels, and outflow of organic acid. Furthermore, it is imperative to replicate the genes responsible for encoding these anion channels. As previously said, it is clear that plants that do not depend on organic acids have developed systems to tolerate aluminium, but our understanding of these mechanisms is still lacking. Further research is necessary to ensure that sufficient levels of aluminium (Al) tolerance are achieved for agricultural use. However, there have been some achievements in genetically modifying plants to enhance their tolerance to Al. One suggested mechanism for plants to tolerate aluminium (Al) involves the secretion of an Al-chelating molecule into the rhizosphere. Two varieties of snapbeans, "Romano" and "Dade" (*Phaseolus vulgaris* L.), which have different levels of resistance to aluminium (Al), were cultivated in a hydroponic system under sterile conditions, both with and without the presence of Al. This experiment was conducted by Miyasaka et al. (1991). Following an 8-day period of development in nutrient solutions, the aliphatic and phenolic organic acids present in the culture solutions were examined using an ion chromatograph and a high pressure liquid chromatograph. The snapbean variety "Dade," known for its tolerance to aluminium (Al), released citric acid into the rhizosphere at a concentration 70 times higher than when it was grown without Al. This concentration was also 10 times higher than that of the "Romano" snapbean variety, regardless of whether it was grown with or without Al. Under Al stress, the "Romano" cultivar released a significantly higher amount of citric acid into the growth medium compared to non-stressed circumstances. One possible mechanism for aluminium tolerance in snapbeans is the release of citric acid into the rhizosphere. This release can be triggered by either high quantities of poisonous aluminium or low levels of phosphorus caused by the formation of insoluble aluminium phosphates. The study did not have the ability to differentiate between these two components; nevertheless, it is crucial for plants growing in Al-toxic soils to have resistance to both primary and secondary Al stress damage (Miyasaka et al. 1991).

An assumed mechanism for aluminium (Al) tolerance involves the binding and removal of Al by organic acids, which can occur either inside the plant (internal tolerance) or in the rhizosphere (exclusion) (Foy 1984, 1988; Taylor 1988). It has been documented that organic acids can bind with aluminium and improve its harmful effects on plants (Foy 1984, 1988; Taylor 1988). In their study, Hue et al. (1986) demonstrated that the inclusion of citric, oxalic, and tartaric acids in the hydroponic solution effectively mitigated the suppressive impact of aluminium on the growth of cotton roots (*Gossypium hirsutum* L.). In a similar vein, Bartlett and Riego (1972) discovered that the presence of citric acid or EDTA complexed Al did not inhibit the growth of maize plants (*Zea mays* L.) in terms of both root and shoot development, unlike ionic Al. These investigations suggest that the organic acids may have detoxified aluminium either by acting externally in the rhizosphere or inside after being absorbed by plant roots. Suhayda and Haug (1984, 1985, 1986) discovered through laboratory experiments that organic acids were able to reverse the conformational change generated by aluminium in the regulatory protein, calmodulin. This reversal restored the protein's ability to stimulate specific enzymes. The organic acids that provided protection to calmodulin from the harmful effects of Al were ranked in the following order: citric acid > oxalic acid > malic acid > tartaric acid (Suhayda and Haug 1984, 1985). The results indicate that organic acids have the ability to internally safeguard enzyme function within the plant, shielding it from the harmful impact of aluminium. Cambraia et al. (1983) provided evidence of internal detoxification through the use of organic acids. Their study demonstrated that when exposed to aluminium, sorghum (Sorghum bicolor Moench) roots exhibited an overall rise in the concentration of organic acids. In addition, the Al-tolerant sorghum cultivar had larger concentrations of organic acids, specifically t-aconitic and malic acids, in its roots compared to the Al-sensitive cultivar (Cambraia et al. 1983). Nevertheless, the snapbean roots exhibited a decrease in organic acid levels in response to Al. However, the snapbean cultivar that was tolerant to Al maintained a higher concentration of organic acids compared to the Al-sensitive cultivar (Lee and Foy 1986). Maize, barley, and peas exhibited similar outcomes, where increased Al tolerance was associated with elevated levels of organic acids in both the shoots and roots of plants exposed to aluminium. This correlation was observed in studies conducted by Suhayda and Haug (1986) for maize, Foy et al. (1987) for barley, and Klimashevskii and Chernysheva (1980) for peas.

4.9.5 *Organic Acids Cleanse Al^{3+} Outside the Root*

Upon being exposed to aluminium (Al), it has been observed that over twelve plant species that are tolerant to Al emit organic acids from their roots (Ma 2000; Ryan et al. 2001). When the Al^{3+} ion is found in the external solution, certain plant species release a greater amount of organic acids, namely near the root apex (Delhaize et al. 1993a, b, c; Zheng et al. 1998). It is not practicable or essential for root exudates to detoxify all the Al^{3+} in the soil. The cations that require detoxification are limited to those in close proximity to the apical root cells, as the root apex is very susceptible to Al^{3+} (Ryan et al. 1993). Restricting the supply of organic acids to the apical zone is a sensible approach as it reduces the metabolic energy required for the Al tolerance process. In order to compensate for the loss of organic acids that disperse from the root or are decomposed by microorganisms, the secretion process must persist as the root tip traverses an acidic soil. The organic acids form a protective barrier around the root apex to shield it from the harmful Al^{3+} cations. There are two clearly identifiable patterns of organic acid excretion. In Pattern I, there is no discernible latency period between the introduction of Al and the initiation of the organic acid release. For example, wheat and buckwheat showed evident production of malate or oxalate within 15–30 min after being exposed to aluminium, as reported by Delhaize et al. (1999) and Zheng et al. (1998). Following exposure to Al^{3+} in Pattern II, the production of organic acid is delayed for several hours. As an example, in the programming language C. Tora exhibits peak citrate efflux four hours after being exposed to aluminium, as seen by Ma et al. (1997a, b, c). In contrast, citrate and malate efflux in rye shows a constant increase throughout a ten-hour duration, as reported by Li et al. (2000). Recent evidence from maize suggests that aluminium (Al) can lead to both an immediate release of citrate and a gradual release that becomes stronger over a period of 48 h (Pellet et al. 1995; Pineros and Kochian 2001). Due of Al's rapid activation of a pre-existing pathway, it is probably unnecessary to stimulate the production of new proteins (Ma 2000). In this scenario, Al could initiate an organic anion efflux by activating a transporter located on the plasma membrane. Conversely, the delay in Pattern II-type secretion may indicate the need for protein induction. These proteins that are artificially produced may have a role in the movement of organic acid anions or in the breakdown of organic acids.

4.9.6 *Ion Channels Control the Release of Organic Acid Anion*

In order to detoxify aluminium in the rhizosphere, organic acids must be transported from the cytosol to the apoplasm. At the nearly neutral pH of the cytoplasm, organic acids exist as organic acid anions, with their protons almost completely separated. It is quite probable that these organic acid anions are the substances transmitted from the root cells. Each species secretes just a small number of specific organic acids in response to Al treatment, even though root cells contain a variety of distinct organic acids (Ma 2000; Ma et al. 1997a, b, c). This demonstrates that the plasma membrane possesses a specialised transport mechanism for the organic acid anions. A transport mechanism in wheat and maize has been discovered as an anion channel (Zhang et al. 2001; Ryan et al. 1997; Pineros and Kochian 2001; Kollmeier et al. 2001). Anions have the ability to move passively along their electrochemical gradient due to the presence of anion channels, which are transport proteins located in the cell membrane. Under most conditions, anions generally go from the cytosol to the apoplasm in plant cells due to the substantial electrical potential difference across the plasma membrane (with the inside being negatively charged). While the precise process by which Al opens these anion channels remains unknown, three ideas have been proposed (Delhaize and Ryan 1995):

(i) The initiation of the channel's formation can be ascribed to Al's direct interaction with the protein constituting the channel.
(ii) Aluminium ions interact with a particular receptor on the membrane's surface or with the membrane itself to initiate the opening of the channel. Subsequently, this cascade of events involving secondary messengers is initiated.

(iii) Aluminium then enters the cytoplasm, where it has the ability to directly or indirectly activate the channel through secondary messengers.

Recent research from maize (Pineros and Kochian 2001) indicates that Al is capable of opening the channel in isolated membrane patches. This suggests that secondary messengers are either unnecessary or are bonded to the membrane. The specific factors that differentiate the Al-tolerant and Al-sensitive genotypes, resulting in a substantial discharge of organic acid from the tolerant genotype but negligible or no discharge from the sensitive genotype, remain unidentified. The number of channel proteins in each genotype's membrane may vary, as well as their permeability to organic anions and their activation by Al. Until the channel proteins or the genes encoding these channels are found, one must depend on physiological methods to find the answers to these questions.

4.9.7 Modifications in the Metabolism of Organic Acids and Their Secretion

Information on the influence of Al on organic acid metabolism is only accessible for a limited number of plant species. While the ability of these plants to produce organic acids is not significantly influenced by the presence of aluminium (Al), the metabolism of organic acids in species that release organic acid anions rapidly when exposed to Al (Pattern I) can be changed due to Al exposure. The results indicate that malate production in wheat is not restricted by a maximum rate, and that the Al-tolerant genotype experiences changes in the function of existing enzymes to replenish internal malate levels due to the loss of malate from the cytoplasm. Certain plant species exhibit a time lag between the introduction of aluminium (Al) and the initiation of organic acid secretion (Pattern II). Scientists have noted that this delay is linked to alterations in the enzyme activity responsible for the production of organic acid. In rye, the presence of aluminium does not change the activities of phosphoenolpyruvate carboxylase, malate dehydrogenase, and NADP-dependent isocitrate dehydrogenase. However, the activity of citrate synthase in the root tip increases by thirty percent (Li et al. 2000). Following 6 h of exposure to Al, there is a notable rise in the activity of citrate synthase immediately before an increase in citrate efflux occurs.

4.9.8 Impact of Aluminium on Maize Plants

Aluminium ions have a remarkably sluggish movement towards the upper regions of plants (Ma et al. 1997a, b, c). Most plants typically contain only 0.2 mg of aluminium per unit of dry mass. Certain plants, known as Al accumulators, have the ability to tolerate far higher levels of Al without experiencing any negative effects. Tea plants are known to have a high capacity for accumulating aluminium (Al), with old leaves carrying up to 30 mg per gram of dry mass of Al (Matsumoto et al. 1976). Baker et al. (2000) discovered that 400 species of land plants from 45 distinct families had the ability to accumulate multiple dangerous metals, making them hyperaccumulators. The main signs of aluminium poisoning are the inhibition of root extension and the simultaneous increase in the synthesis of -1,3-glucan (callose). These effects can be observed even after a short period of exposure. Aluminium-induced root damage leads to insufficient absorption of ions and water (Barcelo and Poschenrieder 2002). According to one idea, the root cap cells detect the presence of aluminium, which initiates signal transduction. This, in turn, results in a physiological response inside the root meristem. Recent research has invalidated the potential impact of the root cap and highlighted the root meristem as the sensitive area. In maize (Zea mays), the distal part of the transition zone has been identified as the specific region that is targeted. It has been found that the main source of aluminium injury is the root tips (Sivaguru and Horst 1998). Root elongation is a result of the proliferation of root cells through cell division. Several plants have been seen to have a decrease in mitotic activity upon exposure to aluminium.

According to Panda (2007), the cell division area in maize is shrinking concomitant with a decrease in the number of proliferating cells. Blancaflor et al. (1998) examined the impact of aluminium on microtubules and actin microfilaments in elongated cells of maize root apices. Maize cultivars that are resistant to aluminium (Al) have been specifically selected for acidic soils (Pandey and Gardner 1992). The genetic diversity of Al resistance in maize has been documented by Jorge and Arruda (1997) and Pintro et al. (1996). Tropical maize populations have enhanced the grain yield of maize by selecting varieties that exhibit tolerance to acidic soils. Most of the breeding endeavours aimed at enhancing productivity in acidic soil focused on cultivating tolerance to aluminium toxicity (Garvin and Carver 2003).

4.9.9 Secondary Characteristics Linked to Resistance to Acidic Soils

Secondary features can be utilised to selectively breed maize with poor soil pH tolerance, based on their correlation with yield. Secondary characteristics can serve as indirect indicators of production under acid soil conditions, however yield is primarily controlled by non-additive gene expression. Secondary features have the potential to assist the genotypic characterisation of plants in response to low soil pH stress. According to the study conducted by Baligar et al. (1997), maize genotypes that were tolerant to aluminium (Al) saturation exhibited significant increases in both shoot and root weights, as well as high nutritional efficiency ratios (NER) for phosphorus (P), calcium (Ca), magnesium (Mg), and iron (Fe) at a 41% Al saturation level. Recent maize hybrids developed in Argentina that have higher tolerance to acidity exhibit increased grain production, mostly attributed to a rise in harvest index (Echarte et al. 2013). A higher harvest index was shown to be more strongly associated with an increase in grain yield components such as kernel number and/or weight, rather than an increase in shoot biomass. The attribute most commonly associated with genetic yield improvement in maize under stressful conditions is the delay in leaf senescence, also known as the ability to "stay green" (Nguyen and Blum 2004). Additional secondary traits associated with tolerance to acidic soil conditions at different stages of maize development have been identified over time. Welcker (2000) discovered significant positive associations between anthesis-silking interval, plant height, number of ears per plant, and yield under conditions of low soil pH. However, the anthesis-silking interval and plant height exhibited only modest to moderately weak correlations with yield, suggesting that these factors are not accurate predictors of performance in low soil pH conditions. Significant correlations were observed between yield and leaf area, photosynthetic rate, stress tolerance index, and stress susceptibility index. These attributes have the potential to serve as reliable indicators of yield due to their expected stability throughout varying environmental circumstances. Certain mineral poisons in the soil possess distinct secondary features that are exclusive to them. Stoyanova et al. (2010) proposed that chlorophyll concentration could serve as a useful secondary characteristic in relation to Mn toxicity.

4.10 Al Tolerance Exclusion Mechanisms

4.10.1 pH Variations in the Rhizosphere

The pH of the rhizosphere increases slightly when metal is excluded from the root apoplast. This decrease in pH reduces the solubility, activity, toxicity, and content of aluminium in plants (Yang et al. 2011). The H^+ -ATPase of the plasma membrane regulates changes in the pH of the rhizosphere and the ability of wheat and squash (Cucurbita pepo) to withstand stress caused by aluminium, as stated by Yang et al. (2011). When other metals are present along with aluminium, organic acids (OA) have the ability to change the pH of the rhizosphere, as observed in studies by Kochian et al. (2004) and Javed et al. (2013). Wan et al. (2012) found that tea plants absorb aluminium, which leads to the release of hydrogen ions from the roots. This process contributes to the acidity of the soil. According to the study conducted by Konishi et al. (1985), aluminium has a crucial role in the growth of tea plants by specifically stimulating the elongation of their roots. Soil acidification could potentially preserve the solubility and absorption of aluminium in tea. On the other hand, species that are sensitive to aluminium (Al) suffer from heightened toxicity and accumulation of metals as a result of the acidic conditions in the rhizosphere (Houman et al. 2009).

4.10.2 Changes in Plasma Membrane Properties and Cell Wall Composition

Aluminium toxicity and exclusion in plants mainly target the root cell wall, as it acts as the principal location for aluminium binding (Kopittke et al. 2015; Liu et al. 2016). The alteration of the cellular wall composition is currently recognised to have a substantial impact on the resistance to aluminium, as demonstrated by physiological, biochemical, and molecular evidence (Che et al. 2016; Zhang et al. 2016). Aluminium can induce an elevation in the polysaccharide content of the cell wall, resulting in a reduction in the cell wall's pliability and its capacity to uptake nutrients and water (Nguyen et al. 2005).

The physicochemical and physiological properties of the plasma membrane also influence the plants' capacity to withstand aluminium. The lipid composition of the plasma membrane (PM) has a considerable impact on Al tolerance (Wagatsuma et al. 2015). The union of the metal with the plasma membrane (PM) results in the phospholipids generating a negative charge on the PM's surface, hence enhancing its sensitivity to aluminium (Al). The resistance to aluminium increased when the proportion of phospholipids in the plasma membrane of root cells decreased in rice and wood species (M. malabathricum and M. cajuputi) (Maejima and Watanabe 2014). Similarly, the sterol content has a major impact on Al's stress tolerance. Many plant species often use a technique for tolerating aluminium that involves having a high amount of sterols and a low concentration of phospholipids (Wagatsuma et al. 2015). A peptide with a short length that attaches to the plasma membrane (PM) has also been suggested to enhance the ability of rice to withstand stress by blocking the entry of aluminium into root cells through its interaction with positively charged ions (Xia et al. 2013).

4.11 Crop Tolerance

Aluminium tolerance

Crops that have evolved in calcareous soils, such as cotton, sorghum, and lucerne, are vulnerable to aluminium saturation levels ranging from 10 to 20%. They should aim for zero aluminium saturation when liming. Rice and cowpeas exhibit higher levels of tolerance compared to other crops. Coffee, rubber, pineapple, and certain pasture plants exhibit limited responsiveness to lime, even in soils with elevated aluminium saturation. Farmers cultivating acid soils have deliberately chosen cultivars that possess aluminium tolerance, similar to how they have chosen cultivars with other desirable agronomic features. Charles Foy is said to be the pioneering scientist who extensively studied this subject. He led a team that examined the ability of several species and varieties to tolerate high amounts of aluminium, as well as some for manganese (Foy and Brown 1964; Foy 1997). The signs of aluminium poisoning are exclusively observed in the roots, although the tops of the plant may exhibit indications of dehydration, nitrogen, phosphorus, calcium, or magnesium shortages. The roots exhibit a compact and truncated appearance. Aluminium poisoning hinders the growth of roots and reduces the absorption of nutrients. Healthy roots may sometimes exhibit resistance to growing into subsoil that is poisonous with aluminium, potentially due to a shortage in calcium. Salinas and Sanchez (1976) discovered a correlation between varietal tolerances to high amounts of exchangeable Al^{3+} and low levels of accessible phosphorus in an Oxisol at the Cerrado Centre. The results were derived from the cultivation of upland rice, maize, wheat, sorghum, and common beans, as reported by Salinas and Sanchez in 1976. Spain et al. (1975) found similar results in the Oxisols of the Llanos Orientales of Colombia.

The review conducted by Bahia-Filho et al. (1997) can be succinctly summarised as follows:

- Numerous maize hybrids and open-pollinated types have been introduced. These plants possess a combination of high capacity for producing crops and consistent crop production, together with the ability to withstand high levels of aluminium toxicity (at 45% aluminium saturation), and improved efficiency in utilising phosphorus.
- A significant amount of maize varieties can be examined in a nutritional solution containing a phosphorus:aluminium ratio of 1:5 by measuring the lengths of the seminal roots. Greater aluminium tolerance was shown by longer seminal roots, which determined the accessions ready for field testing.
- Aluminium tolerance is a trait that is inherited in a quantitative manner and is controlled by dominant genes.
- Aluminium-tolerant cultivars also perform exceptionally well on fertile soils that do not have aluminium issues.
- Additionally, they exhibit a more profound and assertive root system, along with the capacity to efficiently absorb phosphorus under favourable soil moisture conditions.
- The primary process by which aluminium tolerance is achieved in plants is through the release of organic acids (mostly malic and citric acids) into the rhizosphere by the root apex of aluminium-tolerant plants. These organic acids then form complexes with Al^{3+} ions. This phenomenon occurs in aluminium-tolerant varieties of wheat, barley, maize, rice, sorghum, beans, and soybeans. In addition, the combination of these two organic acids, along with oxalic acid, enhances the solubility of iron- and aluminium-bound phosphorus, as well as organic phosphorus (Liao et al. 2006), resulting in a highly beneficial synergy.
- Numerous aluminium-tolerant genes (quantitative trait loci) that are responsible for beginning the synthesis of organic acids and aluminium chelation have been found for many of the species mentioned above.
- Aluminium-tolerant variants exhibit less inhibition of phosphorus absorption and translocation compared to aluminium-susceptible varieties or species. Furthermore, numerous species or varieties that can withstand high levels of aluminium are also capable of tolerating low levels of phosphorus in the soil.

- Aluminium does not hinder the absorption and movement of calcium, magnesium, and potassium in aluminium-tolerant variety when it is present in the roots. However, it does have this effect in sensitive kinds. The tolerance of different varieties of soybeans, wheat, and barley to aluminium is linked to their ability to absorb and transport calcium. In sorghum, tolerance to aluminium is tied to potassium, while in potatoes, it is associated with the translocation of both magnesium and potassium.
- Aluminium-tolerant species of nitrogen-fixing pasture legumes exhibit prolific nodulation at low pH levels, irrespective of calcium concentration.
- Certain endomycorrhizal strains that are resistant to aluminium can enhance phosphorus absorption at pH levels ranging from 3.8 to 4.0.

Among certain species, there exists considerable diversity in Al tolerance. This diversity has proven valuable for breeders in developing Al-tolerant varieties of various crops, as well as for academics studying the physiological and biochemical aspects of A1 tolerance. Wheat has been found to be particularly beneficial in this area because of significant variations in A1 tolerance between different genotypes, which can be as high as tenfold. Near-isogenic lines, which are genetically modified to have a single difference in their Al-tolerant locus, offer efficient platforms for studying A1 tolerance processes. This is advantageous even if several wheat cultivars possess multiple genes, both major and minor, that code for A1 tolerance (Berzonsky 1992). The intentional removal of additional genes during the creation of these lines minimises the possible complication of multiple pathways that support tolerance. Given that wheat has been the primary focus of A1 tolerance research, the subsequent discussion will primarily focus on elucidating recent progress in our understanding of A1 tolerance in this particular species.

The utilisation of marginal soil plays a crucial role in enhancing soybean output in Indonesia. The soybean breeding programme, which focuses on developing high-yielding cultivars that are also tolerant to acid soils, is considered a cost-effective and environmentally benign method for enhancing plant production in acidic soil conditions. The study conducted by Puspitasari et al. (2021) demonstrated that acidic conditions and elevated levels of aluminium impede the growth of soybean seedling roots. Each genotype exhibited a distinct response to both seedling evaluation in hydroponics and plant growth evaluation using acid soil.

Soybean is a fundamental commodity for both the food and industrial sectors in Indonesia. Soybean is the primary source of plant-based protein and is considered to be of utmost importance for food security, following rice and maize. Over the past ten years, there has been a comparatively low level of soybean output at the national level. To enhance soybean output in Indonesia, it is necessary to focus on boosting crop productivity and expanding the cultivation area. Land extensification is feasible, but, it is limited to marginal land with relatively poor productivity. Marginal land in Indonesia encompasses arid, acidic, saline, and peat lands, among others, which exhibit unfavourable physical and chemical conditions for plant growth. Indonesia has a total of 148 million hectares of dry land, with 69.46% of this dry land being classified as acid dry land (Mulyani et al. 2009). Acidic dry ground is distinguished by its low pH levels and poor soil fertility. The pH level of acid dry land is less than 5.5, which enables the dissolution of some harmful elements for plants, such as aluminium (Al) and manganese (Mn). The solubility of aluminium in soil is a significant challenge in crop development. Excessive stress leads to inhibited development of roots, resulting in a decrease in the effectiveness of water and nutrient uptake (Kochian et al. 2004). Plants that are under stress typically exhibit symptoms of deficiency in phosphorus (P), calcium (Ca), magnesium (Mg), boron (B), zinc (Zn), molybdenum (Mo), copper (Cu), potassium (K), and sodium (Na) (Uguru et al. 2012; Luchi et al. 2007). Soybean exhibits limited tolerance to extreme soil acidity, characterised by a pH level below 4.5, and aluminium saturation over 20% (Dierolf et al. 2001). Under these circumstances, the plants exhibit signs such as stunted growth, underdeveloped roots, and yellowing of the leaves. The elevated concentration of aluminium mostly impacts the division and elongation of meristematic cells in the roots, consequently impeding root growth (Purnamaningsih and Hutami 2015). The growth of roots becomes feeble and obscured, resulting in a decrease in the uptake of water and essential nutrients, particularly calcium, magnesium, and phosphorus (Merifio et al. 2010). The soil acidity level significantly affects the development and productivity of soybeans (Uguru et al. 2012). One effective method for addressing acid soil is soil amelioration, such as the application of lime. Applying lime can effectively mitigate aluminium toxicity and enhance pH levels (Zhou et al. 2020). Nevertheless, this method is not economically viable and is unable to address the issue of acidity in lower soil strata. The implementation of a plant breeding programme aimed at developing high-yielding cultivars that are also resistant to acid soils is projected to be a more economically efficient approach to enhancing plant productivity on acid soils. Additionally, this method is environmentally beneficial as it does not involve the release of any chemicals into the environment (Samac and Tesfaye 2003). Nevertheless, the range of soybean genetic variation that is capable of adapting to acid soil remains quite restricted. Screening techniques for acid soil tolerance are crucial in practical breeding. Hydroponic screening is a rapid process that allows for the evaluation of numerous genotypes, although it is typically restricted to the early stage of plant growth. Screening with sand culture is comparable to hydroponics, however it might result in inadequate oxygenation and the buildup of nutrient salts. Screening for acid soil is heavily reliant on the exact type of soil utilised, making it regionally specialised and unsuitable for widespread application.

The main factor that hinders root nodulation in soybean is a low pH. Additionally, it has been hypothesised that aluminium toxicity is the primary constraint in the formation of soybean roots in acidic soil (Eswaran et al. 1997). Root nodule growth is suppressed at an early stage when the root is exposed to acidic medium. This effect would persist even when the plant is transplanted to soil with an optimal pH (Lin et al. 2012). It is evident that an acidic environment inhibits root growth, and the presence of aluminium further restricts root development. Under acidic conditions with a low pH (below 4.5–5.0), plant roots exhibit increased absorption of aluminium (Al), resulting in significant disruption of root elongation and damage to root cells (Hui et al. 2011; Matsumoto 2000). The inhibitory impact of aluminium was not observed at pH 6. However, the average length of roots was greater in the Al treatment compared to the treatment without Al, but this difference was not statistically significant. Aluminium has been discovered to have various impacts on plant growth, which can either be poisonous or beneficial (Bojórquez-Quintal et al. 2017). The presence of aluminium in wheat has a beneficial effect by preventing the harmful effects of excess hydrogen ions (H^+) on the plant's growth and development (Kinraide et al. 1992). Nevertheless, the specific method by which Al operates in a stressful situation is still unknown.

The benefits of acid soil resistant plant roots in high-strength acid soils or in acid soils with macropores, compared to acid soil susceptible genotypes, are not well understood. Haling et al. (2011) conducted a study to determine how the growth and structure of roots in seedlings of five different perennial grass genotypes, which varied in their ability to tolerate acidic soil, were affected by soil acidity, soil strength, and macropores. The researchers also investigated how soil acidity and strength interacted to influence root growth and morphology. The acidity and strength of the soil affected the length and structure of the roots, the growth of root hairs, and caused deformities in the root tip, particularly in genotypes that are susceptible to acidic soil. The length of the roots was partially limited by soil acidity in all genotypes. However, the negative effect of soil acidity on root growth was more pronounced in acid soil resistant genotypes, especially under high soil strength conditions. The roots responded to soil acidity when growing in macropores, but their elongation was enhanced when penetrating high-strength soil. The impact of soil acidity on root growth can be complicated by soil strength, where acid-resistant genotypes show a higher level of sensitivity in soils with greater strength. This emphasises the necessity of choosing genotypes that exhibit resistance to both acidity and high soil strength.

The presence of soil acidity and soil strength poses significant limitations on root growth. However, there is a lack of comprehensive knowledge regarding the potential interplay between these two stresses, despite their frequent co-occurrence. Soil acidity impacts more than 50% of the world's arable land. Aluminium (Al) toxicity in acid soils limits root elongation and, thus, inhibits plant growth. The inclusion of lime and/or the utilisation of acid soil resistant genotypes are essential for the control of acid soils (Scott et al. 2001). However, the growth of roots may still be limited by soil physical conditions. The potential advantage of genotypes with acid soil resistance is not well understood in acid soils that also have high penetrometer resistance (Foy et al. 1999). It has been proposed that genotypes with roots that can withstand multiple soil stresses may be necessary in high-strength acid soils (Foy et al. 1999; Gilker et al. 2002). Robust root development is crucial for the absorption of nutrients and water, which in turn affects the productivity and longevity of plants. This is especially true in environments with little precipitation, where perennial plants depend on developing a deep root system to access moisture in the subsoil in order to survive.

4.11.1 Acid-Resistant Crops and Cultivars That Can Tolerate Soil Acidity

Many plant species that have economic value are commonly believed to have a high tolerance for acidic soil conditions. Given that numerous species have their origins in regions with acidic soil, this suggests that the process of evolution involves adapting to environmental constraints (Somani 1996). Although the species as a whole is not resistant, certain subspecies are capable of tolerating acidic soil. Plant tolerance to acid soil stress can be assessed using quantitative evaluations, which consider tolerance to elevated levels of aluminium (Al) or manganese (Mn), as well as tolerance to deficiencies in calcium (Ca), magnesium (Mg), and phosphorus (P). There is a notable variation in the ability of different species and genetic variations within a species to tolerate Al and Mn, as described by Somani (1996) and Kochian et al. (2004). From a practical standpoint, it is essential to select species or varieties that perform well in conditions with high levels of Al saturation, while requiring minimal lime usage. Over the past decade, several countries in the Sub-Saharan Africa (SSA) region have been conducting research on developing crop varieties that are resistant to acidic conditions. In Cameroon, certain P-efficient cultivars demonstrated a 51% increase in maize production in conditions of low soil pH, while other varieties had a loss in yield by 37% (Tandzi et al. 2018). Similarly, some single cross hybrids in Kenya were found to have a greater capacity to withstand Al toxicity, as demonstrated by Tandzi et al. (2018).

4.11.2 Growing Acid-Tolerant Crop

Aluminium toxicity hampers agricultural yield in acidic soils, and the solution to this problem is soil liming. Nevertheless, due to the substantial amounts of lime required and the related expenses connected with improving these soils, cultivating acid-tolerant crops and cultivars could be a feasible alternative. Blueberries, potatoes, and watermelons exhibit greater acid tolerance compared to crops such as corn, soybean, wheat, alfalfa, and clover. The wide range of Al tolerance observed among different plant species has allowed breeders to create Al-tolerant cultivars and investigate the physiological and biochemical aspects of Al tolerance in germplasm. Wheat has demonstrated its usefulness in this regard, exhibiting a significant variation in aluminium tolerance across its many genetic varieties, up to ten times greater than other cereals (Prochnow 2014). Paddy is an optimal selection due to its ability to counteract acidity and its related adverse consequences in areas with ample water supply. Agroforestry methods, such as multi-storey cropping systems, can also decrease the erosivity of raindrops and the leachability of nutrients. The capacity of the system to decrease soil acidity relies on the specific tree species and the arrangement of the agroforestry system. Baggie et al. (2000) examined the capacity of organic wastes from nitrogen-fixing trees, such as Albizia zygia and Gliricidia sepium, to improve the quality of acidic and infertile rice soils. After a period of four weeks of incubation, it was disclosed that A. Zygia and G. Sepium raised the soil pH from 4.4 to 5.1 and 5.3, respectively, by releasing basic cations into the rhizosphere.

Alfalfa is a highly significant forage crop on a global scale due to its excellent feed quality, substantial yield in many habitats, and capacity to thrive in diverse climatic circumstances (Zhang et al. 2008; Turan et al. 2009). Alfalfa is primarily cultivated for the purpose of producing feed, with seed yield being of secondary importance (Iannucci et al. 2002). The forage can be utilised either for direct grazing or preserved as silage or hay. It serves as a dependable and high-quality supply of nutrients, making it a valuable addition to the agricultural sector (Borreani and Tabacco 2006). The acidity level of soil can be determined by considering both the pH value and the quantities of Al^{3+} present in the soil. Soil acidity is a multifaceted problem that has several impacts on crop productivity, such as nutrient and element deficits and toxicities, hindered plant root growth, and diminished activity of beneficial bacteria (Fageria and Baligar 2003). These factors restrict the plant's ability to absorb nutrients and water. The issue is further complicated due to the multitude of interactions that take place among these components (Foy 1992). Correcting soil acidity decreases the levels of Al, Fe, Mn, Zn, and Cu, while increasing the availability of other necessary nutrients. For most crops, save lucerne, the optimal pH range for achieving maximum plant growth is in the slightly acidic range of 6.0–6.5 (Fageria 2009). Liming is a highly effective and widely used method for addressing soil acidity and enhancing the productivity of lucerne crops. In a study conducted by Moreira and Fageria (2010), the reaction of lucerne cultivated in a greenhouse on a Typic Quartzipsamment soil was evaluated by applying lime. The presence of a positive linear association between the dry matter of roots and nodules and the application of lime suggests that these plant characteristics improve as soil acidity decreases. The assessment of lucerne yield was influenced by several significant parameters, including soil acidity indices such as pH, base saturation, Ca^{2+} concentration, Mg^{2+} concentration, and H + Al (Moreira and Fageria 2010). The impact of various soil acidity markers on yield, as measured by the coefficient of determination (R2), varied and decreased in the following order: base saturation, H + Al, pH, Ca, and Mg concentrations. The study identified the ideal levels of specific soil chemical qualities that result in the highest shoot dry matter. These values can be used as a reference for applying lime to acid soils in order to enhance the yield of lucerne feed.

4.11.3 Soil Acidity Tolerance of Fruit Crops

Slightly acidic (pH 6.0–6.8)

The fruits included are mango (Mangifera indica), citrus (Citrus spp.), banana (Musa spp.), guava (Psidium guajava), papaya (Carica papaya), apple (Malus domestica), peach (Prunus persica), carambola (Averrhoa carambola), and kiwi fruit (Actinidia deliciosa).

Moderatly acidic (pH 5.5–6.8)

The fruits mentioned include pineapple (*Ananas comosus*), orange (*Citrus sinensis*), litchi (*Litchi chinensis*), passion fruit (*Passiflora edulis*), and jackfruit (*Artocarpus heterophyllus*).

Highly acidic (pH range of 5.0–6.8)

The following fruits are: strawberry (Fragaria ananassa), gooseberry (*Emblica officinalis*), bael (*Aegle marmelos*), elephant apple (*Dillenia indica*), and plum (*Prunus domestica*).

Aluminium (Al) toxicity is a significant component that impacts crop yields in highly acidic soils. Aluminium ions with a charge of +3, which are discharged from clay minerals when the pH levels drop below 5, have a detrimental effect on plant roots and hinder their growth and functionality. Subsequent to drought and mineral insufficiency, there is a decrease in agricultural yields. Based on the findings of von Uexkül and Mutert (1995), Wood et al. (2000), and other studies, around 40–50% of the world's potentially cultivable land is characterised by acidity. The bulk of these acidic soils are found in the tropical and subtropical regions. The presence of high levels of aluminium (Al) in the soil limits the ability of crops to grow in locations where food security is already a major concern. This is particularly true in developing countries, as they are often located in areas with a high concentration of acidic soils. Drought is the only abiotic factor that decreases agricultural yield more than aluminium toxicity, according to von Uexkül and Mutert (1995). Aluminium toxicity manifests through many processes, which encompass interactions between aluminium and the symplasm, plasma membrane, and root cell wall, owing to Al's strong reactivity.

4.11.4 Techniques to Enhance Resistance and Tolerance of Plants to Aluminium

When exposed to Al stress, cultivars that are resistant to Al exhibit reduced levels of pectin methylesterase activity and increased quantities of methylated pectin. Consequently, the cell wall now has a reduced net negative charge, which hinders the binding and clustering of Al^{3+} ions. In a separate investigation, Yang et al. (2013) discovered that rice and potato (Solanum tuberosum) roots that exhibited increased levels of pectin methylesterases contained higher quantities of aluminium in the cell wall of the root tip and displayed heightened sensitivity to aluminium. A study conducted by Yang et al. (2011) revealed that hemicelluloses in Arabidopsis play a significant role in binding aluminium, similar to pectins. This finding shows that hemicelluloses may contribute to both aluminium toxicity and tolerance. Zhu et al. (2011) discovered that Al stress hinders the activity of xyloglucan endotransglucosylase (XET), an enzyme located in the cell wall of Arabidopsis. XET is responsible for breaking and reconnecting hemicellulosic xyloglucan polymers during cell development.

4.12 Aluminium Accumulator Plants

The presence of a large amount of aluminium (Al) ions dissolved in acidic soils hampers the growth of plants. Plant adaptation to Al stress can be categorised into two mechanisms: Al exclusion and internal Al inactivation. Plant species that possess systems to exclude aluminium are commonly referred to as "Al excluders," while those that possess mechanisms to accumulate aluminium are known as "Al accumulators." Aluminium accumulators are commonly found in acid soils, especially in regions with high humidity, such as the wet tropics. Al accumulators exhibit a broad phylogenetic distribution in plants, in addition to their geographical distribution (Watanabe 2022).

The presence of a high concentration of aluminium ions in soil solutions is widely recognised as the primary factor that limits plant growth in acidic soils. In soils with a pH over 5.5, the naturally existing forms of aluminium are often stable and not harmful. However, when the pH drops below 5.5, aluminium ions that can dissolve in soil solutions become accessible. Aluminium toxicity is predominantly an issue in the roots of the majority of plant species. According to reports, Al enhances the permeability of the plasma membrane by binding to phospholipids (Matsumoto et al. 1992) and membrane proteins (Caldwell 1989), or through lipid peroxidation (Ikegawa et al. 2000). Aluminium (Al) enhances the stiffness of the cell wall by interacting with pectin and hemicelluloses, as demonstrated by Wehr et al. (2004), Horst et al. (2010), and Yang et al. (2011). In general, the processes by which plants tolerate aluminium can be divided into two categories: aluminium exclusion and internal aluminium inactivation. Many plant species have evolved the capacity to prevent the entry of aluminium into their roots as a means of adjusting to acidic soils (Kochian et al. 2015). The primary mechanism by which Al is prevented from entering the plant is by the release of organic acid anions from the root. Certain organic acid anions have the ability to form stable complexes with aluminium, leading to the deactivation of aluminium in the rhizosphere. The effectiveness of Al inactivation relies on the chelating capacity of the organic acid anions that are secreted, which can be correlated with the stability constants of the Al-organic acid anion chelate. The main organic acids that roots release to detoxify aluminium (Al) are citrate, found in maize, Cassia tora, soybean, snapbean, rye, and triticale; malate, found in wheat, rye, and triticale; and oxalate, found in buckwheat and taro. Anion transporter is believed to regulate the exudation of organic acid anions. Anion channel inhibitors were shown to suppress the release of oxalate and reduce the tolerance to aluminium in buckwheat, as reported by Zheng et al. (1998).

Therefore, the release of organic acid anions is a crucial mechanism for plants to tolerate aluminium. Nevertheless, it is important to note that the tolerance of plants to aluminium cannot be solely attributed to the release of organic acid anions. No link was found between Al tolerance and organic acid anion (citrate) exudation under Al stress in seven different plant species, as observed by Wagatsuma et al. (2001). For instance, rice exhibits a comparatively better tolerance to aluminium (Al), yet it does not release substantial quantities of organic acid anions from its roots to enhance its Al tolerance, as observed by Wagatsuma et al. (2001). Multiple studies have been conducted to investigate the methods by which rice tolerates aluminium. According to Watanabe and Okada (2005), the variations in aluminium tolerance between japonica (tolerant) and indica (sensitive) types can be attributed to distinct electrochemical properties of root tip cells. In their study, Khan et al. (2009) found that variations in the lipid content of the plasma membrane could account for the disparity in aluminium tolerance observed in rice.

Another crucial mechanism for Al tolerance is the internal Al tolerance. In general, the majority of Al that is absorbed by roots cannot be transferred to shoots. However, certain plant species that have acclimated to acidic soils are capable of transporting and accumulating significant quantities of aluminium in their shoots. Plant species that exhibit a high degree of internal aluminium tolerance are sometimes referred to as "Al accumulators" and are predominantly woody plants. Accumulator plants, such as those mentioned by Memon and Schröder (2009), can also be found in the presence of other harmful components, such as heavy metals. This suggests that plants may have a shared adaption mechanism to deal with hazardous materials. Phytoremediation has utilised heavy metal accumulator plants to extract harmful substances from polluted soil (Memon and Schröder 2009). How do aluminium accumulators withstand the elevated levels of aluminium in their tissues? Studies have demonstrated that certain species that accumulate substances have persistent complexes of aluminium with organic or inorganic ligands in their tissues. This occurs, for example, in the rhizosphere of plants that reject aluminium and release organic acid ligands from their roots. Melastoma malabathricum, a woody plant found in Southeast Asia, has the ability to store about 10,000 mg of aluminium per kilogram of dry weight in its leaves, primarily in the form of monomeric aluminium and aluminium-oxalate complexes (Watanabe et al. 1998a, b). Furthermore, it was documented that the primary form of aluminium in the leaves of Hydrangea macrophylla is an aluminium citrate complex (Ma et al. 1997a; b, c). Faramea marginata, a woody plant that primarily grows on acidic soils, exhibits high levels of both aluminium (Al) and silicon (Si) accumulation. The chemical compound Si is in the form of F. The accumulation of silicon in marginata, unlike in rice which is a typical silicon accumulator, is distinct and is likely to form a complex with aluminium (Britez et al. 2002). What is the reason for their ability to amass such elevated levels of aluminium in their shoots? In highly acidic soils, the levels of aluminium activity in the rhizosphere are exceptionally elevated. When it comes to Al excluders, significant quantities of organic acid anions, produced by photosynthate, are required to deactivate rhizospheric Al in these types of soils, leading to an increase in carbon loss. In addition, organic acid anions will disperse in the soil solution and undergo decomposition by soil microbes. In contrast, Al accumulators have the ability to concentrate organic acid anions within the plant. This system has a high energy preference for survival in highly acidic soils. In addition, certain Al accumulators employ Al for their development. Despite the first letter "M." Malabathricum is a prevalent species in highly acidic soils that contain high levels of both aluminium (Al) and iron (Fe), such as acid sulphate soils. However, it is extremely susceptible to iron poisoning. Iron (Fe) is a crucial element for plants. However, an excessive amount of Fe generally leads to an overproduction of reactive oxygen species. According to Watanabe et al. (2006), their research shows that the absorption of aluminium (Al) in M competitively reduces the absorption of iron (Fe). Malabathricum. M. causes a reduction in the buildup of Fe. Malabathricum is known to thrive as a dominant plant species in highly acidic soils.

4.12.1 Distribution of Plants That Accumulate Aluminium

Al accumulators are categorised as plant species that possess a minimum of 1000 mg kg^{-1} of aluminium in their leaves or shoots, as stated by Chenery (1948). Jansen et al. (2002) conducted a comprehensive analysis of the literature's data, utilising modern molecular phylogenies to evaluate the phylogenetic and systematic implications of the Al-hyperaccumulation characteristic. It was found that most plants that accumulate aluminium are eudicots, and they are especially common in the lower, woody branches of more advanced groups, including rosids and asterids. Nevertheless, it is probable that the characteristic has been eliminated in more advanced, non-woody plant species (Jansen et al. 2002). The research conducted by Schmitt et al. (2017) also showed that pteridophytes have a wide spread of plants that accumulate high levels of aluminium (Al), and that the occurrence of Al hyperaccumulation is significantly more common in pteridophytes compared to angiosperms.

4.12.2 Phylogenetic Variation of Aluminium Forms in Al Accumulator Species

Oxalate has been identified as the predominant ligand for a specific form of aluminium in the tissue of aluminium accumulator species in different groups of flowering plants (Ma et al. 1997a, b, c; Maejima et al. 2014; Watanabe et al. 1998a, b). Additionally, a significant amount of non-chelated monomeric aluminium was found in the leaves of woody eudicots (Maejima et al. 2014; Watanabe et al. 1998a, b). Oxalate is commonly considered as a final product that is not further metabolised or metabolised at a sluggish rate. The primary role of oxalate is to regulate calcium levels and provide protection against herbivory by generating calcium oxalate crystals (Franceschi and Nakata 2005). Furthermore, oxalate is inherently poisonous when it is in a soluble state. Thus, it is plausible that an abundance of oxalate is neutralised by forming a complex with calcium in plants. The detoxification of aluminium by oxalate in plants may be attributed to the development of calcium oxalate crystals. In addition, silicon (Si) has been proposed as another ligand for aluminium (Al) in certain species that accumulate Al, such as Dicranopteris linearis, a type of fern (Malta et al. 2016; Liu et al. 2019).

4.13 Manganese Toxicity

Aluminium, manganese, and hydrogen exhibit toxicity under acidic conditions. Plants are intolerant of large amounts of these elements. Phosphorus and molybdenum react with liberated aluminium and iron in acidic soils, resulting in reduced availability to plants. Aluminium and manganese are released into the soil solution in excessive amounts when the pH ($CaCl_2$) decreases below 5.0. Aluminium (Al) toxicity occurs when soil extractable Al levels exceed 2 mg/kg or when exchangeable Al levels exceed 5%. Elevated concentrations of aluminium and manganese have a detrimental impact on the growth of roots and the organisms living in the soil. The diminished soil coverage and heightened runoff result in an elevated susceptibility to soil erosion, leading to subsequent water pollution and sedimentation in streams (Mcfarland et al. 2001). Aluminium exists in soils in several forms and is attached to soil components, especially clay particles and organic materials. When the pH of the soil decreases, aluminium becomes soluble and the concentration of aluminium in the soil solution rises. Typically, soil aluminium content ranging from 2 to 5 ppm (mg/kg) is harmful to the roots of plant species that are sensitive, whereas concentrations over 5 ppm are harmful to species that are tolerant. Typically, the top layer of soil includes a sufficient amount of organic matter, which prevents aluminium from becoming detrimental to plant roots. However, laboratory studies have shown that aluminium may still be extracted from the soil. Excessive concentrations of aluminium in the soil solution hinder root cell development and impede the elongation of the root. The apical regions exhibit deformities and fragility. There is a decrease in root development and branching. Inadequate water and nutrition lead to poor crop and pasture development, reduced yield, and smaller grain size. The manifestation of aluminium poisoning is especially pronounced during arid seasons. Roots are unable to properly penetrate acidic subsurface soil, which acts as a barrier and restricts access to stored subsoil water for grain filling. The presentation was given by Gazey and Davies (2009).

Manganese (Mn) is an essential trace element for all phases of plant development, yet it can be toxic in high concentrations. Maize plants are considered deficient in manganese (Mn) when the concentration is below 15 ppm. If the concentration is between 15 and 25 ppm, it is considered low. A concentration between 26 and 150 ppm is considered sufficient. When the concentration is between 151 and 200 ppm, it is considered high. Concentrations over 200 ppm are considered excessive or dangerous (Schulte and Kelling 1999). The hydrolysis of aluminium (Al) and iron (Fe), which are the main mechanisms responsible for soil acidity, are linked to manganese (Mn) toxicity. Soil acidification in natural and agricultural systems enhances the solubility of manganese (Mn) and increases its bioavailability to levels that are fatal (Hong et al. 1991). The increased solubility of Mn due to a decrease in soil pH exacerbates the negative effects of Mn toxicity in susceptible plants (Hong et al. 1991). Chlorosis, a condition characterised by the yellowing of leaves, is an initial indication of manganese poisoning in plants. Over time, this condition progresses to necrosis, which is the death of plant tissue. This phenomenon primarily affects the oldest leaves of the plants (Snowball and Robson 1991). Until now, there have been only a limited number of published articles on the topic of manganese toxicity in plants. Hence, further investigation is required in this area of inquiry.

Manganese is the second most significant hazardous element present in acidic soils, predominantly found in acidic soils located in Africa and certain regions of Asia (Foy et al. 1988). Manganese in its divalent form (Mn^{2+}) is detrimental to plant growth, and under conditions of low soil pH, MnO_2 is converted into this harmful form of manganese. Manganese toxicity typically occurs in soil with a pH of 5.5 or lower, assuming that the soil has an enough amount of manganese present. However, it can also happen in soils with a higher pH, in poorly drained or compacted soils where reducing circumstances promote the creation of divalent manganese that plants can absorb.

4.13.1 Toxicity of Aluminium and Manganese

Soils with a pH below 4.8 often contain high concentrations of aluminium and manganese, which can be hazardous (Table 4.2). Aluminium toxicity hinders the movement of phosphorus, harms plant roots, and limits the ability to search for additional nutrients and water. This phenomenon is particularly pronounced on soils that have undergone extensive weathering, such as podzolic soils. Plants exposed to aluminium typically exhibit stunted growth, while clovers display a pale appearance and have little leaves. Highly delicate plants, particularly lucerne and to a lesser degree certain types of phalaris, exhibit poor health and an extremely slender growth pattern. They may perish within two years of planting in highly acidic soils. Manganese toxicity is prevalent in soils that have undergone minimal weathering, such as red, black clay loam, and self-mulching soils. Manganese toxicity is more likely to occur in hot and dry settings or in warm soil that is saturated. Manganese poisoning is less prevalent than aluminium toxicity. It results in the discoloration of the edges of leaves and overall stunted growth of plants that are vulnerable to it. The incidence of it is occasional, contingent upon the levels of soil moisture and temperature.

Yields are limited by the ineffective root systems that arise from insufficient uptake of nutrients and water. Significantly, aluminium poisoning significantly amplifies the consequences of drought during seasons with lower rainfall. The situation regarding moisture supply is made worse in cases where subsoils are also acidic because plants are unable to fully utilise the moisture stores in the subsoil due to limited root penetration into the lower soil layers (Photos 1a and b). Even on moderately moist soils, it is not unusual to observe plants under these conditions exhibiting signs of drought stress. There are many soils with high Mn that are harmful to plants. The greatest growth-limiting component in acid soils globally, after Al, is excess Mn (Manyowa and Miller 1991). According to Sparrow and Uren (1987), wet soils are linked to decreased and poorly drained settings, which also have high concentrations of Mn in

Table 4.2 Tolerance to manganese (Mn), and critical levels* of manganese for pasture plants

Tolerance to Mn	Plant	Critical level* of manganese (mg/kg)
Highly sensitive	Lucerne, barrel, and burr medics	200–400
Sensitive	White and strawberry clovers	400–700
Tolerant	Subclove	700–1000
Highly tolerant	Most pasture grasses	>1000

*The critical levels of manganese in this table are the levels (in the youngest fully developed leaf) that are sufficient to cause a 10% decline in growth
Source NSW Agriculture Agfact AC.19, Soil Acidity and Liming

the soil solution. Certain environmental conditions (such as high temperatures; Marsh et al. 1989) can also raise Mn availability, which in turn influences crop development. In southeast Australia, high Mn concentrations in the soil (caused by moderate acidification) constitute a common nutritional issue (Culvenor et al. 1986; Weir 1988). In northern Victoria, waterlogging was found to be the cause of elevated soil Mn concentrations (Sparrow and Uren 1987). Similarly, after waterlogging an acidic soil, wheat shoot tissues in western Australia showed high to hazardous quantities of Mn (Khabaz-Saberi et al. 2006), providing the first clue of a possible Mn toxicity issue. Additional proof was supplied by a pot study conducted in waterlogging-prone regions of the western Australian wheatbelt utilising acidic soils (Setter et al. 2008). Australia has vast tracts of agricultural land that are prone to waterlogging, including 1.8 million ha in western Australia (which receives more than 400 mm of rainfall annually) and another 4.1 million ha in Victoria, which largely has duplex soils (Setter and Waters 2003). Raised beds and drainage are two tried-and-true technical options for dealing with waterlogging. However, because to issues with draining drainage water (particularly if acidic) in Australia's agricultural environment, they are not only costly (Dennis et al. 2000) but also usually impractical. The selection of genotypes tolerant to Mn toxicity as a significant consequence of waterlogging represents an alternative approach to drainage. In fact, resistance to excess Mn coincided with resistance to waterlogging in certain plant species (Mn-tolerant Trifolium subterraneum cultivar vs. Mn-sensitive Medicago truncatula) (Robson and Loneragan 1970). This suggests that genotypes tolerant to Mn toxicity may also be tolerant to waterlogging in other agricultural species (such as wheat, which is Australia's most important crop). This highlights the possibility of using the variation in Mn tolerance found in the genotypes of wheat that are currently in use to breed for Mn tolerance.

Researchers have discovered that the amount of bases (mostly calcium and magnesium) and the amount of aluminium influence each other more than the amount of aluminium alone when determining how poisonous aluminium is to plant roots. Because soil acid saturation index is a significantly more reliable indicator of acidity hazard than soil pH, it is used to depict the aluminium toxicity threat on soil test reports. Figure 4.2 shows the relative crop tolerance to acidity in the soil.

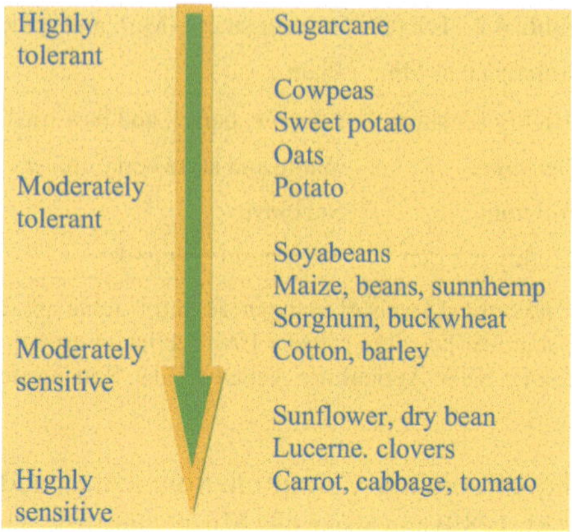

Fig. 4.2 Relative tolerance of crops to soil acidity

4.13.2 Manganese Tolerance

Plant species differ considerably in tolerance of high levels of soil Mn. A review by Kamprath and Foy (1985) suggests that maize and rice are more tolerant of high soil Mn than soybean and barley, and alsike clover and oats are more tolerant than cowpea, lespedeza, and sweet clover. Among flower crops, carnation, poinsettia, and rose were tolerant while snapdragon and calendula were sensitive. Vegetables were ranked in tolerance (from most tolerant to sensitive) as follows: tomato > lettuce > barley and bean > clover > potato (Yost 2000). Many tropical legumes have been reported to be Mn-tolerant. Although critical levels of Mn depend widely on growth conditions, suggested levels for various crops have been established. As in the case of Al tolerance, there also appear to be differences in Mn tolerance within plant species. Varietal tolerance has been reported in soybean, wheat, apple, and cotton. Tolerance of Mn in forage legumes appears to be a combination of prevention of Mn entry into the plant and greater internal tolerance. Superior Mn tolerance in corn (compared to peanut) is believed to be due to reduced transport of Mn to the leaves. On the other hand, Brown and Devine (1980) found that Mn-tolerant and Mn-sensitive soybean cultivars both contained about the same amounts of Mn in their tops. Foy also described differences among plants in their requirements for Ca, Mg, and P that can help confer tolerance when growing in acid soils, which are often low in these nutrients. The ability to grow well with low levels of Ca, Mg, and P is therefore another type of tolerance of (or ability to grow in) soils with low pH. While there are species and varietal differences in tolerance of Al and Mn, it is often assumed that tolerance of high Al might be correlated with tolerance of high Mn. This does not always occur. There are, indeed, many species with considerable tolerance of high soil Al and high soil Mn, and even within species there are often varieties that are tolerant of such adverse conditions. Tropical biodiversity has contributed to the development of this crop management alternative by contributing many of the tolerant species and cultivars.

4.14 Crop Vulnerability

High concentrations of manganese (Mn), iron (Fe), and aluminium (Al) induced in waterlogged acid soils are a potential constraint for growing sensitive wheat cultivars in waterlogged-prone areas of western Australian wheatbelt. Tackling induced ion toxicities by a genetic approach requires a good understanding of the existing variability in ion toxicity tolerance of the current wheat germplasm. A bioassay for tolerance to high concentration of Mn in wheat was developed using Norquay (Mn-tolerant), Columbus (Mn-intolerant), and Cascades (moderately tolerant) as control genotypes and a range of $MnCl_2$ concentrations (2, 250, 500, 750, 1000, 2000, and 3000 lM Mn) at pH 4.8 in a nutrient solution (Khabaz-Saberi et al. 2010). Increasing solution of Mn concentration decreased shoot and root dry weight and intensified the development of toxicity symptoms more in the Mn-intolerant cv. Columbus than in Norquay and Cascades. The genotypic discrimination was based on relative shoot (54–79%) and root dry weight (17–76%), the development of toxicity symptoms (scores 2–4) and the shoot Mn concentration (1428–2960 mg kg^{-1}) was most pronounced at 750 lM Mn. Using this concentration to screen 60 Australian and 6 wheat genotypes from other sources, a wide variation in relative root dry weight (11–95%), relative shoot dry weight (31–91%), toxicity symptoms (1.5–4.5), and shoot Mn concentration (901 to 2695 mg kg^{-1}) was observed. Evidence suggests that Mn tolerance has been introduced into Australian wheat through CIMMYT germplasm having "LERMO-ROJO" within their parentage, preserved either through a co-tolerance to Mn deficiency or a process of passive selection for Mn tolerance. Cultivars Westonia and Krichauff expressed a high level of tolerance to both Mn toxicity and deficiency, whereas Trident and Janz (reputed to be tolerant to Mn deficiency) were intolerant to Mn toxicity, suggesting that tolerance to excess and shortage of Mn are different, but not mutually exclusive traits (Khabaz-Saberi et al. 2010).

In Australia, Scott et al. (2001) reported no yield advantage associated with Mn tolerance in six pairs of Al-tolerant lines with contrasting Mn tolerance growing on an acidic soil with high Mn in southern New South Wales. However, the study site (dry summer and below-average rainfall during the growing season) did not favour Mn toxicity during that particular growing season; moreover, the range of Mn tolerance was relatively narrow. Recognising these limitations, the authors concluded it was premature to consider that Mn tolerance in wheat does not impact on grain yield (Scott et al. 2001). Hence, for evaluation of the importance of Mn tolerance, it is critical to use genotypes with vastly different Mn tolerance grown in soil where reduced conditions exacerbate Mn toxicity. Variation of Mn tolerance has been reported within Brazilian (Camargo 1988), American (Foy et al. 1988), and Canadian wheat varieties (Moroni et al. 1991) as well as in seven Australian wheat varieties in a pot assay (Scott et al. 1998). Clearly, evaluation of a wider range of Australian locally adapted wheat germplasm is needed to assess potential presence of sufficient variation of Mn tolerance to justify the breeding efforts.

Such knowledge would also represent a basis for further characterisation of genetic and molecular aspects of Mn tolerance in wheat, the work that has already been done in some other crops such as rice (Wang et al. 2002) and soybean (Kassem et al. 2006). Therefore, the present study was designed to (1) develop a screening system for tolerance to high concentration of Mn in wheat using genotypes with previously known differential Mn tolerance, and (2) use that screening system to evaluate genotypic variation for Mn tolerance in Australian wheat varieties and advanced breeding lines. The source of Mn tolerance in hexaploid Canadian wheat has been suggested to be Brazilian landraces introduced either directly or indirectly through CIMMYT's germplasm containing Brazilian parentage (Moroni et al. 1991). Similarly, the Brazilian Al-tolerant germplasm with or without co-tolerance to Mn was also introduced into Australia in an attempt to improve Al tolerance, resulting in Al-tolerant varieties without (Tammarin Rock) or with Mn tolerance (Westonia, for Al tolerance: Tang et al. 2003; for Mn tolerance) developed by the major western Australian breeding programme with an active selection for Al tolerance (Carver et al. 1995; Tang et al. 2003).

4.15 Climate and Crops

Climate change has severe consequences on the natural environment and agricultural production, leading to poverty and food insecurity (Arraiza et al. 2014). The indirect impact of such alterations include a shift in cropland distribution, while a direct effect involves, among others, sulphidic wetland drying (because of climate change-induced drought). As a result, climate change has a catalytic effect on the changes in soil quality (Schoenholtz et al. 2000), specifically soil acidification, which is a worldwide problem, but especially so in Africa and Europe. Africa contributes a higher proportion (16.7%) of global acidic soils compared to Europe (9.9%) (von Uexkül and Mutert 1995). In parts of South Africa, for example, the KwaZulu-Natal province, 85% of soils have a pH of less than five (Roberts and Smeda 2001), which has a significant impact on crop production, leading to yield losses of up to 70% (Dewi-Hayati et al. 2014). The extent of yield reduction depends on the level of hydrogen ions (which affects plant growth and development) (Anderson and Bell 2019), climatic conditions and genetic background of the cultivar. In addition, the combination of H^+-toxicity, nutrient deficiency, and reduced water uptake influences growth negatively (Long et al. 2017). Many studies associate soil acidification with nitrogen fertiliser over-application (Oertel et al. 2016), acid rain, changes in soil physical properties (Matsumoto et al. 2018), plant residues (Meda et al. 2001), and changes in soil chemistry (Blake et al. 1999) as the main common reasons worldwide.

Acid soil affects phosphorus uptake, root length, and mean root diameter (Robles-Aguilar et al. 2019). Furthermore, low soil pH could influence the levels of reactive oxygen species (ROS) (Song et al. 2011) such as singlet oxygen ($^1O_2$), superoxide anion (O_2^-), hydrogen peroxide (H_2O_2), and hydroxyl radical (OH). These radicals are able to oxidise vital cellular ultrastructure and give rise to oxidative damage and devastation of cellular organelles (Sharma et al. 2012).

The presence of aluminium, zinc, manganese, and copper in acid soils develop stressful conditions for plants, and result in poor yields of different crops. Therefore strategies have to be developed to make the plants adaptable to the unconducive environments. Better management practices could be used successfully for raising crops in the areas affected. Adaptation of some species to high salt concentrations provides evidence of the genetic potential that existed in plants to cope with the unfavourable conditions (Khan et al. 2001). Variation, both between and within the species, and even at intra-varietal level does exist in different crops. Development of maize varieties tolerant to acidic soils are the elegant examples in the domain of plant breeding. The studies on the genetic basis of tolerance in plants for salinity and toxicity due to aluminium, manganese, copper and zinc revealed that variation in tolerance was genetically controlled and thus heritable in nature. Physiological mechanisms involved in controlling the responses to stress may aid in selecting successful plants (Khan et al. 2001).

The term stress is derived from the Latin word stringere, which means a constraining or impelling force. From crop production point of view, Grime (1979) defined stress as "the external constraints which limit the rate of dry matter production of all or part of the vegetation below its genetic potential." However, Jones and Jones (1989) used economic yield, rather than dry matter production, to measure the response of plants to stress. After planting a seed in soil, development and productivity of plant is subjected to numerous environmental stresses. Generally, plants are considered to be under stress when they experience a relatively severe shortage of an essential constituent or an excess of potentially toxic or damaging substance, and the plants usually face both the situations simultaneously. These abiotic soil stresses constitute a major limiting factor, which hamper plant productivity throughout the world (Clark and Duncan 1993). Therefore strategies have to be developed to make use of these resources for food production. Several plant species are known to have evolved on contaminated soils, and thus developed the potential to grow successfully on affected soils, where non-tolerant individuals of the same species showed complete death (Bradshaw and McNeilly 1981). This tolerance is specific to the individual metal, and is highly heritable (Gartside and McNeilly 1974a, b).

4.15.1 Crop Production on Problem Soils

Development and adoption of management methods/practices is also an important aspect of crop science for getting higher crop yields their adaptation to salt-stressed conditions. In contrast to salinity, soils affected by deposition of aluminium and manganese in the root zone cannot be amended using soil amendments. Although application of lime can reduce the toxic effects of aluminium and manganese in the acid soils by increasing their soil pH, the application of lime is not universally possible due to non-availability of abundant supplies of lime to reclaim huge acreage.

Moreover, liming the soil surface does little to correct chemical impediments in subsoil root penetration, and thus has to be repeated. Therefore the best way of utilising these wasteland is through growing plant species which can survive under these toxic conditions.

4.16 Positive Impact of Aluminium in Plants

Since Maze (1915) and Stoklasa (1922) first discussed the potential function of aluminium in plant growth and development, research on the advantageous effects of aluminium on plants has drawn a lot of interest. Although the precise mechanism underlying the advantageous effect of Al is yet unknown, a few explanations have been put up. Aluminium is beneficial and even potentially essential for some plant species (Bojórquez-Quintal et al. 2017), because of the Al-induced stimulation of nutrient uptake (Watanabe and Osaki 2002). Aluminium supply was shown to stimulate N uptake by several plant species adapted to acid soils (Osaki et al. 1997), and Al treatments increased shoot N contents in wheat and rye (Dinev and Stancheva 1993). In contrast, Al reduced root N uptake and its upward translocation to shoots in sorghum and corn (Pintro et al. 1996). Aluminium promoted the growth of plants supplied with ammonium (NH_4^+) but inhibited that of plants supplied with NO_3^- (Zhao et al. 2014). Nitrogen is a metabolic element involved in the synthesis of amino acids and proteins within plants. Knowledge about Al–N interactions may supply new information to explain instances where Al benefits plant growth. Several reviews have focused on the interactions between Al and phosphorus (Chen et al. 2012), calcium (Meriño-Gergichevich et al. 2010), magnesium (Chen and Ma 2013), boron, and silicon (Horst et al. 2010). Aluminium is a metal and a toxic element to many plants, while N is a non-metal and is an essential element for all plants.

4.16.1 Al Promotes Plant Growth in Acidic Soil

As early as 1928 (McLean and Gilbert) reported that in cereal crops and citrus plants, low concentrations of Al stimulated growth. Improved plant growth induced by Al has been found mostly in plants native or adapted to acidic soils, such as *Melaleuca cajuputi, Arnica montana, Deschampsia flexuosa*, and *Hydrangea paniculata*, when Al is mainly available at low concentrations (Osaki et al. 1998; Pilon-Smith et al. 2009). *Melastoma malabathricum* L. is a dominant woody species that accumulates Al and grows well in tropical acidic and nutrient-deficient soil in Australia, Polynesia, and tropical Asia (Osaki et al. 1998). Al stimulated an increase in dry matter, nutrient uptake, and number of white, fine, and metabolically active roots, but reduced the lignin content in roots (Osaki et al. 1998; Watanabe et al. 2005). Furthermore, abnormal growth, such as rolling of leaves was found in *M. malabathricum* when grown for long periods in the absence of Al. *Miscanthus sinensis* is a dominant species of grass in temperate climates in acidic sulphate soil and its growth is improved by Al supply, although this species is not an Al accumulator (Kayama 2001). Cowpea exhibited growth enhancement when Al was applied at pH 3.8 when compared with pH 5.5 (Malkanthi et al. 1995), indicating that Al is beneficial to plant growth to some extent. In acidic conditions, the growth of *Eucalyptus gummifera* increased by five-fold over 5 months when Al was added in a quartz culture (Mullette 1975). Numerous studies found that Al could cause morphological changes, such as white coloration and elongation in the roots of *Camellia sinensis, M. sinensis, Quercus acutissima, M. malabathricum, Eucalyptus viminalis,* and *Symplocos paniculata* (Ghanati et al. 2005; Watanabe et al. 2006). Similar benefits of Al on growth were also seen in other plants, including *Betula pendula* (Kidd and Proctor 2000), *Quercus serrata* (Tomioka and Takenaka 2007), and *C. sinensis* (Mukhopadyay et al. 2012; Xu et al. 2016). Furthermore, Al elongated central cap cells, resulting in enhanced root elongation in canola (Clune and Copeland 1999). Some recent reports observed that Al application has positive influence on important economic crops, such as rice (Famoso et al. 2011) and corn (Wang et al. 2015a, b, c, d).

Al's temporary effects, which last only a brief time, include the prevention of H^+ toxicity and an increase in root elongation (Kinraide 1998). Root biomass and the quantity of lateral roots both increased after the Al treatment. In Mexico and Central America, *C. xalapensis* is a shrub that hyperaccumulates aluminium and inhabits polluted environments. This species has been proposed as having a potential function as an indicator for the toxicity and pollution of Al (González-Santana et al. 2012). In the presence of 160 and 200 M of Al, an increase in root growth has been noted in rice cultivars that are tolerant to Al (Moreno-Alvarado et al. 2017). Low doses of Al were administered to maize, which hindered root growth but promoted leaf growth (Wang et al. 2015b). Similar to this, horticultural species like turnip (Brassica rapa) and the legume Glycine max exhibit increased root activity and elongation with increased Al concentration (Yu et al. 2011). Other creatures can also develop more quickly when exposed to aluminium. Al promotes the growth and biomass accumulation of phytoplankton, mostly diatoms, in marine environments (Golding et al. 2015; Zhou et al. 2016). With the addition of 500 M of Al, it was possible to see an increase in growth (in vitro) in microorganisms like Frankia (a nitrogen-fixing bacteria), and it was also possible to see that Al avoided the inhibitory effect of the acidic pH (Igual and Dawson 1999). Animal cell bone formation is also influenced by aluminium (Quarles et al. 1989).

4.16.2 Positive Interaction of Al with Other Nutrients and Heavy Metals

Acidic soil has various syndromes including nutrient deficiency and mineral toxicity. In some case, Al maintains nutrient balance in plants by inhibiting toxic minerals and stimulating deficient nutrients.

Al enhances phosphorous efficiency under P deficient condition

The growth of most plants in temperate and tropical climates is induced by Al application in acidic condition, which could be attributed to the improved uptake of N, P, and K. However, increased P concentration in roots might be due to Al-P precipitation on the root surface and/or in the Donnan free space. It was reported that Al application increased the availability of P in *Betula pendula* seedlings in low P soil (Clegg and Gobran 1995). Numerous mechanisms on P availability have been reported in different plants under low P condition, viz., secretion of solutes from roots, including phosphatases, organic acids, phenolic and flavonoids compound that mobilize P from the unavailable form to available. P would bind Fe or Al, which is unavailable to plants, while root exudates may substitute P from these compounds, resulting in increased P efficiency. Organic acids secretion was reported under deficient P soil in white lupine (Johnson et al. 1996), purple lupine (Ligaba et al. 2004), and soybean (Liao et al. 2006). Al application accelerated organic acids exudation, which promoted root growth and ultimately increased the availability and uptake of P in P deficient conditions (Ma et al. 2001; Ligaba et al. 2004). These results were confirmed by Liao et al. (2006), who investigated the mechanism of Al-P interaction. Their results showed no significant differences among tolerant and sensitive genotypes of soybean in homogenous Al and P experiments, while in the heterogeneous nutrient experiments with low P and Al, the P-efficient genotypes performed better than P inefficient genotypes. Whereas the same three organic acids (malate, citrate and oxalate) were secreted from roots, the pattern of secretion was not the same as the homogenous solution system.

Al alleviates H+ toxicity under acidic condition

Acidic soil causes much yield loss for numerous crops. The syndrome includes toxicity of excess ions such as Mn, Al, and H^+, and a deficiency of essential nutrients such as Ca, Mg, and P (Ishitani et al. 2004). Much work has been done to find the genes responsible for Al tolerance, while little information is available about genes that are responsible for H^+ tolerance. It was reported that H^+ toxicity caused severe root growth inhibition in wheat, Arabidopsis, and spinach (Koyama et al. 2001). Furthermore H^+ toxicity existed in acidic sulphate soil (Rosicky et al. 2004).

Aluminium toxicity is one of the major soil factors limiting root growth in acidic soils. Because of the increase in organic matter content in the upper few centimetres of soils under no-till systems (NTS), most Al in soil solution may be complexed to dissolved organic carbon (DOC), thus decreasing its bioavailability. The effects of surface liming on soil chemical attributes and on the composition of the soil solution were dependent on weather conditions, time under NTS, and soil weathering. Most Al in soil solution was complexed to DOC, representing about 70 to 80% of the total Al at pH < 5.0, and about 30 to 40% at pH > 5.0. Under pH 5.5, the results were closely correlated with the solubility line for amorphous Al. Organic complexes may control Al^{3+} release into soil solution at pH < 5.5. Results suggest that in areas under NTS for a long period of time, Al toxicity might decrease due to its complexation to high-molecular-weight organic compounds (Alleoni et al. 2010).

The major beneficial effects of Al are:

- Stimulate plant growth (roots and shoots).
- Promote nutrients uptake.
- Increases defence pathogens.
- Alleviation of abiotic stress.
- Increased metabolism and antioxidant activity.
- Modulate colour of flowers.

4.17 Adaptation

Profound differences in adaptation to local edaphic parameters (Al content, acidity, organic matter, altitude) between cultivated and wild species were observed. Comparison of the distribution patterns of the cultivated and wild species may indicate that the anthropogenic activities played a pivotal role in the development of enhanced Al resistance among the cultivated species, while edaphic conditions seem to be a secondary factor (Gananҫa et al. 2007).

The presence of high amounts of naturally occurring metals in soil can limit plant distribution and colonisation of local ecosystems (Bradshaw and McNeilly 1981; Gravin and Carver 2003). Low soil pH creates favourable conditions for the release of metals into the soil solution at concentrations that can be toxic to growing plants (Kidd and Proctor 2001). These metals in ionic form can affect the growth and reproductive fitness of many plant species (Bradshaw and McNeilly 1981). As a result of selective pressure, inter- and intra-species differences in response to toxic concentrations of metals are widely observed in the plant kingdom (Stanton et al. 2000). Speciation processes are particularly common in volcanic acid soils where high amounts of exchangeable metals may originate both from the natural level of metal or the intensification of the releasing processes into soil solution (Pintro et al. 1999).

In order to have strategies (Thakuria et al. 2016) of plant adaptation to Al toxicity, there is a need in change in root architecture and possible root association with mycorrhizal fungi, and symbiotic and mutualistic association with plant growth promoting bacteria. For search of P-efficient and Al toxicity-tolerant plant genotypes based on molecular uptake mechanisms, advancement in understanding on phosphate transporters, Ca-ATPases, vegetative storage proteins, âglucosidase, PEPCase, MDH, PEP phosphatase mechanisms has been made with respect to Puptake efficiency and Al toxicity tolerance in various crop plants (Bhalerao and Prabhu 2013).

4.17.1 Adaptation Strategies

Acidity tolerance in crop plants is a complex trait, involving multiple genes. Our current knowledge indicates that plants have evolved several mechanisms for combating Al toxicity and limited P supply. Methods for dealing with acidity tolerance could be internal or external. Internal mechanisms include chelation of Al in cytosol, compartmentalisation in vacuole, Al-tolerant isozymes, Al-binding proteins and elevated enzyme activity, increased P-recycling and scavenging for phosphate from intra- and extra-cellular organic pools, and higher number and activity of P-transporters and improved P economy in metabolic pathways. External mechanisms include selective permeability of plasma membrane, differential binding of Al to cell wall, formation of a plant-induced pH barrier in the rhizosphere and exudation of chelating compounds like organic acids or phenolic compounds, changes in root architecture and increased number of root hairs. Many plants adapt to acid soils and their adaptation mechanisms can be broadly categorised into 3 aspects:

(1) Physiological responses of root hairs.
(2) Biochemical responses of plants.
(3) Molecular responses of plants.

References

Adams F, Moore BL (1983) Chemical factors affecting root growth in subsoil horizons of Coastal Plain soils. Soil Sci Soc Am J 47:99–102

Akaya M, Takenaka C (2001) Effects of aluminum stress on photosynthesis of *Quercus glauca* Thumb. Plant Soil 237:137–146

Alleoni LRF, Cambri MA, Caires EF, Garbuio FJ (2010) Acidity and aluminum speciation as affected by surface liming in tropical no-till soils. Soil Sci Soc Am J 74(3):1010–1017

Alva AK, Sumner ME (1988) Effects of phosphogypsum or calcium sulfate on reactive aluminum in solutions at varying pH. Commun Soil Sci Plan 19:1715–1730

Alva AK, Edwards DG, Asher CJ, Blamey FP (1986) Relationships between root length of soybean and calculated activities of aluminium monomers in nutrient solution. Soil Sci Soc Am J 50:959–962

Amjad M (2013) Can we predict the best wheat varieties for low ph, high Aluminium soilsin WA? Agribusiness Crop Updates, Perth

Anderson G, Bell R (2019) Wheat grain-yield response to lime application: relationships with soil pH and aluminium in Western Australia. Crop Pasture Sci 70:295–305

Aniol A, Gustafson JP (1984) Chromosome location of genes controlling aluminum tolerance in wheat, rye, and triticale. Can J Genet Cytol 26:701–705

Arraiza MP, Santamarta JC, Ioras F, García-Rodríguez JL, Abrudan IV, Korjas H, Borála G (eds) Climate change and restoration of degraded land. Colegio de Ingenieros de Montes, Madrid, Spain

Arshad MA and Gill KS (1996). Field pea response to liming of an acid soil under two tillage systems. Can J Soil Sci 76:549–555

Ashenef A (2014) Essential and toxic metals in tea (*Camellia sinensis*) imported and produced in Ethiopia. Food Addit Contam Part B 7:30–36. https://doi.org/10.1080/19393210.2013.831951

Baggie I, Zapata F, Sanginga N, Danso SKA (2000) Ameliorating acid infertile rice soil with organic residue from nitrogen fixing trees. Nutr Cycling Agroecosyst 57:183–190

Bahia-Filho AFC, Magnavaca R, Schaffert RE, Alves VMC (1997) Identification, utilization, and economic impact of maize germplasm tolerant to low levels of phosphorus and toxic levels of exchangeable aluminum in Brazilian soils. In: Moniz AC, Fulani AMC, Schaffert RE, Fageria

NK, Rosolem CA, Cantarella H (eds) Plant–soil interactions at low pH. Sustainable agriculture and forestry production. Brazilian Soil Science Society, Campinas, pp 59–70
Baker DE, Jarrell AE, Marshall LE, Thomas WI (1970) Phosphorus uptake from soils by corn hybrids selected for high and low phosphorus accumulation. Agron J 62:103–106
Baker AJM, McGrath SP, Reeves RD, Smith JAC (2000) Metal hyperaccumulator plants: a review of the ecology and physiology of a biological resource for phytoremediation of metal–polluted soils. In: Terry N, Banuelos G (eds) Phytoremediation of contaminated soil and water. Lewis Publisher, Boca Raton, pp 85–107
Baligar VC, Pitta GVE, Schaffert RE, Bahia Filho AFDC, Clark RB (1997) Soil acidity effects on nutrient use efficiency in exotic maize genotypes. Plant Soil 192:9–13
Barabasz W, Albinska D, Jaskowska M, Lipiec J (2002) Ecotoxicology of aluminium. Pol J Environ Stud 11:199–203
Barber SA (1978) Soil plant interactions in P nutrition of plants. In: Khasawneh FE, Sample EC (eds) The role of phosphorus in agriculture. Soil Science Society of America, Madison, Wisconsion, USA, pp 591–616
Barcelo J, Poschenrieder C (2002) Fast root growth responses, root exudates and internal detoxification as clues to the mechanisms of aluminium toxicity and resistance: a review. Env Exp Bot 48:75–92
Bartlett RJ, Riego DC (1972) Effect of chelation on the toxicity of aluminum. Plant Soil 37:419–423
Basu A, Basu U, Taylor GJ (1994) Induction of microsomal membrane proteins in roots of an aluminum-resistant cultivar of *Triticum aestivum* under conditions of aluminum stress. Plant Physiol 104:1007–1013
Basu U, McDonald JL, Archambault DJ, Good AG, Briggs KG, Aung T, Taylor GJ (1997) Genetic and physiological analysis of doubled-haploid, aluminium-resistant lines of wheat provide evidence for the involvement of a 23 kD, root exudate polypeptide in mediating resistance. Plant Soil 196:283–288
Bates TR, Lynch JP (2000) Plant growth and P accumulation of wild type and two hair mutants of Arabidopsis thaliana. Am J Bot 87:958–963
Becker M, Asch F (2005) Iron toxicity in rice-conditions and management concepts. Plant Nutr Soil Sci 168:558–573
Bengtsson B, Asp H, Jensén P, Berggren D (1988) Influence of aluminium on phosphate and calcium uptake in beech (Fagus sylvatica) grown in nutrient and soil solution. Physiol Plant 74:299–305
Bennet RJ, Breen CM (1991) The aluminum signal: new dimensions to mechanisms of aluminum tolerance. Plant Soil 134:153–166
Berggren D, Mulder J (1995) The role of organic matter in controlling aluminum solubility in acidic mineral soil horizons. Geochim Cosmochim Acta 59:4167–4180
Berzonsky WA (1992) The genomic inheritance of aluminum tolerance in "Atlas 66" wheat. Genome 35:689–693
Bezdicek DF, Beaver T, Granatstein D (2003) Subsoil ridge tillage and lime effects on soil microbial activity, soil pH, erosion, and wheat and pea yield in the Pacifi c Northwest, USA. Soil Tillage Res 74:55–63
Bhalerao SA, Prabhu DV (2013) Toxicity in plants—a review. J Appl Chem 2(3):447–474
Bhalla PL (2006) Genetic engineering of wheat: current challenges and opportunities. Trends Biotechnol 24:305–311
Biana M, Zhoua M, Sunb D, Lic C (2013) Molecular approaches unravel the mechanism of acid soil tolerance in plants. Crop J 1:91–104. https://doi.org/10.1016/j.cj.2013.08.002
Black CA (1968) Soil plant relationships. Wiley, New York
Blake L, Goulding KWT, Mott CJB, Johnston AE (1999) Changes in soil chemistry accompanying acidification over more than 100 years under woodland and grass at Rothamsted Experimental Station, UK. Eur J Soil Sci 50:401–412
Blamey FPC, Ostatek-Boczynski Z, Kerven GL (1997) Ligand effects on aluminium sorption by calcium pectate. Plant Soil 192:269–275

Blamey FPC, Edwards DG, Asher CJ (1983) Effects of aluminium, OH: Al and P: Al molar ratios, and ionic strength on soybean root elongation in solution culture. Soil Sci 136:197–207

Blancaflor EB, Jones DL, Gilroy S (1998) Alterations in the cytoskeleton accompany aluminum-induced growth inhibition and morphological changes in primary roots of maize. Plant Physiol 118:159–172

Bojórquez-Quintal E, Escalante-Magaña C, Echevarría-Machado I, MartínezEstévez M (2017) Aluminum, a friend or foe of higher plants in acid soils. Front Plant Sci 8:1767

Borreani G, Tabacco E (2006) The effect of a baler chopping system of fermentation and losses of wrapped big bales of alfalfa. Agron J 98:1–7

Bose J, Babourina O, Rengel Z (2011) Role of magnesium in alleviation of aluminium toxicity in plants. J Exp Bot 62:2251–2264

Bradshaw AD, McNeilly T (1981) Evolution and pollution. Stud Biol 130:1–77

Britez RM, Watanabe T, Jansen S, Reissmann CB, Osaki M (2002) The relationship between aluminium and silicon accumulation in leaves of Faramea marginata (Rubiaceae). New Phytol 156:437–444

Brown JC, Devine TE (1980) Inheritance of tolerance or resistance to manganese toxicity in soybeans. Agron J 72:898–904

Brown TT, Koenig RT, Huggins DR, Harsh JB, Rossi RE (2007) Lime effects on soil acidity, crop yield, and aluminum chemistry in direct seeded cropping systems. Soil Sci Soc Am J 72:634–640. https://doi.org/10.2136/sssaj2007

Brunner I, Sperisen C (2013) Aluminum exclusion and aluminum tolerance in woody plants. Front Plant Sci 4:172. https://doi.org/10.3389/fpls.2013.00172

Cakmak I, Horst WJ (1991) Effect of aluminum on lipid peroxidation, superoxide dismutase, catalase, and peroxidase activities in root tips of soybean (Glycine max L.). Physiologica Plantarum 834:463–468

Caldwell CR (1989) Analysis of aluminum and divalent cation binding to wheat root plasma membrane proteins using Terbium phosphorescence. Plant Physiol 91:233–241

Camargo CEO (1988) Tolerance of wheat cultivars to different levels of manganese in nutrient solution. In: Webb MJ et al (eds) International symposium on manganese in soils and plants. The University of Adelaide, Australia, pp 109–110

Cambraia J, Galvani FR, Estevao MM (1983) Effects of aluminum on organic acid, sugar and amino acid composition of the root system of sorghum (Sorghum bicolor L. Moench). J Plant Nutr 6:313–322

Campbell CG, Garrido F, Illera V, García-González MT (2006) Transport of Cd, Cu and Pb in an acid soil amended with phosphogypsum, sugar foam and phosphoric rock. Appl Geochem 21:1030–1043

Care DA (1995) The effect of aluminium concentration on root hairs in white clover (Trifolium repens L.) Plant Soil, 171: 159–162.

Carvalho MCS, van Raij B (1997) Calcium sulphate, phosphogypsum and calcium carbonate in the amelioration of acid subsoils for root growth. Plant Soil 192:37–48

Carver BF, Ownby JD (1995) Acid soil tolerance in wheat, in Sparks, DL (ed.): Advances in Agronomy. Vol. 54, Academic Press, Inc., New York, pp. 117–172

Carver et al (1995)

Celis-Arámburo TDJ, Carrillo-Pech M, Castro-Concha LA, Miranda-Ham MDL, Martínez-Estévez M, EchevarríaMachado I (2011) Exogenous nitrate induces root branching and inhibits primary root growth in Capsicum chinense Jacq. Plant Physiol Biochem 49:1456–1464

Chaney RI (1989) Toxic element accumulation in soil and crop, protecting soil fertility and agricultural food-chains. In: Yosef B, Barrow NJ, Goldshmid J (eds) Inorganics contaminants in the vadose zone. Springer-Verlag, Berlin, pp 140–158

Che J, Yamaji N, Shen RF, Ma JF (2016) An Al-inducible expansin gene, OsEXPA10 is involved in root cell elongation of rice. Plant J 88:132–142. https://doi.org/10.1111/tpj.13237

Chen ZC, Ma JF (2013) Magnesium transporters and their role in Al tolerance in plants. Plant Soil 368:51–56

Chen RF, Zhang FL, Zhang QM, Sun QB, Dong XY, Shen RF (2012) Aluminium–phosphorus interactions in plants growing on acid soils: does phosphorus always alleviate aluminium toxicity? J Sci Food Agric 92:995–1000

Chenery EM (1948) Aluminum in the plant world. Part I, general survey in dicotyledons. Kew Bull 3:173–183

Cheng S, Willmann MR, Chen H, Sheen J (2002) Calcium signaling through protein kinases. The Arabidopsis calcium-dependent protein kinase gene family. Plant Physiol 129:469–485

Choudhury S, Sharma P (2014) Aluminum stress inhibits root growth and alters physiological and metabolic responses in chickpea (Cicer arietinum L.). Plant Physiol Biochem 85:63–70

Chu XQ, Wang JG, Li MZ, Zhang SJ, Gao YY, Fan M, Han C, Xiang FN, Li GY, Wang Y et al (2021) HBI transcription factor-mediated ROS homeostasis regulates nitrate signal transduction. Plant Cell 33:3004–3021

Ciamporová M (2002) Morphological and structure responces of plant roots to aluminium at organ, tissue, and cellular levels. Biol Pl 45:161–171

Ciotta MN, Bayer C, Ernani PR, Fontoura SMV et al (2002) Acidificação de um latossolo sob plantio direto. Rev Bras Cienc Solo 26:1055–1064

Clark RB, Duncan RR (1993) Selection of plants to tolerate soil salinity, acidity, and mineral deficiencies. Int Crop Sci 1:371–379

Clarkson DT, Scattergood CB (1982) Growth and phosphate transport in barley and tomato plants during development and recovery from phosphorus stress. J Exp Bot 33:865–875

Clegg S, Gobran GR (1995) Effects of aluminium on growth and root reactions of phosphorus stressed Betula pendula seedlings. Plant Soil 168:173–178

Clune TS, Copeland L (1999) Effects of aluminium on canola roots. Plant Soil 216:27–33

Culvenor RA, Oram RN, David DJ (1986) Genetic variability for manganese concentration in Phalaris aquatica growing in acid soil. Aust J Agric Res 37:409–416

Dale JE, Sutcliffe JF (1986) Water relations of plant cells. https://doi.org/10.1016/B978-0-12-668609-8.50008-9

de la Fuente JM, Ramírez-Rodríguez V, Cabrera-Ponce JL, Herrera-Estrella L (1997) Aluminum tolerance in transgenic plants by alteration of citrate synthesis. Science 276:1566–1568

Delhaize E, Ryan PR (1995) Aluminum toxicity and tolerance in plants. Plant Physiol 107:315–321

Delhaize E, Higgins TJV, Randall PJ (1991) Aluminum tolerance in wheat: analysis of polypeptides in the root apices of tolerant and sensitive genotypes. In: Wright RJ et al (eds) Plant-soil interactions at low pH. Kluwer Academic Publishers, Dordrecht, The Netherlands, pp 1071–1079

Delhaize E, Craig S, Beaton CD, Bennet RJ et al (1993a) Aluminum tolerance in wheat (*Triticum aestivum* L.) (I. uptake and distribution of aluminum in root apices). Plant Physiol 103:685–693

Delhaize E, Hebb DM, Richards KD, Lin J-M, Ryan PR, Gardner RC (1999) Cloning and expression of a wheat (Triticum aestivum L.) phosphatidylserine synthase cDNA. J Biol Chem 274:7082–7088

Delhaize E, Hebb DM, Ryan PR (2001) Expression of a *Pseudomonas aeruginosa* citrate synthase gene in tobacco is not associated with either enhanced citrate accumulation or efflux. Plant Physiol 125:2059–2067

Dengenhardt J, Larsen PB, Howell SH, Kochian LV (1998) Aluminum resistance in the Arabidopsis mutant alr-104 is caused by an aluminum-induced increase in rhizosphere pH. Plant Physiol 117:19–27

Dennis ES, Dolferus R, Ellis M, Rahman M, Wu Y, Hoeren FU, Grover A, Ismond KP, Good AG, Peacock WJ (2000) Molecular strategies for improving waterlogging tolerance in plants. J Exp Bot 51:89–97

Dewi-Hayati PK, Sutoyo A, Syarif A, Prasetyo T (2014) Performance of maize single-cross hybrids evaluated on acidic soils. Int J Adv Sci Eng Inf Technol 4:30–33

Dierolf T, Fairhurst T, Mutert E (2001) A toolkit for acid, upland soil fertility management in southeast Asia, 150. Handbook Series, PPIC-Canada

Dinev N, Stancheva I (1993) Changes in nitrate reductase activity, plastid pigment content, and plant mineral composition of wheat, rye, and triticale grown in the presence of aluminum. J Plant Nutr 16:2397–2409

Dugalic G, Krstic D, Jelic M, Nikezic D, Milenkovic B, Pucarevic M, Zeremski-Skoric T (2010a) Heavy metals, organics and radioactivity in soil of western Serbia. J Hazard Mater 177:697–770

Duncan M (1999) Acid soil management, prepared for the New South Wales Acid Soil Action Program, Armidale, Australia

Echarte L, Nagore L, Di Matteo J, Di J, Robles M, Cambareri M, Maggiora AD (2013) Grain yield determination and resource use efficiency in maize hybrids released in different decades: Chapter 2; IITECH/Agricultural Chemistry, London, UK, pp 1–36

Eswaran H, Reich P, Beinroth F (1997) Global distribution of soils with acidity. In Moniz AC, Furlani AMC, Schaffert RE, Fageria NK, Rosolem CA

Evans and Kamprath (1970)

Exley C (2003) A biogeochemical cycle for aluminium? J Inorg Biochem 97:1–7

Fageria NK (2009) The use of nutrients in crop plants. CRC Press, Boca Raton, p 430p

Fageria NK, Baligar VC (2003) Fertility management of tropical acid soil for sustainable crop production. In: Rengel Z (ed) Handbook of soil acidity. Marcel Dekker, New York, pp 359–385

Famoso AN, Zhao K, Clark RT, Tung CW, Wright MH, Bustamante C, Kochian LV, McCouch SR (2011) Genetic architecture of aluminum tolerance in rice (Oryza sativa) determined through genome-wide association analysis and QTL mapping. PLoS Genet 7:1002221

Foy CD (1974) Effects of aluminum on plant growth. In: Carson EW (ed) The plant root and its environment. University Press of Virginia, Charlottesville, VA, pp 601–642

Foy CD (1983) Plant adaptation to mineral stress in problem soils. Iowa J Res 57:339–354

Foy CD (1984) Physiological effects of hydrogen, aluminum, and manganese toxicities in acid soil. In: Adams F (ed) Soil acidity and liming. American Society of Agronomy, Madison, pp 57–97

Foy CD (1988) Plant adaptation to acid, aluminum-toxic soils. Commun Soil Sci Plant Anal 19:959–987

Foy CD (1992) Soil chemical factors limiting plant root growth. Adv Soil Sci 19:97–149

Foy CD (1997) Tailoring plants to fit problem soils-progress and problems for future research. In: Moniz AC et al (eds) Plant-soil Interactions at low pH: sustainable agriculture and forestry production. Brazilian Soil Science Society, pp 55–57

Foy CD, Brown JC (1964) Toxic factors in acid soils. II. Differential aluminum tolerance of plant species. Soil Sci Soc Am Proc 28:27–32

Foy CD, Fleming AL, Schwartz JW (1973) Opposite alluminum and manganese tolerances of two wheat varieties. Agron J 65:123–126

Foy CD, Chaney RL, White MC (1978) The physiology of metal toxicity in plants Annu. Rev Plant Physiol 29:511–566

Foy CD, Lee EH, Wilding SB (1987) Differential aluminum tolerances of two barley cultivars related to organic acids in their roots. J Plant Nutr 10:1089–1101

Foy CD, Scott BJ, Fisher JA (1988) Genetic differences in plant tolerance to manganese toxicity. In: Graham RD et al (eds) Manganese in soils and plants. Kluwer Academic Publishers, Dordrecht, The Netherlands, pp 293–307

Foy et al (1992)

Foy CD, Sadeghi AM, Ritchie JC, Krizek DT, Davis JR, Kemper WD (1999) Aluminum toxicity and high bulk density: role in limiting shoot and root growth of selected aluminum indicator plants and eastern gamagrass in an acid soil. J Plant Nutr 22:1551–1566

Franceschi VR, Nakata PA (2005) Calcium oxalate in plants: Formation and function. Annu Rev Plant Biol 5:641–671

Frantzios G, Galatis B, Apostolakos P (2001) Aluminium effects on microtubule organization in dividing root-tip cells of Triticum turgidum II. Cytokinetic cells. J Plant Res 114(1114):157–170

Ganança JFT, Abreu I, Sousa NF, Paz RF et al (2007) Soil conditions and evolution of aluminium resistance among cultivated and wild plant species on the Island of Madeira. Plant Soil Environ 53:239–246

Gardner WK, Parbery DG (1983) Acquisition of P by Lupinus albus L. I. Some characteristics of soil/root interface. Plant Soil 68:19–32

Gartside DW, McNeilly T (1974a) Genetic studies in heavy metal tolerant plants. I. Genetics of zinc tolerance in Anthoxanthum odoratum. Heredity 32:287–299

Gartside DW, McNeilly T (1974b) Genetic studies in heavy metal tolerant plants. II. Zinc tolerance in Agrostis tenuis. Heredity 33:303–308

Garvin DF, Carver BF (2003) Role of the genotype in tolerance to acidity and aluminum toxicity. In: Rengel Z (ed) Handbook of soil acidity. Marcel Dekker, NewYork, pp 387–406

Gazey C, Davies S (2009) Soil acidity: a guide for WA farmers and consultants Bulletin 4784. The Department of Agriculture and Food, Western Australia, 52 p. ISSN 1833–7246

Ghanati F, Morita A, Yokota H (2005) Effects of aluminium on the growth of tea plant and activation of antioxidant system. Plant Soil 276:133–141. https://doi.org/10.1007/s11104-005-3697-y

Gilker RE, Weil RR, Krizek DT, Momen B (2002) Eastern gamagrass root penetration in adverse subsoil conditions. Soil Sci Soc Am J 66:931–938

Gilroy S, Bethke PC, Jones RL (1993) Calcium homeostasis in plants. J Cell Sci 106:453–462

Golding LA, Angel BM, Batley GE, Apte SC, Krassoi R, Doyle CJ (2015) Derivation of a water quality guideline for aluminium in marine waters. Environ Toxicol Chem 34:141–151. https://doi.org/10.1002/etc.2771

Gonzalez A, Steffen KL, Lynch JP (1998) Light and excess manganese. Implications for oxidative stress in common bean. Plant Physiol 118:493–504

González-Santana IH, Márquez-Guzmán J, Cram-Heydrich S, Cruz-Ortega R (2012) *Conostegia xalapensis* (Melastomataceae): an aluminum accumulator plant. Physiol Plant 144:134–145. https://doi.org/10.1111/j.1399-3054.2011.01527.x

Grabski S, Schindler M (1995) Aluminum induces rigor within the actin network of soybean cells. Plant Physiol 108:897–901

Grauer VE, Horst WJ (1992) Modelling cation amelioration of A1 phytotoxicity. Soil Sci Soc Am J 134:166–172

Gravin DF, Carver BF (2003) Role of the genotype in tolerance to acidity and aluminum toxicity. In: Rengel Z (ed) Handbook of soil acidity. Marcel Dekker, New York, pp 387–406

Grime JP (1979) Plant strategies and vegetation processes. Wiley, Chichester

Guo T, Chen Y, Zhang Y, Jin Y (2006) Alleviation of Al-toxicity in barley by addition of calcium. Agr Sci China 5:828–833

Guo T, Zhang G, Zhou M, Wu F, Chen J (2004) Effects of aluminum and cadmium toxicity on growth and antioxidant enzyme activities of two barley genotypes with different Al resistance. Plant Soil 258(1):241–248

Haling RE, Simpson RJ, Culvenor RA, Lambers H, Richardson AE (2011) Effect of soil acidity, soil strength and macropores on root growth and morphology of perennial grass species differing in acid-soil resistance. Plant Cell Environ 2011(34):444–456. https://doi.org/10.1111/j.1365-3040.2010.02254.x

Hargrove WL, Thomas GW (1981) Effect of organic matter on exchangeable aluminum and plant growth in acid soils. Chem Soil Environ 40:151–166

Harper SM, Edwards DG, Kerven GL, Asher CJ (1995) Effects of organic acid fractions extracted from Eucalyptus camaldulensis leaves on root elongation of maize (Zea mays) in the presence and absence of aluminium. Plant Soil 171:189–192

Haug A (1984) Molecular aspects of aluminium toxicity. Critical Rev Pl Sci 1:345–373

Haug A, Foy CE (1984) Molecular aspects of aluminum toxicity. Crit Rev Plant Sci 1:345–373

Haug A, Shi B, Vitorello V (1994) Aluminum interaction with phosphoinositide-associated signal transduction. Arch Toxicol 68:1–7

Havlin JL, Tisdale SL, Nelson WL, Beaton JD (2005) Soil fertility and fertilizers: an introduction to nutrient management. Pearson Education, Upper Saddle River, New Jersey

Havlin JL, Tisdale SL, Nelson WL, Beaton JD (2009) Soil fertility and fertilizers: an introduction to nutrient management, sixth edition, provides a thorough understanding of the biological, chemical, and physical properties affecting soil fertility and plant nutrition

Haynes RJ (1982) Effects of liming on phosphate availability. Plant Soil 68:289–308
Haynes RJ (1984) Lime and phosphate in the soil–plant system. Adv Agron 37:249–315
Hecht-Buchholz C, Foy C (1981) Effect of aluminium toxicity on root morphology of barley. Plant Soil, 63:93–95
Hede AR, Skovmand B, López-Cesati J (2001) Acid soil and aluminum toxicity. In: Reynolds MP, Ortiz-Monasterio JJ, Mchab A (eds) Application of physiology in wheat breeding. International Maize and Wheat Improvement Center, pp 172–182
Hepler PK (2005) Calcium: a central regulator of plant growth and development. Plant Cell 17:2142–2155
Hirschi KD, Korenkov VD, Wilganowski NL, Wagner GJ (2000) Expression of arabidopsis CAX2 in tobacco. Altered metal accumulation and increased manganese tolerance. Plant Physiol 124:125–133
Hoekenga OA, Maron LG, Piñeros MA, Cancado GM et al (2006) AtALMT1, which encodes a malate transporter, is identified as one of several genes critical for aluminum tolerance in *Arabidopsis*. Proc Natl Acad Sci U S A 103:9738–9743
Hoisington D (2002) Opportunities for nutritionally enhancing maize and wheat varieties to combat protein and micronutrient malnutrition. Food Nutr Bull 23:376–377
Hong E, Ketterings Q, Mcbride M (2010) Manganese agronomy fact sheet series. In: Field crop extension, vol 49. College of Agriculture and Life Sciences, Ithaca, NY, USA
Hong PYP, Polanin JR, Pigott TD (2012) Validation of the employment hope scale: measuring psychological self-sufficiency among low-income jobseekers. Res Soc Work Pract 22:323–332
Horst WJ (1995) The role of the apoplast in aluminum toxicity and resistance of higher plants: a review. Z Pflanzenernahr Bodenk 158:419–428
Horst W, Asher C, Cakmak I, Szulkiewicz P, Wissemeier AH (1992) Short-term responses of soybean roots to aluminium. J Plant Physiol 140:174–178
Horst W, Fecht M, Naumann A, Wissemeier A, Maier P (1999) Physiology of manganese toxicity and tolerance in *Vigna unguiculata* (L.) Walp. J Plant Nutr 162:263–274
Horst WJ, Wang Y, Eticha D (2010) The role of the root apoplast in aluminum-induced inhibition of root elongation and in aluminum resistance of plants: a review. Ann Bot 106:185–197. https://doi.org/10.1093/aob/mcq053
Houman Y, Tahara K, Shinmachi F, Noguchi A, Satohiko S, Hasegawa I (2009). Differences of growth response to aluminum excess of two Melaleuca trees differing in aluminum resistance. In: Proceedings of the international plant nutrition colloquium XVI. University of California, Davis, CA
Huang JW, Grunes DL, Kochian LV (1992) Aluminium effects on the kinetics of calcium uptake into cells of the wheat root apex. Planta 188:414–421
Hue NY, Craddock GR, Adams F (1986) Effect of organic acids on aluminium toxicity in subsoils. Soil Sci Soc Am J 50:28–34
Hui NY, Liu P, Wang ZY, Chen WR, Xu GD (2011) The effect of aluminum treatments on the root growth and cell ultrastructure of two soybean genotypes. Crop Prot 30(3):323–328
Iannucci A, Di fonzo N, Martiniello P (2002) Alfalfa (*Medicago sativa* L.) seed yield and quality under different forage management systems and irrigation treatments in a Mediterranean environment. Field Crop Res 78:65–7874
Igual JM, Dawson JO (1999) Stimulatory effects of aluminum on *in vitro* growth of *Frankia*. Can J Bot 77:1321–1326. https://doi.org/10.1139/b99-061
Ikeda H, Tadano T (1993) Ultrastructural changes of the root tip cells in barley induced by comparatively low concentration of aluminum. Soil Sci Plant Nutr 39:109–117
Ikegawa H, Yamamoto Y, Matsumoto H (2000) Responses to aluminium of suspension-cultured tobacco cells in a simple calcium solution. Soil Sci Plant Nutr 46:503–514
Inostroza-Blancheteau C, Rengel Z, Alberdi M, Mora MDLL, Aquea F, Arce-Johnson P, Reyes-Díaz M (2012) Molecular and physiological strategies to increase aluminum resistance in plants. Mol Biol Rep 39:2069–2079

Ishitani M, Rao I, Wenzl P, Beebe S, Tohme J (2004) Integration of genomics approach with traditional breeding towards improving abiotic stress adaptation: drought and aluminum toxicity as case studies. Field Crop Res 90:35–45

Itoh S, Barber SA (1983) Phosphorus uptake by six plant species as related to root hairs. Agron J 75:457–461

Iyamuremye F, Dick RP, Baham J (1996) Organic amendments and phosphorus dynamics: I. Phosphorus Chem Sorption Soil Sci 161:426–435

Jansen S, Broadley MR, Robbrecht E, Smets E (2002) Aluminum hyperaccumulation in angiosperms: a review of its phylogenetic significance. Bot Rev 68:235–269

Javed MT, Stoltz E, Lindberg S, Greger M (2013) Changes in pH and organic acids in mucilage of *Eriophorum angustifolium* roots after exposure to elevated concentrations of toxic elements. Environ Sci Pollut Res Int 20:1876–1880. https://doi.org/10.1007/s11356-012-1413-z

Jayasundara HPS, Thomson BD, Tang C (1998) Responses of cool season grain legumes to soil abiotic stresses. Adv Agron 63:77–151

Jelic M, Djalovic I, Milivojevic J, Krstic D (2010) Mobile aluminium content of vertisols as dependent upon fertilization system and small grains genotypes. In: Proceedings of 3nd international scientific/professional conference agriculture in nature and environment protection, Vukovar, Croatia, May 31–June 2, pp 137–142. ISBN 978–953–7693–008

Jiang Y, Huang B (2001) Effects of calcium on antioxidant activities and water relations associated with heat tolerance in two cool-season grasses. J Exp Bot 52:341–349

Jones HG, Jones MB (1989) Some terminology and common mechanisms. In: Jones HG et al (eds) Plants under stress. Cambridge University Press, Cambridge

Jones DL, Kochian LV (1995) Aluminum inhibition of the inositol 1,4,5-triphosphate signal transduction pathway in wheat roots: a role in aluminum toxicity? Plant Cell 7:1913–1922

Jones DL, Kochian LV (1997) Aluminum interaction with plasma membrane lipids and enzyme metal binding sites and its potential role in Al cytotoxicity. FEBS Lett 400(1):51–57

Johnson JF, Vance CP, Allan DL (1996) Phosphorus deficiency in Lupinus albus (altered lateral root development and enhanced expression of phosphoenolpyruvate carboxylase). Plant Physiol 112:31–41

Johnson JP, Carver BF, Baligar VC (1997) Expression of aluminum tolerance transferred from Atlas 66 to hard winter wheat. Crop Sci 37:103–108

Jones DL, Kochian LV, Gilroy S (1998) Aluminum induces a decrease in cytosolic calcium concentration inBY-2 tobacco cell cultures. Plant Physiol 116:81–89

Jorge RA, Arruda P (1997) Aluminum–induced organic acid exudation by roots of aluminum-tolerant tropical maize. Phytochemistry 45:675–681

Jovanovic Z, Djalovic I, Komljenovic I, Kovacevic V, Cvijovic M (2006) Influences of liming on vertisol properties and yields of the field crops. Cereal Res Commun 34:517–520

Jovanovic Z, Djalovic I, Tolimir M, Cvijovic M (2007) Influence of growing sistem and NPK fertilization on maize yield on pseudogley of Central Serbia. Cereal Res Commun 35:1325–1329

Kamprath EJ (1978) Lime in relation to aluminium toxicity in tropical soils. In: Andrew CS, Kamprath EJ (eds) Mineral nutrition of legumes in tropical and subtropical soils. CSIRO, East Melbourne, pp 233–245

Kamprath EJ, Foy CD (1985) Lime-fertilizer-plant interactions in acid soils. In: Englestad O (ed) Fertilizer technology and use, 3rd edn. Soil Science Society of America, Madison, Wisconsin, USA

Kassem MYA, Schultz J, Meksem CH, Cho Y, Wood AJ, Iqbal MJ, Lightfoot DA (2006) An updated 'Essex' by 'Forrest' linkage map and Wrst composite interval map of QTL underlying six soybean traits. Theor Appl Genet 113:1015–1026

Kayama M (2001) Comparison of the aluminum tolerance of Miscanthus sinensis Anderss and Miscanthus sacchariflorus Bentham in hydroculture. Int J Plant Sci 162:1025–1031

Kennedy IR (1992) Acid soil and acid rain, 2nd edn. Research Studies Press Ltd., Taunton, Somerset, England

Kerridge PC, Kronstad WE (1968) Evidence of genetic resistance to aluminum toxicity in wheat (*Triticum aestivum* Vill., Host.). Agron J 60:710–711
Khabaz-Saberi H, Setter TL, Waters I (2006) Waterlogging induces high to toxic concentrations of iron, aluminium and manganese in wheat varieties on acidic soil. J Plant Nutr 29:899–911
Khabaz-Saberi H, Rengel Z, Wilson R, Setter TL (2010) Variation of tolerance to manganese toxicity in Australian hexaploid wheat. J Plant Nutr Soil Sci 173:103–112. https://doi.org/10.1002/jpln.200900063103
Khan MSH, Tawaraya K, Sekimoto H, Koyama H, Kobayashi Y, Murayama T, Chuba M, Kambayashi M, Shiono Y, Uemura M, Ishikawa S, Wagatsuma T (2009) Relative abundance of Δ5-sterols in plasma membrane lipids of root-tip cells correlates with aluminum tolerance of rice. Physiol Plant 135:73–83
Khan AA, McNeilly T, Azhar FM (2001) Stress tolerance in crop plants. Int J Agricult Biol 3:50–57. 1560–8530/2001/03–2–250–255. http://www.ijab.orgReview
Khan NA, Samiullah S, Singh S, Nazar R (2007) Activities of antioxidative enzymes, sulphur assimilation, photosynthetic activity and growth of wheat (*Triticum aestivum*) cultivars differing in yield potential under cadmium stress. J Agron Crop Sci 193:435–444
Kidd C (2001) Satellite rainfall climatology: a review. Int J Climatol 21:1041–1066. https://doi.org/10.1002/joc.635
Kidd PS, Proctor J (2000) Effects of aluminium on the growth and mineral composition of Betula pendula Roth. J Exp Bot 51:1057–1066
Kidd PS, Proctor J (2001) Why plants grow poorly on very acid soils: are ecologists missing the obvious. J Exp Bot 52:791–799
Kidd PS, Llugany M, Poschenrieder C, Gunse B, Barcelo J (2001) The role of root exudates in aluminium resistance and silicon-induced amelioration of aluminium toxicity in three varieties of maize (Zea mays L.). J Exp Bot 52:1339–1352
Kim YS, Park W, Nian H, Sasaki T, Ezaki B, Jang YS, Chung GC, Bae HJ, Ahn SJ (2010) Aluminum tolerance associated with enhancement of plasma membrane H plus-ATPase in the root apex of soybean. Soil Sci Plant Nutr 56:140–149
Kinraide TB (1991) Identity of the rhizotoxic aluminium species. Plant Soil 134:167–178
Kinraide TB (1997) Reconsidering the rhizotoxicity of hydroxyl, sulphate, and fl uoride complexes of aluminum. J Exp Bot 48:1115–1124
Kinraide (1998) Three mechanisms for the calcium alleviation of mineral toxicities. Plant Physiol
Kinraide TB, Parker DR (1990) Apparent phytotoxicity of mononuclear hydroxy-aluminium to four dicotyledonous species. Physiol Plant 79:283–288
Kinraide TB, Ryan PR, Kochian LV (1992) Interactive effects of Al, H+, and other cations on root elongation considered in terms of cell-surface electrical potential. Plant Physiol 99:1461–1468
Kinraide TB, Ryan PR, Kochian LV (1994) A13+-Caz+ interactions in aluminum rhizotoxicity. 11. Evaluating the Ca'+-displacement hypothesis. Planta 192:104–109
Klimashevskii EL, Chernysheva NF (1980) Content of organic acids and physiologically active compounds in plants differing in their susceptibility to the toxicity of Al3. Sov Agri Sci 2:5–7
Kisinyo PO, Othieno CO, Gudu SO, Okalebo JR, Opala PA, Maghanga JK, Ng'etich WK, Agalo JJ, Opile RW, Kisinyo JA, et al (2013) Phosphorus sorption and lime requirements of maize growing acids soil of Kenya. Sustain Agric Res 2:116–123
Kochian (1975)
Kochian LV (1995) Cellular mechanism of aluminum toxicity and resistance in plants. Annu Rev Plant Physiol 46:237–260
Kochian LV, Jones DL (1996) Aluminium toxicity and resistance in plants. In: Yokel R, Golub MS (eds) Research issues in aluminium toxicity. Francis Publishers, Washington, DC, pp 69–89
Kochian LV, Hoekenga OA, Pineros MA (2004) How do crop plants tolerate acid soils? Mechanisms of aluminium tolerance and phosphorous efficiency. Annu Rev Plant Biol 55:459–493. https://doi.org/10.1146/annurev.arplant.55.031903.141655

Kochian LV, Pineros MA, Hoekenga OA (2005) The physiology, genetics and molecular biology of plant aluminum resistance and toxicity. Plant Soil 274:175–195. https://doi.org/10.1007/s11104-004-1158-7

Kochian LV, Piñeros MA, Liu J, Magalhaes JV (2015) Plant adaptation to acid soils: the molecular basis for crop aluminum resistance. Annu Rev Plant Biol 66:571–598. https://doi.org/10.1146/annurev-arplant-043014-114822

Kollmeier M, Dietrich P, Bauer CS, Horst WJ, Hedrich R (2001) Aluminum activates a citrate-permeable anion channel in the Al-sensitive zone of the maize root apex: a comparison between an Al-sensitive and an Al-tolerant cultivar. Plant Physiol 126:397–410

Konarska A (2010) Effects of aluminum on growth and structure of red pepper (Capsicum annuum L.) leaves. Acta Physiol Plant 32:145–151

Kopittke PM, Moore KL, Lombi E, Gianoncelli A, Ferguson BJ, Blamey FPC et al (2015) Identification of the primary lesion of toxic aluminum in plant roots. Plant Physiol 167:1402–1411. https://doi.org/10.1104/pp.114.253229

Kopittke PM, Menzies NW, Wang P, Blamey FP (2016) Kinetic and nature of aluminium rhizotoxic effects: a review. J Exp Bot 67:4451–4467

Koyama H, Toda T, Hara T (2001) Brief exposure to low-pH stress causes irreversible damage to the growing root in Arabidopsis thaliana: pectin–Ca interaction may play an important role in proton rhizotoxicity. J Exp Bot 52:361–368

Krstic D, Stevanovic N, Milivojevic J, Nikezic D (2007) Determination of the soil-to grass transfer of 137Cs and its relation to several soil properties at various locations in Serbia. Isotopes Environ Health St 43:65–73

Krstic D, Djalovic I, Nikezic D and Bjelic D (2012) Aluminium in acid soils: chemistry, toxicity and impact on maize plants. In: Anna Aladjadjiyan (ed) Food production—approaches, challenges and tasks, pp 231–242. ISBN: 978–953–307–887–8

Larsen PB, Kochian LV, Howell SH (1997) Al inhibits both shoot development and root growth in als3, an Al-sensitive Arabidopsis mutant. Plant Physiol 114:1207–1214

Lazof DB, Goldsmith JG, Rufty TW, Linton RW (1994) Rapid uptake of aluminum into cells of intact soybean root tips. A microanalytical study using secondary ion mass spectroscopy. Plant Physiol 106:1107–1114

Lee EH, Foy CD (1986) Aluminum tolerance of two snapbean cultivars related to organic acid content evaluated by highperformance liquid chromatography. J Plant Nutr 9: 1481–1498

Li XF et al (2000) Pattern of Al-induced secretion of organic acids differ between rye and wheat. Plant Physiol 123:1537–1543

Li Y, Yang GX, Luo LT et al (2008) Aluminium sensitivity and tolerance in model and elite wheat varieties. Cereal Res Commun 36(2):257–267

Liao H, Wan H, Shaff J, Wang X, Yan X, Kochian LV (2006) Phosphorus and aluminum interactions in soybean in relation to aluminum tolerance. Exudation of specific organic acids from different regions of the intact root system. Plant Physiol 141:674–684

Ligaba A, Katsuhara M, Ryan PR, Shibasaka M et al (2006) The BnALMT1 and BnALMT2 genes from rape encode aluminum-activated malate transporters that enhance the aluminum resistance of plant cells. Plant Physiol 142:1294–1303

Lin MH, Gresshoff PM, Ferguson BJ (2012) Systemic regulation of soybean nodulation by acidic growth conditions. Plant Physiol 160:2028–2039

Lindsay WL (1979) Chemical equilibria in soils. Wiley, New York

Lipton DS, Balanchar RW, Bleuius DG (1987) Citrate, malate and succinate concentration in exudates from P sufficient and P stressed Medicago sativa L. seedlings. Plant Physiol 85:315–317

Liu Q, Yang JL, He LS, Li YY, Zheng SJ (2008) Effect of aluminum on cell wall, plasma membrane, antioxidants and root elongation in triticale. Biol Plant 52(1):87–92

Liu J, Piñeros MA, Kochian LV (2014) The role of aluminum sensing and signaling in plant aluminum resistance. J Integr Plant Biol 56:221–230

Liu S, Gao H, Wu X, Fang Q, Chen L, Zhao F-J et al (2016) Isolation and characterization of an aluminum-resistant mutant in rice. Rice 9:60. https://doi.org/10.1186/s12284-016-0132-3

Liu WJ, Xu FJ, Lv T, Zhou WW, Chen Y, Jin CW, Lu LL, Lin XY (2018) Spatial responses of antioxidative system to aluminum stress in roots of wheat (Triticum aestivum L.) plants. Sci Total Environ 627:462–469

Liu W-S, Zheng H-X, Guo M-N, Liu C, Huot H, Morel JL, Ent A, Tang Y-T, Qiu R-L (2019) Co-deposition of silicon with rare earth elements (REEs) and aluminum in the fern dicranopteris linearis from China. Plant and Soil 437:427–437

Long A, Zhang J, Yang LT, Ye X, Lai NW, Tan LL, Lin D, Chen LS (2017) Effects of low pH on photosynthesis, related physiological parameters and nutrient profile of *Citrus*. Front Plant Sci 8:185

Luchi S, Koyama H, Luchi A, Kobayashi Y, Kitabayashi S, Ikka T, Hirayama T, Shinozaki K, Kobayashi M (2007) Zinc finger protein STOP1 is critical for proton tolerance in Arabidopsis and coregulates a key gene in aluminium tolerance. PNAS 104(23):9900–9905.18

Lundstrom US (1993) The role of organic acids in the soil solution chemistry of a podzolized soil. J Soil Sci 44:121–133

Ma JF (2000) Role of organic acids in detoxification of aluminum in higher plants. Plant Cell Physiol 41:383–390

Ma JF (2007) Syndrome of aluminum toxicity and diversity of aluminum resistance in higher plants. Survey of Cell Biology 264:225–252

Ma Z, Miyasaka SC (1998) Oxalate exudation by taro in response to Al. Plant Physiol 118:861–865

Ma JF, Hiradate S (2000) Form of aluminium for uptake and translocation in buckwheat (*Fagopyrum esculentum* Moench). Planta 211:355–360

Ma JF, Hiradate S, Nomoto K, Iwashita T, Matsumoto H (1997a) Internal detoxification mechanism of Al in hydrangea: identification of Al form in the leaves. Plant Physiol 113:1033–1039

Ma JF, Hiradate S, Matsumoto H (1998) High aluminum resistance in buckwheat: II. Oxalic acid detoxifies aluminum internally. Plant Physiol 117:753–759. https://doi.org/10.1104/pp.117.3.753

Ma JF, Taketa S, Yang ZM (2000) Aluminum tolerance genes on the short arm of chromosome 3R are linked to organic acid release in triticale. Plant Physiol 122:687–694

Ma JF, Ryan PR, Delhaize E (2001) Aluminum tolerance in plants and the complexing role of organic acids. Trends Plant Sci 6:273–278

Ma JF, Shen R, Nagao S and Tanimoto E (2004a) Aluminum targets elongating cells by reducing cell wall extensibility in wheat roots. Plant Cell Physiol 45: 583–589

Ma JF, Nagao S, Sato K, Ito H, Furukawa J, Takeda K (2004) Molecular mapping of a gene responsible for Al-activated secretion of citrate in barley. J Exp Bot 55:1335–1341

Ma HX, Bai GH, Carver BF, Zhou LL (2005) Molecular mapping of a quantitative trait locus for aluminum tolerance in wheat cultivar Atlas 66. Theor Appl Genet 112:51–57

Ma Y, Bai C, Zhang X, Ding Y (2022) Nitrate increases aluminum toxicity and accumulation in root of wheat. Agriculture 12:1946. https://doi.org/10.3390/agriculture12111946

Maejima E, Watanabe T (2014) Proportion of phospholipids in the plasma membrane is an important factor in Al tolerance. Plant Signal Behav 9:e29277

Maejima E, Hiradate S, Jansen S, Osaki M, Watanabe T (2014) Comparative analysis of aluminum accumulation in leaves of three angiosperm species. Botany 92:327–331

Magalhaes JV, Liu J, Guimaraes CT, Lana UGP, Alves VMC et al (2007) A gene in the multidrug and toxic compound extrusion (MATE) family confers aluminum tolerance in sorghum. Nat Genet 39:1156–1161

Mahanta M, Konjengbam NS, Lyngdoh AA, Pheirim R (2022) Aluminium stress tolerance in legumes under acidic soils: mechanisms and methods—a review. Biol Forum Int J 14(1):1538–1548

Malkanthi DR, Yokoyama K, Yoshida T, Moritsugu M, Matsushita K (1995) Effects of low pH and AI on growth and nutrient uptake of several plants. Soil Sci Plant Nutr 41:161–165

Malta PG, Arcanjo-Silva S, Ribeiro C, Campos NV, Azevedo AA (2016) Rudgea viburnoides (Rubiaceae) overcomes the low soil fertility of the Brazilian Cerrado and hyperaccumulates aluminum in cell walls and chloroplasts. Plant Soil 408:369–384

Manyowa NM, Miller TE (1991) The genetics of tolerance to high mineral concentrations in the tribe Triticeae—a review and update. Euphytica 57:175–185

Maranville JW, Duncan RR, Yohe JM (1994) Adaptation of plants to soil stresses. In: Proceedings of a workshop, University of Nebraska, Lincoln, Nebraska, USA

Maron LG, Piñeros MA, Guimarães CT, Magalhaes JV et al (2010) Two functionally distinct members of the MATE (multi-drug and toxic compound extrusion) family of transporters potentially underlie two major aluminum tolerance QTLs in maize. Plant J 61:728–740

Marschner H (1991) Mechanisms of adaptation of plants to acid soils. In: Wright RJ, Baligar VC, Moorman RP (eds) Plant-soil interactions at low pH. Proceedings of the second international symposium on plant-soil interactions at low pH, June 24–29, 1990, Beckley, West Virginia, USA. Kluwer Academic Publisher, pp 683–702

Marsh KB, Peterson LA, McCown BH (1989) A microculture method for assessing nutrient uptake. II. The effect of temperature on manganese uptake and toxicity in potato shoots. J Plant Nutr 12:219–232

Martin R (1988) Bioinorganic chemistry of aluminum. In: Sigel H, Sigel A (eds) Metal ions in biological systems, vol 24. Marcel Dekker, New York, pp 1–57

Martins PR, Parentoni SN, Lopes MA, Paiva E (1999) Eficiĕncia de indices fenotĭpicos de comprimento de raiz seminal na avaliaĉão de plantas individuais de milho quanto ătolerăncia ao aluminio. Pesquisa Agropecuăria Brasileira 34:1897–1904

Matsumoto H (2000) Cell biology of aluminum toxicity and tolerance in higher plants. Int Rev Cytol 200:1–46

Matsumoto H, Hirasawa E, Torikai H, Takahashi E (1976) Localization of absorbed aluminum in pea root and its binding to nucleic acids. Plant Cell Physiol 17:127–137

Matsumoto H, Yamamoto Y, Kasai M (1992) Changes of some properties of the plasma membrane-enriched fraction of barley roots related to aluminum stress: membrane-associated ATPase, aluminum and calcium. Soil Sci Plant Nutr 38:411–419

Matsumoto H, Yamamoto Y, Rama Devi S (2001) Aluminum toxicity in acid soils. Plant responses to aluminum. In: Prasad M (ed) Metals in the environment. Marcel Dekker, New York, pp 289–319

Matsumoto S, Ogata S, Shimada H, Sasaoka T, Hamanaka A, Kusuma GJ (2018) Effects of pH-induced changes in soil physical characteristics on the development of soil water erosion. Geoscience 8:134

Maze P (1915) Determination des elements inineraux rates ilfcessair'es an developpement du maïs. Conlpt Fend Acad Sci Paris 160:211

McCray JM, Sumner ME (1990) Assessing and modifying Ca and Al levels in acid subsoils. Adv Soil Sci 14:45–75

Mcfarland ML, Haby VA, Redmon L, Bade D (2001) Managing soil acidity. Bulletin No. B-1720. http://overton.tamu.edu/soils

McLean FT, Gilbert BE (1928) Aluminum toxicity. Plant Physiol 3:293

McNeilly T (1994) Metal toxicity. In: Yeo AR, Flowers TJ (eds) Monographs on theoretical and applied genetics, 21. Springer-Verlag, Berlin, Heidelberg, pp 145–174

Meda AR, Cassidato ME, Pavan MA, Miyazawa M (2001) Alleviating soil acidity through plant organic compounds. Braz Arch Biol Technol 44:185–189

Mehra A, Baker CL (2007) Leaching and bioavailability of aluminium, copper and manganese from tea (*Camellia sinensis*). Food Chem 100:1456–1463. https://doi.org/10.1016/j.foodchem.2005.11.038

Memon AR, Schröder P (2009) Implications of metal accumulation mechanisms to phytoremediation. Environ Sci Pollut Res 16:162–175

Merifio CG, Alberdi M, Ivanov AG, Reyes-Diaz M (2010) Al3+ -Ca2+ interaction in plants growing in acid soils: Al phytotoxicity response to calcareous amendments. J Soil Sci Plant Nutr 10(3):217–243

Meriño-Gergichevich C, Alberdi IM, Ivanov AG, Reyes-Díaz M (2010) Al^{3+}–Ca^{2+} interaction in plants growing in acid soils: al-phytotoxicity response to calcareous amendments. J Soil Sci Plant Nutr 10:217–243

Milivojevic J, Nikezic D, Krstic D, Jelic M, Djalovic I (2011) Influence of physical-chemical characteristics of soil on zinc distribution and availability for plants in vertisols of Serbia. Pol J Environ Stud 20:993–1000

Milla MA, Butler E, Huete AR, Wilson CF, Anderson O, Gustafson JP (2002) Expressed sequence tag-based gene expression analysis under aluminum stress in rye. Plant Physiol 130:1706–1716

Mittler R, Vanderauwera S, Gallery M, van Breusegem F (2004) Reactive oxygen gene network of plants. Trends Plant Sci 9:490–498

Miyasaka SC, Buta JG, Howell RK, Foy CD (1991) Mechanism of aluminum tolerance in snapbeans' root exudation of citric acid. Plant Physiol 96:737–743

Mora ML, Schnettler B, Demanet R (1999) Effect of liming and gypsum on soil chemistry, yield, and mineral composition of ryegrass grown in an acidic Andisol. Commun Soil Sci Plant Anal 30:1251–1266

Mora ML, Cartes P, Demanet R, Cornforth IS (2002) Effects of lime and gypsum on pasture growth and composition on an acid Andisol in Chile, South America. Commun Soil Sci Plant Anal 33:2069–2081

Mora ML, Alfaro MA, Jarvis SC, Demanet R, Cartes P (2006) Soil aluminum availability in Andisols of Southern Chile and its effect on forage production and animal metabolism. Soil Use Manage 22:95–101

Moreira A, Fageria NK (2010) Liming influence on soil chemical properties, nutritional status and yield of alfalfa grown in acid soil. R Bras Ci Solo 34:1231–1239

Moreno-Alvarado M, García-Morales S, Trejo-Téllez LI, Hidalgo-Contreras JV, Gómez-Merino FC (2017) Aluminum enhances growth and sugar concentration, alters macronutrient status and regulates the expression of NAC transcription factors in rice. Front Plant Sci 8:73. https://doi.org/10.3389/fpls.2017.00073

Moriyama U, Tomioka R, Kojima M, Sakakibara H, Takenaka C (2016) Aluminum effect on starch, soluble sugar, and phytohormone in roots of *Quercus serrata* Thunb seedlings. Trees 30:405–413. https://doi.org/10.1007/s00468-015-1252-x

Moroni JS, Briggs KG, Taylor GJ (1991) Pedigree analysis of the origin of manganese tolerance in Canadian spring wheat (Triticum aestivum L.). Euphytica 56:107–120

Mossor-Pietraszewska T (2001) Effect of aluminium on plant growth and metabolism. Acta Biochim Pol 3:673–686

Mugwira LM (1980) Growth and Ca, Mg, K and P uptake by triticale, wheat and rye at four Al levels. J Plant Nutr 2:591–606

Mugwira LM, Elgawhary SM, Patel SU (1978) Aluminium tolerance in triticale, wheat and rye as measured by root growth characteristics and aluminium concentrations. Plant Soil 50:681–690

Muhrizal S, Shamshuddin J, Husni MHA, Fauziah I (2003) Alleviation of aluminum toxicity in an acid sulfate soil in Malaysia using organic materials. Commun Soil Sci Plant Anal 34:2993–3012

Mukhopadyay M, Bantawa P, Das A, Sarkar B, Bera B, Ghosh P, Mondal TK (2012) Changes of growth, photosynthesis and alteration of leaf antioxidative defence system of tea [Camellia sinensis (L.) O. Kuntze] seedlings under aluminum stress. Biometals 25:1141–1154

Mullette KJ (1975) Stimulation of growth in eucalyptus due to aluminium. Plant Soil 42:495–499

Mulungu K, Tembo G, Kabwe S (2013) An economic analysis of precision application of lime at reduced rates. J Econ Sust Dev 4(17):211–214

Mulyani A, Rachman A, Dairah A (2009) Penyebaran lahan masam, potensi dan ketersediaannya untuk pengembangan pertanian. Dalam Fosfat Alam. Balai Penelitian Tanah, Bogor

Naidoo G, McD SJ, Lewis RJ (1978) Accumulation sites of Al in snapbean and cotton roots. Agron J 70:489–492

Nguyen HT, Blum A (2004) Physiology and biotechnology integration for plant breeding. Marcel Dekker Inc., New York, NY, USA, 648p

Nguyen NT, Dudzinski M, Mohapatra PK, Fujita K (2005) Distribution of accumulated aluminum and changes in cell wall polysaccharides in *Eucalyptus camaldulensis* and *Melaleuca cajuputi* under aluminum stress. Soil Sci Plant Nutr 51:737–740. https://doi.org/10.1111/j.1747-0765.2005.tb00103.x

Nordstrom DK, May HM (1996) Aqueous equilibrium data for mononuclear aluminum species. In: Sposito G (ed) The environmental chemistry of aluminium. CRC Press Inc, New York, pp 39–80

Oertel C, Matschullat J, Zurba K, Zimmermann F, Erasmi S (2016) Greenhouse gas emissions from soils—a review. Geochemistry 76:327–352

Ono TA, Rompel A, Mino H, Chiba N (2001) Ca2+ function in photosynthetic oxygen evolution studied by alkali metal cations substitution. Biophysical J 81:1831–1840

Osaki MT, Watanabe T, Tadano T (1997) Beneficial effect of aluminum on growth of plants adapted to low pH soils. Soil Sci Plant Nutr 43:551–563. https://doi.org/10.1080/00380768.1997.10414782

Osaki M, Watanabe T, Ishizawa T, Nilnond C, Nuyim T, Sittibush C, Tadano T (1998) Nutritional characteristics in leaves of native plants grown in acid sulfate, peat, sandy podzolic, and saline soils distributed in Peninsular Thailand. Plant Soil 201:175–182

Osawa H, Endo I, Hara Y, Matsushima Y, Tange T (2011) Transient proliferation of proanthocyanidinaccumulating cells on the epidermal apex contributes to highly aluminum-resistant root elongation in camphor tree. Plant Physiol 155:433–46

Ouma E, Ligeyo D, Matonyei T, Were B, Agalo J, Emily T, Onkware A, Gudu S, Kisinnyo P, Okalebo J et al (2012) Development of maize single cross hybrids for tolerance to low phosphorus. Afr J Plant Sci 6:394–402

Ownby JD, Popham HR (1989) Citrate reverses the inhibition of wheat root growth caused by aluminum. J Plant Physiol 135:588–591

Pan JW, Zhu MY, Chen H (2001) Aluminum-induced cell death in root-tip cells of barley. Env Exp Bot 46:71–79

Panda (2007)

Panda SK, Matsumoto H (2007) Molecular physiology of aluminum toxicity and tolerance in plants. Bot Rev 73:326–347

Panda SK, Singha LB, Khan MH (2003) Does aluminium phytotoxicity induce oxidative stress in greengram (*Vigna radiata*)? Bulg J Plant Physiol 29:77–86

Pandey S, Gardner CO (1992) Recurrent selection for population, variety and hybrid improvement in tropical maize. Adv Agron 48:1–87

Papernik LA, Bethea AS, Singleton TE, Magalhaes JV, Garvin DF, Kochian LV (2001) Physiological basis of reduced Al tolerance in ditelosomic lines of Chinese Spring wheat. Planta 212:829–834

Parker DR, Bertsch EM (1992) Formation of the „Al13“ tridecameric polycation under diverse synthesis conditions. Environ Sci Technol 26:914–921

Parker DR, Kinraide TB, Zelazny LW (1989) On the phytotoxicity of polynuclear hydroxy-aluminium complexes in soil. Sci Soc Amer J 53:789–796

Peixoto HP, da Matta FM, da Matta JC (2002) Responses of the photosynthetic apparatus to aluminum stress in two sorghum cultivars. J Plant Nutr 25:821–832

Pellet DM, Grunes DL, Kochian LV (1994) Organic acid exudation as an aluminium tolerance mechanism in maize (Zea mays L.). Planta 196:788–795

Pellet DM, Grunes DL, Kochian LV (1995) Organic acid exudation as an aluminum tolerance mechanism in maize *(Zea* mays L.). Planta 196:788–795

Pellet DM, Papernik LA, Kochian LV (1996) Multiple aluminum-resistance mechanisms in wheat—roles of root apical phosphate and malate exudation. Plant Physiol 112:591–597

Pereira JF, Zhou G, Delhaize E, Richardson T, Ryan PR (2010) Engineering greater aluminium resistance in wheat by over-expressing TaALMT1. Ann Bot 106:205–214

Picton SJ, Richards KD, Gardner RC (1991) Protein profiles in root tips of two wheat (*Triticum aestivum* L.) cultivars with differential tolerance to aluminum. In: Wright RJ et al (eds) Plant soil interactions at low pH. Kluwer Academic Publishers, Dordecht, The Netherlands, pp 1063–1070

Piñeros MA, Magalhaes JV, Carvalho Alves VM, Kochian LV (2002) The physiology and biophysics of an aluminum tolerance mechanism based on root citrate exudation in maize. Plant Physiol 129:1194–1206

Piñeros MA, Shaff JE, Manslank HS, Alves VM et al (2005) Aluminum resistance in maize cannot be solely explained by root organic acid exudation. A comporative physiological study. Plant Physiol 137:231–241

Pilon-Smits EAH, Quinn CF, Tapken W, Malagoli M, Schiavon M (2009) Physiological functions of beneficial elements. Curr Opin Plant Biol 12:267–274. https://doi.org/10.1016/j.pbi.2009.04.009

Pineros M, Kochian LV (2001) A patch clamp study on the physiology of aluminum toxicity and aluminum tolerance in *Zea mays*: identification and characterization of Al^{3+}-induced anion channels. Plant Physiol 125:292–305

Pintro J, Barloy J, Fallavier P (1996) Aluminium effects on the growth and mineral composition of corn plants cultivated in nutrient solution at low aluminum activity. J Plant Nutr 19:729–741

Pintro J, Inoue TT, Tescaro MD (1999) Influence of the ionic strength of nutrient solutions and tropical acid soil solutions on aluminum activity. J Plant Nutr 22:1211–1221

Plieth C (2005) Calcium: just another regulator in the machinery of life? Ann Bot-London 96:1–8

Polle E, Konzak CF, Kittrick JA (1978) Visual detection of aluminum tolerance levels in wheat by hematoxylin staining. Crop Sci 18:823–827

Poschenrieder C, Gunsé B, Corrales I, Barceló J (2008) A glance into aluminum toxicity and resistance in plants. Sci Total Environ 400:356–368

Prijambada ID, Proklamasiningsih E (2010) Effect of organic acids amendment on the growth and yield of soybean (Glycine max) in Ultisol. Int J Agric Biol 12:566–570

Prochnow BL (2014) Soil acidity evaluation and management. Better Crops 98(1):22–25

Purnamaningsih R, Hutami S (2015) Testing of sugarcane mutan somaclone derived from chemical mutation and in vitro selection. In: Prosiding Seminar Nasional Sistem Pertanian-Bioindustri Berkelanjutan, hal 439–447. Yogyakarta, 11 Desember 2014

Quarles LD, Gitelman HJ, Drezner MK (1989) Aluminum-induced de novo bone formation in the beagle. A parathyroid hormone-dependent event. J Clin Invest 83:1644–1650. https://doi.org/10.1172/JCI114063

Rao IM, Zeigler RS, Vera R, Sarkarung S (1993) Selection and breeding for acid-soil tolerance in crops. Bioscience 43:454–465

Reddy KJ, Wang L, Gloss SP (1995) Potential solid phases controlling dissolved aluminum and iron in acidic soils, pp 35–40. In: Date RA et al (ed) Plant soil interactions at low pH: principles and management. Kluwer Acad. Publ., Dordrecht, The Netherlands

Reich PB, Oleksyn J, Tjoelker MG (1994) Relationship of aluminum and calcium to net CO2 exchange among diverse scots pine provenances under pollution stress in Poland. Oecologia 97:82–92

Rengel Z (1992a) Role of calcium in aluminium toxicity. New Phytol 121:499–513

Rengel Z, Zhang WH (2003) Role of dynamics of intracellular calcium in aluminium-toxicity syndrome. New Phytol 159:295–314

Richards KD, Snowden KC, Gardner RC (1994) *Wali6 and wali7*. Genes induced by aluminum in wheat (*Triticum aestivum* L.) roots. Plant Physiol 105:1455–1456

Richardson AE, Barea JM, McNeill AM, Prigent-Combaret C (2009) Acquisition of phosphorus and nitrogen in the rhizosphere and plant growth promotion by microorganisms. Plant Soil 321:305–339

Ritchie GSP (1995) Soluble aluminium in acidic soils: principles and practicalities. Plant Soil 171:17–27

Ritchie GSP (1989) The chemical behaviour of aluminium, hydrogen and manganese. In: Robson AD (ed) Soil acidity and plant growth. Academic Press, New York, USA, pp 1–60

Ritchie GSP (1994) The role of dissolution and precipitation of minerals controlling soluble aluminium in acidic soils. Adv Agron 53:47–83

Ritchie GSP, Posner AM, Ritchie IM (1982) The polarographic study of the equilibrium between humic acid and aluminium in solution. J Soil Sci 33:671–677

Roberts VG, Smeda Z (2001 March 12–16) The distribution of soil fertility constraints in KwaZulu-Natal, South Africa. In Proceedings of the 5th International Symposium on Plant Soil Interactions at low pH, Bergville, South Africa, p.12

Robles-Aguilar AA, Pang J, Postma JA, Schrey SD, Lambers H, Jablonowski ND (2019) The effect of pH on morphological and physiological root traits of *Lupinus angustifolius* treated with struvite as a recycled phosphorus source. Plant Soil 434:65–78

Robson AD, Loneragan JF (1970) Sensitivity of annual Medicago species to manganese toxicity as affected by calcium and pH. Aust J Agric Res 21:223–232

Rosicky MA, Sullivan LA, Slavich PG, Hughes M (2004) Factors contributing to the acid sulfate soil scalding process in the coastal floodplains of New South Wales, Australia. Soil Res 42:587–594

Rounds MA, Larsen PB (2008) Aluminum-dependent root-growth inhibition in Arabidopsis results from AtATR-regulated cell-cycle arrest. Curr Biol 18:1495–1500

Rout GR, Samantaray S, Das P (2001) Aluminium toxicity in plants: a review. Agronomie 21:3–21

Rovira AD, McDougall B (1967) Microbiological and biochemical aspects of the rhizosphere. In: McLaren AD, Peterson GF (eds) Soil biochemistry. Marcel Dekker, New York, pp 417–463

Roy A, Sharma GT (1988) Some aspects of aluminum toxicity in plants Bot. Rev 54:145–178

Ryan PR, Delhaize E (2010) The convergent evolution of aluminium resistance in plants exploits a convenient currency. Funct Plant Biol 37:275–284

Ryan PR, DiTomaso JM, Kochian LV (1993) Aluminium toxicity in roots: an investigation of spatial sensitivity and the role of the root cap. J Exp Bot 44:437–446

Ryan PR, Kinraide TB, Kochian LV (1994) A13+-Caz+ interactions in aluminum rhizotoxicity. I. Inhibition of root growth is not caused by reduction of calcium uptake. Planta 192:98–103

Ryan PR, Delhaize E, Randall PJ (1995a) Characterization of Al-stimulated efflux of malate from the apices of Al-tolerant wheat roots. Planta 196:103–110

Ryan PR, Delhaize E, Randall PJ (1995b) Malate efflux from root apices and tolerance to aluminum are highly correlated in wheat. Aust J Plant Physiol 22:531–536

Ryan PR, Skerrett M, Findlay GP, Delhaize E, Tyerman S (1997) Aluminum activates an anion channel in the apical cells of wheat roots. PNAS 94:6547–6552

Ryan PR, Delhaize E, Jones DL (2001) Function and mechanism of organic anion exudation from plant roots. Annu Rev Plant Physiol Plant Mol Biol 52:527–560

Ryan PR, Tyerman SD, Sasaki T, Furuichi T, Yamamoto Y, Zhang WH, Delhaize E (2011) The identification of aluminium-resistance genes provides opportunities for enhancing crop production on acid soils. J Exp Bot 62:9–20

Salinas JG, Sanchez PA (1976) Soil–plant relationships affecting varietal and species differences in tolerance to low available soil phosphorus. Ciência e Cultura 28:156–168

Samac DA, Tesfaye M (2003) Plant improvement for tolerance to aluminum in acid soils: a review. Plant Cell Tissue Organ Cult 75:189–207

Sanchez PA, Salinas JG (1981) Low-input technology for managing oxisols and ultisols in tropical America. Adv Agron 34(279–406):1982

Scandalios JG (2002) The rise of ROS. Trends Biochem Sci 27:483–486

Schmitt M, Mehltreter K, Sundue M, Testo W, Watanabe T, Jansen S (2017) The evolution of aluminum accumulation in ferns and lycophytes. Am J Bot 104:573–583

Schulte EE, Kelling KA (1999) Understanding Plant Nutrients/Soil and Applied Manganese; University of Wisconsin System Board of Regents and University of Wisconsin-Extension, Cooperative Extension: Madison, WI, USA, pp 4

Schoenholtz S, Miegroet HV, Burger J (2000) A review of chemical and physiological properties as indicators of forest soil quality. Challenges and opportunities. Ecol Manag 138:335–356

Schulte EE, Kelling KA (1999) Understanding plant nutrients/soil and applied manganese; University of Wisconsin System Board of Regents and University of Wisconsin-Extension, Cooperative Extension, Madison, WI, USA, pp 4

Scott BJ, Conyers MK, Poile GJ, Cullis BR (1997) Subsurface acidity and liming affectyield of cereals. Aust J Agric Res 48:843–854

Scott BJ, Fisher JA, Cullis BR (2001) Aluminium tolerance and lime increase wheat yield on the acidic soils of central and southern New South Wales. Aust J Exp Agric 41:523–532

Setter TL, Waters I (2003) Review of prospects for germplasm improvement for waterlogging tolerance in wheat, barley and oats. Plant Soil 253:1–33

Setter TL, Waters I, Sharma SK, Singh KN, Kulshreshtha N, Yaduvanshi NPS, Ram PC, Singh BN, Rane J, McDonald G, Khabaz-Saberi H, Biddulph TB, Wilson R, Barclay I, McLean R, Cakir M (2008) Review of wheat improvement for waterlogging tolerance in Australia and India: the importance of anaerobiosis and element toxicities associated with different soils. Ann Bot 137:1–15

Shaffert RE (1994) Discipline interactions in the quest to adapt plants to soil stresses through genetic improvement. In: Maranville JW, Duncan RR, Yohe JM (eds) Adaptation of plants to soil stresses. Proceedings of a workshop, University of Nebraska, Lincoln, Nebraska, USA

Shann JR, Bertsch PM (1993) Differential cultivar response to polynuclear hydroxo-aluminum complexes. Soil Sci Soc Am J 57:116–120

Sharma P, Jha AB, Dubey RS, Pessarakli M (2012) Reactive oxygen species, oxidative damage, and antioxidative defence mechanisms in plants under stressful conditions. J Bot 217037

Shukry WM, Khattab HKI, El-Bassiouny HMS (2007) Physiological and biochemical studies on flax plant grew in calcareous soil amended with water hyacinth dry manure. J Appl Sci Res 3:64–72

Silva AR (1976) Application of the genetic approach to wheat culture in Brazil. In: Wright MJ (ed) Plant adaptation to mineral stress in problem soils. Cornell University, Ithaca, New York, USA

Silva IR, Ferrufino A, Sanzonowicz C, Smyth TJ, Israel DW, Carter jr, T. E. (2005) Interactions between magnesium, calcium, and aluminum on soybean root elongation. R Bras Ci Solo 29:747–754

Simões CC, Melo JO, Magalhaes JV, Guimarães CT (2012) Genetic and molecular mechanisms of aluminum tolerance in plants. Genet Mol Res 11(3):1949–1957. https://doi.org/10.4238/2012.July.19.14

Singh VP, Tripathi DK, Kumar D, Chauhan DK (2011) Influence of exogenous silicon addition on aluminium tolerance in rice seedlings. Biol Trace Elem Res 144:1260–1274. https://doi.org/10.1007/s12011-011-9118-6

Singh D, Dikshit HK, Singh R (2012) Variation of aluminium tolerance in lentil (Lens culinaris Medik.) Plant Breeding

Singh D, Pal M, Singh CK, Taunk J, Jain P, Chaturvedi AK, Maurya S, Karwa S, Singh R, Tomar RSS, Nongthombam R, Chongtham N, Singh MP (2016) Molecular scanning and morpho-physiological dissection of component mechanism in lens species in response to aluminium stress. PLOS ONE. https://doi.org/10.1371/journal.pone.0160073

Sivaguru M, Horst W (1998) The distal part of the transition zone is the most aluminum-sensitive apical root zone of maize. Plant Physiol 116:155–163

Slootmaker ALJ (1974) Tolerance to high soil acidity in wheat related species, rye and triticale. Euphytica 23:505–513

Snowball K, Robson AD (1991) Nutrient deficiencies and toxicities in wheat: a guide for field identification. CIMMYT, Texcoco, Mexico, 82p

Snowden KC, Gardner RC (1993) Five genes induced by aluminum in wheat (*Triticum aestivum* L.) roots. Plant Physiol 103:855–861

Somani LL (1996) Crop production in acid soils. Agrotech Publishing Academy, New Delhi, p 224

Somers DJ, Briggs KG, Gustafson JP (1996) Aluminum stress and protein synthesis in near isogenic lines of *Triticum aestivum* differing in aluminum tolerance. Physiol Plant 97:694–700

Song H, Xu X, Wang H, Tao Y (2011) Protein carbonylation in barley seedling roots caused by aluminum and proton toxicity is suppressed by salicylic acid. Russ J Plant Physiol 58:653–659

Spain JM, Francis CA, Howeler RH, Cabio F (1975) Differential species and varietal tolerance to soil acidity in tropical crops and pastures. In: Bornemiza E, Alvarado A (eds) Soil management in tropical America. North Carolina State University, USA

Sparrow LA, Uren NC (1987) The role of manganese toxicity in crop yellowing on seasonally waterlogged and strongly acidic soils in north-eastern Victoria. Aust J Exp Agric 27:303–307

Stanton ML, Roy BA, Thiede DA (2000) Evolution in stressful environments. Phenotypic variability, phenotypic selection, and response to selection in five distinct environmental stresses. Evolution 54:93–111

Steiner F, Zoz T, Junior ASP, Castagnara DD, Dranski JAL (2012) Effects of aluminium on plant growth and nutrient uptake in young physic nut plants. Semin Ciênc Agrar 33:1779–1788

Stevenson FJ (1967) Organic acids in soil. In: McLaren AD, Peterson GH (eds) Soil biochemistry, vol 1. Marcel Dekker, New York, pp 119–146

Stevenson FJ (1994) Humus chemistry: genesis, composition, reactions. Wiley, New York

Stevenson FJ, Vance GF (1989) Naturally occurring aluminium—organic complexes. In: Sposito G (ed) The environmental chemistry of aluminium. CRC Press, Boca Raton, pp 117–146

Stoklasa J (1922) Über die Verbreitung des Aluminiums in der Natur und Seine Bedeutung Beim bau- und Betriebsstoffwechsel der Pflanzen. Fisher Scientific, Jena

Stole RJ, Van Helden AK, de Bruyn PL (1976) Hydrolysis precipitation studies of aluminium (III) solutions. 2. A kinetic study and model. J Colloid Interface Sci 57:115–131

Stoyanova Z, Poschenrieder C, Tzvetkova N, Doncheva S (2010) Characterization of the tolerance to excess manganese in four maize varieties. Soil Sci Plant Nutr 55:747–753

Stumm W, Wollast R (1990) Coordination chemistry of weathering: kinetics of the surface-controlled dissolution of oxide minerals. Rev Geophys 28:53–69

Suhayda CG, Haug A (1984) Organic acids prevent aluminuminduced conformational changes in calmodulin. Biochem Biophys Res Commun 119:376–381

Suhayda CG, Haug A (1985) Citrate chelation as a potential mechanism against aluminum toxicity in cells: the role of calmodulin. Can J Biochem Cell Biol 63:1167–1175

Suhayda CG, Haug A (1986) Organic acids reduce aluminum toxicity in maize root membranes. Physiol Plant 68:189–195

Sumner ME, Fey MV, Noble AD (1991) Nutrient status and toxicity problems in acid soils. In: Ulrich B, Sumner ME (eds) Soil acidity. Springer-Verlag, Berlin, pp 149–182

Sun CL, Lu LL, Yu Y, Liu LJ, Hu Y, Ye YQ, Jin CW, Lin XY (2016) Decreasing methylation of pectin caused by nitric oxide leads to higher aluminum binding in cell walls and greater aluminum sensitivity of wheat roots. J Exp Bot 67:97

Sun CL, Lv T, Huang L, Liu XX, Jin CW, Lin XY (2020) Melatonin ameliorates aluminum toxicity through enhancing aluminum exclusion and reestablishing redox homeostasis in roots of wheat. J Pineal Res 68:e12642

Suthipradit S, Edwards DG, Asher CJ (1990) Effects of aluminium on tap-root elongation of soybean (Glycine max), cowpea (Vigna unguiculata) and green gram (Vigna radiata) grown in the presence of organic acids. Plant Soil 124:233–237

Tadano T, Sakai H (1991) Secretion of acid phosphatase by the roots of several crop sp. under P deficient conditions. Soil Sci Plant Nutr 37:129–140

Takahashi T, Ikeda T, Fujita K, Nanzyo M (2006a) Effect of liming on organically complexed aluminum of nonallophanic Andosols from northeastern Japan. Geoderma 130:26–34

Takahashi T, Ikeda Y, Nakamura H, Nanzyo M (2006b) Efficiency of gypsum application to acid Andosols estimated using aluminum release rates and plant root growth. J Soil Sci Plant Nutr 52:584–592

Tamas L, Simonovicova M, Huttova J, Mistrik I (2003) Changes in the composition of cell wall proteins in barley roots during germination and growth in aluminium presence. Plant Soil Environ 49:327–331

Tamás L, Budíková S, Huttovà J, Mistrík I, Simonovicová M, Siroka B (2005) Aluminum-induced cell death of barley-root border cells is correlated with peroxidase- and oxalate oxidase-mediated hydrogen peroxide production. Plant Cell Rep 24:189–194

Tan KH, Binger A (1986) Effect of humic acid on aluminium toxicity in corn plants. Soil Sci 141:20–25

Tandzi NL, Mutengwa C, Ngonkeu E, Gracen V (2018) Breeding maize for tolerance to acidic soils: a review. Agronomy 8:84

Tang C, Nuruzzaman M, Rengel Z (2003) Screening wheat genotypes for tolerance of soil acidity. Aust J Agric Res 54:445–452

Tarafdar JC, Jungk A (1987) Phosphatase activity in the rhizosphere and its relation to the depletion of soil organic phosphorus. Biol Fertil Soil 3:199–204

Taylor GJ (1988) The physiology of aluminum tolerance. Commun Soil Sci Plant Anal 19:7–12. **1957** ©FUNPEC-RP www.funpecrp.com.br Genetics and molecular research 11(3):1949–1957 (2012) Mechanisms of aluminum tolerance in plants

Taylor GJ (1991) Current views of the aluminum stress response: the physiological basis of tolerance. Curr Top Plant Biochem Physiol 10:57–93

Thakuria D, Hazarika S, Krishnappa R (2016) Soil Acidity and management options. Ind J Fertilisers 12:40–56

Thomas GW (1976) The relationship between organic matter content and exchangeable aluminium in acid soils. Soil Sci Soc Am J 39:591

Thornton FC, Schaedle M, Ryanal DJ (1986) Effect of aluminum on the growth of sugar maple in solution culture. Can J for Res 16:892–896. https://doi.org/10.1139/x86-159

Tice KR, Parker DR, McMason DA (1992) Operationally defined apoplastic and symplastic aluminum fractions in root tips of aluminum-intoxicated wheat. Plant Physiol 100:309–318

Tolra R, Barcelo J, Poschenrieder C (2009) Constitutive and aluminium-induced patterns of phenolic compounds in two maize varieties differing in aluminium tolerance. J Inorg Biochem 103:1486–1490

Toma M, Saigusa M (1997) Effects of phosphogypsum on ameliorating strongly acid nonallophanic Andosols. Plant Soil 192:49–55

Tomioka R, Takenaka C (2007) Enhancement of root respiration and photosynthesis in *Quercus serrata* Thunb. seedlings by long-term aluminum treatment. Environ Sci 14:141–148

Turan M, Kiziloglu FM, Ketterings QM (2009) Phosphorus management of lucerne grown on calcareous soil in Turkey. J Plant Nutr 32:516–535

Uguru MI, Oyiga B, Jandong EA (2012) Responses of some soybean genotypes to different soil pH regimes in two planting seasons. Afr J Plant Sci Biotechnol 6(1):26–37

Ulrich B, Mayer R, Khanna PK (1980) Chemical changes due to acid precipitation in a loess-derived soil in central Europe. Soil Sci 130:193–199

Venkatesan S, Jayaganesh S (2010) Characterisation of magnesium toxicity, its influence on amino acid synthesis pathway and biochemical parameters of tea. Res J Phytochem 4:67–77

von Uexkül HR, Mutert E (1995) Global extent, development and economic impact of acid soils. Global extent, development and economic impact of acid soils. Plant Soil 171:1–15

Wagatsuma T, Ishikawa S, Ofei-Manu P (2001) The role of the outer surface of the plasma membrane in aluminum tolerance. In: Ae N, Arihara J, Okada K, Srinivasan A (eds) Plant nutrient acquisition. Springer-Verlag, Tokyo, pp 159–184

Wagatsuma T, Khan MS, Watanabe T, Maejima E, Sekimoto H, Yokota T et al (2015) Higher sterol content regulated by *CYP51* with concomitant lower phospholipid contents in membranes is a common strategy for aluminium tolerance in several plant species. J Exp Bot 66:907–918. https://doi.org/10.1093/jxb/eru455

Wallace SU, Anderson IC (1984) Aluminum toxicity and DNA synthesis in wheat roots. Agron J 76:5–8

Wan Q, Xu R-K, Li X-H (2012) Proton release from tea plant (*Camellia sinensis* L.) roots induced by Al(III) under hydroponic conditions. Soil Res 50:482–488

Wang P, Bi S, Ma L, Han W (2000) Aluminum tolerance of two wheat cultivars (Brevor and Atlas66) in relation to their rhizosphere pH and organic acids exuded from roots. J Agri Food Chem 54:10033–10039

Wang YX, Wu P, Wu YR, Yan XL (2002) Molecular marker analysis of manganese toxicity tolerance in rice under greenhouse conditions. Plant Soil 238:227–233

Wang J, Raman H, Zhang G, Mendham N, Zhou M (2006a) Aluminium tolerance in barley (Hordeum vulgare L.): physiological mechanisms, genetics and screening methods. J Zhejiang Univ Sci B 7:769–787

Wang L, Fan X-W, Pan J-L, Huang Z-B, Li Y-Z (2015b) Physiological characterization of maize tolerance to low dose of aluminum, highlighted by promoted leaf growth. Planta 242:1391–1403. https://doi.org/10.1007/s00425-015-2376-3

Watanabe T, Osaki M, Yoshihara T, et al (1998) Distribution and chemical speciation of aluminum in the Al accumulator plant, Melastoma malabathricum L.. Plant and Soil 201:165–173. https://doi.org/10.1023/A:1004341415878

Watanabe T (2022) Basic understanding of aluminum accumulator plants. Eurasian J Res 22(59–62):2022. https://doi.org/10.14943/EJFR.22.59

Watanabe T, Osaki M (2001) Influence of aluminum and phosphorus on growth and xylem sap composition in Melastoma malabathricum L. Plant Soil 237:63–70

Watanabe T, Osaki M (2002) Mechanisms of adaptation to high aluminum condition in native plant species growing in acid soils: a review. Commun Soil Sci Plant Anal 33:1247–1260

Watanabe T, Okada K (2005) Interactive effects of Al, Ca and other cations on root elongation of rice cultivars under low pH. Ann Bot 95:379–385

Watanabe T, Osaki! M, Toshiaki Tadano T (1998a) Effects of nitrogen source and aluminum on growth of tropical tree seedlings adapted to low pH soils. Soil Sci Plant Nu tr 44:655–666

Watanabe T, Osaki M, Yoshihara T, Tadano T (1998b) Distribution and chemical speciation of aluminum in the Al accumulator plant, *Melastoma malabathricum* L. Plant Soil 201:165–173

Watanabe T, Jansen S, Osaki M (2005) The beneficial effect of aluminium and the role of citrate in Al accumulation in *Melastoma malabathricum.* New Phytol 165:773–780. https://doi.org/10.1111/j.1469-8137.2004.01261.x

Watanabe T, Jansen S, Osaki M (2006) Al-Fe interactions and growth enhancement in Melastoma malabathricum and Miscanthus sinensis dominating acid sulphate soils. Plant, Cell Environ 29:2124–2132

Watt DA (2003) Aluminium-responsive genes in sugarcane: identification and analysis of expression under oxidative stress. J Exp Bot 54:1163–1174

Wehr et al (2004)

Wehr JB, Blamey FPC, Smith TE, Menzies NW (2017) Growth and physiological responses of teak (Tectona grandis Linn. f.) clones to Ca, H and Al stresses in solution and acid soils. New Forests 48:137–152

Weir RG (1988) The occurrence and diagnosis of manganese toxicities and deficiencies in crops and pastures in New South Wales. In: Graham RD et al (eds) Manganese in soils and plants. Kluwer Academic Publishers, Dordrecht, The Netherlands, pp 99–100

Welcker C (2000) Fitting maize into cropping systems on acid soils of the tropics. Final progress report 01/10/1996–30/09/2000; INRA Guadeloupe, Paris, France, pp 89–108

Wenzl P, Patino GM, Chaves AL, Mayer JE, Rao IM (2001) The high level of aluminum resistance in signalgrass is not associated with known mechanisms of aluminum detoxification in root apices. Plant Physiol 125:1473–1484

Werner D, Newton WE (eds) (2005) Nitrogen fixation in agriculture, forestry, ecology, and the environment, vol 4. Springer Science & Business Media

Wheeler BD, Al-Farraj MM, Cook RED (1985) Iron toxicity to plants in base-rich wetlands: comparative effects on the distribution and growth of Epilobium hirsutum and Juncus subnodulosus Schrank. New Phytol 100:653–669

White P, Broadley MR (2003) Calcium in plants. Ann Bot-London 92:487–511

Wild A (1988) Soil acidity and alkalinity. In: Wild A (ed) Russell's soil conditions and plant growth. Longman, Harlow, pp 844–889

Wildey TI (2003) The influence of seed placed lime to reduce the acidifying effects of nitrogen fertilizers in direct seeding systems. M.S. thesis, Washington State Univ., Pullman

Wong MTF, Swift RS (1995) Amelioration of aluminium phytotoxicity with organic matter. In: Date RA, Grundon NJ, Rayment GE, Probert ME (eds) Plant-soil interactions at low ph: principles and management. Kluwer Academic Publishers, Dordrecht, The Netherlands, pp 41–45

Wong MTF, Akyeampong E, Nortcliff S, Rao MR, Swift RS (1995) Initial responses of maize and beans to decreased concentrations of monomeric inorganic aluminium with application of manure or tree prunings to an Oxisol in Burund. Plant Soil 171:275–282

Wood S, Sebastian K, Scherr SJ (2000) Pilot analysis of global ecosystems: agroecosystems. Tech. Rep., Int. Food Policy Res. Inst. and World Resour. Inst., Washington, DC

Woolhouse H (1983) Toxicity and tolerance in the response of plants to metals. In: Lange OL et al (eds) Physiological plant ecology. III. Response to the chemical and biological environment. Springer, Berlin, Heidelberg, p. 254

Wright RJ (1989) Soil aluminium toxicity and plant growth. Commun Soil Sci Plant Anal 20:1479–1497

Wright RJ, Baligar VC, Ritcheyand KD, Wright SF (1989) Influence of soil solution aluminium on root elongation of wheat seedlings. Plant Soil 113:294–298

Xia J, Yamaji N, Ma JF (2013) A plasma membrane-localized small peptide is involved in rice aluminum tolerance. Plant J 76:345–355. https://doi.org/10.1111/tpj.12296

Xu S (1991) Use of solubility method for determining Al-controlling solid phases in acid soils. Ph.D. diss., Washington State Univ., Pullman, WA

Xu Q, Wang Y, Ding Z, Song L, Li Y, Ma D et al (2016) Aluminum induced metabolic responses in two tea cultivars. Plant Physiol Biochem 101:162–172. https://doi.org/10.1016/j.plaphy.2016.02.001

Yamamoto Y, Kobayashi Y, Matsumoto H (2001) Lipid peroxidation is an early symptom triggered by aluminum, but not the primary cause of elongation inhibition in Pea roots. Plant Physiol 125(1):199–208

Yamamoto Y, Kobayashi Y, Devi SR, Rikiishi S, Matsumoto H (2002) Aluminum toxicity is associated with mitochondrial dysfunction and the production of reactive oxygen species in plant cells. Plant Physiol 128:63–72

Yamamoto Y, Kobayashi Y, Devi SR, Rikiishi S, Matsumoto H (2003) Oxidative stress triggered by aluminum in plant roots. Plant Soil 255(1):239–243

Yaman and Kumar P (2021) Physiological stress in agricultural crops: an overview. Biol Forum Int J 13(1):277–292

Yang ZM, Nian H, Sivaguru M, Tanakamaru S, Matsumoto H (2001) Characterization of aluminium-induced citrate secretion in aluminium-tolerant soybean (Glycine max) plants. Physiol Plant 113:64–71

Yang JL, Li YY, Zhang YJ, Zhang SS, Wu YR, Wu P et al (2008) Cell wall polysaccharides are specifically involved in the exclusion of aluminum from the rice root apex. Plant Physiol 146:602–611. https://doi.org/10.1104/pp.107.111989

Yang JL, Zhu XF, Peng YX, Zheng C, Li GX, Liu Y et al (2011) Cell wall hemicellulose contributes significantly to aluminum adsorption and root growth in *Arabidopsis*. Plant Physiol 155:1885–1892. https://doi.org/10.1104/pp.111.172221

Yang XY, Zeng ZH, Yan JY, Fan W, Bian HW et al (2013) Association of specific pectin methylesterases with Al-induced root elongation inhibition in rice. Physiol Plant 148:502–511

Yocum CF (2008) The calcium and chloride requirements of the O2 evolving complex. Coordin Chem Rev 252:296–305

Yost RS (2000) Plant tolerance of low soil pH, soil aluminum, and soil manganese. In: Silva JA, Uchida R (ed) Plant nutrient management in Hawaii's soils, approaches for tropical and subtropical agriculture, College of Tropical Agriculture and Human Resources, University of Hawaii at Manoa

Yu HN, Liu P, Wang ZY, Chen WR, Xu GD (2011) The effect of aluminum treatments on the root growth and cell ultrastructure of two soybean genotypes. Crop Prot 30:323–328

Zang LL, Paven MCML, Clochard T, Porcher A, Satour P, Mojovi´c M, Vidovi´c M, Limami AM, Montrichard F (2020) Nitrate inhibits primary root growth by reducing accumulation of reactive oxygen species in the root tip in Medicago truncatula. Plant Physiol Biochem 146:363–373

Zeigler RS, Pandey S, Miles J, Gourley LM, Sarkarung S (1995) Advances in the selection and breeding of acid-tolerant plants: rice, maize, sorghum and tropical forages. In: Date RA (ed) Plant soil interactions at low pH. Kluwer Acad. Publishers, The Netherlands, pp 391–406

Zhang W-H, Ryan PR, Tyerman S (2001) Malate-permeable channels and cation channels activated by aluminum in the apical cells of wheat roots. Plant Physiol 125:1459–1472

Zhang B, Liu P, Yang YS, Xu GD (2007) Effect of Al in soil on photosynthesis and related morphological and physiological characteristics of two soybean genotypes. Bot Stud 48:435–444

Zhang HM, Wang BR, Xu MG (2008) Effects of inorganic fertilizer inputs on grain yields and soil properties in a long-term wheat-corn cropping system in South China. Commun Soil Sci Plan 39:1583–1599

Zhang H, Jiang Z, Qin R, Zhang H, Zou J, Jiang W et al (2014) Accumulation and cellular toxicity of aluminum in seedling of *Pinus massoniana*. BMC Plant Biol 14:264. https://doi.org/10.1186/s12870-014-0264-9

Zhang M, Ma Y, Horst WJ, Yang Z-B (2016) Spatial-temporal analysis of polyethylene glycol-reduced aluminium accumulation and xyloglucan endotransglucosylase action in root tips of common bean (*Phaseolus vulgaris*). Ann Bot 118:1–9. https://doi.org/10.1093/aob/mcw062

Zhao XQ, Chen RF, Shen RF (2014) Coadaptation of plants to multiple stresses in acidic soils. Soil Sci 179:503–513

Zheng SJ et al (1998) Continuous secretion of organic acids is related to aluminum resistance during relatively long-term exposure to aluminum stress. Physiol Plant 103:209–214

Zhou L, Tan Y, Huang L, Wang W-X (2016) Enhanced utilization of organic phosphorus in a marine diatom *Thalassiosira weissflogii*: a possible mechanism for aluminum effect under P limitation. J Exp Mar Biol Ecol 478:77–85. https://doi.org/10.1016/j.jembe.2016.02.009

Zhou J, Li Z, Zhou T, Xin Z, Wu L, Luo Y, Christie P (2020) Aluminum toxicity decreases the phytoextraction capability by cadmium/zinc hyperaccumulator Sedum plumbizincicola in acid soils. Sci Total Environ 711:134591

Zhu Y, Weindorf DC, Zhang W (2011) Characterizing soils using a portable X-ray fluorescence spectrometer: 1. Soil texture. Geoderma 167:167–177. https://doi.org/10.1016/j.geoderma.2011.08.010

Zysset M, Blaser P, Luster J, Gehring A (1999) Aluminum solubility control in different horizons of a Podzol. Soil Sci Soc Am J 63:1106–1115

Chapter 5
Managing Soil Acidity

Abstract Rainwater substitutes hydrogen for basic cations such as calcium and magnesium, causing the soil to become acidic. The primary leaching base that most affects soil pH is calcium. Carbonates combine in low-rainfall conditions to generate caliche. The depth to caliche is determined by leaching depth, which is affected by soil structure, weather patterns, and annual rainfall. Rainfall and sand content both raise the depth to caliche. Erosion and land levelling can expose caliche, and these carbonate-rich layers can cause pH problems. By decreasing the availability of some essential plant nutrients and increasing the availability of harmful elements like manganese, iron, and aluminium in the soil solution, acidic soils cause problems with production. Acidic soils are known for their poor agricultural performance and crop failure, which is largely due to these harmful substances. While different approaches exist, liming is the most commonly recommended management practice for raising low soil pH. For the P and N fertiliser recovery efficiencies necessary for healthy maize development on acidic soils, lime and P fertiliser application are essential. However, using lime and/or fertiliser is not always cost-effective for small-scale farmers and is not environmentally friendly. Although liming alters the topsoil, it does not completely remove the acidity in the subsoil, which poses a major problem for newly sprouting roots. If a soil is too acidic for a certain crop, it might be balanced with lime or dolomite. The amount of lime or dolomite required to balance an acidic pH will depend on the crop and the soil. Soils with high levels of organic matter and clay will need higher rates of application since they are more resilient to pH changes. Generally speaking, a pH of 5.5 is the ideal (in water). After the desired pH of the soil is reached, further lime or dolomite may be required, depending on the crop. Acidic soils can yield crops much more efficiently by raising the pH level to almost neutral. Even though liming is commonly used to reduce soil acidity, studies have shown that adding animal dung supplements can boost the pH of acidic soils.

Keywords Managing soil acidity · Soil amendments · Liming · Caliche · Farmyard manure · Lime requirement

U. C. Sharma et al., *Soil Acidity*, Progress in Soil Science,
https://doi.org/10.1007/978-3-031-76357-1_5

5.1 General Description

"A country's soils are its most valuable resource because all life on the planet depends entirely on them for survival." For agricultural crops, soils are a substrate that may both physically support plants and serve as a reservoir for nutrients and water necessary for plant growth. When trying to increase agricultural output, a solid grasp of soil is crucial. Plants may not grow at all or produce unsatisfactory harvests occasionally. Then, we want to know what might be the issue: too much water, too little water, a lack of nutrients in the soil, or perhaps too much heat or cold. To ensure that the plants have enough nutrients to develop in poor soil, fertilisers must be used. Because nutrients are taken out of the soil when we harvest crops, it is frequently a good idea to fertilise a field (Bekker and Yapa 1994). The fertiliser will replenish the soil's nutrients. Compost, chicken manure, and cow dung are examples of organic fertilisers that include all the various nutrients (approximately 16 in all). Only a small number of the nutrients that plants require are included in the typical commercial fertilisers. These fertilisers most frequently contain the nutrients nitrogen (N), phosphorus (P), and potassium (K). We may frequently purchase fertilisers that combine all three nutrients—N, P, and K. However, in addition to these macronutrients, plants also require micronutrients including iron, manganese, zinc, and copper as well as the trace minerals calcium (Ca), magnesium (Mg), and sulphur (S). When applying fertilisers—organic or synthetic—doesn't fix the issue, it is preferable to test the soil to determine its fertility. The following factors may be evaluated during a soil test (Bekker and Yapa 1994): how sandy or clayey is the soil?

- Are there enough nutrients in the soil?
- Is there enough OM in the soil?
- How about the soil's ACIDITY?

In general, the acid highland soils (Ultisols and Oxisols) of tropical Asia are AI or Mn poisonous and essentially infertile. Particularly, the Ultisols' thin topsoils are prone to erosion to a great extent. Conventional land-clearing techniques can readily deteriorate and destroy the organic materials including sunburn, rain, and direct sun exposure if after cleansing, appropriately managed. These soils are prone to losing a lot of their initial fertility and advantageous physical characteristics quickly. These soils demand inputs that are far above the means of subsistence farmers in order to sustain output. Sadly, there is no "low-input" technology for the poor, acidic soils of the tropics. If these soils are poorly managed, with no or little input to replenish nutrients lost through cropping or erosion, they will quickly lose their ability to produce a good crop and become infested with Lmperata cylindrica. Eventually, if the degradation is not stopped, they can become barren. According to von Uexkull and Bosshart (1989), primary or secondary forest still covers the majority of acid soils today. Before any forested area is made available for agricultural use, both short- and long-term costs and benefits, as well as potential environmental harm, should be taken into account. In long term, we might find that we depend less on the production of food and more on the acid soils as a supply of timber and a carbon dioxide sink. By bringing the pH level close to neutrality, acidic soils can produce crops with significantly greater efficiency. Animal manure supplements have been shown to raise the pH of acidic soils, despite the fact that liming is frequently used to reduce soil acidity.

The process of soil acidification causes soil pH to drop over time, and the issue frequently goes undetected. Acidification can happen naturally over thousands of years, particularly affecting locations with high rainfall. However, with intense agricultural practices, fast acidification can happen over a few years. Either the subsurface alone or the surface soil can be affected by acidification. Surface acidity can be effectively treated and has a significant positive impact on plant development and output. Correcting subsurface acidity is challenging and expensive.

The need to enhance food crop output is critical due to the growing global population. This can be done by expanding the area under cultivation or by intensifying, diversifying, and specialising agricultural production methods on already-cultivated land. Only 11% of the earth's surface is currently under cultivation, according to estimates from the FAO (1406 Mha), while about 24% (3900 Mha) is potentially arable, the majority of which, 2500 Mha, is made up of acidic soils, with 1700 Mha situated in the humid tropics (IAEA 2006). Therefore, the tropical forest and savannah zones, which are characterised by highly worn, acidic, infertile soils, primarily Oxisols and Ultisols, have the greatest potential for extending agricultural acreage. With growing CO_2 levels in the atmosphere, ongoing use of nitrogenous fertilisers based on ammonium, removal of nutrients from agricultural products without replacement, and nitrate leaching, problems with soil acidification are also anticipated to get worse. The majority of the savannahs are found in humid and subhumid tropical regions, which are ideal for rainfed crop conditions. They also include the largely human-populated savannahs of tropical Asia and a sizeable portion of the agricultural land in several nations in Africa and Latin America. Due to their innate poor fertility and great susceptibility to rapid deterioration, acid savannah soils are typically regarded as marginal (IAEA 2006). The degradation of the natural resource base has been increased as a result of the cultivation of these soils without effective soil management and conservation techniques. In order to prevent further degradation of the resource base and to maintain crop production in tropical acid soils, management practices must be established or enhanced.

In Bangladesh's Old Himalayan Piedmont Plain (OHPP), declining soil fertility and soil acidity are two major problems preventing better crop output. The greater values were found with larger rates of lime application, and the increase in soil pH ranged from 0.5 to 1.11 units. The application of manure resulted in a small increase in soil organic matter (SOM). Following liming, the availability of soil phosphorus increased, that of zinc and boron dropped, while that of potassium and sulphur nearly held steady (Sultana et al. 2019). The output of wheat was significantly increased by the use of lime and manure, and mungbean and T continued to benefit rice, Aman. As yields of wheat, mungbean, and rice increased, so did their ability to absorb nutrients; hence, crops with more nutrient uptake displayed better yields (Sultana et al. 2019).

Acidity has a negative impact on crops both directly through the acidic response and indirectly by decreasing the availability of nutrients. In Bangladesh, more than

30% of the land has acidic soil, which limits crop productivity. The soils in the Old Himalayan Piedmont Plain range in acidity from 4.6 to 6.5 (FRG 2012). Acidic soils have hazardous levels of Al^{3+}, Fe^{3+}, and Mn^{2+}, are lacking in P, and have poorer base availability, all of which reduce crop productivity. In piedmont locations, common crops including potatoes, rice, wheat, and mungbeans are negatively impacted by the acidity of the soil. Application of lime and organic manure changes the characteristics of crops that contribute to yield, which increases crop yields as seen in wheat (Bodruzzaman et al. 2009) and maize (Bodruzzaman 2010).

Even with the finest care, soils can never, if ever, regain their former production in less than ten years after losing their fertile surface layer. Furthermore, if agricultural soil productivity is to be maintained, even with large input, it is necessary to safeguard them against erosion and leaching losses. There is already a lot of technology that can be used to make acid infertile upland soils productive and lucrative over the long run, but it needs to be implemented locally. It will take a lot more research and extension instruction than we currently have to get farmers to apply. Although there is some optimism for smallholder Asian farmers, it is thought that long-term effective and profitable management of acid highland soils will require a great deal of knowledge and money that they do not now have. There are no quick, simple, or inexpensive ways to preserve the current soils while continuing to benefit from them and make a good life now and in the future. However, acid infertile soils, many of which have subpar physical and chemical qualities, are the main cause for concern. In Malaysia, acidic highland soils make up 72% of the land (IBSRAM 1985).

Numerous acidic soils were thought to be unsuitable for growing crops. However, since trees are a fairly natural successor to the rain forest, those soils' original cover, tree crops have been successfully produced on them for many years. In general, the acid highland soils (Ultisols and Oxisols) of tropical Asia are AI or Mn poisonous and essentially infertile. Particularly, the Ultisols' thin topsoils tend to erode easily (Craswell and Pushparajah 1989). Conventional land-clearing techniques can readily deteriorate and destroy the organic materials including sunburn, rain, and direct sun exposure if after cleansing, appropriately managed. These soils are prone to losing a lot of their initial fertility and advantageous physical characteristics quickly. These soils demand inputs that are far above the means of subsistence farmers in order to sustain output. Sadly, there is no "low-input" technology for the poor, acidic soils of the tropics. If these soils are poorly managed, with no or little input to replenish nutrients lost through cropping or erosion, they will quickly lose their ability to produce a good crop and become infested with Lmperata cylindrica. Eventually, if the degradation is not stopped, they can become barren. Right now, primary or secondary forests still cover the majority of acid soils. Before any forested area is made available for agricultural use, both short- and long-term costs and benefits, as well as potential environmental harm, should be taken into account. In long term, humans may require acid soils less for food production and more as a supply of lumber and a carbon dioxide sink (Craswell and Pushparajah 1989).

5.1.1 Diagnosis of the Problem

Because it is difficult to determine soil acidity from symptoms in an above-ground crop, soil analysis is still the sole reliable method for doing so and for figuring out how much lime and gypsum is needed. Soil samples should be taken well before planting because lime and gypsum inclusion is only possible before planting. This will give enough time for the laboratory processing of the samples and for ordering and incorporating the items into the soil. In terms of the actual soil sampling techniques, topsoil samples ought to be collected using a Beater auger, while subsoil samples ought to be taken from the faces of soil pits or using a screw-in type auger. It is important to note that the first signs of excessive soil acidity frequently appear as spots of poor crop development in a stand that is otherwise healthy (excellent). Naturally, it is crucial to address the acidity issues in the weak areas if production is to be optimised. In the light of this, soil samples from these weak areas ought to be collected while the crop is still in its early phases of growth and the poor growth is still readily apparent. After the crop has been removed, samples obtained from the entire field may be biassed due to the better soil conditions in the "good" regions. The "poor" parts can be handled when the issue has been accurately identified, either by liming them separately or liming the field as a whole at the rate advised for the "poor" sections.

Problem soils are those that, due to the features of the land or the soil, cannot be economically utilised for the growth of crops without the adoption of appropriate management techniques. Moreover 50% of the 143 million hectares (ha^{-1}) of net farmed land fall under the category of problematic soils. More than 90% of the land in the northeastern states alone is impacted to varied degrees by soil acidity, which has limited the choice of crops in this area to cereals, especially rice, accounting for more than 70% of the total cropped area. According to Jehangir et al. (2013), the improvement of 25 million hectares will result in an additional 25 million tonnes of food grains. Additionally, the application of fertiliser and its effectiveness are both ineffective because of the unfavourable pH circumstances. The priority should be focused on resolving these soil issues by implementing a variety of management strategies, such as the addition of various nutrients, the selection of crop, irrigation and drainage management, and so on. This will allow for an increase in production as well as productivity on a base that is ecologically sound. The degradation of the soil that occurred as a result of issues with salinity and alkalinity had an impact on a sizeable portion of fertile tracts, encompassing an area of around 67.27 lac hectares.

5.1.2 Sources of Acidity in Soil

- Leaching brought on by a lot of rain.
- Acid parent substance.
- Soluble salts and fertilisers that produce acid.

- Organic acids and humus.
- Minerals made of aluminosilicate.
- Carbon monoxide.
- Hydrogen oxides.
- Polymers made of iron and aluminium.

5.1.3 Acidification and Agriculture

Some agricultural practices hasten the acidity of the soil. In the Burnett district, studies were carried out on a soil that was vulnerable to soil acidification. The following are some factors that hasten acidification:

- Using ammonium-based nitrogen fertilisers at rates higher than what plants need on naturally acidic soils.
- Leaching of nitrate nitrogen from the root zone, which was initially applied as fertilisers based on ammonium.
- Routine removal from the paddock of waste materials and products produced by plants and animals. Alkalinity is removed from the soil by growing plants. The soil turns more acidic as harvested goods are removed. Where a significant amount of material is removed, as in the production of silage, hay, and sugarcane, the impact is highest.

5.1.4 Impacts

Other than decreases in crop and pasture output, which may be dramatic in acute situations, there are no outward signs of soil acidity. Some nutrients may become less accessible as soils become more acidic, while other components may accumulate to dangerous levels. The following issues may arise in acidic soils, some or all of them:

- A decrease in the amount of nutrients recycled by soil microorganisms (perhaps a decrease in the availability of nitrogen).
- The amount of phosphorus accessible to plants in the soil could decrease.
- Brought on calcium, magnesium, and molybdenum deficiency.
- Plants' capacity to utilise subsurface moisture can be constrained.
- The soil may emit aluminium, which is poisonous to plants and microbes.
- Manganese levels could become poisonous.
- Cadmium (a heavy metal pollutant) uptake by pastures and crops may rise.

High soil pH inhibits plant development and nutrient availability. Assessing the long- and short-term economics of soil acidification can be challenging; thus, each site should be taken into account individually. Unfortunately, no one soil test is sufficient to indicate what kind of soil amendment is required or how much to apply when soil acidification is essential. This document serves as a scientific and useful manual for commercial fields' soil acidification. It describes the chemistry of the soil involved, how to decide whether pH adjustment is feasible, and how to acidify the soil. The following five sections make up this manual:

- Knowledge of soil pH.
- The issue is iron chlorosis.
- Iron chlorosis causes.
- Remedies for iron chlorosis.
- Inland Pacific Northwest soil acidification techniques.

5.1.5 Recognising Soil pH

Soil pH is used to measure and express the acidity of soil. This formula generates a pH scale from 0 to 14, with 7 representing neutrality (not being either acidic nor alkaline). When the amounts of hydrogen (H^+) and hydroxide (OH^-) are equal, the pH is neutral ($H^+ = OH^-$). An alkaline soil with hydroxide activity more than hydrogen activity ($OH^- > H^+$) has a pH above 7. The pH of the soil rises as hydroxide levels rise and hydrogen ion activity falls. An acidic soil with hydrogen activity larger than hydroxide ($H^+ > OH^-$) is one with a pH below 7. The pH of the soil lowers as hydroxide levels drop and hydrogen ion activity rises. Parent material and the environment both influence the pH of the soil. Rainwater gently liquefies soil minerals with exchangeable bases and removes them by leaching. Acidification of the soil results from rainwater replacing basic cations like calcium and magnesium with hydrogen. Calcium is the main base that leaches and has the biggest impact on soil pH. The calcium either moves into surface streams and eventually makes its way to the ocean, generating "hard" well water, or it is leached into groundwater. These processes lead to the acidic nature of the soils in high-rainfall regions. Caliche is formed when carbonates assemble in low-rainfall environments. Leaching depth, which is influenced by soil texture, rainfall season, and yearly rainfall, determines the depth to caliche. The depth to caliche increases with rainfall and sand content. Caliche is frequently exposed by land levelling or erosion, and these carbonate-rich layers can lead to pH-related issues. These soils are lighter in colour, have less soil organic matter, have an alkaline pH, and are more challenging to manage when compared to nearby soils. Table 5.1 provides the amount of lime needed to balance the acidity of some nitrogen fertilisers.

5.1.6 Seasonal Changes in Soil pH

Seasonal pH variations in the soil. Before fertilisation, in late winter or early spring, higher soil pH or less acidity is seen. In sandy, low-organic matter soils, the pH of the ground can fall by more than one unit from spring to fall. Soil pH is lowest or most acidic at the start of the growing season or shortly thereafter in low-rainfall

Table 5.1 Nitrogen fertilisers' lime requirements

N sourcs	Lime requirement (kg lime kg^{-1} N)	N source	Lime requirement (kg lime kg^{-1} N)
$(NH^4)^2SO^4$	5.4	Urea	1.8
Ammonium nitrate	1.8	$Ca(NO_3)_2$	1.4
Anhydrous ammonia	1.8	Mono-ammonium phosphate	0.0

Manure, compost, and other organic sources varies. Adapted from Tisdale et al. (1985)

locations. Application of fertiliser and an increase in soil biological activity as the soil warms both reduce the pH of the soil. The magnitude of seasonal pH fluctuations is determined by the soil buffering capacity. Clay and/or organic matter-rich soils are buffered against pH shift and are more likely to experience a small seasonal pH variation of about 0.3 units. Soils with a pH below 7.5 and a sizable amount of silt, clay, or organic matter can shift 1 to 2 units. When soil pH is below 7.0 and there are no carbonates in the soil, large seasonal pH changes are most likely to occur. Large seasonal pH variations are considered unlikely in soils with a carbonate buffer. When evaluating the effectiveness of a soil acidification programme, seasonal pH variations are a crucial factor to take into account.

5.1.7 Conversion of Urea to Bicarbonate

$$(NH_2)_2CO + H_2O \rightarrow 2NH_3 + CO_2$$

(urea + water → ammonia + carbon dioxide)

$$CO_2 + H_2O \rightarrow H_2CO_3 \rightarrow H^+ + HCO_3^-$$

(carbon dioxide + water → carbonic acid → hydrogen + bicarbonate)

The release of carbon dioxide (CO_2) into the soil solution during the conversion of urea [$(NH_2)_2CO$) to ammonia raises the concentrations of carbonates. Rhododendrons, azaleas, and other Vaccinium species, including blueberries, are the plants where this effect is most frequently seen. Apply nitrogen to these plants as ammonium phosphate, ammonium sulphate, or another form that is comparable; try to stay away from urea. When using urea, it is preferable to apply tiny amounts of the substance frequently rather than all at once in big quantities.

5.1.8 Chlorosis Brought on by Lime in Irrigation Water

Bicarbonate overproduction in the soil solution and high pH might result from irrigation with water that is high in carbonate (uncommon) or/and bicarbonate (common).

Some crops that receive irrigation from high-bicarbonate water show chlorosis caused by lime. Lime-induced chlorosis manifests differently depending on drainage, climate, soil pH, and plant type. Monitoring the chemistry of the water, the pH of the soil, and crop performance are the best ways to anticipate future iron chlorosis issues caused by irrigation water. (See below, "Calculating Calcium Carbonate in Irrigation Water," for more information.) When soils with a pH higher than 8.0 are irrigated with high-pH, high-bicarbonate water, chlorosis is most likely to happen. Injecting an acid is a typical method for reducing carbonates in irrigation water. (See below, "Acidifying Irrigation Water with Sulphuric Acid.") Typically, irrigation water is acidified to a pH of 6.5 or below. The most cost-effective way to employ acidifi cation funds is by far to acidify high-pH irrigation water.

5.1.9 Sulphuric Acid for Acidifying Irrigation Water

By changing the pH of the water, acidify irrigation water. When the pH of the water is between 6.0 and 7.0, add acid. In irrigation systems, water with a pH below 6.0 will dissolve metal, whereas water with a pH above 7.0 will leave too much bicarbonate in the water. Using paper pH sticks, it is simple to check the pH of water. The precision is sufficient. By changing the pH of the water, acidify irrigation water. When the pH of the water is between 6.0 and 7.0, add acid. In irrigation systems, water with a pH below 6.0 will dissolve metal, whereas water with a pH above 7.0 will leave too much bicarbonate in the water. Using paper pH sticks, it is simple to check the pH of water. The precision is sufficient.

5.1.10 Elemental S Reactions in Acidification of Soil

Bacteria for the produciton of sulphuric acid from the oxidation of sulphur

$$SO + O_2 + H_2O \rightarrow H_2SO_4$$

(elemental S + oxygen + water + thiobacillus + time → sulfuric acid)

5.1.11 Gypsum Production and Sodium Leaching

During the production of gypsum, elemental sulphur is first converted to sulphuric acid and then reacts with lime found in the soil. This reaction happens rather quickly. There are five tonnes of gypsum that are equivalent to one tonne of elemental sulphur.

$$H_2SO_4 + CaCO_3 \rightarrow CaSO_4 + CO_2 + H_2O$$

Gypsum reacts with sodic soil to produce sodium sulphate, which is removed by leaching.

$$CaSO_4 + NaHCO_3 \rightarrow CaCO_3 + Na_2SO_4 \rightarrow 2Na^+ + SO_4^{-2}$$

Ammonium sulphate [$(NH_4)_2SO_4$], a typical nitrogen fertiliser linked to soil acidification, also contains sulphate. Acidification in fertilisers is caused by ammonium, not sulphate. Acidity is only produced by reduced sulphur substances, as ammonium thiosulphate, or by elements of sulphur. When bacteria oxidise elemental S to produce sulphuric acid, acidification utilising elemental S happens. The duration of this reaction, which is temperature-dependent, is several years. The rate of conversion to sulphuric acid will increase with the addition of elemental S. It can take many years of adding elemental S to reduce soil pH. Periodically check the pH of the soil to confirm the necessity for further acidification. For the majority of agronomic crops, continuous elemental S inputs are typically only cost-effective as sulphur fertiliser. Some high-value horticultural crops may require significant inputs of elemental S to acidify the soil. Keep in mind that the OSU fertiliser guidelines suggest using little more than 100 lb of elemental S for S feeding. Even though the plant will receive S at this rate, your soil won't become acidic. But you won't need any additional S for plant nourishment if you apply enough elemental S or gypsum to acidify the soil. Soil is supplemented with soluble salt (sulphate) by elemental S reactions. If salts are an issue in your soil, then this salt will need to be leached. Salt issues can arise from high elemental S rates over a short time, so it is important to proceed with caution. Economic considerations over the quantity required for complete pH correction typically dictate the rates at which elemental S and gypsum are used.

5.1.12 Fertilisers as Compounds that Cause Acidity

Fertilisers containing ammonium can make soil more acidic. When contrast to acidification by elemental S, this process is sluggish. Ammonium sulphate was used at a rate of 135 lb N/acre in a long-term research on dryland wheat in Pendleton, Oregon. According to Rasmussen and Dick (1995) and Rasmussen and Rohde (1989), the pH of soil fell at a rate of 0.03–0.05 unit each year. In this illustration, it took ten years' worth of ammonium-containing fertiliser applications to reach the same level of acidification as elemental S. This study's soil had an acidic pH of 6.3 and very no carbonate in it. Significant acidification with fertiliser containing ammonium is not practical when carbonates are present. Sand-based Columbia Basin soils have little natural carbonate. Over the past 50 years, farming and nitrogen treatments have caused many irrigated fields to become acidic.

5.1.13 *Amendments to the Soil and Fertilisers*

Compost and manures are two soil additives that may be harmful if their pH, salt, or sodium levels are too high. Before using soil amendments, always do a pH, EC, and sodium analysis. Reclaiming sodic soils can be aided by soil additives with a pH balance of seven and little sodium content. They can aid in aeration and water filtration, which makes it easier for extra salt to be leached. Both crop performance and acidification rate can be enhanced by amendments. Remember that fertilisers containing urea and lime might contribute to lime-induced chlorosis. For further information and solutions, see page 6.

5.2 Crop Preference for Soil pH

Crop preference for soil pH range has been shown in Table 5.2. Acidic soils produce production issues by reducing the availability of some crucial plant nutrients and increasing the availability of the poisonous elements in the soil solution, such as manganese, iron, and aluminium. These toxic elements are a major contributor to the poor crop performance and failure that are characteristic of acidic soils. Strongly to severely acidic soils (pH 5.5) are found in 73% of the Sikkim state of India's farmed land in the East district, 62% in the North district, 44% in the West district, and 37% in the South district. There is a tonne of technology that can make acid infertile soils viable and lucrative over the long run, but it needs to be used locally. Agricultural limestone is the most affordable and comparatively simple source, and it is also the most often utilised liming material. Landowners engaged in organic agriculture have a unique chance to control soil acidity by using locally accessible weed biomass to make biochar. Due to its economic viability and environmental friendliness, agronomical approaches are now receiving more attention for managing soil acidity (Das and Avasthe 2018).

5.2.1 *Crop Selection and Diversification*

The best way to deal with this soil issue is to choose crops that are tolerant of acidity, and breeding such varieties is particularly important for achieving higher production, especially in places where liming is not cost-effective. Crop rotations can assist maximise the nutrient and water usage by crops, boost soil fertility in general and keep it close to neutrality in particular, and improve our soil resources. Due to the production of certain acids that aid in the solubilisation of various kinds of phosphorus, the inclusion of legumes in cropping systems not only reduces the cropping system's need for nitrogen but also aids in the efficient utilisation of native phosphorus (Fenton and Helyar 2002). Because of this ability, legumes are effective

Table 5.2 Preference of crops to different soil reaction

Crop	pH range	Crop	pH range
Maize (*Zea mays*)	5.0–6.5	Sugarcane (*Saccharum officinarum* L.)	6.0–7.5
Barley (*Hordeum vulgare* L.)	6.0–7.5	Sorghum (*Sorghum vulgare* L.)	6.0–7.5
Sugar beet (*Beta vulgaris*)	6.5–8.0	Wheat (*Triticum aestivum*)	6.0–7.5
Lentil, gram etc.	5.5–7.0	Oats (*Avena sativa* L.)	5.0–7.5
Rice (*Oryza sativa* L.)	4.0–6.5	Tobacco (*Nicotiana tabacum*)	5.5–7.5
Cotton (*Gossypium hirsutum*)	5.0–5.5	Rye (*Secale cereale*)	5.0–7.0
Potato (*Solanum tuberosum* L.)	5.0–5.5	Sunflower (*Helianthus annus*)	6.0–7.5
Pea (*Pisum sativum* L.)	5.5–7.0	Barseem (*Trifolium alexandrinum*)	6.0–7.5
Millets	4.0–6.0	Tea (*Camellia sinensis* L.)	4.0–6.0
Cowpea (*Vigna sinensis*)	5.0–6.5	Groundnut (*Arachis hypogaea* L.)	5.0–6.5
French bean (*Phaseolus vulgaris* L.)	5.5–7.0	Soybean (*Glycine max* Merr.)	5.5–7.0

at utilising the phosphorus that is naturally available in various forms. P is obtained from insoluble phosphates through root exudates, increasing the amount of P that is readily available.

The soil in that area becomes more acidic as NO_3–N is leached away. There may be an increase in pHCa at the point of uptake if the NO_3–N is taken up farther down the profile. However, the soil profile becomes more acidic when the NO_3–N is leached below the root zone. It describes each factor's impact and explains how it might be affected.

5.2.2 Needs for Policy

The advantages of managing soil acidity in agriculture include bettering soil health, increasing crop yield and productivity, and increasing input use effectiveness. To help with making decisions at the policy level, it is necessary to conduct policy-related research to quantify the advantages under various circumstances. In order to manage soil acidity in agriculture, some policies are necessary.

- Creating a state-wide policy for managing soil acidity that clearly identifies the numerous competing uses.
- Managing soil health by utilising all available resources.
- Appropriate orientation and training sessions should be held to educate the farming community on the advantages of managing soil acidity.
- Encourage the production of soil acidity management inputs on farms. Routine soil testing.
- To determine the pH level.

Long-term intense crop production leads to the progressive process of soil acidity. Through control, the pace of soil acidity may be slowed. Input costs may be decreased while acidity is decreased by matching applied sulphur and nitrogen with crop demands (Wortmann 2015). Soil acidity has an impact on plant growth and soil microbial activity, which lowers crop yields. The direct impact of acidity on the plants is typically less significant than indirect impacts on nutrient availability and the presence of harmful ions, such as exchangeable aluminium and manganese. As soils get more acidic, some nutrients become less accessible while others become more so. The eastern parts of South Africa frequently experience acid soil conditions that limit crop development. Soils in areas with more rainfall are frequently acidic naturally; however, human activity may hasten acidification. It is important to note that issues with acidic soil are not exclusive to this nation: globally, about 30% of the land that is suitable for cultivation is acidic. Producers typically struggle to understand the many soil acidity characteristics mentioned in soil test reports, and they could also receive inconsistent recommendations for using items like lime and gypsum.

5.3 Soil Acidity Causes

In general, soil acidity rises in the United States from northwest to southeast. Calcium (Ca^{2+}), magnesium (Mg^{2+}), potassium (K^+), and sodium (Na^+), along with other bases (ions with a positive charge), are washed out of the soil and replaced by hydrogen (H^+) as the amount of precipitation that falls on the land increases, resulting in low pH in the high-rainfall regions of India, such as the northeastern states. The following natural processes and management practices can cause short-term pH changes:

i. Rainfall.
ii. Plants eliminating bases (such as Ca^{2+}).
iii. Fertilisers that produce acids, like as ammonia nitrate (NH_4NO_3).
iv. Organic acids produced by decaying plants.
v. CO_2 produced by microbial and root respiration.

It is necessary to discuss some fundamental soil chemistry in order to have a practical grasp of soil acidity. The negative charge in the soil is carried by the organic materials and clays. This negative charge is counterbalanced in non-acidic soils by the positive charges on several plant nutrients, particularly calcium (Ca^{++}), magnesium (Mg^{++}), and potassium (K^+). Other non-nutrient elements, in particular hydrogen (H^+) and aluminium (Al^{+++}), are more abundant in acidifying soils and replace nutrients like calcium and magnesium on the clays and organic matter. The clay particles hold the aluminium under neutral conditions, but when the acidity rises, the clay edges start to dissolve and release soluble aluminium into the soil. Importantly, the most significant growth-limiting component in acid soils—from the perspective of controlling soil acidity—is soluble aluminium. Additionally, it should be remembered that pH only assesses the hydrogen content in the soil and not the aluminium concentration. In order to create recommendations for the treatment of acidity issues that are economically viable, these factors are of utmost significance.

Why do soils get more acidic? Although acidic soils are common in nature, as was previously mentioned, the following human actions may significantly hasten acidification.

Acid rain is a product of industrial air pollution. It has been demonstrated that in some Highveld locations, this is a significant contributing element. The application of nitrogenous fertilisers, especially when done in excess of what the crop needs, is harmful rather than beneficial and may cause decline in productivity. Removing calcium, magnesium, and potassium, which are essential elements, from harvested crops and animal products, is responsible for low productivity and production.

Toxicities of aluminium and manganese, low pH in general, and shortages or low availability of several critical minerals, such as calcium, magnesium, phosphorus, and molybdenum, have all been linked to acid soil infertility as the main growth-limiting factors (Foy 1984). These elements may inhibit the growth of plants directly or indirectly by interfering with the formation and operation of symbiotic relationships with rhizobia, mycorrhizas, and actinomycetes. Water stress that occurs from the limiting of root expansion into acid subsurface layers may also be a barrier to plant growth in extremely acidic soils (Adams 1984).

5.3.1 Lack of Calcium

In acid soils in southeast Queensland (Lee and Wilson 1972; Jones 1982); on solodic soils in southern Queensland (Russell 1966); and on a yellow earth in north Queensland (Probert 1980), pasture legumes have been found to be deficient in calcium. When exchangeable calcium is: 0.65 cmol(+)lkg, sugarcane yield responds to lime (Ridge 1984). Although some surface soils and subsoils in Queensland have been shown to have exchangeable calcium values significantly lower than those mentioned above, there aren't enough local correlations with plant response to allow for reliable conclusions (Bruce et al. 1989).

In agricultural acid soils, aluminium phytotoxicity is thought to be the primary limiting factor for crop productivity. Liming is a widespread practice used to improve the qualities of acidic soils, but choosing the right liming material is crucial for environmental sustainability and agricultural output (Oliver et al. 2021). Parent materials deficient in bases and temperatures that support significant leaching are two soil formation elements that naturally lead to the development of soil acidity (Olego et al. 2016). High-rainfall regions are more likely to experience natural soil acidification, which promotes the leaching of exchangeable calcium (Ca) and magnesium (Mg) and displays high aluminium (Al) saturation because it becomes the major cation in the exchange complex (Olego et al. 2021). One of the nine essential macronutrients that plants use in large amounts is magnesium. It is the key atom in chlorophyll and functions in pigment-protein complexes to gather photons in both photosystem I and II. Magnesium is also involved in CO_2 assimilation reactions in the chloroplast, as well as photophosphorylation and the partitioning of photoassimilates. Its general physiological advantages for active development frequently conceal the precise mechanisms involved in disease resistance, which are also a significant factor in plant health overall (Huber and Jones 2013). Mg is abundant in the environment, but because of the acidity of the soil, it is frequently deficient in the field (Tanoi and Kobayashi 2015).

5.3.2 Administration of Acidic Soils

Liming is the most frequently advised management practice for restoring low soil pH; however, there are other methods as well (Goulding 2016). The application of both lime and P fertiliser is critical for the P and N fertiliser recovery efficiencies required for good maize development on acidic soils, according to Kisinyo (2016). Lime and/or fertiliser application, however, is not always economical for small-scale farmers and is not eco-friendly (Tandzi et al. 2015). Liming also changes the topsoil but does not eliminate acidity in the subsoil, which causes serious issues for emerging roots (Sierra et al. 2006). According to Mwangi et al. (Mwangi et al. 2000), farm yard manure has a stronger buffering capacity and contains both soil acidity and alkalinity, making it a better amendment for adjusting soil pH. However, according to van Averbeke and Yoganathan (2003), the standard guidelines are extremely high (10 tonnes per hectare), and manure is not always readily available. Crop wastes can raise soil pH when added to the environment (Noble et al. 1996). When lucerne meal was fed to acidic soil, Hoyt and Turner (1975) discovered a rise in pH of roughly 0.5 pH units, but they also noticed a fall in pH 20 days later. The addition of residues often results in an initial elevation in soil pH, which is then followed by a pH drop (Wong et al. 1998). An effective and long-lasting alternative to boost yields in acidic soils is the introduction of acid-tolerant maize cultivars (Horst et al. 2017).

A challenge for agriculturalists has been managing acidic soils. Slash-and-burn methods have been used to control acid soils for the majority of human history. These methods make sure that plant biomass is converted into nutrients, some of which lessen the acidity for at least one cropping season. This method is still in use throughout the tropical world since it is both effective and more cost-effective than using correctives like lime, which are frequently out of reach for the typical farmer (Moran 1995). The most popular economic tactic for dealing with acid soils is to burn biomass, while alternative methods have gotten much less attention. A big component of the approach that farmers have taken has been to select, if it has been possible to do so, soils that are less constrained by acidity and to locate plant types that are capable of surviving with low pH on soils that are more acidic. Acid soils are abundant; however, they tend to be found in isolated areas that are adjacent to other soils with higher pH.

The early treatment of soil acidity is crucial. In the event that acidity penetrates the subsoil, yields may be seriously impacted. Controlling subsoil acidity is challenging and expensive. Diagnosing any increase in acidity is the first step in regulating soil acidity. For the plough layer (zero to 10 cm) and the subsurface to 50 or 60 cm, this calls for accurate soil tests for pH, aluminium, and manganese levels.

The following farming methods are suggested for reducing acidification:

- Coordinating crop demand with nitrogen fertiliser additions.
- Using nitrogen fertilisers that don't contribute to acidity.
- Effectively watering the region to reduce leaching
- Early sowing following fallow to facilitate faster use of available nitrogen.
- Cultivating perennial plants with deep roots to better absorb nitrogen.
- Using lime regularly to combat the acidification that the agricultural system is prone to.
- Raising crops that can tolerate acidic soils, such as acid-tolerant crops.

5.3.3 Application of Lime or Dolomite

Lime or dolomite can be used to raise the pH of soils when they are too acidic for a particular crop. The crop and the soil will determine how much lime or dolomite is needed to balance an acidic pH. High organic matter and clay content soils will require higher application rates since they are more resistant to pH fluctuations. A soil test termed "lime requirement" or "buffer pH" should be performed to estimate the amount of lime needed to address an existing soil acidity issue. This soil test is provided by the majority of commercial soil testing labs. The test is used to prescribe lime to increase the soil pH on a hectare of soil's top 10 cm to a target pH that won't reduce crop output. A goal pH (in water) of 5.5 is generally advised. Depending on the crop, more lime or dolomite may be needed once the target soil pH is obtained. The amount of material removed at harvest, the quantity and type of fertilisers typically employed, and the amount of leaching that takes place all affect how acidifying a cropping system is. Table 5.3 compares the pH of a few common things.

Table 5.3 pH of some common items

Item	pH	Item	pH	Item	pH
Most acid soils	4.0–5.0	Lemon juice	2.2–2.4	Clean rain water	5.5–5.7
Vinegar	4.0–4.5	Fresh milk	6.3–6.6	Mild soap solution	8.5–10.0
Orange juice	3.4–4.0	Acid rain	3.0–5.0	Blood plasma	7.0–7.2

5.3.4 *Measurement of Soil Acidity*

When determining if a soil requires an addition of limestone, soil pH, a measurement of hydrogen ion (H^+) activity in the soil solution, is used. The amount of lime a soil needs can be calculated using the soil texture and pH. To more precisely estimate the quantity of limestone required to elevate the soil pH in the plough layer to a desirable level, a lime requirement test (buffer-pH test) is advised. Because the pH scale ranges from 0 to 14, soil with a pH between 6.6 and 7.3 (around 7.0) is classified as neutral. However, unless the soil has been chemically altered or contaminated, the pH of soil only typically ranges from about 3.5–8.5 and is classified as slightly acidic between pH 6.5 and 6.1, moderately acidic between 6.0 and 5.5, strongly acidic between 5.5 and 5.1, very strongly acidic between 5.0 and 4.5, and extremely acidic below pH 4.4. Alkaline soils are those that have pH values greater than 7.4.

5.3.5 *Increasing Crop Yields*

By bringing the pH level close to neutrality, acidic soils can produce crops with significantly greater efficiency. Animal manure supplements have been shown to raise the pH of acidic soils, despite the fact that liming is frequently used to reduce soil acidity. In manure-amended and unamended soils, available S concentrations did not differ noticeably (Whalen et al. 2000). After manure application, extractable Al and Fe decreased marginally, but after incubation, there was no difference between manure-amended and unamended soils. This study found no difference in the cation exchange capacity (CEC) of manure-amended soils compared to unamended soils, and it seems that manure application does not cause significant changes in Al, Fe, and CEC during the short time (weeks). Some findings suggest that adding bovine dung to acid soils can temporarily raise pH and the amount of P and K that is available to plants (Whalen et al. 2000).

5.4 Global Acidity Issues

Australia and New Zealand

Modelling of nutrient cycles, especially the nitrogen cycle, in pasture soils reveals that soil acidification in New Zealand pastures based on legumes is modest but ongoing (Bolan et al. 1991). The natural high buffering capacity of New Zealand soils, however, means that changes in soil pH in the majority of soils over a decade are minimal. The short-term effects of increased soil acidity are seen in Australian soils, nevertheless, because they have poorer buffering capacities (Williams 1980). The situation has gotten so bad that in many places, pasture productivity has been significantly hindered or failures in cereal growth in rotations of cereal and pasture have occurred (Lee 1980). Applying lime to the soil can quickly reduce soil acidity in pastures on flat terrain. Lime use is frequently unprofitable due to the lower productivity and profit margins of hill country soils, and phosphatic and sulphur fertilisers may be the only profitable soil additions. To assist regulate soil acidity on these hill country soils, it would be viable to use phosphatic fertiliser in a less acidulated form than completely acidulated superphosphate. Due to the carbonate found in apatite (francolite) and accessory minerals (calcite, dolomite), the majority of reactive phosphate rocks (RPR) have strong liming action. This is because RPR dissolves using 2 mol of H^+ for every mole of dissolved P.

$$Ca_{10}(PO_4)_5CO_3F_2 + 12H^+ \leftrightarrow 10Ca^{2+} + 5H_2PO_4^- + 2F^- + H_2O + CO_2$$

The application of very high rates of RPR to cultivated soils with inadequate buffering capacity has dramatically reduced soil acidity, according to studies by Easterwood et al. (1989) and Wright et al. (1991).

According to a study of the literature on plant development restrictions in acid soils of tropical Australia, calcium and molybdenum deficiencies are more frequent than aluminium or manganese toxicity. However, adding soluble fertilisers has the potential to make many acidic soils hazardous for aluminium. There are thorough techniques for conducting soil tests that were used to evaluate the nutritional status of acidic soils in Queensland; however, there is no reliable soil test for molybdenum. The use of lime to reduce acid soil's infertile restrictions on the growth of agriculturally significant plant species is the cornerstone of managing soil acidity in tropical Australia. Thus, compared to tropical America, tropical Africa, or tropical Asia, a far smaller percentage of the Australian tropics are dominated by Oxisols and Ultisols. In Australia, Ultisols have a much larger surface area than Oxisols; a similar situation can be seen in tropical Asia, but not in tropical America or Africa (Sanchez 1987). The principal crop grown in the Australian wet tropics is sugarcane, which is primarily grown in coastal lowlands and takes up roughly 126,000 acres. According to Hetherton et al. (1988), sugarcane has a long history of receiving large fertiliser inputs and is a crop that is very tolerant of acidic soil. The Australian Tropics stand out from many other tropical regions due to its reliance on low labour and high horsepower inputs.

There were little soil pH tests taken, and the majority of the early reactions were noted visually without any supporting information on plant output. Accordingly, depending on the species and soil pH, the reaction to liming typically ranged from none to substantial, with established pastures of underground clover frequently producing no worthwhile outcome (Leeper 1964). A body of opinion emerged opposing the application of lime on Australian soils, possibly as a result of these factors and the finding that modest amounts of banded lime were particularly successful in producing underground clover on newly cleared acid soils (Ellington 1984). Since accelerating soil acidification and legume-based agriculture have been linked to issues with plant productivity (Coventry 1985; Porter and Wilson 1984), the use of lime on crops and pastures is once again being thoroughly examined.

Indonesia

Large tracts of arable land outside of Java, particularly in Kalimantan, Sumatra, and Sulawesi, have highly leached soils, are very acidic in response, are very Al-saturated, have poor macro- and micronutrient status naturally, have very low levels of wetland minerals, etc. Low activity clays, which compact particularly in the subsurface and are prone to erosion, make up the majority of the soil. When no lime or fertiliser is used, the production per hectare is relatively low.

Native and immigrant farmers

Both types of farmers are indigent and use various farming methods, which are typically the result of variations in the available land resources and other variables. The farmers who migrate from richer soil regions have some familiarity with more effective farming techniques on permanent fields. To deal with soil issues, indigenous farmers use shifting cropping, or ladang, acid-tolerant crops, and little tillage in traditional farming practices. They just rely on the soil's inherent fertility. Slash-and-bum is the traditional method of clearing land in Indonesia, as it is in other wet tropical areas. After drying for a while, trees are cut down, their branches are cut, and they are burned. Slash-and-bum has a favourable impact on pH, exchangeable acidity, and exchangeable bases, which enhances plant development. Additionally, they never dig the ground deeply to avoid the top 10–15 cm of soil from mingling with the subsoil, which is more barren and has a higher Al saturation. Little physical harm is done by these methods, and the rooting environment is slightly improved. They only grow upland rice (local kinds), which can grow in acidic environments, as a food crop.

Malaysia

According to Tessens and Shamsuddin (1983), the highland soils of Malaysia are often quite worn and are members of the orders Ultisol and Oxisol, which are typified by high acidity, low CEC, poor base saturation, and high AI concentration. Due to their acid tolerance, most of these soils are being used for plantation crops, particularly rubber and oil palm. Due to severe restrictions brought on by soil acidity and limited P availability, which limit development and production, these soils offer little possibility of success for the majority of other crops, most of which are not acid tolerant. According to earlier research by MARDI (1979), the main soil requirements for maize, soybean, peanut, and mungbean produced on mineral soils are P and liming, whereas pasture legumes need P, Mo, K, and liming. When AI saturation exceeded 65, 45, 15, and 25%, respectively, yields of peanuts, soybeans, and maize were negatively impacted (MARDI 1979, 1980). According to Foster et al. (1980), the soil needs to be limed so that AI saturation is below the corresponding levels specified in order for these crops to thrive reasonably. According to Sharifuddin and Dynoodt (1985), in order to get a relative dry matter yield of more than 80% for sorghum grown on four Ultisols and two Oxisols, the Al saturation must be decreased to less than 40%. According to Shamsuddin and Tessens (1983), At predominates in the acidity of Malaysian soils, and the amount of lime these soils need can be calculated by multiplying the amount of base required to elevate the pH to 5.5 by a ratio of 1.3.

Philippines

The Philippines has an acid soil cover that covers up to 58% of its territory, much of which is categorised as hilly country. The majority of the soils are Oxisols and Ultisols, which were previously cultivated by slash and dung but are now abandoned in many places (Afienza 1989). Despite extensive study on liming, few farmers actually use it since it is too expensive. There are now known crop varieties that can withstand acidity, and use of phosphate rock has proven to be a successful method of reversing the widespread phosphorus deficit. The Philippines has a hot, humid climate, similar to many other tropical nations in Southeast Asia, with two distinct seasons: wet or rainy from June to October and dry from November to May. 29 degrees is the usual temperature. During the warm months of March to May, it is 3 °C, whereas from December to February, it is 25 °C. It rains more than 2000 mm annually. There are many different types of soils in the nation as a result of the interaction of these components with the diverse parent materials. With almost 60% of all exports coming from agriculture, the Philippines' economy is dominated by this sector. According to Fernando and de Jesus (1980), soils with a pH of 5.5 or lower are widespread and present serious issues not only in the Philippines but also in other Southeast Asian nations. A significant barrier to increasing and sustained agricultural production is these soils. An estimated 17 million hectares of acid soils are revealed by survey data comparable to about 58% of the nation's total land area. These soils are found widely across the nation, mostly in hilly areas (National Land Use Committee 1985).

Acid soils are defined as typically deep, well-drained, and low in fertility in the Philippines. They are widely distributed as Ultisols and Oxisols (Soil Taxonomy Classification), which are classed as Haplustults and Haplustox at the great group level respectively preliminary findings of a study conducted by the University of the Philippines at Los Banos in Palawan and the Bureau of Soils and Water Management in 1988. The Philippines' fourth-largest province revealed that over 26 320 ha of its entire area—almost 1.5 million ha—is acidic. The thin bed lenses of sandstone and siltstone that make up the sedimentary rocks, which are transgressive and poorly consolidated conglomerate and interconnected, are what gave rise to the soils. Additionally, they are composed of poorly sorted ultramafic rock pebbles, cobbles, and boulders that are found on highly fractured hills and knobs with nearly uniformly high summits. Low levels of almost all macro- and micronutrients lead to low productivity. The local population farms using a slash-and-burn method. When they can no longer receive a significant amount of production, they leave the farms.

South Asia

Agriculture in Sri Lanka makes considerable use of soils that react acidically. Numerous excellent acid soil groups with pH values between 5.5 and 5.6 are maintained for a variety of crops, including annual food crops, roots, and tubers numerous horticultural crops as well as perennial tree crops. The type of soil qualities, how the soils are generally arranged in the landscape, the climate, and their interactions typically influence management options and crop selection. A resource basis for the adaptation of this broad variety of agricultural uses and management techniques is provided by the occurrence and distribution of these soil groups as well as the vast range of climates as identified by the pattern of rainfall distribution and temperature conditions. There are variances in management over short distances due to the location-specificity of the hydrological conditions resulting from the type of the physiography at any site. Serious agricultural research in the field of food crops has been conducted over the past several decades as a result of the government's ongoing policy to achieve self-sufficiency in food crops. Despite significant improvements in our understanding of how to manage acidic soils, it is increasingly clear that further research is needed to maintain the productivity of these soils.

Within the area of acid soils in Sri Lanka, five of the ten soil orders (USDA 1975) have been discovered and mapped. The Ultisols are the most widely utilised soil order in agriculture, however Entisols, Inceptisols, and Histosols are also widely used soil orders for growing food crops. The majority of the Oxisols occur outside the primary acid soil zone, where they are not particularly significant in terms of extent or for the production of food crops.

The management of Sri Lanka's acid soils must take into account the fact that they are quite diverse in terms of their occurrence, characteristics, and potential. Additionally, the climate and hydrological conditions are not constant throughout the region of acid soil. The Department of Agriculture's Land and Water Use Division created a map of agroecological regions in 1977 that took into account the soil characteristics, climate, and land use. This map served as the foundation for regionalising the research effort so that regional research could address the issues and limitations to agricultural development. Local issues that are significant in a regional context can be better identified through regional research. As a result, the following review is primarily centred on biologically uniform natural areas, with the main crop or cropping system receiving the most attention. Rice, root and tuber crops, and horticultural crops are the most significant food crops in the acid soil region. On Sri Lanka's acidic soils, rice will continue to be the most significant food crop. The low country's wet and semi-wet intermediate zones, as well as the mid-country's wet zone, are largely occupied by rice-growing regions. In the upland wet and intermediate zones, potato is the most significant tuber crop. In up-country wet and intermediate zones, vegetables are once more predominate, although in other areas of the acid soil region, vegetables and fruit crops are primarily grown as home garden crops. Fruit farming hasn't become a successful business.

According to meq per unit weight, these soils have a high cation exchange capacity (Dimantha 1977). Phosphorus is scarcely available. Soluble iron and manganese ions are high enough due to severe reduction to result in physiological issues. These soils' potassium levels are normally appropriate. According to investigations by Panabokke and Nagarajah (1964), Ponnamperuma (1960), and Rodrigo (1964), silica content is minimal, and its addition significantly boosts rice productivity. Fertiliser recommendations for low-lying ricelands have been made based on more current research (Nagarajah 1986).

Thailand

45% of Thailand's 51.3 million hectares are taken up by Ultisols, while 0.26% (0.13 million ha) are by Oxisols. Acid soils can also include certain upland Alfisols and Inceptisols (Dystropepts). In a general sense, acid sulphate soils are likewise considered acid soils, albeit management techniques may differ slightly from those used on other well-developed acid soils (Kheoruenromn and Kesawapitak 1989). A wide variety of crops can grow on these acidic soils. Among the important food crops planted were rice, cassava, corn, sorghum, mungbean, peanuts, soybeans, vegetable crops and fruit. As of right now, it appears that the best approach to deal with these acid soils for the production of fruit trees is through high-tech cultural practices. Because most areas of acid soils in Thailand, with the exception of those in Peninsular Thailand and some areas of the southeast coast, are under a tropical savanna climatic regime, field crop practices frequently experience a lack of moisture content due to intraseasonal drought, weed control issues, and proper fertiliser application. Even though farming practices have improved, they are still unable to handle these unfavourable circumstances (Kheoruenromn and Kesawapitak 1989).

Despite having a comparatively small cultivated area in comparison to other food-producing nations, agricultural products make up the majority of Thailand's exports, accounting for around 70% of the overall value of exports (Moncharoen et al. 1987). Despite this, Thailand's agricultural productivity is far from the finest in the world. Crop yield per unit area and farm production rates are still low and not sustainably sufficient. Unexpectedly, agricultural output per unit area in farmers' fields has either decreased or remained constant in recent years (FAO 1984; TDRI 1987). In 1946, the first comprehensive soil map of the nation was created (Changprai 1973; Kheoruenromne 1983). When the Department of Land Development was founded in 1963, the more active phase of the project began. Nine orders of soils are present in Thailand according to taxonomic classification (Soil Survey Staff 1975, 1987); the areas of each are displayed in Table 5.4. Ultisols, Alfisols, and Inceptisols are the three principal orders. Additionally, a research found that Thailand has issue soils covering a total area of about 10.88 million ha (panichapong 1982).

Acid soils are often well developed, strongly leached, with low activity clays and variable-charge properties (Buol et al. 1980; Theng 1980). The low level of accessible nutrients is one of the most fundamental restrictions on crop productivity

Table 5.4 Area under different soil order in Thailand (Million ha)

Soil order	Area	%	Soil order	Area	%
Alfisols	4.70	9.17	Oxisols	0.13	0.26
Entisols	2.28	4.44	Spodosols	0.08	0.15
Histisols	0.08	0.16	Ultisols	22.99	44.81
Inceptisols	4.98	9.70	Vertisols	0.28	0.55
Mollisols	0.69	1.34	Unclassified	15.09	29.42

Source Moncharuen et al. (1987)

in acid soils. These characteristics indicate that farming practices on them require special attention in order to avoid excessive loss of plant nutrients and toxicity of specific elements. Many of Thailand's soils (particularly the somewhat sandy ones) have a low buffering capacity, making fertiliser management challenging. Weeds and moisture deficits exacerbate the situation. The intraseasonal drought period is frequently too extended, and moisture deficiencies become severe. Drought-related crop loss is not rare. Local weeds are more drought tolerant, and soil fertility and weed control issues frequently cause crop loss.

Vietnam

Sloping soils cover three-quarters of Vietnam's total land area. Soils are prone to major erosion and leaching issues, which degrade the soil and reduce crop output. To improve food crop production in Vietnam, there is an urgent need to conserve and increase soil fertility. Recent study on terracing and intercropping has shown increased yields of cassava and legumes in Nghe Tinh's yellow–red soils (Vy and Thi 1989).

Sloping soils in Vietnam can be categorised into three types, accounting for three-quarters of the country's total natural land area:

i. 4.5 million acres of gently sloping soils (3–150).
ii. 3.S million acres of moderately sloping soils (S-25°).
iii. 17.0 million hectares of strong sloping soils (>25°)

Sloping soils in Vietnam are influenced by various unfavourable soil processes, such as erosion and leaching, as a result of natural circumstances and incorrect land use and management by humans. The majority of the soils in the mountainous regions of Vietnam have a yellow–red colour, and they exhibit signs of a dominant ferralitic weathering process, which can be seen in leaching bases and a buildup of iron and aluminium. As a consequence of this, the soils have a low concentration of alkaline and alkali-earth cations, a low base saturation (25–40%), an acidic pH (4.5–5.5 or lower), and a low accessible phosphorus content (due to phosphate fixation produced by a high iron and aluminium content). It has a humid tropical climate with two distinct dry and wet seasons, with heavy and concentrated rainfall (averaging between 1700 and 2000 mm/year in the period from May to October). The climate is characterised by two distinct dry and wet seasons. Strong soil erosion occurred, resulting in plant nutrition loss. The magnitude of soil erosion is enormous. The nutritional content of erosion products is significantly higher than that of surface soil. The characteristics of the yellow–red soils atop granite and gneiss that were once rich in potassium have been completely transformed as a result of erosion. These soils, like the yellow–red soils that are found on basalt and porphyrite, have become potassium-deficient as a result of farming practices that have been carried out on sloped soils where erosion control has not been implemented. Another important type of leaching is vertical leaching, which results in a leachate that contains clay particles, organic matter, nitrogen, and a large amount of calcium, magnesium, and potassium. The results of the soil layer study suggest that the content of alkaline and alkali-earth cations has been decreasing over time. The level of these cations in the soil's bottom layer is three to four times higher than that of the soil's surface layer.

In terms of clay mineral composition, the majority of slope soils in Vietnam include primarily kaolinite. Goethite and gibbsite are two minerals that are widely spread in soils. Iron included in these minerals has several advantages in the soil conditions of Vietnam. It is the interaction of organomineral compounds that results in the production of granule structure, a desirable physical attribute of soils. Because of these minerals, phosphate ions have been chemically absorbed and hence shielded from leaching; also, phosphorus is easily released under reduction conditions. The humus content of soil varies widely and is influenced by latitude and, in particular, the presence of plant cover. The strength of the organic matter mineralisation process is quite high in humid tropical settings. It is greater than the production of humus. Furthermore, the sloping relief and the adoption of improper farming methods, such as excessive cultivation, monoculture, burning and destroying trees, and so on, have accelerated soil erosion and leaching processes, resulting in a significant loss in organic matter content. Fanning techniques have a significant impact on soil organic matter.

Ethopia

Approximately, 43 percent of the arable land in Ethiopia's humid and subhumid highlands is negatively impacted by soil acidity. The health and fertility of the soil, along with the ever-growing need for food and raw materials, have remained the most essential elements in terms of improving and maintaining agricultural production. According to Sivaguru and Horst (1998), excessive amounts of Al in acidic soils mostly cause harm to the root apex and restrict root growth. Crops grown in acidic soils have limited access to both water and nutrients as a result of poor root growth, which in turn reduces the amount of water and nutrients that can be absorbed by the plant. The overall effect is a retardation in the expansion and production of the crop (Marschner 2011; Wang et al. 2006; Sharma and Singh 2002). The application of nitrogen-containing fertilisers or manures, particularly those with high ammonium or urea contents, is the primary factor in the acidification of soil. Nitrification is the process by which hydrogen ions are released. According to Pagani (2011), the pH of soil decreases as acidity increases since pH is a reflection of acidity and is calculated as the negative logarithm of H^+ concentration. The highlands of Ethiopia are one of the most important areas on the African continent in terms of both the fight to preserve the natural resource base and the production of food.

India

In the Northeastern Region of India, 22.26 million ha of land is acidic (Majumdar et al. 2022). The northeastern region of India is distinguished by a variety of agro-climatic and geographical conditions. The limited area under agricultural crops is mostly owing to the region's natural physiographic features, which include more than 15% slope, undulating topography, extensively eroded and degraded soils, and inaccessible terrain. The limited area under agricultural crops is mostly owing to the region's natural physiographic features, which include more than 15% slope, undulating topography, extensively eroded and degraded soils, and inaccessible terrain (Datta and Sharma 2006). The use of local varieties, low agrochemical consumption, low moisture retention capacity of upland soil, undulating topography, no drainage in valley during monsoon, lack of irrigation facilities during winter months, and low sunshine hours, among other factors, all contribute to lower crop productivity in this region. Furthermore, high rainfall in the region causes soil acidity, exacerbating the region's low agricultural output; as a result, NER produces only 1.5% of national food grain production. Soil acidity is a significant abiotic stress that has a negative impact on crop growth and output. Soil acidity affects around 50% of the world's potentially arable land in the humid tropics. Acid soils cover around 90 million hectares (mha) in India. According to estimates, around 25 mha of farmed land in India is strongly to moderately acidic (pH 5.5), whereas 23 mha is somewhat acidic (pH 5.5–6.5). The majority of these soils (54%) are located in NER India, where soil acidity affects more than 90% of the land, with roughly 65% of the region experiencing extreme types of soil acidity (pH < 5.5). Acid soils' production potential is often quite low because to their low fertility. In acidic soils in Northeast India, a number of studies were carried out to determine the impact of applying organic sources of nutrients and amendments on changes in soil pH and nutrient release, especially P, for increased crop productivity (Sharma 1990, 1991a, b, 1993; Shrma et al. 2006). The application of farmyard manure enhanced the soil pH and its stability as well as improved physical condition of the soil (Sharma 1991c).

China

The majority of China's acid soils are located in the country's tropical and subtropical regions, which are also home to some of the country's highest average temperatures and rainfall totals. Actively growing plants, particularly forests, are responsible for the annual return of a sizeable quantity of litter to the soil when conditions are left to their natural state. As a direct consequence of this, the proportion of organic matter present in the surface layers of the soil can reach 5–7%. However, the decline in organic matter content can be attributed to two key contributing processes. Erosion of the soil is a significant problem in various geographic regions. Loss of organic matter is a natural consequence of soil erosion near the surface, which also removes topsoil. The end result of this is that the organic matter content of eroded soils is significantly lower than that of noneroded soils, and this is true even when the natural vegetation is present. The issue worsens if the soil is cultivated for highland crops. Under extreme conditions, soil erosion is so severe that subsoils are exposed, resulting in an organic matter concentration of just around 0.5% in the cultivated layer. Another reason for the fall in soil organic matter content during upland crop cultivation is that soil organic matter decomposes faster in upland circumstances than in submerged conditions.

Farmers use a variety of organic resources to provide plant nutrients and to augment organic matter in the soil. Pig excreta, human excreta, green manures, cattle excreta, plant leftovers, and rape seed cake are the most important, in order of importance.

5.5 Methods for Managing Acidity

Previously, farmers might choose to leave their land either temporarily or permanently, but the demand from an expanding population forces them to control soil acidity in order to retain production and their livelihood. Techniques that treat the issue include amending the soil with the addition of organic materials, lime, mineral fertilisers, etc. (Myers and Pauw 1995). Techniques that partially avoid the issue include adopting alternative land uses and/or crop plants. Since yields on various acid soils are poor for various reasons, techniques that try to change the plant or the soil should rely on accurate diagnosis of the main limits. External elements need to be taken into account when converting such tactics into effective plans. Accordingly, alternative plants must be dependent on the availability of markets, and soil inputs must be practical by being sourced locally or having a sufficient cash flow (Myers and Pauw 1995). Other difficulties with viability include whether subsoil acidity can be controlled, whether the level of control needed is appropriate for the farming system used, and whether the acidity control is temporary or long-term sustainable. Acid soils can technically be made productive. However, the creation of appropriate strategies—that is, the fusion of technical variables with social and economic aspects—is essential for successful implementation (Myers and Pauw 1995).

The management of soil acidity in the past, in a world with fewer people, sometimes involved ceasing to utilise the land for agriculture, either permanently or temporarily. As a result, large areas of land, particularly in the tropics, were undeveloped until recently. Other large regions were utilised sporadically, with infrequently short-term crop production breaks in extended fallows. Man has been obliged to use the less productive lands for farming because of the rapid rise in human population around the world, especially in the tropics, and a large portion of this less productive land has acidic soil. Southeast Asia, Africa, Central America, and South America are experiencing a major crisis in the tropics. Farms that use "slash-and-burn" rotations have been forced to abbreviate their fallow periods due to the acidity of their soils. Population pressures do exist in temperate zones, but more significant is the development of soil acidity as a result of management side effects (Bolan et al. 1991). Farming is the principal industry affected by soil acidity management methods. Except for governments in their capacity as administrators of state forests, other land managers are not very interested in managing soil acidity.

The management of acid soils should aim to increase crop yield, either by modifying agronomic practices in accordance with the climatic and edaphic conditions as listed below, or by adding amendments to remedy the soil abnormalities.

(i) Improving the soil

Because it increases base saturation, P and Mo availability in acidic soils while decreasing Al, Fe, and Mn toxicity, lime has been recognised as an efficient soil ameliorant. Liming also increases the fixation of nitrogen in the atmosphere and the mineralisation of nitrogen in acidic soils.

Following is how soil improvement responds:

$$\text{Colloid in soil} + CaCO_3 = \text{Colloid in soil} + Ca^{++} + H_2O + CO_2$$

Liming, which raises pH, base status, and CEC, is a straightforward solution for managing common issues with acid soils' chemical characteristics, such as low pH, low CEC, and poor base saturation (Sahu and Patnaik 1990).

(ii) Crop selection

The best way to deal with this soil issue is to choose crops that are tolerant of acidity, and breeding such varieties is particularly important for achieving higher production, especially in places where liming is not cost-effective. According to Mandal et al. (1975), rice and tea are the most acid resistant crops and may thus be effectively produced in locations with acidic soils.

(iii) Soil acidity-tolerant varieties

Another method to boost the productivity of acid soils is to grow crop genotypes and cultivars that are resistant to soil acidity.

5.5.1 The Makeup of the Soil

Both the USDA Soil Taxonomy and the FAO Soil Map of the World Legend recognise some significant differences between Ultisols (Acrisols) and Oxisols (Ferralsols) in terms of management features. However, there are several instances of overlap in management qualities that can only be distinguished at the individual profile level or even at the lower levels of classification. Acid soils have unique characteristics, and depending on these characteristics, different management practices may be necessary. Acid soil management strategies either need to be all-inclusive to cover any potential issues or customised to the specific issue(s) that each soil presents. Since not all acid soils have management issues, accurate identification is crucial to prevent the use of pointless inputs. Another significant category of acid soils included those that weren't already acid but turned into them as a result of the application of mineral fertilisers or other inputs (Myers and Pauw 1995). Due to the prolonged use of nitrogenous fertilisers and the presence of pastures that have been fertilised with phosphate, these soils are common throughout the developed, temperate regions. Due to aerial inputs of acidifying substances, acidic soil has also developed in or near industrial areas. According to Pichot et al. (1981), these soils are considered deteriorated and are found outside of temperate zones in the Sudano-Sahel region of West Africa and in some areas of East Africa. Low CEC soils are especially susceptible to this acidification. Numerous scientists have discussed in detail the technical methods intended to meet the limitations of acid soils (Adams 1984; Cochrane 1991; von Uexkiill and Bosshart 1989; Yaman and Kumar 2021; Mahanta et al. 2022; Ma et al. 2022).

One management strategy that gets around the issue is to use the property for purposes other than raising profitable crops. Urban usage is one such strategy, but humanity has not been particularly successful in situating cities and towns on acid soil land. Another management strategy used to treat degraded areas with little to no economic value is the preservation of unproductive land as national parks. Along with steepness, stoniness, and a scarcity of water, acid soil areas may readily meet the requirements for inclusion as national parks. Since many tree species thrive on acidic soils, policymakers have been more active in designating areas for forestry, which is a well-defined technique for managing soil acidity. Such a strategy is applied on three levels: the level of the government, which sets policies and manages the forests; the level of business, which manages the forests; and the level of the individual farmer. Crops adapted to acid soils are planted on the upper portions of the landscape, while more nutrient-demanding and acid soil sensitive crops are grown on the lower slopes' less acidic, higher nutrient status soils.

5.5.2 Traditional Methods in Underdeveloped Nations

Traditional shifting cultivation methods rely on fire's ability to quickly reintroduce nutrients from the surrounding plants into the soil. Both fertiliser and a lime amendment are produced by the ashes. To raise crops, however, all "slash-and-burn" strategies only provide a small window of time, often 1–3 years. Therefore, the systems engaging in "slash and burn" are still low intensity "shifting systems." According to FAO's classification (FAO/SIDA 1974), cropping would occur in around 33% of the years during the fallow-cultivation cycle. To return natural fertility to its pre-industrial levels, long fallows are necessary. The majority of these "slash-and-burn" techniques are only acceptable in situations with unlimited land.

5.5.3 Techniques that Make Use of More Suitable Flora

These methods are intended to alter the plant or switch it from one type to another. Other, more suited crops should be sought out if a soil gets so acidic that the conventional crop no longer performs properly. Farmers have been farming these crops for ages and are familiar with them. However, due to the possibility that the adaptation is limited to one aspect of the soil acidity complex, such as AI, H or Mn, or low P, caution is advised because the adapted plant may thrive on one acidic soil but not another. On the other hand, there exist cultivars adapted simultaneously to low pH and to AI. Wheeler et al. (1992) demonstrated Al tolerance in Medicago sativa, but noted that it was sensitive to low pH. Additionally, a species' genetic diversity may be sufficient to classify particular types as either well adapted or tolerant to acidic soils. While sugarcane and soybean are sometimes thought to be intolerant to acidic soils, they can thrive in other environments.

The fact that growth on acidic soils is subject to a variety of restrictions is one issue with this strategy. The two coping strategies that are required to deal with these limits are tolerance and avoidance, according to Marschner (1991). Low internal mineral nutrient demand and tolerance of hazardous elements are examples of tolerance mechanisms, whereas the ability of roots to alter the rhizosphere, exclude toxic materials from uptake locations, and promote uptake of limiting nutrients are examples of avoidance mechanisms. None of these traits are possessed by any species or cultivar. The capacity to adapt to other harmful elements or to overcome low concentrations of necessary elements is not accompanied by the capacity to tolerate high Al quantities. As a result, a plant or cultivar chosen for its prowess in one acid environment may do badly in another acid soil. For each case, it is either necessary to discover and screen many adapted varieties, or it is necessary to understand the nature of the adaptation and the soil in order to choose the appropriate plant. Farmers have traditionally used species that have been modified. This may be observed frequently in the utilisation of perennial crops like rubber, oil palm, coffee, and other tree crops, which have had some significant triumphs in the tropics. Pineapple and cassava are among the best examples of short-term crops. A downside to such success is overproduction, which leads to low commodity prices, as was the case with coffee and rubber and may currently be the case with pineapple.

There is currently a lot of research being done in the tropics to find cultivars of important crop species that are adapted to acid soils. Accordingly, ICRISAT had a programme to identify cultivars of peanuts that were suited to acid soil; IRRI was interested in rice that was adapted to acid soil; Australia is looking for clovers that are adapted to acid soil; and CIMMYT had a programme to create wheat and maize that are adapted to acid soil (Granados et al. 1993). Rhizobium strains with acid soil adaptability have been found, and some have demonstrated these adaptations in the field (Aarons and Graham 1991; Coventry and Evans 1989). Bottomlands are advantageous for the cultivation of various upland crops during the dry season and for the production of rice during the rainy season due to their hydrological properties and widespread soil acidity (Izac et al. 1991). The lowland rice system is an example of a plant that has been adapted to flourish in an environment where the effects of acidity are limited. As a result, these bottomlands in Asia have nearly reached their full potential as wetland rice areas, but 130 million ha (Raunet 1985) of "inland valleys" suited for lowland rice remain uninhabited in Africa. Health risks and cultural obstacles have made it difficult to develop, but given their high potential for sustainable production—especially given how quickly the upland soils nearby are deteriorating—the bottomlands are likely to be put to use.

5.5.4 Treatment Methods for the Soil Problem

These methods try to alter the soil. This was formerly accomplished in the tropics by staking the land, as shown in conventional slash-and-burn agriculture. Following 1–3 years of cropping, the land returns to being a forest, during which time the soil rebuilds its organic matter and can be successfully farmed once more after a while. This technique has lost viability because to increased land pressure and shorter rotations, especially on steep terrain where it also has detrimental downstream impacts. At least one Southeast Asian nation anticipates a significant decline in hydroelectric capacity as a result of dam siltation from acidic upland soil erosion. Currently, farmers and scientists are attempting to keep up productivity without reverting to a fallow state for a forest or grassland (Sanchez et al. 1982). In these situations, it is necessary to address the root reasons of low output by importing or cultivating inputs that can address toxicities and nutrient deficits.

Enhanced conventional methods

Composting has only been widely utilised in China's traditional communities as a way to maintain fertility. In the tropics, ley systems—a number of years of arable crops that are followed by one or more years of grazing on grass and legumes or restorative fallow—are uncommon (Ruthenberg 1980). Even though there are potential green manure species that are well adapted to acid soils, small farmers in the tropics often employ them sparingly. In general, farmers are reluctant to cultivate a crop that takes up the entire or a large portion of the growing season, provides no direct return, and necessitates effort for planting and digging-in during periods when labour is needed for food crops (Webster and Wilson 1980).

5.5.5 Standards for Liming

In addition to being impractical, liming to raise pH to close to neutrality is often disregarded in tropical soils. In low buffering soils, overliming is a possibility, and lime deficiency in phosphorus and/or other micronutrients can also occur. In order to achieve a target toxicity threshold for specific crops, liming can be based on the exchangeable Al content of the soil (Cochrane et al. 1980; Kamprath 1970), where 1.65 t $CaCO_3$ equivalent ha^{-1} is required for every cmole exchangeable Al kg^{-1}. The method's flaw is that it makes the assumption that Al toxicity is the main issue that needs to be resolved. The approach might underestimate the need for lime if other variables are limiting crop performance. Due to this, liming at a target pH that varies depending on the soil type and the crop species' acid sensitivity may have benefits (Cregan et al. 1989).

Liming substance

Although marl, slag, and other materials are also utilised, limestone is the most frequently used. They include different amounts of silica and magnesium; in certain cases, the silica may have been added by dishonest producers. Dolomite contains 12–13% magnesium, while high Ca limestone only contains 1% magnesium. In Mg-deficient conditions, the presence of Mg becomes significant.

Fineness

Particularly in the developing world, there is a wide variety in particle sizes within commercially produced liming materials. The effectiveness of limestone is influenced by particle size, with material with a particle size of 0.83 mm or more being considered to be relatively ineffective and material with at least 50% passing a 0.25 mm filter being considered to be effective (Barber 1984). The substance operates more quickly the finer it is. However, it could be challenging to apply uniformly if the material is very fine. In terms of cost, response time, and residual effect, 0.25–0.50 mm mesh size is deemed adequate for most soils (von Uexkiill 1986).

Placement

The depth of placement controls the depth over which the liming influences root dispersion, water and nutrient uptake because Ca transport from the lime particles is often slow. Utilising cultivation is common. Alternately, adding anions simultaneously, such as ammonium sulphate and/or potassium chloride, can encourage downward migration (von Uexkiill 1986). According to research by Pavan et al. (1984), nitrate, chloride, and sulphate ions all aid in the leaching of strong base cations into the subsoil. Experimental usage of gypsum ($CaSO_4$) has been made (Ritchey et al. 1980), but it might not be affordable in many tropical areas. Potash or nitrogenous fertilisers may help transfer calcium by nitrate, produced during nitrification, or chloride flowing alongside calcium when applied with lime. Therefore, both the quantity of water and the concentration of anions in the water will affect how well subsoil acidity may be reduced. Rock phosphates can have a liming impact in acidic soils because they elevate pH, give Ca at a low cost, and perhaps reduce exchangeable Al. Rock phosphate's reactivity varies greatly depending on the source, and fine grinding might increase it. The control of organic matter is the third tenet of contemporary acid soil management. There have been attempts to control acidity by attempting to stop the loss of organic matter from soil since it was known that this loss was a contributing element to acidification. While this has happened in the temperate pasture systems of Australia, in certain systems, the accumulation of organic matter is accompanied by nitrogen fixation, which results in higher nitrification rates and acidification (Helyar et al. 1988). It has been shown that green manures, composts, and other organic residues increase crop yields and lessen the harmful effects of soil aluminium (Hoyt and Turner 1975; Hue and Amien 1989).

5.5.6 The Requirement for Interim Systems

Providing a growth path from a traditional system rather than attempting to entirely replace it in one step is a key consideration when establishing management techniques for farmers in emerging nations. The old system serves as the foundation, and better systems emerge as a result of identifying and fixing the traditional system's major flaws. Transitional systems would seek to gradually enhance land use without removing the fallow periods required for fertility restoration in the case of shifting farming. In terms of agronomic and economic performance, such transitional systems are not ideal, but because they are affordable, they are still feasible. Sanchez et al. (1987) outline a transitional system for the Peruvian Amazon region that makes use of a variety of management practices. This system is intended to help the region adapt to changing conditions. This method involves the employment of conventional slash-and-burn clearing techniques, the cultivation of adapted crop cultivars, rotations of adapted crop species, the use of crop residues to their maximum potential, the elimination of the application of fertiliser, and the planting of legume fallow at the end of the rotation in order to smother weeds and restart the cycle.

5.5.7 Techniques for Developing Nations

Let's look at a smallholder farmer's scenario in an acidic highland area of a Southeast Asian nation. In general, there is enough rainfall (1500–2500 men) for this farmer to raise two crops per year on 0.5–5 ha of land. The tiny home garden, which produces a mix of annual and tree crops for grain, vegetables, and fruit, receives the most work. Most mineral fertiliser will be utilised on the farmer's backyard garden if he or she has access to animal manure or can afford it. The amount of workers available will influence how much of the remaining land is cultivated, albeit less intensely. With just little workers and 2 ha available, an additional 0.5 ha may be cultivated. Thus, the cassava plant is grown by the typical subsistence farmer without the use of fertiliser, lime, or organic materials. The plan is to continue producing cassava on the same plot of land for one, two, or three years, or until the yield becomes unprofitable. Then, a different plot of land that has been fallow for a while is utilised. What needs to be done in order to enhance this tactic? It is possible to cultivate alternate crops like maize, peanuts, or upland rice, although probably for a shorter duration. The farmer will need to purchase or grow organic materials, lime, fertiliser, or both in order to properly cultivate these crops. A risky investment like that raises concerns about return on investment. The inputs need to be accurate.

5.5.8 Strategies for Industrialised Nations

The remedy for acidic soil in wealthy nations has been liming. The opportunity cost of soil acidity, the amount and timing of lime application, and government action are the three primary concerns that Cregan et al. (1989) cite as the most important factors. Farmers in this region have access to a broader range of choices, particularly given the fact that soil acidification is frequently occurring and they are required to choose when and to what extent they should take corrective action in order to raise the pH of their fields. If soil acidity limits a farmer's possibilities for business ventures (by, for example, limiting the types of crops that may be grown), there is a large cost that is a combination of direct expenses (due to yield decrease) and opportunity costs. The requirement to take into account the financial effects of various farming practices on acidification rate further complicates this. Therefore, detailed economic models are required as tools for decision-making. The choice of liming tactics is determined by both biological and economic factors, such as the shapes of crop response curves, risk tolerance, governmental regulations, system inputs and outputs of acid and alkali, and the farmer's general knowledge. Assuming that liming in order to maintain a pH near neutral is not justifiable, the choice of liming tactics is determined by both biological and economic factors. Inputs and outputs of acid and alkali, as well as the forms of crop response curves, are all examples of these types of factors. Liming should begin in a biological sense as soon as the most sensitive species begin to react to lime, but this will be regulated by economic worries over the repercussions of not liming when the time comes. In other words, liming should begin as soon as the most sensitive species begin to react to lime. According to Cregan et al. (1989), liming the soil to reduce the toxicity of aluminium, as recommended by Cochrane et al. (1980) for tropical soils, is a rational and realistic strategy for the circumstances that exist in Australia.

5.5.9 Liming Choices

A variety of lime requirement tests have been developed as a result of the necessity to monitor acidification, decide when to lime, and then provide accurate liming recommendations. First, it has been discovered that significantly less lime must be used in tropical locations than was originally advised. Cochrane et al. (1980) created an additional approach that took into account crop tolerance and Ca and Mg levels in addition to Kamprath's (1970) method for liming according to the level of Al saturation. Second, there are regions of the tropics where farmers find it difficult or expensive to get lime. Therefore, the goal of a liming technique must be to apply the proper quantity. Many of the suggested remedies involve amending the soils with substances, such as lime, fertilisers, and organic matter. Only the portion of the soil profile that comes into touch with the additional material will benefit from these treatments' attempts to minimise toxicity. Soil modification to the desired depth constitutes the second element of the liming technique. Although tillage is a useful technique, farmers who use solely hand-operated tools would be hesitant or unable to plough deeper than roughly 15 cm. It is impossible to manage acid soil with minimum input and high output, notwithstanding some outlandish claims (von Uexkiill and Bosshart 1989). Only when nutritional intakes and outputs are in balance can production be sustained. This can be accomplished either with external nutrient sources or with effective residue recycling.

5.5.10 Are Strategies for Managing Soil Acidity Appropriate for Various Environments?

Around the world, acid soil settings are incredibly diverse in terms of climate, human social and economic conditions, preferred farming methods, market and population limits, and the soil itself. Therefore, it is unlikely that straightforward global methods for managing soil acidity can be devised. Instead, tactics have been and will continue to be modified to fit unique local circumstances. Should the strategy be designed to meet a precise but expensive demand, or should it be modified to meet the unique requirements of the circumstance in order to lower the strategy's cost and raise its profitability? It is up to you to decide if you want to pursue a low-cost, low-input (but also low production) solution that would be practical for a resource-poor farmer in a developing country, or you want to pursue maximum profits when there is no restriction on the level of inputs. The farmer will always have the final say. The farmer is the one taking the risk and is the one who is most familiar with the costs of borrowing money as well as the opportunity costs of doing so. Given that reactions to treatments vary over time and geography, the quality of the research material presented to the farmer that suggests the anticipated benefits could vary. Enhancing the quality of the information is one way to improve the strategy selection.

5.5.11 Informational and Research Demands (Myers and Pauw 1995)

A coordinated programme could simultaneously pursue both directions. Inputs of liming material, organic materials, and mineral fertilisers might be tested across important soil types in an area as part of such a programme, which would also collect data on the types of soil, crops, and weather that would allow results to be as broadly applicable as feasible. Some of these researches on highland soils should be carried out for an extended period of time, especially in tropical areas where they are uncommon and sustainability is crucial. A thorough examination of how organic compounds interact with Al as part of the experimental programme would help determine the kinds of organic materials that could be used successfully on acidic soils.

For the purpose of maximising the value of past, present, and future research, coordination and collaboration are required. Too much research is either limited to the local area or unable to be expanded due to the lack of further measures, or it unintentionally replicates work already being done elsewhere. Simple extension to similar soils and climates elsewhere, regression analysis, simulation modelling, GIS, and expert system technologies are a few examples of the methods for expanding the results. Utilising the most economical research instruments is crucial due to the declining availability of research resources. The focus must be on raising production quality and the efficiency with which information is delivered to farmers. Current examples of decision support systems to help with liming decisions include at least two (Hochman et al. 1989; Yost et al. 1988).

5.5.12 Requirement to Address the Issue

Crop production is severely hampered by acidic soils, which make up around 50% of all arable soils worldwide today. Because of the leaching action of carbonic acid, which is CO_2 dissolved in rainwater, soils frequently become acidic. As the pH drops, more micronutrients become available, including aluminium (Al), manganese (Mn), and iron (Fe). However, while basic cations like calcium (Ca) and magnesium (Mg) are lost through leaching and crop harvest, these bases are also restored as a result of the breakdown of organic matter and the weathering of naturally occurring minerals. Acidification of soils is mostly brought on by heavy rainfall, leaching, acidic parent material, organic matter degradation, and the harvest of crops with high yields. Lime, organic matter, crop systems, and the use of plants that are tolerant of soil acidity are all effective ways to regulate soil acidity (Getaneh and Kidanemariam 2021). It is crucial to address the issue producing the main yield restriction. In one instance, even though the soils may be acidic, a P deficit is the main issue. The soils in the tropics are frequently quite acidic, but the main issues are P (and frequently Ca) deficiencies. Rock phosphates may function well in warm, humid situations with acidic soils and may be more cost-effective than superphosphates (Barnes and Kamprath 1975). There are many different types of soil in acid soils; therefore, there may be a variety of actual mechanisms limiting crop performance. Acidic soils can have chemical limitations on growth that include the toxicity of AI, Mn, and other metal ions. The impact of these variables varies depending on the plant (differences in tolerance) and the soil (differences in Al and Mn solubility with pH changes), or it could be due to the impact on related microbes like rhizobia and mycorrhizal organisms. Nevertheless, these soils might also have other growth restrictions unrelated to those brought on by soil acidity. Some might be prone to erosion, experience drought, or have poor drainage. However, effective management of soil acidity may rely on concurrent management of other factors (such as P deficit), or the treatment of soil acidity may have an impact on these other factors (such as induced B deficiency). Treatment of the incorrect issue, such as liming when P deficiency is the main issue or fertilising when the main growth inhibitor is poor soil physical structure, could render a strategy ineffective. The acid sulphate soils provide an additional example. A method of reducing acidity and enabling sustained agricultural production of crops adapted to flourish in certain conditions is to maintain a high water table, which has been commonly practised for centuries. Any alternative approach is pricey. The strategy should also take into account the social and economic limitations. It goes without saying that the proposed level of management needs to work with the current farming system. A low-input subsistence farmer would find little use for an expensive high-tech solution. The farmer also cannot cultivate organic matter at the expense of food. If a farmer cannot get the seed and sell the crop, he cannot switch to an adapted crop.

The main restrictions on acid soils are hazardous concentrations of aluminium (Al) and manganese (Mn), as well as inadequate concentrations of phosphorus (P) (Kochian et al. 2004). According to geology, soil acidity rises as precipitation does. As the pH drops, more micronutrients, including iron (Fe), manganese (Mn), and aluminium (Al), become available. The main contributors to soil acidification are heavy rainfall and leaching, acidic parent material, degradation of organic waste, and the harvest of crops with large yields (Eswaran et al. 1997b; von Uexküll and Mutert 1995; Sharma and Singh 2002).

5.5.13 Controlling Soil Acidity

According to Agegnehu et al. (2019), the management of acid soils should attempt to increase the production potential through the inclusion of amendments to correct the acidity and manipulation of agricultural practices to get the best crop yields.

Using lime

Liming is a significant and efficient method for overcoming acidity limitations and enhancing crop productivity in acid soils. Lime is the "workhorse" or the basis of crop production in acidic soil (Fageria and Baligar 2008). Acidity in the soil can be easily remedied by adding basic minerals or by liming the soil. Agricultural limestone that is calcitic or dolomitic is the most affordable liming material and is also relatively simple to handle (Rengel 2011). Modern agriculture production necessitates the use of effective, long-term, and environmentally responsible management techniques. Liming is a crucial technique for achieving the best yields for all crops produced on acidic soils (Fageria et al. 2007). The most popular long-term technique for reducing soil acidity is liming (Fageria and Baligar 2008). Liming is the most practical way to reduce soil acidity. The soil's pH profile, the type of lime used, the type of soil, the farming technique, and rainfall all affect how much lime is required (Goulding 2016). The primary sources of agricultural lime are dolomitic limestone, crushed limestone, and lime sand from coastal dunes. All of these sources contain carbonate, which neutralises acidic soil. It is composed of calcium carbonate and magnesium carbonate. The neutralising value and particle size are the main determinants of lime quality. When liming to recover acidic soil, lime with a higher proportion of tiny particles will react faster to neutralise acid in the soil (Anderson et al. 2013). Liming has several benefits, including lowering the risk of Mn^{2+} and Al^{3+} toxicity, increasing microbial activity, physical condition, symbiotic legume nitrogen fixation, and forage palatability. It also offers an affordable source of Ca^{2+} and Mg^{2+} when these nutrients are deficient at lower pH levels and increases the abundance of nutrients.

The soil's need for lime will vary depending on:

(1) The necessary pH change.

(2) Clay content and CEC both affect the soil's buffer capacity, or resistance to pH shift. The more the needed lime level, the higher the CEC and clay content.
(3) The liming material's fineness.

When choosing a liming material, the degree of fineness is equally crucial because the surface area in contact with the soil determines how quickly different materials will react. The reaction would be minimal if the soil were coarse, but substantial if the soil were fine. Therefore, there is no fineness issue with minerals like calcium oxide and calcium hydroxide that are powdery by nature. Considering that lime particles do not move around in soils very easily, for both direct and indirect effects on soils and plants. In order to achieve even distribution, it must be applied where it is needed and thoroughly mixed with the soil. When lime is placed to the surface of an acidic subsoil, for example, the anticipated soil response change for increased fertility may not occur as intended because the lime does not quickly and significantly travel. In order to achieve thorough blending with the soil, deeper ploughing would be required (Taye Bekele 2008).

(4) The liming material's chemical makeup.

5.5.14 The Liming Material's Fineness

When choosing a liming material, the degree of fineness is equally crucial because the surface area in contact with the soil determines how quickly different materials will react. The reaction would be minimal if the soil were coarse, but substantial if the soil were fine. Therefore, there is no fineness issue with minerals like calcium oxide and calcium hydroxide that are powdery by nature. Limestone, on the other hand, is a completely different situation because of how its behaviour is affected by particle size (Taye Bekele 2008). Considering that lime particles do not move around in soils very easily, for both direct and indirect effects on soils and plants. In order to guarantee even distribution, it must be added where needed and thoroughly blended with the soil.

5.5.15 Utilising Plant Acid Soil Tolerance

If the soil pH is low, using tolerant species or kinds of pasture and crops will lessen the effects of soil acidity (Rorison 1972). Because the soil will continue to get more acidic without liming treatment, this is not a long-term fix. Through a range of management practices, the rate of soil acidity can be reduced (Goulding 2016). The ability of various crops to withstand acidic soil varies. As a result, one approach is to choose and grow species and varieties that can thrive in acidic soils (Scott et al. 1997). It may be more pertinent to use plant breeding or biotechnology to increase the tolerance of plants and microorganisms to acidity as a last resort when soil pH has dropped to values below six. One of the main obstacles to crop production is the acidity of the soil and the resulting inadequate availability of nutrients. For acid soils to provide the best crop growth, the practice of liming is used to lessen soil acidity and lower phytotoxic levels of Al and Mn. In general, liming ought to be viewed as a soil supplement that increases soil pH to a level suited for maximum nutrient availability, plant growth, and crop output. Integrated soil and crop management can help slow the rate of soil acidification. Organic sources' residual impacts on crop output and soil characteristics may persist for a number of years, making it difficult to accurately quantify their profitability in the short term. Instead, their effects become apparent over an extended period of time. Application of organic wastes enhances the physical and biological characteristics of soils as well as their nutritional content. According to Chairiyah et al. (2021), 2 tonnes of dolomite per hectare are the ideal dosage for acidic soils. When dolomite is applied to acidic soil, it can be limed to bring the pH level from acidic to neutral and improve maize output by 1.91 times. The outcomes demonstrated that adding dolomite to acidic soils can raise the pH level. In comparison to the control treatment (dolomite-free) and normal dolomite treatment, the dolomite provision increased plant height and maize yield (Chairiyah et al. 2021).

5.5.16 IASM (Integrated Acid Soil Management)

IASM is one of the elements involved in managing acid soils. To significantly increase yields, P, K, and N fertilisers should typically be applied in conjunction with lime and other better management techniques. A single lime application also increases productivity over five to seven years. Low efficacy of organic fertilisers used to raise and sequester C in soils of the humid tropics is caused by rapid turnover of organic matter. According to reports, the high levels of soil organic matter and soil fertility found in central Amazonian anthropogenic soils (Terra Preta) are due to charcoal. Following the addition of charcoal to the soil, higher exchange capacity, surface area, and direct nutrient inputs were seen, which were connected to higher nutrient availability and retention. Higher charring temperatures typically enhanced the charcoal's surface area and exchange characteristics. In addition, charcoal is relatively resistant, making it a viable option for a long-term CO_2 sink. The role of microorganisms in oxidising charcoal surfaces and releasing nutrients, as well as the potential for improving the qualities of charcoal as it is being produced in the field, are some of the components of a charcoal management system that are still unknown (Glaser et al. 2002). There are many unmet research needs, including testing charcoal production in the field in tropical agroecosystems, examining the surface characteristics of carbonised materials in the soil environment, and assessing the agronomic and financial viability of using charcoal to manage soil. Leaching with acid rain, which results from industrial pollution, and nitrification after the use of nitrogenous fertilisers are two common anthropogenic sources of acidification (Wild 1988).

5.5.17 Deficiency of P

One of the main barriers to crop production in acidic soils is phosphorus shortage. The main issue with severely weathered acid soil is low P, which necessitates applying substantial amounts of P to boost concentrations of accessible soil P to a sufficient level. To overcome this crop-limiting factor application of lime and alternative liming materials, inorganic and organic sources of nutrients and management of rhizosphere activities as components of integrated nutrient management (INM) practice, and genetic improvement of plants for enhanced P acquisition can be the viable options for P management in acid soils. About 80% of the applied P fertilisers are lost due to low acquisition and use efficiency of plant that results in significant loss of applied P causing serious environmental problems. So, application of inorganic P at right rate, time, and type and at right place can increase P-use efficiency in crops. A few promising management strategies of acid soils are discussed hereunder.

5.5.18 Organic Additions and Fertile Acidic Soil

The utilisation of organic waste products has proven to be a successful substitution for lime (Noble et al. 1996). In order to treat acid soil infertility, uncomposted plant litter, composts, manures, peats, and coal products are typically employed. This is in addition to providing crops with readily available nitrogen, phosphorus, and potassium (Whalen et al. 2000). Due to uric acid's quick conversion to NH_4–N in most soils, poultry manure includes high levels of NH_4–N and can convert a significant amount of N to nitrate-N (NO_3–N) within a week (Sims and Wolf 1994). The types and quantity of organic and inorganic amendments can alter the soil's capacity to bind phosphorus. Whether it is in liquid or solid form, organic manure enhances the soil's texture, water-holding capacity, and microbial activity, ultimately promoting healthy plant growth and yield (Russel and Marsah 1997). Additionally, there is more potential for using chicken manure, which has a high concentration of nutrients and can be used in integrated nutrition management, to make up for a lack of FYM. According to reports, decomposed poultry manure is a very nutrient-dense organic fertiliser that fosters healthy plant growth, provides natural organic plant food, and extends the life of the humus that builds up in the soil and promotes earthworm and microbial activity. In acid soils, amendments such animal dung, grass clippings, and dolomitic lime significantly boosted maize yield as well as the concentrations of exchangeable K, Ca, and Mg as well as extractable P (Naramabuye et al. 2008). The technique not only increased the pH and nutrients that could be extracted from row soil as compared to inter-row soil, but it also increased the size and activity of the soil microbial population. Additionally, the addition of inorganic fertiliser to these organic residues considerably enhanced yields, particularly in the lime, grass residue, and control treatments. After three days of incubation, the addition of lignite and compost to acid soil increased orthophosphate and total P measured in soil solution, with the compost treatments having the most beneficial effects (Schefe et al. 2008). Both untreated and modified soils had higher soil solution P concentrations after a longer incubation period of 26 days, with total P concentrations being most affected. In their 2011 study, Adeleye et al. examined the impact of various organic manures and NPK fertilisers on the chemical characteristics of acid soil with maize. They demonstrated that the chemical characteristics of soil were enhanced when compost, poultry manure, and cane rat droppings were applied in quantities of 5 t ha^{-1} each, together with 100 kg ha^{-1} of NPK (15-15-15) fertiliser. Different kinds of organic manures were applied, and this decreased the soil's acidity. Application of cow dung produced a pH of 6.37, while NPK fertiliser produced a pH of 5.28. When compared to soil fertilised with NPK fertiliser, the application of several forms of organic manures improved the soil's organic C, total N, accessible P, exchangeable K, and CEC. The starting pH of the compost and the soil will likely have an impact on how compost affects soil pH. There are reports in the literature of composts raising, reducing, or not changing soil pH. Numerous studies have shown that cattle manures can raise the pH of acidic soils (Hue 1992). In acid soils treated with pig slurry and solid cattle dung, the populations of bacteria, fungus, actinomycetes, ammonifiers, and nitrifiers (soil microflora) were considerably increased (Ndayegamiye and Cote 1989). Inorganic pyrophosphatase, dehydrogenase, and alkaline phosphomonoesterase activities were all significantly higher in soil treated with cattle manure, according to a study on the effects of long-term application of cow manure on microbial biomass C and enzyme activities in an acid soil (Parham et al. 2002). Due to increased soil microbial activity, applying organic manure promotes the transformation of nutrients in the soil and their availability to plants (Vishwakarma et al. 2012). It was discovered that the phosphomonoesterase activity was boosted by the inclusion of C and N sources through organic materials as opposed to the incorporation of inorganic P (Nannipieri 1994). In other instances, soils with high P concentrations showed decreased phosphomonoesterase activity. According to Diacono and Montemurro (2010), applying organic fertilisers over time repeatedly boosted the soil's organic matter and fertility while also improving crop yields and soil microbial processes as compared to inorganic fertilisation.

5.5.19 Approach to Managing the Rhizosphere

When co-infected with arbuscular mycorrhizal (AM) fungi, which create a beneficial relationship with the plant roots, the phosphate uptake by plants and subsequent growth promotion in plant-soil systems inoculated with P-solubilising microorganisms are more pronounced (Smith and Read 1997). Plant development and health are enhanced by the mycorrhizal fungi's absorption of nutrients. After using the ammonium ion by the hyphae, the AM fungus are also capable of sparsely mobilising soluble inorganic phosphate by excreting H^+ (Richardson et al. 2009). The application of PSB in conjunction with rock phosphate (RP) has been shown to improve yield and P uptake in a variety of crop plants, including rice, according to numerous earlier research (Thakuria et al. 2009). It was discovered that bacterial inoculation with RP produced the desired effect more prominently than when the bacterium was applied in combination with SSP. PSB (e.g. Bacillus firmus) with RP and SSP on yield of paddy and uptake in acidic soils increased significantly due to inoculation. By using the root-dip method, rice seedlings treated with orthophosphate slurry (SSP:soil:water as 3.5:40:100 w/v) produced 690 mg more tissue P than seedlings treated with soil and water slurry (Talukdar et al. 2001). These rhizosphere-based P-management strategies have already been acknowledged as an effective way to increase P-use effectiveness. The key components of rhizosphere-based P-management strategies are: (1) ensuring the availability of highly effective P-solubilisers that are able to function in acidic environments; (2) P-solubilisers need to have a variety of traits that encourage plant growth; (3) enhancing the genetic potential of bioinoculants in terms of P solubilisation in the rhizosphere without compromising the ability of the rhizosphere to colonise; and (4) optimising microbial-based bioinoculum for improving plant growth and soil health. The effectiveness of using the water-soluble inorganic P fertiliser; i.e. in strongly acidic soils of NE India, a single superphosphate exhibits hardly 15–20% of the applied quantity in a crop season. Increased inorganic P fertiliser application is a waste of money for underprivileged farmers and may also damage the environment. Because there are no substitutes for P in agriculture and a potential global P crisis is clearly a distinct possibility in the near future, the rhizosphere-based P-management approach appears to be pertinent in acid soils for enhancing P-use efficiency and yield of crops. In an acid Inceptisol, the root dipping in SSP-MC slurry approach of P-management has recently been shown to be helpful (Kalidas-Singh 2013).

5.5.20 Management of Acid Soils: An Integrated Strategy

Pulses, oilseeds, and coarse cereals are primarily produced on acidic soils, especially on acidic upland. Liming raises the pH of the soil, enhances the availability of nutrients for plants and crop growth, boosts nutrient uptake, activates biological processes, lowers the amount of 3+ extractable Al, and lessens the toxicity of specific elements. Since it is not cost-effective to lime soils with pH levels below 4.5, integrated techniques (combining chemical, engineering, and agronomic measures) are needed to increase the productivity of acid soils in these circumstances.

A. *Chemical techniques*

Simple soil improvement with liming materials, which increase pH, base status, CEC, inactivate Al, Fe, and Mn in soil solution, reduce acidity and P fixation in soil significantly, may effectively manage common problems in acid soils in terms of chemical characteristics.

B. *Technical measures*

- Conservation tillage, ridge-and-furrow tillage.

  Bench terracing, contour, and strip cropping

- Building field bunds to collect rainwater on-site.
- Harvesting and reusing runoff water.
- Surface layer compaction using powerful tractors and rollers.
- Chiesseling of soils with surface compact layers at intervals of 30–45 cm and 60–120 cm.
- *Agricultural practices*
- Increasing aggregate stability through the use of compost, paddy straw, green manure, and rotational planting of legumes.
- Two weeks prior to sowing, rice husks and groundnut shell powder can be added to seed lines to prevent crust formation.
- Adding clay and silt to the tank to boost water retention.
- The use of organic matter, legumes in the cropping system, etc., can result in a decrease in bulk density, a noticeable increase in% of pore space, improvement in soil structure, moisture holding capacity and hydraulic conductivity, etc.
- Use of appropriate crops and cropping systems together with a scientific set of procedures.

5.5.21 Farmers' Expertise

Farmers all throughout the world have created land classification systems that offer helpful insights into their farming practices. These classification systems might be a valuable place to start for agronomic study and the creation of pertinent technology. This was noted by Conklin (1957) approximately 40 years ago. Contrary to soil classification schemes that are popular in western nations, which prioritise soil genesis and subsoil characteristics in classification, folk soil classifications used by the majority of people around the world place a greater emphasis on management and production of the soil. A traditional soil classification essentially identifies which traits are important for growing crops in a specific area. Although there may be similarities within a region if environmental conditions, soils, and crops are comparable, the majority of folk classification schemes are local in scope. Again, this should not come as a surprise because much of the confusion surrounding soil classification today is due to agronomic communities' disagreement with how another group has categorised the variations in their local soils. Additionally, being aware of the local folk classification can assist prevent pointless land use arguments. Some of the soils of Meru, Kenya, are known as Yumba soils, which are used to make clay pots, and Kithaayo soils, which are utilised by livestock and other animals as salt licks. Technicians frequently lament that farmers do not recognise the importance of reducing the acidity of their soils. While there are undoubtedly some farmers who do not comprehend, research on this topic has shown that the majority of farmers are aware of the benefits of fertilisers as well as their drawbacks and high cost (Wilken 1987). Farmers typically use more chemical fertilisers as their disposable income rises. The vast majority of impoverished farmers, however, value their money too highly to spend it on an input that often only lasts one crop season and needs to be applied again the following year. They choose to use other options because of the fluctuating pricing of commodities and the unpredictability of the climate. Manuring is a popular choice in locations with a large cattle population. Manure lasts longer than chemical fertiliser, boosts a soil's ability to retain water, and reduces the danger of crop damage from improper application (Tabor et al. 1990). Fertilisers are frequently used in either insufficient or excessive amounts, according to a large body of literature. Local classification systems show how keenly aware farmers are of the variations in soil quality in their immediate surroundings. According to Dvorak (1988), "The farmers are familiar with the native taxonomies; they are limited to a particular locality; and information on the assignment of plots to suitable categories is easily accessible." According to Conklin (1957), the Hanunh people of the Philippines employed ten basic and thirty derived classifications of soil. These were based on parameters that were pertinent to their agricultural system, including four degrees of firmness, nine colours, five topographic features, and three slopes. Dvorak (1988) examined three communities in India's semi-arid tropics to ascertain the effectiveness of indigenous taxonomies in agricultural production research. The farmers' perspective on their soil and soil fertility was supported by later analyses of soil samples (Dvorak 1988). As previously said, long-term experience within a local ecosystem is a key factor in determining how accurate farmer knowledge is (Conklin 1957; Moran 1981). Particularly where there has been a significant change in cropping patterns or significant switches from polyculture to monoculture, such knowledge will typically be stronger about soils and plant types than about pests and plant diseases (Bentley 1989, 1990). In recent years, not only have extension workers and agronomists tended to disregard farmers' traditional knowledge of soils and soil–plant interactions, but also other scientists as well. It is vital to document these regional systems of land use and classification and to work with farmers to translate them into regional soil maps suitable for management and study. It will be important to establish the degree to which extrapolation is feasible and the dimensions along which local systems experience classification change.

On acidic, nutrient-deficient soils, farmers have a variety of options for achieving high levels of production per unit of land. The extent to which they may achieve adequate amounts of income is largely dependent on economic and social structures that are outside the control of both farmers and scientists. A strong incentive that must be steady and predictable in order to encourage farmers to apply their expertise is the expectation of receiving a fair return for their work. Numerous farmers work hard to improve the growth environment for their crops in acidic soils, yet a much larger proportion are unable to make the same soils profitable. Inadequate local social and political organisations that can mobilise household efforts in controlling common property resources, like water or canal muck, and in resolving the inevitable disputes over access to resources may be the cause for some, who may lack local knowledge as a result of resettlement (Guillet 1992; Netting 1993; Wilken 1987). Prices and transportation systems that guarantee gratifying returns for production labour invested are equally significant. By using slash-and-bum land preparation techniques, choosing crops and variety capable of relatively high production in acid, low-nutrient soils, and making sophisticated use of their folk systems of soil classification and use, native farmers have adapted to the acid soils within their territory. They have also actively developed labour-intensive techniques like manuring, ridging, mounding, and mulching that improve soil conditions and mitigate the effects of initial conditions like low pH. None of these techniques are likely to be used if the farmers lack the social networks necessary to resolve conflicts, their land rights are insecure, and the value of their products is below the market rate (Guillet 1992; Netting 1993).

5.6 Acid Soil Amelioration

5.6.1 Soil Amendments

It is critical to address soil acidity as soon as possible. If acidity seeps into the subsoil, yields may suffer significantly. Controlling subsoil acidity is challenging and expensive. The first step in regulating soil acidity is to identify any acidification. This entails repeatable soil tests for pH, aluminium, and manganese levels in the plough layer (zero to ten centimetres) and the subsurface to 50 or 60 cm. The following farming practices are recommended to reduce acidification:

- Coordinating nitrogen fertiliser inputs with crop demand.
- Using nitrogen fertiliser types that generate less acidity.
- Effectively irrigating the area to reduce leaching
- Planting early after fallow to ensure more rapid nitrogen utilisation.
- Cultivating deep-rooted perennial species to absorb nitrogen at larger depths.
- Using lime on a regular basis to combat the acidity inherent in the agricultural system.
- Cultivating acid-tolerant crops or crop varieties that are more tolerant of acidic soils. Acid-tolerant crops include sugar cane and bananas.

Problem soils may require amendment before cropping and making the best use of the plant nutrients given. Liming of acidic soils and reclamation of alkaline soils are two examples.

In general, calcium carbonate is chosen where it is easily available and reasonably priced. Dolomite limestone is preferred where soils also require Mg treatment. Liming is also accomplished using basic slag and sugar-factory press mud from carbonation facilities. Liming ahead of planting is preferred for some crops, such as potatoes, due to their susceptibility to high soil pH following recent liming, which may produce scab due to Mn insufficiency. To minimise leaf scorch, only carbonate lime should be used for "top-liming" growing plants. In the spring, lime is put on top of short grass, allowed to breakdown, and then rinsed into the topsoil. To ensure a uniform rise in soil reactivity, the liming material should be equally spread over the bare soil and well blended into the topsoil layer. The optimum method is to apply it after harvest but before tillage and sowing (also known as "stubble liming"). Following this practice, the soil layer beneath the topsoil can be improved to some amount. Liming a field may be essential once every three to five years or once during the rotation of crops. Liming is recommended for increasing the pH closer to neutral and, in the course of doing so, improving the availability of numerous nutrients for crops such as sugar beetroot and oilseed rape, which do not grow well under acidic conditions and favour higher soil pH. This is because acidic conditions inhibit the growth of sugar beetroot and oilseed rape. Liming is advised for bringing the pH closer to neutral and, in the process, boosting the availability of various nutrients, with the exception of crops like tea, which require an acidic environment to thrive. Liming is recommended for bringing the pH closer to neutral.

The lime requirement cannot be computed directly from the pH value since reserve acidity must also be neutralised, which is not indicated in the pH value. However, knowing the pH and texture of the soil might help you estimate how much limestone you'll need. In general, the goal is to lime an acidic soil to a pH of 6.5.

5.6.2 *Product Standards for Soil Acidityliming*

Size and other parameters for selling agricultural liming products to assure a quality product are as follows:

(i) Agricultural liming materials must be crushed so that at least 90% pass through a standard 20-mesh screen in the United States (with a tolerance of plus or minus 5%).
(ii) For dolomitic limestone, at least 35% of materials must pass through a standard 100-mesh screen in the United States; for calcitic limestone, at least 25% must pass through the same screen (both with a tolerance of plus or minus 5%).
(iii) To be categorised as dolomitic limestone, a product must contain at least 6% magnesium in the carbonate form.
(iv) The product must be labelled with the amount required to equal that given by a liming material with a CCE of 90%. A product with an 80% CCE, for example, might be labelled "2250 pounds of this material equals 1 tonne of standard agricultural liming material."
(v) When exposed to moisture, pelted lime must slake down to the fineness standards indicated above.

5.6.3 Correction of Soil Acidity

Soil acidity can only be remedied by neutralising the acid present, which is accomplished by the addition of a basic substance. While there are several basic materials that can neutralise acids, the majority are either too expensive or too complicated to maintain. Agricultural limestone (aglime) is the most regularly utilised material. It is employed because it is affordable and simple to manage. Limestone is simple to work with because it is not highly soluble, which means it does not dissolve quickly in water. As a result, it is not extremely corrosive to equipment, and more significantly, its pH at equilibrium (after it has dissolved all of the lime and some remains in the water) is only about 8.2. This is critical since even an excessive application of lime will not reduce crop yields in most cases. Calcium (Ca) from the lime goes to the surface of soil particles and replaces the acidity as it dissolves in the soil. Carbon dioxide (CO_2) and water (H_2O) are formed when acidity combines with carbonate (CO_3). The eventual effect is soil acidification neutralisation.

5.6.4 Rates for Lime

Whether wheat is grown continuously or in rotation with a legume, the amount of lime that should be applied to the soil is dependent on this factor. If only wheat is planted on the field year after year, it is only necessary to apply a rate of lime to raise the pH to slightly above 5.5 in order to avoid disease problems that occur at higher pH; on the other hand, the pH should be raised to at least 6.5 if legumes are grown on the field sometimes. As a consequence of this, the minimum amount of lime that should be applied to continuous wheat is either 0.5 tonne of ECCE lime or fifty percent of the soil test deficiency amount that is required to raise the pH to 6.5 (Zhang et al. 2017a, b). Whichever number is greater, this should be the amount of lime that is applied.

5.6.5 Estimating Costs and Fees

The conclusion reached by their research facility is split into two halves. The first step in determining the lime's quality is to conduct a chemical analysis. The lime material is subjected to this test, which looks for $CaCO_3$ or its equivalent in the substance. The second step in the process involves determining the degree of fineness to which the lime particles are pulverised. When the ECCE is higher, there is more $CaCO_3$ present in the material, and the particle size is more refined. A lime of higher grade will have an ECCE value that is greater than 50%. Due to the fact that aglime does not always have a 100% ECCE, the quantity required to provide a particular level of

100% ECCE must be estimated. The following calculations are those that need to be used.

Amount of aglime required in tonnes of ECCE multiplied by the percentage of ECCE required assume that the aglime that was available included 65% ECCE and that the soil test revealed that 1.5 tonnes of ECCE were necessary in order to raise the pH of the soil to an acceptable level. 1.5 times 100/65 is 2.3 tonnes of aglime, according to the calculations. According to Zhang et al. (2017a, b), in order to obtain the 1.5 tonnes of 100% ECCE lime that are necessary for the assignment, 2.3 tonnes per acre of the 65% ECCE lime would need to be applied. Because it is difficult for lime to dissolve in water, it must be handled in a manner comparable to that of fertilisers, which are used to supply immobile nutrients like phosphorus to the ground. Lime must be thoroughly incorporated into the ground before it can successfully reduce the acidity of the soil. As a consequence of this, gardeners who intend to cultivate their land using a method known as no-till wheat production are highly urged to correct any pH problems that may exist in the soil before beginning no-till wheat production. As a result of the fact that neutralisation involves a process that takes place between soil particles and lime particles, the acidity can be neutralised more effectively if the lime is coupled with the soil more effectively. Because of this, damp materials (such as those obtained from water treatment facilities) that are unable to be adequately mixed with the soil have a tendency to have a lower level of effectiveness. Similar to the soil particles, the particles of pelletised lime are too large to properly combine with the smaller soil particles. When these components are combined with soil, the acidity of the soil is frequently neutralised just around the lime aggregates (or pellets), while the acidity between aggregates is not altered. After determining the correct application rate for the lime and making sure that it is spread evenly across the entire field, it is best to combine the application of the lime with a light tillage activity such as disking. After disking, you can proceed with ploughing, but you need to be careful not to include the soil too deeply. If you do, the lime will be diluted by the subsoil and its effectiveness will decrease. According to Zhang et al. (2017a, b), lime rates are determined by looking at the top six inches of soil. Due to the requirement of water for the lime reaction, the effects of lime on dry soil will take a very long time to manifest. Even if everything is done precisely and the soil is kept wet, it can take anywhere from three to six months for there to be a substantial shift in the soil pH. As a consequence of this, liming the land where wheat is grown should take place as soon after harvest as is practically possible. On the other hand, when the pH of the soil is extremely low, significant change can take place in a matter of weeks, and this can mean the difference between successfully cultivating wheat and not being able to do so.

5.6.6 Where Soil Acidity Issue Happens

There is no simple rule for determining which soils are excessively acidic for plants. There are, nevertheless, some guidelines:

(i) Very acid soils can be found in regions with a lot of rain, such as on top of mountains or on the side of a mountain facing the trade winds.
(ii) It may arise from the parent rock in some situations.

5.6.7 Improving Acid Soils

Fertilisers are required to improve plant growth. Commercial fertilisers (N–P–K) are usually the most convenient to use. However, in highly acidic soils, this may not be sufficient. These soils frequently require additional fertilisers before the plants can develop normally again. Lime is a high-pH fertiliser that contains a lot of calcium (and possibly magnesium as well). When we add lime to an acidic soil, several things can happen at once:

- The pH of the soil will rise.
- When the pH rises over 5.5, the aluminium and manganese toxicity disappears.
- Because lime contains a high concentration of calcium, it helps correct a soil's calcium deficit.
- If we employ magnesium-containing lime, we will also address magnesium deficit.
- Increases soil phosphorus availability.
- Improves soil physical qualities such as soil structure.

5.6.8 Liming the Soil

When we apply lime to soil, it must dissolve before it can be used. Lime has the property of dissolving very slowly. Fine lime particles dissolve fast, whereas coarse lime particles are too large and will remain in the soil for an extended period of time. Even if we use very fine powdered lime, it takes at least a month of rainy weather for the lime to dissolve sufficiently. We may help the lime breakdown by working it into the soil after it has been applied.

5.6.9 Liming Substances

Commercial lime is ground limestone that has been processed. If the soil is sufficiently moist, all of the particles will normally dissolve within 1–2 months.

Dolomite lime contains magnesium in addition to calcium. Dolomite lime is preferable if the soil is poor in magnesium. Coralline lime is a white or grey substance that breaks off from coral reefs that encircle several Pacific islands. It can be found on beaches or in lagoons. It can be used to liming acidic soils and is as effective as commercial lime when handled properly. Because coralline lime is rarely pulverised, it contains a high concentration of coarse particles. These larger particles dissolve considerably more slowly than the small lime. It can take 2–4 years for them to disintegrate. As a result, we require more coralline lime than commercial lime. The size of coralline lime particles may vary depending on location. The number of tiny particles is usually determined by the location of the coralline lime. When we look at white sand on a beach, we can observe that it is not particularly fine. We'll need extra coarse sand if we're going to use it. In addition, we will have to wait longer for the lime to dissolve. When we examine the coralline material found in the lagoon between the shore and the reef, we can see that it has more small particles.

Liming materials are often calcium and magnesium carbonates, oxides, and hydroxides. Liming materials work best when they:

- Eliminate H^+ and Al^+ from exchange sites (potential acidity).
- Remove H^+ from solution (active acidity).
- Are cost-effective.

To be cost-effective, liming material must generally be a salt. Strong bases and acids, such as NaCl (sodium chloride) and $CaSO_4$ (gypsum), are ineffective in raising pH levels because no hydroxyl (OH^-) is released to neutralise H^+ by forming water (H_2O). The equation below shows that a salt of a strong base and a weak acid is required to elevate pH.

$$CaCO_3 + H_2OCa^{2+} + HCO_3{}^- + OHCaCO_3 + H_2O$$

On exchange sites, Ca^{2+} displaces H^+ and Al^{3+}, whereas OH^- neutralises the H^+ in solution.

The efficiency of liming material is determined by two factors:

(i) Calcium carbonate equivalent (CCE).
(ii) Material fineness.

When these two components are combined, the result is known as the effective calcium carbonate (ECC).

The following are examples of common liming materials:

- Carbonate of calcium. It typically comprises 75–95% $CaCO_3$ and 42–53% CaO. More than 5% magnesium carbonate ($MgCO_3$) concentration is beneficial. Hard limestone particle size must be less than 1 mm, and soft material (chalk) particle size must be less than 4 mm.
- Dolomite (calcium magnesium carbonate). Its various kinds contain 15–40% $MgCO_3$ and 60–80% $CaCO_3$. These products are appropriate for acidic soils that are also weak in Mg.
- Slaked lime $Ca(OH)_2$ and quicklime (CaO). These are quick-acting additives for soil acidification; however, they are often more expensive than natural limes. They play an important function in several applications, such as establishing a well-structured soil surface layer for spreading sugar beetroot seeds.

The most common liming material is pulverised natural limestone ($CaCO_3$) of varying fineness depending on the rock's hardness. The finer the grinding required to achieve equivalent efficiency, the harder the rock. Because carbonate limes are only marginally soluble in water and must be dissolved into neutralising forms, they act slowly. In recent years, the degree to which substances are soluble in dilute hydrochloric acid (HCl) has come to be utilised as a criterion for determining how reactive they are. Some of them have a lot of magnesium, which is beneficial for soils that are lacking in magnesium, while others have very little manganese. Lime additives not only lessen the acidity of the soil but also confer a number of other advantages. A form of lime amendment known as marl (sometimes called "lime earth") is described as "lime earth." This was something that was used extensively in Greek and Roman agriculture. Marl is a type of soil that can be found in the uppermost few metres of glacial soils. It has calcium carbonate concentrations ranging from 10 to 30%. It was rediscovered in Europe in the eighteenth century, and by that time, it was being used extensively to remediate the acidic soils that existed at the time. In both the mining and distribution of marl, significant amounts of money are spent on labour. Lime that has been created from red sea algae is very pliable and contains some vitamin B.

When too much lime is given to the soil, overliming results. This will significantly raise the pH of the soil. Our goal is to keep the pH of the soil between 6.0 and 6.5. There are numerous negative consequences of increased pH. When a field is fertilised every year, especially with commercial nitrogen fertilisers, the soil pH can rapidly decline. A soil test on such farms should be performed at least once every five years to determine whether the pH is rapidly falling owing to fertiliser use. The capacity of a soil to collect, store, and transmit energy in a manner that is conducive to the growth of plants is referred to as its fertility. Fertile soils are those that have a healthy balance of carbon, nitrogen, and other elements. It is imperative that the soil be in healthy condition for any of these endeavours to be fruitful. The biological, chemical, and physical components of healthy soils are all operating as they should and are in a state of balance. Soils are living, self-organising systems that have all of these components. This delicate equilibrium can be thrown off by the continued use of acidic or salty synthetic fertilisers, pesticides, fungicides, and herbicides, all of which have the potential to be harmful to the ecosystem. Despite the fact that organic farming as a practice is aware of this fact, there is still work to be done in order to actively manage soil fertility in accordance with the practices of organic farming's pioneers. The preservation of carbon in particular is of the utmost significance and must be accomplished to the greatest extent feasible through the utilisation of solar energy in the process of photosynthesis carried out by green plants, as well as through the storage and utilisation of carbon in the soil that is as effective as possible. In order to actively manage soil biology in order to establish and preserve soil fertility and to attain more sustainable, healthy, and productive farming systems, it is necessary to understand (1) why they are the way that they are and (2) how research and practices might assist in making active management of soil biology more effective. This comprehension will be essential moving forward.

5.6.10 Advantages of Liming

Liming has the following advantages:

- The risk of Mn^{2+} and Al^{3+} toxicity can be mitigated.
- Increase in microbial movement.
- Soil structure improved.
- Legumes exhibit increased symbiotic nitrogen fixation.
- Forage becomes more palatable.
- When these minerals are inadequate at lower pH, it provides an affordable source of Ca^{2+} and Mg^{2+}.
- Improves nutrient availability.

calculating CCE

The CCE is a method of comparing all liming materials to $CaCO_3$ as a benchmark. $CaCO_3$ has a molecular weight of 100 (Ca = 40, C = 12, and O = 16 × 3). $CaCO_3$'s CCE has been hypothetically determined to be 100. When employing materials other than $CaCO_3$, the molecular weight of $CaCO_3$ is divided by the molecular weight of the other material. For example, the following is the calculation for using wood ash as a liming material:

The molecular weight of wood ashes (K_2CO_3) is 138.

$$CaCO_3 = 100$$

When compared to $CaCO_3$, 100/138 = 0.72 (CCE) or 72% effective.

So, if a soil guideline called for 1000 lbs. of agricultural lime ($CaCO_3$), then divide the CCE of 0.72 (K_2CO_3) by the rate required to calculate the amount of K_2CO_3 required.

1000 lbs $CaCO_3$ divided by 0.72 equals 1389 lbs of K_2CO_3.

Because reaction rate is proportional to the size of the material (surface area) in contact with the soil, the efficacy of agricultural limestone is proportional to its fineness. This is because finer limestone has a greater surface area in contact with the soil. Both coarse and tiny particles are included in the composition of agricultural limestone. In a number of states, 75–100% of the limestone must pass through a screen with a mesh size of 8–10, and 25% of it must pass through a screen with a mesh size of 60. Lime laws in some countries require that at least 50% of the material must pass through a 60-mesh screen before it can be considered reactive. These laws also stipulate that no more than 10% of the material may remain on a 10-mesh screen (indicating that the material is too coarse to react).

5.7 Application of Amendments

5.7.1 *Infertility Caused by Acidic Soil*

Acidity can cause adverse plant growth circumstances in soils with pH (pHwater) levels below 5.5 (von Uexküll and Mutert 1995). Acidity is common in temperate temperature areas with considerable rainfall, such as northern Europe (Austrheim et al. 2005; Fystro and Bakken 2005). In the light of these issues, agronomically, surface liming of acid grasslands is highly recommended; nevertheless, the effects on fodder production are controversial. Surface liming, on the other hand, has little or only a limited influence on yield on fertilised grasslands with a complex semi-natural flora (Pinto et al. 1995). Because of their acidity and deficits in a few important nutrients such as nitrogen, phosphate, calcium, zinc, boron, and molybdenum, red and lateritic soil productivity is low.

According to Sehgal (1998), one of the primary reasons for low productivity in red and lateritic soils—which account for over 70 million hectares of cultivable land in India—is the high level of soil acidity. Red and lateritic soils comprise roughly 170.0 million hectares of cultivable land around the world. In addition to that, the P fixation in these soils is rather high (Hocking 2001), but the availability of Si and Zn is quite low (Mandal et al. 1992). According to Fageria and Baligar (2008), one typical method for increasing the productivity of acidic soil is to use lime to lower the soil's acidity. Conyers (2006) describes calcite and dolomite as excellent liming materials. However, due of how expensive they are, resource-poor farmers in India's red and lateritic soil zones are unable to afford to use them in their farming practices. Under these circumstances, a locally accessible low-cost materials such as basic slag, a byproduct of the steel industry, could be utilised as an alternative liming material (Bhat et al. 2007; Khan et al. 2007). This was suggested by Bhat and Colleagues (2007) and Khan and Colleagues (2007). It contains a high concentration of calcium in the form of calcium silicate, which is sparingly soluble and ranges from 21.0 to 38.6% of the total (Collings 1955). Calcium silicate, just like calcite and dolomite, has the ability to reduce soil acidity and increase crop yield. Basic slag also contains P, Si, Zn, Cu, and other elements that can be advantageous in acidic red and lateritic soils that are generally weak in these nutrients (Peregrina et al. 2008). These soils are often acidic and red in colour. Lateritic soils are acidic and red in colour. In point of fact, a number of researchers (Rodriguez et al. 1994; Khan et al. 2008) found that adding basic slag to acid soils increased crop output in addition to increasing the concentrations of nitrogen, phosphorus, potassium, and calcium in plant tissues. Other examples of organic amendments include compost and biochar. These organic additives, when mixed with basic slag, may help to increase the solubility of the basic slag, hence increasing its effectiveness in neutralising the acidity of the soil. According to Tisdale et al. (1997), the decomposition of farmyard manure and PM in soil results in the release of nitrogen and the production of carbon dioxide, which then reacts with the moisture in the soil to form HCO_3. This helps in the process of dissolving $CaSiO_4$ and $Ca_3(PO_4)_2$ in basic slag (Collings 1955), which leads to increased availability of nitrogen, phosphorus, calcium, and silicon in the soil. As a result, mustard is better able to take in these nutrients, which leads to increased crop development. When compared to the use of basic slag by itself, the incorporation of organic amendments into basic slag did, in fact, results in a slight elevation of the soil's pH as well as a reduction in a number of distinct manifestations of soil acidity. Composting boosted both the direct and residual effect of basic slag in improving crop production and phosphorus uptake, as demonstrated by Debnath and Basak (1986) in their studies with two cropping systems, rice-wheat-green grame and green gram-rice-wheat in terai acid soils (Haplaqoll) of West Bengal, India. The terai acid soils were located in Haplaqoll. When it came to increasing the yield of rice grown on residual soils after mustard, basic slag performed significantly better (20.1%) than calcite (12.6%). According to Ma et al. (2006), basic slag treatments led to an increase in the amount of silicon and zinc that were easily accessible in the normally poor soils. This may have contributed to the growth of rice, which requires a high level of silicon and is responsive to zinc. As a result, the production of rice may have improved. The theory was supported by an appreciable rise in the concentrations of silicon and zinc, in addition to an increase in the levels of nitrogen, potassium, and calcium, in rice plants that had been treated with basic slag. On the other hand, Gascho and Korndorfer (1999) found that basic slag performed poorly as a source of silicon in some acidic soils found in the Cerrado region of Brazil. Following the application of basic slag, Khan et al. (1996) found that there was a lower concentration of zinc in the rice. Again, if the solubility of zinc at the accompanying (with basic slag) soil pH of 5.7 is taken into consideration, sufficient zinc would be available for uptake by rice plants provided other counter ions, particularly Fe, did not interfere (Dutta et al. 1989). This conclusion was reached by Dutta and his colleagues. In addition to having a high concentration of zinc and copper, as was mentioned before, basic slag also contained considerable amounts of numerous other heavy metals, including cadmium, chromium, and nickel. It is highly likely that subsequent applications will result in an increase in the concentrations of these metals in the soil, which may have unfavourable repercussions for both the natural environment and agricultural output. Heavy metals can be effectively removed from the environment by Fe and Al oxides, which are found in high concentrations in red and lateritic soil, respectively. They have a powerful binding effect on the potentially dangerous heavy metals, preventing the metals from remaining in the soil in a form that is absorbable by plants. There is also a low probability that these metal ions will leak into the solutions of the soil, seep through the soil profile, and contaminate the aquifers that are located below the surface.

5.7.2 Lime or Dolomite Application

When soil pH is too low for a specific crop, lime or dolomite can be added to raise the pH to the correct level. The amount of lime or dolomite needed to rectify an acidic pH will vary according to the soil and crop. Soils with a high organic matter and clay

Table 5.5 Average nutrient content of bulky organic manures and composts

Type of manure	N (%)	P_2O_5 (%)	K_2O (%)
Dung cattle	0.25–0.35	0.10–0.12	0.12–0.18
Dung of sheep/goat	0.60–0.70	0.45–0.55	0.02–0.04
Human excreta	1.30–1.40	0.75–0.85	0.45–0.55
Hair and wool waste	12.0–12.50	0.08–0.12	0.25–0.35
Farmyard manure	0.45–0.55	0.12–0.18	0.45–0.55
Poultry manure	2.75–3.00	2.80–2.90	2.25–2.40
Town compost	1.40–1.55	0.09–0.15	1.40–1.55
Rural compost	0.45–0.55	0.18–0.22	0.45–0.55

content are more resistant to pH fluctuations and necessitate higher treatment rates. A soil test called "lime requirement" or "buffer pH" should be requested to provide an estimate of how much lime is needed to address an existing soil acidity problem. This soil test is available at the majority of commercial soil testing laboratories. The test is used to prescribe lime to increase the soil pH of one hectare's surface 10 cm to a target pH that will not affect crop yield. In general, a pH (in water) aim of 5.5 is recommended. Depending on the crop, extra lime or dolomite may be required once the target soil pH is reached. The amount of material removed at harvest, the amount and kind of fertiliser generally applied, and the amount of leaching that happens all contribute to cropping systems' acidifying effect.

5.7.3 Animal Slurry and Farmyard Manure

Manures generated from plants, animals, and human waste are included on the list. Organic manure is partially degraded during storage by fermentation, which also produces useful humic compounds. Some nitrogen is lost as ammonia, but this can be decreased by adding around 2% water-soluble phosphate. Fermented wet FYM (25% dry matter) nutrient concentrations are dependent on 1 N provided through 4–5 tonnes of biomass. Nutrient conyent of some organic manures is given in Table 5.5.

5.7.4 Green Manure Crops

According to Marshaller (2011), the pH of the soil has an effect on both the solubility and availability of essential nutrients to plants. According to Yao et al. (2010), liming raises the pH of the soil, which in turn makes other nutrients more accessible. It also prevents aluminium and manganese from being toxic to plant growth. The utilisation of 3.75 t ha^{-1} of lime resulted in an elevation of the pH of the ground from 5.03 to 6.72, as stated by Buni (2014). According to the findings of another study that was carried out by Moges et al. (2018), the pH altered when lime and P fertiliser were applied; however, the application of mineral P fertiliser had no influence on the pH when compared to the control. Dessalegn et al. (2018) came to a similar conclusion, finding that the use of a variety of limes led to an increase in the pH of the soil when compared to the control. These ions and cations are able to remove H+ from exchange sites and form H_2O and CO_2 when they do so. According to Fageria et al. (2007) research, a higher pH results from cations moving into the vacant space left by H^+ during the exchange. Kisinyo et al. (2012) discovered another way in which lime treatment increased soil pH. They found that lime treatment displaced H^+ and Al^{3+} ions from soil adsorption sites and replaced them with Ca^{2+} ions. The process of raising the pH of acidic soils using lime releases phosphate ions precipitated with aluminium and iron ions and makes it possible for plants to absorb phosphorus (Achalu 2012).

5.7.5 *Organic Matter Application*

As a result, more practicable management options must be examined. Organic stuff, which is relatively cheap and widely available, is one such alternative. Farmers have employed plant organic matter for thousands of years to improve soil fertility, recycle nutrients, manage weeds and pests, and regulate moisture content. Several studies have also found that adding plant organic matter to slightly acidic soil causes moderate changes in pH (Reid and Butcher 2011). The overall cation concentration has been linked to the alkalising action of additional organic matter (Pocknee and Sumner 1997). In contrast, the addition of organic compounds such as sodium malate, sodium citrate, calcium oxalate, or calcium gluconate induced a rise in soil pH when tested in other studies (Yan et al. 1996b). This was found to be the case despite the fact that adding inorganic salts (such as Ca_2SO_4) had no effect. Pocknee and Summer (1997) suggested that the biological decarboxylation of carboxylic groups of organic anions was the mechanism responsible for the expenditure of protons that occurred throughout the decarboxylation process. This was because the protons were used up in the process. In contrast, the findings of other researchers (Ritchie and Dolling 1985; Tang and Yu 1999) suggest that the release of protons connected with organic anions is the mechanism by which organic matter contributes to the acidification of soil. Even while it has been demonstrated that adding crude organic matter and organic salts to acidic soils can improve the quality of the soil, very little is known about the process by which this occurs or the material that is the most effective to utilise. In a recent study, Michael et al. (2015) discovered that integrating organic mulch in the form of leaves from Phragmites, which have a high nitrogen content, successfully ameliorates acid sulphate soil. In contrast, simply applying leaf material to the surface of the soil was only somewhat beneficial to the soil's condition.

5.7.6 Byproducts

Numerous byproducts are put to use as soil supplements, and the majority of these byproducts contain alkaline compounds. These chemicals have the ability to reduce the acidity of the soil while also contributing calcium and magnesium. When alkaline compound additions are added to soils, the CCE values of those amendments should be measured in order to evaluate their contribution to the lime requirements of the soil, and the pH of the soil should be monitored in order to avoid the soil from reaching levels of pH that are too high. An excessively high soil pH can result in a deficiency of some micronutrients, most notably manganese and zinc in the coastal plain region, whereas the piedmont and valley regions are deficient in zinc.

Dolomitic lime includes significant amounts of magnesium and calcium, both of which are necessary nutrients for plant growth; calcitic lime contains predominantly calcium. Their CCE values show that they have basically the same potential for neutralising soil acidity.

5.7.7 Marl

Marl is a naturally occurring, chalky mineral that is sometimes employed in the process of lime production. The term “marl” refers to an unconsolidated mixture of calcium carbonate, shards of shells, and various impurities, the most common of which are sand, silt, and clay. Marl is found in beds. Marl deposits are abundant in the eastern part of the state of Virginia, sometimes known as the coastal plain, as well as in the limestone valleys of the Appalachian area and other states along the Atlantic Coast. The CCE of marl, which can range anywhere from 70 to 90, and the amount of money it costs to prepare it are the two primary factors that decide how useful marl is as a liming material. Marl is typically plastic and lumpy, and before it can be used, it needs to be dried out and ground into a powder. The rate of reaction for marl is the same as that for calcitic aglime.

5.7.8 Aglime Pelletised

Pelletised Aglime is made from finely powdered limestone that has been shaped into round pellets using a binding agent (about 0.16–0.20 in. in diameter). When compared to traditional Aglime, the pellets spread more easily and remove dust while handling. Pelletised limestone is also expected to degrade quickly following contact with soil moisture and to react quickly to neutralise soil acidity in the vicinity of the pellet. This is owing to the finely ground particles employed in the pellet manufacturing process. Pelletised limestone is widely used on lawns and gardens by homeowners. Because of the higher processing costs, these materials are often substantially more expensive than traditional Aglime sources. Pelletised dolomitic limestone is similarly successful in preserving soil pH as typical limestone products, according to research conducted in Virginia. However, because the granules dissolve and react in a localised area, these materials are not as successful in adjusting the pH of severely acidic soils. Because the materials are efficient for surface treatments to keep soil pH values in an optimal range, they should be used as part of a regular liming programme that prevents soil pH from falling below 6.0.

5.7.9 Lime Suspension or Fluid (Liquid)

Fluid lime is produced by combining extremely finely ground lime with a high CCE value (normally 100% passing a 100-mesh sieve and normally 80–90% passing a 200-mesh screen) with water (normally 50%) and a suspending agent, which is typically attapulgite clay. This process results in a lime product that is very finely powdered and has a high CCE value. When applying the suspension, a suspension fertiliser applicator is utilised, which ultimately leads to an outstanding distribution pattern. Because of the fineness of the grind, the lime reacts very fast in the soil that has been treated, despite the fact that only moderate amounts of Aglime may be administered at one time using suspension lime (500–1000 pounds/acre of Aglime material). Due to the low application rates, annual applications of suspension lime may be necessary in order to maintain a pH level within the desirable range in the soil. When compared to the price of dry Aglime, the price of lime in suspension or fluid form is often two to four times higher. Due to the high cost of producing and distributing this product, its usage is currently restricted; however, it has been found to be an effective product for liming particular areas, such as in apple orchards, where only the area surrounding the trees is limed.

In order to eradicate pathogens, biosolids at some municipal wastewater treatment plants are subjected to treatment with calcium oxide or hydroxide, which raises the pH of the biosolids to a level that is more than pH 12. These biosolids, which contain calcium hydroxide that has not been reacted, have the potential to be valuable liming materials because they contain unreacted calcium hydroxide. According to the laws, biosolids have to be assessed to determine their unique CCE value, and then they have to be applied at rates that do not exceed the amount of lime that is required for the soil to which they are applied. The CCE value of these chemicals is directly proportional to how effective they are at neutralising the acidity of the soil. The beneficial benefits of chicken litter, mature dairy compost, sheep manure, and lucerne pellets on wheat development in two acid soils were expected to be stronger with increased application rate, but their relative effectiveness with lime was not additive. Butterly and Tang (2018) found that poultry litter was the most effective.

Subsoil acidity (pH5.5 $CaCl_2$) restricts agricultural production primarily through harmful aluminium (Al^{3+}) and manganese (Mn^{2+}) concentrations. Surface lime has limitations in correcting soil acidity at depth. Deep ripping to integrate lime and other materials is costly; therefore, cost-effective ameliorators and/or their mixtures must be discovered. Because of the combined amelioration and fertilisation effects, using lower rates of lime amendments may be helpful to crop development.

5.7.10 Liming Agent: Wood Ash

Most crops thrive in slightly acidic soil with a pH of 6.0–6.8. Acid soils must be limed on a regular basis to sustain agricultural yield. The availability of most plant nutrients is diminished in soils with an acidic pH of less than 6.0, which can also lower the amount of biological activity in the soil. Ash from wood is both more soluble and reactive than ground limestone, and it also has a faster effect on the pH level of soil than lime does. The full effects of lime can take anywhere from six months to a whole year to manifest themselves. Ash from wood contains significant amounts of potassium and calcium, but only traces amounts of phosphorus and magnesium due to its low overall concentration. Because wood ash is not as concentrated in calcium carbonate equivalent (CCE) as lime is, a greater quantity of product is necessary to provide the same effect on the pH of the soil. Because the CCE of wood ash can range from 25 to 59%, and the CCE of lime can be anywhere between 90 and 95%, you will need to use two to four times as much wood ash as you would lime in order to get the same result. There is a wide range of possibility for how effectively wood ash can raise the pH of soil. The recommendations are determined by the current pH of the soil, the desired pH, and the buffering capacity of the soil. Buffering capacity is a measurement of how quickly pH changes in the soil and is based on the amount of clay and organic matter present. A soil pH that has been raised above what is considered acceptable by plants might be lowered by an excessive amount of wood ash. After applying the material, incorporate it into the top two to four inches, and there should be no need for further wood ash throughout that year.

Never mix ash with nitrogen fertilisers like ammonium sulphate, urea, or ammonium nitrate. Ash and nitrogen fertilisers can be toxic to plants. These fertilisers give off ammonia gas if they come into contact with materials that have a high pH, such as wood ash. Ash from wood contains potassium (K) at a concentration of 5%, calcium (Ca) at a concentration of 25%, phosphorus (P) at a concentration of 2%, and magnesium (Mg) at a concentration of 1%. It has been demonstrated that K and P derived from wood ash have a beneficial effect on crop growth. In addition to zinc, boron, copper, molybdenum, and sulphur, wood ash contains a variety of other essential micronutrients. Never use wood ashes that have been generated from waste wood, pressure-treated wood, or wood that has been painted, as these types of wood ashes may contain heavy metals and chemicals that are harmful to plants. The majority of wood ash does include traces of heavy metals, so it is important to exercise extreme caution when working with the product.

5.7.11 Soil Texture Improvement Amendments

The absence of fine clay particles is characteristic of soils with a sandy texture, whereas the absence of coarse clay particles is characteristic of soils with a clay texture. When the particle sizes are either coarse or extremely fine, the chances of natural structure formation are significantly reduced. Clay particles should be added to light soils and sand particles should be added to heavy soils if one wants to achieve the most evident alteration in the particle size composition of soils. The question of how much should be used and whether or not it can actually be done is of the utmost importance. The addition of one percent of a mineral component in the 0–20 cm layer of a topsoil that has a volume of three thousand tonnes per hectare is comparable to the addition of thirty tonnes per hectare of material. In order to transform sandy soil into loamy sand, it is necessary to add 630 or 180 tonnes of clay material per hectare in order to increase the soil's clay content from 4 to 10%. Even if such resources were available without cost in the general area surrounding the field, the cost of transporting them would be prohibitively expensive. If there is sufficient sand or clay material in the subsurface, it may be possible to ameliorate this problem by bringing it to the surface through deep cultivation. Utilising all of the organic material and crop waste that is available can help reduce the negative effects that are linked with extremes in soil texture to a certain extent.

The utilisation of the right tillage techniques is an essential component in the process of improving the structure and exposing the subsoil. However, this only gives a temporary benefit, and other steps need be taken to create favourable conditions for the processes involved in the formation of soil structure. There are a few different additives that have been developed specifically to enhance the structure of the soil. Soil conditioners are substances that are added to soil in order to improve its water-holding capacity (WHC) and its resistance to erosion. When working with fine-textured, heavy soils, these are used to produce a crumb structure, mostly for the purpose of better aeration. The structure of the soil can be indirectly improved by the application of a large number of materials, such as lime and organic manures.

5.7.12 Utilising the Press Mud

Compressed sugar industry trash is known as press mud. Lot of research work on the behaviour, use, and efficacy of press mud has been done by Datta and Gupta (1983a, 1984, 1985; b, c), Datta and Sharma (2006) and Datta et al. (1983) on acid soils. According to Ranjhan (1990), 3% of the waste generated in sulphonation facilities and 7% of the waste generated in carbonation plants comes from the filtration of cane juice. According to Yadav (1992) and Joshi and Sharma (2010), this material is amorphous, dark brown, soft, and sponge-like. It also contains varying concentrations of sugar and fibre, as well as coagulated colloids such as cane wax, albuminoids, inorganic salts, soil particles, and mineral elements. It helps prevent soil erosion, crusting, and cracking, improves drainage, and stimulates the proper growth of bacteria and microorganisms in the soil. In addition to that, it can be exploited as a reclamant and soil conditioner (Tandon 1995). Raw press mud is a fantastic material for the production of biogas due to the fact that it contains 5–15% sugar. According to Gupta et al. (2011), dry press mud can be used to create energy since it contains a higher proportion of combustibles than wet press mud. In recent years, there has been a significant uptick in the practice of biocomposting, which involves the treatment of press mud with wasted distillery wash (Padmanabhan et al. 1993). According to Taylor (2000), carnauba wax, which may be quite pricey, is commonly used in the cosmetics, food, and pharmaceutical industries. Press mud wax could serve as a cheaper alternative to carnauba wax. According to Saleh-e-In et al. (2012), crude press mud comprised between 67.95 and 76.53% moisture, 1.63–2.29% nitrogen, 19.28–30.76% ash, 12.10 to 13.29% sugar, and 6.70–11.01% crude wax. The mean elemental composition of crude press mud was found to be; Ca (21.30–29.97%), Si (9.53–9.94%), P (8.40–9.52%), S (3.77–7.36%), and K (2.51–4.08%).

5.7.13 Reducing the Toxicity of Metals

Phosphorus in the soil reacts with aluminium and/or manganese, both of which are more soluble in acidic soils. It has been known for a long time that phosphorus in some extremely acidic soils is less accessible to crops as a result of these reactions. The reaction of phosphate with certain metals results in the formation of a solid that is almost completely insoluble. One example of this is aluminium phosphate. As a direct consequence of this, not only is there a deficiency in phosphate, but also in aluminium and manganese. When phosphate fertilisers are paired with the seed at the time of planting, the affects of toxic aluminium and manganese are significantly minimised, and it is possible to achieve yields that are close to normal (Zhang et al. 2017a, b).

5.7.14 Use of Biochar

Using a process known as pyrolysis, biomass is heated in an atmosphere with a low level of oxygen to form biochar, which is a substance similar to charcoal that is rich in carbon. It is anticipated that India will have access to approximately 500 million tonnes/year of biomass in the current year (2010–2011). There is an annual surplus of around 313.62 million tonnes of agro-residues produced in India, which amounts to approximately 435.98 million tonnes. These residues are either partially or entirely unutilised as a result of the different constraints that apply to them. A considerable amount of the biomass required for the production of biochar can come from weeds, provided that they are readily available in the area and do not result in a significant loss of crop. Near the ICAR-National Organic Farming Research Institute in Tadong, Gangtok, weeds including Ageratum spp., Lantana spp., Artemisia vulgaris, Chromolaena odorata, Bidens spp., and *Neyraridia spp.* were found, and the biomass from these plants was used to make charcoal. The effect of adding weed biochar to clay loam soil was studied to determine whether or not biochar has the ability to act as a lime substitute. Biochar is naturally basic (pH > 7.0). It can react in the same way that agricultural lime does, by elevating soil pH (up to 1 point). Biochar should be used at the same rate as other inputs like as compost or manure every year to reap the full benefits. Because biochar contains nutrients, the application rates of these organic inputs can be lowered (Das et al. 2014). It was discovered that rates of 5–10 t/ha (0.5–1 kg sqm) have a positive influence on soil characteristics and crop productivity. Biochar has been shown in studies to increase/improve soil moisture retention by up to 20%, nutrient utilisation efficiency by up to 15%, CEC by up to 35%, and crop production by up to 30%. Biochar boosts grain yields (15%) by decreasing leaf SPAD [The SPAD metre measures the difference between the transmittance of a red (650 nm) and an infrared (940 nm) light through the leaf, generating a three-digit SPAD value (Uddling et al. 2007)] value in highland rice at low P availability areas (Das and Avasthe 2016). Thus, biochar can be investigated as an example of utilising a minor waste substance produced as a byproduct of fuel combustion to help the agricultural system through scientific technology.

Properties and benefits of biochar

Lehmann and Joseph (2009) define biochar as the carbonaceous byproduct that is formed when biomass is pyrolysed in a low-oxygen environment, and the charred product that is then applied as a soil supplement. Pyrolysis, which removes oxygen from the reaction, is the process by which biomass is thermally depolymerised at high temperatures. In 2010 Joseph et al., the byproducts of pyrolysis are syngas, bio-oil, and char. Not only can the char, also known as biochar, be added to soil, but it can also be utilised as an energy source. Chan and Xu (2009) state that a variety of organic resources, such as paper mill sludge, crop and forestry waste, and chicken manure, can be used to make it. Applying biochar is becoming more and more popular as a sustainable method of repairing degraded and worn soils (Lehmann and Rondon 2006). According to Yamato et al. (2006), it enhances the microbial population, earthworm population, enzyme activity, bulk density, water-holding capacity, permeability, and other physical, chemical, and biological qualities of the soil that support plant growth and development. For an extended period, soil fertility will be enhanced due to its resilience to microbial destruction (Steiner et al. 2007). Furthermore, as noted by Asai et al. (2009), it raises the xylem sap and saturated hydraulic conductivity of the topsoil in rice fields, resulting in increased crop yields and improved responsiveness to chemical fertiliser treatments, such as N and PA. Because they contain very little in the form of heavy metals or dangerous compounds like As, Cd, Pb, and polycyclic aromatic hydrocarbons, the risk of contamination is relatively minimal. According to Lehmann et al. (2006), Glaser et al. (2006), Singh et al. (2010), they can improve soil fertility, agricultural output, fertiliser and water consumption efficiency, and mitigate N_2O emissions. The increasing amount of CO_2 in the atmosphere can be lowered via the long-term storage of carbon in soil. As a result, biochar has emerged as a viable option for soil carbon storage (Lehmann et al. 2006).

According to Singh et al. (2010), Mukherjee and Zimmerman (2013), the pyrolysis conditions and the feedstock source both significantly affect the properties of the biochar. When comparing wood biochar to manure-based biochar, it is often found to have higher total carbon (C), less ash, lower total N, P, K, S, calcium, magnesium, aluminium, and copper (Al, Na, and Cu) concentrations, as well as lower exchangeable cations and potential cation exchange capacity (CEC). Surface acidity decreased as the pyrolysis temperature rose, while ash content, pH, and surface basicity all increased (Mukherjee and Zimmerman 2013). The main products of slow pyrolysis, which is heating biomass gradually to the necessary peak temperature, are syngas and biochar. Quick pyrolysis is heating biomass quickly, between 400 and 550 degrees Celsius, with the main outcome being bio-oil (Berek 2014). Major et al. (2010a, b) identified several critical physicochemical properties of biochar, including increased surface area and porosity, low bulk density, improved cation exchange capacity (CEC), neutral to high pH, and higher carbon content. According to Jindo et al. (2014), there is also the presence of N, P, and basic cations including Ca, Mg, and K. These are essential plant nutrients for crop growth and development. More biochar is produced during low-temperature pyrolysis, whereas higher-temperature pyrolysis yields biochar with more stable carbon, a larger surface area, robust adsorption capabilities, and more C (Jindo et al. 2014). The functional groups eventually disappear at high pyrolysis temperatures (> 600 C), leaving the material with a polycyclic aromatic structure that is highly condensed and more resistant (Major et al. 2010a, b; Jindo et al. 2014). According to Schulz and Glaser (2012), a higher proportion of aromatic structures, which in turn offers higher resistance against chemical and biological breakdown, is what gives biochar its ability to last hundreds to thousands of years in the soil. According to Singh et al. (2010), whereas the concentrations of C and N in feedstocks rich in minerals decrease as pyrolysis temperature increases, they increase in plant-based biochar.

Significance of biochar

It's well known that soil biochar increases soil capacities and secures carbon. Biochar immediately interacts with soils, microbes, and plant roots after being assimilated into the soil (Lehmann and Joseph 2009). The type of interactions can be influenced by the following factors: (i) feedstock composition, particularly the total percentage and specific composition of the mineral fraction; (ii) conditions of the pyrolysis process; (iii) biochar particle size and delivery system; and (iv) properties of the soil and local environmental conditions. When biochar is introduced to soil, the rate at which it ages starts before it does, and it is somewhat influenced by the soil's moisture content and temperature (Nguyen and Lehmann 2009). According to Singh and Cowie (2008), the amount of carbon generated from biochar decreases significantly over time, peaking during the first two weeks following the application of the biochar amendment. Most surface soils still cause the biomass-derived biochar to remain thermodynamically unstable, according to Macías and Camps Arbestain (2010). Since it includes a large amount of non-aromatic C, low-temperature biochar is more vulnerable to microbial attack (McBeath and Smernik 2009) and subsequent oxidation than high-temperature biochar (Nguyen et al. 2009). With O_2 serving as the most common electron-acceptor species, aromatic C is highly stable but also demonstrates redox activity and functions as a reducing agent. How organic matter and clay mineral surfaces interact with the biochar that has a higher mineral content depends on a number of factors, including the distribution of functional groups on the clays (siloxane, OH), organic matter (COOH, C=O, C–O, CN), the polarity of these compounds, and the makeup and concentration of cations and anions in solution (Sun et al. 2012). The fauna that inhabits soil, including termites, worms, larvae, and other insects, breaks down or covers biochar with organic chemicals, which they then consume or dwell inside of. Earthworm bioturbation helps the soil profile and the biochar physically merge together. According to Major et al. (2010a, b), biochar eventually settles deeper into the soil profile, where microbial activity is lower.

5.7.15 Application of Organic Manures

Most people understand the significance of animal waste as a plant nutrition source or soil amendment, but few are aware of the ability of manure, particularly poultry litter, to neutralise soil acidity and raise soil pH. When manure is put to soil, nitrification and decomposition result in the formation of different acids. It causes H+ to be released into the soil solution (Das et al. 2014). The organic matter in manure serves as a pH buffer, releasing H^+ when alkaline materials are supplied and accepting H^+ when acidic materials are added. Some solid manure also contains high levels of calcium, magnesium, and lime. The organic matter and lime-like elements in solid manure help to balance out the acidity produced when the manure's NH_4 is transformed to NO_3. Solid bovine manure application raises soil pH in acidic soils to neutrality (Benke et al. 2009), boosting nutrient availability, particularly for P and micronutrients (Das 2014). The researchers discovered that poultry manure is just as effective as lime at raising soil pH and lowering aluminium toxicity. It is not feasible for farmers with limited access to resources to use significant quantities of lime to amend acidic soils due to the economic and logistical challenges this presents. As a direct consequence of this, we have begun looking into other options. It has been demonstrated that the addition of organic materials that originate from plants and animals can increase soil fertility, structure, and biological properties, while also decreasing soil acidity and the related Al saturation. The kind of organic manure that is used, the rate at which it is administered, and the capacity of the soil to act as a buffer all influence the magnitude of the pH increase that occurs in the soil. In acidic soils, Hue (1992) discovered that boosting crop yields and decreasing Al toxicity occurred simultaneously when green manure and animal dung were added. In addition to these effects, there is a possibility of an increase in the fertility of the soil, as well as an improvement in its physical qualities and microbiological activity. In a similar vein, biogas slurry, agricultural waste, and organic substances like biochar might be suitable for the management of acidic soils. According to Steiner et al. (2007) and

Chan et al. (2008), pyrolytic biochar can be added to soil as an amendment in order to boost soil fertility and decrease soil acidity. According to research by Xu et al. (2006a, b), the positive impacts of directly integrating plant components into soils are temporary due to the rapid degradation of these components by microbes. It has been demonstrated that biochar is resistant, and it might continue to exist in soils for hundreds of years (Rebecca 2007). Coal in its natural form and coal extracts have both been shown to stimulate root growth in acidic soils by Yazawa et al. (2000).

5.8 The Presence of Lime and Nutrients in the Environment

The presence of excessive levels of the cations hydrogen (H^+), aluminium (Al^{3+}), and manganese (Mn^{2+}) in soils is the primary cause of soil acidity. These cations all prevent plants from developing normally. Crops suffer from the adverse effects of soil acidity mostly as a result of the unavailability of phosphorus (P), which is produced by P fixation in soils. Significant amounts of P are fixed in soils by the soil components iron and aluminium sequioxides. Root elongation limitation and overall crop growth retardation are both indicators that excessive Al^{3+} ions caused by soil acidity tend to collect in plant roots (Kochian et al. 2004; Ligeyo and Gudu 2003–2005). This accumulation of Al^{3+} ions in plant roots prevents P, Mo, and other ions from translocating to the tops of the plant from the roots. Although the negative effect of H^+ ions is not as evident as that of Al^{3+} cations, excess H^+ ions in acid soils influence plant root membrane permeability and therefore interfere with ion transport (Ligeyo and Gudu 2003, 2005). This is according to research conducted by Ligeyo and Gudu. In acidic and phosphorus-deficient tropical soils, where the plant's ability to successfully scavenge native or introduced phosphorus is key, correcting the soil acidity and providing phosphorus fertiliser are both essential practices (Swift et al. 1994). Lime is widely recognised as an efficient method of rectifying soil acidity or as an input recommended for acid soil amelioration. Its immediate effect is a rise in soil pH (The et al. 2006).

Most field crops like soil pH levels between 6.0 and 7.0. The buffer method developed by Shoemaker et al. (1961) is used to determine the lime needs of an acid soil. Table 5.6 shows the amount of lime necessary in terms of pure calcium carbonate to get the soil pH to the appropriate level (6.0, 6.4, and 6.8).

5.8.1 Plant Nursery Raising Method

According to Goulding (2016), the practice of liming is the most popular and successful approach of lowering soil acidity. Liming is often advised for producing field crops on acidic soils, either alone or in conjunction with mineral and organic fertiliser (Ayalew 2011). Liming can be done alone or in conjunction with either of these types of fertilisers. Liming increases the amount of available phosphorus while

Table 5.6 Lime requirement of an acid soil

pH of soil buffer suspension	Lime pH to (tones of required desired $CaCO_3$ to bring level per ha)			pH of soil buffer suspension	Lime pH to (tones of required desired $CaCO_3$ to bring level per ha)		
	6.0	6.4	6.8		6.0	6.4	6.8
6.7	2.4	2.9	3.4	5.7	12.9	15.2	17.1
6.6	3.4	4.2	4.7	5.6	13.9	16.4	18.5
6.5	4.4	5.6	6.1	5.5	15.0	17.6	20.0
6.4	5.4	7.0	7.5	5.4	16.0	18.8	21.4
6.3	6.5	8.5	9.0	5.3	17.1	20.1	22.8
6.2	7.6	10.0	10.4	5.2	18.1	21.3	24.3
6.1	8.6	11.4	11.8	5.1	19.2	22.5	25.8
6.0	9.7	12.8	13.2	5.0	20.2	23.7	27.2
5.9	10.8	14.2	14.7	4.9	21.3	24.9	29.2
5.8	11.8	15.7	16.1	4.8	22.4	26.1	30.6

simultaneously lowering the amount of aluminium present in acidic soils (Jafer and Hailu 2017). Liming also enhances the soil responsiveness, calcium concentration, and cation exchange capacity (CEC). According to Holland et al. (2018), increased root development leads to increased water and nutrient uptake, grain yields, and agricultural sustainability. This is because all of these chemical changes, as long as they remain below a safe range, boost root development. Lime application in alkaline soil creates an unfavourable environment that discourages microbial activity.

In areas with acidic soil, a treatment that combines the application of lime and organic manure can prove to be the most effective method for increasing soil fertility. The addition of organic manure to the soil improves its status in terms of its organic matter content, which is an essential factor in ensuring the fertility of the soil over the long term since it regulates biological processes that have an effect on the availability of nutrients. An alternate source of plant nutrients to artificial fertilisers is provided by the addition of cow dung (CD) and poultry manure (PM), as stated in the research conducted by Banik et al. (2006). This contributes to the preservation of soil fertility. This is of utmost significance with regard to the cultivation of rice. According to the findings of Siavoshi et al. (2011), the impacts of CD and PM on the physicochemical and biological properties of the soil result in an increase in the amount of nutrients that are available for plant growth.

Phytotoxicity from aluminium (Al) is a primary cause of acid soil infertility (Blamey and Edwards 1989). Toxicity severity is determined by the activity of inorganic Al species in soil solution (Blamey et al. 1983). Al species that have been organically complexed are less hazardous (Hue et al. 1986; Suthipradit et al. 1990). The negative logarithm of the activity of monomeric inorganic Al (pAl) in soil solution is proportional to soil pH in most soils (Conyers 1990; Helyar et al. 1993). In the regulation of Al^{3+} phytotoxicity, two soil remedial techniques are thus possible: (i) increasing soil pH and (ii) decreasing Al solubility by altering the connection between soil pH and pAl.

5.9 Liming

Liming with calcium carbonate (calcite) or magnesium carbonate (dolomite) has been used to improve acid soils for ages. The fundamental goal of modern applications is to neutralise exchangeable aluminium and boost the availability of numerous nutrients, primarily calcium, magnesium, and phosphorus. This is normally accomplished by increasing the pH to 5.5–6.0. When manganese poisoning is suspected, elevate the pII to 6.0. Thc parameters to consider are (1) the amount of lime required to reduce the percent aluminium saturation to a level where the specific crop and variety will grow well, (2) the fineness of the lime, and (3) the manner of placement. Given the significance of genetic plant tolerance to aluminium, liming must be employed in conjunction with the aluminium saturation level that the plant can withstand in order to thrive, or just to give calcium and/or magnesium to tolerant species.

Liming

Although growing acid-tolerant crops is a viable alternative for coping with acid soils, liming has long been used to reduce soil acidity and increase soil productivity. When lime ($CaCO_3$) is added to wet soil, the following reactions occur (Hue and Ikawa 2006).

(1) Lime is dissolved slowly by moisture in the soil to produce Ca^{2+} and OH^-

$$CaCO_3 + H_2O \text{ (in soil)} \Rightarrow Ca^{2+} + 2OH^- + CO_2 \text{ (gas)}$$

(2) Newly produced Ca^{2+} will exchange with Al^{3+} and H^+ on the surface of acid soils

$$\begin{aligned} &2Ca^{2+} + \text{soil-Al} \Rightarrow \text{soil-Ca} + Al^{3+} \\ &+ \text{soil-H} - \text{soil-Ca} + H^+ \end{aligned}$$

(3) Lime-produced OH^- will react with Al^{3+} to form $Al(OH)_3$ solid and with H^+ to form water

$$\begin{aligned} &Al^{3+} + 3OH^- \Rightarrow Al(OH)_3 \text{ (solid)} \\ &H^+ + OH^- \Rightarrow H_2O \end{aligned}$$

Through interactions with OH^-, liming reduces harmful Al^{3+} and H_+. The most noticeable consequence of liming is the increase in soil pH caused by excess OH^-

from lime. Another advantage of liming is the availability of Ca^{2+} and Mg^{2+} if dolomite [Ca,Mg$(CO_3)_2$] is utilised.

Because soils range greatly in mineralogy, organic matter, and clay content, varied amounts of lime are required to elevate soil pH to a particular value. Thus, lime need curves for individual soils should be developed for use in lime estimate.

5.9.1 Lime Requirement

A large amount of effort has been devoted to finding the best methods for estimating lime needs in the tropics, with contrasting results. Part of the confusion was caused by attempts to lime soils to neutrality. Kamprath (1970) in North Carolina, and Reeve and Sumner (1970) in South Africa, working independently, suggested that lime recommendations should be based on neutralising exchangeable aluminium. This is calculated by multiplying the centimoles of charge per kilogram of exchangeable Al3+ in the topsoil by a factor of 1.5 (Kamprath 1970). The result is the amount of calcium (in cmolc/kg) needed to be applied as lime, which is equivalent to 1.65 t/ha of pure $CaCO_3$, or 100% $CaCO_3$ equivalent. Lime rates calculated by this method neutralise 85–90% of the exchangeable Al^{3+} in soils containing 2–7% soil organic carbon (SOC) (Kamprath 1970). The reason for 1.5 as a factor is the need to neutralise the Al3+ ions released by organic matter–Al^{3+} complexes or aluminium hydroxides as the pH increases (Juo and Kamprath 1979). This method has been used successfully in most tropical Latin American and Southeast Asian countries and the southern United States, as well as in the aluminium-toxic regions of central Africa (Yamoah et al. 1992; Rufyikiri and Hennebert 1995) and South Africa. This approach also eliminates the need for time-consuming neutralisation tests in the laboratory. The use of the exchangeable Al3+ method has reduced the rates of liming substantially, particularly in acid Oxisol and Ultisols low in ECEC (van Raij 1991). In Puerto Rico, high coffee yields are produced without lime in soils at pH 3.8, but liming is necessary in certain soils high in manganese to prevent manganese toxicity (Abruña et al. 1970). Much of the response to 0.2–0.5 t/ha of lime by aluminium-tolerant crops in the Colombian Llanos Oxisols may simply represent the response to calcium and magnesium fertilisation (Spain et al. 1975). Given the progress in genetic tolerance to aluminium toxicity, lime is no longer needed to neutralise all exchangeable Al3+, but to adjust it to the desired level of aluminium saturation. Therefore, the liming rate formulas are:

Cmol$_c$Ca/kg soil = 1:5 cmol$_c$Al/kg soil.
Tonnes $CaCO_3$ equivalent/ha = 1:65 × cmol$_c$Al/kg soil (Sanchez 2019).

The factor of 1.65 could be increased to 2 or 4 depending on local experience, which is usually associated with the reactive aluminium that is extracted by $CuCl_2$, or, as some others prefer, with the non-exchangeable hydrogen, extracted by buffered solutions. Approximate amount of lime required to raise soil pH and some common liming materials are given in Tables 5.7 and 5.8.

Table 5.7 Approximate amount of limestone needed to raise pH of a 18 cm surface soil layer

	Lime requirement	($t\ ha^{-1}$)
Soil texture	From pH 4.5 to 5.5	From pH 5.5 to 6.5
Sandy and loamy sand	1.2	1.5
Sandy loam	2.0	3.2
Loam	3.0	4.2
Silt loam	3.7	5.0
Clay loam	4.7	5.7
Muck	9.4	10.7

Table 5.8 Calcium carbonate equivalent (CCE) values of some important liming materials

Liming material	Calcium carbonate equivalent (%)	Liming material	Calcium carbonate equivalent (%)
Calcium oxide	179	Dolomie	109
Magnesium oxide	250	Calcium carbonate	100
Calcium hydroxide	136	Basic slag	86
Magnesium carbonate	119	Paper mill sludge	80

Source Das (2014)

5.9.2 Calcium Carbonate Equivalent of Lime Materials

The calcium carbonate equivalent of $CaCO_3$ (shell material), $CaCO_3$ (limestone), $Ca(OH)_2$, CaO (burnt lime), $CaCO_3$–$MgCO_3$ (dolomite), $CaCO_3$ (sugar beet lime), and $CaSiO_3$ (calcium silicate) is 95, 100, 120–135, 150–175, 110, 80–90 and 60–80, respectively. The sources of these lime materials are natural shell deposits, pure form, steam burnt, kiln buernt, natural deposits, sugar beet byproduct lime and slag, respectively. Table gives the CCE of some lime materials.

5.9.3 Overliming

Kamprath (1971) reviewed the reasons for negative lime responses when Oxisols and Ultisols are limed to neutrality. The consequences of such overliming are yield reduction and decreased availability of phosphorus, iron, zinc and boron. Overliming induces phosphorus deficiency in soils with high phosphorus-fixation capacity. When a Gibbsihumox in Hawaii was limed from pH 5.3–6.1, the uptake of phosphorus by sorghum increased dramatically (Fox et al. 1964). When this soil was limed to pH 7, the phosphorus uptake decreased and phosphorus deficiency was observed, apparently because of the formation of insoluble calcium phosphates. Overliming soils that are high in oxidic minerals greatly increases the adsorption of boron by clays and reduces the availability of boron. Kamprath (1971) cites several examples of such problems in the southeastern United States when the soil pH increases beyond 6.5 in Ultisols. Lime-induced manganese deficiencies can also be observed at pH levels above 6.2 because the availability of manganese decreases at this pH level. Overliming may also induce zinc deficiencies. The solubility of zinc decreases rapidly at a pH of 7 and above. In soils that are naturally low in zinc, excessive liming does decrease the availability of zinc when an Oxisol was limed to pH 7.0. The plots around it were limed to about pH 5.5. The bulk of the evidence suggests that variable-charge soils should not be limed to pH values greater than 6.0. Beyond that level, yield decreases can occur.

5.9.4 Residual Effects of Liming

The duration of the residual effects of liming depends on how fast exchangeable Ca^{2+} and Mg^{2+} are being depleted by crop uptake, or how fast acidity-generating processes (primary or secondary) occur. For example, liming usually increases the rate of SOM decomposition because of higher activity by soil biota, which produces H^+ ions. The length of the residual effects can be estimated by decreases in aluminium saturation, which will determine when additional lime should be applied, while the amount of exchangeable Al^{3+} indicates how much should be applied. The high lime application rates of 6 and 8 t/ha maintained maximum maize yields for 8 years. Shorter residual effects were shown with lower rates. Residual effect of liming and press mud application on crop yield in acid soils has been reported by Datta et al. (2001). The lower rates recommended for aluminium-tolerant cultivars would have a residual effect of 2–3 years Another study on residual effects for 6 years in a Typic Paleudult of Malaysia (pH 4.8, 67% aluminium saturation) led Sharifuddin et al. (1995) to conclude that with 2 t/ha of dolomitic lime, liming was not needed again until year 3, when aluminium saturation went back to the pre-lime level. With a 4 t/ha application, the residual effect lasted for four years before going back close to the original aluminium-saturation level (59%). An 8 t/ha application had increased the aluminium saturation from 0 to 23% after 4 years, but caused declines in sweet corn yields due to manganese deficiency, which was triggered by overliming. In an Oxic Dystrand in Costa Rica, Morelli et al. (1971) found residual effects for four years with liming rates of 3.6–7.3 t/ha. Similar durations were found in Oxisols and Ultisols of Puerto Rico (Abruña et al. 1964), and in a Hydrudand of Hawaii (Mahilum et al. 1970). Sharma and Sinha (1989) reported significant residual effect of lime application to rice on the subsequent chick pea (*Cicer arietinum*) crop productivity.

5.9.5 Benefits of Liming

- Decreases acidity of soil while simultaneously improving pH, base saturation, and CEC levels.
- Increases the availability of nutrients.
- Converts insoluble soil complexes of phosphorus and sulphur into forms that are more readily available to plants.
- Biological activity gets a boost.
- Improvement in fixation of N by legume crops.
- Improvement in soil physical structure.
- Reduction in Fe, Al and Mn toxicities.
- Improvement in effectiveness of certain herbicides.
- Reduction in fungal infections.
- Crop yields increase substantially on acid and Ca deficient soils.

According to Sharma and Singh (2002) and Kalidas-Singh (2013), phosphorus (P) management that is based on the rhizosphere may be advantageous in boosting the efficiency with which acidic soils use phosphorous. This involves limiting phosphorous fixation in the rhizosphere, boosting tissue phosphorous content for enhanced root development during the early stages of crop growth, and synchronising the rate of phosphorus mineralisation in the rhizosphere with the rate at which the plant absorbs phosphorus over the various growth phases of the plant. Increasing the population of phosphate solubilizing microorganisms (PSM) in the rhizosphere, delaying the release of phosphorus over time by combining PSM and rhizobial phosphatase (RP), and root dipping seedlings in an orthophosphate solution are all viable options for achieving these goals. Phosphate-solubilising bacteria (PSB) are microorganisms that thrive in soil and have the ability to convert bound forms of phosphates into monocalcium phosphate. According to Sekiya and Yano (2010), seed priming is an important strategy for reducing the amount of fertiliser that is required, improving crop establishment, and boosting crop yields. Seed priming involves the application of water and nutrient solutions such as zinc and phosphorus. Because nutrients or sources of nutrients are delivered directly to seeds, several undesirable interactions between the applied nutrient and the soil matrix, such as P fixation, can be avoided. These interactions can be avoided because nutrients or sources of nutrients are supplied directly to seeds.

5.9.6 The Influence of Applying Manure on the Environment

According to McDowell et al. (2001), routinely applying high rates of manure can result in the accumulation of nutrients in the soil, which can then surpass the crop's capability of absorbing those nutrients and its requirements for them. According to Ceretta et al. (2003), the application of high rates of organic waste together with continuous cultivation that does not involve soil tillage can result in long-term changes in the chemical properties of the soil that are connected to acidity. As a consequence of this, the response to the question of whether or not liming is required, and at what pace it should be done, can change in soils that have been treated with several applications of pig slurry. Cattle dung application can improve soil pH in acidic soils (Whalen et al. 2000). These researchers believe that the presence of $CaCO_3$ in the animal's diet is to blame. $CaCO_3$ is partially absorbed by the animal, while the remaining portion is lost through the animal's faeces.

In their study on the liming capacity of various crop wastes and biochars on an acid Ultisol, Yuan et al. (2011a, b) used incubation experiments to conduct their research. In terms of liming potential, rice hulls performed better than rice hull biochar, whereas the liming potential of soybean and pea straws performed better than the liming potential of their respective biochars. As a result of their increased alkalinity, biochars produced from legume materials had a much greater impact on the soil pH than biochars produced from sources other than legumes. The alkalinity of biochars was an essential factor that determined their capacity for liming, and research found that biochars with higher alkalinity led to significant reductions in the acidity of the soil. The addition of biochar improved soil fertility by decreasing the amount of soil exchangeable acidity while simultaneously raising the amount of soil exchangeable base cations and base saturation.

5.9.7 Challenges of Lime Use

- Lime use challenges for smallholders (Esilaba et al. 2018).
- Limited understanding and awareness of the significance of lime.
- A scarcity of funds to acquire lime and other inorganic inputs.
- For application per unit area, large volumes are necessary.
- Because lime is in bulk, few agro-dealers stock it.
- Transporting the bulk amount of lime necessary per unit area is costly.
- The powder form is difficult to apply since it is blown away by the wind.
- Lime application is time-consuming.
- A scarcity of enhanced crop types that are acid-tolerant.
- Inadequate organic material content.

5.9.8 Factors that Influence Soil Acidity

Several variables influence soil acidity, including:

- Organic matter decomposition, parent material, precipitation, native vegetation, crops cultivated, soil depth, N fertilisation, and flooding.
- At the farm level, a number of factors contribute to soil acidity.

- Hydrogen replaces basic nutrients (calcium, magnesium, and potassium) due to soil erosion, leaching, and crop loss.
- Use of acid-forming fertilisers; nitrification (the conversion of ammonium (NH^{4+}) nitrogen to nitrate (NO_3^-) nitrogen in the soil) produces considerable amounts of acid (H^+). Acid soils cover around 13% of Kenya's total land area.
- Soil acidity is one of the most yield-limiting variables in tropical crop cultivation.
- Soils with a pH of 5.0 have a high exchangeable aluminium content and are hazardous to most crops.
- Excessive exchangeable aluminium harms roots, reduces water and nutrient intake, and increases drought risk.

Methods for lowering soil acidity in such soils include:

- Lime application.
- Organic manure addition.

According to the findings, amending the soil with lime and organic manure before planting the first crop had a significant effect on the yield of that crop. On the other hand, the effects of lime and organic manure were more noticeable in the second and third crops after being applied to the soil. According to Nair et al. (2011) and Ryan et al. (2011), acidic soils have dangerously high concentrations of $A1^{3+}$, Fe^{3+}, and Mn^{2+}, reduced concentrations of P, and low availability of bases, all of which contribute to a decrease in crop output. The acidity of the soil causes significant problems for legumes. The acidity of the soil has a negative impact on the ability of nodule bacteria to live and thrive in the soil, which in turn slows down the nodulation process. According to Rossel and McBratney (2011), if appropriate mitigation strategies are not implemented, soil acidification may become more severe and negatively impact crop production. Materials used for liming include carbonates, oxides, hydroxides, as well as calcium and magnesium silicates. The two types of liming agents that are most commonly used are calcite ($CaCO_3$) and dolomite ($CaCO_3$–$MgCO_3$). The neutralisation of H^+ in soil solution by OH^- or HCO_3 produced from liming components is the first step in the liming reactions that take place. Dolomite, also known as dololime, is a mineral that is being put to widespread use in Bangladesh right now. The application of lime is the procedure that reduces the acidity of soil at the lowest cost. The pH profile of the soil, the quality of the lime, the soil type, the farming system, and the amount of rainfall all play a role in determining the amount of lime that is required. The solubility of nutrients is also improved, which makes it possible for plants to take in more nutrients. According to Jafer and Hailu (2017), lime has the added effect of improving soil quality by elevating its pH, P availability, CEC, and base saturation while simultaneously lowering Al concentrations. According to Kunhikrishnan et al. (2016), lime has the ability to boost the availability of calcium and magnesium in soils. The process of nodulation in legumes speeds up, which results in an increase in nitrogen fixation. Liming also helps to improve the pH of the soil and modifies soil attributes such as OM and plant nutrient availability, all of which contribute to the preservation of high yield (Halim et al. 2014). Increases in soil pH that are produced as a result of sufficient liming help facilitate the release of phosphate anions from aluminium and iron hydroxide surfaces (Barrow 2017). Because it enhances microbial activity, liming helps the mineralisation of agricultural wastes and organic manure in the soil (Aye et al. 2016). This, in turn, can improve the amount of plant nutrients that are easily available in the soil, notably phosphorus. On the other hand, liming has the potential to lower the amount of phosphorus that is available in the soil because it increases the amount of phosphorus that precipitates out as calcium phosphate at higher pH (Penn and Camberato 2019). In field experiments that Rahman et al. (2005) carried out using a wheat-rice cropping pattern, they found that applying 2.4 t of lime per hectare was sufficient to boost crop yields. Lime and organic matter need to be given to acidic soils in order to optimise crop development and the effectiveness of fertiliser use, as well as to avoid the detrimental effects of iron, aluminium, manganese, and manganese. Lime and organic matter need to be added to the soil in order to achieve a level of soil pH that allows for the presence of non-toxic forms of iron, aluminium, and manganese. Because it increases the soil's potential to act as a buffer, the consistent delivery of organic matter that has been well decomposed in acidic soils is an effective method for minimising rapid swings in pH. In addition to this, it improves the availability of phosphorus in acidic soils while simultaneously lowering the toxicity of iron and aluminium. Compost, poultry manure, cow dung, and lime are all viable options for amending soil in order to increase crop yield, maintain soil fertility, and neutralise excess acidity. It is of the utmost importance to calculate the exact quantity of manure that must be applied to acidic soils in order to achieve the desired changes in pH, fertility, and production. When developing crops on acidic soils over the long term, it is recommended that lime be used in conjunction with both organic and chemical fertilisers. This is viewed as the most effective technique. Soil organic matter (SOM) is an essential component in guaranteeing the continued fertility of soil over the long term because it directs biological processes that have an impact on the availability of nutrients. Organic additions like CD and PM help to maintain soil fertility by acting as alternative plant nutrient sources to commercial fertilisers (Siavoshi et al. 2011). This is especially true in rice production where organic additions like CD and PM are utilised. According to Swain et al. (2011), both CD and PM improve the soil's physical, chemical, and biological features which lead to an increase in the amount of nutrients that are available.

5.9.9 Low Productivity

Low crop yields on acidic soils can be attributed to a number of factors, including the toxicity of aluminium, manganese, and hydrogen, as well as deficiencies in nitrogen, phosphorus, potassium, calcium, magnesium, zinc, copper, and boron. The majority of soils in Brazil have an acidic pH, and liming is the procedure that has proven to be the most effective and popular way to neutralise acidity in soil. According to Fageria and Santana (1998), liming promotes crop productivity by increasing pH, Ca, and Mg levels, as well as base saturation. The perfect pH for upland rice was 5.6, whereas the ideal pH for common bean was 6.2, the ideal pH for maize was 6.4, and the ideal pH for soybean was 6.8. Similarly, the optimal degree of base saturation for upland rice was 40%, while the optimal level for common bean was 53%, maize was 59%, and soybeans were 64%. Because of this, upland rice proved to be the most resistant to the acidity of the soil, whereas soybean proved to be the most sensitive. According to Fageria and Stone (1999), the optimal pH range for the majority of annual crops that can be grown in Brazilian Oxisols is between 5.5 and 6.7. Likewise, the optimum values for base saturation for the majority of annual crops grown in Brazilian Oxisols varied from 50 to 60% (Raij and Quaggio 1997).

Lime treatment and integrated nutrient management are two practices that are regularly recommended for acidic soils in order to increase crop output. In Meghalaya, Northeast India, a field experiment was carried out on an acid Alfisol (pH 4.6) in order to investigate the individual and synergistic impacts of lime, NPK, and farmyard manure (FYM) treatment on maize productivity (Manoj Kumar et al. 2012). This research was published in the journal Manoj Kumar et al. When compared to the control, the yield was improved by 53% when the recommended dose of NPK was applied, and it was enhanced by 32% when liming was applied at a rate of 300 kg/ha (furrow application). When compared to the control, yield was improved by 147% through the application of NPK and lime together, while yield was increased by 291% by the application of FYM at 5 t/ha in addition to NPK and lime. According to the results of this research, liming, when combined with other integrated nutrient management practices, can increase maize yield by more than three times on acidic soils in Meghalaya and other northeastern Indian states that have soils that are comparable to Meghalaya's.

According to von Uexküll and Mutert (1995), soil acidity affects almost half of the potentially arable land around the globe, with the largest impact being felt in the wet tropics. According to Mandal (1997), soil acidity affects around one-third of India's agricultural land. According to Sharma and Singh (2002), the majority of these soils can be found in India's northeastern area, which also has approximately 65% of the country's territory that has extremely acidic soil (pH of less than 5.5). These problems were first identified by Patiram (1991). According to Oh and Richter (2004), it is anticipated that the levels of soil acidity, as well as the consequences on soil fertility and agricultural productivity, would continue to rise as the climate continues to change. The control of soil acidity and the enhancement of crop yield on such soils are of the utmost importance for enhancing both global and regional levels of food security. In order to improve the phytoavailability of key nutrients and to alleviate other acidity-induced fertility constraints on such soils, lime treatment is widely proposed in conjunction with integrated nutrient management (Haynes 1984; Patiram 1991).

5.9.10 Particle Size

The standard sieve mesh is used to evaluate the particle size, which is the most essential component in determining the rate of reactivity of lime after it has been applied to soil. The finer the grind, the more surface area the Aglime has available to react with the acidity of the soil. The size of a sieve's mesh is determined by the number of openings that are packed into one linear inch. A sieve with a mesh size of 20 has 20 openings per linear inch, which is equivalent to 400 openings per square inch. On the other hand, a sieve with a mesh size of 100 has 100 openings per linear inch, which is equivalent to 10,000 openings per square inch. The particles of crushed limestone that are able to pass through a sieve with a mesh size of 100 are smaller, have a bigger surface area, and as a result react with the acidity of the soil more quickly than material with a mesh size of 20.

It is necessary for a certain percentage of the particles in liming material to be tiny enough to pass through a 100-mesh filter in order to react swiftly with the acidity of the soil and give the nutritive elements calcium and magnesium. A further percentage (those that pass a sieve with a mesh size of 60–100) should react within one to two years, while the remainder (those that pass a sieve with a mesh size of 20) should react within two to three years. It is possible for a high-quality agricultural limestone to react with the acidity of the soil both quickly (within two weeks) and over the long term (within three years) since it has a range of particle sizes. According to the findings of research, limestone that has been pulverised to a finer consistency (at least 100 mesh) responds rapidly to neutralise the acidity of the soil. However, due to the fact that limestone with a mesh size of 10 to 20 dissolves slowly, it is only moderately effective at increasing the soil pH (reducing the acidity of the soil). Larger particles with a mesh size of 8 to 20 mesh or less were inefficient at reducing the acidity of the soil, while smaller particles with a mesh size of 20 to 30 mesh or above were effective. The particle size of the lime has a major influence on the rate at which the lime degrades under field settings. The fact that large particles (four- to eight-mesh) dissolve very little over a period of one to four years, exemplifies the ineffectiveness of course lime materials for the purpose of reducing the acidity of the soil. It is anticipated that the smaller particles, with a mesh size of 80–100, will completely react in one to four years. Particles of a size greater than 20 mesh are of limited use as an aglime material due to the slow reaction rate they exhibit. The rate at which limestone reacts depends on the pH of the soil, the amount of lime that is mixed in with the soil, and the particle size of the lime. The soil will become more acidic (have a lower pH) in proportion to how quickly the lime reaction occurs. In addition, the pace of the lime reaction is increased when the limestone is coupled with the soil in a more effective manner. The solubility of the liming component is another factor that contributes to the pace of the reaction. When compared to ground Aglime, burnt and hydrated lime has a higher solubility in water, which enables them to neutralise the acidity of soil more quickly. Despite the fact that these lime products are more difficult to handle and distribute than ground Aglime, the quick-acting qualities that they possess can be beneficial in some circumstances.

5.9.11 An Effective Value for Balancing

The effective neutralising value, often known as ENV, is a measurement of the effectiveness of aglime that takes into consideration the combined impacts of the particle size or fineness of grind, as well as the chemical composition. Liming products sold in Virginia are required to bear an ENV label with the relevant values. Calculating the ENV for a specific liming material involves multiplying the CCE by a number of factors that are determined by the degree of fineness of grind of the limestone. These factors are as follows:

$$ENV = ((A + B + C) \times CCE)/100$$

where CCE = calcium carbonate equivalent (%)

A is equal to (the percentage by weight that passes a 20-mesh sieve minus the percentage that passes a 60-mesh sieve) × 0.4, B is equal to (the percentage by weight that passes a 60-mesh sieve minus the percentage that passes a 100-mesh sieve) × 0.8, and C is equal to (the percentage by weight that passes a 100-mesh sieve) × 1.0.

If the same amount of ENV is applied to each acre of land using various liming agents, there should be no difference in the effect on crop output.

5.9.12 Liming's Material Choices for the Process

In many cases, a substantial amount of work is put into determining the appropriate quantity of lime that should be applied to a field based on the results of soil testing; yet, the task of selecting and applying lime is frequently performed inadequately. The majority of the ground agricultural limestone that is used for liming in Virginia is known as Aglime. This has been the case for a number of years. However, there is a wide selection of merchandise available that may be utilised to raise the pH of the soil. When choosing a material for liming, it is important to take into account the following factors: the availability of lime sources in the region; the amount of time that will pass between the application of lime and the planting of crops; the degree to which the soil is acidic; the requirement for magnesium; the value and intensity of the crop.

When it comes to liming, making the decision between using a slow-acting or quick-acting medium can be challenging. This topic can be partially answered by considering the amount of time that passes between the application of lime and the planting of the crop. In the event that lime treatment is performed three to six months prior to planting, more coarsely ground limestone might be utilised. The most effective application of alim is one like this. However, the fine particles in Aglime (the percentage that passes through 100-mesh) react quickly, and yield improvements can be observed even if lime is sprayed at the time that planting is taking place.

5.9.13 Utilisation of Lime as an Asset

Because it moves slowly through the soil, lime is most effective when it is mixed to the depth of the tillage. In addition, as the pH of the soil rises, the rate at which lime particles dissolve slows down. A good liming regimen will benefit from Aglime being applied consistently and integrated completely as much as possible. Incorporating the proposed rate of Aglime to tillage depth during the final primary tillage operation prior to the conversion to no-till would be an example of an appropriate management strategy for transitioning from conventional tillage to no-till. This would take place before the transition.

When it is possible, lime ought to be mixed to the depth of the tilling. A single application of lime, either before or after tilling, typically produces adequate results when applied to soils with a pH ranging from 5.2 to 5.7. When dealing with extremely acidic soils (pH 5.0 or below) that have a high demand for lime, it is possible that it is more beneficial to apply half of the lime before tillage and the other half after. A disc harrow is the tool of choice for incorporating lime into the soil. This method has a better chance of succeeding with high-value crops or crops that are particularly susceptible to the effects of acidic soil conditions in areas with a smaller geographic footprint. If you have a large area that needs a lot of lime, such as three to four tonnes per acre, it is most likely a good idea to apply only half of the required lime in the first year and the other half in the second year. This will allow you to spread the lime out over a longer period of time. The majority of the benefits will be realised in the first year of crop development, while the expense of liming can be spread out over the course of two years.

It is equally as important to apply lime in a uniform manner as it is to select the appropriate kind and quantity of lime to use in any specific circumstance. The application of lime incorrectly can lead to a loss of income as well as other issues, such as specific areas of the field being under-limed while other areas are over-limed. Even when soils have been limed to neutralise the acidity that has built up through time, it is possible to lime them too much. When applications overlap one another to an extreme degree during the spreading operation or when there is unequal dispersion for any other cause, this usually occurs. The pattern of distribution that is created by bulk lime spreaders is determined by a number of parameters, including the rate of application per acre, the driving speed, the consistency of the speed, and the even distribution of lime by the spreader.

5.9.14 Timing

When a lime requirement has been established, you should lime as soon as possible for best results. The application of aglime can take place at any point in time during the interval between harvesting one crop and planting the next. In most cases, lime is first scattered on the top layer of soil, and then it is worked into the ground during the tilling process. Surface applications can be done in conservation tillage systems, pastures, and hay fields whenever the soil conditions are suitable for spreaders to penetrate the fields. Surface applications can also be done in fields of hay. Lime can be purchased at any time of the year.

In the beginning of spring, lime can be sprinkled on any and all crops. Applications made in the spring are highly recommended for autumn harvests. This is because there will be sufficient time for considerable pH adjustment of the soil. However, because to the rapid nature with which small particles react with soil acidity, crops that are sown in the spring will also benefit from lime treatments made in the spring. Because of the delay in the limestone response, it is imperative that applications of limestone never be postponed. Aglime that is up to standard in Virginia will immediately react with the acidity of the soil, which will result in increased crop yields. The summer is an ideal time to lime pastures, particularly those that are going to be fertilised and repaired in the autumn. Applications of Aglime in the fall are advantageous for spring-planted crops such as maize, cotton and soybeans. This is because the lime has more time to reach the acidity in the soil and neutralise it when it is applied in the fall. In forty-five to sixty days, an Aglime treatment of sufficient quality can adequately restore the pH of the soil.

5.9.15 Significance of Liming Acid Soils

The most notable advantage of liming acid soils is that it reduces the solubility of potentially harmful components such as hydrogen, aluminium (Al^{3+}), and manganese (Mn^{2+}). The quantity of hydrogen ions increases as pH falls below 7.0, although hydrogen only becomes hazardous to plants in severely acidic soils (pH 4.0) with very low calcium levels. When the pH falls below 5.5, the concentration of soluble aluminium rises, becoming hazardous to plant root growth when it hits 1.0 part per million (ppm). Manganese concentrations increase below pH 5.2 and can become hazardous. Greatest crops absorb the greatest nutrients when the soil pH is near 7.0. As soil pH drops, the availability of fertiliser nutrients such as nitrogen, phosphorous, and potassium diminishes. Phosphorus is extremely pH sensitive and can become a limiting nutrient in highly acidic soils. When soil acidity is not appropriately regulated, crop performance and fertiliser use efficiency suffer. The species, concentrations, and activity of soil microorganisms are also affected by soil pH. When the pH falls below 5.5, the population of soil microorganisms alters and decreases due to aluminium and manganese toxicity and decreased nutritional availability. Because fungi withstand acidity better than bacteria, they become more common in acidic soils.

5.9.16 Equivalence of Calcium Carbonate

A limestone's ability to neutralise soil acidity is determined by its calcium carbonate equivalency (CCE), which is stated as a percentage. The benchmark is pure calcium carbonate or calcitic limestone, which has a CCE of 100%. This standard is used to compare all other liming materials. The provider should be able to provide the CCE of commercial limestone goods. Some governments and/or vendors use the term neutralising value (NV) instead of CCE. The majority of agricultural limestones utilised in crop production systems are ground calcium carbonate. Some limestones contain impurities that reduce the neutralising value. Other materials are byproducts or byproducts of manufacturing processes, such as oxides and hydroxides. Some materials with strong neutralising values, such as hydrated lime, are difficult and unpleasant to handle in agricultural production systems.

5.10 The Effect of Applied Fertilisers

Farmers are advised to add alkaline materials such as lime (mainly calcium carbonate) to increase soil pH and thus decrease Al toxicity, as well as P fertiliser to increase accessible P in soil, in order to achieve a higher crop yield on acid soils. The most feasible management strategy for most acid soils is to add lime to maintain existing soil pH status or improve surface soil pH; consider liming after the pH goes below pHCa 5.0 for a better chance of successfully growing acid-sensitive species. The application of lime to correct soil acidity is the foundation of an effective soil fertility programme. Lime does more than only balance the pH of the soil. If dolomitic lime is applied, it also provides critical plant nutrients, Ca and Mg. Acidic soil can be turned to neutral soil by adding lime to natural soil. Lime's alkaline nature neutralises the acidity of the soil. Quick lime, commonly known as calcium oxide, can be used to neutralise acidic soil. Because calcium oxide is basic, it neutralises the soil.

5.10.1 Fertiliser Overapplication

The current intensification of agricultural productivity has been made possible in large part by the use of synthetic fertilisers on soils. Historically, the application of N fertilisers has been estimated to account for 30–50% of the increase in crop output (Stewart et al. 2005). N is the most significant fertiliser applied to soils, followed by P and K. Worldwide N fertiliser production is now 123 Tg N per year, a 9.5-fold increase from production in 1961 (FAO 2018). While the use of N fertiliser feeds a large portion of the human population, its efficiency has deteriorated significantly as the size of its use has expanded. When world grain output is compared to global N fertiliser production, the most efficient yield gains are gained when N application rates are lowest, with efficiency declining as N fertiliser application rates increase. Similarly, N consumption efficiency has declined from 68% in 1961 to 47% in 2010 (Luis et al. 2014), implying that less than half of the N fertiliser produced is actually used by plants, with the remainder lost to the environment.

5.10.2 Acidification

There are a number of human-made processes that can cause soils to become more acidic. These processes include the overapplication of nitrogen fertilisers (which was discussed before), the removal of crop products, acid rain, and exposure to potentially acidic sulphate soils. The development of extremely acidic soils, in every circumstance, results in reduced plant growth and crop yield. The acid soil infertility complex is responsible for this stunting of plant growth, which was generated by a combination of the mechanisms described in the previous sentence. The ever-increasing concentration of severely phytotoxic Al is the most important factor among these growth-limiting variables. Al significantly inhibits the growth of plant roots, making Al the most vital of these growth-limiting variables. Similar to other forms of deterioration, the acidification of soil typically occurs over time. According to predictions made by FAO and ITPS (2015), up to thirty percent of the ice-free topsoils around the world are acidic, while up to seventy-five percent of the subsoils are acidic.

Since the 1980s, China's soils have experienced an average pH decrease of between 0.13 and 0.80 units, with the majority of this acidification being linked to the use of excessive amounts of nitrogen fertilisers (Guo et al. 2010). Lime is one method that can be used to reduce the acidity of topsoil; however, this method is not always practicable from a financial standpoint. It is significantly more difficult to neutralise the acidity in the subsurface since lime does not move very easily through the soil profile and there are challenges involved in installing it directly in the subsoil. For instance, it is estimated that only liming 4% of the total acidic land in Australia is economically viable (NLWRA 2002). This is the case in many other countries as well.

5.10.3 Problems

The long-recommended use of synthetic fertilisers, pesticides, and other chemicals to address concerns about agricultural output has resulted in poor soil health as well as resistance among insects, diseases, and weeds. Fertiliser with a higher soluble nitrogen content makes plants more susceptible to disease and pest infestation, and it also makes the weed problem worse. As a consequence of this, both the soil and the plants become dependent on these inputs. These methods of production cannot be maintained indefinitely, and we are already experiencing the fallout from the oversimplification of fertilisation and so-called plant protection practices. The use of soluble acidic-based NPK fertilisers has become an addiction for agricultural systems, and this addiction, when combined with the required use of pesticides and herbicides, leads to the loss of soil quality. According to Pettit (2006), humic compounds, which are necessary for maintaining soil fertility and providing adequate nutrition to plants, have been steadily depleted. Since farming began, there has been a significant decrease in the amount of organic carbon in the soil, which includes humus, which is the interaction between the living and non-living components of the soil. In annual cropping, farming methods are shifting away from heavy cultivation and towards minimum tillage and no-till systems, which are not only profitable but also beneficial to the environment. This is being done to improve the quality of the soil. These so-called sustainable systems, on the other hand, are empirical in the sense that they are built without a firm understanding of the long-term repercussions of their decisions. The prevalent practices continue to make use of harmful chemicals while paying no attention to the delicate equilibrium of the humus, bacteria, trace minerals, and nutrients in the soil. This type of management has led to significant losses of soil organic carbon (including humus), as well as a significant reduction in the diversity and number of microbes (algae, bacteria, fungi, nematodes, and protozoa) and larger creatures (such as mites, ants, beetles, and worms) that are a part of the soil foodweb (Ingham 2006). This makes the roots more sensitive to saline and acid conditions, makes the entire plant more susceptible to pests and diseases, and makes it necessary to provide the plant with supplemental fertiliser and chemical protection (Anderson 2000). This also reduces the capacity of the soil to feed the plants, which reduces the ability of the soil to feed the plants. When the biological and chemical processes of the soil are messed with, the effect is often a deterioration of the soil's physical properties, such as decreased infiltration, increased compaction, and increased erosion. Traditional agriculture is increasingly exploring for strategies to enhance the amount of organic matter and microbial biomass in the soil (Bell 2005; Fisher 2005; Kirkby et al. 2006). This is a reaction to the challenges that traditional agriculture faces. According to FAO (2006), enhanced biodiversity in paddocks and catchments is necessary for maintaining healthy soil. This is because higher biodiversity helps to boost natural predation within a sustainable soil foodweb. This can be accomplished by increasing the amount of organic carbon in the soil by twofold (organic carbon is the basis for a living soil), decreasing the amount of chemicals used, and erecting shelterbelts to improve the microclimate of the soil surface and offer a "home" for an important component of the soil foodweb.

References

Aarons SR, Graham PH (1991) Response of Rhizobium/eguminosarum bv phaseoli to acidity. Plant Soil 134:145–151

Abruña F, Vicente-Chandler J, Pearson RW (1970) Crop response to soil acidity factors in Ultisols and Oxisols. Soil Sci Soc Am Proc 34:629–635

Achalu C (2012) Response of barley to liming of acid soils collected from different land use systems of Western Oromia, Ethiopia. J Biodivers Environ Sci 2:1–13. http://www.innspub.net

Adams F (1984) Crop response to lime in the southern United States. In: Adams F (ed) Soil acidity and liming, 2nd edn. American Society of Agronomy, Madison, Wisconsin, USA, pp 211–265
Afienza RN (1989) Management and utilisation of acid soils in the Philippines. In: Craswell ET, Pushparajah E (eds) Management of acid soils in the humid tropics of Asia. ACIAR Monograph No.13 (IBSRAM Monograph No.1), x + p 118
Agegnehu G, Yirga C, Erkossa T (2019) Soil acidity management. Ethiopian Institute of Agricultural Research (EIAR), Addis Ababa, Ethiopia. **URI:** http://hdl.handle.net/123456789/3151
Anderson AB (2000) Science in agriculture. ACRES, Austin
Anderson NP, Hart JM, Sullivan DM, Horneck DA, Pirelli GJ, Christensen NW (2013) Applying lime to raise soil pH for crop production (Western Oregon)
Asai H, Samson BK, Stephan HM, Songyikhangsuthor K, Homma K, Kiyono Y, Inoue Y, Shiraiwa T, Horie T (2009) Biochar amendment techniques for upland rice production in Northern Laos: 1. Soil physical properties, leaf SPAD and grain yield. Field Crop Res 111:81–84
Austrheim G, Evju M, Mysterud A (2005) Herb abundance and life-history traits in two contrasting alpine habitats in southern Norway. Plant Ecol 179:217–229
van Averbeke W, Yoganathan S (2003) Using kraal manure as a fertilizer. Department of Agriculture and the Agricultural and Rural Development Research Institute, Fort Hare, Alice, South Africa
Ayalew A (2011) The influence of applying lime and NPK fertilizers on yield of maize and soil properties on acid soil of Areka, southern region of Ethiopia. Innov Syst Des Eng 2:33–42
Aye NS, Sale PWG, Tang CX (2016) The impact of long-term liming on soil organic carbon and aggregate stability in low-input acid soils. Biol Fert Soils 52:697–709
Banik P, Ghosal PK, Sasmal TK, Bhattacharya S, Sarkar BK, Bagchi DK (2006) Effect of organic and inorganic nutrients for soil quality conservation and yield of rainfed low land rice in sub-tropical plateau region. J Agron Crop Sci 192:331–343
Barber SA (1984) Liming materials and practices. In: Adams F (ed) Soil acidity and liming, 2nd edn. American Society of Agronomy, Madison, Wisconsin, USA, pp 171–209
Barnes JS, Kamprath EJ (1975) Availability of North Carolina rock phosphate applied to soils. Technical Bulletin 229, North Carolina Agricultural Experiment Station, Raleigh, North Carolina, USA
Barrow NJ (2017) The effects of pH on phosphate uptake from the soil. Plant Soil 410:401–410
Bartlett RJ, Riego DC (1972) Effect of chelation on the toxicity of aluminum. Plant Soil 37:419–423
Bekele T (2008) Estimation of lime requirement. Training manual for regional soil testing laboratory heads and technicians. National soil Testing Center, Ministry of Agriculture and Rural Development
Bekker AW, Yapa LGG (1994) Improving Pacific acid soils using coralline lime. Agro-facts, Soil Science, Institute for Research, Extension and Training in Agriculture. Publications (edited by Emil Adams), Agdex#530
Bell MJ (2005) Chemical fertility and soil health in northern systems. GRDC Ground Cover 56. http://www.grdc.com.au/growers/gc/gc56/supplement/chemfert.htm
Benke MB, Hao XO, Donovan JT, Clayton GW, Lupwayi NZ, Caffyn P, Hall M (2009) Livestock manure improves acid soil productivity under a cold northern Alberta climate. Can J Soil Sci 90:685–697
Bentley JW (1989) What farmers don't how can't help them: the strengths and weaknesses of indigenous technical knowledge in Honduras. Agric Hum Values 6:25–32
Bentley JW (1990) Conocimiento y Experimentos Espontheos de Campesinos Hondwefios sobre el maiz muen9. Manejo Integrado De Plagas (Costa Rica) 17:1626
Berek AK (2014) Exploring the potential roles of biochars on land degradation mitigation. J Degrad Min Lands Manage 1:149–158
Bessho T, Bell CL (1992) Soil solid and solution phase changes and mung bean response during amelioration of aluminium toxicity with organic matter. Plant Soil 140:183–196
Bhat JA, Mandal Biswapati, Hazra GC (2007) Basic slag as a liming material to ameliorate soil acidity in Alfisols of sub-tropical India. Am Eurasian J Agric Environ Sci 2:321–327

Bhat JA, Kundu MC, Hazra GC, Santra GH, Mandal B (2010) Rehabilitating acid soils for increasing crop productivity through low-cost liming material. Sci Total Environ 408:4346–4353

Blamey FPC, Edwards DG (1989) Limitations to food crop production in tropical acid soils. In: Jvan DE (ed) Nutrient management for food crop production in tropical farming systems, pp 73–94

Blamey FPC, Edwards DG, Asher CJ (1983) Effects of aluminium, OH: Al and P: Al molar ratios, and ionic strength on soybean root elongation in solution culture. Soil Sci 136:197–207

Bodruzzaman M (2010) Lime requirement of acid soils for sustainable crop production. PhD thesis, Dept. of Soil Science, BAU, Mymensingh

Bolan NS, Hedley MJ, White RE (1991) Nitrogen cycling and soil acidification. Plant Soil 134:53–63

Bruce RC, Warrell LA, Bell LC, Edwards DG (1989a) Chemical attributes of some Queensland acid soils. 1. Solid and solution phase compositions. Aust J Soil Res 21

Buni A (2014) Effects of liming acidic soils on improving soil properties and yield of haricot bean. J Environ Anal Toxicol 5:1–4

Buol SW, Hol FD, McCracken El (1980) Soil genesis and classification, 2nd edn. Ames, The Iowa State University Press, p 404

Butterly C, Tang C (2018) Evaluating rates of organic amendments with lime for treating acid soils. GRDC. Department of Primary Industries, NSW

Ceretta CA, Durigon R, Basso CJ, Barcellos LAR, Vieira FCB (2003) Características químicas de solo sob aplicação de esterco líquido de suínos em pastagem natural. Pesq Agropec Bras 38:729–735

Chairiyah RR, Ramija KE, Batubara SF (2021) Liming of acid soil and the interaction with soil pH and corn productivity. Earth Environ Sci 807:042071 (IOP Publishing). https://doi.org/10.1088/1755-1315/807/4/042071

Chan KY, Xu Z (2009) Biochar: nutrient properties and their enhancement. Biochar Environ Manage Sci Technol 1:67–84

Chan KY, Van ZL, Meszaros I, Downie A, Joseph S (2008) Using poultry litter biochars as soil amendments. Aust J Soil Res 46:437–444

Changprai C (1973) Soils of Thailand. Proceedings of the fourth international forum on soil taxonomy and agrotechnology transfer. Department of Land Development, Bangkok, pp 29–50

Chew QH, Wei KC, Vasoo S, Chau HC, Sim K (2020) Narrative synthesis of psychological and coping responses towards emerging infectious disease outbreaks in the general population: practical considerations for the COVID-19 pandemic. Singap Med J 61(7):350–356. https://doi.org/10.11622/smedj.2020046

Clare A, Shackley S, Joseph S, Hammond J, Pan G, Bloom A (2015) Competing uses for China's straw: the economic and carbon abatement potential of biochar. GCB Bioenergy 7:1272–1282. https://doi.org/10.1111/gcbb.12220

Cochrane TT (1991) Understanding and managing acid soils oftropical South America. In: Deturck P, Ponnamperurna FN (eds) Rice production on acid soils of the tropics. Institute of Fundamental Studies, Kandy, Sri Lanka, pp 113–122

Cochrane TT, Salinas JG, Sanchez PA (1980) An equation for liming acid mineral soils to compensate soil aluminum tolerance. Trop Agric (Tein) 57:133–140

Collard F-X, Blin J (2014) A review on pyrolysis of biomass constituents: mechanisms and composition of the products obtained from the conversion of cellulose, hemicelluloses and lignin. Renew Sustain Energy Rev 38:594–608

Collings GH (1955) Commercial fertilizers: their sources and use, fifth edition. Tata McGrow-Hill Publishing Company Ltd., Bombay, New Delhi, p 617

Conklin HC (1957) Hanunb agriculture. FAO, Rome, Italy

Conyers M (1990) The control of aluminium solubility in some acidic Australian soils. J Soil Sci 41:147–156

Conyers MK (2006) Liming and lime materials. In: Lal R (ed) Encyclopedia of soil science. Taylor and Francis, Abingdon, UK, pp 1034–1036

Cooper JM, Warman PR (1997) Effects of three fertility amendments on soil dehydrogenase activity, organic C and pH. Can J Soil Sci 77:281–283

Costa A, Rosolem CA (2007) Liming in the transition to no-till under a wheat–soybean rotation. Soil Tillage Res 97:207–217

Coventry DR (1985) Changes in agricultural systems on acid soils in Southern Australia. In: Proceedings third Australian agronomy conference, Hobart, pp 126–145

Coventry DR, Evans J (1989) Symbiotic nitrogen fixation and soil acidity. In: Robson AD (ed) Soil acidity and plant growth. Academic Press, Sydney, Australia, pp 103–137

Craswell ET, Pushparajah E (1989) Management of acid soils in the humid tropics of Asia. ACIAR Monograph No.13 (IBSRAM Monograph No.1), x + p 118

Cregan PD, Hirth JR, Conyers MK (1989) Amelioration of soil acidity by liming and other amendments. In: Robson AD (ed) Soil acidity and plant growth. Academic Press, Sydney, Australia, pp 205–264

Das DK (2014) Soil acidity. In: Introductory soil science, 3rd edn. Kalyani publishers, Ludhiana, India, pp 218–244

Das SK, Avasthe RK (2018) Managing soil acidity for crop production in Sikkim: a policy options. Indian J Hill Farming 31:155–162

Das SK, Avasthe RK (2016) Soil acidity management for higher yields in organic farming. In: Avasthe RK, Pradhan Y, Bhutia K (eds) Transforming rural Sikkim: year round organic crop production. Sikkim Organic Mission, Govt. of Sikkim and ICAR Research Complex, Sikkim Centre, Tadong, Gangtok, Sikkim, pp 219–222

Das SK, Avasthe RK, Reza SK (2014) Management of soil acidity. In: Avasthe RK, Pradhan Y, Bhutia K (eds) Handbook on organic crop production in Sikkim. Sikkim Organic Mission, Govt. of Sikkim and ICAR Research Complex, Sikkim Centre, Tadong, Gangtok, Sikkim, pp 307–310

Das SK, Ghosh GK, Avasthe RK (2021a) Conversion of crop, weed and tree biomass into biochar for heavy metal removal and wastewater treatment. Biomass Conv Bioref. https://doi.org/10.1007/s13399-021-01334-y

Das SK, Ghosh GK, Mishra VK, Choudhury BU et al (2021bc) Utilizing dissimilar feedstocks derived biochar amendments to alter soil biological indicators in acidic soil of Northeast India. Biomass Convers Biorefnery. https://doi.org/10.1007/s13399-021-01670-

Datta M, Gupta RK (1983a) Utilization of pressmud as amendment of acid soils in Nagaland. J Indian Soc Soil Sci 31:511–516

Datta M, Gupta RK (1983b) Response of leguminous crops to pressmud and lime in Nagaland. J Indian Soc Soil Sci 31:535–537

Datta M, Gupta RK (1983c) Utilization of pressmud as amendment of acid soils in Nagaland. J Ind Soc Soil Sci 31:511–516

Datta M, Gupta RK (1984) Studies on P., Ca, Zn and Fe contents in soil and plant tissues as influenced by pressmud and lime. J Indian Soc Soil Sci 32:450–458

Datta M, Gupta RK (1985) Effect of pressmud on manganese contents in soil and plant. J Indian Soc Soil Sci 33:715–717

Datta M, Sharma UC (2006) Soil acidity and amelioration. In: Sharma UC, Datta M, Samra JS (eds) Soils and their management in North East India. ICAR research Complex for NEH Region, Umroi, Meghalaya, India, pp 220–241

Datta M, Gupta RK, Prasad RN (1983) Response of wheat and maize to lime in acid soils of Nagaland. J Ind Soc Soil Sci 31:236–240

Datta M, Bhattacharya BK, Saikh H (2001) Soil fertility—a case study of shifting cultivation sites in Tripura. J Indian Soc Soil Sci 49(1):104–109

Debnath NC, Basak RK (1986) Evaluation of fertilizer value of some indigenous rock phosphate and basic slag of India. Fert Res 8:241–248

Dessalegn T, Bekele A, Getahun D, Bogale W (2018) Response of sorghum and soybean to split application of lime. EIAR, Addis ababa ISBN: 978-99944-66-52-8, p138

Dutta D, Biswapati M, Mandal LN (1989) Decrease in availability of zinc and copper in acidic to near neutral soil on submergence. Soil Sci 147:187–195

Dvorak KA (1988) Indigenous soil classification in semi-arid tropical India. Resource Management Program, Economics Group, Rognss Report 84. ICRISAT

Easterwood GW, Sartain JB, Street JJ (1989) Fertilizer effectiveness of three carbonate apatites on an acid ultisol. Comm Soil Sci Plant Anal 20:789–800

Edmeades DC, Ridley AM (2003) Using lime to ameliorate topsoil and subsoil acidity. In: Rengel Z (ed) Handbook of soil acidity. Marcel Dekker Inc., New York, Basel, pp 297–336

Elad Y, Cytryn E, Meller Harel Y, Lew B, Graber E (2011) The biochar effect: plant resistance to biotic stresses. Phytopathol Mediterr 50:335–349. https://doi.org/10.14601/Phytopathol_Mediterr-9807

Ellington A (1984) Soil acidity—effects and remedies: a review. Department of Agriculture, Victoria. Technical Report Series No. 102

Eswaran H, Reich P, Beinroth F (1997b) Global distribution of soils with acidity. In: Moniz AC (ed) Plant-soil interactions at low pH. Brazilian Soil Science Society, Sao Paulo, Brazil, pp 159–164

Fageria NK, Santana EVP (1998) Fertility problems and management of highly depleted Brazilian soils for optimization of crop production. Paper presented at the first international agronomy congress: agronomy, environment and food security for 21st century, New Delhi, India, 23–27 November 1998

Fageria NK, Stone LF (1999) Acidity management of cerrado and varzea soils of Brazil. Embrapa Arroz e Feija˜o, Document No.92, Santo Antonio de Goias, Brazil

Fageria NK, Baligar VC (2008) Ameliorating soil acidity of tropical oxisols by liming for sustainable crop production. Adv Agron 99:345–399

Fageria NK, Baligar VC, Zobel RW (2007) Yield, nutrient uptake and soil chemical properties as influenced by liming and boron application in a no-tillage system. Commun Soil Sci Plant Anal 38:1637–1652

Fageria NK, Barbosa Filho MP (2007) Dry-matter and grain yield, nutrient uptake, and phosphorus useefficiency of lowland rice as influenced by phosphorus fertilization. Communications in Soil Science and Plant Analysis 38:1289–1297

FAO (1984) Investigation of lands with stagnating and declining productivity. (OCP/INS/l07/JPN). Interim Report. Bangkok

FAO (2006) Soil biodiversity portal. http://www.fao.org/ag/agl/agll/soilbiod/index_en.stm

FAO (2018) FAO statistical databases. http://apps.fao.org/Food and Agriculture Organization of the United Nations (FAO)

FAO and ITPS (2015) Status of the World's Soil Resources (SWSR) – Main Report. Food and Agriculture Organization of the United Nations and Intergovernmental Technical Panel on Soils, Rome, Italy.

FAOISIDA (1974) Report on regional seminar on shifting cultivation and soil conservation in Africa. FAO, Rome, Italy

Fisher P (2005) Using organic matter to maintain the productivity of soils under intensive cropping. GRDC Updates. http://www.grdc.com.au/growers/res_upd/irrigation/i05/fisher.htm

Foster HL, Abdul Rashid A, Abas AG (1980) The correction of acidity in Malaysian soils. In: Pushparajah E, Chin SL (eds) Proceedings of conference on soil science and agricultural development in Malaysia. Malaysian Society of Soil Science, Kuala Lumpur, Malaysia, pp 151–171

Fox RL, DeDatta SK, Wang JM (1964) Phosphorus and aluminum uptake by Latosols in relation to liming. Transactions 8th international congress of soil science (Bucharest), vol 4, pp 595–603

Foy CD (1984) Physiological effects of hydrogen, aluminum, and manganese toxicities in acid soil. In: Adams F (ed) Soil acidity and liming. American Society of Agronomy, Madison, WI, pp 57–97

FRG (Fertilizer Recommendation Guide) (2012) Bangladesh Agricultural Research Council, Farmgate, Dhaka 1215, Bangladesh

Fystro G, Bakken AK (2005) Soil reaction, yields and herbage element content as affected by lime applied on established leys in a multi-site field trial. J Agric Sci Cambridge 143:407–420

Gaskin JW, Steiner C, Harris K, Das KC, Bibens B (2008) Effect of low-temperature pyrolysis conditions on biochar for agricultural use. T ASABE 51:2061–2069

Gerendás J, Führs H (2013) The significance of magnesium for crop quality. Plant Soil 368:101–128
Getaneh S, Kidanemariam W (2021) Soil acidity and its managements: a review. Int J Adv Res Biol Sci 8:70–79. https://doi.org/10.22192/ijarbs.2021.08.03.008
Girmay G, Singh BR, Mitiku H, Borresen T, Lal R (2008) Carbon stocks in Ethiopian soils in relation to land use and soil management. Land Degrad Dev 19(4):351–367. https://doi.org/10.1002/ldr.844
Glaser B, Lehmann J, Zech W (2002) Ameliorating physical and chemical properties of highly weathered soils in the tropics with charcoal—a review. Biol Fertil Soils 35:219–230 https://doi.org/10.1007/s00374-002-0466-4
Goulding KWT (2016) Soil acidification and the importance of liming agricultural soils with particular reference to the United Kingdom. Soil Use Manage 32:390–399
Granados G, Pandey S, Ceballos H (1993) Response to selection for tolerance to acid soils in a tropical maize population. Crop Sci 33:936–940
Gregan PD, Hirth JR and Conyers MK (1989). Amelioration of Soil Acidity by Liming and Other Amendments In: Robson AD (ed) Soil acidity and plant growth. Academic Press Australia, 30–52 Smidmore Street, Marrickville, NSW 2204, pp 231–264
Guillet D (1992) Covering ground: communal water management and the State in the Peruvian Highlands. University of Michigan Press, AM Arbor. USA
Guo H, Ingolia NT, Weissman JS, Bartel DP (2010) Mammalian microRNAs predominantly act to decrease target mRNA levels. Nature 466(7308):835–840
Gupta N, Tripathi S, Balomajumder C (2011) Characterization of press mud: a sugar industry waste. Fuel 90(1):389–394
Halim A, Siddique MNEA, Sarker BC, Islam MJ, Hossain MF, Kamaruzzaman M (2014) Assessment of nutrient dynamics affected by different levels of lime in a mungbean field of the old Himalayan piedmont soil in Bangladesh. J Agric Vet Sci 7:101–112
Hargrove WL, Thomas GW (1981) Effect of organic matter on exchangeable aluminum and plant growth in acid soils. Chem Soil Environ 40:151–166
Havlin JL, Tisdale SL, Nelson WL, Beaton JD (2016) Soil fertility and fertilizers. Pearson Education, Chennai, India
Haynes RJ (1984) Lime and phosphate in the soilñplant system. Adv Agron 37:249ñ315
Helyar KR, Hochman Z, Brennan JP (1988) The problem of acidity in temperate area soils and its management. In: Loveday J (ed) National soils conference 1988 review papers. Australian Society of Soil Science, Canberra, Australia, pp 22–54
Helyar KR, Conyers MK, Munns DN (1993) Soil solution aluminium activity related to theoretical Al mineral solubilities in four Australian soils. J Soil Sci 44:317–333
Hetherington SJ, Asher CJ, Blarney FPC (1988) Comparative tolerance of sugarcane, navybean, soybean and maize to aluminium toxicity. Aust J Agric Res 38:171–176
Hochman Z, Godyn DL, Scott BJ (1989) The integration of data on lime use by modelling. In: Robson AD (ed) Soil acidity and plant growth. Academic Press, Sydney, Australia, pp 265–301
Hocking PJ (2001) Organic acids exuded from roots in phosphorus uptake and aluminium tolerance of plants in acid soils. Adv Agron 74:63–97
Holland JE, Bennett AE, Newton AC, White PJ, McKenzie BM, George TS, Pakeman RJ, Bailey JS, Fornara DA, Hayes RC (2018) Liming impacts on soils, crops and biodiversity in the UK: a review. Sci Total Environ 610–611:316–332
Horst WJ, Wang Y, Eticha D (2010) The role of the root apoplast in aluminium-induced inhibition of root elongation and in aluminium resistance of plants: a review. Ann Bot 106:185–197. https://doi.org/10.1093/aob/mcq053
Horst M, McClintock N, Hoey L (2017) The intersection of planning, urban agriculture, and food justice: a review of the literature. J Am Plann Assoc 83(3):277–295. https://doi.org/10.1080/01944363.2017.1322914
Hoyt PB, Turner RC (1975) Effects of organic materials added to very acid soils on pH, aluminum, exchangeable NIti, and crop yields. Soil Sci 119:227–237
Huber DM, Jones JB (2013) The role of magnesium in plant disease. Plant Soil 368:73–85

Hue NV (1992) Correcting soil acidity of a highly weathered Ultisol with chicken manure and sewage sludge. Commun Soil Sci Plant Anal 23:241–264

Hue NV, Amien I (1989) Aluminum detoxification with green manures. Commun Soil Sci Plant Anal 20:1499–1511

Hue NV, Craddock GR, Adams F (1986) Effect of organic acids on aluminum toxicity in subsoils. Soil Sci Soc Am J 50:28–34

Hue NV, Ikawa H (2006) Acid soils in Hawaii: Problems and management. Department of Agronomy and Soil Science, College of Tropical Agriculture and Human Resources, University of Hawii, Manoa

IAEA (2006) Management practices for improving sustainable crop production in tropical acid soils. Internatinal Atomic Energy Agency, Vienna, Austria

IBSRAM (1985) Acid tropical soils management network. IBSRAM, Bangkok

Ingham ER (2006) Understanding the soil foodweb. First of twelve sub-points. http://www.soilfoodweb.com.au/index.php?pageid=274

Izac A-M, Swift MJ, Andriesse W (1991) A strategy for inland valley agroecosysterns research in West and Central Africa. RCMP Research Monograph 5. IITA, Ibadan, Nigeria

Jafer DG, Hailu G (2017) Application of lime for acid soil amelioration and better soybean performance in South Western Ethiopia. J Biol Agric Healthc 7:95–100

Jehangir IA, Khan MH, Bhat ZA (2013) Strategies of increasing crop production and productivity in problem soils. Int J for Soil Erosion 3:73–78

Jena D (2008) Management of acid soils for sustainable crop production. NAE-acid soil Bulletin No. 4. OUAT, Bhubaneswar, Odisha

Jindo K, Mizumoto H, Sawada Y, Sanchez-Monedero MA, Sonoki T (2014) Physical and chemical characterization of biochars derived from different agricultural residues. Biogeosciences 11:6613–6621

Jones RM (1982) White clover *(Trifolium repens)* in subtropical southeast Queensland. IT. Effects of lime application on nutrient concentrations in soil and white clover. Trop Grasslands 16:127–130

Jones DL, Cross P, Withers PJ, DeLuca TH, Robinson DA, Quilliam RS, Harris IM, Chadwick DR, Edwards-Jones G (2013) Nutrient stripping: the global disparity between food security and soil nutrient stocks. J Appl Ecol 50(4):851–862. https://doi.org/10.1111/1365-2664.12089

Joseph SD, Camps-Arbestain M, Lin Y, Munroe P, Chia CH, Hook J, Van Zwieten L, Kimber S, Cowie A, Singh BP et al (2010) An investigation into the reactions of biochar in soil. Soil Res 48:501–515

Joshi N, Sharma S (2010) Physico-chemical characterization of sulphidation press mud composted press mud and vermicomposted pressmud. Rep Opin 2(3):79–82

Juo ASR, Kamprath EJ (1979) Copper chloride as an extractant for estimating the potentially reactive aluminum pool in acid soils. Soil Sci Soc Am J 43:35–38

Kalidas-Singh S (2013) Effect of phosphorus sources on yield and P uptake of rice in acidic soil. Master's thesis submitted to the Central Agricultural University, Imphal, Manipur

Kalidas-Singh S, Thakuria D (2018) Seedling root-dip in phosphorus and biofertilizer added soil slurry method of nutrient management for transplanted rice in acid soil. J Soil Sci Plant Nutr 18(4):921–938

Kamprath EJ (1970) Exchangeable aluminum as a criteria for liming leached mineral soils. Soil Sci Soc Am Proc 34:252–254

Kamprath EJ (1971) Potential detrimental effects from liming highly weathered soils to neutrality. Proc Soil Sci Soc Fla 31:200–203

Khan MA, Ungar IA (1996) Comparative study of chloride, calcium, magnesium, potassium, and sodium content of seeds in temperate and tropical halophytes. J Plant Nut 19(3–4):517–525

Khan MHR, Bhuiyan MMA, Kabir SM, Blume HP, Oki Y, Aach T (2007) Consequences of basic slag on soil pH, calcium, magnesium status in acid sulphate soils under various water contents. J Biol Sci 7:896–903

Khan MHR, Kabir SM, Bhuiyan MMA, Blume HP, Oki Y, Adachi T (2008) Reclamation of a Badarkhal hot spot of acid sulphate soil in relation to rice production by basic slag and aggregate size treatments under modified plain-ridge ditch techniques. Soil Sci Plant Nutr 54:574–586

Kheoruenromne I (1983) Soil survey. Department of Soils, Faculty of Agriculture, Kasetsart University, Bangkok, p 773

Kheoruenromn I, Kesawapitak P (1989) Management of acid soils for food crop production in Thailand. In: Craswell ET, Pushparajah E (eds) Management of acid soils in the humid tropics of Asia. ACIAR Monograph No. 13 (IBSRAM Monograph No.1), x + p 118

Kirkby C, Fattore A, Smith D, Meyer M (2006) Life cycle assessment of greenhouse gas emissions from irrigated maize. Stubble treatments and plant/soil responses. In: Proceedings 6th Triennial Conference of Maize Association ofAustralia. Griffith, NSW (February 2006)

Kisinyo PO (2016) Effect of lime and phosphorus fertilizer on soil chemistry and maize seedlings performance on Kenyan acid soils. Sky J Agric Res 5:097–104

Kisinyo PO, Gudu SO, Othieno CO, Okalebo JR, Opala PA, Maghanga JK, Agalo JJ, Ng'etich WK, Kisinyo JA, Osiyo RJ, Nekesa AO, Makatiani ET, Odee DW, Ogola BO (2012) Effects of lime, phosphorus and rhizobia on sesbania sesban performance in a Western Kenyan acid soil. Afr J Agric Res 7(18):2800–2809

Kochian LV, Hoekenga OA, Pineros MA (2004) How do crop plants tolerate acid soils? Mechanisms of aluminum tolerance and phosphorous efficiency. Annu Rev Plant Biol 55:459–493. https://doi.org/10.1146/annurev.arplant.55.031903.141655. PMID: 15377228

Kunhikrishnan A, Thangarajan R, Bolan N, Xu Y, Mandal S, Gleeson D, Seshadri B, Zaman M, Barton L, Tang C et al (2016) Functional relationships of soil acidification, liming, and greenhouse gas flux. Adv Agron 139:1–71

Layek J, Narzari R, Hazarika S, Das A, Rangappa K, Devi S, Balusamy A, Saha S, Mandal S, Idapuganti RG et al (2022) Prospects of biochar for sustainable agriculture and carbon sequestration: an overview for eastern Himalayas. Sustainability 2022(14):6684. https://doi.org/10.3390/su14116684

Lee B (1980) Farming brings acid soils. Rur Res 106:4–9

Lee MT, Wilson GL (1972) The calcium and pH components of lime response in tropical legumes. Aust J Agric Res 23:257–265

Leeper GW (1964) 'Introduction to Soil Science'. Melbourne University Press, Carlton

Lehmann J, Rondon M (2006) Bio-char soil management on highly weathered soils in the humid tropics. Biol Approaches Sustain Soil Syst 113:530

Lehmann J, Joseph S (eds) (2009) Biochar for environmental management, vol 1. Earthscan, London, UK

Lehmann J, Gaunt J, Rondon M (2006) Bio-char sequestration in terrestrial ecosystems—a review. Mitig Adapt Strateg Glob Change 11:403–427

Ligeyo DO, Gudu SO (2003–2005) Further laboratory screening of more Kenya maize inbred lines for tolerance to aluminum In: Third year progress report, (March 2004 to 28 February 2005). Mcknight Foundation USA Project, EMBBRAPA, Purdue and Cornell University (USA) and Moi University (Kenya) Phase1

Luis L, Gilles B, Bruna G, Juliette A, Josette G (2014) 50 year trends in nitrogen use efficiency of world cropping systems: the relationship between yield and nitrogen input to cropland. Environ Res Lett 9:105011

Ma JF, Tamai K, Yamaji N, Mitani N, Konishi S, Katsuhara M et al (2006) A silicon transporter in rice. Nature 440:688–691

Ma Y, Bai C, Zhang X, Ding Y (2022) Nitrate increases aluminum toxicity and accumulation in root of wheat. Agriculture 12:1946. https://doi.org/10.3390/agriculture12111946

Macías F, Camps Arbestain M (2010) Soil carbon sequestration in a changing global environment. Mitig Adapt Strateg Glob Change 15:511–529

Mahanta M, Konjengbam NS, Lyngdoh AA, Pheirim R (2022) Aluminium stress tolerance in legumes under acidic soils: mechanisms and methods—a review. Biol Forum Int J 14(1):1538–1548

Mahilum BC, Fox RL, Silva JA (1970) Residual effects of liming volcanic ash soils in the tropics. Soil Sci 109:102–109
Maji AK, Reddy OGP, Meshram S (2008) Acid soil map of India, Annual report. ICAR-National Bureau of Soil Survey and Land Use Planning (ICAR-NBSS&LUP), Nagpur
Major J, Lehmann J, Rondon M, Goodale C (2010a) Fate of soil applied black carbon: downward migration, leaching and soil respiration. Glob Change Biol 16:1366–1379
Major J, Rondon M, Molina D, Riha SJ, Lehmann J (2010b) Maize yield and nutrition during 4 years after biochar application to a Colombian savanna oxisol. Plant Soil 333:117–128
Majumdar S, Somborlong, Wanniang K, Behera UK (2022) Acid soil management in Northeastern region of India. Indian Farming 72:72–77
Mandal SC, Sinha MK, Sinha H (1975) Technical Bulletin No. 51, ICAR New Delhi
Mandal B, Chatterjee J, Hazra GC, Mandal LN (1992) Effect of preflooding on transformation of applied zinc and its uptake by rice in lateritic soils. Soil Sci 153:250–257
Mandal SC (1997) Introduction and historical overview. In: Mahapatra IC, Mandal SC, Misra C, Mitra GN, Panda N (eds) Acidic soils of India. ICAR, New Delhi, pp 3–24
Manoj-Kumar, Hazarika S, Choudhury BU, Ramesh T, Verma BC, Bordoloi LJ (2012) Liming and integrated nutrient management for enhancing maize productivity on acidic soils of Northeast India. Indian J Hill Farming 25:35–37
MARDI (1979) Annual report 1979. Malaysian Agriculture Research and Development Institute, Kuala Lumpur, Malaysia, 1980. Annual Report 1980. Malaysian Agriculture Research and Development Institute, Kuala Lumpur, Malaysia
MARDI (1980) Annual report 1980. Malaysian Agriculture Research and Development Institute, Kuala Lumpur, Malaysia
Marschner H (1986) Mineral nutrition of higher plants. Academic Press Inc., London, England, pp 514–517
Marschner H (1991) Mechanisms of adaptation of plants to acid soils. Plant Soil 134:1–20
Marschner H (1995) Mineral nutrition of higher plants, 2nd. Academic Press Ltd., London
Marschner H (2011) Marschner's mineral nutrition of higher plants. Academic Press, London
Mate CJ, Mukherjee I, Das SK (2015) Persistence of spiromesifen in soil: influence of moisture, light, pH and organic amendment. Environ Monit Assess 187(2):7. https://doi.org/10.1007/s10661-014-4207-625
McBeath AV, Smernik RJ (2009) Variation in the degree of aromatic condensation of chars. Org Geochem 40:1161–1168
McDowell R, Sharpley A, Folmar G (2001) Phosphorus export from an agricultural watershed: linking source and transport mechanisms. J Environ Qual 30:1587–1595
Méndez-Lay J (1973) Effects of lime on phosphorus fixation and plant growth in various soils of Panama. MS thesis, North Carolina State University, Raleigh, NC
Mensah AK, Frimpong KA (2018) Biochar and/or compost applications improve soil properties, growth and yield of maize grown in acidic rainforest and coastal savannah soils in Ghana. Int J Agron. https://doi.org/10.1155/2018/6837404
Michael PS, Rob F, Rob R (2015) The role of organimatter in amelioration of acid sulfate soils. Geoderma 255–256:42–49
Mishra A, Nayak RK, Chander G, Reddy M, Choudhary P (2020) Management of acidic soils. https://www.researchgate.net/publication/346671625
Moges T, Melese A, Tadesse G (2018) Effects of lime and phosphorus fertilizer levels on growth and yield components of malt barley (Hordeum distichum L.) in Angolelana Tera District, North Shewa Zone, Ethiopia. J Adv Agric Res 8(6):582–589
Mokolobate MS, Haynes RJ (2002) Comparative liming effect of four organic residues applied to an acid soil. Biol Fertil Soils 35:79–85
Moncharoen L, Vearasilp T, Eswaran H (1987) Benchmark soils of Thailand Department of Land Development, Thailand and Soil Management Support Services, USA, p 77 with 99 appendices
Moran EF (1981) Developing the Amazon. Indiana University Press, Bloominton, USA

Moran EF (1995) Socio-economic aspect of acid soil management. In: Date R et al (eds) Plant soil interactions at low pH, pp 663–669

Morelli M, Igue K, Fuentes R (1971) Effect of liming on the exchange complex and on the movement of calcium and magnesium. Turrialba 21:317–322

Mukherjee A, Zimmerman AR (2013) Organic carbon and nutrient release from a range of laboratory-produced biochars and biochar—soil mixtures. Geoderma 193–194:122–130

Mwangi TJ, Ngeny JM, Wekesa F, Mulati J (2000) Acidic soil amendment for Maize production in Uasin Gishu District, North Rift Kenya; Kenya Agricultural Research Institute, National Agricultural Research Centre, Kitale, Kenya

Myers RJK, Pauw ED (1995) Strategies for the management of soil acidity. In: Date RA et al (eds) Plant soil interactions at low pH, pp 729–741

Nagarajah S (1986) Fertilizer. Recommendation for rice in Sri Lanka. A historical review. J Soil Sci Soc Sri Lanka 3:1–25

Nair KM, Anilkumar KS, Srinivas S, Sujatha K, Venkatesh DH, Naidu LGK, Sarkar D, Rajasekharan P (2011) Agro-Ecology of Kerala. NBSS Publ. No. 1038; National Bureau of Soil Survey and Land Use Planning: Nagpur, India, p. 408

Nannipieri P (1994) The potential use of soil enzymes as indicators of productivity, sustainability and pollution. In: Pankhurst CE, Doube BM, Gupta VVSR, Grace RR (eds) Soil biota: management in sustainable farming systems. CSIRO, Australia, pp 238–244

Naramabuye FX, Haynes RJ, Modi AT (2008) Cattle manure and grass residues as liming materials in a semi-subsistence farming system. Agr Ecosyst Environ 124:136–141

National Land Use Committee (NLUC) (1985) Guide to the national and regional thematic maps. NLUC, National Economic Development Authority, Philippines

Ndayegamiye A, Cote D (1989) Effect of long-term pig slurry and solid cattle manure application on soil chemical and biological properties. Can J Soil Sci 69:39–47

Netting R (1993) Smallholders, householders: farm families and the ecology of intensive. Sustainable agriculture. Stanford University Press, Stanford, USA

Nguyen BT, Lehmann J (2009) Black carbon decomposition under varying water regimes. Org Geochem 40:846–853

Nguyen BT, Lehmann J, Hockaday WC, Joseph S, Masiello CA (2010) Temperature sensitivity of black carbon decomposition and oxidation. Environ Sci Technol 44:3324–3331

Nguyen TTN, Wallace HM, Xu C-Y, Van Zwieten L, Weng ZH, Xu Z, Che R, Tahmasbian I, Hu H-W, Bai SH (2018) The effects of short term, long term and reapplication of biochar on soil bacteria. Sci Total Environ 636:142–151. https://doi.org/10.1016/j.scitotenv.2018.04.278

NLWRA (2002) Australian agriculture assessment 2001, vol 1. National Land and Water Resources Audit, Canberra

Noble AD, Zenneck I, Randall PJ (1996) Leaf litter ash alkalinity and neutralisation of soil acidity. Plant Soil 179:293–302

Oh NH, Richter DD (2004) Soil acidification induced by elevated atmospheric CO2. Glob Change Biol 10:1936ñ 1946

Olego MA, Visconti F, Quiroga MJ, de Paz JM, Garzón-Jimeno E (2016) Assessing the effects of soil liming with dolomitic limestone and sugar foam on soil acidity, leaf nutrient contents, grape yield and must quality in a mediterranean vineyard. Spanish J Agric Res 14:e1102

Olego MÁ, Quiroga MJ, Mendaña-Cuervo C, Cara-Jiménez J, López R, Garzón-Jimeno E (2021) Long-term effects of calciumbased liming materials on soil fertility sustainability and rye production as soil quality indicators on a typic palexerult. Processes 9:1181

Oliver EE, Houlton BZ, Lipson DA (2021) Controls on soil microbial carbon use efciency over long-term ecosystem development. Biogeochemistry 152:309–325. https://doi.org/10.1007/s10533-021-00758-y

Padmanabhan P, Rakkiyappan P, Alexander KC (1993) Enrichment of pressmud by micro-organism. Sugarcane Breeding Institute. (Extension Publication no. 39)

Pagani A (2011) Soil pH and lime management for corn and soybean, pp 1–112

Panabokke CR, Nagarajah S (1964) The fertility characteristics of the rice growing soils of Ceylon. Trop Agric 120:3–30
Panichapong S (1982) Distribution, characteristics and utilization of problem soils in Thailand. Trop Agric Res Ser 15:83–96
Parham JA, Deng SP, Raun WR, Johnson GV (2002) Long-term cattle manure application in soil I. Effect on soil phosphorus levels, microbial biomass C, and dehydrogenase and phosphatase activities. Biol Fertil Soils 35:328–337
Patiram (1991) Liming of acid soils and crop production in Sikkim. J Hill Res 4:6ñ12
Pauw EFD (1994) The management of acid soils in Africa. Outlook Agric 23:11–16
Pavan MA, Bingham FT, Pratt PF (1984) Redistribution of exchangeable calcium, magnesium and aluminum following lime or gypsum applications to a Brazilian Oxisol. Soil Sci Soc Am 1(48):33–38
Penn CJ, Camberato JJ (2019) A critical review on soil chemical processes that control how soil pH affects phosphorus availability to plants. Agriculture 9:120
Peregrina F, Mariscal I, Ordóñez R, González P, Terefe T, Espejo R (2008) Agronomic implications of converter basic slag as a magnesium source on acid soils. Soil Sci Soc Am J 72:402–411
Pettit RE (2006) The wonderful world of humus and carbon. http://humusandcarbon.blogspot.com/
Pichot J, Sedogo MP, Poulain JF, Arrivets J (1981) Evolution de la fertilite d'un sol ferrugineux tropical sous l'inftuence de fumures minerales et organiques. Agron Trop 36:122–133
Pinto M, Rodriguez M, Besga G, Balcazar N, Lopez FA (1995) Effects of Linz-Donawitz (LD) slag on soil properties and pasture production in the Basque country (Northern Spain). N Z J Agric Res 38:143–155
Pocknee S, Sumner ME (1997) Cation and nitrogen contents of organic matter determine its soil liming potential. Soil Sci Soc Am J 61:86–92
Ponnamperuma FN (1960) Lime as an amendment for the acid lateritic soils of Ceylon. Trop Agric 115:243–252
Porter WM, Wilson IR (1984) Soil acidity in the eastern wheatbelt. J Agric W Aust 25(4):132–135
Probert ME (1980) Growth responses to various calcium sources in a yellow earth soil with low calcium status. Aust J Exp Agric Anim Hus 20:240–246
Rahman MA, Chikushi J, Duxbury JM, Meisner CA, Lauren JG, Yasunaga E (2005) Chemical control of soil environment by lime and nutrients to improve the productivity of acidic alluvial soils under rice-wheat cropping system in Bangladesh. Environ Control Biol 43:259–266
Raij BV, Quaggio JA (1997) Methods used for diagnosis and correction of soil acidity in Brazil: an overview. In: Moniz AC et al (eds) Plant soil interaction at low pH. Barzilian Soil Science Society, Campinas, Brazil, pp 205–214
Ranjhan SK (1990) Agro-industrial by-products and non-conventional feeds for livestock feeding, 127. I.C.A.R., India Publication
Rasmussen PE, Rohde CR (1989) Soil acidifi cation from ammonium-nitrogen fertilization in moldboard plow and stubble- mulch wheat-fallow tillage. Soil Sci Soc Am J 53(1):119–122
Rasmussen PE, Dick RP (1995) Long-term management effects on soil characteristics and productivity. In: Proceedings of the western nutrient management conference, Salt Lake City, UT, March 9–10, 1995. Potash and Phosphate Institute
Raunet M (1985) Les bas-fonds en Afrique et à Madagascar. Géomorphologie-géochimie-péd ologie-hydrologie. Z Geomorphol 52:25–62
Rebecca R (2007) Rethinking biochar. Environ Sci Technol 41:5932–5933
Reeve NG, Sumner ME (1970) Effects of aluminum toxicity and phosphorus fi xation on crop growth on oxisols in natal. Soil Sci Soc Am Proc 34:263–267
Reid RJ, Butcher CS (2011) Positive and negative impacts of plants on acid production in exposed acid sulphate soils. Plant Soil 349:183–190
Rengel Z (2011) Soil pH, soil health and climate change. In: Soil health and climate change. Protecting the root apex from soil acidity: a theoretical framework. Plant Soil 278

Richardson AE, Barea JM, McNeill AM, Prigent-Combaret C (2009) Acquisition of phosphorus and nitrogen in the rhizosphere and plant growth promotion by microorganisms. Plant Soil 321:305–339
Ridge DR (1984) Review of research in sugarcane nutrition. Report on BSES Project 102.30.000. Bureau of Sugar Experiment Stations, Brisbane, Australia
Ritchey K, Souza DMG, Lobato E, Correa O (1980) Calcium leaching to increase rooting depth in a Brazilian savannah Oxisol. Agron J 72:40–44
Ritchie GSP, Dolling PJ (1985) The role of organic matter in soil acidification. Soil Res 23:569–576
Rodrigo DM (1964) Response of rice to silica. Trop Agric 120:219–226
Rodriguez M, Lopez FA, Pinto M, Balcazar N, Besga G (1994) Basic linz-donawitz slag as a liming agent for pastureland. Agron J 86:904–909
Rorison IH (1972) The effect of extreme soil acidity on the nutrient uptake and physiology of plants. AGRIS, pp 223–254, ISSN: 0074-6452
Rossel RV, McBratney AB (2011) A response-surface calibration model for rapid and versatile site-specific lime-requirement predictions in southeastern Australia. Soil Res 39:185–201
Rufyikiri G, Hennebert P (1995) Gestion des Sols Acides en Régions d 'Altitude à Occupation Intense: Le Kirimiro au Burundi. Faculte de Sciences Agronomiques, University du Burundi, Bujumbura
Russel A, Marsah EJ (1997) Soil conditions for plant growth, 10th edn. Longman group limited, London and New York, pp 264–300
Russell IS (1966) Plant growth on a low calcium status solodic soil in a subtropical environment 1. Legume species. Calcium carbonate, zinc, and other minor element interactions. Aust J Agric Res 17:673–686
Ruthenberg H (1980) Farming systems in the tropics, SECOND edition. Clarendon Press, Oxford, UK, p 366. Sanchez P A 1976 Properties and Management of Soils in the Tropics. John Wiley and Sons, New York, USA. p 618
Ryan PK, Delhaize E (2010). The convergent evolution of Al resistence in plants exploit a convenient currency. Funct Plant Biol 37: 275–284
Ryan PR, Tyerman SD, Sasaki T, Furuichi T, Yamamoto Y, Zhang WH, Delhaize E (2011) The identification of aluminiumresistance genes provides opportunities for enhancing crop production on acid soils. J Exp Bot 62:9–20
Sahu GC, Patnaik S (1990) Influence of lime or organic matter on pH and ion exchange phenomenon of some alfisol. J Indian Soc Soil Sci 38:154–157
Saleh-e-In MM, Yeasmin S, Paul BK, Ahsan M, Rahman MZ, Roy SK (2012) Chemical studies on press mud: a sugar industries waste in Bangladesh. Sugar Tech (Int J Sugar Crops Relat Ind). https://doi.org/10.1007/s12355-012-0139-z
Sanchez PA, Bandy DE, Villachica JH, Nicholaides JJ (1982) Amazon basin soils: management for continuous crop production. Science 216:821–827
Sanchez PA, Benites J, Bandy D (1987) Low-input systems and managed fallows for acid soils in the humid tropics. In: Soil management under humid conditions in Asia and Pacific. IBSRAM, Bangkok, Thailand, pp 353–360
Sanchez PA (1987) Management of acid soils in the humid tropics of Latin America. In: IBSRAM management of acid tropical soils for sustainable agriculture: proceedings of an IBSRAM inaugural workshop, Bangkok, Thailand, pp 63–107
Sanchez PA (2019) Properties and Management of Soils in the Tropics (Second Edition), Cambridge University Press, University Printing House, Cambridge CB2 8BS, United Kingdom
Schefe CR, Patti AF, Clune TS, Jackson WR (2008) Organic amendment addition enhances phosphate fertiliser uptake and wheat growth in an acid soil. Aust J Soil Res 46:686–693
Schulz H, Glaser B (2012) Effects of biochar compared to organic and inorganic fertilizers on soil quality and plant growth in a greenhouse experiment. J Plant Nutr Soil Sci 175:410–422
Scott BJ, Conyers MK, Poile GJ, Cullis BR (1997) Subsurface acidity and liming affectyield of cereals. Aust J Agric Res 48:843–854

Sehgal J (1998) Red and lateritic soils—an overview. In: Sehgal J, Blum WE, Gajbhiye KS (eds) Red and lateritic soils, vol 1. Oxford and IBH Pub. Pvt. Ltd., New Delhi, pp 3–10
Sekiya N, Yano K (2010) Seed P-enrichment as an effective P supply to wheat. Plant Soil 327:347–354
Shainberg I, Sumner ME, Miller WP, Farina MPW, Pavan MA, Fey MV (1989) In: Stewart BA (ed) Use of gypsum on soils: a review. Advances in Soil Science, vol 9. Springer-Verlag, New York, pp 1–111
Sharifuddin HAH, Dynoodt RFP (1985) Evaluation of nutritional status of selected Malaysian soils. In: Mokhtaruddin AM et al (eds) Advances in soil research in Malaysia. Penerbit Universiti Pertanian Malaysia
Sharifuddin HAH, Shamshuddin J, Grundon NJ, Edwards DG, Zaharah R (1995) Residual effects of dolomitic limestone on corn grown on an acidic Malaysian ultisol. In: Date RA, Grundon NJ, Rayment GE, Probert ME (eds) Plant–soil interactions at low pH: principles and management. Kluwer, Dordrecht, pp 511–517
Sharma UC (1990) Reaction of potato (Solanum tuberosum) cultivars to phosphorus stress and screening of phosphorus-efficient genotypes on acid soils. Indian J Agric Sci 60:523–528
Sharma UC (1991a) Effect of P and FYM on the availability of P in acidic soil of Meghalaya state. J Hill Res 4:100–102
Sharma UC (1991b) Response of potato to phosphorus and farmyard manure on acid soils of Khasi hills of Meghalaya. J Indian Soc Soil Sci 39:306–307
Sharma UC, Patel CS (1991c) Relative efficacy of nitrogen sources and amendments on acidic soil of Meghalaya with potato as test crop. J Indian Society of Soil Science 39: 463-467
Sharma UC (1993) Effect of N and FYM and their interactions on the yield of rice in acidic soil of Nagaland. J Hill Res 6:231–232
Sharma UC, Singh RP (2002) Acid soils of India: their distribution, management and future strategies for higher productivity. Fertiliser News 47(3):45–52
Sharma PK, Verma T, Gupta J (1990) Ameliorating effects of phosphorus, lime and animal manure on wheat yield and root cation exchange capacity in degraded Alfisols of north-west Himalayas. Fert Res 23:7–13
Sharma UC, Sinha H (1989) Residual effect of phosphatic sources in a rice—gram rotation on acid soil
Sharma UC, Datta M, Samra JS (2006) Soils and their management in North East India. ICAR Research Complex for NEH Region, Umroi Road, Umroi, Meghalaya, India, p 502
Shoemaker HE, McLean EO, Pratt PF (1961) Buffer methods of determining lime requirements of soils
Siavoshi M, Nasiri A, Laware SL (2011) Effect of organic fertilizer on growth and yield component in rice. J Agric Sci 3:217–224
Sierra J, Ozier-Lafontaine H, Dufour L, Meunier A, Bonhomme R, Welcker C (2006) Nutrient and assimilate partitioning in two tropical maize cultivars in relation to their tolerance to soil acidity. Field Crop Res 95:234–249
Sims JT, Wolf DC (1994) Poultry waste management: agricultural and environmental issues. Adv Agron 52:1–83
Singh B, Cowie A (2008) A novel approach, using 13C natural abundance, for measuring decomposition of biochars in soil. In: Carbon and nutrient management in agriculture. Fertilizer and Lime Research Centre, Massey University, Palmerston North, New Zealand
Singh B, Singh BP, Cowie AL (2010) Characterisation and evaluation of biochars for their application as a soil amendment. Soil Res 48:516–525
Sivaguru M, Horst WJ (1998) The distal part of the transition zones is the most aluminum-sensitive apical root zone of maize. Plant Physiol 116:155–163
Smith SE, Read DJ (1997) Mycorrhizal symbiosis, 2nd edn. Academic Press, San Diego
Soil Survey Staff (1975) Soil taxonomy, a basic system of soil classification for making and interpreting soil surveys. U.S. Department of Agriculture. Washington D.C., U.S. Government Printing Office, p 754

Soil Survey Staff (1987) Keys to soil taxonomy (third printing). SMSS Technical Monograph No. 6. Ithaca, New York, p 280
Song W, Guo M (2012) Quality variations of poultry litter biochar generated at different pyrolysis temperatures. J Anal Appl Pyrolysis 94:138–145
Spain JM, Francis CA, Howeler RH, Calvo F (1975) Differential species and varietal tolerance to soil acidity in tropical crops and pastures. In: Bornemisza E, Alvarado A (eds) Soil management in tropical America. North Carolina State University, Raleigh, NC, pp 308–324
Steiner C, Teixeira WG, Lehmann J, Nehls T, Macedo JLVD, Blum WEH, Zech W (2007a) Long-term effects of manure, charcoal, and mineral fertilization on crop production and fertility on a highly weathered central Amazonian upland soil. Plant Soil 291:275–290
Stewart WM, Dibb DW, Johnston AE, Smyth TJ (2005) The contribution of commercial fertilizer nutrients to food production. Agron J 97:1–6
Sultana BS, Mian MH, Jahiruddin M, Rahman MM et al (2019) Amendment of acidic soil with lime and manure for enhancing fertility, nutrient uptake and yield of wheat-mungbean-monsoon rice in the Old Himalayan Piedmont Plain. Asian J Soil Sci Plant Nutr 4(2):1–26
Sun K, Jin J, Keiluweit M, Kleber M, Wang Z, Pan Z, Xing B (2012) Polar and aliphatic domains regulate sorption of phthalic acid esters(PAEs) to biochars. Bioresour Technol 118:120–127
Suthipradit S, Edwards DG, Asher CJ (1990) Effects of aluminium on taproot elongation of soybean, cowpea and green gram grown in the presence of organic acids. Plant Soil 124:233–237
Swain MD, Anderson GP, Serrano-González J, Liu JL, Zabetakis D, Goldman ER (2011) Immunodiagnostic reagents using llama single domain antibody–alkaline phosphatase fusion proteins. Anal Biochem 417:188–194
Swain MR, Laxminarayana K, Ray RC (2012) Phosphorus solubilization by thermotolerant Bacillus subtilis isolated from cowdung microflora. Agric Res 1:273–279
Swift MJ, Dvorak KA, Mulongoy K, Musoka M, Sanginga N, Tian G (1994) The role of soil organism in the sustainability of tropical cropping systems In: Syers SK, Rimmer DI (eds) Soil science and sustainable land management in tropics. CAB International, Cambridge University Press U.K., pp 155–170
Tabor JA. Kilambya DW, Kibe JM (1990) Reconnaissance survey of the ethno-;dology in Embu. Meru. Machakos and ~iki districts of Kenya's eastern province. Unpublished manuscript. International Agricultural hgnm, University of Missouri. Uo Dr Bryce Ratchford, Columbia. MO 652 I I. USA
Talukdar NC, Thakuria D, Bordoloi L, Goswami C (2001) Half a decade research on utilization of soil biological agents as components of nutrient management for quality and resilience of agro- and forest ecosystem. In: Yadav AK, Roychoudhury S (eds) Three decades of research in biofertilizers and organic farming in Northeast India. Assam Agricultural University, Jorhat, Assam, India and Regional Biofertilizer Development Centre, Imphal, Manipur, pp 7–33
Tandon HLS (1995) Waste recycling in agriculture an introduction. Fertilizer Development Consultation Organization, New Delhi, India
Tandzi NL, Ngonkeu ELM, Youmbi E, Nartey E, Yeboah M, Gracen V, Ngeve J, Mafouasson H (2015) Agronomic performance of maze hybrids under acid and control soil conditions. Int J Agron Agric Res 6:275–291
Tang C, Yu Q (1999) Impact of chemical composition of legume residues and initial soil pH on pH changes of a soil after residue incorporation. Plant Soil 215:29–38
Tanoi K, Kobayashi NI (2015) Leaf senescence by magnesium deficiency. Plants 4:756–772
Taylor AK (2000) From raw sugar to raw materials. Chemical Innovation, November, pp 45–48
TDRI (Thailand Development Research Institute) (1987) Land resources: natural resources management Part IT. In: Thailand natural resources profile—is the resource base for Thailand's development sustainable'? National Environment Board, Department of Technical and Economic Cooperation and United States Agency for International Development, Bangkok, pp 19–45
Tessens E, Sharnsuddin J (1983) Quantitative relationship between mineralogy and properties of tropical soils. UPM Press, Serdang, Selangor, Malaysia

Thakuria D, Talukdar NC, Goswami C, Hazarika S, Kalita MC, Bending GD (2009) Evaluation of rice–legume–rice cropping system on grain yield, nutrient uptake, nitrogen fixation, and chemical, physical, and biological properties of soil. Biol Fertil Soils 45:237–251

The C, Calba H, Zonkeng C, Ngonkeu EIM, Adetimirin V (2006) Response of maize grain to changes in acid soils characteristics after soil amendment. Plant Soil 28:45–57

Theng BKG (1980) Soils with variable charge. New Zealand Society of Soil Science, Palmerston North, New Zealand, Offset Publications, p 448

Tisdale SL, Nelson W, Beaton J (1985) Soil fertility and fertilizers, 4th edn. Macmillan Publishers, New York

Tisdale SL, Nelson WL, Beaton JD, Havlin JL (1997) Soil fertility and fertilizers. Second Indian reprint, 5th edn. Prentice Hall of India Pvt. Ltd., New Delhi

U.S. Department of Agriculture (1975) Soil taxonomy, agriculture handbook No. 436. U.S. Government Printing Office, Washington. D.C., p 754

van Raij B (1991) Fertilidade do Solo e Adubação. Ceres, São Paulo. 2008. Gesso na Agricultura. Instituto Agronômico de Campinas, São Paulo (cited in Sanchez, 2019)

von Uexkiill HR (1986) Efficient fertiliser use in acid upland soils of the humid tropics. FAO Fertiliser and Plant Nutrition Bulletin, p 10

von Uexkiill HR, Bosshart RP (1989) Management of acid upland soils in Asia. In: Craswell ET, Pushparajah E (ed) Management of acid soils in the humid tropics of Asia. ACIAR Monograph No.13 (IBSRAM Monograph No.1), x + p 118

Vishwakarma AK, Brajendra, Pathak KA, Ramakrishna Y (2012) Effect of different sources of nutrient application on productivity, nutrient uptake and economics of groundnut (Arachis hypogaea L.) in Kolasib district of Mizoram. Indian J Soil Conserv 40:152–157

von Uexküll HR, Mutert E (1995) Global extent, development and economic impact of acid soils. In: Date RA, Grundon NJ, Rayment GE, Probert ME (eds) Plant-soil interactions at low pH: principles and management. Kluwer Academic Publishers, Dordrecht, The Netherlands, pp 5–19

Vy N, Thi NT (1989) Management of sloping soils for food production in Vietnam

Wang J, Raman H, Zhang G, Mendham N, Zhou M (2006) Aluminium tolerance in barley (Hordeum vulgare L.): physiological mechanisms, genetics and screening methods. Zhejiang Univ Sci 7:769–787

Webster CC, Wilson PN (1980) Agriculture in the tropics, second edition. Longman tropical agriculture series. Longman, London, UK, p 640

Whalen JK, Chang C, Clayton GW, Carefoot JP (2000) Cattle manure amendments can increase the pH of acid soils. Soil Sci Soc Am J 64:962–966

Wheeler DM, Edmeades DC, Christie RA, Gardner R (1992) Effect of aluminium on the growth of 34 plant species: a summary of results obtained in low ionic strength solution culture. Plant Soil 146:61–66

Wild A (1988) Soil acidity and alkalinity. In: Wild A (ed) Russell's soil conditions and plant growth. Longman, Harlow, pp 844–889

Wilken GC (1987) Good farmers: traditional agricultural resource management in Mexico and Central America. University of California Press, Berkeley, USA

Williams CH (1980) Soil acidification under clover pasture. Aust J Exp Agric Anim Husb 20:561–567

Wilson DO, Boswell FC, Ohki K, Parker MB, Shuman LM, Jellum MD (1982) Changes in soybean seed oil and protein as influenced by manganese nutrition. Crop Sci 22:948–952

Wong MTF, Nortcliff S, Swift RS (1998) Method for determining the acid ameliorating capacity of plant residue compost, urban waste compost, farm yard manure and peat applied to tropical soils. Commun Soil Sci Plant Anal 29:2927–2937

Wortmann CS (2015) Management strategies to reduce the rate of soil acidification. University of Nebraska-Lincoln Extension, Institute of Agriculture and Natural Resources, NebGuide

Wright RJ, Baligar VC, Belesky DP, Snuffer JD (1991) The effect of phosphate rock dissolution on soil chemical properties and wheat seedling root elongation. Plant Soil 134:21–30

Xu JM, Tang C, Chen ZL (2006a) The role of plant residues in pH change of acid soils differing in initial pH. Soil Biol Biochem 38:709–719

Yadav DV (1992) Utilization of press mud cakes in Indian agriculture. Indian J Sugarcane Technol 7:1–16

Yaman, Kumar P (2021) Physiological stress in agricultural crops: an overview

Yamato M, Okimori Y, Wibowo IF, Anshori S, Ogawa M (2006) Effects of the application of charred bark of Acacia mangium on the yield of maize, cowpea and peanut, and soil chemical properties in South Sumatra, Indonesia. Soil Sci Plant Nutr 52:489–495

Yamoah CF, Burleigh JR, Eylands VJ (1992) Correction of acid infertility in Rwandan oxisols with lime from an indigenous source for sustainable cropping. Exp Agric 28:417–424

Yan F, Schubert S, Mengel K (1996) Soil pH changes during legume growth and application of plant materials. Biol Fertil Soils 23:236–242

Yan F, H¨utsch BW, Schubert S (2006) Soil-pH dynamics after incorporation of fresh and oven-dried plant shoot materials of faba bean and wheat. J Plant Nutr Soil Sci 169:506–508

Yang T, Siddique KHM, Liu K (2020) Cropping systems in agriculture and their impact on soil health-a review. Glob Ecol Conserv 23. https://doi.org/10.1016/j.gecco.2020.e01118

Yao MK, Angui PK, Konaté S, Tondoh JE, Tano Y, Abbadie L, Benest L (2010) Effects of land use types on soil organic carbon and nitrogen dynamics in mid-west Côte d'Ivoire. Eur J Sci Res 40:211–222

Yao C, Joseph S, Li L, Pan G, Lin Y, Munroe P, Pace B, Taherymoosavi S, Van Zwieten L, Thomas T, Nielsen S, Ye J, Donne S (2015) Developing more effective enhanced biocharfertilisersforim-provement of pepper yield and quality. Pedosphere 25:703–712. https://doi.org/10.1016/S1002-0160(15)30051-5

Yazawa Y, Wong MTF, Gilkes RJ, Yamaguchi T (2000) Effect of additions of brown coal and peat on soil solution composition and root growth in acid soil from wheat belt of Western Australia. Commun Soil Sci Plant Anal 31:743–758

Yost R, Itoga S, Hansen J, Arnieni LI (1988) Expert systems for managing acid soils. In: Demeterio JL, De Guzman B (eds) Management and utilisation of acid soils of Oceania. University of Guam, Guam, Agricultural Experiment Station, pp 180–185

Yuan JH, Xu RK, Wang N, Li JY (2011a) Amendment of acid soils with crop residues and biochars. Pedosphere 21(3):302–308

Zhang A, Cheng G, Hussain Q, Zhang M, Feng H, Dyck M, Wang X (2017a) Contrasting efects of straw and straw–derived biochar application on net global warming potential in the Loess Plateau of China. Field Crop Res 205:45–54

Chapter 6
Recent Advances and Future Challenges

Abstract Because of a combination of poor water absorption, nutritional shortages, and mineral toxicity, acidic soils are ubiquitous and present a serious challenge to agriculture. Conventional restoration methods, such as enhancing the soil with lime, magnesium, or calcium, are expensive and detrimental to the ecosystem. The most effective technique to lessen soil acidity is to cultivate cultivars that are acid-tolerant. The ability of plants to tolerate acidic soils varies greatly, and a major factor influencing this ability is the toxicity of aluminium (Al). Understanding the physiological, biochemical, and genetic bases of Al tolerance is therefore essential to the effective development of crops that are acid-tolerant. The amount of human demand on soil resources is getting worse even though soils are a basic necessity for supporting human needs. Anthropogenic effects such as pollution, acidity, salinity, and/or sodicity, erosion, and the loss of soil organic matter will pose major obstacles to future developments in soil science. Reversing soil deterioration through the buildup of soil organic matter and sustainable soil management offers a substantial chance to reduce climate change since soil stores atmospheric carbon. Indirect effects of human activities on soils include acid deposition and heavy metal contamination. Global agricultural decision-making is being optimised by breakthroughs in analytics and modelling expertise, as well as greater capacity for data processing, storage, and sharing. These developments anticipate the many complexities that impact agricultural production processes. The use of interactive decision support technologies inside the Agronomic Response Unit architecture is expected to boost resource consumption efficiency when combined with socioeconomic variables, such as farmers' ability to invest. Soil science research has not only produced highly qualified scientists in the field, but it has also played a pivotal role in generating novel insights and solutions that have helped address a multitude of social problems. However, a number of challenges stand in the way of further advancements in soil science. Soil science education in the future needs to adjust to new socioeconomic concerns and take into account soils in contexts other than agriculture. The education must ensure that the many roles that soils play in an ecosystem are properly understood and that soil continues to rank among the most important strategic resources that are necessary to solving societal challenges.

U. C. Sharma et al., *Soil Acidity*, Progress in Soil Science,
https://doi.org/10.1007/978-3-031-76357-1_6

Keywords Recent advances in acid soils research · Future challenges and advancements · New data generation · Framework with socio-economic conditions · Analytics and modelling proficiency

6.1 General Description

A key component of agricultural sustainability is the preservation of soil quality and the use of sustainable soil management techniques. One of the main obstacles to agricultural productivity is the acidity of the soil and the resulting reduced availability of nutrients. One of the physiological properties of the soil solution that is expressed in terms of pH is soil response, which shows whether the soil is neutral, acidic, or alkaline. It has a major influence on a variety of soil characteristics, such as the availability of nutrients, biological activity, and physical state of the soil. For the best crop performance in acid soils, liming has been shown to be essential in reducing soil acidity and phytotoxic levels of Al, Mn, and Fe. However, lime treatment should not be viewed as a means of achieving potential yield on its own; rather, it should be viewed as a strategy to raise soil pH in order to maximise nutrient availability for optimal plant growth and yield (Agegnehu et al. 2019). Applying the recommended amounts of organic and inorganic fertilisers, especially P and K fertilisers, should go hand in hand with liming. Furthermore, it's important to pinpoint the regions where applying lime improves crop output significantly. Integrated crop and soil management is another way to slow down the rate of soil acidification. Crop output and soil productivity can both be increased using an integrated approach to soil fertility management. It is clear that applying organic wastes promotes the accumulation of nutrients in the soil, and following years of application, the amount of nutrients to be applied in either organic or inorganic form will progressively decline. Since the long-term effects of organic sources on soil characteristics and crop productivity can persist for several years, it is not possible to accurately estimate their profitability in the near term; nonetheless, their long-term benefits are evident. When organic wastes are applied to soils, their biological and physical qualities are enhanced as well as their nutritional content is raised. The tolerance of acidic soil conditions varies greatly amongst crops. In rice, corn, wheat, beans, and soybeans, there are notable varietal variations with regard to Al tolerance. These variations offer a strong chance for plant selection and breeding for increased acidity tolerance, a cheap and efficient substitute for the expensive process of amending acid soil with lime (Agegnehu et al. 2019). Generally speaking, it is extremely important from a practical standpoint to utilise all of the resources at hand, especially crop species and acid-tolerant varieties that enhance and maintain soil and agricultural output. In general, acid soil management must prioritise strategic research that combines improved crop varieties with soil and water management to create prototype and environmentally friendly technologies for sustained food production within the proper socioeconomic and policy framework.

Acidic soils are common and pose a significant problem for agriculture because they reduce productivity due to a combination of poor water absorption, nutrient deficiencies, and mineral toxicity. Traditional restoration techniques, including adding lime, magnesium, or calcium to the soil, are costly and harmful to the environment. Cultivating acid-tolerant cultivars is the most efficient way to reduce soil acidity. There is considerable variation in plants' capacity to withstand acidic soils, and aluminium (Al) toxicity plays a major role in determining this capacity. Thus, for the purpose of successfully developing acid-tolerant crops, it is imperative to comprehend the physiological, molecular, and genetic foundations of Al tolerance (Zhu and Shen 2023). Molecular markers have been designed to promote marker-assisted selection in the breeding of Al-tolerant cultivars and to facilitate gene cloning. There are cultivars mainly present in local germplasms of various crops having capability to withstand P-deficiency mostly prevailing in acid soils. There are cultivars mainly present in local germplasms of various crops having capability to withstand P-deficiency mostly prevailing in acid soils. Different genes and quantitative trait loci in different plant species govern different tolerance processes. In order to cultivate Al-resistant varieties through backcrossing, intercrossing, and molecular marker-assisted selection, it is anticipated that new Al-resistant genes will be discovered through the use of genome-wide association methods (Zhu and Shen 2023). This will support the sustainable use of acidic soils.

Although soils are a basic requirement for supporting human requirements, the amount of human strain on soil resources is becoming increasingly extreme. Future advancements in soil science will face significant hurdles from anthropogenic consequences such as pollution, acidity, salinity and/or sodicity, erosion, and the loss of soil organic matter. By storing atmospheric carbon in the soil, the reversal of soil degradation via the accumulation of soil organic matter and sustainable soil management presents a significant opportunity to mitigate climate change (Beyene et al. 2023). Additionally, human activity has indirect effects on soils, such as acid deposition and heavy metal pollution. By anticipating the numerous complexity that affect agricultural production processes, advances in analytics and modelling proficiency, along with increased capacity for data processing, storage, and sharing, are optimising agricultural decision-making on a global scale. It is thought that resource usage efficiency can be increased by combining the deployment of interactive decision support tools inside the Agronomic Response Unit framework with the socioeconomic conditions, such as the farmers' ability to invest. In addition to producing competent scientists in the area, soil science research has been crucial in producing new information and technologies that have assisted in resolving numerous societal issues (Beyene et al. 2023). But there are a number of obstacles in the way of more developments in soil research. Future soil science education must therefore consider soils in contexts other than agriculture and adapt to new socioeconomic challenges. The education must guarantee that soil remains one of the most crucial strategic resources that are essential to resolving societal problems and that the many roles that soils play in an ecosystem are fully comprehended.

To ensure food security and agricultural sustainability, it will be essential to increase crop production in acid soils. Widespread crop growth and yield are severely hindered by the combination of high soil acidity, aluminium (Al) toxicity, and phosphorus (P) deficiency. In addition to searching the soil for the nutrients and water needed for plant growth and development, roots are also sensitive to pressures below ground and react accordingly. Plants, especially roots, have developed a range of strategies for adjusting to these challenges in the terrestrial environment of widespread P deficiency and Al toxicity demands (Li et al. 2023). Research on legumes is crucial and should continue in the future due to recent developments in our knowledge of root-based mechanisms connected to modifications of root architecture, exudation, and symbiosis, as well as related genetic and molecular strategies in adaptation to individual and/or interacting P and Al conditions in acid soils (Li et al. 2023). Using genetic selection to breed legume cultivars with enhanced nutrient efficiency and/or Al tolerance traits could prove to be a potent tactic for creating crop varieties that can sustain or increase yields under more demanding environmental conditions and stressed soil conditions.

The primary cause of the detrimental effects of soil acidity on plant growth and yield is phosphorus deficiency, which is brought on by P adhering to colloidal fractions, converting to insoluble Al and/or Fe compounds, and being toxic to iron, manganese, and aluminium (Hocking 2001; Brady and Weil 2008). Crop productivity in acidic soils has also been shown to be limited by deficiencies of calcium, magnesium, potassium, and molybdenum (Sumner 2000). Applying lime is a popular solution for problems with acidity in soil. In general, surface liming takes a while to reduce subsurface acidity but quickly reduces topsoil acidity (Ernani et al. 2004). To achieve acceptable yields in acidic regions, it is essential to reduce the acidity of the soil by surface liming in order to bring the pH range into compliance with greater crop production.

Global agriculture faces a major issue from acidic soils (pH < 5.5) (Che et al. 2022). Based on Kochian et al. (2015), there are over 2.5 billion hectares of potentially arable land that are acidic, which makes up 40–50% of all arable land in the globe. However, crop productivity is limited and the full agricultural potential of acidic soils is not exploited due to mineral nutrient shortages, such as phosphate (P) insufficiency, aluminium (Al), manganese (Mn), and proton (H^+) toxicities, and other stress factors (von Uexküll and Mutert 1995). Due to overapplication of nitrogen (N) fertilisers, paddy soil acidification has grown to be a significant issue (Zhao et al. 2014). Extensive fertilisation studies conducted on red soil in Hu'nan Province have shown that crop growth is significantly impeded or even fails as a result of soil acidity (Wang et al. 2017). Thus, minimising the detrimental impacts of acidic soils on crop growth and encouraging their sustainable use are crucial if we are to maximise agricultural production potential. Tea (*Camellia sinensis*) really loves acid soils (pH 4.0–5.5), in contrast to most crops where soil acidity significantly inhibits yield. Particularly, tea is highly resistant to aluminium (Al) toxicity caused by acidity, which is a significant problem limiting the production of the majority of other crops. In fact, tea really needs Al for optimal growth (Ding et al. 2021). Thus, it may be essential to comprehend tea's Al tolerance and Al-stimulatory pathways in order to design crops that are eventually suited for acid soils.

6.1.1 Acid Rain (AR)

Around the world, acid rain (AR) has complicated detrimental effects on plants and creates a number of environmental issues. The direct effects of AR or its depositional compounds on plant damage and performance have been the subject of numerous investigations in the past. The indirect effects of AR on plants, which are mediated by soil microbes and the abiotic soil rhizosphere environment, have not, however, been extensively studied. It is important to pay more attention to the indirect effects of acidic deposition because it can have a negative impact on plants by altering the dynamics of certain substances in the soil and affecting plant-associated microorganisms' distribution, composition, abundance, function, and activity (Zhang et al. 2023a, b). The AR-soil microbes-plants route was assessed by:

- combining knowledge about the possible benefits of soil microbes for plants in reducing acidity stress and the negative impacts of AR on soil microorganisms connected with plants;
- investigating the ways in which different types of plant mycorrhizal affect the detection of AR effect on plants.

It was demonstrated by the meta-analysis that plant mycorrhizal types influenced the effects of AR-induced pH on leaf chlorophyll content, plant height, and root biomass. Different types of mycorrhizal symbiotic relationships with plants were mentioned, along with some potential explanations for them. According to Zhang et al. (2023a, b), in order to assess the overall impact of AR on plant performance, future studies pertaining to the effects of AR on plants should concentrate on the combined direct and indirect effects.

6.1.2 Soil Organic Matter

The goal of modern agricultural research and development is to encourage farmers to better use organic resources in order to improve fertility, reduce elemental toxicity, and safeguard the soil. One important strategy for doing this is the enhancement of soil organic matter (SOM) (Amede et al. 2021). Enhancing soil organic matter and recycling nutrients back into the soil can be achieved primarily through the use of green manuring, composting, farmyard manure (FYM), biochar, and crop residue retention. According to studies, organic matter increases water-holding capacity, aeration, microbial activity, and soil structure (Agegnehu and Amede 2017; Amede et al. 2021). This is supported by data from several studies. While green manures both protect the soil from erosion and provide a significant quantity of nitrogen to succeeding crops, crop leftovers have the ability to recycle nutrients that crops have withdrawn from the soil (Kumar and Sukul 2020). By removing harmful components from the soil solution and integrating them into organic compounds, the use of organic matter in the form of crop residues, green manures, FYM, compost, or biochar may also help lessen their impacts (Cornelissen et al. 2018).

6.1.3 Integrated Acid Soil Management (IASM)

Acid soil management includes IASM among its components. P availability and usage efficiency are increased in acid soils by the release of a variety of organic acids by FYM treatment, which can then form stable complexes with Al and Fe to block the P retention sites (Prasad and Power 1997; Agegnehu and Amede 2017). Partially or completely, lime can be replaced with organic sources such agricultural residues, manures, compost, and biochar (Agegnehu and Amede 2017). Acid soils have shown improvement in yields when organic amendments are added, primarily due to the reduction of Al's phytotoxic levels (Haynes and Mokolobate 2001).

6.1.4 Sustainable Cropping Systems

The method used to cultivate various crops while taking socioeconomic and biophysical factors into account in order to sustainably improve crop production and soil quality is known as a sustainable cropping system. The market, agroecology, soil types, farming system, and crop diversification all influence cropping system practices. Major cropping system components, crop rotation and intercropping, are implemented in different ways by farmers throughout Sub-Saharan Africa. A better strategy of intercropping could optimise the complementarity between annual crops and perennials or between cereal and legume crops (Pauw 1994). It has long been known that agricultural yield sustainability and land productivity can both be increased by the systematic rotation of crop species (Xu et al. 2002). Legumes can harbour N-fixing bacteria in their roots, which helps plants develop to their full potential while emitting fewer greenhouse gases. Crop rotations increase soil fertility, which is essential for growing crops that are robust and fruitful.

6.1.5 Acid Soils

Spodosols present in cool and wet climate, are also acidic soils having a strongly leached surface layer (Albic horizon) and a subsoil accumulation of organo-Al/Fe oxides complexes (Spodic horizon), mainly present in Northern Europe, Russia and North America. In the humid tropics, the three most prevalent acid soil types are oxisols, ultisols, and acid sulphate soils. Because of their acidity, cationic toxicity, and low nutrient availability, these soils present challenges for agriculture and the environment, and management is required. Over the last two decades, a body of

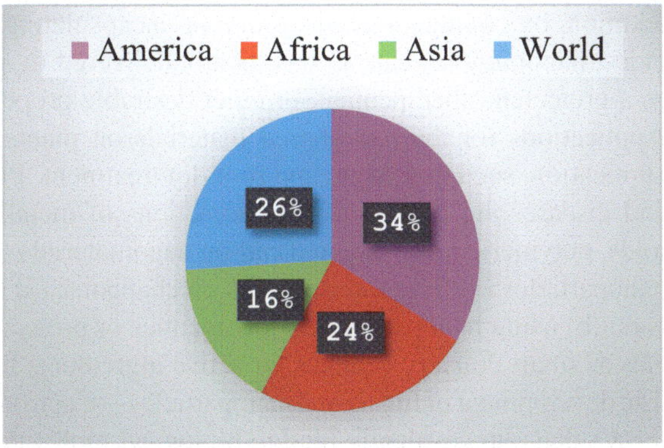

Fig. 6.1 Distribution of oxisols and ultisols in the world

research has identified the causes, mechanisms, and effects of acid soil formation (Michaela 2023). In the light of climate change, research is required to increase farm output and safeguard the different ecological services that the environment offers to humans. In order to increase agricultural yield and profitability, these research needs were compiled and presented as necessary future research on acid soils. The % distribution of Oxisols and Ultisols is displayed in Fig. 6.1.

Applying organic matter or revegetation is one of the more user-friendly and sustainable approaches needed in the poor economies. Research in many humid trophic areas, including Australia and Thailand, Malaysia, Indonesia, and Singapore, is gaining traction, especially in the area of ASS management. Management of forest cover, application of organic matter, and sustainable farming methods are a few examples of integrated soil fertility management techniques. In agricultural settings, it is practically important to introduce legumes to increase soil productivity and crop varieties that can withstand acidity and acid minerals. Research linking soil, organic matter, and water management with tolerant and high-yielding crops that are environmentally and agriculturally sustainable must be prioritised as the main approach to managing acid soils in the humid tropics sustainably. To create affordable and long-lasting soil nutrient management, an integrated reclamation strategy addressing the interaction of soil nutrients and lime fertiliser is essential (Mossisa 2018). In several developing nations that depend on agriculture to feed their expanding populations, the principal soil-related barriers to agricultural development include soil acidity linked to Al toxicities, soil erosion, and soil nutrient depletion (Tolera Abera et al. 2006).

6.1.6 Nanotechnology in Agriculture

According to Hayles et al. (2017), nanoscience is being investigated in the areas of controlled release of agrochemicals, water management, plant hormone administration, seed germination, transfer of genes of interest, and nano-biosensors. In addition, by conjugating, adsorbing, or encapsulating active substances like fungicides, nanoparticles can be specifically delivered to a particular site while serving as a protectant, therapeutant, or other desirable property (Khandelwal et al. 2016). Applications for the nano-based materials on plants include foliar spraying, soil application, seed treatment, and root dip treatment. Plant development is promoted and diseases are suppressed by the actions of metallic oxides, nonmetals, metalloids, polymeric materials, and carbon nanomaterials (Elmer et al. 2018). The use of nanoparticles involves three primary mechanisms: (a) using nanoparticles as biosensors; (b) using nanoparticles as therapeutics or protectants; and (c) using nanoparticles as smart delivery vehicles for active ingredients like target genes or fungicides. The development of functional nanoparticles—electronic, optical, magnetic, or structural—that can covalently bind to biological molecules like proteins, peptides, and nucleic acids is the result of recent advances in nanotechnology. As one of the most promising nanomaterials, quantum dots (QD) have been widely used in many bio-related applications, such as the rapid and accurate identification of a particular biological marker (Kashyap et al. 2015).

6.1.7 Molecular Approaches

Many places have acid soils, which reduce plant productivity worldwide. On acid soils, the main factors limiting plant development and productivity are phosphate (P) deficiency, manganese (Mn), Fe and Al toxicity. However, plants have developed a variety of coping mechanisms to deal with these pressures. These tactics include controlling Mn absorption, translocation, and distribution to prevent Mn toxicity, utilising both internal and external mechanisms to adapt to Al toxicity, and coordinating a number of P transport pathways to effectively take up P from the soil. The current understanding of the molecular mechanisms by which plants adjust to these limitations in acid soils, with a particular emphasis on the functions of transporters related to P efficiency and Al/Mn tolerance has been studied. Using gene editing in conjunction with other biotechnologies will help create new approaches to enhance plant adaption to acidic soils. Additional research is necessary as the molecular processes underlying plants' co-adaptation to various stressors in acid soils are mainly unexplored.

6.2 Advances in Soil Acidity Research

The study of Al resistance mechanisms in plants and the sustainable usage of acidic soils have advanced significantly in recent years. Al is one of the most common metals in the Earth's crust and is abundant in acidic soils. The necessity to lessen aluminium toxicity in acidic soils in order to increase plant productivity has increased due to the growing world population. While there has been success with a variety of approaches for introducing exogenous medications, the greatest potential lies in creating plants that can withstand acidic soil. In recent decades, significant progress has been achieved in understanding genetic pathways through the use of molecular technologies. Plants may have similar tolerance mechanisms even though their tolerance to Al varies greatly. The importance of external processes, especially the production of organic acids, in Al detoxification has been shown in numerous investigations. To track pertinent genes and provide helpful resources for molecular marker-assisted selection, several markers, including SNP (single-nucleotide polymorphism), RFLP (restriction fragment length polymorphism), and SSR (simple sequence repeats) markers, have been established. The use of marker-assisted selection to guide gene tracing developed from its association with gene-specific molecular markers. It has been possible to create gene-specific markers for marker-assisted selection using HvAACT1 and TaALMT1 (Bian et al. 2013; Raman et al. 2006).

It is anticipated that the effects of marker-assisted selection in breeding will be amplified by recent advancements in marker development. Although marker-assisted selection for Al tolerance is effective, markers are currently far from Al-tolerant genes (Bian et al. 2013). However, closer markers, gene-specific markers, and newly developed genome-wide association study (GWAS) analysis can improve the effectiveness of selection for Al tolerance. As several methods of tolerance can be coupled to generate greater tolerance, the finding of new genes will remain the main emphasis for enhancing Al tolerance in crop plants. It is possible that in the future, GWAS research could uncover novel Al-resistant genes, and molecular marker-assisted selection and intercrossing will yield Al-resistant cultivable cultivars that do well in acidic soils. For this reason, it is not necessary to enhance the acidic soil in order to attain yield and a sustainable use of acid soils.

6.2.1 Harmful Effects of Acidic Soils

The toxicity of acidic soils is largely caused by heavy metal toxicity, nutritional shortages, and soil acidity. High concentrations of H^+ ions in soil cause a detrimental impact on nutrient availability as pH falls (Wang et al. 2023). This reduces nutrient availability. Furthermore, low pH levels cause heavy metals like zinc (Zn), magnesium (Mn), copper (Cu), aluminium (Al), and iron (Fe) to become more soluble. This can be detrimental to plant growth because these heavy metals don't make up a large amount of what plants need to grow, and many of them can be toxic or hazardous (Ginocchio et al. 2009).

6.2.2 Sustainable Use of Acidic Soil

Agronomic techniques to mitigate the toxicity of acidic soils have historically centred on raising soil pH, with lime application being the most often used method. In order to enhance cell defences and decrease the absorption of Al in the soil, lime is supplied (Guo et al. 2006; Kinraide 1998). Although this method has been successful in improving plant development, it has a number of drawbacks. Lime can raise the pH of the topsoil, but it has little effect on the subsoil, which is typically contaminated with Al and inhibits plant roots' capacity to spread widely and take up water and nutrients. Furthermore, the expense of applying large amounts of lime to acidic soils may be a problem in developing nations. Lime can lessen the toxicities of H^+, Al, and Mn, but deficiencies in other minerals, including N and P, do not go away (Rao et al. 1993; Zhao et al. 2014). As such, care must be used while applying lime to reduce soil acidity.

6.2.3 Soil Remediation of Heavy Metals

Heavy metal-contaminated sites have become more noticeable as industry and urbanisation grow, especially due to excessive emissions from electroplating, smelting nonferrous metals, mining tailings, etc. The technology for cleaning up contaminated sites with heavy metals have advanced quickly in the last few years. More successful real-world applications have surfaced, and new and efficient remediation methods have been developed one after the other (Song et al. 2022). As such, a methodical summary of the existing state of affairs is important. For sites contaminated with heavy metals, certain commonplace soil remediation techniques are required. These include physical remediation, which involves replacing and heating soil, bioremediation, which involves phytoremediation and microbial remediation, chemical remediation, which involves chemical leaching, chemical stabilisation, electrokinetic remediation-permeable reactive barrier, and chemical oxidation/reduction, and various combined remediation methods that are thoroughly studied. A thorough analysis is also conducted of the contributing variables, benefits, drawbacks, remediation process, and real-world applications. According to Jiang et al. (2020), metals and metalloids having densities higher than 5 g cm^{-3}, such as As, Hg, Zn, Cu, and Ni, are classified as heavy metals (HMs). With the rapid growth of industry and urbanisation, heavy metal-contaminated soil has become more commonplace due to its high toxicity and durability (Bi et al. 2020; Men et al. 2020). Both anthropogenic interference inputs and natural sources contribute to the buildup of heavy metals in the soil. Volcanic eruptions, weathering of metal-rich rocks, and atmospheric volcanic emissions are examples of natural sources (Khan et al. 2021). Heavy metal buildup is a result of anthropogenic interference inputs such mining, smelting, burning coal, and other activities (Rajendran et al. 2022). The biological structure and function of the soil are deteriorated by heavy metal contamination, which also negatively impacts agricultural output (Mao et al. 2012). Secondary ecological and environmental problems might arise from the easy discharge of contaminated soil into the atmosphere and water. Furthermore, the food chain can introduce heavy metals into the bodies of humans or other animals. Human organs are severely and fatally affected by excessive heavy metal deposition, which can result in cancer, deformity, and mutation (Jan et al. 2015).

Low-cost technologies that are integrated into continuous agricultural production are particularly desirable. Research has been conducted in response to these demands to lower the levels of heavy metals in soil and crops. $Ca(NO_3)_2$ or NH_4HCO_3, Ca-Mg-phosphate, and K_2SO_4 were suggested as chemical fertilisers for soils contaminated with cadmium. Tests revealed that the leftover $CaCO_3$ from the sugar industry might be used to lower the bioavailability and uptake of Cd, Hg, and Pb by leafy greens. Compared to utilising regular cultivars, employing these cultivars could reduce the content of metal in the crop by almost 50% (Wu et al. 2018). Several remediation strategies, such as soil thermal desorption, soil washing, phytoremediation, and others, have been developed in response to the severity and prevalence of heavy metal-contaminated sites (Shah and Daverey 2020; Zhao et al. 2019a). Remediation technology for heavy metal-contaminated soils has advanced quickly, leading to the maturity of certain methods and more advanced studies on the mechanisms and influencing elements of remediation. Furthermore, a number of novel and efficient combination remediation solutions have gradually surfaced.

6.2.4 Mechanism of Tea Plant Al Tolerance

Remarkably, considering the general toxicity of aluminium, the tea plant needs aluminium for proper growth in addition to being tolerant of it (Hajiboland et al. 2013). For instance, it is commonly recognised that Al^{3+} directly promotes the growth of tea plants and other acidophiles, such as *Melaleuca cajuputi, Hydrangea macrophylla, and Melastoma malabathricum*, but not low pH (Osaki et al. 1997). Al resistance is generally enabled by two different higher plant mechanistic mechanisms. The first tactic is called Al exclusion, in which organic acid anions or other Al chelators are secreted into the rhizosphere to block Al^{3+} ions from reaching the root apex (Ma et al. 2001). According to Kochian et al. (2015), the second mechanism is Al tolerance, which entails changing the root cell wall's ability to bind Al and sequestering or compartmentalising Al^{3+} ions once they enter the root cytosol (also known as apoplastic detoxification). The main way that tea deals with Al is probably through Al tolerance, as it is an Al hyper-accumulator. According to studies, phenolics, oxalate, fluoride, and citrate are among the substances that help tea roots and shoots detoxify by chelating Al^{3+} ions (Bojórquez-Quintal et al. 2017).

6.2.5 *Molecular Mechanisms of Plant Adaptation to Acid Soils*

Many places have acid soils, which reduce plant productivity worldwide. Phosphorus (P) shortage and aluminium (Al)/manganese (Mn) toxicity are the main factors limiting plant growth and production in acid soils. However, plants have developed a variety of coping mechanisms to deal with these pressures. These tactics include controlling Mn absorption, translocation, and distribution to prevent Mn toxicity, utilising both internal and external mechanisms to adapt to Al toxicity, and coordinating a number of P transport pathways to effectively take up P from the soil. Here, we summarise our present understanding of the molecular processes via which plants adjust to these limitations in acid soils, with a particular emphasis on the functions of transporters related to P efficiency and Al/Mn tolerance (Che et al. 2023). Using gene editing in conjunction with other biotechnologies will help create new approaches to enhance plant adaption to acidic soils. More research is needed to fully understand the molecular mechanisms behind plants' coadaptation to various stressors in acid soils, as they are currently mostly understood (Che et al. 2023). Acid soils, which are characterised by a pH of less than 5.5, are among the most challenging soil types for global agricultural output. The humid northern temperate zone and the humid high rainfall tropical area are the two primary regions on Earth where acid soils are found (von Uexküll and Mutert 1995). Al is usually insoluble in neutral or slightly acidic soils. On the other hand, it is quickly dissolved into solution as poisonous Al^{3+} in acidic soils, which prevents root elongation and lowers nutrient and water intake (Kochian et al. 2004, 2015). Even though micromolar Al is thought to be harmful to agricultural plants, some native plants have a high tolerance for Al's toxicity, and Al can even promote the growth of these plants. Determining the mechanisms of Al tolerance in various plant species is crucial. Mn toxicity is a significant additional limiting factor in acid soils (Li et al. 2015). Mn is regarded as a crucial element for plant growth since it is involved in several metabolic processes. However, because the concentration of plant-bioavailable Mn^{2+} in soils increases 100-fold when soil pH lowers by one unit, Mn toxicity frequently occurs in acidic soils. While Mn toxicity mostly affects plant shoots and causes leaf chlorosis and necrosis, Al toxicity largely affects root growth. Despite being a crucial macronutrient for the growth and development of plants, most soils lack P. The form of P that is available to plants is orthophosphate, and because Al and Fe oxides fix inorganic phosphate (Pi), the concentration of available P is low in acid soils (López-Arredondo et al. 2014). P is hence a significant stressor for plant growth in acidic soils.

6.2.6 *Plant Adaptation to Multiple Stresses in Acid Soils*

In order to cope with the various pressures that coexist in acidic soils, plants have evolved defence systems. For instance, to chelate Al^{3+} and boost P acquisition, both

P deficit and Al toxicity promote organic acid production (Chen and Liao 2016). However, different plant species emit organic acids in different ways. Citrate exudation in rice is triggered by both P deficit and Al toxicity (Yokosho et al. 2011, 2016). Significant advancements in our knowledge of the molecular mechanisms underlying plant adaptation to acid soils have been accomplished within the past ten years. The absorption, translocation, distribution, and sequestration of Al, Mn, and P by transporters regulates plant adaptation to Al/Mn toxicity and P deficit in acid soils, and they thus play a significant role in plant adaptation to limitations. The transporters responsible for hyper-accumulators' resistance to Mn and Al, however, are still unknown.

6.2.7 Acid Rain Impacts

The term "acid rain" (AR) refers to a combination of dry and wet atmospheric deposition that decreases precipitation to a pH of less than 5.6. It is caused by higher-than-normal amounts of nitric (HNO_3), sulphuric (H_2SO_4), and acidifying chemicals. Smith (1872) undertook a thorough investigation of the substance in AR and documented some potentially dangerous side effects. Natural and artificial activities can provide several antecedents and causes of AR development. According to Zhang et al. (2007a, b), human-induced sources include burning coal, natural gas, oil, and agricultural emissions from intensive paddy farming, stubble burning, fertiliser and pesticide use, and lightning and other biogenic activities. Natural sources include volcanic eruptions, vegetation decay, and lightning.

Because acid rain has large-scale, cross-national negative effects on the ecosystem, scientists and policymakers have taken an interest in it. Over the course of the last century, acid rain (AR) has grown to be a significant environmental issue. Precipitation is referred to be AR when it reduces pH to less than 5.6. The primary causes of air pollution (AR) include the presence of sulphur dioxide (SO_2), nitrogen oxides (NO_x), ozone (O_3), and organic acids in air that is created by both natural and man-made sources. The leading SO_2 emitter, India, also exhibits a steady rise in NO_2 levels, which is what causes AR formation (Prakash et al. 2022). Due to their immobility, plants are inevitably exposed to AR, which has a detrimental effect on the surrounding environment. AR either directly damages photosynthetic systems and reproductive organs, resulting in reduced growth, productivity, and yield, or indirectly affects subterranean components including the soil and root system. In reaction to acid rain, genes that are crucial for abiotic stress defence in plants are also altered. AR causes soil acidification and alters the metabolism of carbon and nitrogen in balance, as well as the characteristics of litter and the activities of microorganisms and enzymes (Prakash et al. 2022).

Since the start of the industrial revolution, acid rain (AR) pollution has frequently resulted in natural disasters and economic loss globally (Grennfelt et al. 2019). Carbon dioxide (CO_2) and nitrogen oxides (NOx) are examples of acidic compounds that are typically produced by human activity-related emissions, including those from burning coal, industrial production, and vehicle exhaust (Vahedpour and Zolfaghari 2011; Yun et al. 2018). According to Ruiz-López et al. (2019), sulphur oxides (SO_x), concentrated sulphuric acid (H_2SO_4), and other acidic chemicals are also produced by naturally occurring occurrences including ocean foam, volcanic eruptions, and lightning-caused forest fires. The development of acid rain, or AR ($pH < 5.6$), is the result of acidic compounds reacting with atmospheric moisture. These chemicals can travel great distances and eventually settle as acid deposition (Shukla et al. 2012; Warren et al. 2017). According to Singh and Agrawal (2008), acid rain has the potential to deplete agricultural output, acidify soils and aquatic ecosystems, and erode infrastructure. Ultimately, this could lead to a decrease in hardwood habitats.

In terrestrial ecosystems, plants are crucial for both production and diversity. They also act as the primary interceptors and sensors of AR contaminants. Thus, the ways in which plants respond to AR can offer scholars a theoretical framework within which to investigate associated problems and assist land managers in creating technical countermeasures for the restoration and reconstruction of damaged ecosystems in AR-affected areas (Fu et al. 2019). There is a direct correlation between the acidity of AR and plant performance (Du et al. 2017; Shi et al. 2021). A reduction in photosynthesis and growth has been shown by simulation trials to occur in simulated AR (SAR) treatments, as the growth and development of plants' above- and below-ground organs (leaf and root, respectively) are suppressed (Shi et al. 2021; Du et al. 2017; Pham et al. 2021). Both the direct impacts of AR's acidity and the indirect effects of soil chemical properties mediated by AR on plants have been thoroughly investigated (Singh and Agrawal 2008; Shi et al. 2021).

In nations like North America, Europe, and East Asia, AR has emerged as one of the biggest environmental catastrophes (Singh and Agrawal 2007). Acid deposition incidents occurred often in China (Zhou et al. 2019). According to Li et al. (2017a, b) and Andrade et al. (2020), emissions of SO_2 and NOx, the two main drivers of AR, are predicted to increase until at least 2030, with India being the second-known emitter of SO_2. Over the previous forty years, there has been a growing trend in the incidence of acid rain ($pH < 5.6$) in India (Bhaskar and Rao 2017). Further increasing the acidity of depositions are emissions from agricultural operations caused by overuse of pesticides and fertilisers, which release ammonia (NH_3) and reactive nitrogen (Nr) species into the atmosphere (Sutton et al. 2017). The global occurrence of AR may have a negative impact on ecosystem components, changing the characteristics of litter and accelerating soil acidification (Fei et al. 2020), which may result in a drop in soil microbial communities (Wei et al. 2020) and forest declines (Zheng et al. 2019). AR pollution also adversely affects plant productivity, one of the most important elements of terrestrial ecosystems (Liu et al. 2018a, b). This results in visible symptoms such as necrotic spots, premature defoliation or ageing, growth inhibition, leaf loss, and other visible symptoms (Bobbink et al. 2010; Du et al. 2017). According to Shu et al. (2019), AR directly affects the sections above ground, which inhibits the functions of wax biosynthesis, intracellular H^+ ion buildup, and other detrimental ions in mesophyll cells. Owing to the production of reactive oxygen species (ROS), an excessive buildup of intracellular H^+ ions can cause oxidative stress

(Neves et al. 2009). Acidification of water bodies puts local animals and plants at risk of extinction by making their habitat inhospitable (Singh and Agrawal 2007).

Effect of acid rain on plant growth and yield

Crop development and productivity are negatively impacted by acid rain, which has a detrimental influence on the agricultural ecology (Singh and Agrawal 2004). It is generally known that herbaceous plants are more vulnerable to direct AR harm than woody plants are (Heck et al. 1986). Dicotyledons are more responsive to AR than monocotyledons are (Evans 1988; Knittel and Pell 1991). AR causes changes to the cuticle's thickness, trichome loss in the epidermis, cellular deformation, mesophyll cell collapse, stomatal cell occlusion, and the development of scar tissue, among other anatomical changes (Cape 1986; Da Silva et al. 2005). Chlorosis, necrosis, dehydration, wilting, early senescence, stunting, pathogen infection, and mortality are among the adverse consequences of simulated acid rain (SAR) on morphology (Milton and Abigail 2015).

Effect of acid rain on soil

Because soil is naturally dynamic and complex, it is easily impacted by AR. This leads to acidification of the soil, an increase in the exchange of H^+ with nutritional cations (Mg, K, and Ca), and leaching (Breemen et al. 1984). According to Mishima et al. (2013), a shortage of essential nutrients has an indirect impact on plant growth and soil fertility. By preventing legumes from transmitting signals that draw rhizobia, nutrient deficiencies prevent nodulation in plants (Sullivan et al. 2017). In addition, they indirectly prevent ectomycorrhizal fungal interaction with plants (Maltz et al. 2019), which lowers plant vigour and production. According to Ma et al. (2020), under Chinese cabbage farming, AR affected the chemical qualities of the soil. Spraying pH 3.5 SAR was found to lower soil pH by 0.21, 0.19, and 0.15 units at depths of 0, 4, and 8 cm in comparison to pH 7.0 (control). Nonetheless, there was no discernible change in soil pH between the pH 4.5, 5.5, and 7.0 treatments. Similar findings were made by Zhou et al. (2020), who discovered that SAR decreased the pH of the soil in both the rhizosphere and the non-rhizosphere when it affected Masson pine (*Pinus massoniana*) seedlings.

The industrialised world has addressed acid rain's complications to some degree by enacting emission standards for the gases that cause acid rain. The emerging world must adopt strict regulating regulations and conduct thorough and efficient emissions monitoring in order to prevent such issues. Moreover, the chemical makeup of AR is altered by rising NOx pollution globally. Given the paucity of research on these topics, a thorough evaluation and prediction of the effects of various forms of acid rain on plant development and function, biodiversity, and soil qualities are required. In order to get important insights into how diverse plant responses to acid rain stress differ, further research is required to evaluate the potential future effects of acid rain on various aspects of plants and ecosystems in India and around the world in an environment that is changing rapidly. The biomass, diversity, and activity of soil microbes are all greatly reduced by high AR frequency (Liu et al. 2021). Microorganisms in the rhizosphere have the potential to influence how soil-deposited AR pollutants are transformed (Fujii 2014). They may also cause plants to release hormones, like indoleacetic acid, which can control growth and reproduction in response to AR stress (Lu et al. 2018). An integral part of soils, microorganisms are crucial in balancing the impacts of AR on the soil–plant system. Unfortunately, prior research on the AR-soil microbe-plant route was lacking. According to Soudzilovskaia et al. (2019), the majority of plant species produce root symbioses with arbuscular mycorrhizal fungi (AMF), followed by ectomycorrhizal fungi (EcMF). These symbioses have a significant impact on plant growth and response to environmental stresses, such as AR. However, symbiotic microbes in plants can mitigate a number of the negative impacts of AR on plants, such as heavy metal toxicity. For example, Xia et al. (2021), and He et al. (2022), describe the inoculation of AMF and EcMF. Previous meta-analyses of the literature have well-documented the detrimental effects of acidic stress or nitrogen deposition on plant-associated soil microorganisms (Meng et al. 2019; Han et al. 2020). However, the response of different plant mycorrhizal types to AR stress has received less attention.

Effects on leaf micromorphology

As a result of AR's impacts on leaf blade micromorphology, plant photosynthesis may be negatively impacted in a number of ways, including the loss of epidermal cells and the waxy structure on the surface. As per Andrade and Silva (2017), AR has the potential to cause partial collapse of spongy tissue, deformation of stomata, and collapse of the epidermis.

6.2.8 Rhizosphere Management

Within 24 h of application, P delivered as water soluble Single Super Phosphate (SSP) is frequently converted into iron- and aluminium-bound complexes and may no longer be available for plant uptake, depending on the pH, clay, organic matter, sesquioxides, and phosphorous fixing capacity of acidic soils. According to Bhattacharya and Singh (1990), there appears to be less of this fixation in the case of rock phosphate (RP). In these conditions, phosphorous usage efficiency in acidic soils may be improved through rhizosphere-based P management (Kalidas-Singh 2013). This entails minimising phosphorous fixing in the rhizosphere and raising tissue phosphorous concentration for improved root formation during the early stages of crop growth. It also entails synchronising the rate of P mineralization in the rhizosphere with P uptake by the plant during different growth phases. These can be accomplished by increasing the number of Phosphate Solubilizing Microorganisms (PSM) in the rhizosphere, applying RP and PSM together to release P gradually over an extended period of time, and immersing seedling roots in an orthophosphate solution. The bound forms of phosphates can be dissolved into accessible monocalcium phosphate in soil by Phosphate Solubilizing Bacteria (PSB). This happens as a result of the release of metabolites that suppress pathogens, such as phytohormones, lytic enzymes, and organic acids like gluconic acid, as well as an increase in phosphatases' activity in the roots, which hydrolyses organic P compounds to enhance plant uptake of P (Richardson et al. 2009). Another effective way to manage acidic soils is to lime the wetted zone or use fertigation to supply calcium nitrate as a nitrogen source. Fallow period is necessary for the establishment of related soil chemical and physical conditions as well as for the climax vegetation to acquire the necessary nutrients. Sekiya and Yano (2010) state that a key tactic for lowering the need for fertiliser, bolstering crop establishment, and raising crop yields is seed priming using water and nutrient solutions, such as P and Zn. One might prevent a lot of unwanted interactions (such fixation of applied P) between the applied nutrient and the soil matrix by directly applying nutrients or nutrient sources to seeds. The combination application of lime and P fertiliser on acidic Kenyan soils has been shown in numerous research to improve soil accessible P and its use (Kisinyo 2011; Opala et al. 2014). For this reason, managing P-deficient acid soils for the long run as well as in the short term requires the simultaneous application of lime and fertiliser.

6.2.9 Use of Organic Materials (Organic Materials)

In order to lessen Al toxicity in acid soils, the use of OMs has been suggested as an alternative to liming (Lungu 1993). Al solubility is reduced as a result of the release and synthesis of organic compounds during the breakdown of OMs, which mixes with Al to produce a solid-organic material phase (Tang et al. 2007; Haynes and Mokolobate 2001). Additionally, there are several ways in which organic compounds interact with P in soils, which may have an impact on P sorption and release processes. Guppy et al. (2005) have suggested both direct and indirect processes for the rise in soil accessible P caused by the addition of OMs. In acidic soils, OMs have been shown to decrease soil acidity, decrease Al toxicity, and increase soil accessible P. Tithonia diversifolia green manure has been shown by Opala et al. (2010) to be effective in raising maize production in western Kenya because it can lower exchangeable Al levels in soils without necessarily raising soil pH. The ability of an organic substance to complex Al through organic acids released during its decomposition process was shown to be connected to the material's ability to lessen Al toxicity. In two acid soils, Opala et al. (2014) investigated the impact of a variety of organic elements with varying compositions that are frequently found on small-holder farms on the generation of dry matter in maize. They were contrasted with triple superphosphate (TSP) and lime.

In contrast to organic amendments, such as crop residue or legume-based green manure, which add nutrients as well as organic matter, organic fertilisers (OFs) are materials of plant, animal, or microbiological origin that have partially or fully decomposed and add plant nutrients (N, P, K, Ca, S, Mg, etc.) (Melero et al. 2007). The availability of OFs varies greatly between locations and even different land use circumstances. Industrial processing wastes, deadstock, and manures are the most frequent OFs with animal origins. Green manure, leftovers (vermicompost), and industrial and municipal wastes are examples of OFs with a plant origin (Bernal et al. 1998). OFs in plant litter and root exudates biofertilize soil under natural soil conditions (e.g. Bauer and Black 1994). Even while plants and animals are the primary sources, OFs with a microbiological origin are also significant, despite this fact being little known. Numerous microorganisms (bacteria, viruses, and fungi) can be found in soil, living there freely or adhering to animals and plants to form hosts. These organisms gather in surface settings and offer various ecological services, such as the breakdown of organic materials. These bacteria represent a significant source of OFs when they perish due to resource depletion or harmful changes in the soil ecosystems (Albiach et al. 2000). This is the primary reason why the first OFs in a barren soil originate from pioneer microorganisms and microscopic plants. The kind of microbial ecology that the different types of OFs are able to establish determines the regional variation in the microbe composition. Just like with OFs of animal and plant origin, the quantity of OFs of microbial origin is influenced by host soil types, soil conditions, and climatic variability (Belyaeva and Haynes 2009). For example, an alkaline soil may contain more bacterial than other microbe-derived organic fertiliser. Similar to this, there might be more fungal biomass (matter) in an acidic soil. According to Larney et al. (2011), the many sources of organic matter (OFs) that are accessible are crucial in establishing a suitable microbial ecology. This, in turn, influences the composition and availability of nutrients, moisture content, and the biochemical and characteristics of soils. OFs play a significant role in problem soils, especially acid soils where toxicity, low nutrient availability, and acidity are issues for overall soil usage and management (e.g. Michael et al. 2015; Materrechera and Mkhabela 2002). The regulation of acidity and toxicity—two significant factors that affect soil productivity—as well as their roles as reservoirs of soil nutrients underscore the significance of OFs (Pichtel et al. 1994).

Plant and animal organic fertilisers

As previously mentioned, plants, animals, and soil microorganisms (bacteria, fungus, and algae) are the primary sources of OFs. The animal-based organic fertilisers are fish emulsion, milk, urea (urine), bone meal, blood meal, manure, and manure tea. Cottonseed meal, molasses, legume cover crops, green manure cover crops, kelp weed, and compost tea are common sources of OFs of a plant origin. Depending on the biological and nutritional state of the plants and animals as well as the habitat in which the organic fertiliser sources were discovered, the fertiliser or nutrient content of all these sources of OFs may change. For example, an animal that is well-nourished and in good health may have a higher nutrient content than one that is poorly fed and has biological illnesses. In a similar vein, plant kinds and production methods have an impact on nutrient contents. A plant or animal's ability to interact with its surroundings influences the kinds and quantities of nutrients that can be obtained from them. The nutrients that are available to plants are greatly influenced by the parent materials and soil-forming elements of a soil that was used to grow plants or cultivate pastures. Given that OFs are only used for plant nutrition, the sources of the OFs, the current state of the soil microbial ecology, and the surrounding environment all have a significant role in determining what happens to the sources. When compared to other plant and animal debris, some are quite large suppliers of OFs.

6.2.10 Use of Acid-Tolerant Crops

Plant breeding programmes have produced germplasms tolerant to Al toxicity and/or low soil accessible P to address issues connected to soil acidity. Through improved microbial colonisation and symbiotic interaction with P solubilizing microorganisms in the rhizosphere, low soil availability tolerant genotypes can receive enough P even from sparingly soluble P (Oliveira et al. 2006). Low grain yields can be attributed to the sensitive crop germplasms' inability to express this gene, which prevents them from making optimal use of administered fertilisers and/or fixed P. Crop yields in acid soils could be significantly increased by adding this gene to cultivars that are sensitive to P deficiencies. The main factor limiting crop productivity in extremely acidic soils ($pH < 5.0$) is aluminium toxicity. As mentioned, liming is not a sustainable way to lessen its effects, which has led to the identification and application of genotypes that are tolerant of Al. The activation of membrane transporters by aluminium facilitates the release of organic acids from the root apex, the site of aluminium phytotoxicity. The released organic acids then form stable, nontoxic complexes with Al^{3+} in the rhizosphere, which is a major physiological mechanism of plant tolerance to aluminium (Magalhaes et al. 2007). Recent research has found Kenyan germplasms for sorghum and maize that are efficient at using P and/or tolerant of Al toxicity (Ouma et al. 2013; Matonyei 2010; Too 2011). These superior materials offer a solid basis for cultivar breeding that is resistant to Al toxicity and/or maximises P uptake.

6.2.11 Organic Matter Management

In the humid tropics, where shifting cultivation (in a small area) or plantation agriculture are common practices, organic matter management is crucial to preventing loss, particularly when the natural forest is cut and burned. Organic matter is critical for soil health and production in these areas. First, regardless of the kind of organic matter, mineralization and decomposition—which are comprised of four primary processes—supply cations for CEC, recycle nutrients, and enhance nutrient retention and micronutrient leaching in the soil (Michael 2021). There are several forms of organic matter accessible, ranging from green manure from cover crops and agricultural residues (estimated at 3.8 billion tonnes annually) to farm yard manure from animal operations, in contrast to the organic litter from forest trees (Lal 2005). Acid soils require careful consideration and use of these various forms of organic matter (Michael 2021). Significant amounts of CO_2 are evolved during the microbially assisted breakdown process, and this CO_2 dissolves in water to create H_2CO_3. According to Bolan and Hedley (2003), acid soils are unable of storing much CO_2, so the pH cannot drop below 5. Organic litter decomposes and returns nutrients to the soil, a process that keeps the soil fertile even in acidic soils. This process occurs naturally in forests, where nutrients are taken up by trees and organic litter falls back into the soil. When a rain forest is cleared and the fallen trees or debris are burned, the organic litter is reduced, microbial ecology imbalances cause lower decomposition, soil moisture is lost, and as a result, soil nutrients are lost (Michael 2021).

One way to control pore space stability and distribution, which encourages water losses, in cases where forest land has been cleared is to manage the stumps, logs, and other debris in a way that permits them to decompose and return to the soil (Chan et al. 2003). Due to the low water-holding capacity of sandy soils and the availability of negative charges on them, organic matter is important for both water retention and the leaching of cationic nutrients (Walker et al. 2004; Michael 2020). The effects of organic matter on the pH of soil vary and are dependent on the starting pH, base cations, organic anions, and the C:N ratio. The addition of organic matter raises, reduces, or has no effect on the pH of the soil under general land use systems (Xu et al. 2006). For instance, organic matter raises pH and neutralises acidity (H^+) in acidic soils (Michael et al. 2015). The pH rises as protons are used by the decarboxylation of organic anions and the release of hydroxyl groups during decomposition. As organic matter breaks down, soil bacteria utilise H^+ as an electron acceptor and lower its content (Mitchael 2015a, b). Conversely, nitrification of ammonium, ammonification of residue-N, and nitrate leaching reduce soil pH (Xu et al. 2002). On the other hand, adding organic matter may raise low pH and decrease alkaline pH. The sorts of organic matter applied and the type of soil in which it was applied are the sources of these conflicting outcomes (Michael 2013). The majority of our ASS investigations showed an elevation in pH with the addition of organic materials, for example (Michael et al. 2015, 2016, 2017. The application of organic matter through farmyard manure not only increased the productivity of rice and potato in acid soils of northeastern region of India (Sharma 1991; Sharma and Tripathy 1991; Sharma and Sharma 2009, 2011) but, also enhanced the N use efficiency (Sharma and Arora 1986, 1987).

6.2.12 Molecular Basis for Crop Al Resistance

We now know much more about the diversity of Al resistance genes and mechanisms, how Al and Al-induced signals regulate and trigger the expression of resistance genes, and how the proteins these genes encode function and are regulated (Kochian et al. 2015). This is an impressive amount of progress made over that time. By examining the novel Al resistance genes and mechanisms that have been discovered over the past ten years, it is necessary to assess the current state of our understanding of the physiological, genetic, and molecular grounds for crop Al tolerance.

Plant breeders, particularly in tropical breeding programmes, have been successfully breeding for increased crop Al resistance for more than 50 years by taking advantage of the substantial genetic variability in Al resistance seen in many crop species. The molecular biology and physiology of crop Al resistance have been extensively studied since the last Annual Review of Plant Biology article on acid soil resistance (Kochian et al. 2004) in order to identify Al resistance genes and their corresponding mechanisms. This is due to the agronomic significance of acid soils. As a result, the molecular, genetic, and physiological underpinnings of crop Al resistance have advanced significantly, and a multitude of published materials has been produced. Researchers have started to apply their fundamental knowledge of the role and interaction of Al resistance genes and their networks, as the field of crop Al resistance research has grown, to enable further advancements in Al resistance through molecular-assisted breeding and biotechnology methods, which can be utilised to build on the gains made by earlier crop breeding initiatives.

6.2.13 Aluminium Exclusion Through Root Organic Acid Exudation

Organic acid (OA) anions are exuded by roots of plants into the rhizosphere by Al-mediated root exudation. Here, the anions chelate Al^{3+} ions to produce harmless compounds that do not enter the root, making this the most thoroughly studied Al exclusion mechanism. Early in the 1990s, groundbreaking physiological research on Al resistance in wheat (Triticum aestivum) and snap beans (Phaseolus vulgaris) revealed a relationship between rising levels of Al-dependent exudation of Al-chelating OAs and mounting Al resistance (Delhaize et al. 1993). The Al-activated exudation of malate, citrate, and (in certain species) oxalate from roots was found to correlate similarly with variance in Al resistance, as evidenced by many investigations conducted on a variety of monocot and dicot species. Plant species that use this Al resistance mechanism differ primarily in three ways: (a) in which OAs (malate, citrate, or oxalate) are produced; (b) in how much of an exudate occurs; and (c) in how long it takes for Al to induce and activate the exudation. While many plant species only release one OA in response to Al stress, it is not uncommon to find plant species that release multiple OAs; such as, wheat and Arabidopsis release both citrate and malate in response to Al stress (Liu et al. 2009), suggesting that multiple release mechanisms or transporters are working in tandem in these species. The first insights into the molecular mechanisms mediating OA transport came from electrophysiological studies of root Al resistance mechanisms in the late 1990s and early 2000s. These studies showed that certain plasma membrane anion-permeable channels existed in protoplasts isolated from the root tips of Al-resistant wheat (Zhang et al. 2001) and maize (Zea mays) (Kollmeier et al. 2001; Pineros ad Kochian 2001), and that these channels were activated by extracellular Al^{3+}. It is not surprising that the first genes found encoded malate and citrate transporters in the roots of wheat, sorghum (Sorghum bicolor), and barley (Hordeum vulgare). The first genes driving Al resistance mechanisms were cloned more recently (Sasaki et al. 2004).

6.2.14 Aluminium Exclusion Through Other Compounds

Numerous organic chemicals, including phenolics, are released by roots and have the ability to chelate Al^{3+} ions in the rhizosphere. Although phenolic compounds are not as effective as OAs at chelating Al^{3+}, they nevertheless have a good capability because of the electrophilic nature of the oxygen atoms in the ^{-}OH groups of phenolic rings. Kidd et al. (2001) found that the roots of maize plants carrying an Al-resistant genotype emit catechol, catechin, and quercetin in response to Al. This finding led the researchers to hypothesise that root phenolic exudation, in conjunction with citrate release, may be involved in maize root Al exclusion. No further published reports of phenolics being exuded from the roots as a mechanism for Al resistance have been found, but the same research team discovered that the roots of an Al-resistant maize variety had significantly higher internal levels of caffeic acid, catechol, and catechin in response to Al stress than the roots of an Al-sensitive variety. This suggests that an internal mechanism for Al tolerance functions in conjunction with root tip Al exclusion mediated by citrate exudation (Tolra et al. 2009).

6.2.15 Root Organic Acid Efflux

A vital metabolic centre for all organisms, the tricarboxylic acid cycle includes the OAs malate and citrate, which are utilised by numerous plant species for Al resistance based on root Al exclusion (Berg et al. 2002). In the light of this, the plant's Al resistance mechanism is probably going to have a large carbon cost, therefore it should come as no surprise that it is highly restricted in both space and Al response. Since only the root apex needs to be shielded from harm owing to Al toxicity, it is well-documented that Al-activated root malate and citrate exudation is spatially localised to the root tips, lowering the carbon cost (Sivaguru and Horst 1998; Sivaguru et al. 2013). Sivaguru et al. (2013) demonstrated that Al-induced SbMATE gene expression was specifically localised to the epidermal and outer cortical cell layers of the sorghum root distal transition zone (DTZ, the root region between cell division and cell elongation), which they also showed to be the site of greatest Al-induced damage. They achieved this by using laser capture microdissection coupled to a high-sensitivity quantitative real-time polymerase chain reaction (qRT-PCR) assay. Numerous publications have shown that exposure to root Al not only causes the activation of the ALMT and MATE genes in many species, but also post-translationally activates the OA transport proteins, causing the release of OA anions. This adds another layer of regulation. Hence, the root releases malate and citrate only when there are sufficient levels of Al^{3+} ions in the rhizosphere to require root protection from Al toxicity. This is important because acid soils are usually quite heterogeneous spatially with regard to pH and Al saturation (Wong et al. 2008), and acid soil pH can vary by more than 1 unit over relatively small distances. Particularly in systems that use conservation agriculture, surface applications of liming materials can produce abrupt vertical fertility gradients and increase the vulnerability of plants to water stress (Butterly et al. 2022). Thus, the most efficient approach is probably to directly incorporate lime and other amendments into the subsoil layers; however, this might be expensive and time-consuming, expose the soil to erosion, and encourage the mineralization of organic matter.

Numerous studies and efforts have been directed into determining the best substances and techniques to treat acid subsoils with little to no soil disturbance. Tests have been conducted on lime, gypsum, calcium oxide, organic additions, and biological solutions. The best approach would be combining lime with organic materials since it improves the physical and nutritional qualities of the soil while also offering ameliorative and nutritional benefits. In Brazil, the establishment of cover crops on modified soils is recommended to enhance the physical state of the soil profile. According to Butterly et al. (2022), the continuous biopores created by roots or macrofauna enhance macroporosity and maximise water infiltration, both of which are essential for improving subsurface layers. It is preferable to take steps to slow down the rate at which subsurface acidity develops rather than trying to ameliorate acid subsoils, as this may not always be feasible or lucrative. It is still necessary to develop cost-effective and integrated ways in the future.

6.2.16 Biochar for Amending Soil Acidity

The organic resource can be returned back to the farm with the help of compost and biochar (Agegnehu et al. 2019). Reducing the amount of organic materials means using less labour and money to bring them back to the field. This can be achieved by composting crop leftovers and manure together or by partially carbonising them to create biochar. f resistant organic matter and nutrients have not been leached out or denitrified throughout the composting process, farmers may be able to benefit from applying biochar to their soils; however, the labile carbon may be lost to the atmosphere. Because it can nourish soil organisms, which are in charge of several advantageous processes in the soil, the labile carbon would be advantageous to the soil (Agegnehu et al. 2017). Studies on liming acid soils have shown that biochar can raise pH at a pace of roughly one-third that of lime, raise calcium levels, and lessen the toxicity of aluminium on acid soils (Lehmann et al. 2003; Steiner 2007). To help with the implementation of integrated soil fertility management (ISFM), appropriate policies for sustainable land use intensification as well as the organisations and systems required to put them into practice and assess their effectiveness are crucial. While agricultural yield has increased dramatically due to the use of improved fertilisers and seeds, ISFM treatment has the potential to greatly enhance poor agronomic efficiency (Vanlauwe and Zingore 2011), with all the associated financial benefits for farmers. Although there have been initiatives to advance seed and fertiliser technology, other initiatives, such as the creation of site-specific decision guides that facilitate the resolution of more complicated problems, can be launched to direct farming communities towards comprehensive ISFM, which includes appropriate organic matter management and localised technology adaptation. All things considered, ISFM adoption in the face of climate change has the potential to help tropical agro-ecosystems' environment in addition to increasing farm output and farmers' revenue.

Based on recent studies, biochar has the potential to be added to acid soils as an amendment due to its liming properties and other advantageous qualities. Being alkaline by nature, biochar has the potential to improve upland acid soil productivity due to its liming action, ability to retain water and nutrients, highly recalcitrant nature, and carbon sequestration capacity, as suggested by recent research (Berek 2019). The most common amendment techniques for acid soils are liming the soil or adding organic ingredients. But there is a chance that the soil will be overlimed, which could have negative consequences on the distribution of liming materials, trace element deficiencies, and cation imbalances, among other things. It has been demonstrated that biochar, a rich carbon byproduct of biomass pyrolysis with a restricted oxygen supply, works well as a soil amendment and carbon sequestration agent. It can be made at the high heat pyrolysis temperature of 300–700 °C using a single biomass source or a combination of biomass materials. Recent research has shown that biochar has a good pH buffering capacity and liming potential, which suggests that it may be able to reduce soil acidity. Additionally, biochar is being recognised as an adsorbent for the removal of contaminants or heavy metals. Because of its high porosity, biochar may also provide a safe haven for the soil microbial community. Researchers from all around the world have been interested in the positive effects of biochar on acid soils, and in certain nations, the results of these studies have been incorporated into official regulations. Using biochar presents a significant chance to address both soil acidity and climate change issues at the same time. Nevertheless, there is a dearth of studies on the use of biochar to improve severely worn acid soils.

Effects of biochar

The nature of biochar is alkaline. Both the organic and inorganic components of biochar are the source of its alkalinity (Fidel 2012). More precisely, according to Fidel et al. (2017), there are four divisions in the biochar alkalinity method: carbonates, other inorganics, low-pKa organic structures, and other organics. Alkalinity in biochar is caused by carbonate content (Mukome et al. 2013), especially in high temperature pyrolysis biochar (Yuan and Xu 2011). According to Fidel et al. (2017), there is a strong link between the quantity of basic cations in the biochars and their alkalinity, while Berek and Hue (2016) found a strong correlation between the quantity of basic cations and its carbonate content. The alkalinity of biochar produced at low pyrolysis temperatures may also be attributed to oxidative functional groups such phenolics and carboxylic acids (Wang et al. 2014; Yuan and Xu 2011). The primary source of the biochars' strong cation exchange capacity and excellent buffering ability is pH change resistance. For instance, biochars have a high buffering capacity because of the negative charge produced by the deprotonation of their surface functional groups and the carbonates (Dai et al. 2014).

Benefits of biochar

According to a meta-analysis by Biederman and Harpole (2013), biochar increased plant yield and productivity. For instance, it was demonstrated that the short-term benefits of biochar included increasing its capacity to store water and adding nutrients like K and P, whilst the long-term benefits included liming and nutrient retention. It has been demonstrated that alkaline biochar is more successful than acidic biochar at raising soil pH, lowering exchangeable Al and Fe, and increasing P availability in acidic soils. According to a global-meta study by Jeffery et al. (2017), applying biochar to improve crop output in the Tropics proved to be more successful than liming. Incorporating biochar into acid soil is anticipated to yield several positive outcomes, the principal ones being raising pH, reducing P fixation, and reducing Al toxicity and, to a lesser extent, Mn or Fe toxicity. The oxygenic surface functional groups connected to the surface of biochar, as well as its alkalinity, which is derived from the basic cations in the ash impurity, are what primarily contribute to its favourable effects (Chintala et al. 2013; Smider and Singh 2014; Wang et al. 2014; Yuan and Xu 2011). Carbonates and oxides are the basic cations found in biochar. The alkalinity of biochar is derived from these cations, which will take the role of lime in creating the ion OH^-, which will raise the pH of the soil by neutralising excess ion H^+ (Berek and Hue 2016). The reduced activity of monomeric Al^{3+} and other species by precipitation as a result of the elevated soil pH, a similar mechanism by which lime neutralises pH and detoxifies Al phytotoxicity, may be responsible for alleviating Al toxicity. Fixing P will then be released into a form that plants can use by raising the pH of the soil and precipitating Al and Fe. Biochar is mostly composed of carbonates and oxides as cations. According to Berek and Hue (2016), the alkalinity of biochar is a result of these cations, which act as lime to create the ion OH^-, which raises the pH of the soil by neutralising excess protons. The method by which lime neutralises pH and detoxifies Al phytotoxicity may also be responsible for reducing the activity of monomeric Al^{3+} and other species by precipitation as a result of the higher soil pH. This mechanism could explain the diminished Al toxicity. Elevating the pH of the soil and precipitating Al and Fe will then release fixed P into a form that plants can use. After the soil's acidity decreases, micronutrients like zinc and molybdenum will become more easily accessible, and the activity of helpful microbes like rhizobium, mycorrhizal fungus, and phosphate-solubilised bacteria will also intensify. As a result, it appears possible that biochar will take the place of lime in solving soil acidity issues. To achieve the best results, it is necessary to have a thorough understanding of the interactions between acid soil and biochar. This is because the properties of biochar vary due to the wide diversity of biomass used in its production and the variable conditions under which it is produced. As a result, applying biochar to different soils can have positive, neutral, or negative effects (Tusar et al. 2023).

Fertilisation, acid rain, nutrient leaching, removal of agricultural leftovers, and other agricultural methods are some of the many elements that influence soil acidity (Kalkhoran et al. 2019). According to Goulding (2016), acidic soils have a range of detrimental effects on soil biology and plant performance, such as base cation leaching, aggregate structure instability, increased metal toxicity, and decreased nutrient availability. Because it prevents cell division and elongation, a high Al concentration in acidic soils limits root growth. Crop productivity and output are ultimately negatively impacted by this. Seguel and others (2013). Because acid soil is so common and getting worse, it is critical to create long-term strategies to stop acidification in the future and to restore acid soil that already exists in order to preserve food security. To reduce soil acidity and boost agricultural output, researchers and agricultural experts have suggested a variety of techniques, such as liming, integrated nutrient management, and adding organic manure to acid soils (Thakuria et al. 2016). Due to lime's provision of basic cations like calcium and magnesium, the pH of the soil rises and Al toxicity decreases (Behera and Shukla 2015). Long-term lime treatment in agricultural soil can lead to re-acidification and promote physical hardness, but it can also leach out mineral nutrients like Mg^{2+} and NO^{3-} (Mosharrof et al. 2021). An attractive substitute for the existing methods is biochar, a type of pyrogenic carbon-rich substance that may be added to a variety of soils, including acidic soils, to boost agricultural production and yield. According to Qambrani et al. (2017), biochar is regarded as a natural and environmentally beneficial soil amendment due to its resistance to microbial degradation, which may allow it to lower or stabilise soil pH over an extended length of time. According to Steiner (2016), the constituents of biochar include moisture, ash components, labile carbon, and other volatile chemicals.

Biochar for soil acidity reclamation

Pyrolysis temperature is one of the driving parameters that may have a big influence on the properties of biochar. The CEC and volatile matter were negatively correlated with the rising pyrolysis temperature, indicating that biochar produced at higher temperatures would have less CEC and volatile carbon. According to Shen et al. (2008), there is a correlation between the decline in CEC and the production of basic functional groups as well as a reduction in the amount of acidic functional groups, especially carboxylic functional groups at higher temperatures. The later effect may be shortlived, while the earlier effect will increase with the passage of time (Mosharrof et al. 2021). In addition, biochar can improve soil microbial functions, including symbiotic association with mycorrhiza that helps to acquire nutrients under acidic conditions (Singh et al 2022; Mrunalini et al. 2022). Higher temperatures result in more ash, which raises the pH of the biochar and gives it more liming properties. Conversely, lower temperatures result in higher levels of carboxylic and phenolic functional groups, which help buffer the pH during deprotonation (Tomezyk et al. 2020). In addition to pyrolysis temperature, other parameters that affect the synthesis of biochar include the biomass's moisture content, lignin, and cellulose content (Tripathi et al. 2016). The intrinsic basic cations in the biochar can also aid in lowering soil pH, as can the carboxylic and phenolic functional groups in the surface of the biochar.

Interactions between biochar and acid soils

Following its application to the soil, biochar interacts with organic matter and soil minerals. These interactions can be rather intense, especially in acidic soils where the relative quantity of Fe/Al-bearing minerals is high (Weng et al. 2022). Among the various ways that biochar interacts with soil minerals are: (a) the electrostatic and ligand exchange between negatively charged biochar functional groups and positively charged mineral surfaces; (b) the electrostatic and negatively charged biochar, multivalent cations, and negatively charged soil minerals; (c) the ligand exchange and electrostatic interactions between negatively charged soil minerals and positively charged biochar surfaces; (d) electrostatic interactions between negatively charged functional groups of biochar, multivalent ions, and positively charged mineral surfaces; (e) ligand/electrostatic interactions between positively charged soil minerals and soil organic matter. In addition to these chemical interactions, soil microorganisms (bacterial biomass and fungal hyphae) may promote soil aggregation, and biochar may have physical interactions with soil minerals and organic matter. These interactions might vary greatly depending on the qualities of the biochar (fresh and old), with aged biochar and soil minerals interacting more. However, the interactions between minerals and biochar may boost soil carbon stabilisation and contribute to nutrient cycling (Zhang et al. 2023a, b).

The pH of the soil will rise due to the decarboxylation of organic anions and the negatively charged functional groups consuming protons (Wang et al. 2014). Al adsorption may be facilitated by the oxygenic functional groups in biochar, such as carboxylic acids and phenolics. Al in the soil solution can also form complexes with carboxylic acids, which lessens the toxicity of Al to plant growth (Qian et al. 2013). These processes are comparable to those via which organic amendments reduce the toxicity of aluminium in acidic soils (Hue 2011). Therefore, it is anticipated that biochar will outperform traditional acid soil ameliorants.

One sustainable material that has emerged as a possible solution for soil carbon sequestration is biochar, which is enhanced in carbon through biomass pyrolysis. The use of biochar as a useful technique for soil C sequestration appears promising. It is obvious that it has the ability to increase C storage, improve soil qualities, and reduce greenhouse gas emissions (Li and Tasnady 2023). To fully achieve this potential, though, more study is required to comprehend the intricate relationships between soil processes and biochar, as well as to create sustainable methods for producing and using it. Biochar improves soil fertility, nitrogen retention, and water-holding capacity when added to soils. Because of its porous nature, beneficial microbes can live there, enhancing soil health and encouraging plant growth (Liao and Thomas 2019). Moreover, biochars have an amazing molecular structure with strong chemical and microbiological stability.

Challenges of biochar application

The temperature of pyrolysis, residence time, heating rate (rapid or slow), degree of oxygen limitation, and feedstock characteristics (such as cellulose or lignin concentration, moisture status, and particle size) are some of the variables that affect the properties of biochar. Many of the techniques for altering biochar, such as chemical oxidation and mineral impregnation, are costly, time-consuming, and labour-intensive. Taking into account the purchasing perspective of farmers, it can be challenging for them to determine the type of biochar they require because biochar packet labels lack a uniform classification system. Nonetheless, the International Biochar Initiatives are working to create a classification system for biochar. To maintain and enhance soil health, clear guidelines would reinforce the use of biochar as a soil addition (Loehrlein 2020). Potential problems that could limit the use of biochar as an amendment include the kind of biochar, the production conditions, the properties of the soil, and the amount of biochar utilised (Jatav et al. 2021). Establishing standardised practices and controls for quality assurance is essential. The high expense of mass-producing and utilising biochar may hinder its widespread adoption. Cost-effective manufacturing processes and application strategies must be developed in order to increase the availability of biochar to farmers and land managers.

6.2.17 Soil Acidity and Land Degradation

A decline in the productivity and quality of the soil is referred to as soil degradation, which is a part of the process of land degradation. The term "land degradation" describes the loss or decrease in the land's production, which results from a variety of natural processes that are frequently sped up by human disturbance (Lal 1993). Land usage, climate change, overpopulation, and urbanisation are the main factors contributing to land degradation. Reduced soil quality and a diminished chance of future life forms surviving are the results of land degradation (Fitzpatrick 2002). There are three main types of this worldwide threat: desertification, deterioration caused by humans, and natural degradation. According to Fitzpatrick (2002), induced degradation happens more quickly than natural deterioration as a result of improper land use and management.

Desertification, which affects drylands that make up around 40% of the planet's land surface, is the most severe type of degradation (UNEP 1992). There are repercussions of land degradation both on and off site. While off-site consequences—such as flooding, sedimentation, and a decline in water quality—occur in the surrounding areas, on-site effects are felt directly where degradation occurs (a drop in the productive capacity of land) (FAO 2011; Stringer 2017). Compared to soil degradation, which is defined as a reduction in the quality and productivity of the soil, land degradation has a larger scope. The properties of the soil, the climate, and the flora, as well as human influences like land use, soil management, and farming/cropping systems, all influence how quickly soil degradation occurs (Lal 2001). Globally, almost 75% of land is degraded (Gibbs and Salmon 2015). According to the five worldwide evaluations of soil deterioration conducted between 1977 and 2003, the range for global degradation is estimated to be between 15 and 63%, whereas the range for dryland degradation is between 4 and 74% (Safriel 2007). According to the Food and Agriculture Organisation and the Intergovernmental Technical Panel on Soils

(2015), land degradation is primarily caused by salinization, organic carbon losses, and water erosion, with Latin America and the Caribbean accounting for 14% of the world's land degradation.

The potential for agricultural land production is decreased by acidification, a severe type of deterioration similar to salinization. Increased nutrient leaching, erosion, and the permanent degradation of silicate minerals in the soil are all consequences of the acidification process, which stops plants from using water (Fitzpatrick 2002). Naturally occurring in regions with lower precipitation levels, soil acidity can be exacerbated by human activity. Degradation can be used to identify the following morphological and visual findings that were noted in the field:

- Topsoil erosion, as shown by gullies and rills on the exposed soil surface.
- Water erosion, as shown by gullies, rills, exposed roots, etc.
- Terrain change brought on by mass movement.
- Physical degradation, as seen by low agricultural yields and Acidification.
- Chemical and Biological Deterioration.
- Predominance of unproductive grassland vegetation.

According to Lal (1999), there is a direct connection between nutrition, food security, environmental quality, and soil deterioration.

6.2.18 Nanotechnology

Growing technical expertise has significantly improved scientific communities' ability to reason logically and form theoretical ideas into research that is almost exclusively focussed on products. "There's plenty of room at the bottom" was one of Feynman's imaginative thoughts that led to the identification and rise of nanotechnology as one of the major emerging technologies of the next decades (Feynman 1960). The application of nanotechnology in several fields has greatly improved the qualities of applications (Nikalje 2015). The assessment of nanoparticles is one of the key facets of nanotechnology. A new age known as the "nano-revolution," which involves with particles with at least one dimension at the nanoscale, has been brought about by the use of nanomaterials (Syed et al. 2016). Nanomaterials have garnered significant attention and are regarded as the particles of the century (Zhao et al. 2014). Because of their physicochemical qualities that are dependent on size and structure, these nanomaterials have advanced in many different fields and are now considered the most inspiring materials in the technological world (Porchezhiyan and Noorjahan 2016). Due to their huge surface plasmon resonance, tiny nanomaterials have strong reactivity, which greatly enhances their analytical capabilities (Baker et al. 2015). In the world of agriculture, sustainable contests persist despite the tremendous growth of nanotechnology. Given the current circumstances, the agriculture industry needs to ensure that a fast growing global population receives adequate nourishment (Mba et al. 2012). Regretfully, it is believed that every year, around one-third of the world's crop production is lost to waste during the production to consuming phase. A variety of technological solutions are emerging to address this and raise post-harvest standards. One such development is the application of goods based on nanomaterials to modern agriculture reform (Sekhon 2014). Recent years have seen a significant focus on the use of nanoparticles in agriculture, which is predicted to revolutionise current agricultural methods (Fraceto et al. 2016). As of right now, the agricultural community is dealing with a wide range of issues, such as declining yield, crops being destroyed by pests, losing soil fertility, not having enough water, unpredictable weather patterns, and soil contamination from pollutants in the environment that have caused significant loss (Pouratashi and Iravani 2012). Creative thinking and thoughtful design are crucial for overcoming these obstacles. According to Parisi et al. (2015), nanoagroparticles are nanosized particles intended to mitigate issues associated with the agricultural industry. In addition to being effective pesticides, herbicides, insecticides, and fungicides, these nanoagroparticles can also stimulate plant growth. However, one of the main issues is how to produce simple nanoparticles without using any hazardous materials in the synthesis process. According to Kavitha et al. (2013), the majority of conventional procedures are frequently associated with a number of drawbacks, including the production of excessive heat, the need for hazardous materials for synthesis, the need for complex facilities, etc. Therefore, scientific groups are working to create simple, easy-to-use methods for creating nanoparticles with the needed characteristics. Using biogenic sources to synthesise nanoparticles is one such topic. The biogenic sources are microorganisms and plants, or their byproducts, which mediate the synthesis and stabilise nanoparticles while acting as reductive agents.

Because of its endless applicability in a wide range of sectors, nanotechnology has become an essential component of evolving societies (Ramsden 2016). The creation, processing, and utilisation of sophisticated nanomaterials for a wide range of applications to address present and future issues encountered by different industries is the primary driver of the nano-based industry's expansion (Shang et al. 2019). Thousands of commercial goods based on nanoparticles are presently on the market, and this number is likely to rise in the future (Hansen et al. 2020), according to various estimations. Engineered nanomaterials (ENMs) are nanoparticles that are purposely made, as opposed to naturally occurring nanomaterials like montmorillonite, kaolinite, saponite, etc. Materials that have at least one dimension of 100 nm or less are considered nanomaterials (Kreyling et al. 2010). The specialised structural and functional components, either internal or surface-based, of these materials typically result in unique and extraordinary physicochemical and biological features when compared to their bulk equivalents (Ji et al. 2021). Even though nanoparticles have unparalleled possibilities and applications, there is a significant risk to the environment and public health (Chen et al. 2017).

Guidelines and directives are needed to assess potential hazards and provide recommendations for the safe use of nanotechnology (Coles and Frewer 2013). Researchers should characterise the toxicological characteristics of nanoparticles and save the information they collect in an easily accessible database. To remove workplace dangers, safety measures should be given top priority in nanoparticle manufacturing facilities. Additionally, labs should establish safety protocols for the safe handling, application, and disposal of waste materials connected to nanoparticles. In order to ensure that there are no negative effects from the use of nanoparticles in agriculture, caution must be exercised when marketing their advantages. In addition to validating these technologies against current technologies, more study is required to ascertain the viability, sustainability, efficiency, application, and releasability of goods based on nanotechnology in real-world settings. Additionally, more field and in vivo experiments involving nanomaterials laden with pesticides, or nanopesticides, are needed. The current limitation is the lack of long-term trial data, which is necessary to obtain a complete understanding of the efficacy of these pesticides. In contrast to traditional pesticides, regulatory bodies have not clearly defined nanomaterials and nanopesticides. In contrast to conventional pesticides, the effects of nanopesticides may depend on the absorption, bioavailability, concentration, and toxicity of the nanoparticles as well as the ratio of the active linked to them, as Kookana et al. (2014) described in great depth. Furthermore, little is known about the issue of pesticide resistance and possible strategies for reducing its prevalence through the addition of nanoparticles. It is impossible to develop regulatory criteria for risk assessment without employing sophisticated analytical methods. A wide variety of nanoformulations must be accessible for usage in the future since regulatory bodies often rotate and revise the ways that pesticides are applied in the field, restricting or outlawing the use of chemical pesticides.

There is an accumulation of nanomaterials in the soil as a result of their specialised development for use in agriculture as nano-pesticides, nano-fertilisers, and nano-based biosensors (Khan et al. 2021). The variety and abundance of microorganisms in the soil microbiome are significantly impacted by the presence of engineered nanomaterials (ENMs). They also have an impact on important microbial functions like mineralization, nitrogen fixation, and activities that stimulate plant growth. The behaviour of ENMs in soil is usually influenced by a variety of soil and ENM characteristics. Silver and zinc oxide nanoparticles have been produced and investigated in great detail because of their outstanding industrial qualities and well-known antibacterial activity. Thus, it is crucial to comprehend how these ENMs affect the soil microbiome and associated activities at this point. These studies could contribute to boosting agricultural output by providing the data required to control the use of ENMs in sustainable agriculture (Khan et al. 2021).

Many ENMs typically reach the soil during their lifecycle, either directly or indirectly, under a variety of conditions, and in significant quantities. For this reason, it is critical to look into the effects of these ENMs on the soil ecosystem as well as the pathways by which they enter the soil (Maghsoodi et al. 2019). Studies must be done on ENM behaviour and persistence in the soil in addition to their release pathways. Studying the life cycle of ENMs in soil and how the characteristics of the soil affect their activity can also help to better understand the function of nanomaterials in soil pollution (Aslani et al. 2014). For example, the chemical form, mobility, bioavailability, and toxicity of ENMs may be influenced by the physicochemical characteristics of the soil, including its pH, temperature, hydrostatic pressure, and ionic composition (Schlich and Hund-Rinke 2015).

Plant health and microbiome

While there are currently a number of commercial products based on ENMs on the market, such as fertilisers that enhance plant growth and yield, pesticides that control plant diseases, and sophisticated sensors that track plant health and soil quality, the widespread use of ENMs in agriculture is still in its early stages (Iavicoli et al. 2017). While nanopesticides have a low amount of active ingredients and great efficacy, enhanced solubility, and durability, ENM-based nanofertilisers are known to improve the efficiency with which nutrients are used (Pirzadah et al. 2020). Furthermore, a number of other advantages of agricultural products based on ENMs have been well documented (Hofmann et al. 2020). As a result, the chance of ENM contamination of soil is rising dramatically along with the growing usage of ENMs in agriculture (Iavicoli et al. 2017). In fact, because of their increased interactions with soil and plant microbiome, huge quantities of ENMs in the soil may have a significant impact on the plant rhizosphere (Kibbey and Strevett 2019). In particular, if ENMs are discharged into the subsoil with the intention of achieving immediate advantages, the plants may store them and they may have adverse effects over an extended period of time (Mousavi Kouhi et al. 2015). In the rhizosphere, plants release a variety of exudates that promote microbial growth. Meanwhile, microorganisms work in concert with plant roots to support plant growth by facilitating a variety of nutrient cycles (Zhang et al. 2014). On the other hand, exudates from plants, extracellular materials made by microorganisms, and rhizosphere microbial communities are all greatly impacted by the presence of ENMs (Gao et al. 2018). Consequently, either directly or indirectly through altering the plant and soil microbiome, the presence of ENMs in the soil has a significant impact on plant health. We go into more depth about these effects below.

Soil remediation through nanotechnology

Globally, there is growing concern about the use of nanotechnology for soil remediation in order to remove toxins from soil environments (Kristani et al. 2021). Diffuse pollution and point sources pollution are the main sources of pollutants in soil (Petruzzelli et al. 2010). According to Cachada et al. (2018), the main methods for transferring contaminants into the soil include volatilization, leaching, and suspension. Pollutants released into the soil are carried via its pores and subsequently deposited there, where they are also held on the surfaces of organic and mineral materials. A greater quantity of nanoparticles will enter the environment as nanotechnology is used more often for soil remediation. An unexpected risk to the ecosystem and soil environment as well as an indirect harm to human health could result from the high concentration of nanomaterials. According to Guerra et al. (2018), the primary worry is that nanomaterials transported through the soil environment could enter the groundwater system, contaminate drinking water supplies, and have a negative effect on terrestrial life and human health. Nanoparticles' detrimental effects on microorganisms are one of the main obstacles facing nanotechnology. Nanomaterials' toxicity will have an impact on microorganisms and prevent the soil environment's enzymes from working. A number of investigations have been conducted to lessen and prevent the toxicity of nanomaterials to soil organisms.

6.3 Future Investigations

The three primary challenges with all of the acid soil types that have been studied are generally acidity, toxicity, and nutrient deficiencies. This suggests that the problems are complex and call for specialised and methodical management (Agegnehu and Amede 2017). More advanced research in soil acidity management, restoration, and nutrient management is needed for these three major issue of acid soils (Tully et al. 2015). In the humid tropics, certain study needs to be location specific, while other research might be more general and broadly regional in scope. According to Agegnehu et al. (2014), research on managing acid soils must therefore be directed towards key stakeholders, such as small farmers, land managers, and curators, or simply controlling the acid environment for general protection and usage. Researchers need to concentrate on multiple topics in certain places related to significant acid soils, as demonstrated by the three primary problems with acid soils in the humid tropics. Farms, the overall environment, and site-specific locations that require extra care or attention—like protected areas and Ramsar wetlands—are often considered to be these areas (Agoume and Birang 2009). The inherent acidity of humid tropical soils is mostly caused by the leaching of cations during periods of excessive rainfall, which makes management of these areas essential.

6.3.1 Future Issues

Without a doubt, the demands of society should guide the development of all essential soil issues, which are crucial to the field of soil sciences and its advancement (El-Ramady et al. 2019). As a result, soil scientists need to play a highly proactive role in maximising their nation's potential and minimising its constraints. According to Terribile et al. (2013), this position may enable soil scientists to make real contributions to overcoming social restrictions. As proposed by Merino et al. (2016), the following scientific challenges can be incorporated in the future soil sciences concerns:

- Carbon sequestration in decommissioned soils;
- Pollution and gas balances from soils contributing to the greenhouse effect;
- Nutrient imbalances in agricultural and forest soils;
- Efficiency of soil and water in rangelands, and in intensive agriculture;
- Soil acidification and impoverishment in soils submitted to percolation;

- Soil contamination in areas supporting intensive agriculture (greenhouses); and
- Soil reclamation and decontamination of contaminated areas;
- Use of Artificial Intelligence (AI) protocols addressing farming system options for soil acidity identification and amelioration.

We feel that the following aspects of acid soils need further comprehensive studies:

- Use of nanotechnology in solving the problems related to soil acidity;
- The role of soils in global climate changes'
- Breeding cultivars to withstand soil acidity;
- Impact of organic nutritional sources on soil acidity stress;
- Reclamation of degraded acidic soils;
- Agroforestry on acid soils;
- Search of Arbuscular mycorrhizal associations and efficient P- solubilizers.

6.3.2 Impact of Climate Change on Soils and Vice Versa

Because of their unique characteristics, soils are impacted by climate change and may even be a contributing factor in it. Changes in rainfall patterns brought about by climate change may have an impact on soils through increased soil erosion and runoff, decreased organic matter supply, depletion of chemicals, and degradation of the soil's structure and porosity. The provision of various essential ecosystem services and sustainable development objectives, such as the production of food, feed, fibre, fuel (or biomass), biodiversity, carbon storage, water supply, and additional cultural and social functions, depend heavily on soil. Consequently, soils may experience a loss of vital functionalities due to a depletion of essential components. Although soils are the second-largest carbon storage after oceans, they also have the potential to release significant amounts of greenhouse gases into the atmosphere, such as carbon dioxide, methane, and nitrous oxide. When it comes to how climate change affects soils, variations in temperature or precipitation patterns have an impact on agricultural yields, and an increase in extreme weather events has the potential to completely destroy harvests on vast tracts of land (Streck and Gay 2017). There is a substantial correlation between soil and climate change. Thus, soil deterioration, conservation, and management are now influenced by changes in the global climate. These issues, and climate changes in particular, have become research hotspots globally (Adams et al. 2017; Chilkoti et al. 2017; Gao et al. 2017; Liu et al. 2017; Zhang et al. 2017). The adaptation and mitigation of climate change depend heavily on soils. When N fertiliser is applied to fields, drained peatlands and manure on fields release carbon dioxide and methane into the atmosphere. Nitrous oxide emissions from these sources are the most frequent gas emissions from soil that are linked to agricultural operations. Mitigation measures that are anticipated include reducing the gas emission from the soil. Techniques for mitigating or lowering soil emissions consist of the

- Improvement of rice-growing techniques.
- The restoration of degraded areas and cultivated organic or peaty soils.

- Improved management practices for livestock and manure.
- The creation of N-fertiliser application methods that lower emissions of nitrous oxide (Berga 2016; Streck and Gay 2017).
- Reduction in methane emission from rice growing areas of the globe through use of alternate wetting /drying, aerobic rice production and system of rice intensification (SRI).

The ability of the soil to retain moisture, resist erosion, and enhance ecosystem biodiversity may all be improved by soil carbon sequestration. According to Lozano García et al. (2017) and Zhang et al. (2017), soil carbon sequestration enables soils to take up carbon from the atmosphere and store it in the soil, which helps with both adaptation and mitigation. However, the loss of carbon from soils frequently causes soil deterioration by quickening the processes of erosion and mineralization, changing land use, and creating desertification (Bojko and Kabala 2017; Islam and Nursey-Bray 2017; Streck and Gay 2017). As per the UN General Assembly's presentation of "the 2030 Agenda for Sustainable Development," the concept of "land degradation neutrality" is a reflection of the growing political consciousness regarding the need for enhanced protection of soil (Flasbarth 2017; Boer et al. 2017; Ginzky et al. 2017).

6.3.3 Potential Contribution from Soil Scientists

The capacity of soil scientists to tackle various issues pertaining to agriculture, forestry, and the environment is often cited as a predictor of the future of their field. The dynamic and multipurpose role of soils has been highlighted by these challenges. Yet, Terribile et al. (2013) stated that sophisticated technical tools are unquestionably necessary due to the complexity of these earlier difficulties. In any given country, the demands of the nation's society determine the needs of the land and its potential contribution. This implies that planning should address all soil-related risks, sustainable soil management, and land use. Because soil is so important as a natural resource, for potential productivity, and for all other ecosystem processes, these plans should also prioritise protecting the soil. Emerging soil challenges should address both contemporary and historical societal demands, as well as the possible role that soil scientists might play in addressing these needs. Therefore, it is the great task of soil scientists to improve and advance soil sciences and their challenges, which might be accomplished by addressing the following issues:

- Boost the C-stock in agro-ecosystems;
- Enhance digital soil evaluation;
- Encourage and improve sustainable output;
- To prevent pollution of groundwater;
- To produce biomass on marginal areas;
- To establish a system for food quality and traceability;
- To enhance soil remediation processes; and

- To mitigate and adapt to the anticipated changes in climate;
- Nature of Al-hydroxyl interstratifications in mixed mineralogy with 2:1 minerals such as vermiculite, Chlorite and Mica in the soil matrix.

6.3.4 Solving Emerging Problems

There are many future trends in soil sciences. These trends require soil professionals to diversify their expertise, maintaining and deep understanding of soil science disciplines, e.g. chemistry, physics, and biology, to tackle new needs, technologies and new mathematical and conceptual tools. These trends also should implement more studies in basic and applied soil sciences among soil scientists to create a large synergistic interconnection between soil science branches. The future of soil sciences should be much productive in knowledge relevant to societal needs and should empower soil sciences in society. Future soil scientists must have not only an integrated and comprehensive view but also an improved interaction between the soil sciences and policymakers, as well as stakeholders (Terribile et al. 2013). To aid in the advancement of soil sciences in this approach, the following four major points were identified:

- Raising knowledge of the policy cycle as the foundation for organising soil research,
- Offering a variety of options rather than specific answers when addressing specific sustainability issues,
- When mobilising knowledge, adhere to the knowledge chain; and
- Enhance communication (Bouma 2010; Terribile et al. 2013).

6.3.5 On-Farm Research

It is necessary to conduct research on the application of organic matter, specifically on how to incorporate it into the soil by bulk mixing as opposed to applying it on the surface as surface mulch and under various moisture regimes. Acid sulphate soils (ASS) have been extensively studied, and it has been found that adding organic matter improves ASS more when soil conditions are aerobic (75% field capacity) or anaerobic (flooded, 100% saturation capacity) (Michael et al. 2016).

Farm research should concentrate on managing acidity on farms for farmers' advantage (Agegnehu et al. 2021). It has been mentioned that lime should be applied, however affordability is a problem in many economies due to the high cost of lime (Foley et al. 2011; Goulding 2016). Thus, management requires methodical approaches to the introduction of new crops, pasture, or crop varieties that can withstand acidity (de Sant-Anna et al. 2017). However, research on the best ways for farmers to get and apply this material is still lacking, and is a crucial approach

(Haynes and Mokolobate 2001). Reiterating the local understanding that the addition of organic matter alleviates nutrient deficiencies and reduces acidity and toxicity needs to be done (Haile and Boke 2011). Farmers must determine whether locality-specific sources of organic matter are easily accessible and reasonably priced (Ageg-nehu et al. 2014). While a lot of work has gone into translating our understanding of Al-resistance mechanisms from a variety of crops and model plants that are not acidophiles (Delhaize et al. 2004; Kochian et al. 2015), it's possible that studying species that are fully adapted to acid soils will yield more crucial information that will improve the breeding of crops that are both acid- and Al-tolerant. This strategy works well for tea since it grows best in acidic soil, can withstand high levels of Al, and even needs Al to thrive to its full potential (Hajiboland et al. 2013). If it were possible to help other crops develop a comparable resistance to Al, it would be a significant advancement. Today, tea is grown in more than 60 countries, with 6.3 million hectares of plantations covering 8.57 million tonnes of production globally (FAO 2017). This viewpoint seeks to stimulate additional analysis of the processes by which tea plants use and tolerate Al, as our knowledge of the tea response to Al is currently quite restricted. This will help to prepare the ground for the breeding of crops that will be acid-soil tolerant in the future.

6.3.6 Designing Future Crops for Acid Soils

A significant portion of the world's cultivatable land area is made up of acid soils, which makes a wide range of crop species susceptible to Al stress. These include many tree crops (such as 90% of coffee species, 60% of cocoa, and 100% of oil palm), legumes (such as 35% of soybean), root and tuber crops (such as 60% of white potatoes, 80% of sweet potatoes, and 100% of cassava), and cereals (20% of barley, 20% of maize, and 13% of rice) or Al stress (von Uexküll and Mutert 1995). Soil acidification is becoming a far bigger issue than it was in the past due to rising nitrogenous fertiliser inputs and rising acidity from rain. For instance, over the previous few decades, the pH of the soil has dropped by 0.13–0.8 in the main crop-producing regions of China (Guo et al. 2010). Al stress now affects a lot more crops as a result. Furthermore, Ulrich et al. (1980) have recognised Al toxicity as a significant contributing factor to the decrease of forests. Thus, the creation of species tolerant to Al is crucial for maintaining agricultural output and ensuring the world's food security, but it's also necessary for forest restoration, which enhances the ecological environment on a worldwide scale.

The primary focus of current breeding efforts to produce crops with enhanced Al tolerance is on Al tolerance per se. Tea plants, on the other hand, may offer a better model for future Al-tolerant crops because they are not only extremely tolerant of Al but also actually depend on it for optimal growth. Rather than just exhibiting Al tolerance, these crops may also show Al stimulation of growth and productivity, which would completely adapt them to acid soils. According to this theory, growing crops that are resistant of Al may take three different approaches in the future. First, it is necessary to develop a strong resistance to aluminium, maybe by combining effective compartmentalization with aluminium-exclusion mechanisms. Secondly, crops need to be able to be stimulated to grow by Al. A deeper understanding of how Al is used by tea plants and how it promotes tea growth at the molecular and genetic levels will be crucial to achieving this crucial second phase in the tea research process. Molecular breeding of these systems into other crops will be made possible, in particular, by the eventual identification of the critical genes and pathways mediated by the Al-dependent growth enhancement of tea plants. Third, it is imperative to develop the capacity to limit the translocation of Al into the tissues of food crops. Edible crop portions must have their Al content closely regulated to prevent unsafe levels of accumulation because Al is a possible neurotoxicant that can induce cell injury (Yokel 2000) and has also been connected to Alzheimer's disease (Exley 2007). These factors contribute to the very low Al concentration and, in particular, the very low Al bioavailability in tea infusions (Flaten 2002), which makes tea consumption both safe and healthful. Because tea plants contain significant levels of polyphenols that have germicidal qualities, traditional Agrobacterium-mediated gene transformation has only extremely low effectiveness in tea (Mondal et al. 2004). Novel high-efficiency, species-independent plant transformation technologies have recently been developed by a number of labs. These systems include zygote-based transformation and regeneration (Toda et al. 2019) and nanomaterial-based gene delivery (Kwak et al. 2019). These new techniques could be used to increase tea's transformation efficiency. In spite of these unresolved issues, "Genomics"-based methodologies offer significant prospects for identifying pivotal genes implicated in Al tolerance or Al growth stimulation. Recent endeavours in this regard have been conducted via transcriptome, proteome, and ionomic analyses of complete tea root or leaf specimens (Li et al. 2017a, b; Yamashita et al. 2020).

Research has shown that implementing pasture or fodder crops, or crop types that are tolerant of acid soil, is necessary for a certain kind of agroecosystem and a customised strategy for controlling soil acidity (Wang et al. 2006; Halimi 2011). In addition to the crops' resilience to stress, what matters most is how the different ecological services they provide—such as the cycling of nutrients and moisture retention, the establishment of a particular microbial ecology, and the return of leaf litter and root exudates to the soil—contribute to the administration of acidity. Plants use either hyperaccumulation or tolerance as a means of surviving in environments with high toxic levels and acidity.

It's necessary to conduct research on the best times to grow a given kind of crop. One example of study may be the rotation of hyperaccumulator annuals or perennials with less tolerant crops, or the mixing or intercropping of hyperaccumulator annuals with sensitive crops. According to Agegnehu and Amede (2017), the strategy needs to incorporate techniques like strip cropping and relay such that the yearly accumulator is harvested before the crop. For the purpose of controlling acidity and facilitating the return of carbon and cations to the soil, the hyperaccumulator's harvested materials can be burned (pyrolysed) at high temperatures and then returned to the soil as biochar. Lehmann et al. (2003) and Steiner (2007), two recent studies, demonstrated that applying biochar raised pH and Ca levels and decreased Al toxicity. In addition to

Al toxicity prevailing in acid soils (<PH 4.5), phosphorus deficiency is encountered in acid soils across the globe. As reported by Pan et al (2022), inspite of associated gene (phoA for organic P, ppa for Inorganic P, phoB for P-starvation) known with P-uptake and the mechanism of P-uptake by plant from soil, the molecular translocation of P- in plant cells is still unclear through P-signalling process. It is necessary to develop the system to monitor signals of Input/outcome of P-signalling pathways. With the help of genomic techniques (GWAS and CRISPR/Cas 9), elite plant/crop allelic variation with potential ways of P-efficiency breeding can be unearthed in the near future.

6.3.7 Organic Nutrient Sources

Utilising fully decomposed material from microbial, animal, or plant sources, biofertilisers serve as supplies of nutrients for the soil as well as ameliorants or detoxifiers. Microbes that have died, animals, and plants, or their byproducts, are the primary sources of biofertilisers. What kind of microbial ecology forms in the presence of the sources determines how bioavailable the biofertilisers are. Given everything, further research is required to determine the relative benefits of organic fertilisers in the effort to improve food production on problematic soils in a changing environment. Among these is the fact that, despite the fact that the fundamental mechanisms addressing nutrient deficiencies, acidity, and toxicity are known, information regarding the precise amount of organic fertiliser given by microorganisms, animals, and plants in various surface settings is lacking. Tree leaf sources are having Ca content varying from 4.2 to 19.7 g/kg dry foliage in hedge row tree species such as *Tephrosia* sp., *Crotolaria* sp., *Indogoferra* sp. As reported by Datta and Singh (2005–06), such Ca-rich tree leaf sources can be utilised in acid soil amendment.

Under acidic, alkaline, saline, drought- or flood-prone soil conditions, studies must determine the sorts and kinds of organic fertilisers that various microorganism species contribute to organic soil fertility. Research on live microorganisms and their interactions in biofertilisers is a crucial area of study that should complement efforts being made in the field of organic fertiliser research. How much organic fertiliser is added needs to be investigated under specific farming or cropping systems, including slash-and-burn agriculture, when the original microbial ecology is lost or a new but complicated one is established due to various crops being grown. For example, crop productivity is increased when non-legume crops are interplanted with legumes. It is necessary to determine the kinds and varieties of organic fertilisers that legumes and other agricultural plants, as well as a particular type of microbe, give. Even though these studies are technically challenging, current molecular and genome techniques enable the determination of the kind of organic fertiliser that crop plants and soil microbes contributed at various growth stages and crop production management settings. Research is needed since the majority of plant and animal sources of organic fertiliser do not have their complete nutrient contents (macro- and micronutrients) readily available in the literature.

6.3.8 Current Challenges in Biochar Production

Feedstock availability and sustainability

Biochar manufacturing requires a steady supply of biomass feedstock. However, ensuring a sufficient and sustainable supply of feedstock can be difficult, especially when thinking about large-scale biochar applications. It is crucial to prioritise the use of ecologically sustainable feedstock sources that don't interfere with food production or damage ecosystems. A number of feedstocks, such as grape residues, distiller's grains, sugarcane residues, palm oil residues, apple pomace, and sawdust, have been investigated recently for their feasibility in the manufacture of biochar (Xu et al. 2011). Interestingly, biochar yields from agriculture leftovers were higher than those from forestry residues. Grape residues showed the best biochar yields of all the agricultural residues tested, indicating its potential as a workable biomass source for biochar production. The production of biochar needs a consistent supply of biomass feedstock. It might be challenging to guarantee a steady and adequate supply of feedstock, nevertheless, particularly when considering large-scale biochar applications. Using environmentally sustainable feedstock sources that don't harm ecosystems or impede food production must be given top priority. Many feedstocks have lately been examined for their viability in the production of biochar, including grape residues, distillers grains, sugarcane residues, palm oil residues, apple pomace, and sawdust (Xu et al. 2011). It's interesting to note that biochar outputs from forestry residues were lower than those from leftovers in agriculture. Out of all the agricultural residues studied, grape residues had the highest biochar yields, suggesting that they could be a viable biomass source for biochar production. Given the abundance of agricultural and forest residues (Kan et al. 2020), biochar production exhibits sustainable strength from an economic perspective. This allows for the use of biochars derived from wood residues, which have the potential to have higher heating values and be applied in metallurgical and power generation contexts (Tomczyk et al. 2020), as well as the reduction of CO_2 emissions at the site (Wang et al. 2020). In a similar vein, biochars made from agricultural leftovers have increased plant nutrients and certain physicochemical properties that enable better nutrient recovery, nutrient usage efficiency, and environmental applications (Li and Chan 2022). Moreover, careful study and strategic assurance are required to create a sustainable biomass supply chain that connects the biomass source location to the biomass conversion site. Many inorganic and organic contaminants, such as dioxins, furans, polycyclic aromatic hydrocarbons (PAHs), polychlorinated biphenols (PCBs), and heavy metals, can generally have an impact on the efficiency and applicability of biochar (Freddo et al. 3012). Therefore, when using biochar, questions come up about how emissions from ironmaking and energy production will affect the environment, as well as how potentially toxic elements (PTEs) will move around and be bioavailable, and how these elements will leach when doing tasks like improving soil and reducing pollution (Terzano et al. 2021). The issues regarding toxicity are relatively lower when biochar is created at higher temperatures. This is explained by the fact that PTEs in biochar are less volatile and are concentrated in the residual ash, as opposed to the higher emissions that are frequently linked to the coal or coke conversion processes used in the production of iron (Freddo et al. 2012). Furthermore, it is less likely that harmful trace elements will leach out or become bioavailable in biochar, particularly in higher pH circumstances (Januševicius et al. 2022). Apart from PAH production, vaporisation also affects the concentration of polycyclic aromatic hydrocarbons (PAHs) in biochar. Interestingly, according to Buss et al. (2016), both of these processes—PAH vaporisation and formation—are accelerated by raising the pyrolysis temperature. Because of the more favourable conditions for their generation, the formation of PAHs normally increases as the temperature of pyrolysis rises. Nevertheless, an increase in the vaporisation of PAHs from the biochar may also result from the higher temperature (Buss et al. 2016). The lack of a continuous trend in PAH content in biochar with temperature observed in the literature could be explained by the interaction of these two mechanisms. However, more investigation is required in this area due to the paucity of thorough systematic investigations, especially with regard to the modes of occurrence, total concentrations, behaviour, and fate of dangerous pollutants in biochar as well as their chemical speciation or sequential extraction. Using the right pyrolysis procedures and technologies is essential to reducing the negative effects on the environment and human health caused by the emissions of toxic gases during the synthesis of biochar. Retort kilns are one example of this, since they have been shown to drastically cut down on incomplete combustion product emissions. In order to reduce hazardous emissions, retort kilns work by internally recirculating and burning the pyrolysis gases.

Early publications on biochar research, including those by Martinsen et al. (2015), Islami et al. (2011), and Berek and Hue (2016), suggested that biochar might be useful as an amendment for acid soils. Therefore, more research is required to determine how adding biochar to acid soils can improve soil productivity, increase agricultural output, and mitigate the effects of climate change. Therefore, in order for researchers to proceed, a research framework needs to be created. Mapping the prospective feedstock for biochar is necessary, then the production routes must be determined. The products, biochars, should next be characterised using a local or generic standard appropriate for acidic soils. To evaluate the impact of biochar on soil characteristics, carbon sequestration, greenhouse gas emissions, plant development, and yield, field, pot, and greenhouse experiments are required. Figure 6.2 illustrates the framework and research priorities for conceptual biochar in acid soils.

As a substitute for traditional methods of adjusting soil acidity, biochar holds great promise for bolstering global food and energy security in the future. Biochar development has a lot of prospects since biochar feedstock is widely available, it is inexpensive to produce, there is a lot of interest in research, and the community is familiar with using the result.

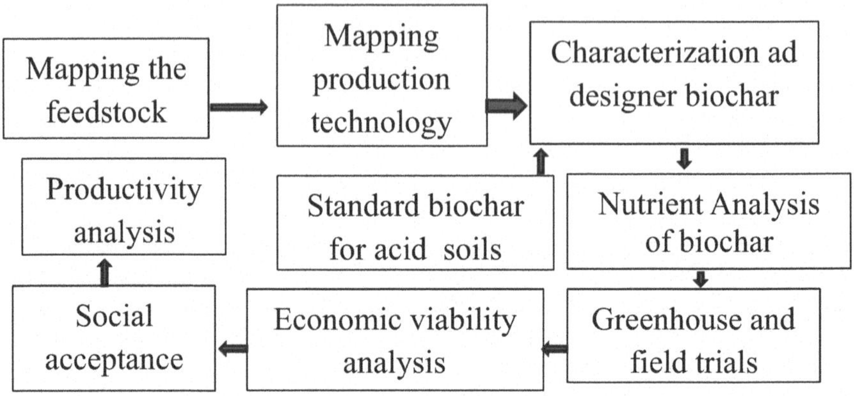

Fig. 6.2 A biochar research priority and a framework for acid soils. *Source* Based on Berek (2019)

6.3.9 Nanotechnology for Soils

A balanced relationship between environmental preservation and agricultural productivity can be achieved by putting integrated water and soil management strategies into practice (Montzka and Bunning 2015). Agriculture could undergo significant disruption in a number of industries thanks to the tremendously promising field of nanotechnology, which manipulates matter at the nanoscale. When it comes to crop yield, soil health, resource management, and sustainable food production, the use of nanotechnology in agriculture provides creative answers. This article discusses the various applications of nanotechnology and gives an outline of its potential in agriculture. A crop's development and yield may be improved by nanoparticles because of their special characteristics at the nanoscale. Growth regulators, insecticides, and nutrients can all be transported to plants specifically by means of engineered nanoparticles. It has been demonstrated, for instance, that critical nutrient-loaded nanoparticles enhance plant nutrient uptake, hence increasing plant growth and yield (Raliya and Tarafdar 2013). In order to minimise their negative effects on the environment and maximise their effectiveness, nanoparticles can also be used as vehicles for the controlled release of agrochemicals (Giraldo et al. 2014). Precision agriculture benefits from nanotechnology since it makes it possible to monitor and evaluate crop health, environmental parameters, and soil conditions in real time.

6.3.10 Nano-enabled Soil Amendments

When it comes to changing soil properties and interactions, nanoparticles (NPs) have shown a great deal of promise due to their distinct physical and chemical characteristics at the nanoscale. NPs affect soil structure, nutrient availability, and water dynamics when they are introduced into soil environments. They do this through interacting with organic matter, soil particles, and aqueous components. Utilising nanoparticles' benefits while reducing potential hazards requires an understanding of the mechanics behind the interactions between nanoparticles and soil. Complex mechanisms involving different physical, chemical, and electrostatic forces interact with soil constituents through nanoparticles. Van der Waals forces, surface charge interactions, and electrostatic attraction all have a big impact on how NPs behave in soil. The functional groups and charges that make up a nanoparticle's surface chemistry determine how attracted it is to organic materials and dirt particles. Because they have an impact on soil aggregation, porosity, and compaction, nanoparticles can change the structure of soil. To improve soil structure and lessen erosion, for example, nanoparticles can act as a bridge between soil particles and encourage aggregate stability. In sandy soils, titanium dioxide nanoparticles increased aggregate stability and water retention, according to Liu et al. (2015). This led to an improvement in soil structure. Furthermore, NPs have the ability to modify the distribution of soil pore sizes, which affects nutrient diffusion, root penetration, and water movement. The availability of nutrients and soil's ability to retain water can both be improved by nanoparticles. Essential nutrients can be transported by engineered nanoparticles, which guarantee regulated release and precise delivery to plant roots. For instance, studies have looked into the possibility of using silver nanoparticles to improve crops' uptake of nutrients. According to Wang et al.'s (2018) research, the addition of silver nanoparticles to agricultural soils boosted the availability of phosphorus, which in turn promoted plant growth.

According to Younis et al. (2021), developing "precision farming" systems through the effective use of nanotechnology is one way that nanotechnology plays a critical role in modern farming. Every barrier to using agri-nanotech products (agri-NPs) in farming has been noted and examined, taking into account the products' life cycle assessment (LCA) in plants and soil. The assessment of the effects of different nanotech goods on crop development with respect to conventional market technologies has received a lot of attention. In order to increase public understanding and acceptance of nanotechnologies in agro-environmental domains, it also highlights rules, legislative policies, and economic outputs.

In order to fully understand the applicability and compatibility of integrated approaches in agri-nanotechnology and their potential for sustainable agriculture, researchers and academics are currently facing significant challenges. In the world of agriculture, sustainable contests persist despite the tremendous growth of nanotechnology. Nowadays, agriculture is a response to the need to feed the world's continuously expanding population (Baker et al. 2017). Roughly one-third of the world's crop yield is thought to be lost each year. Numerous factors, including pest infestation, microbial diseases, weeds, natural disasters, low soil fertility, and many more, are to blame for the loss. Many technological solutions are used to get around these restrictions, but most of them have drawbacks of their own. As a result, research into the application of nanoparticles in agriculture is progressing slowly but has the potential to revolutionise the current agricultural system (Baker et al. 2017). Applications of these nanoparticles can be extremely beneficial given the present worldwide food shortage situation. Adverse impacts from the widespread use of chemical agrochemicals, which are known to produce biomagnification in an ecosystem, can be mitigated by nanotechnology. Since there are few publications on the application of nanotechnology in agriculture, current research points to the development of nanoagroparticles as a promising avenue for the future that could revitalise the current agricultural system.

References

Abera T, Molla A, Feyissa A, Liben M, Woyema A, Admasu L, Bekele A (2006) Barely research and development in Ethiopia: proceedings of the 2nd national barely research and development review workshop. Holetta Agricultural Research Center, Ethiopia 28–30 November 2006

Adams VM, Álvarez-Romero JG, Capon SJ, Crowley GM, Pressey RL (2017) Making time for space: the critical role of spatial planning in adapting natural resource management to climate change. Environ Sci Policy 74:57–67

Agegnehu G, Amede T (2017) Integrated soil fertility and plant nutrient management in tropical agro-ecosystems: a review. Pedosphere 27:662–680

Agegnehu G, Vanbeek C, Bird M (2014) Influence of integrated soil fertility management in wheat and tef productivity and soil chemical properties in the highland tropical environment. J Soil Sci Plant Nutr 14:532–545

Agegnehu G, Srivastava A, Bird MI (2017) The role of biochar and biochar-compost in improving soil quality and crop performance: a review. Appl Soil Ecol 119:156–170

Agegnehu G, Yirga C, Erkossa T (2019) Soil acidity management. Ethiopian Institute of Agricultural Research (EIAR). Addis Ababa, Ethiopia. ISBN: 9789994466597

Agegnehu G, Amede T, Erkossa T et al (2021) Extent and management of acid soils for sustainable crop production system in the tropical agroecosystems: a review. Acta Agric Scand B Soil Plant Sci 71:852–869

Agoumé V, Birang A (2009) Impact of land-use systems on some physical and chemical soil

Albiach R, Canet R, Pomares F, Ingelmo F (2000) Microbial biomass content and enzymatic activities after the application of organic amendments to a horticultural soil. Biores Technol 75:43–48

Amede T, Legesse G, Agegnehu G, Gashaw T, Degefu T, Desta G, Mekonnen K, Schulz S, Thorne P (2021) Short term fallow and partitioning effects of Green manures on wheat systems in east African highlands. Field Crops Res. https://doi.org/10.1016/j.fcr.2021.108175

Amede T, Legesse G, Agegnehu G, Gashaw T, Degefu T, Desta G, Mekonnen K, Schulz S, Thorne P (2021) Short term fallow and partitioning effects of Green manures on wheat systems in east African highlands. Field Crops Res. https://doi.org/10.1016/j.fcr.2021.108175

Andrade GC, Silva LC (2017) Responses of tropical legumes from the Brazilian Atlantic rainforest to simulated acid rain. Protoplasma 254:1639–1649

Andrade GC, Castro LN, da Silva LC (2020) Micromorphological alterations induced by simulated acid rain on the leaf surface of *Joannesia princeps* Vell. (Euphorbiaceae). Ecol Indic 116:106526. https://doi.org/10.1016/j.ecolind.2020.106526

Aslani F, Bagheri S, Muhd Julkapli N, Juraimi AS, Hashemi FSG, Baghdadi A (2014) Effects of engineered nanomaterials on plants growth: an overview. Sci World J 2014

Baker S, Kumar KM, Santosh P, Rakshith D, Satish S (2015) Extracellular synthesis of silver nanoparticles by novel Pseudomonas veronii AS41G inhabiting Annona squamosa L. and their bactericidal activity. Spectrochim Acta Mol Biomol Spectrosc 136:1434–1440

Baker S, Volova T, Prudnikova SV, Satish S, Nagendra PMN (2017) Nanoagroparticles emerging trends and future prospect in modern agriculture system. Environ Toxicol Pharmacol 53:10–17. https://doi.org/10.1016/j.etap.2017.04.012
Bauer A, Black AL (1994) Quantification of the effect of soil organic matter content on soil productivity. Soil Sci Soc Am J 58(185–193):563
Behera SK, Shukla AK (2015) Spatial distribution of surface soil acidity, electrical conductivity, soil organic carbon content and exchangeable potassium, calcium and magnesium in some cropped acid soils of India. Land Degrad Dev 26:71–79
Belyaeva ON, Haynes RJ (2009) Chemical, microbial and physical properties of manufactured soils produced by co-composting municipal green waste with coal fly ash. Bioresour Technol 100:5203–5209
Berek AK (2019) The potential of biochar as an acid soil amendment to support indonesian food and energy security—a review. Pertanika J Trop Agric Sc 42(2):745–759
Berek AK, Hue NV (2016) Characterization of biochars and their use as an amendment to acid soils. Soil Sci 181(9/10):412–426
Berg JM, Tymoczko JL, Stryer L (2002) The citric acid cycle. In: Biochemistry, 5th edn. Freeman, New York, pp 465–487
Berga L (2016) The role of hydropower in climate change mitigation and adaptation: a review. Engineering 2(3):313–318
Bernal MP, Paredes C, Sánchez-Monedero MA, Cegarra J (1998) Maturity and stability parameters of composts prepared with a wide range of organic wastes. Bioresour Technol 63:91–99
Beyene S, Kibret K, Erkossa T (2023) Future/emerging soil issues. In: Beyene S, Regassa A, Mishra BB, Haile M (eds) The soils of Ethiopia. World soils book series. Springer, Cham. https://doi.org/10.1007/978-3-031-17012-6_13
Bhaskar VV, Rao SP (2017) Annual and decadal variation in chemical composition of rain water at all the ten GAW stations in India. J Atmos Chem 74:23–53. https://doi.org/10.1007/s10874-016-9339-3
Bhattacharya NG, Singh B (1990) Transformation of applied phosphorus and its availability in acid soils. Two Bud 37(1):24–30
Bi X, Zhang M, Wu Y, Fu Z et al (2020) Distribution patterns and sources of heavy metals in soils from an industry undeveloped city in Southern China. Ecotoxicol Environ Saf 205. https://doi.org/10.1016/j.ecoenv.2020.111115
Bian M, Zhou M, Sun D, Li C (2013) Molecular approaches unravel the mechanism of acid soil tolerance in plants. Crop J 1(2):91–104
Biederman LA, Harpole WS (2013) Biochar and its effects on plant productivity and nutrient cycling: A meta-analysis. Glob Change Biol Bioenergy 5(2):202–214
Bobbink R, Hicks K, Galloway J, Spranger T, Alkemade R, Ashmore M, De Vries W (2010) Global assessment of nitrogen deposition effects on terrestrial plant diversity: a synthesis. Ecol Appl 20:30–59. https://doi.org/10.1890/08-1140.1
Boer BW, Ginzky H, Heuser IL (2017) International soil protection law: history, concepts and latest developments. In: Ginzky H et al (eds) International yearbook of soil law and policy 2016, International yearbook of soil law and policy. Springer International Publishing AG, pp 49–72. https://doi.org/10.1007/978-3-319-42508-5_7
Bojko O, Kabala C (2017) Organic carbon pools in mountain soils—sources of variability and predicted changes in relation to climate and land use changes. CATENA 149(Part 1):209–220
Bojórquez-Quintal E, Escalante-Magaña C, Echevarría-Machado I, Martínez-Estévez M (2017) Aluminum, a friend or foe of higher plants in acid soils. Front Plant Sci 8:1767
Bolan NS, Hedley MJ (2003) Role of carbon, nitrogen and sulfur cycles in soil acidification. In: Rengel Z (ed) Handbook of acidity. Marcel Dekker, New York, pp 29–52
Bouma J (2010) Implications of the knowledge paradox for soil science. Adv Agron 106. ISSN: 0065–2113. https://doi.org/10.1016/s0065-2113(10)06004-9
Brady NC, Weil RR (2008) Nature and properties of soils, 14th edn. Prentice Hall, Upper Saddle River, p 992

Breemen NV, Driscoll CT, Mulder J (1984) Acidic deposition and internal proton sources in acidification of soils and waters. Nature 307:599–604. https://doi.org/10.1038/307599a0

Buss W, Graham MC, MacKinnon G, Mašek O (2016) Strategies for producing biochars with minimum PAH contamination. J Anal Appl Pyrolysis 119:24–30

Butterly CR, Amado TJC, Tang C (2022) Soil acidity and acidification. In: Oliveira TSD, Bell RW (eds) Subsoil constraints for crop production. Springer, Cham. https://doi.org/10.1007/978-3-031-00317-2_3

Cachada A, Rocha-Santos T, Duarte AC (2018) Chapter 1—soil and pollution: an introduction to main issues. In: Duarte AC, Cachada A, Rocha-Santo T (eds) Soil pollution. Academic press, pp 1–28. https://doi.org/10.1016/B978-0-12-849873-6.00001-7

Cape JN (1986) Effects of air pollution on the chemistry of surface waxes of Scots pine. Wat Air Soil Poll 31:393–399. https://doi.org/10.1007/BF00630856

Chan KY, Heenan DP, So HB (2003) Sequestration of carbon and changes in soil quality under conservation tillage on light-textured soils in Australia: a review. Aust J Exp Agric 43:325–334

Che J, Zhao XQ, Shen RF (2022) Molecular mechanisms of plant adaptation to acid soils. Pedosphere

Che J, Zhao XQ, Shen RF (2023) Molecular mechanisms of plant adaptation to acid soils: a review. Pedosphere 33:14–22. https://doi.org/10.1016/j.pedsph.2022.10.001

Chen C, Liao H (2016) Organic acid anions: an effective defensive weapon for plants against aluminum toxicity and phosphorus deficiency in acidic soils. J Genet Genomics 43:631–638. https://doi.org/10.1016/j.jgg.2016.11.003

Chen M, Qin X, Zeng G (2017) Biodiversity change behind wide applications of nanomaterials? Nano Today 17:11–13

Chilkoti V, Bolisetti T, Balachandar R (2017) Climate change impact assessment on hydropower generation using multi-model climate ensemble. Renew Energy 109:510–517

Chintala R, Mollinedo J, Schumacher TE, Malo DD, Julson JL (2013) Effect of biochars on chemical properties of acidic soil. J Arch Agron Soil Sci 60(3):393–404

Coles D, Frewer LJ (2013) Nanotechnology applied to European food production—a review of ethical and regulatory issues. Trends Food Sci Technol 34:32–43. https://doi.org/10.1016/j.tifs.2013.08.006

Cornelissen G, Nurida NL, Hale SE, Martinsen V, Silvani L, Mulder J (2018) Fading positive effect of biochar on crop yield and soil acidity during five growth seasons in an Indonesian Ultisol. Sci Total Environ 634:561–568

da Silva LC, Azevedo AA, da Silva EAM, Oliva MA (2005) Effects of simulated acid rain on the growth of five Brazilian tree species and anatomy of the most sensitive species (*Joannesia princeps*) Aust J Bot 53:789–796. https://doi.org/10.1071/BT04096

Dai Z, Wang Y, Muhammad N, Yu X, Xiao K, Meng J et al (2014) The effects and mechanisms of soil acidity changes, following incorporation of biochars in three soils differing in initial pH. Soil Sci Soc Am J 78(5):1606–1614

Datta M, Singh NP (2005–06) Annual report, ICAR Research Complex for NEH Region, Meghalaya. Available from https://icarneh.ernet.in and https://kiran.nic.in

de Sant-Anna SA, Jantalia CP, Sá JM, Vilela L, Marchão RL, Alves BJ, Urquiaga S, Boddey RM (2017) Changes in soil organic carbon during 22 years of pastures, cropping or integrated crop/livestock systems in the Brazilian cerrado. Nutri Cycl Agroecosyst 108:101–120

Delhaize E, Craig S, Beaton CD, Bennet RJ, Jagadish VC, Randall PJ (1993) Aluminum tolerance in wheat (Triticum asetivum L.): I. Uptake and distribution of aluminum in root apices. Plant Physiol 103:685–693

Delhaize E, Ryan PR, Hebb DM, Yamamoto Y, Sasaki T, Matsumoto H (2004) Engineering high-level aluminum tolerance in barley with the *ALMT1* gene. Proc Natl Acad Sci U S A 101:15249–15254

Ding ZJ, Shi YZ, Li GX, Harberd NP, Zheng SJ (2021) Tease out the future: how tea research might enable crop breeding for acid soil tolerance. Plant Commun 2(3):100182. https://doi.org/10.1016/j.xplc.2021.100182

Du E, Dong D, Zeng X, Sun Z, Jiang X, de Vries W (2017) Direct effect of acid rain on leaf chlorophyll content of terrestrial plants in China. Sci Total Environ 605–606:764–769. https://doi.org/10.1016/j.scitotenv.2017.06.044

Elmer W, Ma C, White J (2018) Nanoparticles for plant disease management. Curr Opin Environ Sci Health 6:66–70. https://doi.org/10.1016/j.coesh.2018.08.002

El-Ramady H, Alshaal T, Abdelrahman H, El-Hady O (2019) Future soil issues. In: El-Ramady H, Alshaal T, Bakr N, Elbana T, Mohamed E, Belal A (eds) The soils of Egypt. https://doi.org/10.1007/978-3-319-95516-2

Ernani PR, Ribeiro MFS, Bayer C (2004) Chemical modifications caused by liming below the limed layer in a predominantly variable charge acid soil. Commun Soil Sci Plant Anal 35:889–901

Evans LS (1988) Effect of acidic decomposition on vegetation: state of science. In: Rao DN, Yunus M, Ahmad KJ, Singh SN (eds) Perspectives in environmental botany, today and tomorrows. Printers and Publishers, New Delhi, pp 73–119

Exley C (2007) Aluminium, tau and Alzheimer's disease. J Alzheimers Dis 12:313–315

FAO (2011) The state of the world's land and water resources for food and agriculture (SOLAW)—managing systems at risk. Food and Agriculture Organization of the United Nations, Rome and Earthscan, London. Available via: http://www.fao.org/3/a-i1688e.pdf

FAO (2017), Food and Agriculture Organization of the United Nations. http://www.fao.org/faostat/en/#data

Fei J, Ma J, Yang J, Liang Y, Ke Y, Yao L, Min X (2020) Effect of simulated acid rain on stability of arsenic calcium residue in residue field. Environ Geochem Health 42:769–780. https://doi.org/10.1007/s10653-019-00273-y

Feynman RP (1960) There's plenty of room at the bottom. Eng Sci 23:22–36

Fidel RB (2012) Evaluation and implementation of methods for quantifying organic and inorganic component of biochar alkalinity. Master's thesis, Iowa State University, USA. Retrieved 21 Mar 2018, from https://lib.dr.iastate.edu/etd/12752/

Fidel RB, Laird DA, Thompson ML, Lawarinenko M (2017) Characterization and quantification of biochar alkalinity. Chemosphere 167:367–373

Fitzpatrick RW (2002) Land degradation processes. ACIR Monogr 84:119–129

Flasbarth J (2017) Soils need international governance: a European perspective for the first volume of the international yearbook of soil law and policy. In: Ginzky H et al (eds) International yearbook of soil law and policy 2016, international yearbook of soil law and policy. Springer International Publishing AG, pp 15–19. https://doi.org/10.1007/978-3-319-42508-5_3

Flaten TP (2002) Aluminium in tea—concentrations, speciation and bioavailability. Coord Chem Rev 228:385–395

Foley JA, Ramankutty N, Brauman KA, Cassidy ES, Gerber JS, Johnston M et al (2011) Solutions for a cultivated planet. Nature 478:337–342

Fraceto LF, Grillo R, De Medeiros GA, Scognamiglio V, Rea G, Bartolucci C (2016) Nanotechnology in agriculture: which innovation potential does it have? Front Environ Sci 4:20. https://doi.org/10.3389/fenvs.2016.00020/full

Freddo A, Cai C, Reid BJ (2012) Environmental contextualisation of potential toxic elements and polycyclic aromatic hydrocarbons in biochar. Environ Pollut 171:18–24

Fujii K (2014) Kazumichi Fujii soil acidification and adaptations of plants and microorganisms in Bornean tropical forests. Ecol Res 2014(29):371–381. https://doi.org/10.1007/s11284-014-1144-3

Fu Y, Tian Z, Amoozegar A, Heitman JL (2019) Measuring dynamic changes of soil porosity during compaction. Soil Tillage Res

Gao J, Wang E, Ren W, Liu X, Yang Y (2017) Effects of simulated climate change on soil microbial biomass and enzyme activities in young Chinese fir (Cunninghamia lanceolata) in subtropical China. Acta Ecol Sin 37(4):272–278

Gao X, Avellan A, Laughton S, Vaidya R, Rodrigues SM, Casman EA, Lowry GV (2018) CuO nanoparticle dissolution and toxicity to wheat (Triticum aestivum) in rhizosphere soil. Environ Sci Technol 52(5):2888–2897

Ginocchio R, De la Fuente LM, Sánchez P, Bustamante E, Silva Y, Urrestarazu P, Rodríguez PH (2009) Soil acidification as a confounding factor on metal phytotoxicity in soils spiked with copper-rich mine wastes. Environ Toxicol Chem Int J 28(10):2069–2081
Ginzky H, Heuser IL, Qin T, Ruppel OC, Wegerdt P (2017) International yearbook of soil law and policy 2016. In: International yearbook of soil law and policy series, vol 2016. Springer International Publishing AG. https://doi.org/10.1007/978-3-319-42508-5
Giraldo JP, Landry MP, Faltermeier SM, McNicholas TP et al (2014) Plant nanobionics approach to augment photosynthesis and biochemical sensing. Nat Mater 13(4):400–408
Goulding KWT (2016) Soil acidification and the importance of liming agricultural soils with particular reference to the United Kingdom. Soil Use Manage 32:390–399
Grennfelt P, Engleryd A, Forsius M, Hov Ø, Rodhe H, Cowling EB (2019) Acid rain and air pollution: 50 years of progress in environmental science and policy. Ambio 49:849–864
Guerra FD, Attia MF, Whitehead DC, Alexis F (2018) Nanotechnology for environmental remediation: materials and applications. Molecules 23(7):1760. https://doi.org/10.3390/molecules23071760
Guo TR, Ying C, Zhang YH, Jin YF (2006) Alleviation of Al toxicity in barley by addition of calcium. Agric Sci China 5(11):828–833
Guo JH, Liu XJ, Zhang Y, Shen JL, Han WX, Zhang WF, Christie P, Goulding KW, Vitousek PM, Zhang FS (2010) Significant acidification in major Chinese croplands. Science 327:1008–1010
Guppy CN, Menzies NW, Moody PW, Blamey FPC (2005) Competitive sorption reactions between phosphorus and organic matter in soil: a review. Aust J Soil Sci 43:189–202
Haile W, Boke S (2011) Response of Irish potato (Solanum tuberosum) to the application of potassium at acidic soils of chencha, southern Ethiopia. Int J Agric Biol 13:595–598
Hajiboland R, Rad SB, Barceló J, Poschenrieder C (2013a) Mechanisms of aluminum-induced growth stimulation in tea (*Camellia sinensis*). J Plant Nutr Soil Sci 176:616–625. https://doi.org/10.1002/jpln.201200311
Hajiboland R, Bahrami Rad S, Barceló J, Poschenrieder C (2013b) Mechanisms of aluminum-induced growth stimulation in tea (Camellia sinensis). J Plant Nutr Soil Sci. 176:616–625. https://doi.org/10.1002/jpln.201200311
Halimi ES (2011) Development of acid-soil tolerant corn (Zea mays L.) with high-quality protein. Agrivita 33:127–132
Han Y, Feng J, Han M, Zhu B (2020) Responses of arbuscular mycorrhizal fungi to nitrogen addition: a meta-analysis. Glob Change Biol 26:7229–7241. https://doi.org/10.1111/gcb.15369
Hansen SF, Hansen OFH, Nielsen MB (2020) Advances and challenges towards consumerization of nanomaterials. Nat Nanotechnol 15(12):964–965
Hayles J, Johnson L, Worthley C, Losic D (2017) Nanopesticides: a review of current research and perspectives. New Pestic Soil Sens 2017:193–225. https://doi.org/10.1016/B978-0-12-804299-1.00006-0
Haynes R, Mokolobate M (2001) Amelioration of Al toxicity and P deficiency in acid soils by additions of organic residues: a critical review of the phenomenon and the mechanisms involved. Nutri Cycl Agroecosyst 59:47–63
He M, Hua Z, Chen H, Liu Y, Li Y, Zhang Z (2022) Effects of simulated acid rain on rhizosphere microorganisms of invasive *Alternanthera philoxeroides* and native *Alternanthera sessilis*. Front Microbiol 13:993147. https://doi.org/10.3389/fmicb.2022.993147
Heck WW, Heagle AS, Shriner DS (1986) *Hordeum vulgare* exposed to long term fumigation with low concentration of SO_2. Physiol Plant 76:445–450
Hocking PS (2001) Organic acids exuded from roots in phosphorus uptake and aluminum tolerance of plants in acid soils. In: Sparks DL (ed) Advances in agronomy. Academic
Hofmann T, Lowry GV, Ghoshal S, Tufenkji N, Brambilla D, Dutcher JR, Wilkinson KJ et al (2020) Technology readiness and overcoming barriers to sustainably implement nanotechnology-enabled plant agriculture. Nat Food 1(7):416–425
Hue NV (2011) Alleviating soil acidity with crop residues. Soil Sci 176(10):543–549

Iavicoli I, Leso V, Beezhold DH, Shvedova AA (2017) Nanotechnology in agriculture: opportunities, toxicological implications, and occupational risks. Toxicol Appl Pharmacol 329:96–111
Islam MT, Nursey-Bray M (2017) Adaptation to climate change in agriculture in Bangladesh: the role of formal institutions. J Environ Manage 200:347–358
Islami T, Guritno B, Nurbasuki, Suryanto A (2011) Maize yield and associated soil quality changes in cassava + maize intercropping system after 3 years of biochar applications. J Agric Food Technol 1(7):112–115
Jan AT, Azam M, Siddiqui K, Ali A, Choi I, Haq QM (2015) Heavy metals and human health: mechanistic insight into toxicity and counter defense system of antioxidants. Int J Mol Sci 1612:29592–29630
Janušević ius T, Mažeikiene A, Danila V, Paliulis D (2022) The characteristics of sewage sludge pellet biochar prepared using two different pyrolysis methods. Biomass Convers Bioref
Jatav HS, Rajput VD, Minkina T, Singh SK et al (2021) Sustainable approach and safe use of biochar and its possible consequences. Sustainability 13:10362
Jeffery S, Abalos D, Prodana M, Bastos AC, Van Groenigen JM, Hungate BA, Verheijen F (2017) Biochar boosts tropical but not temperate crop yields. Environ Res Lett 12(5):1–6
Ji Z, Guo W, Sakkiah S, Liu J, Patterson TA, Hong H (2021) Nanomaterial databases: data sources for promoting design and risk assessment of nanomaterials. Nanomaterials 11(6):1599
Jiang HH, Cai LM, Wen HH, Hu GC, Lai-Guo Chen LG, Luo J (2020) An integrated approach to quantifying ecological and human health risks from different sources of soil heavy metals. Sci Total Environ 701:134466. https://doi.org/10.1016/j.scitotenv.2019.134466
Kalidas-Singh S (2013) Effect of phosphorus sources on yield and P uptake of rice in acidic soil. Master's Thesis. Central Agricultural University, Imphal, Manipur
Kalkhoran SS, Pannell DJ, Thamo T, White B, Polyakov M (2019) Soil acidity, lime application, nitrogen fertility, and greenhouse gas emissions: optimizing their joint economic management. Agric Syst 176:102684
Kan Z-R, Liu Q-Y, Wu G, Ma S-T, Virk AL, Qi J-Y, Zhao X, Zhang H-L (2020) Temperature and moisture driven changes in soil carbon sequestration and mineralization under biochar addition. J Clean Prod 265:121921
Kashyap PL, Kumar S, Gurjar MS, Singh A et al (2015) Phytopathogenomics in plant disease management: a paradigm shift. In: Prasad D, Ray DP (eds) Biotechnological approaches in crop protection. Biotech Book Publishers, New Delhi, pp 241–262
Kavitha KS, Baker S, Rakshith D, Kavitha HU, Yashwantha Rao HC, Harini BP, Satish S (2013) Plants as green source towards synthesis of nanoparticles. Int Res J Biol Sci 2:66–76
Khan ST, Adil SF, Shaik MR, Alkhathlan HZ, Khan M, Khan M (2021) Engineered nanomaterials in soil: their impact on soil microbiome and plant health. Plants (Basel) 11(1):109. https://doi.org/10.3390/plants11010109
Khandelwal N, Barbole RS, Banerjee SS, Chate GP, Biradar AV, Khandare JJ et al (2016) Budding trends in integrated pest management using advanced micro-and nano-materials: challenges and perspectives. J Environ Manage 184:157–169. https://doi.org/10.1016/j.jenvman.2016.09.071
Kibbey TC, Strevett KA (2019) The effect of nanoparticles on soil and rhizosphere bacteria and plant growth in lettuce seedlings. Chemosphere 221:703–707
Kidd PS, Llugany M, Poschenrieder C, Gunse B, Barcelo J (2001) The role of root exudates in aluminium resistance and silicon-induced amelioration of aluminium toxicity in three varieties of maize (*Zea mays* L.). J Exp Bot 52:1339–1352
Kinraide TB (1998) Three mechanisms for the calcium alleviation of mineral toxicities. Plant Physiol 118(2):513–520
Kisinyo PO (2011) Constraints of soil acidity and nutrient depletion on maize (Zea mays L.) production in Kenya. Ph.D thesis, Moi University, Kenya
Knittel R, Pell EJ (1991) Effects of drought stress and simulated acidic rain on foliar conductance of *Zea mays* L. Environ Exp Bot 31:79–90. https://doi.org/10.1016/0098-8472(91)90010-L

Kochian LV, Hoekenga OA, Pineros MA (2004) How do crop plants tolerate acid soils? Mechanisms of aluminum tolerance and phosphorous efficiency. Annu Rev Plant Biol 55:459–493. https://doi.org/10.1146/annurev.arplant.55.031903.141655

Kochian LV, Piñeros MA, Liu J, Magalhaes JV (2015) Plant adaptation to acid soils: the molecular basis for crop aluminum resistance. Annu Rev Plant Biol 66:571–598

Kollmeier M, Dietrich P, Bauer CS, Horst WJ, Hedrich R (2001) Aluminum activates a citrate-permeable anion channel in the aluminum-sensitive zone of the maize root apex. A comparison between an aluminum-sensitive and an aluminum-resistant cultivar. Plant Physiol 126:397–410

Kookana RS, Boxall ABA, Reeves PT, Ashauer R, Beulke S, Chaudhry Q et al (2014) Nanopesticides: guiding principles for regulatory evaluation of environmental risks. J Agric Food Chem 62:4227–4240. https://doi.org/10.1021/jf500232f

Kreyling WG, Semmler-Behnke M, Chaudhry Q (2010) A complementary definition of nanomaterial. Nano Today 5(3):165–168

Kristanti RA, Liong RMY, Hadibarata T (2021) Soil Remediation applications of nanotechnology. Trop Aquat Soil Pollut 1:35–45

Kumar S, Sukul P (2020) Green manuring and its role in soil health management, soil health. Springer, Switzerland (AG), pp 219–241

Kwak SY, Lew TTS, Sweeney CJ, Koman VB, Wong MH, Bohmert-Tatarev K, Snell KD, Seo JS, Chua NH, Strano MS (2019) Chloroplast-selective gene delivery and expression in planta using chitosan-complexed single-walled carbon nanotube carriers. Nat Biotechnol 14:447–455

Lal R (1993) Tillage effects on soil degradation, soil resilience, soil quality and sustainability. Soil Tillage Res 27:1–8

Lal R (1999) Soil management and restoration for carbon sequestration to mitigate the accelerated greenhouse effect. Prog Environ Sci 1:307–326

Lal R (2001) World cropland soils as a source or sink for atmospheric carbon. Adv Agron. 71. https://doi.org/10.1016/S0065-2113(01)71014-0

Lal R (2005) World crop residues production and implications of its use as a biofuel. Environ Int 31:575–584

Larney FJ, Janzen HH, Olson AF (2011) Residual effects of one-time manu crop residue and fertilizer amendments on a desurfaced soil. Can J Soil Sci 91:1029–1043

Lehmann J, da Silva JP, Steiner C, Nehls T, Zech W, Glaser B (2003) Nutrient availability and leaching in an archaeological anthrosol and a ferralsol of the Central Amazon basin: fertilizer, manure and charcoal amendments. Plant Soil 249:343–357

Li S, Chan CY (2022) Will biochar suppress or stimulate greenhouse gas emissions in agricultural fields? Unveiling the dice game through data syntheses. Soil Syst 2022(6):73

Li S, Tasnady D (2023) Biochar for soil carbon sequestration: current knowledge, mechanisms, and future perspectives. 9:67. https://doi.org/10.3390/c9030067

Li P, Song A, Li ZJ, Fan FL, Liang YC (2015) Silicon ameliorates manganese toxicity by regulating both physiological processes and expression of genes associated with photosynthesis in rice (*Oryza sativa* L.). Plant Soil 397:289–301. https://doi.org/10.1007/s11104-015-2626-y

Li C, McLinden C, Fioletov V, Krotkov N, Carn S, Joiner J, Streets D, He H, Ren X, Li Z, Dickerson RR (2017a) India is overtaking China as the world's largest emitter of anthropogenic sulfur dioxide. Sci Rep 7:1–7. https://doi.org/10.1038/s41598-017-14639-8

Li Y, Huang J, Song X, Zhang Z, Jiang Y, Zhu Y, Zhao H, Ni D (2017b) An RNA-Seq transcriptome analysis revealing novel insights into aluminum tolerance and accumulation in tea plant. Planta 246:91–103

Li X, Zhang X, Zhao Q, Liao H (2023) Genetic improvement of legume roots for adaption to acid soils. Crop J 11. https://doi.org/10.1016/j.cj.2023.04.002

Liao W, Thomas SC (2019) Biochar particle size and post-pyrolysis mechanical processing affect soil pH, water retention capacity, and plant performance. Soil Syst 3:14

Liu J, Magalhaes J V, Shaff J, Kochian LV (2009) Aluminum-activated citrate and malate transporters from the MATE and ALMT families function independently to confer Arabidopsis aluminum tolerance. Plant J 57:389–399. https://doi.org/10.1111/j.1365-313X.2008.03696

Liu X, Li Q, Wang O, Zhao Q et al (2015) Influence of TiO2 nanoparticles on the bioavailability of nutrients in Chinese cabbage. J Environ Sci 30:153–158

Liu Y-R, Delgado-Baquerizo M, Trivedi P, He J-Z, Singh BK (2017) Identity of biocrust species and microbial communities drive the response of soil multi-functionality to simulated global change. Soil Biol Biochem 107:208–217

Liu WJ, Xu FJ, Lv T, Zhou WW, Chen Y, Jin CW, Lu LL, Lin XY (2018) Spatial responses of antioxidative system to aluminum stress in roots of wheat (Triticum aestivum L.) plants. Sci Total Environ 627:462–469

Liu H, Ren X, Zhu J, Wu X, Liang C (2018a) Effect of exogenous abscisic acid on morphology, growth and nutrient uptake of rice (*Oryza sativa*) roots under simulated acid rain stress. Planta 248:647–659. https://doi.org/10.1007/s00425-018-2922-x

Liu X, Fu Z, Zhang B, Zhai L, Meng M, Lin J, Zhang J (2018b) Effects of sulfuric, nitric, and mixed acid rain on Chinese fir sapling growth in Southern China. Ecotoxicol Environ Saf 160:154–161. https://doi.org/10.1016/j.ecoenv.2018.04.071

Liu Z, Wei H, Zhang J, Saleem M et al (2021) Seasonality regulates the effects of acid rain on microbial community in a subtropical agricultural soil of Southern China. Ecotoxicol Environ Saf 224:112681. https://doi.org/10.1016/j.ecoenv.2021.112681

Loehrlein M (2020) Soil health. In: Sustainable landscaping. CRC Press, Boca Raton, FL, USA, pp 169–188

Lopez-Arredondo DL, Leyva-Gonzalez MA, Gonzalez-Morales SI et al (2014) Phosphate nutrition: improving low-phosphate tolerance in crops. Annu Rev Plant Biol 65:95–123

Lozano-García B, Muñoz-Rojas M, Parras-Alcántara L (2017) Climate and land use changes effects on soil organic carbon stocks in a Mediterranean semi-natural area. Sci Total Environ 579:1249–1259

Lü G, Guo S, Zhang H, Geng L, Song F, Fei Z et al (2011) Transcriptional profiling of watermelon during its incompatible interaction *with fusarium oxysporum* f. sp. *niveum*. Eur J Plant Pathol 131:585–601. https://doi.org/10.1007/s10658-011-9833-z

Lungu OI (1993) Effects of lime and farmyard manure on soil acidity and maize growth on an acid alfisol from Zambia. Trop Agric 70:309–313

Ma JF, Ryan PR, Delhaize E (2001) Aluminium tolerance in plants and the complexing role of organic acids. Trends Plant Sci 6:273–278

Ma S, Chen W, Zhang J, Shen H (2020) Influence of simulated acid rain on the physiological response of flowering Chinese cabbage and variation of soil nutrients. Plant Soil Environ 66:648–657. https://doi.org/10.17221/469/2020-PSE

Magalhaes J, Liu J, Gumaraes CT, Lana UGP et al (2007) A gene in the multidrug and toxcic compound extrusion (MATE) family confers aluminium tolerance in sorghum. Nat Genet 39(9):1157–1161. https://doi.org/10.1038/ng2074

Maghsoodi MR, Lajayer BA, Hatami M, Mirjalili MH (2019) Chapter 17—challenges and opportunities of nanotechnology in plant-soil mediated systems: beneficial role, phytotoxicity, and phytoextraction. In: Ghorbanpour M, Wani SH (eds) *Advances* in Phytonanotechnology. Academic Press, Cambridge, MA, USA, pp 379–404

Maltz MR, Chen Z, Cao J, Arogyaswamy K, Shulman H, Aronson EL (2019) Inoculation with *Pisolithus tinctorius* may ameliorate acid rain impacts on soil microbial communities associated with *Pinus massoniana* seedlings. Fungal Ecol 40:50–61. https://doi.org/10.1016/j.funeco.2018.11.011

Mao J, Thornton EPE, Xiaoying S, Maosheng Z, Post WM (2012) Remote sensing evaluation of CLM4 GPP for the period 2000–09. J Clim 25:5327–5342. https://doi.org/10.1175/JCLI-D-11-00401.1

Martinsen V, Alling V, Nurida ML, Mulder J, Hale SE, Ritz C, Cornelissen G et al (2015) pH effects of the addition of three biochars to acidic Indonesian mineral soils. J Soil Sci Plant Nutr 61(5):821–834

Materrechera SA, Mkhabela TS (2002) The effectiveness of lime, chicken manure and leaf litter ash in ameliorating acidity in a soil previously under black wattle (*Acacia mearnsii*) plantation. Biores Technol 85:9–16
Matonyei KT (2010) Acid soil tolerance studies on selected maize breeding lines from Kenya. Master of Philosophy thesis, presented to Moi University, Kenya
Mba C, Guimaraes P, Ghosh K (2012) Re-orienting crop improvement for the changing climatic conditions of the 21st century. Agric Food Secur 1:7
Melero S, Madejón E, Ruiz JC, Herencia JF (2007) Chemical and biochemical properties of a clay soil under dryland agriculture system as affected by organic fertilization. Eur J Agron 26:327–334
Men C, Liu R, Xu L, Wang Q, Guo L, Miao Y, Shen Z (2020) Source-specific ecological risk analysis and critical source identification of heavy metals in road dust in Beijing, China. J Hazard Mater 388:121763
Meng C, Tian D, Zeng H, Li Z, Yi C, Niu S (2019) Global soil acidification impacts on belowground processes. Environ Res Lett. https://doi.org/10.1088/1748-9326/ab239c
Merino A, Moreno G, Navarro FB, Gallardo JF (2016) Future issues. In: Gallardo JF (ed) The soils of Spain, world soils book series. Springer International Publishing Switzerland. https://doi.org/10.1007/978-3-319-20541-0_6
Michael PS (2013) Ecological impacts and management of acid sulphate soil: a review. Asian J Water Environ Pollut 10:13–24
Michael PS (2015a) Effects of organic matter amendments and plants on the chemistry of acid sulfate soils under aerobic and anaerobic conditions. J Proc R Soc NSW 148:185–186
Michael PS (2015b) Effects of alkaline sandy loam on sulfuric soil acidity and sulfidic soil oxidation. Int J Environ 4:42–54
Michael PS (2020) Organic carbon and nitrogen amendment prevents oxidation of sulfidic soil of acid sulfate soils under aerobic conditions. Eurasian Soil Sci 53:1743–1751
Michael PS (2021) Role of organic fertilizers in the management of nutrient deficiency, acidity, and toxicity in acid soils—a review. J Glob Agric Ecol 12:19–30
Michael PS, Fitzpatrick R, Reid R (2015) The importance of organic matter on amelioration of acid sulfate soils with sulfuric horizons. Geoderma 225:42–49
Michael PS, Fitzpatrick R, Reid R (2016) The importance of soil carbon and nitrogen in amelioration of acid sulphate soils. Soil Use Manage 32:97–105
Michaela PS (2023) The importance of sustainable management of acid soils in the humid tropics under climate change and the future research directions. Ecofeminism Clim Change (EFCC) 4:39–50. https://doi.org/10.26480/efcc.01.2023.39.50
Milton MB, Abigael TO (2015) Effects of simulated acid rain on the morphology, phenology and yield of Okra (*Abelmoschus esculentus* (L). Moench). J Environ Sci Comp Sci Eng Technol 4:501–518 [Google Scholar]
Mishima SI, Kimura SD, Eguchi S, Shirato Y (2013) Changes in soil available-nutrient stores and relationships with nutrient balance and crop productivity in Japan. Soil Sci Plant Nutr 59:371–379. https://doi.org/10.1080/00380768.2013.786646
Mondal TK, Bhattacharya A, Laxmikumaran M, Ahuja PS (2004) Recent advances of tea (*Camellia sinensis*) biotechnology. Plant Cell Tissue Organ Cult 76:195–254
Montzka C, Bunning S (2015) Combating soil degradation and water scarcity: the case of a Mediterranean agroecosystem in Morocco. Sustainability 7(7):8668–8688
Mosharrof M, Uddin MK, Sulaiman MF, Mia S, Shamsuzzaman SM, Haque ANA (2021) Combined application of biochar and lime increases maize yield and accelerates carbon loss from an acidic soil. Agronomy 2021(11):1313
Mossisa F (2018) Progress of soil acidity management research in Ethiopia. Greener J Soil Sci Plant Nutr 5:9–22. https://doi.org/10.15580/GJSSPN.2018.2.051518072

Mousavi Kouhi SM, Lahouti M, Ganjeali A, Entezari MH (2015) Comparative effects of ZnO nanoparticles, ZnO bulk particles, and Zn 2+ on Brassica napus after long-term exposure: changes in growth, biochemical compounds, antioxidant enzyme activities, and Zn bioaccumulation. Water Air Soil Pollut 226:1–11

Mrunalini K, Behera B, Jayaraman S, Abhilash PC, Dubey PK, Swamy GN, Prasad JVNS, Rao KV, Krishnan P, Pratibha G (2022) Nature-based solutions in soil restoration for improving agricultural productivity. Land Degrad Dev 33:1269–1289

Mukome FN, Six J, Parikh SJ (2013) The effects of walnut shell and wood feedstock biochar amendments on greenhouse gas emissions from a fertile soil. Geoderma 200:90–98

Neves NR, Oliva MA, da Cruz CD, Costa AC, Ribas RF, Pereira EG (2009) Photosynthesis and oxidative stress in the restinga plant species *Eugenia uniflora* L. exposed to simulated acid rain and iron ore dust deposition: potential use in environmental risk assessment. Sci Total Environ 407:3740–3745. https://doi.org/10.1016/j.scitotenv.2009.02.035

Nikalje A (2015) Nanotechnology and its applications in medicine. Med Chem 15:81–89

Oliveira CA, Sa NMR, Marriel IE, Gomes EA et al (2006) Phosphate solubilizing microorganisms closely associated with maize rhizosphere. In: Alves et al (eds) 3rd international symposium on phosphorus dynamics in soil-plant continuum: integrating marginal lands into productive agricultural systems by means of improving soil and fertilizer phosphorus efficiency. 14–19th May, 2006. Uberlandia, Minas Gerais, Brazil, pp 234–235

Opala PA, Okalebo JR, Othieno CO, Kisinyo P (2010) Effect of organic and inorganic phosphorus sources on maize yields in an acid soil in western Kenya. Nutr Cycl Agroecosyst 86:317–329

Opala PA, Nyambati RO, Kisinyo PO (2014) Response of maize to organic and inorganic sources of nutrients in acid soils of Kericho County, Kenya. Am J Exp Agric 4(6):713–723. https://doi.org/10.9734/AJEA/2014/6415

Osaki M, Watanabe T, Tadano T (1997) Beneficial effect of aluminum on growth of plants adapted to low pH soils. Soil Sci Plant Nutr 43:551–563

Ouma E, Ligeyo D, Matonyei T, Agalo J, Were BE, Too E, Onkware, Gudu S, Kisinyo P, Nyangweso P (2013) Enhancing maize grain yield in acid soils of western Kenya using aluminium tolerant germplasm. J Agric Sci Technol A 3:33–46

Pan Y, Song Y, Zhao L, Chen P, Bu C, Liu P, Zhang D (2022) The genetic basis of phosphorus utilization efficiency in plants provide new insight into woody perennial plant improvement. Int J Mol Sci 23(4):2354. https://doi.org/10.3390/ijms23042353

Parisi C, Vigani M, Rodríguez-Cerezo E (2015) Agricultural nanotechnologies: what are the current possibilities? Nano Today 10:124–127

Pauw EFD (1994) The management of acid soils in Africa. Outlook Agric 23:11–16

Petruzzelli G, Gorini F, Pezzarossa B, Pedron F (2010) The fate of pollutants in soil. In: Bianchi F, Cori L, Moretti PF (eds) CNR environment and health inter-department project. Consiglio Nazionale delle Ricerche, Roma, Italy, pp 1–31. https://www.researchgate.net/publication/326463752_The_fate_of_pollutants_in_soil

Pham H, Thinh N, Do A, Hens L (2021) Impacts of simulated acid rain on the growth and the yield of soybean (Glycine max (L.) Merr.) in the mountains of Northern Vietnam. Sustainability 13:4980. https://doi.org/10.3390/su13094980

Pichtel JR, Dick WA, Sutton P (1994) Comparison of amendments and management practices for long-term reclamation of abandoned mine lands. J Environ Qual 23:766–772

Pineros MA, Kochian LV (2001) A patch-clamp study on the physiology of aluminum toxicity and aluminum tolerance in maize. Identification and characterization of Al3+-induced anion channels. Plant Physiol 125:292–305

Pirzadah B, Pirzadah TB, Jan A, Hakeem KR (2020) Nanofertilizers: a way forward for green economy. In: Hakeem KR, Pirzadah TB (eds) Nanobiotechnology in agriculture: an approach towards sustainability. Springer International Publishing, Cham, Switzerland, pp 99–112

Porchezhiyan V, Noorjahan SE (2016) Fabrication of versatile chitosan nanocomposite hydrogel impregnated with biosynthesized silver nanoparticles using Sapindus mukorossi: characterization and applications. RSC Adv 6:95564–95573

Pouratashi M, Iravani H (2012) Farmers knowledge of integrated pest management and learning style preferences: implications for information delivery. Int J Pest Manage 58:347–353

Prakash J, Agrawal SB, Agrawal M (2022) Global trends of acidity in rainfall and its impact on plants and soil. J Soil Sci Plant Nutr (2023) 23(1):398–419 (Epub 2022 Nov 17). https://doi.org/10.1007/s42729-022-01051-z

Prasad R, Power JF (1997) Soil fertility management for sustainable agriculture. CRC Lewis Publishers, Boca Raton, p 349

Qambrani NA, Rahman MM, Won S, Shim S, Ra C (2017) Biochar properties and eco-friendly applications for climate change mitigation, waste management, and wastewater treatment: a review. Renew Sustain Energy Rev 7(79):255–273

Qian L, Chen B, Hu D (2013) Effective alleviation of aluminum phytotoxicity by manure-derived biochars. Environ Sci Technol 47(6):2737–2745

Rajendran S, Priya AK, Senthil Kumar P, Hoang TK, Sekar K, Chong KY, Khoo KS, Ng HS, Show PL (2022) A critical and recent developments on adsorption technique for removal of heavy metals from wastewater—a review. Chemosphere 135146

Raliya R, Tarafdar JC (2013) Nanoparticles and their potential application as antifertilizers and pesticides. Plant Sci 10(1):61–70

Raman H, Raman R, Wood R, Martin P (2006) Repetitive indel markers within the ALMT1 gene conditioning aluminium tolerance in wheat (Triticum aestivum L.). Mol Breed 18:171–183

Ramsden J (2016) Nanotechnology: an introduction. Elsevier Science, Amsterdam, The Netherlnds

Rao IM, Zeigler RS, Vera R, Sarkarung S (1993) Selection and breeding for acid-soil tolerance in crops. Bioscience 43(7):454–465

Richardson AE, Barea JM, McNeill AM, Prigent-Combaret C (2009) Acquisition of phosphorus and nitrogen in the rhizosphere and plant growth promotion by microorganisms. Plant Soil 321:305–339

Ruiz-López MF, Martins-Costa MT, Anglada JM, Francisco JS (2019) A new mechanism of acid rain generation from HOSO at the air-water interface. J Am Chem Soc

Safriel UN (2007) The assessment of global trends in land degradation. In: Sivakumar MVK, Ndiang'ui N (eds) Climate and land degradation. Springer, Berlin/Heidelberg/New York

Sasaki T, Yamamoto Y, Ezaki B, Katsuhara M, Ahn SJ et al (2004) A wheat gene encoding an aluminum-activated malate transporter. Plant J 37:645–653

Schlich K, Hund-Rinke K (2015) Influence of soil properties on the effect of silver nanomaterials on microbial activity in five soils. Environ Pollut 196:321–330

Sekhon BS (2014) Nanotechnology in agri-food production: an overview. Nanotechnol Sci Appl 7:31–53

Sekiya N, Yano K (2010) Seed P-enrichment as an effective P supply to wheat. Plant Soil 327:347–354

Shah V, Daverey A (2020) Phytoremediation: a multidisciplinary approach to clean up heavy metal contaminated soil. Environ Technol Innov. https://doi.org/10.1016/j.eti.2020.100774

Shang Y, Hasan MK, Ahammed GJ, Li M, Yin H, Zhou J (2019) Applications of nanotechnology in plant growth and crop protection: a review. Molecules 24(14):2558

Sharma UC (1991) Response of potato to phosphorus and farmyard manure on acidic soils of Khasi hills of Meghalaya. J Indian Soc Soil Sci 39:386–388

Sharma UC, Arora BR (1986) Effect of nitrogen, phosphorus potassium and FYM application on the micronutrients content of potato. J Micronutrient Anal (England) (merged in Food Chemistry in 1990) 2:275–282

Sharma UC, Arora BR (1987) The effect of nitrogen, phosphorus, potassium and FYM on the yield of potato. J Agric Sci (Cambridge) 108:321–329

Sharma UC, Tripathy AK (1991) Integrated management of P and FYM in rice in Nagaland. Indian J Agron 36:402–405

Sharma UC, Sharma V (2009) Plant-livestock-soil-hydrology interactions in the north-eastern region of India. Biologia 64(3):460–464

Sharma UC, Sharma V (2011) Soil as a sink for carbon sequestration: how agroforestry can help? Indian J Agrofor 13:65–77

Shen W, Li Z, Liu Y (2008) Surface chemical functional groups modification of porous carbon. Recent Pat Chem Eng 1:27–40

Shi Z, Zhang J, Xiao Z et al (2021) Effects of acid rain on plant growth: a meta-analysis. J Environ Manage 297:113213. https://doi.org/10.1016/j.jenvman.2021.113213

Shu X, Zhang K, Zhang Q, Wang W (2019) Ecophysiological responses of *Jatropha curcas* L. seedlings to simulated acid rain under different soil types. Ecotoxicol Environ Saf 185:1–12. https://doi.org/10.1016/j.ecoenv.2019.109705[PubMed]

Shukla J, Sundar S, Sharma S, Naresh R (2012) Modeling and analysis of the acid rain formation due to precipitation and its effect on plant species. Nat Resour Model. https://doi.org/10.1111/j.1939-7445.2012.00119.x

Singh B, Agrawal M (2004) Impact of simulated acid rain on growth and yield of two cultivars of wheat. Wat Air Soil Poll 152:71–80. https://doi.org/10.1023/B:WATE.0000015331.02874.df

Singh A, Agrawal M (2007) Acid rain and its ecological consequences. J Environ Biol 29:15

Singh A, Agrawal M (2008) Acid rain and its ecological consequences. J Environ Biol 29(1):15–24

Singh H, Northup BK, Rice CW, Prasad PVV (2022) Biochar applications influence soil physical and chemical properties, microbial diversity, and crop productivity: a meta-analysis. Biochar 4:8

Sivaguru M, Horst W (1998) The distal part of the transition zone is the most aluminum-sensitive apical root zone of maize. Plant Physiol 116:155–163

Sivaguru M, Liu J, Kochian LV (2013) Targeted expression of SbMATE in the root distal transition zone is responsible for sorghum aluminum resistance. Plant J 76:297–307

Smider B, Singh B (2014) Agronomic performance of a high ash biochar in two contrasting soils. Agr Ecosyst Environ 191:99–107

Smith RA (1872) Air and rain: the beginnings of a chemical climatology. Longmans, Green, and Company, London, p 600

Song P, Xu D, Yue J, Ma Y, Dong S, Feng J (2022) Recent advances in soil remediation technology for heavy metal contaminated sites: a critical review. Sci Total Environ 838(Part 3). https://doi.org/10.1016/j.scitotenv.2022.156417

Soudzilovskaia NA, van Bodegom PM, Terrer C et al (2019) Global mycorrhizal plant distribution linked to terrestrial carbon stocks. Nat Commun 10:5077. https://doi.org/10.1038/s41467-019-13019-2

Steiner C (2007) Soil charcoal amendments maintain soil fertility and establish carbon sink-research and prospects. In: Li TX (ed) Soil ecology research developments. Nova, New York, pp 1–6

Steiner C (2016) Considerations in biochar characterization. Agric Environ Appl Biochar Adv Barriers 63:87–100

Streck C, Gay A (2017) The role of soils in international climate change policy. In: Ginzky H et al (eds) International yearbook of soil law and policy 2016. Springer International Publishing AG, pp 105–128. https://doi.org/10.1007/978-3-319-42508-5_10

Stringer LC (2017) Land degradation. In: Richardson D, Castree N, Goodchild MF et al (eds) International encyclopedia of geography: people, the earth, environment, and technology. Wiley, New Jersey

Sullivan TS, Barth VP, Lewis RW (2017) Soil acidity impacts beneficial soil microorganisms. Washington State University Extension and the U.S. Department of Agriculture, pp 1–6

Sumner ME (2000) Handbook of soil science. CRC Press, Washington, D.C. USA

Sutton MA, Drewer J, Moring A, Adhya TK, Ahmed A, Bhatia A, Brownlie W, Dragosits U, Ghude SD, Hillier J, Hooda S (2017) The Indian nitrogen challenge in a global perspective. Indian Nitrogen Assess. https://doi.org/10.1016/B978-0-12-811836-8.00002-1

Syed B, Nagendra-Prasad MN, Satish S (2016) Synthesis and characterization of silver nanobactericides produced by Aneurinibacillus migulanus 141, a novel endophyte inhabiting Mimosa pudica L. Arab J Chem. https://doi.org/10.1016/j.arabjc.2016.01.005

Tang Y, Zhang H, Schroder JL, Payton ME, Zhou D (2007) Animal manure reduces aluminium toxicity in acid soils. Sci Soc Am J 71:1699–1707. https://doi.org/10.2136/sssaj2007.0008

Terribile F, Basile A, Bonfante A, Carbone A, Colombo C, Langella G, Iamarino M, Manna P, Minieri L, Vingiani S (2013) Future soil issues. In: Costantini EAC, Dazzi C (eds) The soils of Italy, world soils book series. Springer Science+Business Media Dordrecht. https://doi.org/10.1007/978-94-007-5642-7_11

Terzano R, Rascio I, Allegretta I, Porfido C, Spagnuolo M, Khanghahi MY, Crecchio C, Sakellariadou F, Gattullo CE (2021) Fire effects on the distribution and bioavailability of potentially toxic elements (PTEs) in agricultural soils. Chemosphere 281:130752

Thakuria D, Hazarika S, Krishnappa R (2016) Soil acidity and management options. Indian J Fertil 12:40–56

Toda E, Koiso N, Takebayashi A, Ichikawa M, Kiba T, Osakabe K, Osakabe Y, Sakakibara H, Kato N, Okamoto T (2019) An efficient DNA-and selectable-marker-free genome-editing system using zygotes in rice. Nat Plants 5:363–368

Tolra R, Barcelo J, Poschenrieder C (2009) Constitutive and aluminium-induced patterns of phenolic compounds in two maize varieties differing in aluminium tolerance. J Inorg Biochem 103:1486–1490

Tomczyk A, Sokołowska Z, Boguta P (2020) Biochar physicochemical properties: pyrolysis temperature and feedstock kind effects. Rev Environ Sci Bio/technol 19:191–215

Too EJ (2011) Physiological and molecular characterization of resistance to aluminium stress in selected grain sorghum. PhD thesis Moi University—Eldoret Kenya

Tripathi M, Sahu JN, Ganesan P (2016) Effect of process parameters on production of biochar from biomass waste through pyrolysis: a review. Renew Sustain Energy Rev 55:467–481

Tully K, Sullivan C, Weil R, Sanchez P (2015) The state of soil degradation in Sub-Saharan Africa: baselines, trajectories, and solutions. Sustainability 7:6523–6552

Tusar HM, Uddin MK, Mia S, Suhi AA, Wahid SBA, Kasim S, Sairi NA, Alam Z, Anwar F (2023) Biochar-acid soil interactions—a review. Sustainability 15:13366. https://doi.org/10.3390/su151813366

Ulrich B, Mayer R, Khanna PK (1980) Chemical changes due to acid precipitation in a loess-derived soil in Central Europe. Soil Sci 130:1993–1999

UNEP (1992) World atlas of desertification. United Nations Environment Programme. Edward Arnold, Nairobi/London

Vahedpour M, Zolfaghari F (2011) Mechanistic study on the atmospheric formation of acid rain base on the sulfur dioxide. Struct Chem 22:1331–1338. https://doi.org/10.1007/s11224-011-9823-y

Vanlauwe B, Zingore S (2011) Integrated soil fertility management: an operational definition and consequences for implementation and dissemination. Better Crops 95:4–7

von Uexkull HR, Mutert E (1995) Global extent, development and economic impact of acid soils. Plant Soil 171:1–15

Walker DJ, Clemente R, Bernal MP (2004) Contrasting effects of manure and compost on soil pH, heavy metal availability and growth of Chenopodium album L. in a soil contaminated by pyritic mine waste. Chemosphere 57:215–224

Wang J, Raman H, Zhang G, Mendham N, Zhou M (2006) Aluminium tolerance in barley (Hordeum vulgare L.): physiological mechanisms, genetics and screening methods. J Zhejiang Univ Sci B 7:769–787

Wang Y, Yin R, Liu R (2014) Characterization of biochar from fast pyrolysis and its effect on chemical properties of the tea garden soil. J Anal Appl Pyrol 110:375–381

Wang C, Zheng M, Song W, Wen S, Wang B, Zhu C, Shen R (2017) Impact of 25 years of inorganic fertilization on diazotrophic abundance and community structure in an acidic soil in southern China. Soil Biol Biochem 113:240–249

Wang D, Shi Y, Samake A, Fu S, Chen S, Yang M (2018) Silver nanoparticles enhance plant growth in a saline environment. J Nanopart Res 20(5):141

Wang R, Duan X, Wang S, Ren N-Q, Ho S-H (2020) Production, properties, and catalytic applications of sludge derived biochar for environmental remediation. Water Res 187:116390
Wang X, Ai S, Liao H (2023) Deciphering interactions between phosphorus status and toxic metal exposure in plants and rhizospheres to improve crops reared on acid soil. Cells 12(3):441
Warren D, Kraft CE, Josephson DC (2017) Acid rain recovery may help to mitigate the impacts of climate change on thermally sensitive fish in lakes across eastern North America. Glob Change Biol 23:2149–2153. https://doi.org/10.1111/gcb.13568
Wei H, Liu Y, Xiang H, Zhang J, Li S, Yang J (2020) Soil pH responses to simulated acid rain leaching in three agricultural soils. Sustainability 12:280. https://doi.org/10.3390/su12010280
Weng Z, Van Zwieten L, Tavakkoli E, Rose MT et al (2022) Microspectroscopic visualization of how biochar lifts the soil organic carbon ceiling. Nat Commun 13:5177
Wong MTF, Asseng S, Robertson MJ, Oliver Y (2008) Mapping subsoil acidity and shallow soil across a field with information from yield maps, geophysical sensing and the grower. Precis Agric 9:3–5
Wu Q, Wei Z, Long X, Jiang C (2018) Advances in remediation of acid agricultural soils contaminated by heavy metals in South China. In: Luo Y, Tu C (eds) Twenty years of research and development on soil pollution and remediation in China. Springer, Singapore. https://doi.org/10.1007/978-981-10-6029-8_21
Xia L, Shao C, Zhang N et al (2021) Improved tolerance of mycorrhizal torreya grandis seedlings to sulfuric acid rain related to phosphorus and zinc contents in shoots. J Fungi Open Access Mycol J 7:296. https://doi.org/10.3390/jof7040296
Xu Z, Owens PN (2011) Developments and directions of the journal of soils and sediments: 2011 and beyond. J Soils Sed 11:1–2. https://doi.org/10.1007/s11368-010-0320-5
Xu RK, Coventry DR, Farhoodi A, Schultz JE (2002) Soil acidification as influenced by crop rotations, stubble management and application of nitrogenous fertiliser, Tarlee, South Australia. Aust J Soil Res 40:483–494
Xu JM, Tang C, Chen ZL (2006) The role of plant residues in pH change of acid soils differing in initial pH. Soil Biol Biochem 38:709–719
Yamashita H, Fukuda Y, Yonezawa S, Morita A, Ikka T (2020) Tissue ionome response to rhizosphere pH and aluminum in tea plants (*Camellia sinensis L.*), a species adapted to acidic soils. Plant Environ Interact. https://doi.org/10.1002/pei3.10028
Yokel RA (2000) The toxicology of aluminium in the brain: a review. Neurotoxicology 21:813–828
Yokosho K, Yamaji N, Ma JF (2011) An Al-inducible MATE gene is involved in external detoxification of Al in rice. Plant J 1061–1069. https://doi.org/10.1111/j.1365-313X.2011.04757.x
Yokosho K, Yamaji N, Kashino-Fujii M, Ma JF (2016) Retrotransposon-mediated aluminum tolerance through enhanced expression of the citrate transporter OsFRDL4. Plant Physiol 172:2327–2336. https://doi.org/10.1104/pp.16.01214
Yuan J-H, Xu R-K (2011) The amelioration effects of low temperature biochar generated from nine crop residues on an acidic Ultisol. Soil Use Manage 27(1):110–115
Yun J, Zhu C, Wang Q, Hu Q, Yang G (2018) Catalytic conversions of atmospheric sulfur dioxide and formation of acid rain over mineral dusts: molecular oxygen as the oxygen source. Chemosphere. 217. https://doi.org/10.1016/j.chemosphere.2018.10.201
Zhang WH, Ryan PR, Tyerman SD (2001) Malate-permeable channels and cation channels activated by aluminium in the apical cells of wheat roots. Plant Physiol 125:1459–1472
Zhang JE, Ouyang Y, Ling DJ (2007) Impacts of simulated acid rain on cation leaching from the latosol in south China. Chemosphere 67(11):2131–2137
Zhang JE, Ouyang Y, Ling DJ (2007b) Impacts of simulated acid rain on cation leaching from the latosol in south China. Chemosphere 67(11):2131–2137
Zhang N, Wang D, Liu Y, Li S, Shen Q, Zhang R (2014) Effects of different plant root exudates and their organic acid components on chemotaxis, biofilm formation and colonization by beneficial rhizosphere-associated bacterial strains. Plant Soil 374:689–700

Zhang L, Wang G, Zheng Q, Liu Y, Fan X (2017) Quantifying the impacts of agricultural management and climate change on soil organic carbon changes in the uplands of Eastern China. Soil Tillage Res 174:81–91

Zhang C, Zhao X, Liang A, Li Y, Song Q, Li X, Li D, Hou N (2023a) Insight into the soil aggregate-mediated restoration mechanism of degraded black soil via biochar addition: emphasizing the driving role of core microbial communities and nutrient cycling. Environ Res 228:115895

Zhang Y, Li J, Tan J, Li W, Singh BP, Yang X, Bolan N, Chen X, Xu S, Bao Y, Lv D, Peng A, Zhou Y, Wang H (2023b) An overview of the direct and indirect effects of acid rain on plants: relationships among acid rain, soil, microorganisms, and plants. Sci Total Environ 873:162388. https://doi.org/10.1016/j.scitotenv.2023.162388

Zhao XQ, Chen RF, Shen RF (2014) Coadaptation of plants to multiple stresses in acidic soils. Soil Sci 179(10–11):503–513

Zhao Z, Chang E, Lai P, Dong Y, Xu R, Fang D, Jiang J (2019) Evolution of soil surface charge in a chronosequence of paddy soil derived from alfisol. Soil and Tillage Research,192:144–150. https://doi.org/10.1016/j.still.2019.05.011

Zhao X, Sun Y, Huang J et al. (2020) Effects of soil heavy metal pollution on microbial activities and community diversity in different land use types in mining areas. Environ Sci Pollut Res 27:20215–20226 (2020). https://doi.org/10.1007/s11356-020-08538-1

Zheng W, Li R, Yang Q, Zhang W, Huang K, Guan X, Wang S (2019) Short-term response of soil respiration to simulated acid rain in *Cunninghamia lanceolata* and *Michelia acclurei* plantations. J Soils Sed 19:1239–1249. https://doi.org/10.1007/s11368-018-2148-3

Zhou X, Xu Z, Liu W, Wu Y, Zhao T, Jiang H, Zhang X, Zhang J, Zhou L, Wang Y (2019) Chemical composition of precipitation in Shenzhen, a coastal mega-city in South China: influence of urbanization and anthropogenic activities on acidity and ionic composition. Sci Total Environ 662:218–226. https://doi.org/10.1016/j.scitotenv.2019.01.096

Zhou J, Li Z, Zhou T, Xin Z, Wu L, Luo Y, Christie P (2020) Aluminum toxicity decreases the phytoextraction capability by cadmium/zinc hyperaccumulator Sedum plumbizincicola in acid soils. Science of the Total Environment, 711:134591

Zhu XF, Shen RF (2023) Towards sustainable use of acidic soils: deciphering aluminum-resistant mechanisms in plants. J Fundam Res. https://doi.org/10.1016/j.fmre.2023.03.004

Index

U. C. Sharma et al., *Soil Acidity*, Progress in Soil Science,
https://doi.org/10.1007/978-3-031-76357-1

The manufacturer's authorised representative in the EU is Springer Nature Customer Service Centre GmbH, Europaplatz 3, 69115 Heidelberg, Germany. If you have any concerns regarding our products, please contact ProductSafety@springernature.com

Printed and bound by CPI Group (UK) Ltd, Croydon, CR0 4YY
01/10/2026
02225960-0019